Subcellular Biochemistry

Volume 109

The book series SUBCELLULAR BIOCHEMISTRY is a renowned and well recognized forum for disseminating advances of emerging topics in Cell Biology and related subjects. All volumes are edited by established scientists and the individual chapters are written by experts on the relevant topic. The individual chapters of each volume are fully citable and indexed in Medline/Pubmed to ensure maximum visibility of the work.

Vladimir N. Uversky
Editor

(Macro)Molecular Crowding

Life of the Pottage

Springer

Editor
Vladimir N. Uversky
Department of Molecular Medicine
and USF Health Byrd Alzheimer's Research Institute
Morsani College of Medicine
University of South Florida
Tampa, FL, USA

ISSN 0306-0225 ISSN 2542-8810 (electronic)
Subcellular Biochemistry
ISBN 978-3-032-03369-7 ISBN 978-3-032-03370-3 (eBook)
https://doi.org/10.1007/978-3-032-03370-3

This Springer imprint is published by the registered company Springer Nature Switzerland AG
The registered company address is: Gewerbestrasse 11, 6330 Cham, Switzerland

Contents

Chapter 1
A Brief Historico-Philosophical Overview of Macromolecular Crowding: Making "Physiological Conditions" More Physiological

Vladimir N. Uversky

Abstract Although cellular environment represents a kind of pottage with the increased viscosity and molecular interactions, changed water activity, volume exclusion, and considerably restricted amounts of free water, most of the biomolecular research in vitro is traditionally conducted under the "physiological conditions"; i.e., in warm, slightly salted, neutral or slightly alkaline aqueous solution, where the effects of such crowded environment on the behavior of biological macromolecules are mostly ignored. Furthermore, macromolecular crowding is spatially and temporally heterogeneous, as cell contains numerous biological condensates, overcrowded liquid droplets, which, being formed in response to the changing environment, reflect complexity of the cellular molecular kitchen. This chapter represents a brief historical overview of the developments in the field of macromolecular crowding reflecting the evolution of understanding of what are the appropriate experimental conditions for gaining the most accurate information on the behavior of biological macromolecule in biological systems.

Keywords Physiological conditions · Macromolecular crowding · Excluded volume · Reductionism

It is clear that to get the most accurate observation of how a biological macromolecule would behave in a biological system, one should carefully recreate environmental features, such as the temperature, pH, ionic concentration, and other factors to closely resemble the conditions within a living cell. On the other hand, to understand physico-chemical molecular mechanisms driving the functionality of a given biological macromolecule (e.g., enzymatic activity), one would try finding a way to have an unperturbed look exclusively at a query molecule. As a result, for many

V. N. Uversky (✉)
Department of Molecular Medicine and USF Health Byrd Alzheimer's Research Institute, Morsani College of Medicine, University of South Florida, Tampa, FL, USA
e-mail: vuversky@usf.edu

V. N. Uversky (ed.), *(Macro)Molecular Crowding*, Subcellular Biochemistry 109,
https://doi.org/10.1007/978-3-032-03370-3_1

molecular biologists, molecular biochemists, or molecular biophysicists, "physiological conditions" suitable for the analysis of a biological macromolecule (enzyme) structure and function represent a warm (around 37 °C, which roughly corresponds to human body temperature or room temperature (23 °C) if non-mammalian samples are studied), slightly salted (around 150 mM NaCl, which closely resembles the salt concentration found in the extracellular fluid of most living organisms), neutral or slightly alkaline (pH ~7.0–7.8, which is a typical pH range for most bodily fluids or inside a typical bacterial, plant, or animal cell) aqueous solution containing as low concentration of a highly purified query biomolecule as accurate measurements would allow. It was pointed out that measurements conducted under such relatively ideal thermodynamic conditions would provide the characteristics of an isolated query macromolecule "in a bath of solvent containing salts, substrates, and cofactors as required" (Minton and Wilf 1981). This view is a reflection of a reductionist approach, which is traditionally used in biology, physics, and chemistry, being based on an assumption that the properties and behavior of the complex system can be explained by understanding their smaller, individual components, like molecules or cells, in isolation. In other words, the isolated molecules and their structure have sufficient explanatory power to provide an understanding of the whole complex system (Van Regenmortel 2004). This idea is rooted in the famous claim made in 1966 by Francis Harry Compton Crick (1916–2004) that "the ultimate aim of the modern movement in biology is to explain all biology in terms of physics and chemistry" (Crick 1966). This means that higher-level phenomena are defined by molecular states (Van Regenmortel 2004), which can be understood by looking at isolated molecules. To do so, biophysicist would try to use the simplest system, preferentially containing only the molecules of interest to avoid unwanted effects of additional components, whereas biochemist would try to work with the purest sample to eliminate contributions of impurities. This brings to life an aforementioned warm, slightly salted, neutral aqueous medium as "physiological conditions" suitable for the analysis of highly purified biological macromolecules. According to G Rickey Welch and James S Clegg, "The isolation of individual enzymes and the observation of "saturation kinetics" *in vitro* at the turn of the century, culminating in the Michaelis-Menten and Briggs-Haldane theoretical treatments, strengthened the conviction that the mechanistic understanding of cellular function was to be found in physical chemistry" (Welch and Clegg 2010). Therefore, it is not surprising that the reductionism has pervaded molecular biology, especially in its early days.

On the other hand, almost for two centuries, biologists have known that cellular inside represents a complex system, and cells cannot be considered as simple sacks with water. In fact, the special term "protoplasm" (from the Greek ***protos*** for *first*, and ***plasma*** for *thing formed*) was introduced in the first half of the XIX century to describe the content of the cell. In 1839, a Czech anatomist and physiologist Jan Evangelista Purkyně (also spelled as Johann Evangelist Purkinje) (1787–1869) coined term" protoplasm" for the description of the fluid substance found inside a cell, i.e., a kind of living material within the cell (Purkinje 1839; Bynum et al. 2014). Even at that time it was recognized that the protoplasm, which was also called *Primordialschlauch* (primordial utricle) by a German botanist Hugo von Mohl (1805–1872), represents "tough, slimy, granular, semi-fluid" substance within the plant cells (Larson 2012; Evert 2006; Mohl 1846). A French biologist Félix Dujardin (1801–1860) used the term "sarcode" the formless life substance he found in the *Foraminifera* (which are single-celled organisms, members of a *Rhizarian protists* phylum) (Dujardin 1835; Dujardin 1841). The importance of the substance found inside a cell is reflected in the formulation of the Protoplasm Doctrine proposed in 1861 by a German microscopic anatomist Max Johann Sigismund Schultze (1825–1874), who was the first to recognize that protoplasm represents a fundamental substance found in both plants and animals an concluded that "all living cells are made of a living substance called Protoplasm" (Max Johann Sigismund Schultze; Schultze and Schultze 1863). In 1869, an English biologist and anthropologist Thomas Henry Huxley (1825–1895) defined protoplasm as the "physical basis of life", pointing out that the distribution of molecules within this substance defines the property of life (Huxley 1870).

Therefore, it is not surprising that at early days of cellular biology, the original functionalist view of living cells was rooted not in the "cell theory" but in the so-called "protoplasmic theory of life" formulated in 1874 by John James Drysdale (1817–1892) (Drysdale and Drysdale 1874) and stating that "the physical basis of life (including the foundation of heredity) must be sought in the properties of this universal biological substance" (Welch and Clegg 2010), since by that time, protoplasm was taken as "stuff of life," "locus of life," "living essence," and "Urschleim", among other things (Welch and Clegg 2010). Notably, over the years, the cellular content was described by an impressive set of terms (http://www.enjoyed.today/Protoplasm/#Ref-20) (Battaglia 1985), reflecting constant interest of the researchers to this matter and recognition of the complexity of this system. According to the Wikipedia (https://en.wikipedia.org/wiki/Protoplasm), besides the aforementioned ***protoplasm(a)***, ***Primordialschlauch***, and ***sarcode*** these terms are (in order of their appearance in the scientific literature):

- 1854, 1867—***Hautschicht/Körnerschicht*** (***ectoplasm/endoplasm***, proposed by German botanist Nathanael Pringsheim (1823–1894) (Pringsheim 1854) and a German biologist and botanist Wilhelm Friedrich Benedikt Hofmeister (1824–1877) (Hofmeister 1867);
- 1863—***Cytoplasm*** (proposed by a Swiss anatomist, physiologist, and histologist Albert von Kölliker (born Rudolf Albert Kölliker, 1817–1905) as synonym for "protoplasm" (Kölliker 1867), however, currently, it describes the cell substance and organelles outside the nucleus;
- 1863—***Grundsubstanz*** (***ground substance***, proposed by a Russian botanist, protozoologist, and bacteriologist Lev Semyonovich Tsenkovsky, also Leon Cienkowski (822–1887) (Cienkowski 1863);
- 1868—***metaplasm/protoplasm*** (introduced by a German botanist Johannes Ludwig Emil Robert von Hanstein (1822–1880) to emphasize that "plasma is made up of two distinct portions: one of these is truly living matter, and should

alone be called protoplasm; the other is not alive, and is called metaplasm" (Darwin 1877));

- 1870—***deutoplasm/protoplasm*** (proposed by a Belgian embryologist, cytologist and marine biologist Édouard Joseph Louis Marie Van Beneden (1846–1910)) (Beneden 1870);
- 1872—***bioplasm*** (proposed by a British physician and microscopist Lionel Smith Beale (1828–1906) to describe the "essence of life" found in nucleus) (Beale 1872);
- 1875—***paraplasm/protoplasm*** (proposed by a Baltic German anatomist Karl Wilhelm Ritter von Kupffer (born Karl Wilhelm Kupffer, 1829–1902)) (Kupffer 1875);
- 1876—***inter-filar substance*** (proposed by Wilhelm Velten) (Velten 1876);
- 1877—***Hyaloplasm(a)*** (proposed by a German botanist and plant physiologist Wilhelm Friedrich Philipp Pfeffer (1845–1920) to describe the clear, fluid portion of the cytoplasm) (Pfeffer 1921);
- 1880—***Protoplast*** (coined by a German botanist Johannes Ludwig Emil Robert von Hanstein (1822–1880) to describe the entire cell, excluding the cell wall) (von Hanstein 1880);
- 1880—***Enchylema/Hyaloplasma*** (used by a German botanist Johannes Ludwig Emil Robert von Hanstein (1822–1880)) (von Hanstein 1880);
- 1882—***Kleinkörperchen*** or ***Mikrosomen*** (***small bodies*** or ***microsomes***, used by a German botanist Johannes Ludwig Emil Robert von Hanstein (1822–1880));
- 1882—***paramitome*** (proposed by a German biologist Walther Flemming (1843–1905) to describe the fluid portion of the protoplasm of a cell) (Flemming 1882);
- 1882—***nucleoplasm*** (proposed by a Polish-German plant cytologist Eduard Adolf Strasburger (1844–1912) to describe the fluid-like substance inside the nucleus of a cell where the chromatin and nucleolus are suspended; he also was a first person to proclaim: "New cell nuclei can only arise from the division of other nuclei") (Strasburger 1882);
- 1884—***Idioplasma*** (used by a Swiss botanist Carl Wilhelm von Nägeli (1817–1891) to describe a part of the protoplasm, from which cells receive their hereditary characters) (Battaglia 1985);
- 1892—***Zwischensubstanz*** (***inter-alveolar substance***, proposed by a German zoologist Johann Adam Otto Bütschli (1848–1920) to emphasize that protoplasm is composed of a two-phase system similar to an emulsion, thereby possessing an alveolar structure) (Bütschli 1892);
- 1892—***kinoplasm/trophoplasm*** (proposed by a Polish-German plant cytologist Eduard Adolf Strasburger (1844–1912) to emphasize that one can find structurally different substances in the protoplasm, the active kinoplasm in a form of the fibrils and other structures, and trophoplasm, which is mostly related to nutrition) (Strasburger);
- 1895—***Grundplasma*** (***ground plasma***, introduced by German botanist Franz Schütt (1859–1921)) (Schütt 1895);

- 1897—***ergastoplasm*** (proposed by Charles Garnier (1855–1939) to describe the "working" part of the protoplasm, endoplasmic reticulum) (Garnier 1897);
- 1899—***cytoplasmic matrix*** (described by a Norwegian-American zoologist John P. Munson (1860–1928)) (Munson 1899);
- 1901—***Protoplasma oder Zelleinschlüsse*** (***protoplasmic or cellular inclusions***, or ***deutoplasm***, proposed by a Polish histologist and embryologist of Armenian descent Władysław (Ladislaus) Szymonowicz (1869–1933) to describe various substances of different nature (fat, carbohydrates, pigment granules) found in the protoplasm but not belonging to it) (Szymonowicz 1901, 1902);
- 1931—***phaneroplasm/cryptoplasm*** (introduced by an American biologist William Seifriz (1888–1955) to describe the "visible" and "hidden" parts of protoplasm, respectively) (Seifriz 1931);
- 1956—***zytoplasmatische Matrix*** (used by a German histologist Wolfgang Bargmann (1906–1978)) (Bargmann 1956);
- 1965—***cytosol*** (coined by an American biochemist Henry A. Lardy (August—2010) to describe the aqueous fluid inside cells) (Lardy 1965).

Many researchers also recognized that protoplasm is not a homogeneous substance but have complex structure and composition, potentially containing two major phases, fluid and solid. These conclusions that were reflected in all those terms used to describe protoplasm, were originally based on data generated by light microscopy and cytochemical staining techniques. By the late 1800s it became clear that protoplasm is not homogeneous but has noticeable internal structure, with "solid" phase observed in it being described by several models, such as reticular/fibrillar (protoplasm is made up of a thread-like network and an interstitial substance), foam/alveolar (a protoplasmic network forms a number of spaces closed in on every side (small alveoli or cavities), with all these spaces being filled with fluid), and microsomal/granular (the cells consist of fine granules, which are distributed in the jelly-like intergranular substance) (Butschli 1894; Hall 1969; Welch and Clegg 1986; Seifriz 1936; Welch and Clegg 2010). As per J.G. M'Kendrick, by the 1860s, protoplasm was commonly defined as "a diaphanous semi-liquid, viscous mass, extensible but not elastic, homogenous – that is to say, without structure, without visible organisation, having in it numerous granules, and endowed with irritability and contractility" (M'Kendrick 1888).

In 1899, an American zoologist and geneticist Edmund Beecher Wilson (1856–1939) made an important observation that the researchers of the protoplasm structure often observe a fundamentally different picture of the internal structure of a cell, even when working with the same objects (Wilson 1899). He pointed out that the observed internal structure of the cell can be very different and depends on the differences in methods used for fixation and staining of the studied cells. Based on these observations, the conclusion was made that the protoplasm of the eggs of various creatures, although being differ visually, represents an emulsion consisting of a solid substance and droplets of various sizes and, apparently, of various chemical nature (Wilson 1899). Curiously, Edmund Wilson was the first to show that droplets observed within the oocytes of the sea star species have a liquid nature. Since the microspheres in the protoplasm of the oocytes of one of the sea star species were natively pigmented, he observed that after the destruction of the egg, droplets having the same color merge when colliding with each other, forming a larger droplet (coalescing) (Wilson 1899). The idea of the fluidity of the protoplasm components was also supported by the analysis of the protoplasmic emulsion conducted by a German zoologist and cell physiologist Josef Spek (1895–1964), who revealed that small protoplasmic granules can fuse to form larger liquid globules (JOSEF 1924; Spek 1930). Notably, Edmund Wilson was also the first to propose that the seemingly continuous matter of protoplasm consists of smaller objects that were inaccessible to the eyes of his contemporaries due to the limitations of the resolution of light microscopes of his time. Furthermore, he suggested that it is from these small and invisible elements the alveoli and fibrillar structures visible to the eye are formed, similar to the processes of coagulation of gelatin and albumin *in vitro* (Wilson 1899). Therefore, it was pointed out that Edmund Beecher Wilson was the first to propose considering the phase separation of biopolymers as a way of the internal organization of the cell (Antifeeva et al. 2022).

As evidenced, e.g., by the pioneering work of a Swiss botanist Albert Friedrich Frey-Wyssling (1900–1988), by the early twentieth century, it became clear that the protoplasm is characterized by the "submicroscopic structure", and that protoplasmic structures are dynamic and labile (Frey-Wyssling 1940; Frey-Wyssling 1953). This recognition of the dynamic nature of the protoplasmic structures paralleled earlier observations of Edmund Wilson and Josef Spek and indicated intricate complexity of the protoplasm. This view is supported by a picture generated by numerous modern laboratory techniques, which clearly showed that the cellular interior represents the dynamic, highly responsive, greatly integrated, and convoluted matter with the intricate arrangement of organelles, membranous structures, cytoskeletal elements, and membrane-less organelles. Note that in modern biology, the term protoplasm is rather unpopular, being considered obsolete, whereas "cytoplasm" became a popular descriptor of the cellular content. It was also pointed out that the usage of the term "protoplasm" began to decline partly due to the increasing reductionist approach of cell biologists, who became more and more focused on the analysis of structure and functionality of specific components of the cellular (or protoplasmic) composition (Welch and Clegg 2010).

This completes the circle and brings us back to the widely accepted utilization of the aforementioned highly idealistic "physiological conditions" (i.e., dilute solutions containing, in addition to the molecule under study, a dilute buffer and some salt) in studies of the isolated cellular components and biological macromolecules. Again, undoubtedly, this reductionist approach reflected, e.g., in the biophysical and biochemical studies on proteins *in vitro* was, is, and will be crucial to have highly controlled conditions. Multiple examples of the apparent validity of this approach are given by numerous studies reporting successful use of highly purified proteins for their structural characterization, as well as for the analysis of their functionality (including the famous "lock and key" hypothesis formulated in 1894 by a German chemist Hermann Emil Louis Fischer (1852–1919) (Fischer 1894; Lemieux and Spohr 1994) and claiming that the unique 3-D structure is a prerequisite to protein function, as well as the subsequent extension of this model to induced fit representation, where binding of an enzyme to the appropriate substrate is accompanied by subtle changes in a protein's active site (Koshland 1958)), interactability, folding (including the discovery that polypeptide chains of globular proteins can spontaneously form their spatial structures (Anfinsen et al. 1961; Gutte and Merrifield 1971), indicating that the native protein structure represents the most stable of all the structures of the chain under physiological conditions (Anfinsen et al. 1961; Gutte and Merrifield 1971; Anfinsen 1973b; Anfinsen and Scheraga 1975), or, in other words, that information dictating the native structure of foldable proteins is encoded in their amino acid sequences (Anfinsen 1973a), and, therefore, at optimal conditions (temperature, solvent concentration and composition, etc.), the native structure represents a unique, stable, and kinetically accessible minimum of the free energy, a famous dogma or thermodynamic hypothesis formulated by an American biochemist Christian Boehmer Anfinsen Jr. (1916–1995)) as well as protein misfolding and aggregation.

An important question is how reliably the idealistic "physiological conditions" traditionally used in biophysical and biochemical *in vitro* studies of, say, proteins, mimic the actual environment of a protein inside a cell. As it was already pointed out, intracellular environment is far from being slightly salted neutral aqueous medium, as cells are tightly packed with various biological macromolecules (proteins, protein complexes, nucleic acids, ribonucleoproteins, polysaccharides, etc.), suggesting that the intracellular environment is extremely crowded and has rather limited free space. In fact, estimates showed that the total intracellular concentration of biological macromolecules ranges from 80 to 400 mg/ml (Zimmerman and Trach 1991; van den Berg et al. 1999; Rivas et al. 2004; Ho et al. 2018; Illmer et al. 1999; Model et al. 2021; Yamada and Sgarbieri 2005), that these macromolecules occupy 5–40% of cellular volume (Ellis and Minton 2003), and that all this creates a unique crowded medium with considerably restricted amounts of free water (Zimmerman and Trach 1991; Zimmerman and Minton 1993; Fulton 1982; Minton 1997, 2000c; Ellis 2001). Although intracellular media contain billions of protein molecules (Lodish et al. 2000) and very large quantities of DNA, RNA, ribonucleoprotein, and polysaccharide molecules, typically, no individual macromolecular species is present there at very high concentrations. As a result, intracellular environment is typically described as "crowded" rather than "concentrated" (Minton 1997, 2000b, c). Looking at this very thick intracellular molecular soup, one can clearly find logical grounds of an important claim made as early as in 1877 by a German physiologist Wilhelm Friedrich Kühne (1837–1900) that an enzyme in a test tube is one thing, whereas an enzyme in the cell is another (Kuhne 1877; Gutfreund 1976). In fact, if a structure of a foldable ordered protein (and therefore its function) is determined by both its amino acid sequence and environment, then the dramatic differences between the idealistic *in vitro* modeled "physiological conditions" and the actual very complex intracellular environment could be expected to be at least somehow reflected in the protein structure and function. This is because the solvent properties of water that define the "folding power" of water are affected by the presence of all those crowders (Breydo et al. 2015a, b; Ferreira et al. 2016, 2017).

Note that these considerations are not directly related to the macromolecular crowding concept rightly promoted by an American physical biochemist Allen P. Minton (Minton 1980, 1983, 1997, 2000a, b, c; Minton and Wilf 1981). The concept is rooted in the excluded volume effects analyzed by a British biochemist Alexander George Ogston (1911–1996) in his studies of the connective tissue polysaccharide hyaluronan (Laurent and Ogston 1963; Ogston 1970) and described even earlier (in 1934) by a Swiss physical chemist Werner Kuhn (1899–1963) (Kuhn 1934), who was first to define the excluded volume of an object as the volume around this object into which the center of another similar object is not allowed to enter if overlapping of these two objects is to be avoided (Onsager 1949), and by Sho Asakura and Fumio Oosawa in their depletion theory, also known as the Asakura-Oosawa model, describing the behavior of colloidal particles in the presence of non-adsorbing polymers (Asakura and Oosawa 1954, 1958).

Because of a cell contains from 80 to 400 mg/ml of biological macromolecules and because of these macromolecules occupy very significant fraction of the cellular volume, cell interior represents a crowded milieu. As a result, the volume available to a given protein (or another biological macromolecule) inside the cell is significantly lower than the total cell volume according to the principle of excluded volume, which is the volume that is inaccessible to a query molecule as a result of the presence of other molecules in the system, and which originates from the mutual impenetrability of solute molecules that generates steric repulsion. Therefore, one should at least try to take this complexity into account, since it is clear that "the relatively ideal thermodynamic conditions typically used for the *in vitro* analysis of protein structure and function (where a protein of interest is dissolved at a low concentration in warm slightly salted water) cannot be used even as a first approximation of the "real life" intracellular protein environment with its highly packed milieu" (Kuznetsova et al. 2015).

To understand the effects of excluded volume on the behavior of a macromolecule (protein), crowding conditions should be modeled (Minton 1980, 1983, 1997, 2000c, b, a; Minton and Wilf 1981). It is accepted now that the macromolecular crowding effects on biological macromolecules of interest may be examined experimentally by using concentrated solutions of some model "crowding agents", such as poly(ethylene glycol), Dextran, Ficoll, or inert proteins (Minton 2001; Hatters et al. 2002). "If we consider macromolecular crowders as hard objects, their sheer presence limits the volume available to other macromolecules, with consequences on the conformation and interactions with other molecules" (Alfano et al. 2024). In line with these considerations, the addition of high concentration of water-soluble polymers or other macromolecular crowders was shown to affect protein solubility (Atha and Ingham 1981; Juckes 1971), conformational stability and structural properties of biological macromolecules (Minton 2000b, c; Eggers and Valentine 2001a, b; Bismuto et al. 2002; Samiotakis et al. 2009; Shahid et al. 2017), binding of small molecules (Turoverov et al. 2015), enzymatic activity (Turoverov et al. 2015), protein-protein interactions (Minton 2000b), protein compaction (Morar et al. 2001), self-association (Nichol et al. 1981), folding (Choi et al. 2013; Choi et al. 2012; Zhou 2008), as well as misfolding and aggregation (Breydo et al. 2014,

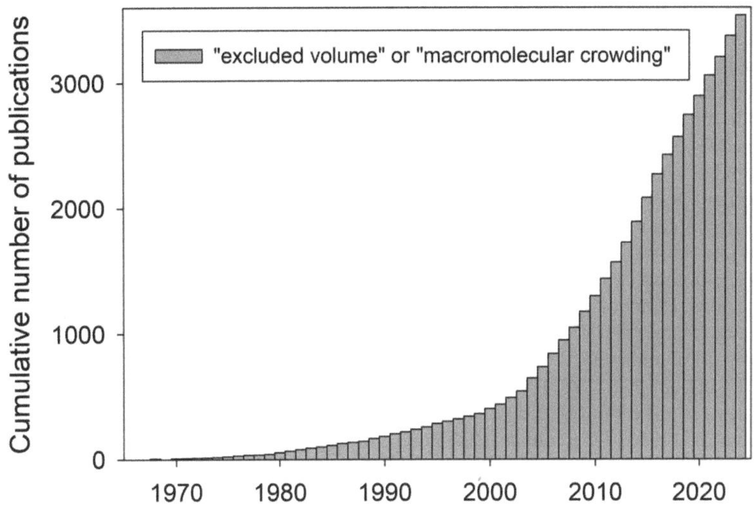

Fig. 1.1 Continuous increase in the interest of researchers to the crowding phenomenon as reflected in steady increase in the number of papers discussing "excluded volume" or "macromolecular crowding"

2015b, 2017; Kuznetsova et al. 2014; Munishkina et al. 2004, 2008; Uversky et al. 2002; Ellis 2001; Ellis and Minton 2006; Ma et al. 2013; Minton 2000b, 2005; Patel et al. 2024; Siddiqui and Naeem 2023).

With introduction of the macromolecular crowding concept paralleled by the elaboration of experimental and computational means for the analysis of corresponding model systems, molecular and cellular biologist, biochemists, and biophysicists are clearly making an important attempt to move their investigation of biological macromolecules a few steps closer to the natural physiological conditions. The current standing of this field of molecular biology is illustrated by Fig. 1.1 showing continuous increase in the number of papers discussing "excluded volume" or "macromolecular crowding". There is no need to further emphasize the importance of the macromolecular crowding concept and related excluded volume effects, as this volume is dedicated to these phenomena.

Compliance with Ethical Standards Authors declare no conflict of interest.

Funding source Not applicable.

References

Alfano C, Fichou Y, Huber K, Weiss M, Spruijt E, Ebbinghaus S, De Luca G, Morando MA, Vetri V, Temussi PA, Pastore A (2024) Molecular crowding: the history and development of a scientific paradigm. Chem Rev 124(6):3186–3219. https://doi.org/10.1021/acs.chemrev.3c00615

Anfinsen CB (1973a) Principles that govern folding of protein chains. Science 181(4096):223–230

Anfinsen CB (1973b) Principles that govern the folding of protein chains. Science 181(96):223–230
Anfinsen CB, Scheraga HA (1975) Experimental and theoretical aspects of protein folding. Adv Protein Chem 29:205–300. doi:https://doi.org/10.1016/s0065-3233(08)60413-1, Elsevier
Anfinsen CB, Haber E, Sela M, White FH Jr (1961) The kinetics of formation of native ribonuclease during oxidation of the reduced polypeptide chain. Proc Natl Acad Sci USA 47:1309–1314. https://doi.org/10.1073/pnas.47.9.1309
Antifeeva IA, Fonin AV, Fefilova AS, Stepanenko OV, Povarova OI, Silonov SA, Kuznetsova IM, Uversky VN, Turoverov KK (2022) Liquid-liquid phase separation as an organizing principle of intracellular space: overview of the evolution of the cell compartmentalization concept. Cell Mol Life Sci 79(5):251. https://doi.org/10.1007/s00018-022-04276-4
Asakura S, Oosawa F (1954) On interaction between two bodies immersed in a solution of macromolecules. J Chem Phys 22(7):1255–1256
Asakura S, Oosawa F (1958) Interaction between particles suspended in solutions of macromolecules. J Polym Sci 33(126):183–192
Atha DH, Ingham KC (1981) Mechanism of precipitation of proteins by polyethylene glycols. Analysis in terms of excluded volume. J Biol Chem 256(23):12108–12117
Bargmann W (1956) Über einige Probleme und Ergebnisse des elektronen-mikroskopischen Studiums der Zelle. DMW-Deutsche Medizinische Wochenschrift 81(28):1109–1113
Battaglia E (1985) Meiosis and mitosis: a terminological criticism. Annali di Botanica (Rome) 43:101–140
Beale LS (1872) Bioplasm: an introduction to the study of physiology&medicine. With numerous illustrations. J. & A. Churchill
Beneden E (1870) van, Recherches sur la composition et la signification de l'ceuf. Mémoires couronnés de l'Académie Royale de Belgique 34:1867–1870
Bismuto E, Martelli PL, De Maio A, Mita DG, Irace G, Casadio R (2002) Effect of molecular confinement on internal enzyme dynamics: frequency domain fluorometry and molecular dynamics simulation studies. Biopolymers 67(2):85–95
Breydo L, Reddy KD, Piai A, Felli IC, Pierattelli R, Uversky VN (2014) The crowd you're in with: effects of different types of crowding agents on protein aggregation. Biochim Biophys Acta 1844 2:346–357. https://doi.org/10.1016/j.bbapap.2013.11.004
Breydo L, Sales AE, Ferreira L, Fedotoff O, Shevelyova MP, Permyakov SE, Kroeck KG, Permyakov EA, Zaslavsky BY, Uversky VN (2015a) Effects of osmolytes on protein-solvent interactions in crowded environment: analyzing the effect of TMAO on proteins in crowded solutions. Arch Biochem Biophys 570:66–74. https://doi.org/10.1016/j.abb.2015.02.021
Breydo L, Sales AE, Frege T, Howell MC, Zaslavsky BY, Uversky VN (2015b) Effects of polymer hydrophobicity on protein structure and aggregation kinetics in crowded milieu. Biochemistry 54(19):2957–2966. https://doi.org/10.1021/acs.biochem.5b00116
Breydo L, Redington JM, Uversky VN (2017) Effects of intrinsic and extrinsic factors on aggregation of physiologically important intrinsically disordered proteins. Int Rev Cell Mol Biol 329:145–185. doi:https://doi.org/10.1016/bs.ircmb.2016.08.011, Elsevier
Bütschli O (1892) Untersuchungen über mikroskopische Schäume und das Protoplasma. Engelmann
Butschli O (1894) Investigations on microscopic foams and on protoplasm
Bynum WF, Browne J, Porter R (2014) Dictionary of the history of science, vol 533. Princeton University Press
Choi SI, Son A, Lim KH, Jeong H, Seong BL (2012) Macromolecule-assisted de novo protein folding. Int J Mol Sci 13(8):10368–10386. https://doi.org/10.3390/ijms130810368
Choi SI, Kwon S, Son A, Jeong H, Kim KH, Seong BL (2013) Protein folding in vivo revisited. Curr Protein Pept Sci 14(8):721–733. https://doi.org/10.2174/1389203711408131227170544
Cienkowski L (1863) Zur Entwicklungsgeschichte der Myxomyceten. Jahrb Wiss Bot 3:325–337
Crick F (1966) Of molecules and men
Darwin F (1877) On the protrusion of protoplasmic filaments from the glandular hairs on the leaves of the common teasel (Dipsacus sjlvestris). J Cell Sci 2(67):245–272

Drysdale JGS, Drysdale JJ (1874) The protoplasmic theory of life. Bailliere, Tindall & Cox
Dujardin F (1835) Recherches sur les organismes inférieurs. Ann Sci Nat 4
Dujardin F (1841) Histoire naturelle des zoophytes: Infusoires, comprenant la physiologie et la classification de ces animaux, et la manière de les étudier à l'aide du microscope: Ouvrage accompagné de planches, vol 2. Roret
Eggers DK, Valentine JS (2001a) Crowding and hydration effects on protein conformation: a study with sol-gel encapsulated proteins. J Mol Biol 314(4):911–922
Eggers DK, Valentine JS (2001b) Molecular confinement influences protein structure and enhances thermal protein stability. Protein Sci 10(2):250–261
Ellis RJ (2001) Macromolecular crowding: obvious but underappreciated. Trends Biochem Sci 26(10):597–604
Ellis RJ, Minton AP (2003) Cell biology: join the crowd. Nature 425(6953):27–28. https://doi.org/10.1038/425027a
Ellis RJ, Minton AP (2006) Protein aggregation in crowded environments. Biol Chem 387(5):485–497. https://doi.org/10.1515/BC.2006.064
Evert RF (2006) Esau's plant anatomy: meristems, cells, and tissues of the plant body: their structure, function, and development. Wiley
Ferreira LA, Madeira PP, Breydo L, Reichardt C, Uversky VN, Zaslavsky BY (2016) Role of solvent properties of aqueous media in macromolecular crowding effects. J Biomol Struct Dyn 34(1):92–103. https://doi.org/10.1080/07391102.2015.1011235
Ferreira LA, Uversky VN, Zaslavsky BY (2017) Role of solvent properties of water in crowding effects induced by macromolecular agents and osmolytes. Mol BioSyst 13(12):2551–2563. https://doi.org/10.1039/c7mb00436b
Fischer E (1894) Einfluss der configuration auf die wirkung der enzyme. Ber Dt Chem Ges 27:2985–2993
Flemming W (1882) Zellsubstanz, kern und zelltheilung. Vogel
Frey-Wyssling A (1940) The submiceoscopic structure of the cytoplasm. J R Microsc Soc 60(3):128–139
Frey-Wyssling A (1953) Submicroscopic morphology of protoplasm
Fulton AB (1982) How crowded is the cytoplasm? Cell 30(2):345–347
Garnier C (1897) Les filaments basaux des cellules glandulaires. Note préliminaire, Berger-Levrault
Gutfreund H (1976) Wilhelm Friedrich Kuhne: an appreciation. FEBS Lett 62 (Suppl):E1–12. https://doi.org/10.1016/0014-5793(76)80846-0
Gutte B, Merrifield RB (1971) The synthesis of ribonuclease a. J Biol Chem 246(6):1922–1941
Hall TS (1969) Ideas of life and matter: studies in the history of general physiology, 600 BC-1900 AD
Hatters DM, Minton AP, Howlett GJ (2002) Macromolecular crowding accelerates amyloid formation by human apolipoprotein C-II. J Biol Chem 277(10):7824–7830
Ho B, Baryshnikova A, Brown GW (2018) Unification of protein abundance datasets yields a quantitative Saccharomyces cerevisiae proteome. Cell Syst 6 (2):192-205 e193. https://doi.org/10.1016/j.cels.2017.12.004
Hofmeister W (1867) Die Lehre von der pflanzenzelle, vol 1. Engelmann, Engelmann
Huxley TH (1870) On the physical basis of life, vol 1. CC Chatfield
Illmer P, Erlebach C, Schinner F (1999) A practicable and accurate method to differentiate between intra- and extracellular water of microbial cells. FEMS Microbiol Lett 178(1):135–139. https://doi.org/10.1111/j.1574-6968.1999.tb13769.x
Josef S (1924) Neue beiträge zum problem der plasmastrukturen. Von Zeitschrift für wissenschaftliche Biologie: Zeitschrift für Zellen-und Gewebelehre Abteilung B 1:278
Juckes IR (1971) Fractionation of proteins and viruses with polyethylene glycol. Biochim Biophys Acta 229(3):535–546
Kölliker A (1867) Handbuch der gewebelehre des menschen. Wilhelm Engelmann
Koshland DE (1958) Application of a theory of enzyme specificity to protein synthesis. Proc Natl Acad Sci USA 44(2):98–104

Kuhn W (1934) Über die gestalt fadenförmiger moleküle in lösungen. Kolloid-Zeitschrift 68:2–15
Kuhne W (1877) Über das Verhalten verschiedener organisirter und sog. ungeformter Fermente. Verhandlungen Naturhistorisch-Med Ver Zu Heidelber 1:190–193
Kupffer KW (1875) Ueber Differenzierung des Protoplasmas an den Zellen tierischer Gewebe. Schriften des Naturwissenschaft Vereins für Schleswig-Holstein 1(3):229–242
Kuznetsova IM, Turoverov KK, Uversky VN (2014) What macromolecular crowding can do to a protein. Int J Mol Sci 15(12):23090–23140. https://doi.org/10.3390/ijms151223090
Kuznetsova IM, Zaslavsky BY, Breydo L, Turoverov KK, Uversky VN (2015) Beyond the excluded volume effects: mechanistic complexity of the crowded milieu. Molecules 20(1):1377–1409. https://doi.org/10.3390/molecules20011377
Lardy HA (1965) On the direction of pyridine nucleotide oxidation-reduction reactions in gluco-neogensis and lipogenesis. In: Control of energy metabolism. Elsevier, pp 245–248
Larson PR (2012) The vascular cambium: development and structure. Springer
Laurent TC, Ogston AG (1963) The interaction between polysaccharides and other macromolecules. 4. The osmotic pressure of mixtures of serum albumin and hyaluronic acid. Biochem J 89(2):249–253. https://doi.org/10.1042/bj0890249
Lemieux UR, Spohr U (1994) How Emil Fischer was led to the lock and key concept for enzyme specificity. Adv Carbohydrate Chem Biochem 50:1–20, Elsevier
Lodish H, Berk A, Zipursky SL, Matsudaira P, Baltimore D, Darnell J (2000) Molecular cell biology, 4th edn. W. H. Freeman, New York
M'Kendrick J (1888) On the modern cell theory. In: Proceedings of the Royal Philosophical Society of Glasgow. pp 1–55
Ma Q, Hu JY, Chen J, Liang Y (2013) The role of crowded physiological environments in prion and prion-like protein aggregation. Int J Mol Sci 14(11):21339–21352. https://doi.org/10.3390/ijms141121339
Max Johann Sigismund Schultze. https://doi.org/10.1093/oi/authority.20110803100447653
Minton AP (1980) Excluded volume as a determinant of protein structure and stability. Biophys J 32(1):77–79. https://doi.org/10.1016/S0006-3495(80)84917-4
Minton AP (1983) The effect of volume occupancy upon the thermodynamic activity of proteins: some biochemical consequences. Mol Cell Biochem 55(2):119–140. https://doi.org/10.1007/BF00673707
Minton AP (1997) Influence of excluded volume upon macromolecular structure and associations in 'crowded' media. Curr Opin Biotechnol 8(1):65–69
Minton AP (2000a) Effect of a concentrated "inert" macromolecular cosolute on the stability of a globular protein with respect to denaturation by heat and by chaotropes: a statistical-thermodynamic model. Biophys J 78(1):101–109. https://doi.org/10.1016/S0006-3495(00)76576-3
Minton AP (2000b) Implications of macromolecular crowding for protein assembly. Curr Opin Struct Biol 10(1):34–39
Minton AP (2000c) Protein folding: thickening the broth. Curr Biol 10(3):R97–R99
Minton AP (2001) The influence of macromolecular crowding and macromolecular confinement on biochemical reactions in physiological media. J Biol Chem 276(14):10577–10580
Minton AP (2005) Influence of macromolecular crowding upon the stability and state of association of proteins: predictions and observations. J Pharm Sci 94(8):1668–1675. https://doi.org/10.1002/jps.20417
Minton AP, Wilf J (1981) Effect of macromolecular crowding upon the structure and function of an enzyme: glyceraldehyde-3-phosphate dehydrogenase. Biochemistry 20(17):4821–4826. https://doi.org/10.1021/bi00520a003
Model MA, Hollembeak JE, Kurokawa M (2021) Macromolecular crowding: a hidden link between cell volume and everything Else. Cell Physiol Biochem 55(S1):25–40. https://doi.org/10.33594/000000319
Mohl H (1846) über die Saftbewegung im Innern der Zellen. Bot Ztg 4:73–78

Morar AS, Olteanu A, Young GB, Pielak GJ (2001) Solvent-induced collapse of alpha-synuclein and acid-denatured cytochrome c. Protein Sci 10(11):2195–2199
Munishkina LA, Cooper EM, Uversky VN, Fink AL (2004) The effect of macromolecular crowding on protein aggregation and amyloid fibril formation. J Mol Recognit 17(5):456–464. https://doi.org/10.1002/jmr.699
Munishkina LA, Ahmad A, Fink AL, Uversky VN (2008) Guiding protein aggregation with macromolecular crowding. Biochemistry 47(34):8993–9006. https://doi.org/10.1021/bi8008399
Munson JP (1899) The ovarian egg of limulus. Athenaeum Press, Boston
Nichol LW, Ogston AG, Wills PR (1981) Effect of inert polymers on protein self-association. FEBS Lett 126(1):18–20. https://doi.org/10.1016/0014-5793(81)81022-8
Ogston A (1970) On the interaction of solute molecules with porous networks. J Phys Chem 74(3):668–669
Onsager L (1949) The effects of shape on the interaction of colloidal particles. Ann N Y Acad Sci 51(4):627–659
Patel SP, Nikam T, Sreepathi B, Karankar VS, Jaiswal A, Vardhan SV, Rana A, Toga V, Srivastava N, Saraf SA, Awasthi S (2024) Unraveling the molecular jam: how crowding shapes protein aggregation in neurodegenerative disorders. ACS Chem Biol 19(10):2118–2130. https://doi.org/10.1021/acschembio.4c00365
Pfeffer W (1921) Osmotische untersuchungen: studien zur zellmechanik. W. Engelmann
Pringsheim N (1854) Untersuchungen über den Bau und die Bildung der Pflanzenzelle, vol 1, Hirschwald
Purkinje J (1839) Uber die Analogien in den Strukturelementen des thierischen and pflanzlichen Organismus. Ȗbersicht der Arbeiten und Verȧnderungen der Schlesischen Gesellschaft fçr Vaterlȧndische Cultur im Jahre:81–82
Rivas G, Ferrone F, Herzfeld J (2004) Life in a crowded world. EMBO Rep 5(1):23–27. https://doi.org/10.1038/sj.embor.7400056
Samiotakis A, Wittung-Stafshede P, Cheung MS (2009) Folding, stability and shape of proteins in crowded environments: experimental and computational approaches. Int J Mol Sci 10(2):572–588. https://doi.org/10.3390/ijms10020572
Schultze M, Schultze MJS (1863) Das Protoplasma der Rhizopoden und der Pflanzenzellen: ein Beitrag zur Theorie der Zelle. W. Engelmann
Schütt F (1895) Die peridineen der plankton-expedition, vol 1. Lipsius & Tischer
Seifriz W (1931) The structure of protoplasm. Science 73(1902):648–649
Seifriz W (1936) Protoplasm
Shahid S, Hassan MI, Islam A, Ahmad F (2017) Size-dependent studies of macromolecular crowding on the thermodynamic stability, structure and functional activity of proteins: in vitro and in silico approaches. Biochim Biophys Acta Gen Subj 1861(2):178–197. https://doi.org/10.1016/j.bbagen.2016.11.014
Siddiqui GA, Naeem A (2023) Connecting the dots: macromolecular crowding and protein aggregation. J Fluoresc 33(1):1–11. https://doi.org/10.1007/s10895-022-03082-2
Spek J (1930) Zustandsänderungen der Plasmakolloide bei Befruchtung und Entwicklung des Nereis-Eies. Protoplasma 9(1):370–427
Strasburger E (1882) Ueber den Theilungsvorgang der Zellkerne und das Verhältniss der Kerntheilung zur Zelltheilung, vol 21. Cohen
Strasburger EA Complete dictionary of scientific biography
Szymonowicz L (1901) Lehrbuch der Histologie und der mikroskopischen Anatomie. Stuber
Szymonowicz L (1902) A text-book of histology and microscopic anatomy of the human body: including microscopic technique. Lea Bros
Turoverov KK, Kuznetsova IM, Uversky VN (2015) What macromolecular crowding can do to a protein. Int J Mol Sci 15:23090–23140
Uversky VN, Elisa MC, Bower KS, Li J, Fink AL (2002) Accelerated alpha-synuclein fibrillation in crowded milieu. FEBS Lett 515(1–3):99–103. https://doi.org/10.1016/s0014-5793(02)02446-8

van den Berg B, Ellis RJ, Dobson CM (1999) Effects of macromolecular crowding on protein folding and aggregation. EMBO J 18(24):6927–6933. https://doi.org/10.1093/emboj/18.24.6927
Van Regenmortel MH (2004) Reductionism and complexity in molecular biology. Scientists now have the tools to unravel biological and overcome the limitations of reductionism. EMBO Rep 5(11):1016–1020. https://doi.org/10.1038/sj.embor.7400284
Velten W (1876) Die physikalische Beschaffenhoit des pflanzlichon Protoplasmas. Sitzber Akad Wiss Wien, Math-Nat Kl 73:131–151
von Hanstein JLER (1880) Das Protoplasma als Träger der pflanzlichen und thierischen Lebensverrichtungen für Laien und Fachgenossen. Selbstverlag
Welch GR, Clegg JS (1986) The organization of cell metabolism: a historical vignette. In: The organization of cell metabolism. Springer, pp 1–5
Welch GR, Clegg JS (2010) From protoplasmic theory to cellular systems biology: a 150-year reflection. Am J Physiol Cell Physiol 298(6):C1280–C1290. https://doi.org/10.1152/ajpcell.00016.2010
Wilson EB (1899) The structure of protoplasm. Science 10(237):33–45
Yamada EA, Sgarbieri VC (2005) Yeast (Saccharomyces cerevisiae) protein concentrate: preparation, chemical composition, and nutritional and functional properties. J Agric Food Chem 53(10):3931–3936. https://doi.org/10.1021/jf0400821
Zhou HX (2008) Protein folding in confined and crowded environments. Arch Biochem Biophys 469(1):76–82. https://doi.org/10.1016/j.abb.2007.07.013
Zimmerman SB, Minton AP (1993) Macromolecular crowding: biochemical, biophysical, and physiological consequences. Annu Rev Biophys Biomol Struct 22:27–65
Zimmerman SB, Trach SO (1991) Estimation of macromolecule concentrations and excluded volume effects for the cytoplasm of Escherichia coli. J Mol Biol 222(3):599–620

Chapter 2
Effects of Molecular Crowding on the Structural, Dynamic, and Functional Properties of Biological Macromolecules: A General Overview

Altijana Hromić-Jahjefendić, Hana Krnjić, and Vladimir N. Uversky

Abstract Intracellular environment is characterized by high content of various biomacromolecules, limited amount of free water, and almost complete lack of unoccupied space. Since such extremely crowded milieu is poorly modeled by slightly salted aqueous solutions of low concentrations of a biomolecule of interest, in the laboratory practice, it is typically mimicked by concentrated solutions of various polymers or inert proteins that serve as model "crowding agents". Analysis of the behavior of different proteins under these conditions that model the effects of macromolecular crowding *in vitro* revealed that protein structure, folding, shape, conformational stability, binding of small molecules, enzymatic activity, protein-protein interactions, protein-nucleic acid interactions, and pathological aggregation are all affected by the presence of crowders. The goal of this chapter is to provide an overview of the variety of effects of macromolecular crowding on a protein molecule.

Keywords Macromolecular crowding · Excluded volume · Protein structure · Protein folding · Protein function · Protein-protein interaction · Intrinsically disordered protein · Protein aggregation

A. Hromić-Jahjefendić (✉) · H. Krnjić
Department of Genetics and Bioengineering, Faculty of Engineering and Natural Sciences, International University of Sarajevo, Sarajevo, Bosnia and Herzegovina
e-mail: ahromic@ius.edu.ba

V. N. Uversky (✉)
Department of Molecular Medicine and USF Health Byrd Alzheimer's Research Institute, Morsani College of Medicine, University of South Florida, Tampa, FL, USA
e-mail: vuversky@usf.edu

V. N. Uversky (ed.), *(Macro)Molecular Crowding*, Subcellular Biochemistry 109,
https://doi.org/10.1007/978-3-032-03370-3_2

The Essence of Macromolecular Crowding

Cells are not simple water-filled membranous sacks, being instead complex systems, whose interior represents an intricate milieu tightly packed with various biological macromolecules, such as proteins, protein complexes, nucleic acids, ribonucleoproteins, polysaccharides, etc., which are estimated to be present in cells at concentrations between 80 to 400 mg/mL (Rivas et al. 2004; van den Berg et al. 1999; Zimmerman and Trach 1991). Furthermore, these macromolecules occupy 5–40% of cellular volume (Ellis and Minton 2003), thereby creating a unique crowded medium with considerably restricted amounts of free water (Ellis 2001; Fulton 1982; Minton 1997, 2000b; Zimmerman and Minton 1993; Zimmerman and Trach 1991). Such congested intracellular media is typically described as "crowded" rather than "concentrated" (Minton 1997, 2000b, a), since despite the fact that cells contain billions of protein molecules (Lodish et al. 2000) and very large quantities of DNA, RNA, ribonucleoprotein, and polysaccharide molecules, none of those individual macromolecular species is present there at very high concentrations. Such crowded medium resembles very thick molecular soup, being characterized by a high volume occupancy, where the average distances between macromolecules could be noticeably smaller than the size of the macromolecules themselves (Homouz et al. 2008). As a result, the excluded volume (i.e., the volume occupied by macromolecules and unavailable to other solutes) emerges, since two (macro) molecules cannot occupy the same space in solution, as steric hindrance or impediment of a macromolecule is expected to exclude other molecules from its neighborhood (Ralston 1990). In solutions with increasing concentrations of such particles, the number of ways that can be used to place added molecules is progressively limited, since the volume of solution available to the new molecules is progressively restricted to the part of space, from which they are not excluded (Ralston 1990). The consequence of this phenomenon is decreased randomness of the particle distribution in the concentrated solutions. This results in the noticeable decrease in the entropy of the crowded solution in comparison with the entropy of ideal solution. Such entropy decrease increases the free energy of the solute and thereby produces the increase in the thermodynamic activity of solute, and therefore is expected to affect various processes determined by the activity (Ralston 1990). Importantly, the excluded volume effects do not only influence the thermodynamic activity of the concentrated solute itself but also affect any other solute present in the solution at low concentrations (Ralston 1990). Therefore, the behavior of a molecule of interest in the non-ideal crowed solutions is expected to be different from the behavior of this molecule in the diluted solutions (Ralston 1990). Consequently, such crowed environments might significantly affect any reaction the course of which is dependent on the available volume (Minton 2000a, Minton 2005; Zimmerman and Minton 1993). Furthermore, such crowded environment is expected to place constraints on the active factors present in the microenvironment (Chen et al. 2011; Ellis 2001; Minton 2001). The thermodynamic consequences of this unavailable volume are described in terms of excluded volume effects (Minton 2001; Zimmerman and

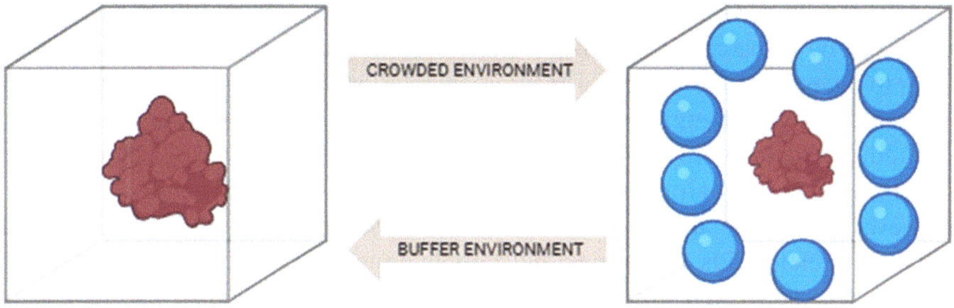

Fig. 2.1 Protein aggregation in crowded environment *vs.* buffer environment (generated using BioRender)

Minton 1993). In fact, volume exclusion was demonstrated to have a significant impact on the structural characteristics and conformational stability of biological macromolecules using a variety of *in vitro* systems under the model crowded settings (Bismuto and Irace 2001; Bismuto et al. 2002). Additionally, different biological equilibria may be impacted by macromolecular crowding. The graphical view of protein behavior in buffer and crowded environment can be seen from Fig. 2.1.

How One Can Model the Macromolecular Crowding and Confinement *In Vitro*?

Based on these considerations, it is obvious that the relatively ideal thermodynamic conditions typically used for the *in vitro* analysis of protein structure and function (where a protein of interest is dissolved at a low concentration in warm slightly salted water) cannot be used even as a first approximation of the "real life" intracellular protein environment with its highly packed milieu, where said biomolecule of interest scrambles and wades through the tightly packed crowd. However, there is an increasing appreciation of this shortcoming, and the view that the macromolecular crowding is an important phenomenon that should be taken into account in biochemical studies is gaining attention (Ellis 2001; Zimmerman and Minton 1993). It is accepted now that the macromolecular crowding effects on biological macromolecules of interest may be examined experimentally by using concentrated solutions of some model "crowding agents", such as poly(ethylene glycol), Dextran, Ficoll, or "inert" proteins (Bismuto and Irace 2001; Bismuto et al. 2002; Hatters et al. 2002). Although concentrated solutions of such polymers can mimic macromolecular crowding characterized by high excluded volume, it is important to remember that the efficiency of the artificial macromolecular crowding is expected to depend on a ratio between the hydrodynamic dimensions (or occupied volumes) of the crowder and the test molecule, with the most effective conditions being those where the volumes occupied by a crowding agent and a test molecule are comparable (Chen et al. 2011; Minton 1993; Tokuriki et al. 2004).

Artificial crowding agents (such as PEG, Dextran or Ficoll) are commonly employed by researchers *in vitro* to simulate intracellular conditions. This enables them to investigate the effects of macromolecular crowding on the protein crystallization that are challenging to crystallize under typical circumstances.

Traditionally, *in vitro* protein folding has been thoroughly investigated in diluted solutions under the aforementioned relatively ideal thermodynamic conditions. However, due to the abundance of soluble and insoluble macromolecules, such as proteins, nucleic acids, ribosomes, and carbohydrates, the intracellular environment is extremely congested. This indicates that other macromolecular species cannot access a sizable portion of the intracellular area. It has been suggested that crowding indirectly stabilizes the folded protein by compacting the denatured states. High concentrations of inert natural or synthetic macromolecules, known as crowding agents, can be added to the systems in vitro to simulate crowding experimentally. Crowding has been shown to have significant influence on the thermodynamics and kinetics of numerous biological processes, such as protein binding, folding, and aggregation, in both experimental and theoretical work (Cheung et al. 2005; Uversky et al. 2002; van den Berg et al. 2000).

Large, complex proteins (i.e., multidomain and/or disulfide containing) and frequently extreme solvent conditions (such as acidic pH) have been the focus of experimental crowding studies in the past. For instance, the impact of crowding agents on the oligomerization of GroEL subunits (Galan et al. 2001), the refolding of reduced denatured lysozyme (van den Berg et al. 2000), the amyloid formation of the human apolipoprotein C-II (Hatters et al. 2002), the self-assembly of the HIV capsid protein and cell division protein FtsZ (del Alamo et al. 2005; Rivas et al. 2001), and the refolding of reduced denatured lysozyme (van den Berg et al. 2000).

The impact of macromolecular crowding on the function of tiny single-domain proteins that fold in diluted liquids by straightforward methods hasn't been thoroughly investigated, though. Stagg et al. (2007) reported a study that combined computational and experimental methods to carefully evaluate the native and denatured structural ensembles, as well as the thermal stability, of a well-behaved α/β model protein (apoflavodoxin) as a function of concentration of an inert spherical macromolecular crowding agent Ficoll 70 (Stagg et al. 2007). This study aimed to provide additional information about this kind of mechanism. Through structural enhancement in both the folded and denatured ensembles of molecules, they showed that crowding improves protein-thermal stability (Stagg et al. 2007).

Ficoll 70 was used as the crowding agent for this project. This drug has a lot of benefits; it is a polar, inert polysaccharide that doesn't interact with proteins (sucrose epichlorohydrin copolymer; average molecular mass of 74 kDa). It exhibits the characteristics of a semirigid sphere with a radius of about 55 °C (Venturoli and Rippe 2005). It is a pleasing imitation of globular macromolecules that could exist in the biological environment where proteins fold naturally.

The authors used far-UV circular dichroism (CD) at 222 nm to investigate the effect of Ficoll 70 on the thermal stability of apoflavodoxin. The thermal midpoint (T_m) of apoflavodoxin increases dramatically from 45 °C to 65 °C in Hepes buffer (pH 7), showing improved thermal stability, as the concentration of Ficoll 70

(0–400 mg/ml) rises. The responses don't depend on scan rate and are cooperative and reversible. Protein stability is also influenced by buffer composition; apoflavodoxin is more stable in phosphate buffer than in Hepes. The protein is continuously stabilized by Ficoll 70, and its effects are stronger in buffer conditions that are less stable (Stagg et al. 2007). Moreover, the effects of Ficoll 70 on apoflavodoxin's native state have been investigated in the study. The protein's secondary structure is enhanced by increasing Ficoll 70 concentrations (0–400 mg/ml), as seen by a 10–15% increase in the negative signal at 222 nm obtained by far-UV CD spectra. Ficoll 70 causes a 20% increase in helical content and a more than 10% decrease in random coil content. The native-state structure of apoflavodoxin in solution is altered by these modifications to more closely resemble the crystal structure. Holoflavodoxin in both folded and unfolded forms by applying Dextran 70 showed similar effect (Stagg et al. 2007). Apoflavodoxin's structural organization and thermal stability have been improved by applying Ficoll 70. The stabilizing impact of apoflavodoxin is highlighted by the fact that its thermal midpoint (T_m) rises with an increase in Ficoll 70 concentration, especially under less stable buffer conditions. These results imply that proteins can be stabilized by crowding agents such as Ficoll 70 and Dextran 70, which improve structural and thermal stability.

In a different study (Gagarskaia et al. 2017) the impact of two crowding agents, Dextran-70 k and PEG-8 k, with distinct chemical compositions and hydrodynamic dimensions on the structural characteristics and conformational stability of actin have been examined. While the Dextran-70 k molecules' dimensions are significantly greater than those of globular actin and are like those of I-actin, the hydrodynamic dimensions of PEG-8 k molecules are comparable to those of globular actin. These two inert polymers in high concentrations were used to mimic the crowded environment of a living cell. The two crowding agents were able to stabilize the native structure of actin by delaying the denaturation process. Compared to Dextran-70 k, PEG-8 k showed a significantly stronger stabilizing impact. By keeping actin stable in its dormant condition, both substances stopped the protein from unfolding completely (Gagarskaia et al. 2017). The summary of this example can be seen in Table 2.1.

Another important side of the intracellular crowded environment is molecular confinement. It was pointed out that there is a principle difference between crowding and confinement phenomena, although the two terms are often used interchangeably. In fact, "confinement refers to a space limitation and the constraints determined by it. Crowding refers to a more dynamic situation, where molecules are restrained by the presence of many others. The two concepts are certainly related but do not entirely overlap: proteins can for instance be crowded both in an organelle but also in the bloodstream, much as in a very wide space people could group in a specific area. On the other hand, molecules are certainly confined in a small organelle, without necessarily being under crowding conditions" (Alfano et al. 2024). Confinement can be modeled by encapsulating a query protein in silica glass using the sol-gel techniques and encapsulating it into the reverse micelles. The sol-gel encapsulation is considered as a reliable model of the molecular confinement despite the fact that the fraction of the total volume excluded by the silica matrix is smaller than the

Table 2.1 Properties and effects of crowding agents, characteristics, and specific impacts on protein stability

Crowding agent	Molecular weight	Key characteristics	Effects on protein stability	Specific findings	References
Ficoll 70	74 kDa	Inert, polar, behaves like a semirigid sphere (≈55 Å radius)	Enhances thermal stability and structural organization of proteins	Increases thermal midpoint (T_m) of apoflavodoxin from 45 °C to 65 °C with concentrations from 0–400 mg/ml Enhances secondary structure, increasing helical content by up to 20%	Stagg et al. (2007)
Dextran 70	70 kDa	Larger than Ficoll 70, used to simulate crowded environments	Stabilizes native structure of proteins; slows down denaturation processes	Similar stabilizing effects observed as with Ficoll 70 Maintains stability of actin in its inactive state, preventing full unfolding	Gagarskaia et al. (2017)
PEG (Polyethylene Glycol)	Varies	Hydrodynamic dimensions comparable to globular proteins	Stabilizes proteins and affects folding kinetics; stronger stabilizing effect than Dextran in some cases	PEG-8 k had a much stronger stabilizing effect on actin compared to Dextran 70 k, maintaining stability in its inactive state while slowing down denaturation	Gagarskaia et al. (2017)

fractional volume occupied by macromolecules in living cell (Eggers and Valentine 2001a, b). However, since the protein itself may dictate the pore size during the gelation process, the sizes of protein-occupied pores in these gels have the same order of magnitude as the diameter of a protein (Lan et al. 1999). Although proteins are not bound covalently to the highly porous silica matrix, and although encapsulated macromolecules are unable to escape the glass under most solvent conditions (Eggers and Valentine 2001b), the sol-gel glass matrix does substantially limit the rotational freedom of the protein (Gottfried et al. 1999) not limiting the exchange of the solvent that freely permeates the silica matrix. Another important advantage of this approach is the optical transparency of the resulting glass products that allows application of the majority of spectroscopic techniques used to monitor the structure of proteins in dilute solutions (Bismuto and Irace 2001; Bismuto et al. 2002;

Brennan 1999; Eggers and Valentine 2001a, b). Using this approach, the solvent effects on the secondary structure of several ordered proteins have been analyzed following their encapsulation in the hydrated pores of a silica glass matrix by the sol-gel method (Eggers and Valentine 2001a, b). For several globular proteins, it has been shown that the conditions resulting from the sol-gel glass encapsulation do not dramatically affect their structure and function (Brennan 1999; Dave et al. 1994; Gill and Ballestros 2000).

Alternative or complementary approach to study the effect of molecular confinement on the dynamics of both a target protein and solvent molecules is the encapsulation of the protein of interest inside a reverse micelle and dispersing the resulting particle in a low viscosity fluid (Babu et al. 2003; Roche et al. 2013; Ehrhardt et al. 1999; Peterson et al. 2005; Pometun et al. 2006). Reverse micelles are unique because they represent nano-size pockets of encapsulated water separated from an organic medium by a surfactant (Davis et al. 1996; Grandi et al. 1981; Pometun et al. 2006). By varying the ratio of water to the concentration of the surfactant in the organic solvent one can tune the size of the reverse micelles and the resulting degree of protein/peptide hydration inside them (Christ and Schurtenberger 1994; Roche et al. 2013). Adding water increases the size of the micelle, permitting an examination of the effect of a systematic enhancement of the number of hydration waters surrounding a confined protein (Grandi et al. 1981; Mukherjee et al. 2006). Similar to the sol-gel glass encapsulation, structural properties of proteins in reverse micelles can be analyzed by various spectroscopic techniques (Babu et al. 2003; Roche et al. 2013; Davis et al. 1996; Ehrhardt et al. 1999; Peterson et al. 2005; Pometun et al. 2006).

Protein and peptides can also be encapsulated in mesoporous silica with tunable pore diameter and easily functionalized surface (Ravindra et al. 2004; Siefker et al. 2014), in silica tubes (Mandal et al. 2014) or carbon nanotubes (O'Brien et al. 2008), inside the cavities of polyacrylamide gels (Bolis et al. 2004; Pastore et al. 2007; Sanfelice et al. 2009, 2013) or weakly hydrophobic pores of methacrylate resin (Radhakrishna et al. 2013), or in self-assembled organic nanotubes (Kameta et al. 2010).

It was indicated that "although all aforementioned encapsulation techniques provide suitable conditions for studying the effects of molecular confinement on different aspect of protein structure and dynamics and on the solvent dynamics and properties, the use of solutions with high concentrations of inert polymers represents the most common approach to model the various effects of macromolecular crowding" (Kuznetsova et al. 2015).

Crowding Effects on Protein Behavior

An oversimplified view of macromolecular crowding effects on a protein molecule is presented in Fig. 2.2, which shows that the major mechanism by which macromolecular crowding is expected to affect proteins is related to the minimization of

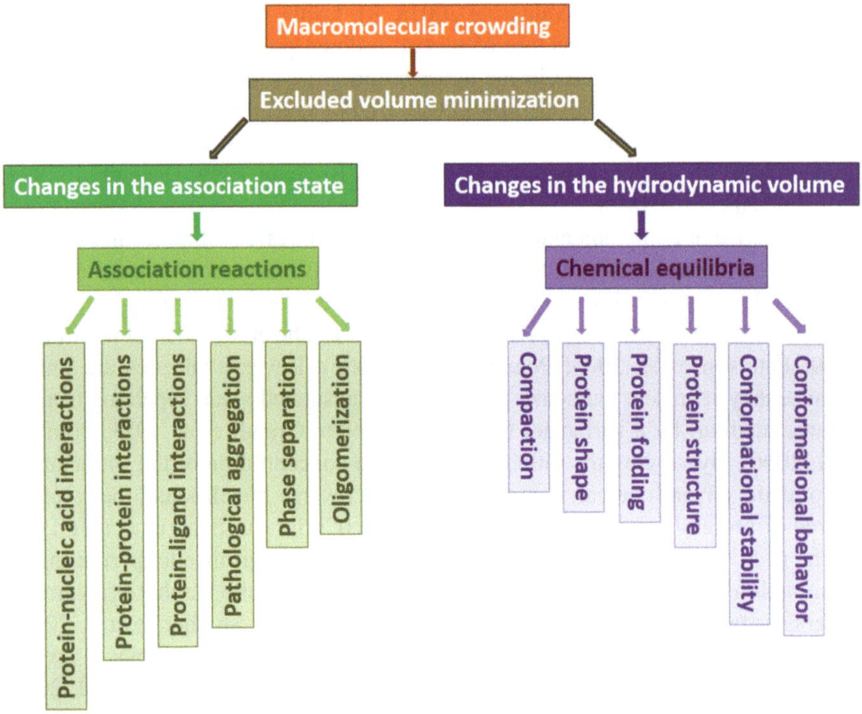

Fig. 2.2 Schematic representation of the potential effects of excluded volume on the behavior of proteins in crowded milieu

excluded volume. This tendency for the excluded volume minimization in crowded environments is driven by the need of a system to maximize its entropy. The excluded volume can be minimize either by changes in the hydrodynamic volume of a protein or via changes in its oligomerization/association state. Changes in hydrodynamic volume are promoted by modulation of the chemical equilibria that would affect protein folding, structure, shape, compaction, conformational stability, and conformational equilibrium. Changes in the association state are controlled by the modulating the association reactions and would have effects on protein-protein interactions, protein-nucleic acid interactions, protein oligomerization, pathologic aggregation, and phase separations (Kuznetsova et al. 2014).

Protein Function in Crowded Environment

By changing the effective reactant concentrations and the structural dynamics of proteins, macromolecular crowding can have a substantial impact on the activity of monomeric enzymes. The size ratio of the crowding agent to the substrate determines the crowding effect. Macromolecular crowding can affect an enzymatic reaction if it involves big substrates or if the enzyme changes its shape (Chen et al. 2011).

According to studies, high concentrations of crowding agents (such as PEG, Ficoll 70, and BSA) cause enzymes like DNA ligases and T4 DNA ligase to exhibit higher activity because they increase the effective concentrations of DNA ends, which improve ligation efficiency (Zimmerman and Pheiffer 1983). Likewise, crowding increased the effectiveness of DNA polymerase I and the protein recA (Lavery and Kowalczykowski 1992), encouraging DNA binding and strand exchange in otherwise non-permissive circumstances (Harrison and Zimmerman 1984).

Through a variety of ways, macromolecular crowding also affects the catalytic efficiency of different enzymes. For instance, by altering substrate binding and internal dynamics, crowding agents boost *Saccharomyces cerevisiae* Fet3p's catalytic efficiency. *Plasmodium falciparum* purine nucleoside phosphorylase and α-chymotrypsin activity are similarly affected, with the effects changing according to the concentration of the crowding agent (Pozdnyakova and Wittung-Stafshede 2010).

Furthermore, through a quasi-processive mechanism, crowding conditions were discovered to increase MEK's phosphorylation of ERK MAP kinase, so bolstering the "distributive phosphorylation model" for cellular signaling (Aoki et al. 2013).

Crowding does not, however, influence all enzymes. Crowding effects are extensively context-dependent and differ with specific enzyme-substrate interactions and environmental conditions, as evidenced by the fact that some enzymes, such as trypsin and some glycolytic enzymes, do not exhibit substantial modifications in catalytic parameters under high crowding conditions (Asaad and Engberts 2003).

Oligomeric proteins are affected by macromolecular crowding in a different way than monomeric enzymes. Because crowding reduces the excluded volume of oligomers, it promotes the production of dimers or higher-order oligomers, which impacts protein functionality. The impact of crowding on oligomer equilibrium was highlighted in early work by Minton and Wilf, which showed that crowding agents caused rabbit muscle glyceraldehyde-3-phosphate dehydrogenase (GAPD) to change from monomers to less active tetramers (Minton and Wilf 1981). Similar results were seen for other enzymes, such as hexokinase and G6P DH, where crowding changed kinetic characteristics and encouraged dimer stability, respectively.

PEG and BSA crowding agents also influenced structural alterations and catalytic activity of several enzymes, such as glucosidase II, lactate dehydrogenase, and the SARS-CoV 3C-like protease (Okamoto et al. 2010). Structural alterations due to the crowding affected activity in some reported cases, while stabilizing active oligomeric forms in others. For example, in Enterobacter synthetase, crowding inhibited the production of side products through structural alterations. Crowding has an influence to stabilize macromolecular complexes, which impacts nuclear activities as well. Furthermore, it has been reported that for example crowding controls the activity of the GroEL–GroES chaperone machinery, preventing substrate loss (Martin and Hartl 1997). It has been also reported that crowding affects small-molecule substrates as well by changing enzymatic activity according to the chemistry of the substrate and how it interacts with crowders. In crowded settings horseradish peroxidase (HRP), for example, reacts differently with hydrophilic OPD and hydrophobic TMB substrates. These results highlight how crucial it is to consider the characteristics of both the enzyme and the substrate when researching enzymatic reactions in crowded settings (Aumiller et al. 2014). Summary of crowder effects on enzyme activity and structural dynamics can be obtained from Table 2.2.

Table 2.2 Effects of different crowding agents on enzyme activity and structural dynamics

Crowding agent	Effects on enzymes	Specific findings	References
PEG (Polyethylene Glycol)	Increases activity of DNA ligases and T4 DNA ligase; promotes dimer stabilization in other enzymes	Enhances effective concentrations of DNA ends, improving ligation efficiency. Crowding promotes dimer stabilization and alters kinetic parameters in enzymes like G6P DH and hexokinase	Zimmerman and Pheiffer (1983) and Okamoto et al. (2010)
Ficoll 70	Increases activity of DNA ligases; enhances phosphorylation of ERK MAP kinase by MEK	High concentrations lead to increased effective concentrations of reactants, promoting DNA binding and strand exchange. Supports the "distributive phosphorylation model" for cellular signaling	Aoki et al. (2013) and Zimmerman and Pheiffer (1983)
BSA (Bovine Serum Albumin)	Enhances activity of DNA ligases; modulates structural changes in various enzymes	Like PEG, it increases effective concentrations of reactants and stabilizes active oligomeric forms while inducing structural changes that influence catalytic activities in enzymes like glucosidase II and lactate dehydrogenase	Okamoto et al. (2010)
Dextran	Affects structural properties and conformational dynamics of IDPs can promote folding in certain proteins	High concentrations (up to 250 g/L) did not induce significant folding in certain IDPs but can shift conformational equilibrium toward folded states in destabilized mutants when present at specific concentrations (e.g., Dextran 20)	Aden and Wittung-Stafshede (2014)
Hydrophobic Substrates (e.g., TMB)	Alters enzymatic activity based on interaction with crowders	Reacts differently with horseradish peroxidase (HRP) under crowded conditions compared to hydrophilic substrates like OPD, emphasizing the importance of substrate chemistry in crowded environments	Aumiller et al. (2014)
Hydrophilic Substrates (e.g., OPD)		Like hydrophobic substrates, it shows distinct reactivity with HRP under crowded conditions, highlighting the variability in enzyme-substrate interactions influenced by crowding agents	Aumiller et al. (2014)

Conformational Changes

It has been discovered that intrinsically disordered proteins (IDPs) and intrinsically disordered regions (IDRs), also referred to as naturally flexible regions, have distinct functions in promoting the formation of supramolecular complexes and controlling the activity of their binding partners (Dunker et al. 2001). Under physiological conditions in vitro, IDPs and IDRs may contain collapsed disorder (i.e., regions where intrinsic disorder is present in a molten globular form) and extended disorder (i.e., regions where intrinsic disorder is present in a form of random coil or pre-molten globule), as their conformational behavior and structural features are like those of non-native states of globular proteins (Dunker et al. 2001). Numerous human diseases are known to be linked to internally displaced people (IDPs) (Uversky et al. 2008).

The impact of elevated crowder concentrations on the structural characteristics of many IDPs was examined. Flaugh and Lumb (2001) demonstrated that two IDPs, FosAD and p27ID, did not exhibit substantial folding in response to molecular crowding approximated by large concentrations (up to 250 g/L) of the Dextrans of average molecular weights 9.5, 37.5, and 77 kDa and Ficoll 70. Whereas p27ID corresponds to the cyclin-dependent kinase inhibition domain of the cell-cycle inhibitor human p27Kip1 (residues 22–97) and is active as a cyclin A-Cdk2 inhibitor (Bienkiewicz et al. 2002), FosAD is the C-terminal activation domain of human c-Fos (residues 216–310) and is functional for interacting with transcription factors in whole-cell extract (Campbell et al. 2000; Flaugh and Lumb 2001). None of these IDPs had any appreciable conformational changes in the presence of macromolecular crowding agents, as evidenced by discernible variations in their fluorescence or circular dichroism spectra. Consequently, ordered structure in IDPs cannot always be induced by molecular crowding effects alone (Flaugh and Lumb 2001).

However, in the presence of 200 mg/mL Dextran 20, this IDP can effectively fold to a structure that resembles the structure of the wild type protein, according to an evaluation of the conformational behavior of a destabilized mutant of the immunoglobin G binding domain of protein L from the mesophile *Streptococcus magnus*, which is primarily unfolded in water but can be folded upon the addition of salt (Aden and Wittung-Stafshede 2014). Additionally, this study demonstrated unequivocally that when a polypeptide is in the transition area between folded and unfolded states, crowding can cause the conformational equilibrium to move toward the folded form (Aden and Wittung-Stafshede 2014).

The calcium binding protein RCL provides a unique chance to examine the identical polypeptide chain in two radically different folding states: as a compact folded structure in the Ca2 + -bound form (holo-form, RH of 2.2 nm) and as an extended IDP in its apo-form (RH of 3.2 nm) (Sotomayor-Perez et al. 2011). It was recently demonstrated that the crowding agent Ficoll 70 significantly enhanced the protein affinity for calcium and thermal stability of the two forms of RCL, despite having no effect on the structural contents of the apo-state and holo-state of the protein (Sotomayor-Perez et al. 2011).

By assessing their NMR spin relaxation values both with and without 160 mg/mL of Ficoll 70, the impact of macromolecular crowding has also been examined on the structural characteristics and conformational dynamics of several IDPs with varying degrees of residual structures (Cino et al. 2012). It was demonstrated that all the proteins from the chosen study maintained their segmental movements on the nanosecond timeframe and were largely disordered in crowded situations (Cino et al. 2012).

According to the results, intrinsically disordered proteins (IDPs) can be divided into two groups based on how they behave in a crowded environment: foldable and non-foldable. In these conditions (and hence inside living cells), foldable IDPs can gain structure, most likely because of crowding-induced hydrophobic core formation. Non-foldable IDPs, on the other hand, continue to be mainly unstructured in busy environments. To act as a scaffold for the creation of their structures, some of these non-foldable IDPs might require additional proteins, DNA, RNA, or other naturally occurring binding partners (Kuznetsova et al. 2014). Summary of crowder effects on IDPs can be seen in Table 2.3.

Table 2.3 Effects of various crowding agents on intrinsically disordered proteins

Crowding agent	Concentration	Effect on IDPs	Specific findings	References
Dextran (9.5 kDa)	Up to 250 g/L	No significant folding in IDPs (FosAD and p27ID)	Molecular crowding did not induce noticeable conformational changes in circular dichroism or fluorescence spectra for FosAD and p27ID	Flaugh and Lumb (2001) and Campbell et al. (2000)
Dextran (37.5 kDa)	Up to 250 g/L			
Dextran (77 kDa)	Up to 250 g/L			
Ficoll 70	Up to 250 g/L	Did not induce significant folding in IDPs (FosAD and p27ID); increased protein affinity for calcium in RCL	Molecular crowding effects were insufficient to induce ordered structures in FosAD and p27ID Increased calcium binding affinity and thermal stability of RCL were observed without affecting structural content	Sotomayor Perez et al. (2011) and Flaugh and Lumb (2001)
Dextran 20	200 mg/mL	Promoted folding of destabilized mutant of protein L	Crowding shifted the conformational equilibrium toward the folded form when the polypeptide was in the transition region between folded and unfolded states	Aden and Wittung-Stafshede (2014)
Ficoll 70	160 mg/mL	Retained disordered state; segmental motions on nanosecond timescale	All selected proteins remained mostly disordered under crowded conditions, retaining their segmental motions despite the presence of Ficoll 70	Cino et al. (2012)

Effect on Shape of the Protein

Multidomain proteins existing in close and open functional states or proteins undergoing transitions between elongated and more compact forms are among other targets that could be influenced by the presence of crowding agents.

Borrelia burgdorferi's aspherical protein VlsE has an elongated football shape with a helical core surrounded by floppy loops at each end. Homouz et al. used *in vitro* and *in silico* methods to examine the impact of macromolecular crowding on the structural changes and shape of this protein (Eicken et al. 2002; Homouz et al. 2008). The scientists demonstrated that high concentrations of Ficoll 70 enhance the folding kinetics of this football-shaped protein (Homouz et al. 2008). Additionally, the helical secondary structure of this protein increased significantly with increasing Ficoll 70 concentrations. When high Ficoll 70 concentrations were added to the VslE solution with urea present, a noticeable non-native β-structure was formed (Homouz et al. 2008). However, *in silico* simulation showed that crowding causes a considerable increase in helical structure along with a change from a football-like shape to a more compact bean-like shape. Because they reveal a concealed antigenic area, these crowding-induced shape alterations of VslE were thought to have biological importance (Homouz et al. 2008).

Calmodulin (CaM), a dumbbell-shaped molecule with four EF hands (two in the N-lobe and two in the C-lobe), is transformed into more compact structures in the presence of crowding agents, according to another study that combined *in silico* (coarse-grained molecular simulations) and experimental (FRET techniques methods) (Homouz et al. 2009). These analyses also showed that crowding was able to stabilize multiple compact conformations of CaM, indicating that the intrinsic plasticity of this protein's structure is what defines it (Homouz et al. 2009).

Another study that used phosphoglycerate kinase (PGK) used a similar combination method. When Ficoll 70 was present, there was a discernible compaction of this protein along with a sharp rise in its enzymatic activity (Dhar et al. 2010).

We may conclude from the previously mentioned studies that macromolecular crowding agents have a major impact on the structure and functionality of multidomain proteins. These results imply that protein structure alterations brought on by crowding may have biological implications.

Effect on Protein Folding

It is expected that the hydrodynamic volume of the unfolded state may be influenced by macromolecular crowding, which could impact protein folding. Therefore, by destabilizing the more stretched and pliable denatured states, macromolecular crowding may indirectly stabilize the folded states of proteins (Cheung et al. 2005). In the presence of Dextran, the unfolded ensemble of S16 was more compact than in the diluted solutions, demonstrating the direct effect of macromolecular crowding on a protein's unfolded state, forcing it to become more compact in crowded conditions. This partially collapsed unfolded form had observable residual structure (Mikaelsson et al. 2013).

To learn how macromolecular crowding can affect the folding capability of short independently foldable structural motifs, the effects of high levels of Dextran 70 (200 g/L) and Ficoll 70 (200 g/L) on the thermal stability and folding-unfolding kinetics of three small folding motifs—a 34-residue α-helix, a 34-residue cross-linked helix-turn-helix, and a 16-residue β-hairpin—were examined (Mukherjee et al. 2009). Crowding was found to have no effect on the folding of helical motifs, but to considerably slow down the folding rate of the shortest β-hairpin peptide (Mukherjee et al. 2009).

The detailed refolding and formation of the tetradecameric GroEL from urea-unfolded monomers demonstrates the need of macromolecular crowding for protein folding. A synergistic impact of crowding agents and nucleotides is seen in this process. In the presence of crowding agents alone, urea-unfolded GroEL is unable to successfully restore its natural oligomeric structure. The reactivation and folding efficiency of GroEL is considerably increased when large concentrations of crowding agents like Ficoll 70, Dextran, or BSA are added in conjunction with ATP, as opposed to ATP alone, even though nucleotides (ATP or ADP) can only partially refold GroEL (Galan et al. 2001).

Additional research on green fluorescent protein, enolase, and dihydrofolate reductase revealed that although these proteins can fold spontaneously in diluted solutions, they lose the capacity to fold in crowded settings. They create soluble, protease-sensitive non-native species under such circumstances, most commonly in the form of soluble oligomers. However, with the help of ATP hydrolysis, these proteins can regain their correct folding capacity when the entire GroEL/GroES chaperonin system is present. This demonstrates that many proteins may need chaperonin support for appropriate folding in congested cellular settings (Martin 2002).

In their study, Ye et al. showed how membrane insertion of outer membrane proteins (OMPs) and macromolecular crowding affect refolding (Ye et al. 2013). OmpA's refolding and membrane insertion rates were considerably slowed down by Ficoll 70, whereas OmpT's refolding and membrane insertion rates were little affected. The authors produced a shortened version of OmpA that only included the transmembrane domain (OmpA171) because the unique characteristic of OmpA is the presence of a substantial periplasmic domain. While OmpA171 and full-length OmpA exhibited comparable folding rates in the absence of a crowding agent, full-length OmpA refolded significantly more slowly than OmpA171 in the presence of 20% Ficoll 70 (Ye et al. 2013). However, when compared to OmpA, the folding efficiency of OmpA171 was noticeably lower. According to these findings, the periplasmic domain has two functions in the refolding-insertion process of OmpA. It slows down the rates of OmpA folding and membrane insertion while increasing their efficiency in the presence of crowders (Ye et al. 2013).

Based on the previously mentioned studies, we may draw the conclusion that macromolecular crowding influences the stability of unfolded states, which frequently results in a more compact form. This may stabilize the folded state by

Table 2.4 Effects of different crowding agents on protein shape and folding

Crowding agent	Effect on protein shape	Effect on protein folding	Specific findings	References
Ficoll 70	Accelerates folding kinetics of VlsE; induces transition to a more compact shape	Enhances reactivation and folding efficiency of GroEL in the presence of ATP	Increases the helical secondary structure of VlsE and leads to exposure of hidden antigenic regions High concentrations enhance reactivation efficiency of GroEL compared to ATP alone	Galan et al. (2001) and Homouz et al. (2008)
Dextran 70	Induces compaction of proteins; promotes folding in some cases	Slows down folding rates of certain motifs; stabilizes unfolded states	Unfolded ensemble of S16 becomes more compact in the presence of Dextran High concentrations impact thermal stability and folding-unfolding kinetics of small motifs, slowing down β-hairpin folding	Mukherjee et al. (2009) and Mikaelsson et al. (2013)
BSA (Bovine Serum Albumin)	Enhances structural stability and promotes folding in complex systems like GroEL	Works synergistically with nucleotides to enhance refolding efficiency	When combined with ATP, BSA significantly improves the reactivation and folding efficiency of GroEL compared to ATP alone	Galan et al. (2001)
Ficoll 70 (20%)	Slows down refolding and membrane insertion rates for OmpA compared to truncated OmpA171	Affects membrane insertion rates differently based on protein structure	Full-length OmpA refolds slower than OmpA171 in the presence of Ficoll 70, indicating that the periplasmic domain influences the refolding-insertion process by slowing down rates while improving efficiency	Ye et al. (2013)

destabilizing the denatured states (Cheung et al. 2005; Mikaelsson et al. 2013). Depending on certain structural characteristics and circumstances, crowding can either accelerate or improve folding in a variety of proteins and folding motifs (Galan et al. 2001; Mukherjee et al. 2009). Detailed effects of various crowders on protein stability and shape are summarized in Table 2.4.

Effects on Protein Assembly

A number of studies support the idea that high concentrations of non-specific polymers might alter different biological equilibria, impacting the different types of processes that result in the formation of proteins and protein complexes (Minton 2001).

It has been shown that macromolecular crowding greatly improves several biological assembly processes. It facilitates the interaction of 70S particles to generate 100S dimers by encouraging the assembly of *E. coli* 60S ribosomal particles from separate 30S and 50S subunits (Zimmerman and Trach 1988). Other interactions are also impacted, including the self-assembly of bacterial proteins like FtsZ (Gonzalez et al. 2003; Naddaf and Sayyed-Ahmad 2014; Rivas et al. 2001), the binding of hepatic aldolase B to hepatocyte matrix (Agius 1996), and the association of rat brain hexokinase with mitochondria from rat liver and yeast (Azoulay-Zohar and Aflalo 2000). Furthermore, crowding improves the processes of binding and assembly for various protein interactions (Alcorlo et al. 2009; Bookchin et al. 1999; Huang et al. 2010; Kim and Yethiraj 2010; Kim et al. 2010; Lindner and Ralston 1997; Wilf and Minton 1981; Snoussi and Halle 2005).

Despite these effects, high concentrations of PEG and Dextrans do not impact some high-affinity heterocomplex forms, such as those between barnase and barstar or TEM1-β-lactamase and β-lactamase inhibitor protein (BLIP) (Phillip et al. 2009). Instead of encouraging dimerization, macromolecular crowding can sometimes prevent weak contacts, such CyPet–YPet aggregation in PEG solutions (Phillip et al. 2009).

During cellular starvation, filament production by metabolic enzymes such as glutamine synthetase has been proposed as a defense mechanism. For instance, higher-order fibrils are formed when glutamine synthetase assembles into filaments, which are reliant on macromolecular crowding (Petrovska et al. 2014).

Additionally, crowding significantly improves bacteriophage T4's DNA replication complex formation. Under crowded conditions, the complex, which is made up of five core proteins, can assemble even in the absence of ATP hydrolysis (Jarvis et al. 1990; Reddy et al. 1995; Reddy et al. 1993). Furthermore, the DNA-tracking state of the T4 replication accessory protein gp45, a "sliding clamp" protein that also functions as a transcriptional activator, is stabilized by macromolecular crowding (Fu et al. 1996; Herendeen et al. 1989; Herendeen et al. 1992).

Actin filament polymerization and interaction with accessory proteins are facilitated by macromolecular crowding in cytoskeletal structure. This affects collagen hydrogel assembly and characteristics, including fiber organization and mechanical stress resistance, as well as the reversible conversion between actin filaments and bundles (Dewavrin et al. 2014; Grazi 1994; Suzuki et al. 1989; Tellam et al. 1983).

As demonstrated by the swelling and shrinking of macrophage nuclei in response to solute size, crowding also impacts membrane-associated activities, affecting cellular volume regulation and ion flux across cell membranes (Minton et al. 1992; Rosania and Swanson 1995). The size of the agents and the interacting proteins determine their capacity to promote assembly; they have different impacts on HIV-1 capsid assembly and viral recognition by short peptides (Rincon et al. 2011).

Furthermore, although macromolecular crowding has little effect on the effectiveness of inhibitors in both crowded and non-crowded settings, it can speed up harmful processes such as the polymerization of deoxyhemoglobin S in sickle cell disease (Bookchin et al. 1999).

Disease and Therapeutic Implications

Neurodegenerative Diseases

Macromolecular crowding overall has an important role in various neurodegenerative diseases, including Alzheimer's disease, Huntington's disease, Parkinson's disease, amyloidosis, and amyotrophic lateral sclerosis. The pathology of these conditions revolves around macromolecular crowding, impacting folding, aggregation, clearance, organelle dynamics, and intracellular transport. Under normal conditions, crowding helps proteins achieve their native structure and stabilizes it. However, it can also have similar effects on misfolded proteins or partially folded intermediates, including alpha-synuclein, amyloid-beta (Ma et al. 2012; Munishkina et al. 2004), huntingtin (Groover et al. 2021), and tau proteins (Wu et al. 2016), increasing the chance of their aggregation. These aggregates can affect cellular functions, become toxic to cells, and result in neuronal death (Cicardi et al. 2021; Mittal et al. 2015; Takahashi et al. 2020). Intrinsically disordered proteins, including huntingtin and tau, are more prone to aggregation induced by the crowding effect, and have been shown to have an increased rate of crowding under specific conditions (Cino et al. 2012; Uversky 2009).

Additionally, stress granules can be affected by macromolecular crowding. They are cytoplasmic granules that isolate mRNAs and RNA-binding proteins during stress when functioning normally. However, problems in macromolecular crowding can result in abnormal stress granules which can worsen cellular dysfunction under stress, and are linked to amyotrophic lateral sclerosis and dementia (Bounedjah et al. 2012).

Crowding can accelerate the nucleation stage of aggregation, which can result in rapid amyloid fibrils formation. Aggregation hotspots can also form, especially in cytoplasmic regions that exhibit impaired proteostasis (Minton 2001; Werner et al. 2021). Moreover, neurodegenerative diseases often have a malfunction in protein quality control systems. In turn, different pathways important for the clearance of misfolded proteins and protein aggregates, including proteasomes and autophagy pathways, become overwhelmed, resulting in further cellular crowding (Amzallag and Hornstein 2022; Zhou et al. 2008).

Therapeutics

Macromolecular crowding, while extremely useful in certain circumstances, can lead to problems and protein aggregation disorders. Understanding and analysis of these pathways can give us insights into potential therapeutic strategies. Currently, there are different approaches to developing therapeutics for disorders caused by macromolecular crowding in cytoplasm.

Osmolytes are small molecules involved in the regulation of a cell's osmotic pressure. They do not directly affect protein function, but they can contribute to the stabilization of proteins' structures and reduce aggregation by preventing and decreasing intermediate protein conformations that can contribute to disorders. These compounds have been found useful in the treatment of diseases like Alzheimer's and Huntington's (Aravindan et al. 2020; Khan et al. 2023; Kushwah et al. 2020).

Proteostasis—the process responsible for the homeostasis of proteins and their quality control—is another approach that has potential benefits for aggregation-related diseases. Molecules involved in the proper folding and clearing of misfolded proteins, including chaperons and proteasome activators, can also counter the negative effects of macromolecular crowding. Targeting these through specific drugs, and focusing on autophagy-inducing pathways as well, could prove beneficial (Amzallag and Hornstein 2022).

Other approaches to the design of the treatment include targeting biochemical pathways dysregulated in crowding, including those responsible for stress granule formation, through stress granule modulators, which could help with amyotrophic lateral sclerosis (Portz et al. 2021; Yang et al. 2023). Additionally, using nanotechnology holds potential as well, and nanoparticles that could modulate crowding, or deliver anti-aggregation agents to aggregation hotspots could be helpful as well (Asefy et al. 2021).

Concluding Remarks

This chapter provides an overview of various effects of macromolecular crowding on different aspects of protein structure and function. Since there is virtually no a single protein-related feature that would not be affected by the addition of high concentrations of macromolecules, the macromolecular crowding phenomenon is expected to have very important consequences. Clearly, proteins are not an exception, and other biological macromolecules can similarly "feel" effects of macromolecular crowding. Therefore, macromolecular crowding cannot be ignored, and *in vitro* studies on the various structural and functional aspects of biological molecules should take this phenomenon into account.

Compliance with Ethical Standards Authors declare no conflict of interest.

Funding source: Not applicable.

References

Aden J, Wittung-Stafshede P (2014) Folding of an unfolded protein by macromolecular crowding in vitro. Biochemistry 53(14):2271–2277. https://doi.org/10.1021/bi500222g

Agius L (1996) Substrate modulation of aldolase B binding in hepatocytes. Biochem J 315 (Pt 2) (Pt 2):651–658. https://doi.org/10.1042/bj3150651

Alcorlo M, Jimenez M, Ortega A, Hermoso JM, Salas M, Minton AP, Rivas G (2009) Analytical ultracentrifugation studies of phage phi29 protein p6 binding to DNA. J Mol Biol 385(5):1616–1629. https://doi.org/10.1016/j.jmb.2008.11.044
Alfano C, Fichou Y, Huber K, Weiss M, Spruijt E, Ebbinghaus S, De Luca G, Morando MA, Vetri V, Temussi PA, Pastore A (2024) Molecular crowding: the history and development of a scientific paradigm. Chem Rev 124(6):3186–3219. https://doi.org/10.1021/acs.chemrev.3c00615
Amzallag E, Hornstein E (2022) Crosstalk between biomolecular condensates and proteostasis. Cells 11(15). https://doi.org/10.3390/cells11152415
Aoki K, Takahashi K, Kaizu K, Matsuda M (2013) A quantitative model of ERK MAP kinase phosphorylation in crowded media. Sci Rep 3:1541. https://doi.org/10.1038/srep01541
Aravindan S, Chen S, Choudhry H, Molfetta C, Chen KY, Liu AYC (2020) Osmolytes dynamically regulate mutant Huntingtin aggregation and CREB function in Huntington's disease cell models. Sci Rep 10(1):15511. https://doi.org/10.1038/s41598-020-72613-3
Asaad N, Engberts JB (2003) Cytosol-mimetic chemistry: kinetics of the trypsin-catalyzed hydrolysis of p-nitrophenyl acetate upon addition of polyethylene glycol and N-tert-butyl acetoacetamide. J Am Chem Soc 125(23):6874–6875. https://doi.org/10.1021/ja034298f
Asefy Z, Hoseinnejhad S, Ceferov Z (2021) Nanoparticles approaches in neurodegenerative diseases diagnosis and treatment. Neurol Sci 42(7):2653–2660. https://doi.org/10.1007/s10072-021-05234-x
Aumiller WM Jr, Davis BW, Hatzakis E, Keating CD (2014) Interactions of macromolecular crowding agents and cosolutes with small-molecule substrates: effect on horseradish peroxidase activity with two different substrates. J Phys Chem B 118(36):10624–10632. https://doi.org/10.1021/jp506594f
Azoulay-Zohar H, Aflalo C (2000) Binding of rat brain hexokinase to recombinant yeast mitochondria: identification of necessary physico-chemical determinants. Eur J Biochem 267(10):2973–2980. https://doi.org/10.1046/j.1432-1033.2000.01313.x
Babu CR, Flynn PF, Wand AJ (2003) Preparation, characterization, and NMR spectroscopy of encapsulated proteins dissolved in low viscosity fluids. J Biomol NMR 25(4):313–323
Bienkiewicz EA, Adkins JN, Lumb KJ (2002) Functional consequences of preorganized helical structure in the intrinsically disordered cell-cycle inhibitor p27(Kip1). Biochemistry 41(3):752–759. https://doi.org/10.1021/bi015763t
Bismuto E, Irace G (2001) The effect of molecular confinement on the conformational dynamics of the native and partly folded state of apomyoglobin. FEBS Lett 509:476–480
Bismuto E, Martelli PL, De Maio A, Mita DG, Irace G, Casadio R (2002) Effect of molecular confinement on internal enzyme dynamics: frequency domain fluorometry and molecular dynamics simulation studies. Biopolymers 67(2):85–95. https://doi.org/10.1002/bip.10058
Bolis D, Politou AS, Kelly G, Pastore A, Temussi PA (2004) Protein stability in nanocages: a novel approach for influencing protein stability by molecular confinement. J Mol Biol 336(1):203–212
Bookchin RM, Balazs T, Wang Z, Josephs R, Lew VL (1999) Polymer structure and solubility of deoxyhemoglobin S in the presence of high concentrations of volume-excluding 70-kDa dextran. Effects of non-s hemoglobins and inhibitors. J Biol Chem 274(10):6689–6697. https://doi.org/10.1074/jbc.274.10.6689
Bounedjah O, Hamon L, Savarin P, Desforges B, Curmi PA, Pastre D (2012) Macromolecular crowding regulates assembly of mRNA stress granules after osmotic stress: new role for compatible osmolytes. J Biol Chem 287(4):2446–2458. https://doi.org/10.1074/jbc.M111.292748
Brennan JD (1999) Using fluorescence to investigate proteins entrapped in sol-gel derived materials. Appl Spec 53:106A–121A
Campbell KM, Terrell AR, Laybourn PJ, Lumb KJ (2000) Intrinsic structural disorder of the C-terminal activation domain from the bZIP transcription factor Fos. Biochemistry 39(10):2708–2713. https://doi.org/10.1021/bi9923555
Chen C, Loe F, Blocki A, Peng Y, Raghunath M (2011) Applying macromolecular crowding to enhance extracellular matrix deposition and its remodeling in vitro for tissue engineering and cell-based therapies. Adv Drug Deliv Rev 63(4–5):277–290. https://doi.org/10.1016/j.addr.2011.03.003

Cheung MS, Klimov D, Thirumalai D (2005) Molecular crowding enhances native state stability and refolding rates of globular proteins. Proc Natl Acad Sci USA 102(13):4753–4758. https://doi.org/10.1073/pnas.0409630102
Christ S, Schurtenberger P (1994) Optical contrast variation experiments in water-in-oil microemulsions: Size distribution and structure of protein-free and protein-containing microemulsions. J Phys Chem 98(48):12708–12714
Cicardi ME, Marrone L, Azzouz M, Trotti D (2021) Proteostatic imbalance and protein spreading in amyotrophic lateral sclerosis. EMBO J 40(10):e106389. https://doi.org/10.15252/embj.2020106389
Cino EA, Karttunen M, Choy WY (2012) Effects of molecular crowding on the dynamics of intrinsically disordered proteins. PLoS One 7(11):e49876. https://doi.org/10.1371/journal.pone.0049876
Dave BC, Dunn B, Valentine JS, Zink JI (1994) Sol-gel encapsulation methods for biosensors. Anal Chem 66:1120–1127
Davis DM, McLoskey D, Birch DJ, Gellert PR, Kittlety RS, Swart RM (1996) The fluorescence and circular dichroism of proteins in reverse micelles: application to the photophysics of human serum albumin and N-acetyl-L-tryptophanamide. Biophys Chem 60(3):63–77
del Alamo M, Rivas G, Mateu MG (2005) Effect of macromolecular crowding agents on human immunodeficiency virus type 1 capsid protein assembly in vitro. J Virol 79(22):14271–14281. https://doi.org/10.1128/JVI.79.22.14271-14281.2005
Dewavrin JY, Hamzavi N, Shim VP, Raghunath M (2014) Tuning the architecture of three-dimensional collagen hydrogels by physiological macromolecular crowding. Acta Biomater 10(10):4351–4359. https://doi.org/10.1016/j.actbio.2014.06.006
Dhar A, Samiotakis A, Ebbinghaus S, Nienhaus L, Homouz D, Gruebele M, Cheung MS (2010) Structure, function, and folding of phosphoglycerate kinase are strongly perturbed by macromolecular crowding. Proc Natl Acad Sci USA 107(41):17586–17591. https://doi.org/10.1073/pnas.1006760107
Dunker AK, Lawson JD, Brown CJ, Williams RM, Romero P, Oh JS, Oldfield CJ, Campen AM, Ratliff CM, Hipps KW, Ausio J, Nissen MS, Reeves R, Kang C, Kissinger CR, Bailey RW, Griswold MD, Chiu W, Garner EC, Obradovic Z (2001) Intrinsically disordered protein. J Mol Graph Model 19(1):26–59. https://doi.org/10.1016/s1093-3263(00)00138-8
Eggers DK, Valentine JS (2001a) Crowding and hydration effects on protein conformation: a study with sol-gel encapsulated proteins. J Mol Biol 314(4):911–922
Eggers DK, Valentine JS (2001b) Molecular confinement influences protein structure and enhances thermal protein stability. Protein Sci 10(2):250–261
Ehrhardt MR, Flynn PF, Wand AJ (1999) Preparation of encapsulated proteins dissolved in low viscosity fluids. J Biomol NMR 14(1):75–78
Eicken C, Sharma V, Klabunde T, Lawrenz MB, Hardham JM, Norris SJ, Sacchettini JC (2002) Crystal structure of Lyme disease variable surface antigen VlsE of Borrelia burgdorferi. J Biol Chem 277(24):21691–21696. https://doi.org/10.1074/jbc.M201547200
Ellis RJ (2001) Macromolecular crowding: obvious but underappreciated. Trends Biochem Sci 26(10):597–604
Ellis RJ, Minton AP (2003) Cell biology: join the crowd. Nature 425(6953):27–28. https://doi.org/10.1038/425027a
Flaugh SL, Lumb KJ (2001) Effects of macromolecular crowding on the intrinsically disordered proteins c-Fos and p27(Kip1). Biomacromolecules 2(2):538–540. https://doi.org/10.1021/bm015502z
Fu TJ, Sanders GM, O'Donnell M, Geiduschek EP (1996) Dynamics of DNA-tracking by two sliding-clamp proteins. EMBO J 15(16):4414–4422
Fulton AB (1982) How crowded is the cytoplasm? Cell 30(2):345–347
Gagarskaia IA, Povarova OI, Uversky VN, Kuznetsova IM, Turoverov KK (2017) The effects of crowding agents Dextran-70k and PEG-8k on actin structure and unfolding reaction. J Mol Struct 1140:46–51

Galan A, Sot B, Llorca O, Carrascosa JL, Valpuesta JM, Muga A (2001) Excluded volume effects on the refolding and assembly of an oligomeric protein. GroEL, a case study. J Biol Chem 276 (2):957-964. https://doi.org/10.1074/jbc.M006861200

Gill I, Ballestros A (2000) Bioencapsulation within synthetic polymers (part 1): Sol-gel encapsulated biologicals. Trends Biotechnol 18:282–296

Gonzalez JM, Jimenez M, Velez M, Mingorance J, Andreu JM, Vicente M, Rivas G (2003) Essential cell division protein FtsZ assembles into one monomer-thick ribbons under conditions resembling the crowded intracellular environment. J Biol Chem 278(39):37664–37671. https://doi.org/10.1074/jbc.M305230200

Gottfried DS, Kagan A, Hoffman BM, Friedman JM (1999) Impeded rotation of a protein in sol-gel matrix. J Phys Chem B 103:2803–2807

Grandi C, Smith RE, Luisi PL (1981) Micellar solubilization of biopolymers in organic solvents. Activity and conformation of lysozyme in isooctane reverse micelles. J Biol Chem 256 (2):837–843

Grazi E (1994) Cytoskeleton, motile structures and macromolecular crowding. Adv Exp Med Biol 358:123–130. doi:https://doi.org/10.1007/978-1-4615-2578-3_12, Springer US

Groover SE, Adegbuyiro A, Fan CK, Hodges BL, Beasley M, Taylor K, Stonebraker AR, Siriwardhana C, Legleiter J (2021) Macromolecular crowding in solution alters huntingtin interaction and aggregation at interfaces. Colloids Surf B Biointerfaces 206:111969. https://doi.org/10.1016/j.colsurfb.2021.111969

Harrison B, Zimmerman SB (1984) Polymer-stimulated ligation: enhanced ligation of oligo- and polynucleotides by T4 RNA ligase in polymer solutions. Nucleic Acids Res 12(21):8235–8251. https://doi.org/10.1093/nar/12.21.8235

Hatters DM, Minton AP, Howlett GJ (2002) Macromolecular crowding accelerates amyloid formation by human apolipoprotein C-II. J Biol Chem 277(10):7824–7830

Herendeen DR, Kassavetis GA, Barry J, Alberts BM, Geiduschek EP (1989) Enhancement of bacteriophage T4 late transcription by components of the T4 DNA replication apparatus. Science 245(4921):952–958. https://doi.org/10.1126/science.2672335

Herendeen DR, Kassavetis GA, Geiduschek EP (1992) A transcriptional enhancer whose function imposes a requirement that proteins track along DNA. Science 256(5061):1298–1303. https://doi.org/10.1126/science.1598572

Homouz D, Perham M, Samiotakis A, Cheung MS, Wittung-Stafshede P (2008) Crowded, cell-like environment induces shape changes in aspherical protein. Proc Natl Acad Sci USA 105(33):11754–11759. https://doi.org/10.1073/pnas.0803672105

Homouz D, Sanabria H, Waxham MN, Cheung MS (2009) Modulation of calmodulin plasticity by the effect of macromolecular crowding. J Mol Biol 391(5):933–943. https://doi.org/10.1016/j.jmb.2009.06.073

Huang L, Jin R, Li J, Luo K, Huang T, Wu D, Wang W, Chen R, Xiao G (2010) Macromolecular crowding converts the human recombinant PrPC to the soluble neurotoxic beta-oligomers. FASEB J 24(9):3536–3543. https://doi.org/10.1096/fj.09-150987

Jarvis TC, Ring DM, Daube SS, von Hippel PH (1990) "Macromolecular crowding": thermodynamic consequences for protein-protein interactions within the T4 DNA replication complex. J Biol Chem 265(25):15160–15167

Kameta N, Minamikawa H, Someya Y, Yui H, Masuda M, Shimizu T (2010) Confinement effect of organic nanotubes toward green fluorescent protein (GFP) depending on the inner diameter size. Chemistry 16(14):4217–4223. https://doi.org/10.1002/chem.200903413

Khan S, Siraj S, Shahid M, Haque MM, Islam A (2023) Osmolytes: wonder molecules to combat protein misfolding against stress conditions. Int J Biol Macromol 234:123662. https://doi.org/10.1016/j.ijbiomac.2023.123662

Kim JS, Yethiraj A (2010) Crowding effects on association reactions at membranes. Biophys J 98(6):951–958. https://doi.org/10.1016/j.bpj.2009.11.022

Kim YC, Best RB, Mittal J (2010) Macromolecular crowding effects on protein–protein binding affinity and specificity. J Chem Phys 133(20):205101

Kushwah N, Jain V, Yadav D (2020) Osmolytes: a possible therapeutic molecule for ameliorating the neurodegeneration caused by protein misfolding and aggregation. Biomolecules 10(1). https://doi.org/10.3390/biom10010132
Kuznetsova IM, Turoverov KK, Uversky VN (2014) What macromolecular crowding can do to a protein. Int J Mol Sci 15(12):23090–23140. https://doi.org/10.3390/ijms151223090
Kuznetsova IM, Zaslavsky BY, Breydo L, Turoverov KK, Uversky VN (2015) Beyond the excluded volume effects: mechanistic complexity of the crowded milieu. Molecules 20(1):1377–1409. https://doi.org/10.3390/molecules20011377
Lan EH, Dave BC, Fukuto JM, Dunn B, Zink JI, Valentine JS (1999) Synthesis of sol-gel encapsulated heme proteins with chemical sensing properties. J Mater Chem 9:45–53
Lavery PE, Kowalczykowski SC (1992) Enhancement of recA protein-promoted DNA strand exchange activity by volume-occupying agents. J Biol Chem 267(13):9307–9314
Lindner RA, Ralston GB (1997) Macromolecular crowding: effects on actin polymerisation. Biophys Chem 66(1):57–66. https://doi.org/10.1016/s0301-4622(97)00011-2
Lodish H, Berk A, Zipursky SL, Matsudaira P, Baltimore D, Darnell J (2000) Molecular cell biology, 4th edn. W. H. Freeman, New York
Ma Q, Fan JB, Zhou Z, Zhou BR, Meng SR, Hu JY, Chen J, Liang Y (2012) The contrasting effect of macromolecular crowding on amyloid fibril formation. PLoS One 7(4):e36288. https://doi.org/10.1371/journal.pone.0036288
Mandal SS, Cristiglio V, Lindner P, Bhattacharyya AJ (2014) Small-angle neutron scattering studies of hemoglobin confined inside silica tubes of varying sizes. Chemphyschem 15(2):302–309. https://doi.org/10.1002/cphc.201300863
Martin J (2002) Requirement for GroEL/GroES-dependent protein folding under nonpermissive conditions of macromolecular crowding. Biochemistry 41(15):5050–5055. https://doi.org/10.1021/bi0159251
Martin J, Hartl FU (1997) The effect of macromolecular crowding on chaperonin-mediated protein folding. Proc Natl Acad Sci USA 94(4):1107–1112. https://doi.org/10.1073/pnas.94.4.1107
Mikaelsson T, Aden J, Johansson LB, Wittung-Stafshede P (2013) Direct observation of protein unfolded state compaction in the presence of macromolecular crowding. Biophys J 104(3):694–704. https://doi.org/10.1016/j.bpj.2012.12.020
Minton AP (1993) Macromolecular crowding and molecular recognition. J Mol Recogn JMR 6(4):211–214. https://doi.org/10.1002/jmr.300060410
Minton AP (1997) Influence of excluded volume upon macromolecular structure and associations in 'crowded' media. Curr Opin Biotechnol 8(1):65–69
Minton AP (2000a) Implications of macromolecular crowding for protein assembly. Curr Opin Struct Biol 10(1):34–39
Minton AP (2000b) Protein folding: Thickening the broth. Curr Biol 10(3):R97–R99
Minton AP (2001) The influence of macromolecular crowding and macromolecular confinement on biochemical reactions in physiological media. J Biol Chem 276(14):10577–10580
Minton AP (2005) Models for excluded volume interaction between an unfolded protein and rigid macromolecular cosolutes: macromolecular crowding and protein stability revisited. Biophys J 88(2):971–985. https://doi.org/10.1529/biophysj.104.050351
Minton AP, Wilf J (1981) Effect of macromolecular crowding upon the structure and function of an enzyme: glyceraldehyde-3-phosphate dehydrogenase. Biochemistry 20(17):4821–4826. https://doi.org/10.1021/bi00520a003
Minton AP, Colclasure GC, Parker JC (1992) Model for the role of macromolecular crowding in regulation of cellular volume. Proc Natl Acad Sci USA 89(21):10504–10506. https://doi.org/10.1073/pnas.89.21.10504
Mittal S, Chowhan RK, Singh LR (2015) Macromolecular crowding: macromolecules friend or foe. Biochim Biophys Acta 1850 (9):1822–1831. https://doi.org/10.1016/j.bbagen.2015.05.002
Mukherjee S, Chowdhury P, Gai F (2006) Tuning the cooperativity of the helix-coil transition by aqueous reverse micelles. J Phys Chem B 110(24):11615–11619. https://doi.org/10.1021/jp062362k

Mukherjee S, Waegele MM, Chowdhury P, Guo L, Gai F (2009) Effect of macromolecular crowding on protein folding dynamics at the secondary structure level. J Mol Biol 393(1):227–236. https://doi.org/10.1016/j.jmb.2009.08.016

Munishkina LA, Cooper EM, Uversky VN, Fink AL (2004) The effect of macromolecular crowding on protein aggregation and amyloid fibril formation. Journal of molecular recognition : JMR 17(5):456–464. https://doi.org/10.1002/jmr.699

Naddaf L, Sayyed-Ahmad A (2014) Intracellular crowding effects on the self-association of the bacterial cell division protein FtsZ. Arch Biochem Biophys 564:12–19. https://doi.org/10.1016/j.abb.2014.08.016

O'Brien EP, Stan G, Thirumalai D, Brooks BR (2008) Factors governing helix formation in peptides confined to carbon nanotubes. Nano Lett 8(11):3702–3708. https://doi.org/10.1021/nl8019328

Okamoto DN, Oliveira LC, Kondo MY, Cezari MH, Szeltner Z, Juhasz T, Juliano MA, Polgar L, Juliano L, Gouvea IE (2010) Increase of SARS-CoV 3CL peptidase activity due to macromolecular crowding effects in the milieu composition. Biol Chem 391(12):1461–1468. https://doi.org/10.1515/BC.2010.145

Pastore A, Salvadori S, Temussi PA (2007) Peptides and proteins in a confined environment: NMR spectra at natural isotopic abundance. J Pept Sci 13(5):342–347. https://doi.org/10.1002/psc.848

Peterson RW, Lefebvre BG, Wand AJ (2005) High-resolution NMR studies of encapsulated proteins in liquid ethane. J Am Chem Soc 127(29):10176–10177. https://doi.org/10.1021/ja0526517

Petrovska I, Nuske E, Munder MC, Kulasegaran G, Malinovska L, Kroschwald S, Richter D, Fahmy K, Gibson K, Verbavatz JM, Alberti S (2014) Filament formation by metabolic enzymes is a specific adaptation to an advanced state of cellular starvation. elife 3. https://doi.org/10.7554/eLife.02409

Phillip Y, Sherman E, Haran G, Schreiber G (2009) Common crowding agents have only a small effect on protein-protein interactions. Biophys J 97(3):875–885. https://doi.org/10.1016/j.bpj.2009.05.026

Pometun MS, Peterson RW, Babu CR, Wand AJ (2006) Cold denaturation of encapsulated ubiquitin. J Am Chem Soc 128(33):10652–10653. https://doi.org/10.1021/ja0628654

Portz B, Lee BL, Shorter J (2021) FUS and TDP-43 phases in health and disease. Trends Biochem Sci 46(7):550–563. https://doi.org/10.1016/j.tibs.2020.12.005

Pozdnyakova I, Wittung-Stafshede P (2010) Non-linear effects of macromolecular crowding on enzymatic activity of multi-copper oxidase. Biochim Biophys Acta 1804 4:740–744. https://doi.org/10.1016/j.bbapap.2009.11.013

Radhakrishna M, Grimaldi J, Belfort G, Kumar SK (2013) Stability of proteins inside a hydrophobic cavity. Langmuir 29(28):8922–8928. https://doi.org/10.1021/la4014784

Ralston G (1990) Effects of "crowding" in protein solutions. J Chem Educ 67(10):857

Ravindra R, Zhao S, Gies H, Winter R (2004) Protein encapsulation in mesoporous silicate: the effects of confinement on protein stability, hydration, and volumetric properties. J Am Chem Soc 126(39):12224–12225. https://doi.org/10.1021/ja046900n

Reddy MK, Weitzel SE, von Hippel PH (1993) Assembly of a functional replication complex without ATP hydrolysis: a direct interaction of bacteriophage T4 gp45 with T4 DNA polymerase. Proc Natl Acad Sci USA 90(8):3211–3215. https://doi.org/10.1073/pnas.90.8.3211

Reddy MK, Weitzel SE, Daube SS, Jarvis TC, von Hippel PH (1995) Using macromolecular crowding agents to identify weak interactions within DNA replication complexes. Methods Enzymol 262:466–476. https://doi.org/10.1016/0076-6879(95)62038-9

Rincon V, Bocanegra R, Rodriguez-Huete A, Rivas G, Mateu MG (2011) Effects of macromolecular crowding on the inhibition of virus assembly and virus-cell receptor recognition. Biophys J 100(3):738–746. https://doi.org/10.1016/j.bpj.2010.12.3714

Rivas G, Fernandez JA, Minton AP (2001) Direct observation of the enhancement of noncooperative protein self-assembly by macromolecular crowding: indefinite linear self-association of bacterial cell division protein FtsZ. Proc Natl Acad Sci USA 98(6):3150–3155. https://doi.org/10.1073/pnas.051634398

Rivas G, Ferrone F, Herzfeld J (2004) Life in a crowded world. EMBO Rep 5(1):23–27. https://doi.org/10.1038/sj.embor.7400056

Roche CJ, Dantsker D, Heller ER, Sabat JE, Friedman JM (2013) Reverse micelles as a tool for probing solvent modulation of protein dynamics: reverse micelle encapsulated hemoglobin. Chem Phys 430:88–97. https://doi.org/10.1016/j.chemphys.2013.04.006

Rosania GR, Swanson JA (1995) Effects of macromolecular crowding on nuclear size. Exp Cell Res 218(1):114–122. https://doi.org/10.1006/excr.1995.1137

Sanfelice D, Tancredi T, Politou A, Pastore A, Temussi PA (2009) Cold denaturation and aggregation: a comparative NMR study of titin I28 in bulk and in a confined environment. J Am Chem Soc 131(33):11662–11663. https://doi.org/10.1021/ja904462n

Sanfelice D, Politou A, Martin SR, De Los RP, Temussi P, Pastore A (2013) The effect of crowding and confinement: a comparison of Yfh1 stability in different environments. Phys Biol 10(4):045002. https://doi.org/10.1088/1478-3975/10/4/045002

Siefker J, Karande P, Coppens MO (2014) Packaging biological cargoes in mesoporous materials: opportunities for drug delivery. Expert Opin Drug Deliv 11(11):1781–1793. https://doi.org/10.1517/17425247.2014.938636

Snoussi K, Halle B (2005) Protein self-association induced by macromolecular crowding: a quantitative analysis by magnetic relaxation dispersion. Biophys J 88(4):2855–2866. https://doi.org/10.1529/biophysj.104.055871

Sotomayor-Perez AC, Ladant D, Chenal A (2011) Calcium-induced folding of intrinsically disordered repeat-in-toxin (RTX) motifs via changes of protein charges and oligomerization states. J Biol Chem 286(19):16997–17004. https://doi.org/10.1074/jbc.M110.210393

Stagg L, Zhang SQ, Cheung MS, Wittung-Stafshede P (2007) Molecular crowding enhances native structure and stability of alpha/beta protein flavodoxin. Proc Natl Acad Sci USA 104(48):18976–18981. https://doi.org/10.1073/pnas.0705127104

Suzuki A, Yamazaki M, Ito T (1989) Osmoelastic coupling in biological structures: formation of parallel bundles of actin filaments in a crystalline-like structure caused by osmotic stress. Biochemistry 28(15):6513–6518. https://doi.org/10.1021/bi00441a052

Takahashi A, Nagao C, Murakami K, Kuroi K, Nakabayashi T (2020) Effects of molecular crowding environment on the acquisition of toxic properties of wild-type SOD1. Biochim Biophys Acta Gen Subj 1864 2:129401. https://doi.org/10.1016/j.bbagen.2019.07.010

Tellam RL, Sculley MJ, Nichol LW, Wills PR (1983) The influence of poly(ethylene glycol) 6000 on the properties of skeletal-muscle actin. Biochem J 213(3):651–659. https://doi.org/10.1042/bj2130651

Tokuriki N, Kinjo M, Negi S, Hoshino M, Goto Y, Urabe I, Yomo T (2004) Protein folding by the effects of macromolecular crowding. Protein Sci 13(1):125–133. https://doi.org/10.1110/ps.03288104

Uversky VN (2009) Intrinsically disordered proteins and their environment: effects of strong denaturants, temperature, pH, counter ions, membranes, binding partners, osmolytes, and macromolecular crowding. Protein J 28(7–8):305–325. https://doi.org/10.1007/s10930-009-9201-4

Uversky VN, Elisa MC, Bower KS, Li J, Fink AL (2002) Accelerated alpha-synuclein fibrillation in crowded milieu. FEBS Lett 515(1–3):99–103. https://doi.org/10.1016/s0014-5793(02)02446-8

Uversky VN, Oldfield CJ, Dunker AK (2008) Intrinsically disordered proteins in human diseases: introducing the D2 concept. Annu Rev Biophys 37:215–246. https://doi.org/10.1146/annurev.biophys.37.032807.125924

van den Berg B, Ellis RJ, Dobson CM (1999) Effects of macromolecular crowding on protein folding and aggregation. EMBO J 18(24):6927–6933. https://doi.org/10.1093/emboj/18.24.6927

van den Berg B, Wain R, Dobson CM, Ellis RJ (2000) Macromolecular crowding perturbs protein refolding kinetics: implications for folding inside the cell. EMBO J 19(15):3870–3875. https://doi.org/10.1093/emboj/19.15.3870

Venturoli D, Rippe B (2005) Ficoll and dextran vs. globular proteins as probes for testing glomerular permselectivity: effects of molecular size, shape, charge, and deformability. Am J Physiol Renal Physiol 288(4):F605-613. doi:https://doi.org/10.1152/ajprenal.00171.2004

Werner TER, Horvath I, Wittung-Stafshede P (2021) Response to crowded conditions reveals compact nucleus for amyloid formation of folded protein. QRB Discov 2:e2. https://doi.org/10.1017/qrd.2020.17
Wilf J, Minton AP (1981) Evidence for protein self-association induced by excluded volume. Myoglobin in the presence of globular proteins. Biochim Biophys Acta 670 (3):316-322. https://doi.org/10.1016/0005-2795(81)90103-3
Wu Y, Teng N, Li S (2016) Effects of macromolecular crowding and osmolyte on human Tau fibrillation. Int J Biol Macromol 90:27–36. https://doi.org/10.1016/j.ijbiomac.2015.11.091
Yang C, Wang Z, Kang Y, Yi Q, Wang T, Bai Y, Liu Y (2023) Stress granule homeostasis is modulated by TRIM21-mediated ubiquitination of G3BP1 and autophagy-dependent elimination of stress granules. Autophagy 19(7):1934–1951. https://doi.org/10.1080/15548627.2022.2164427
Ye C, Chai Q, Zhong M, Wei Y (2013) Effect of crowding by Ficolls on OmpA and OmpT refolding and membrane insertion. Protein Sci 22(2):239–245. https://doi.org/10.1002/pro.2205
Zhou HX, Rivas G, Minton AP (2008) Macromolecular crowding and confinement: biochemical, biophysical, and potential physiological consequences. Annu Rev Biophys 37:375–397. https://doi.org/10.1146/annurev.biophys.37.032807.125817
Zimmerman SB, Minton AP (1993) Macromolecular crowding: biochemical, biophysical, and physiological consequences. Annu Rev Biophys Biomol Struct 22:27–65
Zimmerman SB, Pheiffer BH (1983) Macromolecular crowding allows blunt-end ligation by DNA ligases from rat liver or Escherichia coli. Proc Natl Acad Sci USA 80(19):5852–5856. https://doi.org/10.1073/pnas.80.19.5852
Zimmerman SB, Trach SO (1988) Effects of macromolecular crowding on the association of E. coli ribosomal particles. Nucleic Acids Res 16(14A):6309–6326. https://doi.org/10.1093/nar/16.14.6309
Zimmerman SB, Trach SO (1991) Estimation of macromolecule concentrations and excluded volume effects for the cytoplasm of Escherichia coli. J Mol Biol 222(3):599–620

Chapter 3
(Macro)Molecular Crowding Effects Beyond Volume Exclusion

Boris Y. Zaslavsky and Vladimir N. Uversky

Abstract Macromolecular crowding affects many areas, including protein folding, binding of small molecules, enzymatic activity, interaction with nucleic acids, protein aggregation, and protein-protein interactions. While the common belief has been that the primary impact of crowded environments on the function, structure, thermodynamics, and aggregation of a protein can be described in terms of excluded volume effects, it has now become clear that other factors, which originate from high concentrations of "inert" macromolecules in crowded solution, must be considered in order to get a clearer understanding a protein's behavior in a crowded environment. This review will highlight several important factors that arise in a crowded environment, including perturbed diffusion, viscosity, soft interactions, direct physical interactions between the crowding agents and proteins, and, most importantly, the effects of crowders on solvent properties.

Keywords Macromolecular crowding · Excluded volume effect · Viscosity · Soft interactions · Solvent properties · Aqueous two-phase system · Partition

Crowded Environments Beyond Excluded Volume Effects

While numerous studies have suggested that the excluded volume model offers a sufficient phenomenological description of how protein behavior is affected by crowded media, recent findings show that other aspects of these solutions must be considered as well. One study performed by a calorimetric analysis of ubiquitin

B. Y. Zaslavsky
Cleveland Diagnostics, Cleveland, OH, USA
e-mail: Boris.Zaslavsky@ClevelandDx.com

V. N. Uversky (✉)
Department of Molecular Medicine and USF Health Byrd Alzheimer's Research Institute, Morsani College of Medicine, University of South Florida, Tampa, FL, USA
e-mail: vuversky@usf.edu

V. N. Uversky (ed.), *(Macro)Molecular Crowding*, Subcellular Biochemistry 109,
https://doi.org/10.1007/978-3-032-03370-3_3

thermal unfolding in the presence of different cosolutes, such as KCl, glucose, Dextran, urea, and PEG, and found that, contrary to the excluded volume theory which states that the behavior of a protein in crowded environments is based on hardcore steric repulsions among the macromolecules and should therefore have a clear entropic origin, an entropic destabilization and an enthalpic stabilization for Dextran, glucose, and PEG have been discovered (Senske et al. 2014). Curiously, PEG and Dextran displayed different effects on ubiquitin stability, with Dextran leading to enthalpic stabilization, while the presence of PEG resulted in entropy dominating over enthalpy, causing an overall destabilization (Senske et al. 2014). As such, there are clearly other aspects than excluded volume effects which result in the mechanistic complexity of crowded environments; some of these aspects are outlined in detail in the ensuing sections.

Molecular Confinement

Molecular confinement is another important aspect of an intracellular crowded environment. Two experimental methods have been developed to model the limited available space of this molecular imprisonment: using sol-gel techniques to examine the encapsulation of a studied protein in silica glass, and encapsulating a test protein into reverse micelles. Even though the portion of the total volume excluded by the silica matrix is smaller than the fractional volume occupied by macromolecules in a living cell, sol-gel encapsulation is nonetheless considered as a reliable model of molecular confinement(Eggers and Valentine 2001a, b). However, since pore size during the gelation process may depend on the protein itself, the sizes of protein-occupied pores are on the same scale as the protein diameters (Lan et al. 1999).

Even though proteins are not covalently bound to the highly porous silica matrix, and the encapsulated macromolecules cannot escape the glass under most solvent conditions (Eggers and Valentine 2001a), the sol-gel glass matrix significantly limits the rotational freedom of the protein (Gottfried et al. 1999) without hindering the exchange of the solvent that freely permeates the silica matrix. This approach is also useful due to the optical transparency of the resulting glass products, which allow popular spectroscopic techniques for monitoring protein structure in dilute solutions to be used, such as circular dichroism (CD) (Eggers and Valentine 2001a, b) and fluorescence (Brennan 1999; Bismuto and Irace 2001). Using this approach, CD was used to analyze the solvent effects on the secondary structure of several ordered proteins following encapsulation in the hydrated pores of a silica glass matrix by the sol-gel method (Eggers and Valentine 2001a, b). The findings showed that the conditions resulting from sol-gel glass encapsulation do not dramatically affect the function and structure of several globular proteins (Brennan 1999; Dave et al. 1994; Gill and Ballestros 2000).

Another method for exploring the effect of molecular confinement on the dynamics of both a target protein and solvent molecules is encapsulating the protein within a reverse micelle shell and dispersing the resulting particle in a low viscosity fluid, such as short chain alkanes (Ehrhardt et al. 1999; Babu et al. 2003; Peterson et al. 2005; Pometun et al. 2006; Roche et al. 2013). Reverse micelles are unique since they represent nano-size pockets of encapsulated water that are separated from an organic medium by a surfactant (Grandi et al. 1981; Davis et al. 1996; Pileni 1993). The micelle size and the consequent degree of protein/peptide hydration can be adjusted by changing the water-to-surfactant concentration ratio (surfactants like AOT and sodium bis(2-ethylhexyl) sulfosuccinate) in the organic solvent (Christ and Schurtenberger 1994; Roche et al. 2013). Increasing the amount of water results in a larger micelle, which in turn allows for examining the effect of a systematic enhancement of the number of hydration waters surrounding a confined protein (Grandi et al. 1981; Mukherjee et al. 2006). As with sol-gel glass encapsulation, various spectroscopic techniques can be used to analyze the structural properties of proteins in reverse micelles, such as fluorescence, NMR (Ehrhardt et al. 1999; Babu et al. 2003; Peterson et al. 2005; Pometun et al. 2006; Roche et al. 2013), and CD (Davis et al. 1996).

Both peptides and proteins can also be encapsulated in mesoporous silica with tunable pore diameter and easily functionalized surface (Ravindra et al. 2004; Siefker et al. 2014), in carbon nanotubes (O'Brien et al. 2008) or silica tubes (Mandal et al. 2014), inside the weakly hydrophobic pores of methacrylate resin (Radhakrishna et al. 2013) or the cavities of polyacrylamide gels (Bolis et al. 2004; Pastore et al. 2007; Sanfelice et al. 2009, 2013), or in self-assembled organic nanotubes (Kameta et al. 2010).

While the encapsulation techniques outlined above offer suitable conditions for studying the effects of molecular confinement on protein dynamics and structure, as well as solvent properties and dynamics, using solutions with high concentrations of inert polymers for modelling the various effects of macromolecular crowding remains the most common approach.

Not All Crowding Agents Are Created Equal

The chemical nature of bystander molecules does not play a role in the excluded volume model, given that the phenomenon is solely described by the volume that these molecules occupy. As such, inert molecules with similar hydrodynamic volumes but different chemical natures are likely to affect protein behavior in a similar way. However, some studies have shown that not all crowding agents are created equal, and solutes vary in their effectiveness as crowders (Turoverov et al. 2015). Elcock posited the important question regarding the existence of truly inert crowding agents which could be utilized to provide only a direct readout of excluded volume effects, or if all crowding agents will inevitably cause additional effects that cannot be ignored (Elcock 2010).

It has been mentioned that the presence of site-specific interactions, surface charges, and an intricate surface topology could result in making proteins more complex and biologically relevant crowding agents than any simple polymer (King et al. 2014). A comparative study of the enzymatic activity modulating effects of high concentrations of globular proteins (lysozyme and hemoglobin), PEGs, and various Dextrans showed that these crowding agents had different effects on the reaction rates of various decarboxylating enzymes, such as homotetrameric pyruvate decarboxylase, multisubunit enzyme urease, and monomeric glutamate decarboxylase (Derham and Harding 2006). Globular proteins were shown to drastically increase enzymatic activity, while enzymatic activity decreased in a concentration-dependent manner in the presence of PEGs and Dextrans (Derham and Harding 2006). Dextran-70 and Ficoll-70 affected the aggregation propensity of a β-rich protein bovine carbonic anhydrase (BCA) in different ways (Mittal and Singh 2014). Here, the presence of Dextran-70 resulted in a noticeably lower BCA aggregation relative to the effects of Ficoll-70 (for solutions containing the same proportions of these crowders) (Mittal and Singh 2014).

The inequality of crowding agents can also be illustrated by the observation that mixed macromolecular crowding agents have different effects on protein behavior than solutions of the individual crowders. For instance, while oxidative refolding of unfolded and reduced hen egg white lysozyme was almost entirely abolished at high concentrations of individual crowding agents due to protein aggregation (van den Berg et al. 1999), using a mixed macromolecular crowding agent comprising both a polysaccharide and bovine serum albumin (BSA) caused a noticeable increase in properly folded protein yield, suggesting that mixed crowded media may be more favorable for protein folding (Zhou et al. 2004). The refolding another protein, rabbit muscle creatine kinase (MM-CK), from its guanidinium hydrochloride (GdmHCl)-unfolded state was differently affected by four single macromolecular crowding agents: Dextran-70, Ficoll-70, calf thymus DNA (CT DNA), and PEG-2000, and three mixed crowding agents containing both a polysaccharide (or PEG-2000) and CT DNA, with mixed crowding agents having a better effect on MM-CK refolding (Du et al. 2006).

Finally, a set of flexible, inert, and hydrophilic polymers with largely spherical shape (Dextran, PEG, and Ficoll), which were normally used for modeling macromolecular crowding, was recently expanded to include more rigid polymers with large surface to volume ratio, such as the cellulose derivative hydroxypropyl cellulose (HPC) (Breydo et al. 2014). This expansion was justified since, in addition to spherical biomacromolecules, a cell contains several more rigid (and therefore characterized by rod-like shapes) biopolymers such as protein fibers, DNA, and polysaccharide components of the extracellular matrix. Because these rigid polymers have high exposed surface areas, they could be more available for protein interactions. One study showed that compact, flexible polysaccharides (such as Dextrans) and more rigid HPCs have varying effects on the aggregation and structure of several proteins, with differing degrees of oligomeric state and intrinsic disorder(Breydo et al. 2014). It has therefore been concluded that different biopolymers have varying influence on protein aggregation due to complex effects of different nature, including solution viscosity, excluded volume effect, and protein-polymer interaction (Breydo et al. 2014).

Therefore, there is a lot of evidence to support the idea that, in addition to excluded volume effects, other factors caused by the presence of high concentrations of "inert" macromolecules in crowded solution should be considered. The important players of crowded environments are direct physical interactions between proteins and crowding agents, viscosity, soft interactions, the effects of crowders on solvent properties, and perturbed diffusion.

Roles of Direct Interactions of Proteins with Crowding Agents

If an "inert" bystander molecule possesses some affinity to a particular protein, then direct interactions between the protein and the crowding agent will indubitably occur due to the high concentrations of crowder used. Different affinities of a protein to various crowding agents may explain the non-uniform response of the protein to different crowded environments. For instance, the structural stability of human α-lactalbumin (HLA) and the molten globule state transitions of this protein that can be induced in Ca^{2+}-depleted HLA (apo-HLA) at pH 7.0 and 55 °C (T_m of this process in dilute media is 39.7 °C) or at pH 3.0 and 25.0 °C, were differently affected by varying crowding agents, such as Dextran-70, Ficoll-70, PEG-2000, and PEG-20000 (Zhang et al. 2012). The thermal stability of apo-HLA was decreased noticeably by high concentrations of PEG-2000 and PEG-20000 (the corresponding T_m decreased by 6 and 4 °C, respectively), resulting in the formation of a stable molten globular state at lower temperatures. On the other hand, high concentrations of Dextran-70 and Ficoll-70 drastically increased the thermal stability of apo-HLA (by 12.6 and 3.8 °C, respectively) (Zhang et al. 2012). These differences in the impacts of different crowders on thermal stability of apo-HLA were ascribed to the fact that no noticeable interaction was observed between apo-HLA and Dextran-70 or Ficoll-70, while there was a minor non-specific interaction between the protein and PEG-2000 (Zhang et al. 2012).

Crowding agent interactions with a protein can affect not only its conformational behavior, but its functionality as well. For example, the Ca^{2+}/CaM-regulated formation of a complex between the Ca^{2+}/CaM-dependent protein kinase II (CaMKII) and the NMDA-type glutamate receptor (NMDAR) subunit GluN2B is differently impacted by varying macromolecular crowding agents, where interaction efficiency is reduced by lysozymes but enhanced by BSA (Goodell et al. 2014). As with BSA, high concentrations of Dextran-10, Dextran-70, and poly(vinylpyrrolidone) with molecular weight of 40 kDa (PVP-40) caused a more efficient formation of the CaMKII-GluN2B complex, while adding 100 mg/ml of non-specific rabbit IgG caused a reduction in CaMKII binding to GluN2B (Goodell et al. 2014). The inhibitory effects of IgG and lysozyme in the presence of Ca^{2+} were caused by the binding of the regulatory CaM protein to these protein crowders (Goodell et al. 2014).

The rate of 4-methylumbelliferyl-β-D-N-N′-N″ triacetylchitotrioside ((NAG)(3)-MUF) hydrolysis by hen egg white lysozyme was drastically decreased by the addition of human serum albumin (HSA) (Calderon et al. 2011). Though the reaction followed a Michaelis-Menten mechanism under the specified conditions, the increase in HSA concentration caused a tenfold decrease of the catalytic rate constant, while the Michaelis constant was not dependent on HSA in a wide range of albumin concentration (Calderon et al. 2011). The authors explained that the decrease in lysozyme catalytic efficiency was not caused by HSA-induced macromolecular crowding or substrate association with albumin. The results of the ultracentrifugation analysis instead suggested that the existence of an HSA-lysozyme interaction was established, with the lowered catalytic rate constant with HSA concentration mirroring the drop in free lysozyme concentration, signifying the existence of at least two enzyme populations: the active free enzyme and the inactive HSA-bound lysozyme (Calderon et al. 2011). Analysis of the theoretical HSA-lysozyme complex explained this HSA-induced inhibition, where the active site residues Asp52 and Glu35 of lysozymes were oriented toward the HSA surface (Calderon et al. 2011).

One study offers analysis of oxidative folding of four disulfide bond-containing model peptides with lengths varying from 13 residues (1.4 kDa) to 58-residues (6.5 kDa) (conotoxins GI, PVIIA, and r11a, and bovine pancreatic trypsin inhibitor) in various crowded environments (Buczek et al. 2007). The addition of high concentrations of polysaccharides did not have an effect on the folding rates or equilibria for these peptides, while the presence of protein-based crowding agents (albumin or ovalbumin) drastically accelerated folding reactions, even when the crowder concentrations were lower than those predicted to provide the excluded volume effects (Buczek et al. 2007). Given that both protein-based crowding agents contained cysteins, it was concluded that they might serve as redox-active crowding agents (Buczek et al. 2007).

Another recent study examined the interactability of the globular folded protein, peptidyl-prolyl isomerase Pin1, in *Xenopus laevis* oocytes and in native-like crowded oocyte extract by in-cell NMR spectroscopy, and it was determined that, prior to substrate recognition, active Pin1 is involved in many weak nonspecific attractive interactions with intracellular proteins (Luh et al. 2013). Furthermore, the popular synthetic crowding agent Ficoll-70 was unable to mimic the intracellular environment, since it cannot simulate biologically important weak attractive interactions (Luh et al. 2013). The authors showed that these findings agree with another model (McConkey 1982) where, due to molecular crowding inside the cells, most globular folded proteins with surface charge properties close to neutral under physiological conditions may play a role in forming macromolecular complexes with other sticky proteins (Luh et al. 2013).

Soft Interactions Between Target Proteins and Crowding Agents

In addition to the direct interactions described above, proteins can be involved in non-specific (or soft) interactions with different crowding agents, and these interactions can affect how proteins in a crowded environment behave (Minton 2013).

In order to examine the contributing interactions which may modulate protein response to the addition of cosolutes in aqueous solution, Harries and colleagues explored the effects that adding various cosolutes had on folding a 16 amino-acid peptide model into a β-hairpin (Politi and Harries 2010; Sukenik et al. 2013; Sapir and Harries 2014). The authors compared the folding-inducing and structure-stabilizing effects of naturally occurring stabilizing osmolytes, such as polyols (glycerol, sorbitol, manitol, erythriol, and xylitol), sugar (dissacharide trehalose), and polymers affecting a target peptide via macromolecular crowding effects (such as dextran-20, dextran-40, PEG-100, PEG-400, and PEG-4000).

It was shown that while both classes of solutes had similar, size-dependent stabilization effects on the peptide, there were considerable differences in the underlying molecular mechanisms of this stabilization (Politi and Harries 2010; Sukenik et al. 2013; Sapir and Harries 2014). In fact, osmolytes acted as an enthalpic stabilization mechanism which clearly promoted peptide folding (Politi and Harries 2010), while macromolecular crowding agents acted in a concentration-dependent manner (Politi and Harries 2010; Sukenik et al. 2013; Sapir and Harries 2014). Here, the effects of crowders were primarily entropic in nature at low concentrations, while a significant enthalpic contribution was observed at higher concentrations. These findings show that in solutions of macromolecular crowders, there is an intricate balance between the excluded volume effects (or "depletion interactions") driven by entropy, and some direct "chemical" interactions between the crowder and target protein involving considerable enthalpic contributions (Politi and Harries 2010; Sukenik et al. 2013; Sapir and Harries 2014). The authors theorized that the decomposition of folding free energy into its entropic and enthalpic components offers a glimpse into the methods by which target proteins are stabilized or destabilized by different cosolute families (Politi and Harries 2010; Sukenik et al. 2013; Sapir and Harries 2014).

A recent in-cell NMR study analyzed the capability of the cytoplasm of *Escherichia coli* to promote folding of a destabilized globular protein (Schlesinger et al. 2011).A mutant form of protein L (ProtL), a 7 kDa globular protein from the mesophile *Streptococcus magnus* with seven lysine residues replaced by glutamic acids, was expressed in *E. coli*. While the wild-type protein exhibited relatively high conformational stability (since just 0.1% of its molecules populated the denatured state at room temperature in dilute buffer) 84% of the ProtL mutant populated the unfolded state under the same conditions (Schlesinger et al. 2011). Such a mutation-induced destabilization was believed to be caused by the variant's decreased hydrophobic surface area, rather than an increase in negative charge, and was overcome by adding Na^+ or K^+ salts in dilute solutions (Schlesinger et al. 2011). Mutant ProtL remained mostly unfolded in the cytoplasm, even after an increase of salt concentration inside *E. coli* (Schlesinger et al. 2011). These findings indicated the presence of some nonspecific interactions between the cytoplasmic components and the protein. As such, the stabilizing excluded-volume effects caused by crowded conditions can be improved by nonspecific interactions between cytoplasmic components (Schlesinger et al. 2011).

The enthalpic and entropic contributions that crowding had on the stability of ubiquitin were determined via analysis of temperature dependence of NMR-detected amide proton exchange (Wang et al. 2012b). It was shown that chemical interactions with crowding agents may be a major contributor to protein stability, often overcoming the contribution by hardcore repulsions (excluded volume effect). Consequently, both soft chemical interactions and hard-core repulsions caused by excluded volume have to be taken into consideration when evaluating the effects that crowding has on protein dynamics, structure, and stability in cells (Wang et al. 2012b).

High concentrations of crowding agents are enough to define the appearance of the excluded volume effects, which are entirely entropic hard-core repulsions. Meanwhile, a protein of interest's weak interactions with crowding agents can be attractive or repulsive. These weak interactions can therefore diminish or enhance the excluded volume effects (Sarkar et al. 2013). By employing an NMR-detected amide proton exchange and different concentrations of the lyophilized, dialyzed, and resuspended *E. coli* cytoplasm as a model crowding agent, one study discovered the presence of a noticeable destabilization of chymotrypsin inhibitor 2 (CI2) in these crowded solutions (Sarkar et al. 2013). Increased cytosol concentration resulted in an increase in the destabilization effect, suggesting that the non-specific favorable interaction of the cytoplasm with CI2 managed to overcome the stabilizing hard-core repulsions that excluding volume effects typically cause (Sarkar et al. 2013). Furthermore, the cytosol effects were considerably stronger than the destabilizing effects of homogeneous protein crowders, strongly suggesting the biological significance of weak, nonspecific interactions (Sarkar et al. 2013).

Similarly, after analyzing free energies of the 56-residue-long B1 domain of protein G (GB1) in living *E. coli* cells, it was discovered that, under nonperturbing conditions, the NMR-detected H/D exchange can measure the equilibrium thermodynamic stability of a protein inside the cell on an individual amino acid residue level (Monteith and Pielak 2014). The thermodynamic stability of a protein was calculated using the relative populations of its denatured and native states which are evaluated by NMR via analysis of the efficiency of hydrogen–deuterium exchange at the residue level. These experiments indicated that, unlike dilute solution data (pH 7.6 and 37 °C), the cellular environment causes significant stabilization of the protein (Monteith and Pielak 2014). Meanwhile, GB1 was considerably destabilized when among homogeneous protein crowders (such as lysozymes) *in vitro* due to the prevalence of weak, attractive interactions between positively charged lysozymes (pI = 11.3) and anionic GB1(Monteith and Pielak 2014). The study also showed that these attractive interactions between the crowded molecules (crowdees) and the crowding agents (crowders) are likely to cause destabilization since, being highly extended, an unfolded protein molecule exhibits a larger reactive surface than a folded one, and as such the free energy of the unfolded state ensemble relative to the folded state is lower (Monteith and Pielak 2014). Furthermore, because the cellular environment and concentrated solutions of crowding agents had opposite effects on the stability of GB1, there is clearly a challenge in recreating the cellular interior by simple means (Monteith and Pielak 2014). Given these findings, the use of in-cell NMR and NMR-detected hydrogen–deuterium exchange of quenched cell lysates was suggested for quantifying the stability of globular proteins inside cells and for the identifications of residues, the two most important aspects for intracellular protein folding (Monteith and Pielak 2014).

The effects of nonspecific, weak, chemical interactions between proteins and crowders on the stability and structure of a globular protein are believed to depend on these soft interactions (Sarkar et al. 2014). Steric repulsion's effect on the equilibrium thermodynamic stability of a test protein is thought to be strengthened by repulsive soft interactions, such as interactions between similarly charged species. Meanwhile, a protein's conformational stability is expected to lower if attractive soft interactions occur between the protein and the crowding molecules (Sarkar et al. 2014). The anionic proteins from *Escherichia coli* used as crowding agents' ability to impact the conformational stability of anionic test protein CI2 at pH 7.0 was examined to confirm these hypotheses (Sarkar et al. 2014). Unexpectedly, though both the test protein and the bacterial anionic proteins used as crowding agents exhibited net charges of the same sign, CI2 was destabilized in an anionic crowded environment, signifying that, in addition to the expected charge-charge repulsion, there were also nonspecific, weak, and attractive interactions between the crowding proteins and CI2. As such, these soft attractive interactions can counteract the stabilizing influence of excluded volume effects and surpass the stabilizing charge-charge repulsion (Sarkar et al. 2014).

Gnutt et al., by using a functionalized PEG whose end-groups were labeled by Atto488 and Atto565 to measure the mean end-to-end distance by FRET, recently showed that *in vitro* model crowding conditions (high concentrations of Ficoll-70 or unlabeled PEG-10,000) result in significant compaction of this crowding sensor, while cell lysates or living HeLa cells where the crowding sensor was injected exhibited no contraction (Gnutt et al. 2014). When this sensor was injected directly into the nucleus, it displayed considerably more expansion than in the diluted buffer or the cytosol, signifying the presence of even stronger interactions between the environment and the crowding sensor. The authors concluded that transient, weak, and nonspecific interactions (e.g. electrostatic, van der Waals, or hydrophobic forces) between the crowder and the crowdee are able to counteract the excluded-volume effects of the highly crowded cellular milieu (Gnutt et al. 2014). Given that this study analyzed the behavior of PEG, rather than proteins in artificially and naturally crowded environments, we can conclude that these findings can be applied globally.

The findings in this section, along with numerous other studies emphasizing the non-equivalency of crowding agents and the superiority of natural components over synthetic crowders, help reinforce the recently proposed hypothesis that crowded environments can be divided into two different classes: structured crowding and uniform crowding (Ma and Nussinov 2013). Using this classification, random crowding conditions created by synthetic particles with a narrow size distribution are labeled as 'uniform crowding', while cellular environments that are highly coordinated and possess characteristic organization and clustering of proteins and other macromolecules can be considered as 'structured crowding' (Ma and Nussinov

2013). Structured crowding's notable features are that crowders are not necessarily inert, and the presence of a specific organization of such crowded media may restrict the search for appropriate binding partners (Ma and Nussinov 2013).

Effects of Solvent Viscosity

Solvent viscosity impacts protein diffusion and varies according to the chemical concentration, nature, and molecular mass of a crowder when in a crowded environment. Past a certain concentration of a crowding agent, the diffusion rates of a protein may be considerably altered by the increased solvent viscosity. One study showed that high-mass macromolecular crowding agents not only affect a solution's dynamic (or bulk) n, but can also alter the microviscosity of the protein environment (Erlkamp et al. 2014). These changes could have significant effects on protein behavior in crowded environments. For instance, it was discovered by analysis of fibrillation and aggregation behavior of proteins in crowded media with increasing crowding agent concentrations that, once a crowder's concentration is sufficiently high, the resulting increase in viscosity defines the lowered amyloid species diffusion rate, resulting in a drastic reduction of the overall aggregation rates (Munishkina et al. 2004; Lee et al. 2012; Breydo et al. 2014). Because rigid polymer (such as HPC) solutions have much higher viscosities, they could have bigger impacts on protein behavior than solutions of less rigid, spherical polysaccharides (such as Dextrans). This notion was confirmed in a study where protein diffusion was strongly reduced, and protein aggregation and folding were slowed considerably in solutions with HPCs (Breydo et al. 2014).

The considerable effect that increased solvent viscosity in crowded environments had on protein folding rates was confirmed in a comprehensive study of the effect of Ficoll on the pressure – and temperature-dependent stability diagram and folding reaction of the globular protein Staphylococcal Nuclease (SNase) (Erlkamp et al. 2014). Using small-angle X-ray scattering, FTIR, and calorimetric measurements it was shown that Ficoll caused a noticeable stabilization of SNase, and the pressure- and temperature-induced equilibrium unfolding transitions of this protein exhibited a noticeable shift to higher pressures and temperatures in solutions containing 30 wt% Ficoll (Erlkamp et al. 2014). In the kinetic analysis of SNase refolding, Ficoll's presence caused a dramatic decrease in the folding rate of SNase (Erlkamp et al. 2014). This was contrary to the effect that macromolecular crowding was expected to have, since folding causes a compaction of the polypeptide chain, and the macromolecular presence near a protein was believed to accelerate folding reaction by means of the excluded volume effect. In fact, it was shown that when a protein undergoes a folding transition, the unfolded state is less compact than either the transition state or any partially folded intermediate state. As such, steric repulsion and excluded volume interactions induced by a macromolecular crowding agent likely stabilize the folding transition state or partially folded intermediate state, consequently increasing the folding rate of the protein (Erlkamp et al. 2014).

Following these findings, it was concluded that the increased folding rate in a crowded milieu lends strong credence to the idea that crowding effects on protein folding kinetics are predominately caused by excluded-volume interactions (Erlkamp et al. 2014). Meanwhile, folding kinetics may also undergo the influence of various dynamic effects, such as frictional drag modulation which occurs due to conformational motions along the folding coordinates. As such, crowding might also result in a reduction in translational (and, to a lesser extent, rotational) diffusion constants (Erlkamp et al. 2014). Given that the folding rate of SNase was decreased proportionally to the crowding-induced increase of the microviscosity of the solvent, it was concluded that the primary effect that a macromolecular crowding agent has on folding kinetics must be based on changes in solvent-induced dynamics (Erlkamp et al. 2014).

Analysis of the effects of solvent viscosity and macromolecular crowding on the apparent translational diffusion coefficients (ADCs) of small metabolites with radius of gyration R_G ranging from 0.7 Å to 4.5 Å in solutions with different degrees of molecular crowding and increasing bulk viscosity yielded several important conclusions (Garcia-Perez et al. 1999). This analysis was vital given that small molecule behavior can be affected by both bulk solvent viscosity and by macromolecular crowding, which affects molecular diffusion by creating obstacles. The authors showed that, following the modified Stokes-Einstein relationship, ADC = k/R_G $(1 + 2.5\Phi)$, where k is a constant for a given temperature and Φ is an obstruction factor representing the fraction of a solution's volume that is occupied by cosolutes, a measure of the molecular crowding in the solution, for same viscosity solutions, metabolite ADCs decreased as the concentration of macromolecular crowding agents increased (Garcia-Perez et al. 1999).

The molecular mechanisms of influence of the crowding agents on the viscosity of a solvent are also expected to vary based on crowding agent size. Here, small crowders have an effect on solvent microviscosity, which measures the friction experienced by a single particle undergoing diffusion due to its environmental interaction on a micrometer-length scale. Meanwhile, crowders of a larger size are believed to impact the macroviscosity of solution, which can be viewed as a measure of steric hindrance to a protein's diffusion as a result of having macromolecular obstacles in solution.

These ideas are supported by a detailed analysis of the formation of a complex between the β-lactamase inhibitor protein (BLIP) and β-lactamase (TEM) in crowded solution using crowding agents of both low and high molecular mass, such as PEG-200, PEG-1000, PEG-3350, PEG-8000, ethylene glycol (EG), and Ficoll-70 (Kozer and Schreiber 2004). This study showed that the dissociation rate constant (k_{off}) was almost entirely unaffected by all crowded environments, whereas the protein–protein association rate constant (k_{on}) was considerably lower in all crowded solutions. It is also worth noting that protein association was affected by small and large crowders in fundamentally different ways (Kozer and Schreiber 2004). Solutions containing low mass crowding agents displayed an inverse dependence between solution viscosity and k_{on}, while high mass polymer solutions had only a small effect on k_{on}, even at viscosities 12-fold higher than water (Kozer and Schreiber 2004).

These findings were interpreted in terms of the formation of a structurally heterogeneous or non-uniform media in the presence of large crowders, where, in addition to how much volume the crowding agent occupied, some areas in the solution contained bulk water, proteins were able to associate and diffuse in these areas almost as if they were in diluted environments (Kozer and Schreiber 2004). Another study provided very similar findings for the effect of macro- and micro-molecular crowding on the interaction between cytochrome *f* and plastocyanin from the cyanobacterium *Phormidium laminosum* (Schlarb-Ridley et al. 2005). Here, high molecular mass viscogens (Dextran-70 and Ficoll-70) had almost no effect on protein association up to a relative viscosity of 4, while low molecular mass viscogens (glycerol, ethanediol, and sucrose) strongly decreased the bimolecular association rate constant over a similar viscosity range (Schlarb-Ridley et al. 2005).

Perturbed Diffusion of Target Proteins in Crowded Milieu

Diffusion governs various dynamic biological processes, and is as such considered the main transport process inside the cell (Roosen-Runge et al. 2011). Though diffusion can result in the dispersion of individual molecules, motion caused by diffusion can also mediate reactant encounters, thereby acting as the major driving force behind biochemical reactions and pattern formation (Weiss 2014). Since to a protein that is diffusing in a crowded environment, free diffusion is impeded by bystander molecules, the *in vivo* diffusion coefficients for globular proteins in living cells were found to be considerably lower than the diffusion coefficients of the same proteins in dilute solutions *in vitro* (Wojcieszyn et al. 1981; Arrio-Dupont et al. 2000; Verkman 2002; Ando and Skolnick 2010; Roosen-Runge et al. 2011).

Another study pointed out the existence of a complex dependence of the lowered protein diffusion rate on the combination of a crowding agent and a protein (Muramatsu and Minton 1988; Dix and Verkman 2008; Banks and Fradin 2005; Wang et al. 2010), and, in several instances, macromolecular crowding induced subdiffusive behavior (Weiss et al. 2004; Banks and Fradin 2005). This subdiffusive motion originates from the deviation from the linearity of the mean square displacement of a protein, $v(t)$, which grows in time, suggesting a power law dependence $v(t) \sim t^{\alpha}$ with $\alpha < 1$; this anomaly varies with the conformation and size of the protein, and with the solution's total protein concentration (Weiss et al. 2004). Computer simulations also showed that, unlike to the naive expectation that subdiffusive behavior may impede diffusive transport efficiency and the associated cell sampling processes, subdiffusion in fact increased the likelihood of finding a nearby target, suggesting that subdiffusion may act as a slower but more reliable diffusive search process inside the cell (Guigas and Weiss 2008).

These findings all suggest that proteins diffuse in crowded solutions anomalously. For instance, NMR analysis of the translational and rotational diffusion of a small globular protein, chymotrypsin inhibitor 2 (CI2) showed a drastic difference in the diffusion behavior in solutions of synthetic polymers, glycerol, and proteins/cell lysates (Wang et al. 2010). In fact, while both rotational and translational diffusion decreased with increasing glycerol solution viscosity, and followed the Stokes-Einstein-Debye law, $D_r = \kappa T/8\pi\eta r^3$, and the Stokes-Einstein law, $D_t = \kappa T/6\pi\eta r$ respectively (where D_r is the rotational diffusion coefficient, D_t is the translational diffusion coefficient, κ is the Boltzmann constant, η is the solution viscosity, and r is the radius of the studied protein), macromolecular crowding induced some considerable deviations from the expected behavior (Wang et al. 2010).

In fact, synthetic polymers were shown to have a greater effect on translation than rotation, resulting in a negative deviation from Stokes laws (Wang et al. 2010). Meanwhile, cell lysates and protein crowders had the opposite effect, as they reduced rotational diffusion more than translational diffusion, resulting in a positive deviation from Stokes laws (Wang et al. 2010). This phenomenon was believed to be generalizable given that the mentioned rotational attenuation did not depend on the total charge and size of the crowding agent protein. NMR relaxation experiments concluded that the presence of weak interactions between the proteins and CI2 serves as the primary source of the difference in the diffusion behavior of CI2 in solutions crowded with synthetic polymers and proteins, implying that weak, non-specific protein-protein interactions may have a fundamental impact on protein diffusion in cells, and that synthetic polymers may not be adequate replicas of an intracellular environment (Wang et al. 2010).

Following analysis of several proteins' diffusion behavior in solutions with high concentrations (up to 200 g/liter) of various background or bystander proteins, it was discovered that the fractional reduction of a protein's diffusion coefficient in the presence of a given weight/volume concentration of crowders usually increased when the protein being examined was larger, and when the bystander proteins were smaller (Muramatsu and Minton 1988). This suggests that diffusive transport of larger proteins and aggregates inside cells may be several orders of magnitude slower than in dilute solution (Muramatsu and Minton 1988). This was confirmed by the finding that the diffusional mobility of a tracer particle in the cytoplasm was considerably lowered when the tracked particle's radius increased, leaving particles with a radius > 25–30 nm immobile (Luby-Phelps et al. 1987; Seksek et al. 1997; Arrio-Dupont et al. 2000).

Roosen-Runge et al. used quasielastic neutron backscattering to examine protein self-diffusion in crowded aqueous solutions of BSA that use BSA itself as a crowding agent (Roosen-Runge et al. 2011). Their findings showed that crowding has a considerable effect on protein self-diffusion on a nanosecond time scale, and that, at biological volume fractions, the diffusion coefficient is significantly lower when compared to the dilute limit. This suggests that, in crowded environments, protein dynamics in general and protein diffusion in particular cannot be solely described in terms of confined motions and excluded volume, and that one must also take hydrodynamic interactions into account (Roosen-Runge et al. 2011). Following these conclusions, analysis of the binding of a β-lactamase to its protein inhibitor in living HeLa cells showed that, for a wild-type and an electrostatically optimized mutant, the association rate constants were only 25% and 50% lower than the corresponding *in vitro* values, and a slower binding mutant exhibited no changes in the rate constant (Phillip et al. 2012). The authors of this study also stressed that the detected changes were considerably smaller than those anticipated for the cell's highly crowded environment (Phillip et al. 2012).

A study comparing the peculiarities of the translational diffusion of an intrinsically disordered protein α-synuclein (14 kDa) and a well-folded globular protein, chymotrypsin inhibitor 2 (CI2, 7.4 kDa), revealed markedly different behavior in dilute solution and under crowded conditions (i.e., under high concentrations (300 g/L) of the synthetic polymers PVP-40 and Ficoll-70 and the globular proteins lysozyme (15 kDa), and BSA (67 kDa)) (Wang et al. 2012a). The findings showed that, in dilute solution, CI2 diffused faster than the disordered protein, while α-synuclein diffused faster than CI2 under crowded conditions, despite the hydrodynamic volume of α-synuclein being known to be considerably larger than that of CI2 (Wang et al. 2012a). The authors concluded that protein shape could be one of the key parameters which determines protein diffusion under crowded conditions (Wang et al. 2012a).

Changes in Protein Hydration and Hydration Dynamics

Water molecules in close proximity to a protein molecule form a hydration shell that is crucial for protein structure and folding, and helps define substrate binding, conformational changes, and molecular recognition (Verma et al. 2011). Changes in protein catalytic activity have been shown to cause changes in protein hydration (Partridge et al. 1998; Pocker 2000; Kornblatt and Kornblatt 2002, Yang et al. 2004; Halling 2004) and may have a considerable effect on the conformational stability and dynamics of a protein (Partridge et al. 1998; Halling 2004). When hydration is lowered, it results in a more rigid structure and altered functionality (Partridge et al. 1998; Halling 2004). In severe cases of dehydration, when a solid protein is dispersed in anhydrous organic solvent, thermal stability may drastically increase and protein functionality may be altered in a number of ways (Verma et al. 2011). In fact, in such unnatural circumstances, enzymes were shown to gain remarkable properties, such as radically altered substrate and enantiomeric specificities, molecular memory, greatly enhanced stability, and the ability to catalyze unusual reactions (Klibanov 1989, 2001).

Lipases (whose natural lipolytic reaction in aqueous media is hydrolysis of carboxylic esters), following dispersal in nearly anhydrous organic solvents, acted as efficient catalysts that stimulated the transesterification reaction between heptanol and tributyrin (Zaks and Klibanov 1985). In addition, these anhydrous enzymes, after dispersal in a 99.98% organic medium, gained new and remarkable properties, becoming extremely thermostable and more selective (Zaks and Klibanov 1985). Furthermore, well-known proteases *Bacillus subtilis* protease (subtilisin Carlsberg) and bovine pancreatic α-chymotrypsin, once dispersed as solids in anhydrous organic solvents, acted as strong catalysts, efficiently directing enzymatic transesterification reactions (Zaks and Klibanov 1988b). The transesterification reaction rate was enhanced by chymotrypsin and subtilisin in octane by an order of 100 billion (Zaks and Klibanov 1988b). Additional analysis of the dependence of catalytic activity of three unrelated enzymes (mushroom polyphenol oxidase, yeast alcohol oxidase, and horse liver alcohol dehydrogenase) in several organic solvents as a function of water content showed that while enzymes remained catalytically active in a variety of organic solvents, enzymatic activity greatly increased in response to the increased water content in the solvents (which was always below the solubility limit) (Zaks and Klibanov 1988a). It was concluded that organic solvents affect an enzyme primarily via interactions with the enzyme-bound layer of water, rather than with the enzyme itself (Zaks and Klibanov 1988a).

PEGs are believed to be preferentially excluded from the hydration shells around proteins or other biomolecules. This results in osmotic stress, which draws water out of a polymer-inaccessible crevice or away from a protein's surface (Verma et al. 2011). Osmotic stress has been used to identify changes in the amount of water molecules associated with single macromolecules undergoing conformational changes (Reid and Rand 1997). Such an osmotically stressed enzyme's catalytic activity is expected to change due to the alterations in the number of bound water molecules in substrate, enzyme, and in the enzyme-substrate complex (change in K_M, the Michaelis constant) or between the transition state and the ground state of the complex (change in V_{max}, the maximal reaction rate) (Verma et al. 2011). As confirmation of this theory, the addition of PEGs of differing molecular weights to yeast hexokinase solutions improved this enzyme's affinity for its substrate glucose (Reid and Rand 1997). These PEG-induced changes in hexokinase activity were believed to be due to PEG's indirect action on an enzyme through PEG-modulated changes in water activity (Reid and Rand 1997). In fact, evaluation of ΔN_w, the difference in the number of water molecules between the glucose-free and glucose-bound conformations of hexokinase which can be considered a release of water from around the protein when HK changes conformation in the direction of cleft closure, showed that ΔN_w measured under different conditions was affected drastically by PEGs (Reid and Rand 1997). ΔN_w increased from 50 to about 326 water molecules per hexokinase molecule PEG molecular weight increased from 300 to 1000 Da, and stayed constant for PEGs over a wide range of higher molecular weights (Reid and Rand 1997). Finally, an increase in PEG-1000 concentrations caused a drastic decrease in the ΔN_w value from 326 to 50, suggesting that the unbound form of hexokinase can be significantly dehydrated by high concentrations of PEG (Reid and Rand 1997).

The effects of PEG-400 or high concentrations of lysozyme on protein and hydration dynamics of hen egg white lysozyme labeled with a metal-carbonyl vibrational probe covalently attached to a solvent accessible histidine residue by ultrafast two-dimensional infrared (2D-IR) spectroscopy were examined in a comparative analysis, which clearly showed that the effects of the protein crowder with complex electrostatic profile and the chemically inert polymer and are different (King et al. 2014). This technique is unique in its ability to offer an accurate description of system dynamics via the frequency–frequency correlation function (FFCF), a powerful observable unique to 2D-IR that identifies the equilibrium structural fluctuations that modulate a probe molecule's transition frequency (King et al. 2014). The resulting FFCFs exhibit rapid initial picosecond decays as a result of the hydration water's motion, followed by a noticeable static offset caused by fluctuations that are too slow to be fully sampled within the experimental window, and are as such attributed to slow protein fluctuations (King and Kubarych 2012).

This technique revealed that hydration dynamics occur with a 2.7 ps time constant in pure D_2O, which is twice as slow as the dynamics of bulk D_2O (King et al. 2014). In solutions with high PEG-400 concentrations (80% v/v) the hydration dynamics almost 4 times slower, and protein contribution increased by ~75% relative to the corresponding contribution at pure D_2O. In addition, at around 50% D_2O, a dynamic transition manifested by an abrupt and significant slowing of protein-hydration dynamics was observed (King et al. 2014). Though both protein and hydration dynamics were strongly coupled at all solvent compositions, they were only weakly coupled to the polymer concentration on either side of this transition (King et al. 2014). Analyzing the FFCFs in solutions containing non-labeled lysozymes as a crowding agent showed the existence of a similar dynamic transition that occurs at a higher water content (~70%) than when PEG-400 was used as a crowding agent. This difference was likely caused by the more significant constraining effect that lysozymes have on surrounding water molecules (King et al. 2014). Examination of the inter-protein distance at which the mentioned dynamic transition occurred suggested that collective hydration of protein molecules over distances of 30–40 Å took place, which was believed to suggest the existence of a dynamical influence of hydration water extending upward of 15–20 Å from the surface of each protein (King et al. 2014).

These findings were confirmed by Terahertz spectroscopy studies which revealed that water dynamics in the solvation layer around one protein molecule are distinct from bulk water out to ≈10 Å, and that protein solvation layers overlap an result in the correlated water network motion beyond 20 Å at high protein concentrations (Ebbinghaus et al. 2007). The findings of these studies suggest that, at the expected concentrations of macromolecules inside of cells (300 mg/mL), most of the water within cells is expected to be involved in slow, collective hydration, utilizing minimal amounts of "bulk-like" water, despite being made up of 50–70% water (King et al. 2014).

Effects of Crowding Agents on Solvent Properties

Crowded environments can be very interesting due to the influence that crowding agents can have on solvent properties. The dielectric and thermodynamic properties of water in aqueous solutions of polymers such as Ficoll, Dextran, PVP, and PEG can change significantly relative to those in pure water (Arnold et al. 1985; Zaslavsky 1994). The solvent polarity of aqueous media in Ficoll, Dextran, and PEG solutions also displayed changes based on concentration and polymer type(Zaslavsky 1994). The process of using PEGs as osmotic stressing agents to change water activity and consequently have an effect on protein hydration was described earlier in Section 2.7 (Verma et al. 2011).

Another approach to evaluate solvent changes in crowded environments utilizes aqueous two-phase systems (ATPSs). These systems are formed in aqueous different water-soluble polymer mixtures, or in a solution of a certain salt and a single polymer, where, under the right conditions, at least two distinct aqueous phases emerge with a well-defined interface (Ananthapadmanabhan and Goddard 1987). For instance, when two specific polymers like Ficoll and Dextran are mixed in water above certain concentrations, two immiscible aqueous layers form in the mixture. An interfacial boundary separating two distinct aqueous-based phases clearly exists, each preferentially rich in one of the polymers, with both phases' aqueous solvent being suitable for biological products (Zaslavsky 1994; Albertsson 1986; Walter et al. 1985). These systems are unique since each phase tends to contain well over 80% water on a molal basis, but they remain immiscible and have different solvent properties (Zaslavsky 1994; Madeira et al. 2010, 2011, 2012; Ferreira et al. 2012; Moody et al. 2005; Willauer et al. 2002). Similar phase separations have also been shown to occur in aqueous solutions of a single polymer upon a change in temperature or increase in salt concentration. The aforementioned phase separation in solution of macromolecules varies with their concentration. The typical thresholds of phase separation in aqueous mixtures of proteins, polymers, and polysaccharides fall within a range of several weight percent of each macromolecule (Keating 2012; Tolstoguzov 2000).

One study showed that, for a 1:1 mixture of Dextran and polyvinyl alcohol, even when two polymers are completely compatible (i.e. miscible) in a dry state (i.e., in the absence of water), their mixture may form an ATPS in water (Zaslavsky 1994). This confirms the theory [49] that phase separation in aqueous mixtures of two polymers stems from the different effects that polymers have on water structure, causing the formation of immiscible hydrogen bond networks of the same aqueous nature.

All known ATPSs are able to modulate the partitioning of various solutes. In fact, each phase of ATPS offers a unique solvent environment for proteins, small organic molecules, RNPs, nucleic acids, or other solutes. An unequal solute distribution can often be caused by differences in solute-solvent interactions in the two phases. As such, a new liquid phase may be specifically depleted or enriched in particular solutes. It was initially shown (Zaslavsky 1994) and later confirmed that, in ATPSs, the partitioning of various solutes from small organic compounds to proteins and nucleic acids does not typically involve their interactions with phase-forming polymers (Moody et al. 2005; Madeira et al. 2008b, a, 2011; Silvério et al. 2010; da Silva et al. 2014; Ferreira et al. 2014, 2015). It wastherefore suggested that the unequal distribution of soluble compounds in ATPSs is caused by differences between solute-solvent interactions in the coexisting phases (Zaslavsky 1994).

The partition ratio of an ionizable solute in ATPS can be defined as (Madeira et al. 2011):

$$\log K_s = S_s \Delta\pi^* + A_s \Delta\beta + B_s \Delta\alpha + C_s c \quad (3.1)$$

where K is the solute partition ratio; $\Delta\pi^*$ is the difference between the solvatochromic solvent dipolarity/polarizability of the coexisting phases, $\Delta\alpha$ is the difference between the solvatochromic solvent hydrogen-bond donor acidity of the phases, $\Delta\beta$ is the difference between the solvatochromic solvent hydrogen-bond acceptor basicity of the phases; c is the difference between the electrostatic properties of the phases; S_s, A_s, B_s and C_s are constants (solute-specific coefficients) that describe the susceptibility of the complementary solute-solvent interactions to the changes in the corresponding solvent properties. The solute-specific coefficients are based on the partition ratios for the solute in a set of ATPS with established solvent properties of the phases by multiple linear regression analysis.

A new amino acid scale was proposed after free amino acids were partitioned into ten different polymer/polymer ATPSs containing 0.15 M NaCl in 0.01 M phosphate buffer at pH 7.4 (Madeira et al. 2014). This scale was based on each amino acid's partition ratio values in various ATPSs using multiple linear regression analysis. This is the first amino acid scale defined by their interactions with water. It offers a unique amino acid physico-chemical description that describes their hydrogen bonding, dipole-dipole, and electrostatic interactions with an aqueous environment (Madeira et al. 2014). It was also shown that linear combinations of solute-specific coefficients A_s, S_s, C_s, and B_s were correlated with various structural, physicochemical, and biological amino acid properties (Madeira et al. 2014). Finally, partitioning closely related proteins in ATPSs was shown to be different, and protein partitioning in a set of ATPSs of different compositions was also shown to be used to quantify differences between closely related proteins' 3D structures (Zaslavsky et al. 2013).

Another study showed that protein partitioning in a set of ATPSs of different ionic compositions served as a useful tool for quantifying the structural differences between globular proteins, α-synuclein, and its variants (Breydo et al. 2013). The mechanisms underlying the behavior of protein in ATPSs were studied in a recent analysis, where 15 proteins in Dextran-70-PEG-8000 ATPSs were partitioned in the presence of 0.01 M sodium phosphate buffer, pH 7.4, and various salt additives (CsCl, NaCl, NaClO4, Na_2SO4, and NaSCN) (Ferreira et al. 2014). It was shown that the protein partition behavior is affected by the presence and concentration of salt additives. The differences in protein partition behavior that were observed in these systems cannot be explained by protein size differences, or by differences in the phases' relative hydrophobicity and electrostatic properties.

Examination of roughly 50 different protein structure descriptors suggested that protein partition behavior is defined by their surface peculiarities (e.g., the average hydrophobicity of surface residues and the number of water-filled cavities), and by the intrinsic flexibility of the protein structure expressed in terms of the B-factor (or temperature factor) (Ferreira et al. 2014).In the absence of physical interaction of proteins with polymers in different ATPS phases, the identified protein surface features must obviously define the peculiarities of protein-solvent interactions in different phases. As such, the inequality of distribution of soluble compounds in ATPSs, which is defined by differences between the solute-solvent interactions in the coexisting phases, serves as a direct indication that, in different phases, solvent water has different properties. Given that the concentrations of polymers experiencing phase separation are in the same range used to model molecular crowding (20–50%), there is a strong indication that the properties of water are affected differently in different crowded environments.

For any given solute, the solute-solvent interactions are governed by the solvent's properties, though the contribution offered by each property would depend on the solute's physicochemical features and structure. Cabot and Hunter (2012) mention that most quantitative approaches studying solvation phenomena have used specially designed spectroscopic probes that can easily detect changes in their environment (Reichardt and Welton 2010; Marcus 1998; Reichardt 2007). The most commonly used term for the classification of solvents is 'polarity'. This is a very poorly defined term, currently meaning the sum of all possible specific and non-specific interactions between the solvent and any potential solute, and excludes any interactions that lead to the solute's chemical transformations (Reichardt and Welton 2010; Ab Rani et al. 2011). The solute-solvent interactions are varied, and include dipole-dipole, dipole-induced dipole, electrostatic, hydrogen bonding, and electron pair donor-acceptor interactions. It is particularly important that polarity defines the potential behavior of the solvent in a relationship with the solute, which is not an absolute property that a pure solvent possesses (Ab Rani et al. 2011). There are many different polarity scales that use various probes and spectroscopic techniques (IR, NMR, and UV/Visible absorption and emission spectroscopy to name a few) (Cabot and Hunter 2012). Since all polarity scales are estimates made under varying assumptions, and since different scales can offer differing estimates for the same solvent, there can be no universal measure of solvent polarity (Ab Rani et al. 2011).

Two studies have measured the solvent polarity of aqueous media in solutions of different ATPS-forming polymers, such as Ficoll, Dextran, and PEG of different molecular weights using the water soluble carboxylate-substituted anionic betaine Reichard's solvatochromic dye and thymol blue (Zaslavsky et al. 1990), or a number of different sulphonephthalein dyes, such as eosin and fluorescein (Zaslavsky et al. 1991). The phase solvent polarity of Dextran-Ficoll and Dextran-PEG ATPS was calculated using the same solvatochromic dyes, and was found to differ in the coexisting phases (Zaslavsky et al. 1992). The differences in the polarity of a solvent of aqueous media in the two phases of polymer/polymer ATPSs, as well as the differences in the electrostatic and hydrophobic phase properties described above, were used to posit the theory (Zaslavsky 1992, 1994) that polymer/polymer ATPSs can be considered similar to organic solvent-water biphasic systems, with the distinction of both phases having the same aqueous nature.

Any single-parameter polarity scale is too limited to describe the large number of possible solute-solvent interactions. As such, the Linear Solvation Energy Relationship (LSER) was used to develop three multi-parameter polarity scales, namely hydrogen bond donor acidity (α) (Taft and Kamlet 1976), hydrogen bond acceptor basicity (β) (Kamlet and Taft 1976), and dipolarity/polarizability (π*)

(Kamlet et al. 1979). Together, these three scales help explain a given solvent's ability to participate in solute-solvent interactions; i.e., solvent polarity, much more effectively than any single-parameter polarity scale. This LSER model can be defined as:

$$(XYZ) = (XYZ)_o + s\pi^* + a\alpha + b\beta \quad (3.2)$$

where (XYZ) is the solute property (reaction rate, solubility, the logarithm of a gas/solvent or solvent/solvent partition coefficient, equilibrium constant, etc.) in a given solvent; $(XYZ)_o$ is the same solute property in a reference state; *s*, *a*, and *b* are the solute-dependent coefficients characterizing the respective influence of the π^*, α, and β terms on the (XYZ) property being studied.

In order to test the theory that crowding agents may induce changes in the solvent properties of aqueous media, the solvatochromic comparison method was utilized to calculate the hydrogen-bond donor acidity (α), solvent dipolarity/polarizability (π^*),and hydrogen-bond acceptor basicity (β) of aqueous solutions of different polymers (PEG, Dextran, Ficoll, PVP, and UCON (a copolymer of ethylene glycol and propylene glycol)), with the crowding agents typically used being polymer concentrations up to 40% (Ferreira et al. 2015). Polymer-induced changes in these features were found to depend on polymer type and concentration, and, in the case of PEGs, on molecular mass. It is worth noting that similarly sized polymers UCON and PEG generated drastically different changes in the solvent properties of water in their solutions. The formation of α-synuclein aggregates with very different morphologies was also shown to be induced by these two polymers (Ferreira et al. 2015). Finally, the altered solvent properties in solutions of polymers offer a quantitative definition of the crowding effects on the stability and refolding of a protein that has been described in literature (Ferreira et al. 2015). All these findings strongly suggest the idea that crowding agents do induce changes in solvent properties of aqueous media in a crowded environment, and that any crowding effect analysis should take these changes into consideration.

Conclusions

The excluded volume effect is clearly not the only factor that affects how various biomolecules (including proteins) behave in a crowded environment. In addition to excluded volume effects, which can be viewed as the 'hard' non-specific steric interactions, there are other important factors of crowded environments, such as perturbed diffusion, viscosity, direct physical interactions between the crowding agents and proteins, so-called "soft" interactions (such as hydrophobic, electrostatic, and van der Waals interactions) between a protein and the crowding agent, and, most importantly, the effects that crowders have on solvent properties. It is highly probable that the changes that crowding agents can induce in the solvent properties of aqueous media may be the root of the "soft" interactions between a crowding agent and a studied protein (Ferreira et al. 2015).

Macromolecular crowding agents have been shown to alter the solvent properties of aqueous media in their solutions. The solvent dipolarity/polarizability, hydrogen-bond acceptor basicity, and hydrogen-bond donor acidity of aqueous media evaluated in crowding agent solutions were shown to be crowding agent-specific and to depend on crowder concentration. Additional support of the idea that steric impediment or hindrance that defines the basis of excluded effect is not the sole mechanism that affects the behavior of proteins in crowded environments comes from the important observations that UCON and PEG, despite being the same size, generate drastically different changes in the solvent properties of water in their solutions, and induce morphologically different α-synuclein aggregate forms. This clearly shows that crowding agent-induced changes in the solvent properties of aqueous media play a major role in macromolecular crowding effects. As such, "soft" interactions of a biological macromolecule with crowding agents were proposed to be considered as the interactions between the macromolecule and aqueous media with solvent properties altered via influence of a crowding agent.

However, it should be kept in mind that the complexity of the intracellular environment extends far beyond macromolecular crowding. Selenko and colleagues pointed out that understanding the peculiarities of protein function inside cells relies on understanding the intracellular environment's specific physical properties, where multiple factors are involved (Theillet et al. 2014). The factors that shape protein cellular behavior include the compositions of the eukaryotic cytoplasm and nucleoplasm and the prokaryotic cytoplasm, concentrations of metabolite and ion, dielectric properties of the medium, intracellular viscosity, the presence of numerous membrane surfaces, and, of course, macromolecular confinement and crowding (Theillet et al. 2014).

Compliance with Ethical Standards

Authors declare no conflict of interest.
Funding source: Not applicable.

References

Ab Rani MA, Brant A, Crowhurst L, Dolan A, Lui M, Hassan NH, Hallett JP, Hunt PA, Niedermeyer H, Perez-Arlandis JM, Schrems M, Welton T, Wilding R (2011) Understanding the polarity of ionic liquids. Phys Chem Chem Phys 13(37):16831–16840. https://doi.org/10.1039/C1cp21262a

Albertsson PA (1986) Partition of cell particles and macromolecules, 3rd edn. Wiley, New York

Ananthapadmanabhan KP, Goddard ED (1987) Aqueous biphase formation in polyethylene oxide-inorganic salt systems. Langmuir 3:25–31

Ando T, Skolnick J (2010) Crowding and hydrodynamic interactions likely dominate in vivo macromolecular motion. Proc Natl Acad Sci USA 107(43):18457–18462. https://doi.org/10.1073/pnas.1011354107

Arnold K, Herrmann A, Pratsch L, Gawrisch K (1985) The dielectric properties of aqueous solutions of poly(ethylene glycol) and their influence on membrane structure. Biochim Biophys Acta 815(3):515–518

Arrio-Dupont M, Foucault G, Vacher M, Devaux PF, Cribier S (2000) Translational diffusion of globular proteins in the cytoplasm of cultured muscle cells. Biophys J 78(2):901–907. https://doi.org/10.1016/S0006-3495(00)76647-1

Babu CR, Flynn PF, Wand AJ (2003) Preparation, characterization, and NMR spectroscopy of encapsulated proteins dissolved in low viscosity fluids. J Biomol NMR 25(4):313–323

Banks DS, Fradin C (2005) Anomalous diffusion of proteins due to molecular crowding. Biophys J 89(5):2960–2971. https://doi.org/10.1529/biophysj.104.051078

Bismuto E, Irace G (2001) The effect of molecular confinement on the conformational dynamics of the native and partly folded state of apomyoglobin. FEBS Lett 509:476–480

Bolis D, Politou AS, Kelly G, Pastore A, Temussi PA (2004) Protein stability in nanocages: a novel approach for influencing protein stability by molecular confinement. J Mol Biol 336(1):203–212

Brennan JD (1999) Using fluorescence to investigate proteins entrapped in sol-gel derived materials. Appl Spec 53:106A–121A

Breydo L, Mikheeva LM, Madeira PP, Zaslavsky BY, Uversky VN (2013) Solvent interaction analysis of intrinsically disordered proteins in aqueous two-phase systems. Mol BioSyst 9(12):3068–3079. https://doi.org/10.1039/c3mb70329k

Breydo L, Reddy KD, Piai A, Felli IC, Pierattelli R, Uversky VN (2014) The crowd you're in with: effects of different types of crowding agents on protein aggregation. Biochim Biophys Acta 1844 2:346–357. https://doi.org/10.1016/j.bbapap.2013.11.004

Buczek O, Green BR, Bulaj G (2007) Albumin is a redox-active crowding agent that promotes oxidative folding of cysteine-rich peptides. Biopolymers 88(1):8–19. https://doi.org/10.1002/bip.20615

Cabot R, Hunter CA (2012) Molecular probes of solvation phenomena. Chem Soc Rev 41(9):3485–3492. https://doi.org/10.1039/c2cs15287h

Calderon C, Abuin E, Lissi E, Montecinos R (2011) Effect of human serum albumin on the kinetics of 4-methylumbelliferyl-beta-D-N N′ N″ Triacetylchitotrioside hydrolysis catalyzed by hen egg white lysozyme. Protein J 30(6):367–373. https://doi.org/10.1007/s10930-011-9339-8

Christ S, Schurtenberger P (1994) Optical contrast variation experiments in water-in-oil microemulsions: size distribution and structure of protein-free and protein-containing microemulsions. J Phys Chem 98(48):12708–12714

da Silva NR, Ferreira LA, Mikheeva LM, Teixeira JA, Zaslavsky BY (2014) Origin of salt additive effect on solute partitioning in aqueous polyethylene glycol-8000-sodium sulfate two-phase system. J Chromatogr A 1337:3–8. https://doi.org/10.1016/j.chroma.2014.02.031

Dave BC, Dunn B, Valentine JS, Zink JI (1994) Sol-gel encapsulation methods for biosensors. Anal Chem 66:1120–1127

Davis DM, McLoskey D, Birch DJ, Gellert PR, Kittlety RS, Swart RM (1996) The fluorescence and circular dichroism of proteins in reverse micelles: application to the photophysics of human serum albumin and N-acetyl-L-tryptophanamide. Biophys Chem 60(3):63–77

Derham BK, Harding JJ (2006) The effect of the presence of globular proteins and elongated polymers on enzyme activity. Biochim Biophys Acta 1764(6):1000–1006. https://doi.org/10.1016/j.bbapap.2006.01.005

Dix JA, Verkman AS (2008) Crowding effects on diffusion in solutions and cells. Annu Rev Biophys 37:247–263. https://doi.org/10.1146/annurev.biophys.37.032807.125824

Du F, Zhou Z, Mo ZY, Shi JZ, Chen J, Liang Y (2006) Mixed macromolecular crowding accelerates the refolding of rabbit muscle creatine kinase: implications for protein folding in physiological environments. J Mol Biol 364(3):469–482. https://doi.org/10.1016/j.jmb.2006.09.018

Ebbinghaus S, Kim SJ, Heyden M, Yu X, Heugen U, Gruebele M, Leitner DM, Havenith M (2007) An extended dynamical hydration shell around proteins. Proc Natl Acad Sci USA 104(52):20749–20752. https://doi.org/10.1073/pnas.0709207104
Eggers DK, Valentine JS (2001a) Crowding and hydration effects on protein conformation: a study with sol-gel encapsulated proteins. J Mol Biol 314(4):911–922
Eggers DK, Valentine JS (2001b) Molecular confinement influences protein structure and enhances thermal protein stability. Protein Sci 10(2):250–261
Ehrhardt MR, Flynn PF, Wand AJ (1999) Preparation of encapsulated proteins dissolved in low viscosity fluids. J Biomol NMR 14(1):75–78
Elcock AH (2010) Models of macromolecular crowding effects and the need for quantitative comparisons with experiment. Curr Opin Struct Biol 20(2):196–206. https://doi.org/10.1016/j.sbi.2010.01.008
Erlkamp M, Grobelny S, Winter R (2014) Crowding effects on the temperature and pressure dependent structure, stability and folding kinetics of staphylococcal nuclease. Phys Chem Chem Phys PCCP 16(13):5965–5976. https://doi.org/10.1039/c3cp55040k
Ferreira LA, Parpot P, Teixeira JA, Mikheeva LM, Zaslavsky BY (2012) Effect of NaCl additive on properties of aqueous PEG-sodium sulfate two-phase system. J Chromatogr A 1220:14–20. https://doi.org/10.1016/j.chroma.2011.12.001
Ferreira L, Fan X, Mikheeva LM, Madeira PP, Kurgan L, Uversky VN, Zaslavsky BY (2014) Structural features important for differences in protein partitioning in aqueous dextran-polyethylene glycol two-phase systems of different ionic compositions. Biochim Biophys Acta 1844 3:694–704. https://doi.org/10.1016/j.bbapap.2014.01.016
Ferreira LA, Madeira PP, Breydo L, Reichardt C, Uversky VN, Zaslavsky BY (2015) Role of solvent properties of aqueous media in macromolecular crowding effects. J Biomol Struct Dynam
Garcia-Perez AI, Lopez-Beltran EA, Kluner P, Luque J, Ballesteros P, Cerdan S (1999) Molecular crowding and viscosity as determinants of translational diffusion of metabolites in subcellular organelles. Arch Biochem Biophys 362(2):329–338
Gill I, Ballestros A (2000) Bioencapsulation within synthetic polymers (part 1): sol-gel encapsulated biologicals. Trends Biotechnol 18:282–296
Gnutt D, Gao M, Brylski O, Heyden M, Ebbinghaus S (2014) Excluded-volume effects in living cells. Angew Chem 54. https://doi.org/10.1002/anie.201409847
Goodell DJ, Eliseeva TA, Coultrap SJ, Bayer KU (2014) CaMKII binding to GluN2B is differentially affected by macromolecular crowding reagents. PLoS One 9(5):e96522. https://doi.org/10.1371/journal.pone.0096522
Gottfried DS, Kagan A, Hoffman BM, Friedman JM (1999) Impeded rotation of a protein in sol-gel matrix. J Phys Chem B 103:2803–2807
Grandi C, Smith RE, Luisi PL (1981) Micellar solubilization of biopolymers in organic solvents. Activity and conformation of lysozyme in isooctane reverse micelles. J Biol Chem 256 (2):837–843
Guigas G, Weiss M (2008) Sampling the cell with anomalous diffusion – the discovery of slowness. Biophys J 94(1):90–94. https://doi.org/10.1529/biophysj.107.117044
Halling PJ (2004) What can we learn by studying enzymes in non-aqueous media? Philos Trans Royal Soc London Ser B Biol Sci 359(1448):1287–1296; discussion 1296–1287, 1323–1288. https://doi.org/10.1098/rstb.2004.1505
Kameta N, Minamikawa H, Someya Y, Yui H, Masuda M, Shimizu T (2010) Confinement effect of organic nanotubes toward green fluorescent protein (GFP) depending on the inner diameter size. Chemistry 16(14):4217–4223. https://doi.org/10.1002/chem.200903413
Kamlet MJ, Taft RW (1976) Solvatochromic comparison method .1. Beta-scale of solvent hydrogen-bond acceptor (Hba) basicities. Journal of the American Chemical Society 98 (2):377–383. https://doi.org/10.1021/Ja00418a009
Kamlet MJ, Hall TN, Boykin J, Taft RW (1979) Linear solvation energy relationships .6. Additions to and correlations with the pi* scale of solvent polarities. J Org Chem 44 (15):2599–2604. https://doi.org/10.1021/Jo01329a001

Keating CD (2012) Aqueous phase separation as a possible route to compartmentalization of biological molecules. Acc Chem Res 45(12):2114–2124. https://doi.org/10.1021/ar200294y

King JT, Kubarych KJ (2012) Site-specific coupling of hydration water and protein flexibility studied in solution with ultrafast 2D-IR spectroscopy. J Am Chem Soc 134(45):18705–18712. https://doi.org/10.1021/ja307401r

King JT, Arthur EJ, Brooks CL 3rd, Kubarych KJ (2014) Crowding induced collective hydration of biological macromolecules over extended distances. J Am Chem Soc 136(1):188–194. https://doi.org/10.1021/ja407858c

Klibanov AM (1989) Enzymatic catalysis in anhydrous organic solvents. Trends Biochem Sci 14(4):141–144. https://doi.org/10.1016/0968-0004(89)90146-1

Klibanov AM (2001) Improving enzymes by using them in organic solvents. Nature 409(6817):241–246. https://doi.org/10.1038/35051719

Kornblatt JA, Kornblatt MJ (2002) Water as it applies to the function of enzymes. Int Rev Cytol 215:49–73, Elsevier

Kozer N, Schreiber G (2004) Effect of crowding on protein-protein association rates: fundamental differences between low and high mass crowding agents. J Mol Biol 336(3):763–774. https://doi.org/10.1016/j.jmb.2003.12.008

Lan EH, Dave BC, Fukuto JM, Dunn B, Zink JI, Valentine JS (1999) Synthesis of sol-gel encapsulated heme proteins with chemical sensing properties. J Mater Chem 9:45–53

Lee CF, Bird S, Shaw M, Jean L, Vaux DJ (2012) Combined effects of agitation, macromolecular crowding, and interfaces on amyloidogenesis. J Biol Chem 287(45):38006–38019. https://doi.org/10.1074/jbc.M112.400580

Luby-Phelps K, Castle PE, Taylor DL, Lanni F (1987) Hindered diffusion of inert tracer particles in the cytoplasm of mouse 3T3 cells. Proc Natl Acad Sci USA 84(14):4910–4913

Luh LM, Hansel R, Lohr F, Kirchner DK, Krauskopf K, Pitzius S, Schafer B, Tufar P, Corbeski I, Guntert P, Dotsch V (2013) Molecular crowding drives active Pin1 into nonspecific complexes with endogenous proteins prior to substrate recognition. J Am Chem Soc 135(37):13796–13803. https://doi.org/10.1021/ja405244v

Ma B, Nussinov R (2013) Structured crowding and its effects on enzyme catalysis. Top Curr Chem 337:123–137. https://doi.org/10.1007/128_2012_316

Madeira P, Teixeira JA, Macedo EA, Mikheeva LM, Zaslavsky BY (2008a) Correlations between distribution coefficients of various biomolecules in different polymer/polymer aqueous two phase systems. Fluid Phase Equil 267:150–157

Madeira PP, Teixeira JA, Macedo EA, Mikheeva LM, Zaslavsky BY (2008b) "On the Collander equation": protein partitioning in polymer/polymer aqueous two-phase systems. J Chromatogr A 1190(1–2):39–43. https://doi.org/10.1016/j.chroma.2008.03.003

Madeira PP, Reis CA, Rodrigues AE, Mikheeva LM, Zaslavsky BY (2010) Solvent properties governing solute partitioning in polymer/polymer aqueous two-phase systems: nonionic compounds. J Phys Chem B 114(1):457–462. https://doi.org/10.1021/jp907346s

Madeira PP, Reis CA, Rodrigues AE, Mikheeva LM, Chait A, Zaslavsky BY (2011) Solvent properties governing protein partitioning in polymer/polymer aqueous two-phase systems. J Chromatogr A 1218(10):1379–1384. https://doi.org/10.1016/j.chroma.2011.01.020

Madeira PP, Bessa A, Alvares-Ribeiro L, Aires-Barros MR, Reis CA, Rodrigues AE, Zaslavsky BY (2012) Salt effects on solvent features of coexisting phases in aqueous polymer/polymer two-phase systems. J Chromatogr A 1229:38–47. https://doi.org/10.1016/j.chroma.2012.01.029

Madeira PP, Bessa A, Alvares-Ribeiro L, Raquel Aires-Barros M, Rodrigues AE, Uversky VN, Zaslavsky BY (2014) Amino acid/water interactions study: a new amino acid scale. J Biomol Struct Dyn 32(6):959–968. https://doi.org/10.1080/07391102.2013.800994

Mandal SS, Cristiglio V, Lindner P, Bhattacharyya AJ (2014) Small-angle neutron scattering studies of hemoglobin confined inside silica tubes of varying sizes. Chemphyschem 15(2):302–309. https://doi.org/10.1002/cphc.201300863

Marcus Y (1998) The properties of solvents. Wiley, Chichester

McConkey EH (1982) Molecular evolution, intracellular organization, and the quinary structure of proteins. Proc Natl Acad Sci USA 79(10):3236–3240
Minton AP (2013) Quantitative assessment of the relative contributions of steric repulsion and chemical interactions to macromolecular crowding. Biopolymers 99(4):239–244. https://doi.org/10.1002/bip.22163
Mittal S, Singh LR (2014) Macromolecular crowding decelerates aggregation of a beta-rich protein, bovine carbonic anhydrase: a case study. J Biochem 156(5):273–282. https://doi.org/10.1093/jb/mvu039
Monteith WB, Pielak GJ (2014) Residue level quantification of protein stability in living cells. Proc Natl Acad Sci USA 111(31):11335–11340. https://doi.org/10.1073/pnas.1406845111
Moody ML, Willauer HD, Griffin ST, Huddleston JG, Rogers RD (2005) Solvent property characterization of poly(ethylene glycol)/dextran aqueous biphasic systems using the free energy of transfer of a methylene group and a linear solvation energy relationship. Ind Eng Chem Res 44:3749–3760
Mukherjee S, Chowdhury P, Gai F (2006) Tuning the cooperativity of the helix-coil transition by aqueous reverse micelles. J Phys Chem B 110(24):11615–11619. https://doi.org/10.1021/jp062362k
Munishkina LA, Cooper EM, Uversky VN, Fink AL (2004) The effect of macromolecular crowding on protein aggregation and amyloid fibril formation. J Mol Recog JMR 17(5):456–464. https://doi.org/10.1002/jmr.699
Muramatsu N, Minton AP (1988) Tracer diffusion of globular proteins in concentrated protein solutions. Proc Natl Acad Sci USA 85(9):2984–2988
O'Brien EP, Stan G, Thirumalai D, Brooks BR (2008) Factors governing helix formation in peptides confined to carbon nanotubes. Nano Lett 8(11):3702–3708. https://doi.org/10.1021/nl8019328
Partridge J, Dennison PR, Moore BD, Halling PJ (1998) Activity and mobility of subtilisin in low water organic media: hydration is more important than solvent dielectric. Biochim Biophys Acta 1386(1):79–89
Pastore A, Salvadori S, Temussi PA (2007) Peptides and proteins in a confined environment: NMR spectra at natural isotopic abundance. J Pept Sci 13(5):342–347. https://doi.org/10.1002/psc.848
Peterson RW, Lefebvre BG, Wand AJ (2005) High-resolution NMR studies of encapsulated proteins in liquid ethane. J Am Chem Soc 127(29):10176–10177. https://doi.org/10.1021/ja0526517
Phillip Y, Kiss V, Schreiber G (2012) Protein-binding dynamics imaged in a living cell. Proc Natl Acad Sci USA 109(5):1461–1466. https://doi.org/10.1073/pnas.1112171109
Pileni MP (1993) Reverse micelles as microreactors. J Phys Chem 97(27):6961–6973
Pocker Y (2000) Water in enzyme reactions: biophysical aspects of hydration-dehydration processes. Cell Mol Life Sci CMLS 57(7):1008–1017
Politi R, Harries D (2010) Enthalpically driven peptide stabilization by protective osmolytes. Chem Commun 46(35):6449–6451. https://doi.org/10.1039/c0cc01763a
Pometun MS, Peterson RW, Babu CR, Wand AJ (2006) Cold denaturation of encapsulated ubiquitin. J Am Chem Soc 128(33):10652–10653. https://doi.org/10.1021/ja0628654
Radhakrishna M, Grimaldi J, Belfort G, Kumar SK (2013) Stability of proteins inside a hydrophobic cavity. Langmuir 29(28):8922–8928. https://doi.org/10.1021/la4014784
Ravindra R, Zhao S, Gies H, Winter R (2004) Protein encapsulation in mesoporous silicate: the effects of confinement on protein stability, hydration, and volumetric properties. J Am Chem Soc 126(39):12224–12225. https://doi.org/10.1021/ja046900n
Reichardt C (2007) Solvents and solvent effects: an introduction. Org Process Res Dev 11(1):105–113. https://doi.org/10.1021/Op0680082
Reichardt C, Welton T (2010) Solvents and solvent effects in organic chemistry, 4th edn. Wiley-VCH, Weinheim
Reid C, Rand RP (1997) Probing protein hydration and conformational states in solution. Biophys J 72(3):1022–1030. https://doi.org/10.1016/S0006-3495(97)78754-X

Roche CJ, Dantsker D, Heller ER, Sabat JE, Friedman JM (2013) Reverse micelles as a tool for probing solvent modulation of protein dynamics: reverse micelle encapsulated hemoglobin. Chem Phys 430:88–97. https://doi.org/10.1016/j.chemphys.2013.04.006
Roosen-Runge F, Hennig M, Zhang F, Jacobs RM, Sztucki M, Schober H, Seydel T, Schreiber F (2011) Protein self-diffusion in crowded solutions. Proc Natl Acad Sci USA 108(29):11815–11820. https://doi.org/10.1073/pnas.1107287108
Sanfelice D, Tancredi T, Politou A, Pastore A, Temussi PA (2009) Cold denaturation and aggregation: a comparative NMR study of titin I28 in bulk and in a confined environment. J Am Chem Soc 131(33):11662–11663. https://doi.org/10.1021/ja904462n
Sanfelice D, Politou A, Martin SR, De Los RP, Temussi P, Pastore A (2013) The effect of crowding and confinement: a comparison of Yfh1 stability in different environments. Phys Biol 10(4):045002. https://doi.org/10.1088/1478-3975/10/4/045002
Sapir L, Harries D (2014) Origin of enthalpic depletion forces. J Phys Chem Lett 5(7):1061–1065
Sarkar M, Smith AE, Pielak GJ (2013) Impact of reconstituted cytosol on protein stability. Proc Natl Acad Sci USA 110(48):19342–19347. https://doi.org/10.1073/pnas.1312678110
Sarkar M, Lu J, Pielak GJ (2014) Protein crowder charge and protein stability. Biochemistry 53(10):1601–1606. https://doi.org/10.1021/bi4016346
Schlarb-Ridley BG, Mi H, Teale WD, Meyer VS, Howe CJ, Bendall DS (2005) Implications of the effects of viscosity, macromolecular crowding, and temperature for the transient interaction between cytochrome f and plastocyanin from the cyanobacterium Phormidium laminosum. Biochemistry 44(16):6232–6238. https://doi.org/10.1021/bi047322q
Schlesinger AP, Wang Y, Tadeo X, Millet O, Pielak GJ (2011) Macromolecular crowding fails to fold a globular protein in cells. J Am Chem Soc 133(21):8082–8085. https://doi.org/10.1021/ja201206t
Seksek O, Biwersi J, Verkman AS (1997) Translational diffusion of macromolecule-sized solutes in cytoplasm and nucleus. J Cell Biol 138(1):131–142
Senske M, Tork L, Born B, Havenith M, Herrmann C, Ebbinghaus S (2014) Protein stabilization by macromolecular crowding through enthalpy rather than entropy. J Am Chem Soc 136(25):9036–9041. https://doi.org/10.1021/ja503205y
Siefker J, Karande P, Coppens MO (2014) Packaging biological cargoes in mesoporous materials: opportunities for drug delivery. Expert Opin Drug Deliv 11(11):1781–1793. https://doi.org/10.1517/17425247.2014.938636
Silvério SC, Rodriguez O, Teixeira JA, Macedo EA (2010) Solute partitioning in polymer-salt ATPS: the Collander equation. Fluid Phase Equilib 296:173–177
Sukenik S, Sapir L, Gilman-Politi R, Harries D (2013) Diversity in the mechanisms of cosolute action on biomolecular processes. Faraday Discuss 160:225–237; discussion 311–227
Taft RW, Kamlet MJ (1976) Solvatochromic comparison method .2. Alpha-scale of solvent hydrogen-bond donor (Hbd) acidities. J Am Chem Soc 98(10):2886–2894. https://doi.org/10.1021/Ja00426a036
Theillet FX, Binolfi A, Frembgen-Kesner T, Hingorani K, Sarkar M, Kyne C, Li C, Crowley PB, Gierasch L, Pielak GJ, Elcock AH, Gershenson A, Selenko P (2014) Physicochemical properties of cells and their effects on intrinsically disordered proteins (IDPs). Chem Rev 114(13):6661–6714. https://doi.org/10.1021/cr400695p
Tolstoguzov V (2000) Phase behaviour of macromolecular components in biological and food systems. Nahrung 44(5):299–308. https://doi.org/10.1002/1521-3803(20001001)44:5<299::AID-FOOD299>3.0.CO;2-9
Turoverov KK, Kuznetsova IM, Uversky VN (2015) What macromolecular crowding can do to a protein. Int J Mol Sci 15:23090–23140
van den Berg B, Ellis RJ, Dobson CM (1999) Effects of macromolecular crowding on protein folding and aggregation. EMBO J 18(24):6927–6933. https://doi.org/10.1093/emboj/18.24.6927
Verkman AS (2002) Solute and macromolecule diffusion in cellular aqueous compartments. Trends Biochem Sci 27(1):27–33

Verma PK, Rakshit S, Mitra RK, Pal SK (2011) Role of hydration on the functionality of a proteolytic enzyme alpha-chymotrypsin under crowded environment. Biochimie 93(9):1424–1433. https://doi.org/10.1016/j.biochi.2011.04.017

Walter H, Brooks DE, Fisher D (eds) (1985) Partitioning in aqueous two-phase systems: theory, methods, use, and applications to biotechnology. Academic, Orlando

Wang Y, Li C, Pielak GJ (2010) Effects of proteins on protein diffusion. J Am Chem Soc 132(27):9392–9397. https://doi.org/10.1021/ja102296k

Wang Y, Benton LA, Singh V, Pielak GJ (2012a) Disordered protein diffusion under crowded conditions. J Phys Chem Lett 3(18):2703–2706. https://doi.org/10.1021/jz3010915

Wang Y, Sarkar M, Smith AE, Krois AS, Pielak GJ (2012b) Macromolecular crowding and protein stability. J Am Chem Soc 134(40):16614–16618. https://doi.org/10.1021/ja305300m

Weiss M (2014) Crowding, diffusion, and biochemical reactions. Int Rev Cell Mol Biol 307:383–417. https://doi.org/10.1016/B978-0-12-800046-5.00011-4

Weiss M, Elsner M, Kartberg F, Nilsson T (2004) Anomalous subdiffusion is a measure for cytoplasmic crowding in living cells. Biophys J 87(5):3518–3524. https://doi.org/10.1529/biophysj.104.044263

Willauer HD, Huddleston JG, Rogers RD (2002) Solvent properties of aqueous biphasic systems composed of polyethylene glycol and salt characterized by the free energy of transfer of a methylene group between the phases and by a linear solvation energy relationship. Ind Eng Chem Res 41:2591–2601

Wojcieszyn JW, Schlegel RA, Wu ES, Jacobson KA (1981) Diffusion of injected macromolecules within the cytoplasm of living cells. Proc Natl Acad Sci USA 78(7):4407–4410

Yang L, Dordick JS, Garde S (2004) Hydration of enzyme in nonaqueous media is consistent with solvent dependence of its activity. Biophys J 87(2):812–821. https://doi.org/10.1529/biophysj.104.041269

Zaks A, Klibanov AM (1985) Enzyme-catalyzed processes in organic solvents. Proc Natl Acad Sci USA 82(10):3192–3196

Zaks A, Klibanov AM (1988a) The effect of water on enzyme action in organic media. J Biol Chem 263(17):8017–8021

Zaks A, Klibanov AM (1988b) Enzymatic catalysis in nonaqueous solvents. J Biol Chem 263(7):3194–3201

Zaslavsky BY (1992) Bioanalytical applications of partitioning in aqueous polymer 2-phase systems. Anal Chem 64(15):A765. https://doi.org/10.1021/Ac00039a001

Zaslavsky B (1994) Aqueous two-phase partitioning: physical chemistry and bioanalytical applications. Marcel Dekker, New York

Zaslavsky BY, Miheeva LM, Masimov EA, Djafarov SF, Reichardt C (1990) Solvent polarity of aqueous polymer-solutions as measured by the Solvatochromic technique. J Chem Soc Faraday Trans 86(3):519–524. https://doi.org/10.1039/Ft9908600519

Zaslavsky BY, Miheeva LM, Gulaeva ND, Borovskaya AA, Rubtsov MI, Lukatskaya LL, Mchedlovpetrossyan NO (1991) Influence of nonionic polymers on solvent properties of water as detected by studies of acid-base equilibria of Sulphonephthalein and fluorescein dyes. J Chem Soc Faraday Trans 87(7):931–938. https://doi.org/10.1039/Ft9918700931

Zaslavsky BY, Borvskaya AA, Gulaeva ND, Miheeva LM (1992) Physicochemical features of solvent media in the phases of aqueous polymer 2-phase systems. Biotechnol Bioeng 40(1):1–7. https://doi.org/10.1002/bit.260400102

Zaslavsky A, Madeira P, Breydo L, Uversky VN, Chait A, Zaslavsky B (2013) High throughput characterization of structural differences between closely related proteins in solution. Biochim Biophys Acta 1834 2:583–592. https://doi.org/10.1016/j.bbapap.2012.11.004

Zhang DL, Wu LJ, Chen J, Liang Y (2012) Effects of macromolecular crowding on the structural stability of human alpha-lactalbumin. Acta Biochim Biophys Sin Shanghai 44(8):703–711. https://doi.org/10.1093/abbs/gms052

Zhou BR, Liang Y, Du F, Zhou Z, Chen J (2004) Mixed macromolecular crowding accelerates the oxidative refolding of reduced, denatured lysozyme: implications for protein folding in intracellular environments. J Biol Chem 279(53):55109–55116. https://doi.org/10.1074/jbc.M409086200

Chapter 4
Role of Aqueous Media in Macromolecular Crowding

Pedro P. Madeira, Vladimir N. Uversky, and Boris Y. Zaslavsky

Abstract The behavior of biomolecules in a crowded environment is affected by a number of factors, with the excluded volume being traditionally considered as a major contributor to the macromolecular crowding effects in polymer solutions. In addition to the hard nonspecific steric interactions representing the excluded volume effects, other commonly considered factors are the so-called soft interactions between the crowder and the crowdee, such as electrostatic, hydrophobic, and van der Waals interactions. However, the changes in the solvent properties of aqueous media induced by the crowding agents may represent important but commonly neglected factor affecting behavior of biomolecules in a crowded environment. This chapter systematically represents information in support to this hypothesis. Presented data clearly show that the crowding agents do induce changes in solvent properties of aqueous media in crowded environment. Therefore, these changes should be taken into account for crowding effect analysis.

Keywords Macromolecular crowding · Solvent properties of aqueous media · Aqueous two-phase systems · Solvatochromic dyes · Dipolarity/polarizability · Hydrogen-bond donor acidity · And hydrogen-bond acceptor basicity

It is generally accepted that protein folding, protein-protein interactions, and other biochemically important processes *in vivo* may differ from those in dilute solutions commonly used in laboratory experiments (Alfano et al. 2024; Elcock 2010; Nakano

P. P. Madeira
I3S – Instituto de Investigação e Inovação em Saúde, Universidade do Porto, Porto, Portugal
e-mail: pmadeira@i3s.up.pt

V. N. Uversky (✉)
Department of Molecular Medicine and Byrd Alzheimer's Research Institute, Morsani College of Medicine, University of South Florida, Tampa, FL, USA
e-mail: vuversky@usf.edu

B. Y. Zaslavsky (✉)
Cleveland Diagnostics, Cleveland, OH, USA
e-mail: boris.zaslavsky@clevelanddx.com

V. N. Uversky (ed.), *(Macro)Molecular Crowding*, Subcellular Biochemistry 109,
https://doi.org/10.1007/978-3-032-03370-3_4

et al. 2014; Phillip and Schreiber 2013; Zhou et al. 2008). One of the reasons is believed to be the high overall concentrations of biological macromolecules that may occupy up to 40% of the cellular volume (Alfano et al. 2024; Elcock 2010; Nakano et al. 2014; Phillip and Schreiber 2013; Zhou et al. 2008). The term "macromolecular crowding" stresses that the influence of high macromolecule concentrations results from the steric interactions of crowding agents with the biomolecules of interest. Under the idealistic conditions, the crowding molecules are expected to be inert toward the protein or nucleic acid under study. They physically occupy a significant fraction of the solution volume, leaving only restricted space available to biomolecules, hence the term "excluded volume effect" is often used (Alfano et al. 2024; Elcock 2010; Nakano et al. 2014; Phillip and Schreiber 2013; Zhou et al. 2008).

According to Elcock (2010), there is the question of whether genuinely inert crowding agents exist that could be used in experiments to provide a direct readout of excluded volume effects only, or whether it is inevitable that all crowding agents will also cause additional effects that must be considered.

The experimental data accumulated and reviewed in the literature (Alfano et al. 2024) show that the excluded volume effect is not the only factor affecting the behavior of biomolecules in a crowded environment (Alfano et al. 2024; Elcock 2010; Nakano et al. 2014; Phillip and Schreiber 2013; Zhou et al. 2008). To explain some experimental observations inconsistent with the excluded volume effect, it was suggested that there are "soft" interactions, such as electrostatic, hydrophobic, and van der Waals interactions between the crowding agent and the protein, in addition to hard nonspecific steric repulsion (Nakano et al. 2014; Phillip and Schreiber 2013). This hypothesis allows one to explain the experimental data by a balance of attractive as well as repulsive crowding agent/protein interactions (Nakano et al. 2014; Phillip and Schreiber 2013; Benton et al. 2012; Knowles et al. 2011; Wang et al. 2012). It is generally ignored that there is a third component in all crowded solutions—water, which is important for all the biochemical processes (protein folding, aggregation, protein/protein interactions, etc.) and for life in general (Ben-Naim 2003). The commonly used macromolecular crowding agents include dextran, Ficoll, polyethylene glycol (PEG), and polyvinylpyrrolidone (PVP), though proteins, such as albumin or lysozyme, are sometimes used as well. It was reported recently that under crowding conditions, there is an overlapping of hydration shells for the crowding agent, implying that water in the solution is affected by the agent (King et al. 2014). This finding agrees with the suggested water restructuring in the presence of low molecular weight osmolytes and in the presence of crowding agents as an important factor in the enhancement of protein stability (Canchi and Garcia 2013; Politi and Harries 2010; Sukenik et al. 2013).

It is known that the dielectric and thermodynamic properties of water in aqueous solutions of polymers, such as dextran, Ficoll, PEG, and PVP, change significantly relative to those in pure water (Arnold et al. 1985; Zaslavsky 1994). Furthermore, according to the theoretical analysis, confinement of water molecules in a hydration shell around the hydrophobic interface produces a thin layer of water molecules characterized by low correlation, entropy, dielectric constant, and slow reorientation of their intrinsic molecular dipoles (Despa et al. 2004). Hydrophobe-structured water with the hindered rotational motion of water molecules and decreased dielectric constant can enhance the effective forces between charged groups (Despa et al. 2004). The solvent polarity of aqueous media in solutions of dextran, Ficoll, and PEG changed depending upon the polymer type and concentration (Zaslavsky 1994). Using the solvatochromic comparison method, Kim et al. demonstrated that PEG can also affect water's hydrogen-bond donor (HBD) acidity (Kim et al. 2002). It is well known that all the aforementioned synthetic polymers in different combinations may form aqueous two-phase systems (ATPSs) (Zaslavsky 1994). These systems arise in aqueous solutions of two particular polymers, for example, dextran and PEG or dextran and Ficoll, above certain concentration thresholds. Two partially immiscible phases are formed with one phase containing predominantly one polymer, and the other phase containing predominantly the other, while both containing 70–90% water. It is well established that phase separation occurs because of different effects of the two polymers on the water structure (Zaslavsky 1994). The solvent properties of aqueous media in the two phases are different (Madeira et al. 2010). These differences are determined primarily by the polymer composition of the phases.

The solvent properties of different solvents may be studied by the approach developed by Taft, Kamlet, and others (Taft and Kamlet 1976; Kamlet and Taft 1976; Kamlet et al. 1977). This approach is based on using a set of solvatochromic dyes with the wavelength positions of their UV–visible absorption maximum shifting depending on different solvent properties. This approach was used to quantify the solvent's dipolarity/polarizability, HBD acidity, and hydrogen-bond acceptor (HBA) basicity in the phases of ATPS (Madeira et al. 2010) as well as in aqueous solutions of PEG of different molecular mass (Kim et al. 2002). It was demonstrated that the partitioning of organic compounds and proteins in ATPS can be described and even predicted by a linear combination of different solute/water interactions using solvatochromic solvent features of aqueous media in the phases (Madeira et al. 2010). Therefore, it was suggested (Ferreira et al. 2016b) that the macromolecular crowding effect may be related to the crowding agent's influence on the solvent features of aqueous media. The data on the effects of different polymers on solvent features of water reported in the literature (Ferreira et al. 2016b; Zaslavsky et al. 1990; Ferreira et al. 2017) are summarized in Table 4.1.

It should be kept in mind that the results obtained are to be viewed as relative estimates of the solvent properties in the solutions under study. Some of the data in Table 4.1 are illustrated graphically in Figs. 4.1, 4.2 and 4.3. The solvent dipolarity/polarizability (π^*) of aqueous media increases with the polymer concentration for all polymers examined except Ucon . The polymer effect on the solvent dipolarity/polarizability (π^*) characterizing the interactions of aqueous media with solute dipoles and induced dipoles decreases in the following sequence: PVP-40 > dextran-40 = dextran-75 = Ficoll-70 > PEG-10 K = PEG-4.5 K > PEG-600 > PPG-400 > Ucon > PEGDME-250, at 30 and 40% concentrations.

The data presented in Table 4.1 and illustrated in Fig. 4.2 show that the polymer effect on the solvent HBD acidity (α) of aqueous media decreases with the polymer concentration for all polymers examined. The polymer effect decreases in the sequence: Ucon>

Table 4.1 Solvent features of water in aqueous solutions of different polymers[a]

Polymer	Concentration, %	π*	α	β
None		1.096 ± 0.003	1.237 ± 0.003	0.596 ± 0.001
PEG-300	10	1.107 ± 0.001	1.112 ± 0.001	0.603 ± 0.001
	20	1.111 ± 0.001	1.015 ± 0.001	0.617 ± 0.002
	30	1.115 ± 0.001	0.923 ± 0.002	0.635 ± 0.002
	40	1.120 ± 0.001	0.830 ± 0.001	0.655 ± 0.001
PEG-600	10	1.106 ± 0.005	1.098 ± 0.004	0.636 ± 0.002
	201.109	1.107 ± 0.006	1.009 ± 0.004	0.652 ± 0.002
	30	1.109 ± 0.005	0.899 ± 0.004	0.669 ± 0.001
	40	1.110 ± 0.001	0.790 ± 0.003	0.694 ± 0.001
PEG-1000	10	1.107 ± 0.002	1.085 ± 0.003	0.611 ± 0.002
	20	1.110 ± 0.002	0.987 ± 0.003	0.627 ± 0.002
	30	1.116 ± 0.002	0.889 ± 0.002	0.646 ± 0.002
	40	1.121 ± 0.003	0.789 ± 0.003	0.669 ± 0.003
PEG-4500	10	1.109 ± 0.002	1.060 ± 0.007	0.638 ± 0.002
	20	1.112 ± 0.001	0.963 ± 0.006	0.654 ± 0.001
	30	1.119 ± 0.001	0.872 ± 0.004	0.677 ± 0.002
	40	1.124 ± 0.002	0.762 ± 0.005	0.699 ± 0.003
PEG-8000	10	1.107 ± 0.001	1.072 ± 0.002	0.608 ± 0.002
	20	1.110 ± 0.001	0.974 ± 0.003	0.625 ± 0.002
	30	1.114 ± 0.002	0.875 ± 0.002	0.646 ± 0.001
	40	1.120 ± 0.002	0.772 ± 0.001	0.670 ± 0.001
PEG-10000	10	1.109 ± 0.001	1.070 ± 0.007	0.640 ± 0.002
	20	1.113 ± 0.001	0.972 ± 0.005	0.657 ± 0.002
	30	1.120 ± 0.002	0.874 ± 0.005	0.676 ± 0.002
	40	1.122 ± 0.001	0.774 ± 0.004	0.701 ± 0.002
Ucon-3930	10	1.115 ± 0.002	0.996 ± 0.009	0.656 ± 0.002
	20	1.120 ± 0.002	0.86 ± 0.01	0.681 ± 0.002
	30	1.105 ± 0.003	0.766 ± 0.009	0.709 ± 0.006
	40	1.083 ± 0.001	0.666 ± 0.003	0.733 ± 0.002
Ficoll-70	10	1.128 ± 0.007	1.105 ± 0.009	0.641 ± 0.006
	20	1.144 ± 0.001	1.049 ± 0.003	0.656 ± 0.001
	30	1.159 ± 0.004	0.988 ± 0.006	0.660 ± 0.005
	40	1.174 ± 0.003	0.935 ± 0.005	0.671 ± 0.003
Dextran-70	10	1.119 ± 0.001	1.165 ± 0.002	0.634 ± 0.002
	20	1.136 ± 0.003	1.110 ± 0.003	0.642 ± 0.003
	30	1.157 ± 0.003	1.052 ± 0.004	0.649 ± 0.003
	40	1.175 ± 0.004	1.000 ± 0.005	0.655 ± 0.005
Dextran-40	10	1.127 ± 0.001	1.169 ± 0.002	0.631 ± 0.001
	20	1.146 ± 0.001	1.106 ± 0.003	0.641 ± 0.002
	30	1.161 ± 0.003	1.055 ± 0.003	0.655 ± 0.003
	40	1.178 ± 0.005	1.004 ± 0.006	0.66 ± 0.01

(continued)

Table 4.1 (continued)

Polymer	Concentration, %	π*	α	β
PVP-40	10	1.148 ± 0.005	1.046 ± 0.008	0.670 ± 0.005
	20	1.161 ± 0.007	0.967 ± 0.007	0.699 ± 0.009
	30	1.182 ± 0.008	0.926 ± 0.009	0.718 ± 0.009
	40	1.20 ± 0.01	0.85 ± 0.02	0.74 ± 0.01
PPG-400	10	1.118 ± 0.001	1.052 ± 0.001	0.623 ± 0.001
	20	1.117 ± 0.001	0.934 ± 0.001	0.658 ± 0.001
	30	1.089 ± 0.001	0.916 ± 0.001	0.688 ± 0.001
	40	1.029 ± 0.001	–	0.718 ± 0.001
PEGDME-250	10	1.101 ± 0.001	1.105 ± 0.001	0.610 ± 0.002
	20	1.104 ± 0.001	0.992 ± 0.001	0.628 ± 0.001
	30	1.100 ± 0.002	0.885 ± 0.002	0.653 ± 0.002
	40	1.093 ± 0.001	0.780 ± 0.001	0.680 ± 0.001

[a] π* is the solvent dipolarity/polarizability; α is the solvent hydrogen bond donor (HBD) acidity; β is the solvent hydrogen bond acceptor (HBA) basicity; polymer concentration is expressed in wt.%

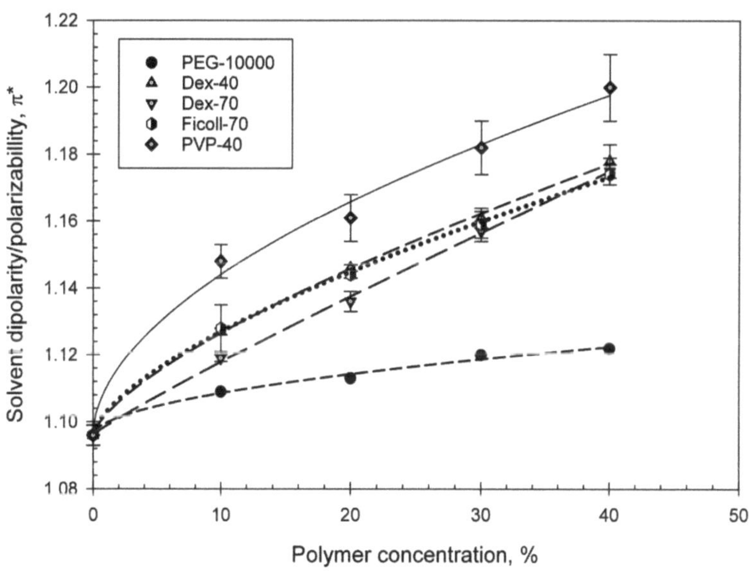

Fig. 4.1 Solvent dipolarity/polarizability, π*, as a function of concentration of polymers indicated in the figure

PEG-4500 ≥ PEG-8000 = PEG-10000 = PEGDME-250 > PEG-1000 = PEG-600 > PEG-300 > PVP-40 > Ficoll-70 > dextran-75 = dextran-40 > PPG-400, at both concentrations of 30 and 40%. It should be noted that the sequence is almost similar to the one determined for the polymer effect on the solvent HBA basicity (β), with only a difference in the PVP position.

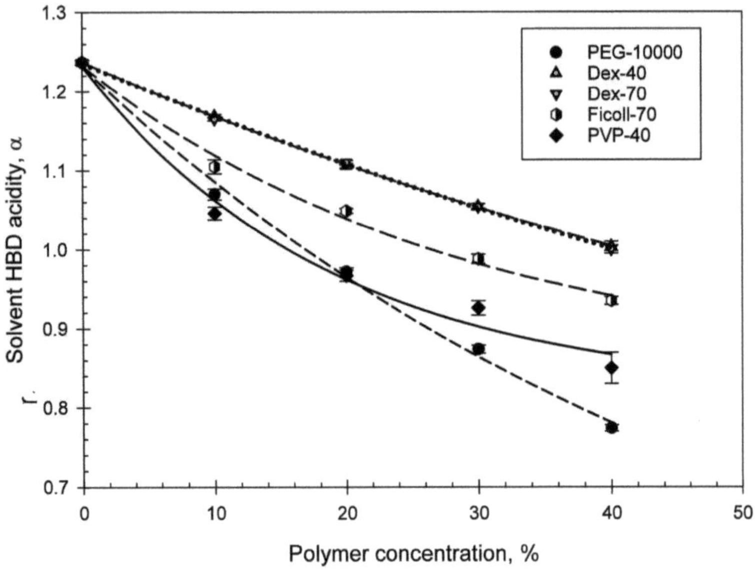

Fig. 4.2 Solvent hydrogen bond donor acidity, α, as a function of concentration of polymers indicated in the figure

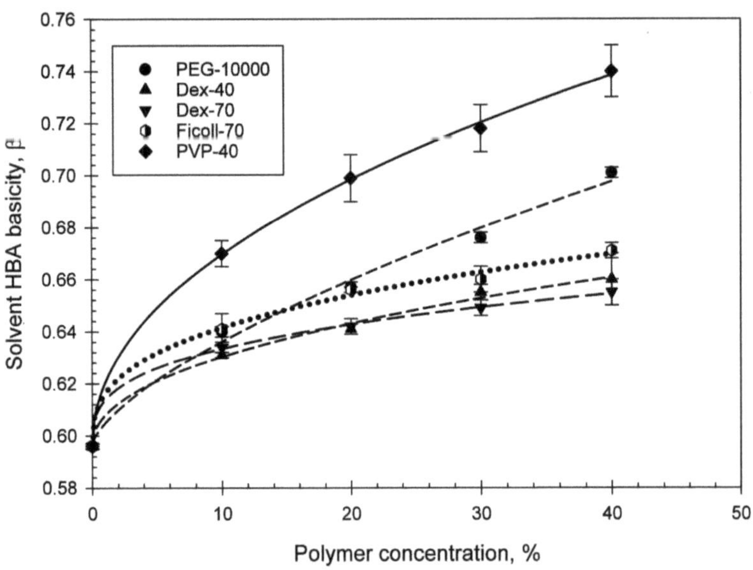

Fig. 4.3 Solvent hydrogen bond acceptor basicity, β, as a function of concentration of polymers indicated in the figure

On the other hand, the data presented in Table 4.1 and illustrated in Fig. 4.3 indicate that the solvent HBA basicity (β) of aqueous media increases with the polymer concentration for all polymers examined. The polymer effect decreases in the sequence: PVP-40 ≥ Ucon > PPG-400 > PEG-10000 = PEG-4500 = PEG-600 > P EGDME-250 > Ficoll-70 = PEG-8000 = PEG-1000 > Dex-40 = PEG-300 = Dex-7 0, at both concentrations of 30 and 40%. It should be noted that this sequence is different from the one found for the polymer effect on the solvent dipolarity/ polarizability.

Proteins, such as albumin, for example (Miklos et al. 2011) are also frequently used as crowding agents. However, it is expected that like other proteins, albumin can affect solvent features of water. In fact, the solvent properties of aqueous media were examined for proteins including plant dehydrins (dehydration proteins) (Close 1997; Hara 2010; Allagulova et al. 2003; Graether and Boddington 2014; Ferreira et al. 2018b), HspB6 (formerly Hsp20) (Ferreira et al. 2018a) with molecular weight of 17.136 kD, and elastin-like polypeptide (ELP) with the $(GVGVP)_{40}$ sequence and molecular weight of 17.063 kD (Ferreira et al. 2015a). The data obtained for proteins effects on the solvent features of aqueous media are listed in Table 4.2.

The data indicating concentration effects of proteins on the solvent properties of aqueous media listed in Table 4.2 are illustrated graphically in Figs. 4.4, 4.5 and 4.6. It should be noted that to show the differences between proteins measured at concentrations up to 15 mg/ml and polymers measured at the concentrations starting with 100 mg/ml only the data obtained at concentrations from 0 up to 110 mg/ml are shown.

Table 4.2 Solvent features of aqueous media in solutions of different proteins

Protein	Conc. mg/ml	Conc. mM	π*	α	β
None	0	0	1.096 ± 0.003	1.237 ± 0.003	0.596 ± 0.002
HspB6	5	0.292	1.105 ± 0.002	1.201 ± 0.003	0.599 ± 0.001
	10	0.584	1.105 ± 0.002	1.165 ± 0.004	0.602 ± 0.001
	15	0.876	1.106 ± 0.002	1.128 ± 0.005	0.605 ± 0.002
Dehydrin K2	5	0.93	1.105 ± 0.001	1.223 ± 0.001	0.597 ± 0.001
	10	1.86	1.106 ± 0.001	1.219 ± 0.001	0.599 ± 0.001
	15	2.79	1.107 ± 0.001	1.210 ± 0.001	0.600 ± 0.001
Dehydrin K4	5	0.38	1.100 ± 0.001	1.226 ± 0.002	0.597 ± 0.001
	10	0.76	1.102 ± 0.001	1.221 ± 0.001	0.599 ± 0.001
	15	1.14	1.104 ± 0.001	1.216 ± 0.001	0.600 ± 0.002
Dehydrin K10	5	0.14	1.104 ± 0.001	1.224 ± 0.002	0.595 ± 0.003
	10	0.29	1.109 ± 0.001	1.216 ± 0.001	0.600 ± 0.002
	15	0.42	1.113 ± 0.002	1.207 ± 0.002	0.604 ± 0.002
ELP	15.0	0.90	1.104 ± 0.002	1.248 ± 0.02	0.596 ± 0.002

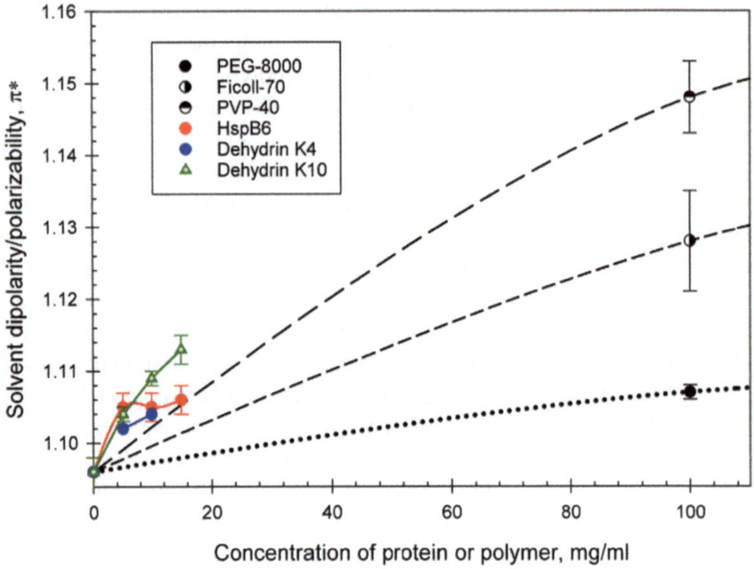

Fig. 4.4 Solvent dipolarity/polarizability as a function of protein concentration. (Data are shown in Table 4.2)

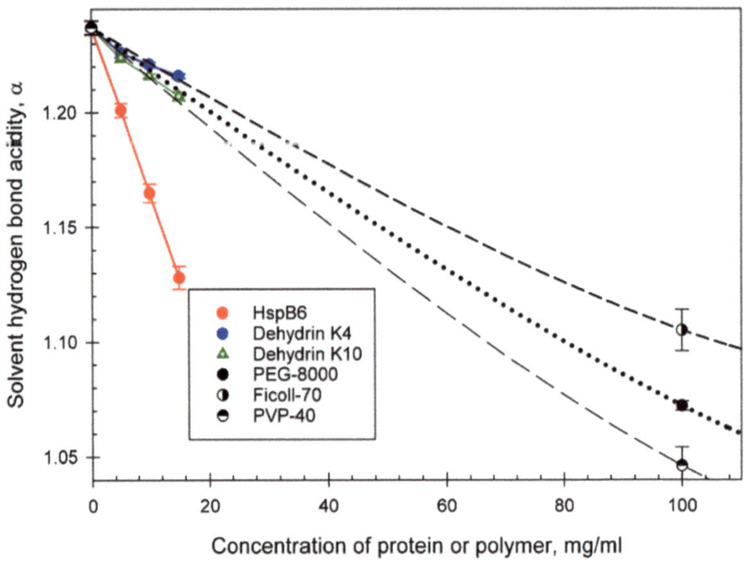

Fig. 4.5 Solvent hydrogen bond donor acidity, α, as a function of protein concentration. (Data see in Table 4.2)

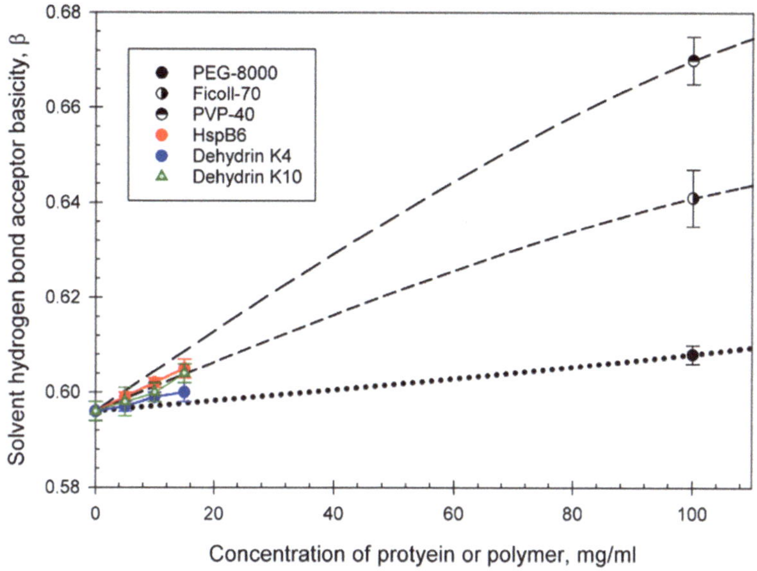

Fig. 4.6 Solvent hydrogen bond acceptor basicity, β, as a function of protein concentration. (Data see in Table 4.2)

The effect of proteins, such as dehydrin K10 and HspB6, on solvent dipolarity/polarizability, π*, characterizing the interactions of aqueous media with solute dipoles and induced dipoles clearly exceed those of nonionic polymers, while Dehydrin K4 effect is like that displayed by PVP-40 (see Fig. 4.4).

Figure 4.5 shows that protein HspB6 influence on the solvent hydrogen bond donor acidity, α, is the strongest in the whole macromolecular crowders family including nonionic polymers and proteins, while the effects of dehydrins K4 and K10 are close to those of Ficoll-70 and PVP-40, correspondingly.

The effect of HspB6 on solvent hydrogen bond acceptor basicity, β, is between those of PVP-40 and Ficoll-70, while the effect of dehydrin K10 is almost identical with that of Ficoll-70 and the effect of dehydrin K4 is just exceeds that of PEG-8000 (see Fig. 4.6).

To analyze the possible role of the established influence of polymers on the solvent properties of aqueous media in the crowding effects of the polymers and proteins, it is important to consider how significant these effects are. The dipolarity/polarizability parameter, π*, of aqueous media in 40% Ucon solution (see Table 4.1) differs from that of the polymer-free media by 0.022. The difference between the π* values for such organic solvents as methanol and ethanol is 0.06 (Kamlet et al. 1977), that is, three times larger.

In the aqueous two-phase system formed by dextran-75 and Ficoll-70 (Zaslavsky 1994; Madeira et al. 2008b; Madeira et al. 2013), however, the difference between the π* values for the coexisting phases amounts only to 0.003, and this difference affects the distribution of small compounds and proteins between the two phases.

Similarly, the polymer influence on the HBA basicity β of aqueous media is rather small. The difference between the β values for aqueous media in 40% dextran-75 solution and polymer-free media is 0.033. The similar difference between methanol and ethanol is 0.15 (Kamlet et al. 1977). On the other hand, the difference between the HBA basicity β for aqueous media in the coexisting phases of dextran-75-PEG-600 is just 0.005 (Zaslavsky 1994; Madeira et al. 2008b; Madeira et al. 2013).

The polymer influence on the HBD acidity α of aqueous media is quite significant. The difference between the α values for aqueous media in 40% dextran-40 solution and for polymer-free media is 0.223, much larger than the difference between the α values for methanol and ethanol of just 0.10 (Kamlet et al. 1977). In polymer-polymer ATPS, the differences between the HBD acidity α of the aqueous media in the coexisting phases vary from 0 in Ficoll-70-PEG-6000 to 0.181 in dextran-75-Ucon ATPS, depending on the polymer and salt composition (Zaslavsky 1994; Madeira et al. 2008b; Madeira et al. 2013).

It follows from the experimental data obtained in (Ferreira et al. 2016b; Zaslavsky et al. 1990; Ferreira et al. 2017) that nonionic polymers used as macromolecular crowding agents change the solvent properties of water in their aqueous solutions. If the data obtained in the studies of polymer-protein interactions are considered regarding the data mentioned earlier, it becomes clear that the conclusions about polymer-protein interactions (Phillip and Schreiber 2013) are commonly based on the deviation of the experimental data from one or the other model chosen by the respective authors.

For example, heats of mixing lysozyme or ovalbumin solutions with PEG were measured calorimetrically (Pico et al. 2007). Heats of corresponding dilutions of these solutions were measured separately, and it was found that the sum of the heats of dilutions was not equal to the heat of mixing. The difference observed was interpreted as evidence of protein-PEG interactions. However, it may readily be explained by protein transfer from water to aqueous media with PEG-induced changes in the solvent properties (Pico et al. 2007). Similarly, the conclusion about PEG-lysozyme interactions was made by Bloustine et al. (2006), based on the deviation of the light scattering data from the water depletion model, while the effect of PEG-induced changes in the solvent properties of water may explain the same data. Same explanation may be applicable to the other reported data (Crowley et al. 2008; Kulkarni et al. 2000). The studies of partition behavior of numerous different proteins in different polymer-polymer ATPSs do not indicate the presence of protein-polymer interactions, though this possibility cannot be excluded for any specific protein.

It should be mentioned that the possible involvement of aqueous media in the crowding effects was discussed in the literature (Canchi and Garcia 2013; Nakano et al. 2014; Politi and Harries 2010; Sukenik et al. 2013; Harada et al. 2012). To test whether the changes of the solvent properties of aqueous media are relevant to the macromolecular crowding effects, the results of the studies, where the numerical experimental data were reported, were examined in (Zhou et al. 2004).

The oxidative refolding of reduced, denatured hen egg white lysozyme was examined in the presence of bovine albumin, dextran-70, and Ficoll-70 (Zhou et al.

2004). The refolding yield of lysozyme reported was examined in terms of solvent properties of aqueous media in dextran-75 and Ficoll-70 solutions (see Table 1) (Zhou et al. 2004). It has been shown previously (Madeira et al. 2010; Madeira et al. 2012; Madeira et al. 2014) that different properties of solutes in aqueous solutions may be expressed as linear combinations of different solvent properties of the aqueous media. Therefore, a similar expression for protein folding/refolding in the presence of crowding agents was applied in (Ferreira et al. 2016b). The data fit with the simplest linear model based on previously reported linear relationships between a variety of solvation-related parameters for various proteins and small organic compounds and the solvent polarity/polarizability, HBD acidity, and HBA basicity of aqueous media. The observed relationship can be described as follows:

$$\text{Yield}(\%) = 4200_{\pm 902} - 3200_{\pm 653}\pi^* - 440_{\pm 152}\alpha \tag{4.1}$$

$$\text{N} = 4; \text{r}^2 = 0.9849; \text{SD} = 4.2; \text{F} = 32.7$$

where yield is indicated above; π^* and α are the solvent polarity/polarizability and solvent HBD acidity in the aqueous polymer solution, respectively; N is the number of experimental data; all the other parameters are defined above.

The number of experimental data is extremely small (the yields reported in (Zhou et al. 2004) in the presence of 10 and 20% dextran, 10% Ficoll, and in polymer-free solution were used), and hence the relationship (Eq. 4.1) cannot be viewed as sufficiently reliable. However, the fact that it exists at all, even though the ionic composition of the refolding media used differs from that used in our solvatochromic measurements is interesting (Zhou et al. 2004).

Similarly, the yield of refolded rabbit muscle creatine kinase can be described in terms of solvent properties for dextran-70 and Ficoll-70 (Du et al. 2006), but not for the PEG-2000 effect, which may be assigned to the different excluded volume effects of PEG-2000 and those of dextran and Ficoll of the same molecular mass. It was established that the effects of PEG-600 and PEG-4500 on the solvent properties of aqueous media are different. Since the PEG-2000 effects were not examined, we could not include the PEG-2000 crowding effect data in our analysis. The relationship can be described as follows:

$$\text{Yield}(\%) = 4000_{\pm 1339} - 2800_{\pm 974}\pi^* - 600_{\pm 218}\alpha \tag{4.2}$$

$$\text{N} = 5; \text{r}^2 = 0.8104; \text{SD} = 6.3; F = 4.2$$

where yield is the yield of refolded rabbit muscle creatine kinase; all the other parameters are defined above. The yields reported in (Du et al. 2006) were used in the presence of 10 and 20% dextran, 10 and 20% Ficoll, and in the polymer-free solution. It should be noted that the protein refolding was analyzed in 0.05 M aqueous Tris–HCl, pH 7.5 (Du et al. 2006), and the ionic composition of the polymer solutions was already shown to affect the solvent properties of aqueous media (Du et al. 2006). Therefore, the observed relationship should only be viewed as a trend and not a reliable correlation.

The relationships described by Eqs. (4.1) and (4.2) do not provide unambiguous experimental evidence, but they support the assumption that polymer-induced changes in the solvent properties of aqueous media may play an important role in macromolecular crowding effects. We suggest that the solvent properties measured here by solvatochromic dyes represent one aspect of the structure of water in the solutions of crowding agents.

Results of the analysis of the interrelationship between the stability of ubiquitin and solvent properties of aqueous solutions of glucose, dextran-40, and PEG-20,000 at different concentrations are summarized in Fig. 4.7. The relationship observed in Fig. 4.7 may be described as:

$$T_m = -72.5_{\pm 16} + 338.8_{\pm 12.6}\pi^* + 16.4_{\pm 4.4}\alpha \quad (4.3)$$

$$N = 10; r^2 = 0.9904; SD = 1.01; F = 359$$

where T_m is the melting temperature of ubiquitin; all the other parameters are as defined above.

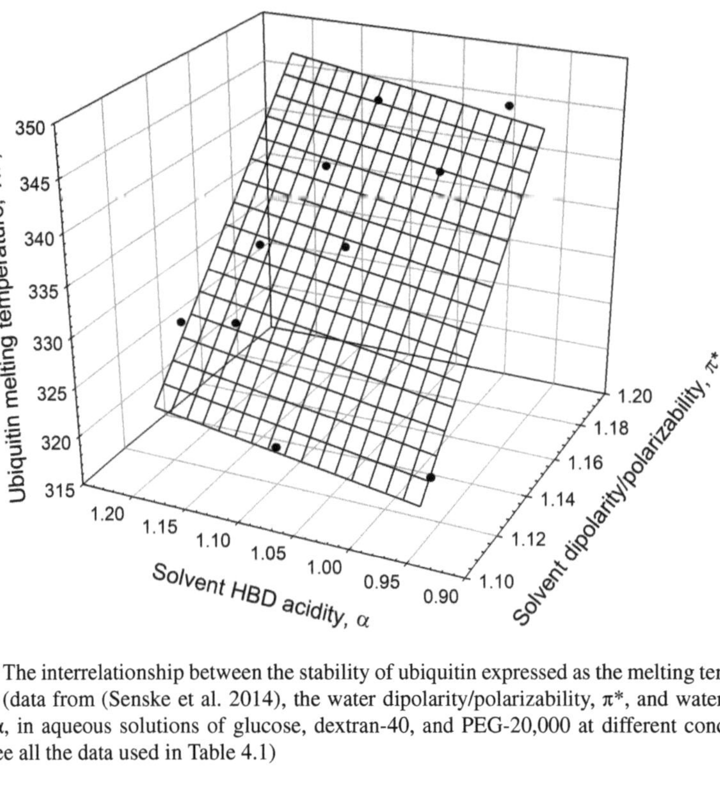

Fig. 4.7 The interrelationship between the stability of ubiquitin expressed as the melting temperature, T_m, (data from (Senske et al. 2014), the water dipolarity/polarizability, π^*, and water HBD acidity, α, in aqueous solutions of glucose, dextran-40, and PEG-20,000 at different concentrations. (See all the data used in Table 4.1)

It should be noted that the T_m values reported in (Senske et al. 2014) for the co-solute free solution, pH 2.0, and the solutions of 0.1 M and 0.2 M KCl do not fit the relationship. It should be mentioned also that solutions of dextran-40 in (Senske et al. 2014) have been prepared in grams per liter of water, and from our experience it is a questionable procedure for the preparation of aqueous solutions of highly hydrated polymers, such as dextran or Ficoll, and should always be followed by checking the final concentration of the solution by gravimetrical analysis using lyophilization of aliquots or measuring optical rotation or other properties of a solution, since the "dry" polymer may contain rather significant amount of water. Surprisingly, the data for dextran-40 fit the above relationship so well.

The relationship described by Eq. 4.3 implies that the ubiquitin stability in aqueous solutions of different crowders, such as dextran and PEG, and small organic osmolytes, such as glucose, is increased due to these compounds' influence on the solvent properties of water.

Similar relationships were also found for chymotrypsin inhibitor 2 in the presence of Ficoll-70 and sucrose (Senske et al. 2014) alkaline phosphatase in aqueous solutions of PEG-4000, Ficoll-70, and dextran-70 (Jia et al. 2017) and globular actin in the presence of Ficoll-70, TMAO, urea (up to 2.0 M) and TMAO-urea mixture (Schummel et al. 2016). An additional example worth mentioning is the effects of Ficoll, sucrose, TMAO, and urea on the polymerization kinetics of α/β-tubulin reported by Schummel et al. (2016) The apparent growth rate characterized the polymerization kinetics, K_{app} value, assayed at 37 °C in the buffer containing 3.26 M glycerol, pH 6.8; (Schummel et al. 2016); i.e., under conditions drastically different from those used to estimate the solvent properties of aqueous media in the polymers and osmolytes solutions (Ferreira et al. 2016b; Ferreira et al. 2017). Nevertheless, the reasonably good relationship with solvent properties of water was established, implying that protein-water interactions are important to the protein aggregation and stability. It should be mentioned, however, that we could not find any relationship between the data reported in (Guseman and Pielak 2017) on the effects of polymers and osmolytes on the equilibrium stability of side-by-side dimerization of A34F variant of B1 domain of protein G and the solvent properties of aqueous media in solutions of TMAO, urea, sucrose, PEG-8000, and Ficoll-70.

Shahid et al. (2019) examined effects of Dextran70, Dextran-40 and Ficoll-70 on stability of α-lactalbumin and lysozyme. The results reported in (Shahid et al. 2019) analyzed in terms of the solvent features of aqueous media in crowders solutions are illustrated graphically in Fig. 4.8. Hong and Gierasch (2010) examined the stability of the cellular retinoic acid-binding protein (CRBP I) in Ficoll-70 solutions. The data from (Hong and Gierasch 2010) analyzed in terms of the solvent properties of Ficoll-70 solutions are illustrated graphically in Fig. 4.9.

The data accumulated so far do not allow one to answer the most important question—whether the macromolecular crowding effect is the effect of agent-induced changes on the properties of aqueous media or a combination of size-exclusion effect and solvent restructuring effects. Our current limited views of the water structure and the complete lack of knowledge of the relationship between specific properties of biological macromolecules and the solvent properties of aqueous media

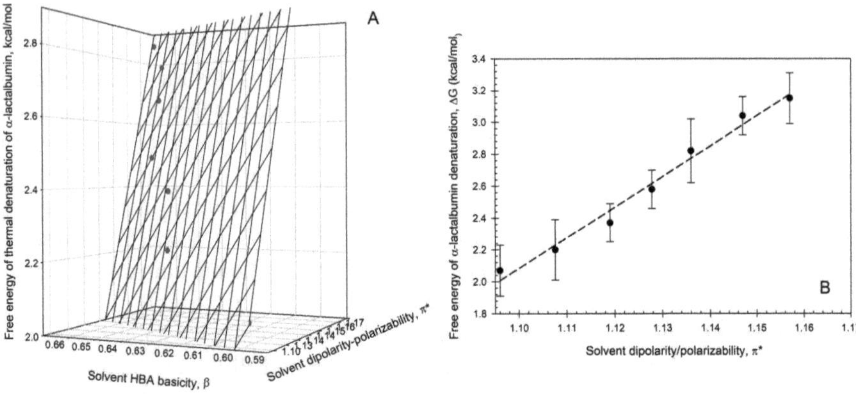

Fig. 4.8 (**a**) The linear relationship between the free energy of denaturation of α-lactalbumin, solvent dipolarity/polarizability, π*, and solvent HBA basicity, β, in aqueous solutions of Ficoll-70. Data from (Shahid et al. 2019). (**b**) The free energy of denaturation of α-lactalbumin as a function of solvent polarity/polarizability, π*, in aqueous solutions of Dextran-70. (Data from Shahid et al. 2019)

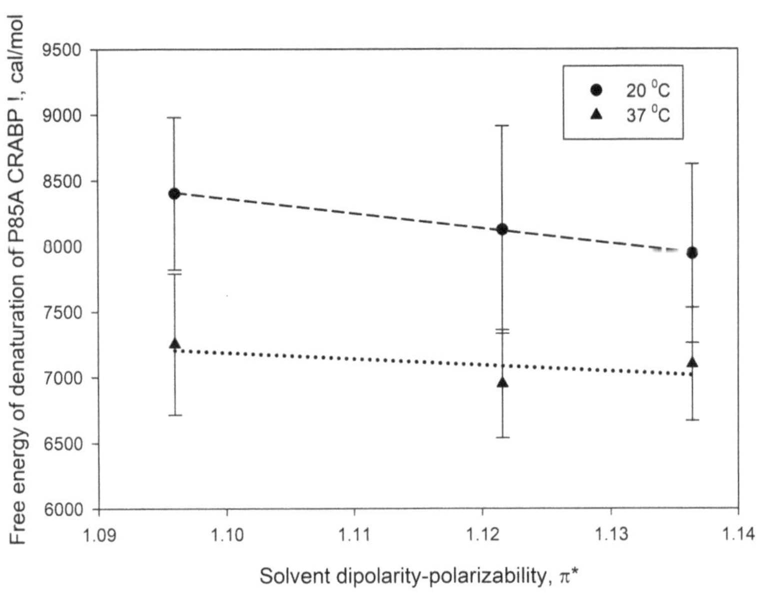

Fig. 4.9 The free energy of denaturation of CRBP I as a function of solvent polarity/polarizability, π*, in aqueous solutions of Ficoll-70 at 20 °C and 37 °C. Data from (Hong and Gierasch 2010)

complicate the issue. At this time, we may suggest combining the two effects results in the experimentally observed changes in biomolecule behavior in crowded solutions. The two effects types of relative importance may be specific to the protein or nucleic acid under analysis. Protein aggregation is very sensitive to environmental conditions. It was suggested that high concentrations of inert polymers that are used to mimic macromolecular crowding in vitro experiments may have a large influence on the behavior of biological macromolecules (Hong and Gierasch 2010; Bismuto et al. 2002; Eggers and Valentine 2001), affecting protein-protein interactions in general (Martin et al. 2014; Minton 2000a, b) and could modulate both the rate and the extent of amyloid formation *in vivo* (Martin et al. 2014; Morar et al. 2001). Accelerated *in vitro* aggregation and fibrillation in the presence of crowding agents have been reported for human apolipoprotein C-II (Hatters et al. 2002), α-synuclein (Breydo et al. 2014; Munishkina et al. 2008a; Munishkina et al. 2004; Munishkina et al. 2008b; Shtilerman et al. 2002; Uversky et al. 2002), β-synuclein (Yamin et al. 2005), amyloid-β peptide (Li and Mehler 2006), human tau protein (Ma et al. 2013b) and its 244–441 fragment (Zhou et al. 2009), human copper, zinc superoxide dismutase (Ma et al. 2013b), islet amyloid polypeptide (Rivera et al. 2009), human prion protein (Zhou et al. 2009), β-lactoglobulin (Ma et al. 2013a), FG nucleoporins (Milles et al. 2013). Crowders of similar chemical nature are known to affect protein aggregation in a concentration-dependent manner (Uversky et al. 2002).

Also, crowders of different chemical nature can modulate protein aggregation in a different manner (Breydo et al. 2014; Uversky et al. 2002; Assarsson et al. 2014). For example, in the case of the $A\beta_{42}$ aggregation, accelerating effects were observed from the positively charged polymers, whereas no aggregation modulating effects were seen from the negative or neutral polymers (Assarsson et al. 2014).

It was also shown that rigid and flexible polysaccharides influence protein aggregation via different mechanisms. Furthermore, it has been suggested that, in addition to excluded volume effects, changes in solution viscosity and nonspecific "soft" protein-polymer interactions might influence the structure and dynamics of proteins in crowded environments (Kuznetsova et al. 2014, 2015). To further clarify factors affecting protein fibrillation in crowded environments, we examined the effects of PEG and Ucon of similar size (~4.0 kDa) on aggregation of α-synuclein into amyloid fibrils under conditions closely resembling physiological. We want to empha size here, once again the analysis of the partition of recombinant human α-synuclein in aqueous dextran-PEG and Ficoll-Ucon two-phase systems (namely, using the Collander solvent regression relationship between the proteins partition coefficients in different ATPSs) suggested that this protein does not specifically bind to the polymers used in the study.

It has been reported that the effects of mixtures of Ficoll-70 and dextran-6 on stability of protein FKBP were non-additive relative to those of individual crowding agents (Senske et al. 2014). Synergistic effects of mixtures of crowding agents, such as Ficoll-70, dextran-70 and PEG-2000, with calf thymus DNA on refolding of rabbit muscle creatinine kinase have been reported as well (Du et al. 2006). The individual crowding effects of dextran and sucrose on refolding of human muscle creatine kinase were reported to cancel each other in the mixture, in contrast to the expected increase of excluded volume effect (Patel et al. 2002). It has been suggested that small compounds may display volume exclusion effects exceeding those of macromolecules (Sharp 2015). This suggestion apparently agrees with the multiple studies of the effects of small organic osmolytes, such as glucose, sucrose, and trimethylamine N-oxide (TMAO), which, being present at high concentrations, may display crowding effects and hence are sometimes called molecular crowders (Bai et al. 2017; Benton et al. 2012; Schummel et al. 2017; Senske et al. 2014). Therefore, osmolytes are often considered as molecular crowders in the literature due to the fact that any compound, including osmolytes, occupies definite volume in the solution, thereby restricting space that might be occupied by biomacromolecule, such as protein or nucleic acid. It is important to emphasize also that the applicability of the excluded volume theory to the description of the effect of osmolytes on protein behavior was considered in (Ferreira et al. 2015b), where the effects of high concentrations of three osmolytes, such as glucose, sucrose, and raffinose on the reversible homodimerization of α-chymotrypsin were investigated by sedimentation equilibrium. The authors used two models to explain the osmolyte-mediated stabilization of the α-chymotrypsin homodimer and considered both binding interactions (transfer-free energy analysis) and steric interactions (excluded volume theory). Their analysis revealed that the observed stabilization of the alpha-chymotrypsin homodimer can be reasonably described by the thermodynamic nonideality arising from osmolyte molecular crowding (Ferreira et al. 2015b). These compounds, including osmolytes, may affect the stability of proteins similarly to common macromolecular crowding agents, and it is well known that many of them do not interact with many proteins directly. It gradually becomes unavoidable to admit that the excluded volume effect, being a cornerstone of the theoretical views on crowding phenomena, is oversimplified, and its role in the crowding effect is likely overestimated. The representation of macromolecular crowders and proteins as hard spheres is a very crude approximation. The view that all proteins interact with typical synthetic macromolecular agents, such as polyethylene glycol (PEG), dextran, polyvinylpyrrolidone, or Ficoll, contradicts all the experimental data reported so far in the studies of protein partitioning in aqueous two-phase systems formed by different pairs of these polymers (Ferreira et al. 2015b; Madeira et al. 2011). It should be also emphasized here that the results show the presence of interactions between the protein and the macromolecular crowding agent are typically based on the models used for the interpretation of the experimental data (this issue is discussed in detail in (Ferreira et al. 2016b)).

It is known that protein conformational stability and peculiarities of the processes of protein folding and aggregation can be modulated by crowding agents (Chebotareva et al. 2004; Eronina et al. 2005; Chebotareva 2007; Kuznetsova et al. 2014; Kuznetsova et al. 2015). To understand how the addition of high concentrations of PEG-8000 and Dextran-70 affects the resistance of the globular proteins toward the urea-induced unfolding, a method of intrinsic fluorescence was used in (Stepanenko et al. 2016). Concentrations of polymers in ATPSs are typically in the range of polymer concentrations used to model crowded environment. These considerations defined the choice of polymer concentrations used in (Stepanenko et al. 2016) (12% of PEG-8000 and 24% of Dextran-70).

Since only intrinsic fluorescence was used for recording protein unfolding in (Stepanenko et al. 2016), the data provided below are not universal and not necessarily reflect gross protein unfolding. In fact, intrinsic fluorescence mostly reports local structural changes occurring in vicinity of Trp or Tyr residues, and it cannot be excluded that other methods (such as microcalorimetry or far- and near-UV CD spectroscopies) will provide different results. The results of these analyses were summarized in the form of the unfolding curves for proteins of interest in the absence and presence of different additives monitored by changes in the intrinsic fluorescence (Stepanenko et al. 2016). Major findings of these analyses are summarized below. Despite being predicted to contain high level of disordered residues, ribonucleases A and B were shown to be highly stable against the urea unfolding. In fact, no sigmoidal transitions were induced in these proteins by urea at all conditions studied. It is likely that during the urea-induced unfolding of RNase A, some local structural changes take place in the pre-transition region. Although conformational stability of RNase B (which is a glycosylated form of a protein) was not affected by crowders, RNase A became a bit more stable when PEG-8000 or Dextran-70 was added. Similarly, analysis revealed that lysozyme is a highly stable protein, since no noticeable structural changes were induced by 6 M urea (Stepanenko et al. 2016). The stability of this protein was minimally perturbed by the additions of high concentrations of PEG-8000 or Dextran-70. Similar to lysozyme and RNases, papain was characterized by high conformational stability, with no noticeable structural changes being induced by 6 M urea (Stepanenko et al. 2016).

It was also established that proteins were stabilized by both crowding agents (Stepanenko et al. 2016). In fact, addition of PEG-8000 or Dextran-70 stabilized β-lactoglobulin B, shifting $C_{1/2}$ by ~.5 M. Concanavalin A illustrates a protein, which is dramatically stabilized by both crowders (Stepanenko et al. 2016). In the presence of PEG-8000 or Dextran-70, protein is not completely unfolded even by 7.5 M urea. At all conditions studied, structural transitions start at >4.0 M urea, illustrating high conformational stability of this protein (Stepanenko et al. 2016). This study also revealed that hemoglobin was another protein which was stabilized by both PEG and Dextran. It seems that the stabilization effects induced by Dextran-70 were a bit greater than those promoted by PEG ($C_{1/2}$ is shifted by 1.5 vs. 1.2 M) (Stepanenko et al. 2016).

On the other hand it was also illustrated that some proteins can be differently affected by different crowding agents (Stepanenko et al. 2016). Here, the unfolding behavior of β-lactoglobulin A was slightly affected by PEG, whereas Dextran stabilized the protein, shifting $C_{1/2}$ by ~0.8 M (Stepanenko et al. 2016). It was also shown that PEG-8000 slightly stabilized α-chimotrypsin (by ~0.1 M), whereas Dextran-75 destabilizes this protein by ~.8 M. PEG does not affect unfolding behavior of α-chimotrypsinogen, whereas Dextran destabilizes this protein, shifting C1/2 of its urea-induced unfolding curve by ~0.5 M (Stepanenko et al. 2016). At all conditions, structural transitions start at >4.0 M urea, indicating that α-chimotrypsinogen is a highly stable protein.

Finally, it was established that trypsinogen occupied a unique niche, since PEG did not affect conformational stability of this protein, whereas Dextran induces very strong perturbations in the unfolding behavior of this protein. In the presence of Dextran, the unfolding curve below 1.5 M urea passed through a minimum in the vicinity of 0.5 M urea, whereas the trend of the curve above 2.0 M urea testifies for the destabilizing effect of this polymer on conformational stability of trypsinogen. Unfortunately, there is no reasonable explanations for this unusual behavior of trypsinogen (Stepanenko et al. 2016).

Therefore, according to their response to the addition of polymers globular proteins can be grouped into three major categories as follows: highly stable proteins with very small effects of the PEG-8000 or Dextran-70 addition (lysozyme, papain, ribonuclease A, and ribonuclease B), proteins noticeably stabilized by at least one of the polymers (concanavalin A, hemoglobin, β-lactoglobulin A, β-lactoglobulin B), and proteins noticeably destabilized by at least one of the polymers (α-chymotrypsin, α-chymotrypsinogen, trypsinogen) (Stepanenko et al. 2016).

In several cases, the protein conformational stability of a given protein was differently affected by different polymers. For example, PEG slightly stabilized α-chymotrypsin (by ~0.1 M), whereas Dextran destabilized this protein by ~0.8 M. The fact that ribonucleases, lysozyme, papain, and β-lactoglobulin were very resistant to urea-induced unfolding can be explained by the fact that these proteins contain three or four disulfide bridges which can strongly improve protein stability and by this means decrease effect of crowding agents (Stepanenko et al. 2016).

Analysis of data shown in Table 4.3 provides further support to the differential effects of different crowders on protein stability. In fact, PEG-8000 either did not affect conformational stability of some proteins (RNase A, RNase B, lysozyme, papain, trypsinogen, α-chymotrypsinogen, and β-lactoglobulin A) or made some proteins more stable (α-chymotrypsin, β-lactoglobulin B, hemoglobin, and concanavalin A). The effects of Dextran-70 were even more diverse, since this additive did not affect unfolding of some proteins (RNase A, RNase B, lysozyme, and papain), increased conformational stability of other proteins (β-lactoglobulin A, β-lactoglobulin B, hemoglobin, and concanavalin A), and destabilized some globular proteins (trypsinogen, α-chymotrypsin, and α-chymotrypsinogen). As far as correlation between the effects of crowding agents on the unfolding behavior of target proteins and structural properties of said proteins is concerned, Table 4.3 shows that some conclusions can be made on the features contributing to the stability of proteins in crowded environments (Stepanenko et al. 2016).

In fact, the data in (Stepanenko et al. 2016) shows the conformational stability of globular proteins in crowded milieu is not obviously related to the level of their predicted intrinsic disorder. On the other hand, the data show that crowding agents do not have much effect on the conformational stability of small, monomeric positively charged proteins (i.e., proteins with high pI values), whereas oligomeric negatively charged proteins (i.e., proteins with low pI values) are typically stabilized by both crowders used in (Stepanenko et al. 2016). It is likely that the ability of crowding agents to stabilize oligomeric proteins against unfolding by urea is due to the crowding-induced shift of the oligomer–monomer equilibrium toward the oligomeric state.

Although the efficiency of macromolecular crowding is believed to be dependent on a ratio between the hydrodynamic dimensions (or occupied volumes) of the crowder and the test molecule, with the most effective conditions being those where the volumes occupied by the crowder and the test molecule are of similar size

Table 4.3 Some structural characteristics of proteins used in (Stepanenko et al. 2016)

Protein (Uniprot for ID)	Abbreviation (oligomerization status)	Mol. Wight, kDa	pI	Struct. class	Stokes radius (Å)	Effects of crowders
Ribonuclease A (P61823, residues 27–150)	RNase A (monomer)	13.7	9.63	α ± β	18.7 ± 1.8	0/0
Lysozyme (P00698; residues 19–147)	HEL (monomer)	14.3	11.0	α ± β	19.0 ± 1.9	0/0
Ribonuclease B	RNase B (glycosylated monomer)	17.0	8.88	α ± β	20.2 ± 2.0	0/0
Papain (P00784), rersidues 134–345)	Pap (monomer)	23.4	8.75–9.55	α ± β	22.7 ± 2.3	0/0
Trypsinogen (P00760)	TRY (monomer)	24.0	8.7; 9.3	Multidomain (α and β	22.9 ± 2.4	0/–
α-Chymotrypsin (P00766, residues 1–13, 16–146, 149–245)	CHY (processed monomer, three chains)	25.0	8.75	All-β	23.5 ± 2.4	+/–
α-Chymotrypsinogen (P00766)	CHTG (monomer)	25.7	8.97	All-β	26.7 ± 2.8	0/–
β-Lactoglobulin A (P02754, residues 17–178 G80D, A134 V)	bLGA	36.8	5.3	All-β	26.7 ± 2.8	0/+
β-Lactoglobulin B (P02754, residues 17–178)	bLGB	36.8	5.1	All-β	26.7 ± 2.8	+/+
Hemoglobin human (P69905 + P68871)	HHb (heterotetramer of two α- and two β-chains)	64.5	6.8	All-α	32.6 ± 3.5	+/+
Concanavalin A (P02866, residues 30–148, 164–281)	ConA (homotetramer, each processed monomer has two chains)	104.0	4.5–5.5	All-β	38.6 ± 4.3	+/+

[a] Molecular weight of β-lactoglobulins correspond to their homodimeric forms; [b] Molecular weight of hemoglobin correspond to the heterotetrameric form;; [c] Molecular weight of Concanavalin A corresponds to the homote;trameric form [d] Struct. Class is defined by SCOP (Andreeva et al. 2004; Murzin et al. 1995); [e] Hydrodynamic radii (Stokes radii) were calculated based on the known molecular weights of proteins and assuming their globular conformation. The following equation was used (Uversky 1993): $\log(R_s) = (0.357_{\pm 0.05})\log(MW)—0.204_{\pm 0.023})$; [f] Effects of macromolecular crowders (PEG/Dextran) on conformational stability of target proteins were evaluated by the ability of PEG-8000 or Dextran-75 to visually increase (+), decrease (−) or not affect (0) the urea- = induces unfolding of the proteins

(Minton 1993; Tokuriki et al. 2004; Chen et al. 2011), the analysis in (Stepanenko et al. 2016) showed that the hydrodynamic dimensions of macromolecular crowding agents used (29.2 and 64.9 Å for the PEG-8000 and Dextran-70, respectively (Armstrong et al. 2004)) did not play noticeable role in the ability of these agents to affect conformational stability of the globular proteins with the hydrodynamic radii ranging from ~19 to ~39 Å.

Since two macromolecules cannot be at the same time in the same place, the excluded volume (i.e., the volume occupied by macromolecules and unavailable to other solutes) is generated. This excluded volume originates from the mutual impenetrability of solute molecules that generates steric repulsion. Therefore, crowed environments might significantly affect any reaction the course of which is dependent on the available volume (Zimmerman and Minton 1993; Minton 2005). In agreement with this hypothesis, previous studies revealed that macromolecular crowing can affect the process of protein folding, influence conformational equilibrium, and thereby modulate the conformational behavior and stability of a protein (Kuznetsova et al. 2014, 2015). In fact, the application of the excluded volume model to a protein folding process (i.e., a transition from a less compact unfolded form to a more compact folded state) suggests that the presence of a crowding agents might significantly affect the protein folding process by favoring a more compact (folded) state over the more extended (unfolded) form, and these folding-promoting effects are expected to increase exponentially with the increase in the concentration of the crowding agent (Ralston 2005). In line of this model, it is expected that macromolecular crowding might affect protein folding by influencing the hydrodynamic volume of the unfolded state. In other words, macromolecular crowding might provide an indirect stabilizing effect to the folded states of proteins due to destabilization of the more extended and malleable denatured states (Zhou 2004; Cheung et al. 2005). Therefore, according to the excluded volume theory, the most probable outcome of the presence of crowders in the solution of a protein undergoing unfolding is the increased conformational stability of this protein.

However, the data in (Stepanenko et al. 2016) revealed that some proteins are in fact destabilized by the addition of high concentrations of crowding agents (see Table 4.3). These observations are in line with the results of some earlier studies which indicated that not all crowders are made equal and therefore can have rather different effects on protein structure, function, and stability (Stepanenko et al. 2016). For example, crowding agents Ficoll-70, Dextran-70, PEG-2000, and PEG-20000 were shown to have different effects on the structural stability of human α-lactalbumin (Zhang et al. 2012). Here, the thermal stability of the apo-form of this protein was significantly decreased by the high concentrations of PEG-2000 and PEG-20000, whereas high concentrations of Dextran-70 and Ficoll-70 dramatically enhanced the thermal stability of this form of α-lactalbumin (Zhang et al. 2012). The difference in the action of different crowders on the thermal stability of this protein was attributed to the fact that no detectable interaction was observed between the protein and Ficoll-70 or Dextran-70, whereas there was a weak, non-specific interaction between the protein and PEG-2000 (Zhang et al. 2012).

Similarly, the unfolding behavior of myoglobin was differently affected by Ficoll-70, Dextran-70, and Dextran-40 (Malik et al. 2012). Here, the analysis of thermal denaturation suggested that all the crowders at the concentration of 200 g/L had a destabilizing effect on Mb as compared to the thermal unfolding of this protein in dilute solutions (Malik et al. 2012). However, when myoglobin was predestabilized by urea, Ficoll-70 showed a destabilizing effect on the thermal unfolding of this protein under all conditions, whereas Dextran-40 and Dextran-70 were destabilizing at 0–2 M urea but did not have such additional destabilizing effects when thermal denaturation was monitored in the presence of 3–5 M urea (Malik et al. 2012).

In order to explain some experimental observations inconsistent with the excluded volume effect, and to explain the differences in the capability of different crowders to differently affect the conformational stability of proteins, the existence was proposed of some "soft" interactions, such as electrostatic, hydrophobic, and van der Waals interactions between the crowding agent and the protein, in addition to hard non-specific steric interactions (Phillip and Schreiber 2013; Nakano et al. 2014). According to this model, when a protein and a crowder do not interact, crowding agent can exercise its "excluded volume" capacity and thereby increases protein stability, whereas this ability to stabilize proteins via the excluded volume effect is diminished when a protein and a crowder do interact (Zhang et al. 2012). One should also keep in mind that the weak interactions of a protein of interest with crowding agents can be repulsive or attractive. Therefore, such weak interactions can either enhance or diminish the excluded volume effects (Sarkar et al. 2013).

However, this explanation is not applicable to proteins studied here, since our analysis of the behavior of these proteins in the PEG-8000-Dextran-70-based ATPSs revealed the lack of detectable interactions between these proteins and polymers used to form ATPSs. An important feature of all known ATPSs is their ability to modulate partitioning of various solutes. In fact, each phase of ATPS provides a distinct solvent environment for small organic molecules, proteins, nucleic acids, and other solutes. Differences in solute–solvent interactions in the two phases commonly lead to unequal solute distribution. As a result, each liquid phase may be specifically enriched or depleted in particular solutes. It has been shown previously that partitioning of various solutes in ATPSs generally does not involve interactions of these solutes with phase forming polymers (Zaslavsky 1994; Moody et al. 2005; Madeira et al. 2008a, b, 2011; Silvério et al. 2010; da Silva et al. 2014; Ferreira et al. 2014, 2016b).

It has been shown that concentrations of water-soluble polymers used to generate ATPSs are in a range of polymer concentrations typically used in crowding experiments. Therefore, when a protein distributes between the phases of an ATPS it exists under macromolecular crowding conditions (Ferreira et al. 2016b, 2017; Albertsson 1986; Zaslavsky et al. 2013). If proteins do not interact with polymers, why they do partition and how, then, crowders differently affect conformational stabilities of proteins? In our view, the answer to these questions is in the ability of the crowding agents to induce changes in the solvent properties of aqueous media (Ferreira et al. 2016b).

Verma et al. used three different IR probes to show that the water dynamics remain unchanged in aqueous solutions of PEG dimethyl ether. At the same time, there are significant perturbations in in the hydrogen-bonded network of water (Verma et al. 2016). Kilburn et al. reported changes in water activity in aqueous solution of RNA induced by polymer crowder PEG-1000 and suggested that these changes might be related to the RNA stabilization (Kilburn et al. 2010). Changes in water activity in the presence of macromolecular crowders are also considered in a work by Nakano and Sugimoto (2017). Results of molecular dynamics simulations demonstrate that different polyols affect the water structure in their aqueous solutions (Politi et al. 2009). It was shown that TMAO makes the water hydrogen-bond network relatively stronger (Sarma and Paul 2013) and that TMAO reorganizes the hydrogen-bond network of water in a specific way (Rezus and Bakker 2009) whereas the data reported in (Ohto et al. 2017) did not confirm this observation. However, it has been shown that TMAO interactions with water lead to a more structured hydrogen-bond network (Sahle et al. 2016). It has also been reported that TMAO and urea, in their aqueous mixtures, cooperatively enhance the hydrogen-bond structure of water (Hunger et al. 2015). These effects of various solutes on water structure in their aqueous solutions, however, are not generally taken into account in studies of crowding effects (Bellissent-Funel et al. 2016). It has been shown that various physicochemical properties of aqueous solutions of polymers, salts, and small organic compounds, such as water activity, osmotic coefficient, surface tension, relative permittivity, and viscosity, may be described.

The thermal unfolding of ubiquitin was studied in aqueous solutions of glucose, dextran-40, and PEG-20,000 at their different concentrations (Senske et al. 2014). The melting temperatures reported in (Senske et al. 2014) at pH 2.0 are plotted in Fig. 4.7 against the solvent dipolarity/polarizability, π^*, and HBD acidity, α, of water in aqueous solutions of the compounds above. It should be mentioned that the π^* and α values at the particular concentrations used in (Senske et al. 2014) were interpolated from the concentration dependences of these properties reported earlier (Ferreira et al. 2016b) and for solutions of PEG-20,000, we used the concentration dependences reported for PEG-10,000 (Ferreira et al. 2016b).

It should be noted that the T_m values reported in (Senske et al. 2014) for the co-solute free solution, pH 2.0 and for the solutions of 0.1 M and 0.2 M KCl do not fit the relationship. It should be mentioned also that solutions of dextran-40 in (Senske et al. 2014) have been prepared in grams per liter of water, and from our experience it is a questionable procedure for preparation of aqueous solutions of highly hydrated polymers, such as dextran or Ficoll, and should always be followed by checking the final concentration of the solution by gravimetrical analysis using lyophilization of aliquots or measuring optical rotation or other properties of solution, since the “dry” polymer may contain rather significant amount of water. It is surprising that the data for Dextran-40 fit the above relationship so well. The relationship described by Eq. 4.3 implies that the ubiquitin stability in aqueous solutions of different crowders, such as dextran and PEG and small organic osmolyte, such as glucose, is increased due to these compounds influence on the solvent properties of water.

Similar relationships were also found for chymotrypsin inhibitor 2 in the presence of Ficoll-70 and sucrose, alkaline phosphatase in aqueous solutions of PEG-4000, Ficoll-70, and dextran-70,58 and globular actin in the presence of Ficoll-70, TMAO, urea (up to 2.0 M) and TMAO-urea mixture. An additional example worth mentioning is the effects of Ficoll, sucrose, TMAO, and urea on the polymerization kinetics of α/β-tubulin reported by Schummel et al. (2017). The polymerization kinetics was characterized by the apparent growth rate, *kapp* value, assayed at 37 °C in the buffer containing 3.26 M glycerol, pH 6.8;33 i.e., under conditions drastically different from those used to estimate the solvent properties of aqueous media in the polymers and osmolytes solutions (Ferreira et al. 2016b; Ferreira et al. 2017). Nevertheless, the reasonably good relationship with coefficients listed in Table 2 was established implying that protein-water interactions are important to the protein aggregation as well as for the protein stability. It should be mentioned, however, that we could not find any relationship between the data reported in ref.60 on the effects of polymers and osmolytes on the equilibrium stability of side-by-side dimerization of A34F variant of B1 domain of protein G and the solvent properties of aqueous media in solutions of TMAO, urea, sucrose, PEG-8000, and Ficoll-70.

One should mention here that cosolutes affects behavior of not only proteins, but DNA as well. For example, the melting temperatures of DNA of two different sequences have been reported in the presence of various concentrations of TMAO and urea (Ueda et al. 2016). The melting temperatures for the dsDNA forming canonical DNA duplex is plotted against the aforementioned properties of water in solutions of urea and TMAO in Fig. 4.10. The solvent properties of water were earlier examined in TMAO solutions up to TMAO concentration of 2.0 M.53 Therefore we plotted only T_m dsDNA values for TMAO solutions not exceeding 15%wt. The relationship shown in Fig. 4.10a may be described as:

$$T_m^{dsDNA} = 222_{\pm 22} - 120_{\pm 10}\pi^* - 19.3_{\pm 10.5}\alpha \quad (4.4a)$$

$$N = 8; r^2 = 0.9710; SD = 0.86; F = 83.6$$

where T_m^{dsDNA} is the melting temperature of dsDNA forming canonical DNA duplex; all the other parameters are as defined above.

Qualitatively similar relationship is shown in Fig. 4.10b for the DNA sequence forming a four-stranded G-quadruplex. This relationship may be described as:

$$Tm^{gqDNA} = 455_{\pm 21} - 245_{\pm 10.3}\pi^* - 95.7_{\pm 10.1}\alpha \quad (4.4b)$$

$$N = 8; r^2 = 0.9918; SD = 0.83; F = 304$$

where $T_m$gqDNA is the melting temperature of the DNA sequence forming a four-stranded G-quadruplex; all the other parameters are as defined above.

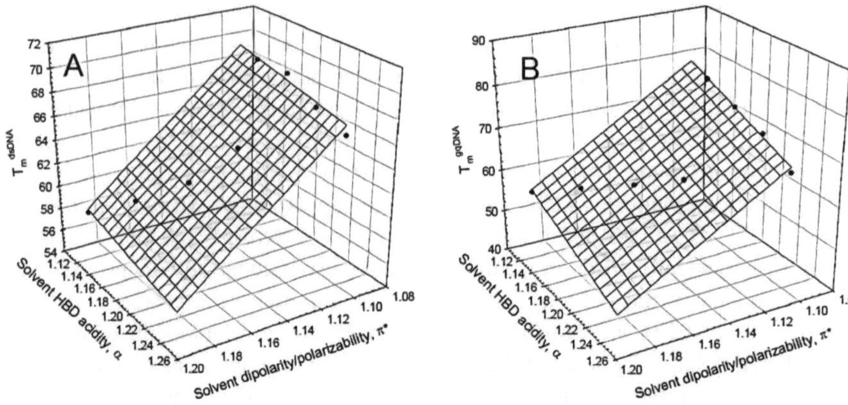

Fig. 4.10 Interrelationship between the stability of DNA of two different sequences expressed as the melting temperature, T_m, (see all the data used in Table S1) and the water dipolarity/polarizability, π^*, and water HBD acidity, α, in aqueous solutions of urea and TMAO at different concentrations (data from (Ueda et al. 2016)) (**a**) dsDNA sequence forming canonical DNA duplex; (**b**) gqDNA sequence forming a four-stranded $G_{quadruplex}$

The quantitative differences between Eq. 4.4a and 4.4b seem to imply that differences between the aforementioned DNA structures may lead to significant differences between various types of DNA-water interactions.

It has been shown that the RNA hairpin stability is directly related to the water activity in aqueous solutions of sucrose, Ficoll-70, and PEG (Gao et al. 2016). It has been reported by us (Ferreira et al. 2016a) that water activity in aqueous solutions of various compounds may be described in terms of solvent dipolarity/polarizability and HBD acidity of water. Therefore, it is unsurprising that the same solvent properties of water describe relatively well the reported in (Ferreira et al. 2017) melting temperatures of the RNA in solutions of sucrose, PEG-6000, PEG-20,000, and Ficoll-70. The observed relationship may be described as:

$$T_m^{RNA} = 189_{\pm 36} - 108 \pm 31.4\pi^* - 20.1_{\pm 4.7}\alpha \quad (4.5a)$$

$$N = 9; r^2 = 0.8170; SD = 1.79; F = 13.4$$

where T_m^{RNA} is the melting temperature of the hairpin structured RNA; all the other parameters are as defined above.

It is important to note that if the data for Ficoll solutions (prepared in gram per liter of water, and hence questionable due to the reasons discussed above) are omitted, the same relationship is dramatically improved and is described as:

$$T_m^{RNA} = 247_{\pm 24.1} - 162_{\pm 21.4}\pi^* - 18.6_{\pm 2.5}\alpha \quad (4.5b)$$

$$N = 6; r^2 = 0.9748; SD = 0.93; F = 58$$

where all the parameters are as defined above.

Different biological macromolecules would respond differently to the same changes in the solvent properties of water. Therefore, if the changes in the solvent properties of aqueous media is the major component of the crowding effect induced by macromolecular and/or low molecular crowders it is readily understandable why the same crowders may affect different proteins or nucleic acids in different ways (Miklos et al. 2011). The solvent properties of water in aqueous mixtures of two different polymers could be assayed only over the limited polymers concentration range because these mixtures separate into two aqueous phases at concentrations exceeding certain thresholds. The experimental data for solutions of individual polymers and their mixtures are presented in Table 4.4 together with the data calculated for the mixtures using the principle of additivity of the individual contributions of each polymer. Each solvent property of the mixture was calculated by addition of the experimentally assayed values in individual solutions of two polymers and subtraction of the experimental value for pure water. The term "calculated value" will be used further in this meaning in each case.

The data in Table 4.4 show that in four pairs of different macromolecular crowders, their overall effects on the solvent HBD acidity, α, is reduced in all cases, while in Ficoll-PEG, and dextran-Ficoll mixtures the overall effects on the solvent dipolarity/polarizability, π^*, are additive within the experimental error limits. The total effects on the solvent HBA basicity, β, are also additive within the experimental error limits in the mixtures of dextran-Ficoll and PEG-Ucon. The reported effects of mixtures of dextran-6 and Ficoll-70 on the stability of the protein FKBP exceed calculated values (Batra et al. 2009). It should be noted that the data presented in Table 4.4 show that the effects of two polymers on the solvent properties of water in their mixtures are typically non-additive, and these effects may affect various proteins in different manners.

These data are graphically illustrated in Fig. 4.11a for changes in the solvent dipolarity/polarizability π^* induced by individual polymers and their mixtures relative to water where π^* amounts to 1.096 ± 0.003, and in Fig. 4.11b for changes in the solvent HBD acidity α relative to water where $\alpha = 1.237 \pm 0.003$. To compare these changes to common organic solvents it may be mentioned that the difference between the π^* values for methanol and ethanol amounts to 0.06, and the difference between the α values in the same solvents is 0.10 (Kamlet et al. 1983). The data presented in Table 4.4 for the mixture of dextran-75 and PEG-8000 may be compared to those reported previously for the coexisting phases in aqueous two-phase system formed by these two polymers in 0.01 M sodium phosphate buffer, pH 7.4 (Madeira et al. 2012). These data are presented in Table 4.4 together with the calculated values, and they show that the overall effects of the two polymers are quite different in the two phases and do not agree with those for mixtures of individual polymers. The solvent dipolarity/polarizability and HBD acidity of water in the upper Ficoll-rich phase containing 28.31%wt. Ficoll and 3.23%wt. dextran-70 are reduced relatively to the calculated values, while the same properties in the lower dextran-rich phase containing 11.57%wt. dextran and 9.03%wt. Ficoll exceed the

Table 4.4 Solvent features of aqueous media in mixtures of macromolecular crowding agents (data for solutions of individual macromolecular crowding agents taken from (Ferreira et al. 2017)

Exp./Calc.	Polymer 1	Polymer 2	π*	α	β
	None	None	1.096 ± 0.003	1.237 ± 0.003	0.596 ± 0.001
	Ficoll-70	PEG-600	π*	α	β
Exp.	10%	0	1.128 ± 0.007	1.105 ± 0.009	0.615 ± 0.006
Exp.	0	10%	1.106 ± 0.001	1.113 ± 0.004	0.611 ± 0.002
Calc.	10%	10%	1.138 ± 0.011	0.981 ± 0.016	0.630 ± 0.009
Exp.	10%	10%	1.126 ± 0.001	1.032 ± 0.001	0.622 ± 0.001
	PEG-8000	Ucon-3930	π*	α	β
Exp.	10%	0	1.107 ± 0.002	1.072 ± 0.002	0.608 ± 0.002
Exp.	0	10%	1.129 ± 0.002	0.996 ± 0.009	0.631 ± 0.002
Calc.	10%	10%	1.140 ± 0.007	0.831 ± 0.014	0.643 ± 0.005
Exp.	10%	10%	1.110 ± 0.001	0.923 ± 0.003	0.642 ± 0.002
	Dextran-75	Ficoll-70	π*	α	β
Exp.	10%	0	1.105 ± 0.001	1.172 ± 0.002	0.609 ± 0.002
Exp.	0	10%	1.128 ± 0.007	1.105 ± 0.009	0.615 ± 0.006
Calc.	10%	10%	1.137 ± 0.011	1.040 ± 0.014	0.628 ± 0.009
Exp.	10%	10%	1.144 ± 0.002	1.075 ± 0.002	0.620 ± 0.002
	PVP-40	Ucon-3930	π*	α	β
Exp.	10%	0	1.140 ± 0.005	1.063 ± 0.008	0.644 ± 0.005
Exp.	0	10%	1.129 ± 0.002	0.996 ± 0.009	0.631 ± 0.002
Calc.	10%	10%	1.173 ± 0.010	0.822 ± 0.020	0.679 ± 0.008
Exp.	10%	10%	1.134 ± 0.001	0.942 ± 0.002	0.662 ± 0.004
	Dextran-75[a]	Ficoll-70 [a]	π*	α	β
Calc.	3.23%	28.31%	1.160 ± 0.012	0.974 ± 0.010	0.640 ± 0.009
Exp.	3.23%	28.31%	1.114 ± 0.003	1.014 ± 0.001	0.689 ± 0.002
Calc.	11.57%	9.03%	1.142 ± 0.010	1.041 ± 0.007	0.629 ± 0.007
Exp.	11.57%	9.03%	1.166 ± 0.002	1.022 ± 0.002	0.636 ± 0.001

[a]Solvent features experimentally assayed in the coexisting phases of aqueous two-phase system formed in the mixture of 12.9%wt. dextran-75 and 18.1%wt. Ficoll-70 in 0.01 M sodium phosphate buffer, pH 7.4. (Data from Ferreira et al. 2017)

calculated values. The solvent HBA acidity of water exceeds the calculated value in the upper phase and agrees with the calculated value for the lower phase.

It should be noted that water exists as a flexible network of hydrogen bonds and changes in the solvent properties of water imply that this network is rearranged. It is currently impossible to particularize structural changes in water corresponding to those in its solvent properties, and it is possible that the identical solvent properties do not imply identical structural arrangement of water. Analysis of effects of temperature, urea, and different salts on phase separation in aqueous mixtures of two nonionic polymers and thermodynamic analysis of phase diagrams indicated that phase separation is due to the incompatibility of the polymer modified water structures (Zaslavsky et al. 1989). The difference between the relative hydrophobicity of the coexisting phases estimated by the free energy of interfacial transfer of a CH_2 group may be described by a linear combination of the differences between the

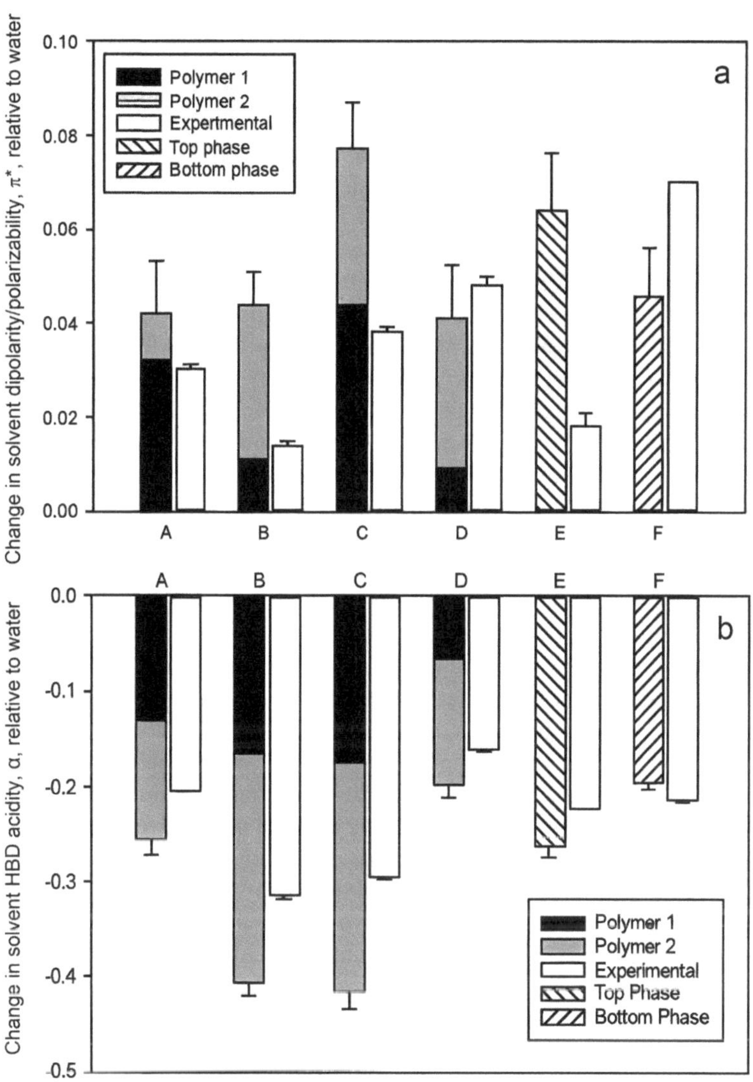

Fig. 4.11 Effects of mixtures of macromolecular crowding agents on (a) the water dipolarity/polarizability, π^*, and (b) water HBD acidity, α, in aqueous solutions containing 10%wt. of each crowding agent. The following mixtures listed in Table 4.4 were used: A: Polymer 1—Ficoll-70, polymer 2—PEG-600; B: Polymer 1—PEG-8000, polymer 2—Ucon-3930; C: Polymer 1—PVP-40, polymer 2—Ucon-3930; D: Polymer 1—Dextran-75, polymer 2—Ficoll-70; E: top Ficoll-rich phase of aqueous two-phase system; F: bottom dextran-rich phase of aqueous two-phase system (composition of both phases see in Table 4.4). Left column represents the overall effect of two polymers calculated based on the principle of additivity of contributions of each polymer; right column represents the experimentally assayed effect. All effects are shown as changes in the solvent property (a) relative to the solvent dipolarity/polarizability, π^*, of water, and (b) solvent water HBD acidity, α, of water. All the numerical data see in Table 4.4

solvent features of aqueous media in the two phases in aqueous two-phase systems formed by various pairs of nonionic polymers (Zaslavsky et al. 1989).

To compare these changes to common organic solvents it may be mentioned that the difference between the π* values for methanol and ethanol amounts to 0.06, and the difference between the α values in the same solvents is 0.10 (Kamlet et al. 1983). The data presented in Table 4.4 for the mixture of dextran-75 and PEG-8000 may be compared to those reported previously for the coexisting phases in aqueous two-phase system formed by these two polymers in 0.01 M sodium phosphate buffer, pH 7.4 (Madeira et al. 2012). These data are presented in Table 3 together with the calculated values, and they show that the overall effects of the two polymers are quite different in the two phases and do not agree with those for mixtures of individual polymers. The solvent dipolarity/polarizability and HBD acidity of water in the upper Ficoll-rich phase containing 28.31%wt. Ficoll and 3.23%wt. dextran-70 are reduced relatively to the calculated values, while the same properties in the lower dextran-rich phase containing 11.57%wt. dextran and 9.03%wt. Ficoll exceed the calculated values. The solvent HBA acidity of water exceeds the calculated value in the upper phase and agrees with the calculated value for the lower phase. It should be noted that water exists as a flexible network of hydrogen bonds and changes in the solvent properties of water imply that this network is rearranged. It is currently impossible to particularize structural changes in water corresponding to those in its solvent properties, and it is possible that the identical solvent properties do not imply identical structural arrangement of water. Analysis of effects of temperature, urea, and different salts on phase separation in aqueous mixtures of two nonionic polymers and thermodynamic analysis of phase diagrams indicated that phase separation is due to the incompatibility of the polymer modified water structures (Zaslavsky et al. 1989). The difference between the relative hydrophobicity of the coexisting phases estimated by the free energy of interfacial transfer of a CH_2 group may be described by a linear combination of the differences between the solvent features of aqueous media in the two phases in aqueous two-phase systems formed by various pairs of nonionic polymers (Madeira et al. 2012).

The solvent features of water in aqueous mixtures of several individual polymers with TMAO were examined at the polymer concentration of 30%wt. and TMAO concentrations varied from 1 up to 1.4 M in the solutions of dextran-75, and up to 2.0 M in solutions of other polymers. The TMAO concentration range was chosen to provide reliably measurable changes in the solvent features of water but was limited by TMAO solubility in polymer solutions examined. The experimental data for solutions of individual polymers and TMAO and their mixtures are presented in Table 4.5 together with the “additive”.

In the mixtures of TMAO and Ficoll-70, the solvent HBD acidity α and HBA basicity β exceed calculated values, while the solvent dipolarity/polarizability π* is either equal or slightly decreases relative to the calculated values. However, in mixtures of TMAO and dextran-75, the experimental values of all the water solvent properties agree with the calculated values within the experimental error’s values. Solvent dipolarity/polarizability in these mixtures agrees with the calculated values, while the experimentally measured solvent HBA basicity exceeds the calculated values over the whole range of TMAO concentrations examined.

Table 4.5 Solvent features of aqueous media in mixtures of macromolecular crowding agents with trimethylamine N-oxide (TMAO)

Exp./Calc.	Polymer	TMAO	π*	α	β
	None	None	1.096 ± 0.003	1.237 ± 0.003	0.596 ± 0.001
	Ficoll-70	TMAO	π*	α	β
Exp.	30.0%	0	1.164 ± 0.001	0.997 ± 0.002	0.642 ± 0.003
Exp.	0	1.0 M	1.096 ± 0.003	1.190 ± 0.003	0.633 ± 0.003
Calc.	30.0%	1.0 M	1.158 ± 0.007	0.950 ± 0.008	0.679 ± 0.007
Exp.	30.0%	1.0 M	1.148 ± 0.003	0.941 ± 0.003	0.670 ± 0.003
Exp.	0	1.5 M	1.096 ± 0.003	1.163 ± 0.005	0.650 ± 0.002
Calc.	30.0%	1.5 M	1.158 ± 0.007	0.923 ± 0.011	0.696 ± 0.006
Exp.	30.0%	1.5 M	1.143 ± 0.002	0.904 ± 0.002	0.705 ± 0.003
Extrap.	0	1.96 M	1.096 ± 0.003	1.140 ± 0.004	0.666 ± 0.004
Calc.	30.0%	1.96 M	1.158 ± 0.007	0.897 ± 0.009	0.713 ± 0.008
Exp.	30.0%	1.96 M	1.140 ± 0.002	0.869 ± 0.003	0.734 ± 0.004
	Dextran-75	TMAO	π*	α	β
Exp.	30.0%	0	1.151 ± 0.001	1.062 ± 0.002	0.630 ± 0.002
Exp.	0	1.0 M	1.096 ± 0.003	1.190 ± 0.003	0.633 ± 0.003
Calc.	30.0%	1.0 M	1.145 ± 0.007	1.015 ± 0.014	0.667 ± 0.005
Exp.	30.0%	1.0 M	1.145 ± 0.003	1.016 ± 0.003	0.664 ± 0.005
Interp.	0	1.4 M	1.096 ± 0.003	1.168 ± 0.003	0.647 ± 0.005
Calc.	30.0%	1.4 M	1.145 ± 0.009	0.993 ± 0.009	0.681 ± 0.011
Exp.	30.0%	1.4 M	1.143 ± 0.002	0.983 ± 0.002	0.676 ± 0.004
	PEG-600	TMAO	π*	α	β
Exp.	30.0%	0	1.109 ± 0.005	0.899 ± 0.004	0.669 ± 0.001
Exp.	0	1.0 M	1.096 ± 0.003	0.996 ± 0.009	0.631 ± 0.002
Calc.	30.0%	1.0 M	1.109 ± 0.005	–	0.679 ± 0.004
Exp.	30.0%	1.0 M	1.113 ± 0.004	Nd	0.704 ± 0.005
Exp.	0	1.5 M	1.096 ± 0.003	1.163 ± 0.005	0.650 ± 0.002
Calc.	30.0%	1.5 M	1.113 ± 0.005	–	0.696 ± 0.008
Exp.	30.0%	1.5 M	1.13 ± 0.004	Nd	0.749 ± 0.003
Exp.	0	2.0 M	1.096 ± 0.003	1.137 ± 0.004	0.667 ± 0.004
Calc.	30.0%	2.0 M	1.113 ± 0.005	–	0.713 ± 0.008
Exp.	10.0%	2.0 M	1.113 ± 0.002	Nd	0.803 ± 0.005
Dextran-rich phase[a]		TMAO	π*	α	β
Exp.	Nd	0	1.101 ± 0.001	1.039 ± 0.003	0.618 ± 0.001
Interp.	0	0.56 M	1.096 ± 0.003	1.210 ± 0.004	0.617 ± 0.002
Calc.	Nd	0.56 M	1.101 ± 0.007	1.012 ± 0.010	0.639 ± 0.007
Exp.	Nd	0.56 M	1.132 ± 0.001	1.072 ± 0.002	0.634 ± 0.001

[a]Dextran-rich bottom phase of ATPS formed by 12.2%wt. Dextran-75, 6.0%wt. PEG-8000, and 0.01 M sodium/potassium phosphate buffer, pH 7.4

Overall, the data in Table 4.5 show that the solvent properties of water depend on the particular composition of the polymerosmolyte mixture. This observation may explain why the overall effect of dextran-70 and sucrose on refolding of human

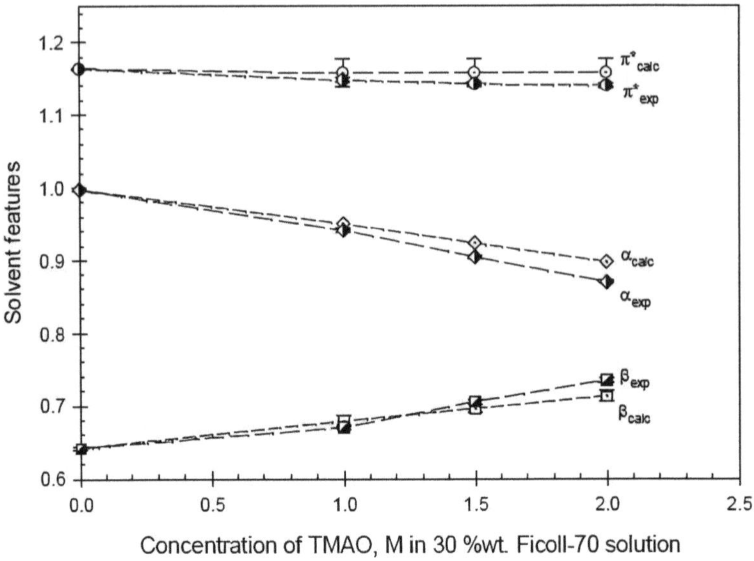

Fig. 4.12 Solvent dipolarity/polarizability, π*, HBD acidity, α, and HBA basicity, β, experimentally assayed (exp.) and calculated based on the assumption of additivity of contributions of the components (calc.) in aqueous mixtures of Ficoll-70 and TMAO as functions of TMAO concentration in the mixtures containing 30%wt. Ficoll-70. All the numerical data are shown in Table 4.5

muscle creatine kinase was shown to be less than the sum of their individual effects (Du et al. 2006) values calculated for the mixtures as described above. The data obtained for mixtures of TMAO and Ficoll are illustrated graphically in Fig. 4.12.

It was found that the solvatochromic band of the Reichardts' dye, due to unclear as of yet reason, was unstable in aqueous mixtures of PEG-600 and TMAO, and hence the water solvent HBD acidity could not be assayed in these mixtures. Solvent dipolarity/polarizability in these mixtures agrees with the calculated values, while the experimentally measured solvent HBA basicity exceeds the calculated values over the whole range of TMAO concentrations examined.

Overall, the data in Table 4.5 show that the solvent properties of water depend on the particular composition of the polymer osmolyte mixture. This observation may explain why the overall effect of dextran-70 and sucrose on refolding of human muscle creatine kinase was shown to be less than the sum of their individual effects (Lin and Li 2010). Furthermore, it was shown that the solvent properties of the dextran-rich bottom phase change with addition of 0.56 M TMAO concentration in the aqueous two-phase system formed by dextran-75 and PEG-8000 and 0.5 M TMAO. In fact, comparison of the solvent properties of water in the dextran-rich phase in the presence of 0.56 M TMAO with those assayed in the same phase in the same TMAO free system shows that the solvent dipolarity/polarizability of water exceeds the calculated value, and the HBD acidity decreases relatively to the calculated value. The HBA basicity agrees with the calculated value. In this case again the overall effect of TMAO and polymer in the phase of aqueous two-phase system differs from the one observed in the aqueous mixtures of individual dextran-75 and TMAO.

Analysis of solvent properties of water in aqueous solutions of individual macromolecular crowders and osmolytes shows that linear combinations of the solvent dipolarity/polarizability and hydrogen bond donor acidity of water determined in these solutions strongly correlate with changes in stability of proteins and nucleic acids reported in the literature. The same properties in mixtures of two macromolecular crowders or mixtures of osmolyte and macromolecular crowder vary differently for various solvent properties, and the overall effects of the two components in the mixture on the solvent properties of water may be additive, reduced or enhanced depending on the particular composition of the mixture. It was hypothesized (Ferreira et al. 2017) that changes in the solvent properties of water are related to changes in the water hydrogen-bonding structuring. It is suggested that the observed crowder-induced changes in the solvent properties of water should be taken into account in theoretical considerations of crowding effects in biological systems.Compliance with Ethical StandardsAuthors declare no conflict of interest. Funding source: Not applicable.

References

Albertsson PA (1986) Partition of cell particles and macromolecules, 3rd edn. Wiley, New York

Alfano C, Fichou Y, Huber K, Weiss M, Spruijt E, Ebbinghaus S, De Luca G, Morando MA, Vetri V, Temussi PA, Pastore A (2024) Molecular crowding: the history and development of a scientific paradigm. Chem Rev 124(6):3186–3219. https://doi.org/10.1021/acs.chemrev.3c00615

Allagulova CR, Gimalov F, Shakirova F, Vakhitov V (2003) The plant dehydrins: structure and putative functions. Biochem Mosc 68:945–951

Andreeva A, Howorth D, Brenner SE, Hubbard TJ, Chothia C, Murzin AG (2004) SCOP database in 2004: refinements integrate structure and sequence family data. Nucleic Acids Res 32 (Database issue):D226–D229. https://doi.org/10.1093/nar/gkh039

Armstrong JK, Wenby RB, Meiselman HJ, Fisher TC (2004) The hydrodynamic radii of macromolecules and their effect on red blood cell aggregation. Biophys J 87(6):4259–4270. https://doi.org/10.1529/biophysj.104.047746

Arnold K, Herrmann A, Pratsch L, Gawrisch K (1985) The dielectric properties of aqueous solutions of poly(ethylene glycol) and their influence on membrane structure. Biochim Biophys Acta 815(3):515–518. https://doi.org/10.1016/0005-2736(85)90381-5

Assarsson A, Hellstrand E, Cabaleiro-Lago C, Linse S (2014) Charge dependent retardation of amyloid beta aggregation by hydrophilic proteins. ACS Chem Neurosci 5(4):266–274. https://doi.org/10.1021/cn400124r

Bai J, Liu M, Pielak GJ, Li C (2017) Macromolecular and small molecular crowding have similar effects on α-synuclein structure. ChemPhysChem 18(1):55–58

Batra J, Xu K, Zhou HX (2009) Nonadditive effects of mixed crowding on protein stability. Prot Struct Funct Bioinform 77 (1):133–138

Bellissent-Funel M-C, Hassanali A, Havenith M, Henchman R, Pohl P, Sterpone F, Van Der Spoel D, Xu Y, Garcia AE (2016) Water determines the structure and dynamics of proteins. Chem Rev 116(13):7673–7697

Ben-Naim A (2003) Hydrophobic hydrophilic phenomena in biochemical processes. Biophys Chem 105(2–3):183–193. https://doi.org/10.1016/s0301-4622(03)00088-7

Benton LA, Smith AE, Young GB, Pielak GJ (2012) Unexpected effects of macromolecular crowding on protein stability. Biochemistry 51(49):9773–9775. https://doi.org/10.1021/bi300909q

Bismuto E, Martelli PL, De Maio A, Mita DG, Irace G, Casadio R (2002) Effect of molecular confinement on internal enzyme dynamics: frequency domain fluorometry and molecular dynamics simulation studies. Biopolymers 67(2):85–95. https://doi.org/10.1002/bip.10058

Bloustine J, Virmani T, Thurston G, Fraden S (2006) Light scattering and phase behavior of lysozyme-poly (ethylene glycol) mixtures. Phys Rev Lett 96(8):087803

Breydo L, Reddy KD, Piai A, Felli IC, Pierattelli R, Uversky VN (2014) The crowd you're in with: effects of different types of crowding agents on protein aggregation. Biochim Biophys Acta 1844 2:346–357. https://doi.org/10.1016/j.bbapap.2013.11.004

Canchi DR, Garcia AE (2013) Cosolvent effects on protein stability. Annu Rev Phys Chem 64:273–293. https://doi.org/10.1146/annurev-physchem-040412-110156

Chebotareva NA (2007) Effect of molecular crowding on the enzymes of glycogenolysis. Biochemistry (Mosc) 72(13):1478–1490. https://doi.org/10.1134/S0006297907130056

Chebotareva NA, Kurganov BI, Livanova NB (2004) Biochemical effects of molecular crowding. Biochemistry (Mosc) 69(11):1239–1251

Chen C, Loe F, Blocki A, Peng Y, Raghunath M (2011) Applying macromolecular crowding to enhance extracellular matrix deposition and its remodeling in vitro for tissue engineering and cell-based therapies. Adv Drug Deliv Rev 63(4–5):277–290. https://doi.org/10.1016/j.addr.2011.03.003

Cheung MS, Klimov D, Thirumalai D (2005) Molecular crowding enhances native state stability and refolding rates of globular proteins. Proc Natl Acad Sci USA 102(13):4753–4758. https://doi.org/10.1073/pnas.0409630102

Close TJ (1997) Dehydrins: a commonalty in the response of plants to dehydration and low temperature. Physiol Plant 100(2):291–296

Crowley PB, Brett K, Muldoon J (2008) NMR spectroscopy reveals cytochrome c–poly (ethylene glycol) interactions. Chembiochem 9(5):685–688

da Silva NR, Ferreira LA, Mikheeva LM, Teixeira JA, Zaslavsky BY (2014) Origin of salt additive effect on solute partitioning in aqueous polyethylene glycol-8000-sodium sulfate two-phase system. J Chromatogr A 1337:3–8. https://doi.org/10.1016/j.chroma.2014.02.031

Despa F, Fernandez A, Berry RS (2004) Dielectric modulation of biological water. Phys Rev Lett 93(22):228104. https://doi.org/10.1103/PhysRevLett.93.228104

Du F, Zhou Z, Mo Z-Y, Shi J-Z, Chen J, Liang Y (2006) Mixed macromolecular crowding accelerates the refolding of rabbit muscle creatine kinase: implications for protein folding in physiological environments. J Mol Biol 364(3):469–482

Eggers DK, Valentine JS (2001) Molecular confinement influences protein structure and enhances thermal protein stability. Protein Sci 10(2):250–261. https://doi.org/10.1110/ps.36201

Elcock AH (2010) Models of macromolecular crowding effects and the need for quantitative comparisons with experiment. Curr Opin Struct Biol 20(2):196–206. https://doi.org/10.1016/j.sbi.2010.01.008

Eronina TB, Chebotareva NA, Kurganov BI (2005) Influence of osmolytes on inactivation and aggregation of muscle glycogen phosphorylase b by guanidine hydrochloride. Stimulation of protein aggregation under crowding conditions. Biochemistry (Mosc) 70(9):1020–1026

Ferreira L, Fan X, Mikheeva LM, Madeira PP, Kurgan L, Uversky VN, Zaslavsky BY (2014) Structural features important for differences in protein partitioning in aqueous dextran-polyethylene glycol two-phase systems of different ionic compositions. Biochim Biophys Acta 1844 3:694–704. https://doi.org/10.1016/j.bbapap.2014.01.016

Ferreira LA, Cole JT, Reichardt C, Holland NB, Uversky VN, Zaslavsky BY (2015a) Solvent properties of water in aqueous solutions of elastin-like polypeptide. Int J Mol Sci 16(6):13528–13547

Ferreira LA, Fan X, Madeira PP, Kurgan L, Uversky VN, Zaslavsky BY (2015b) Analyzing the effects of protecting osmolytes on solute–water interactions by solvatochromic comparison method: II. Glob Prot Rsc Adv 5(73):59780–59791

Ferreira LA, Loureiro JA, Gomes J, Uversky VN, Madeira PP, Zaslavsky BY (2016a) Why physicochemical properties of aqueous solutions of various compounds are linearly interrelated. J Mol Liq 221:116–123
Ferreira LA, Madeira PP, Breydo L, Reichardt C, Uversky VN, Zaslavsky BY (2016b) Role of solvent properties of aqueous media in macromolecular crowding effects. J Biomol Struct Dyn 34(1):92–103. https://doi.org/10.1080/07391102.2015.1011235
Ferreira LA, Uversky VN, Zaslavsky BY (2017) Role of solvent properties of water in crowding effects induced by macromolecular agents and osmolytes. Mol BioSyst 13(12):2551–2563. https://doi.org/10.1039/c7mb00436b
Ferreira LA, Gusev NB, Uversky VN, Zaslavsky BY (2018a) Effect of human heat shock protein HspB6 on the solvent features of water in aqueous solutions. J Biomol Struct Dyn 36(6):1520–1528. https://doi.org/10.1080/07391102.2017.1328316
Ferreira LA, Walczyk Mooradally A, Zaslavsky B, Uversky VN, Graether SP (2018b) Effect of an intrinsically disordered plant stress protein on the properties of water. Biophys J 115(9):1696–1706. https://doi.org/10.1016/j.bpj.2018.09.014
Gao M, Gnutt D, Orban A, Appel B, Righetti F, Winter R, Narberhaus F, Müller S, Ebbinghaus S (2016) RNA hairpin folding in the crowded cell. Angew Chem Int Ed 55(9):3224–3228
Graether SP, Boddington KF (2014) Disorder and function: a review of the dehydrin protein family. Front Plant Sci 5:576. https://doi.org/10.3389/fpls.2014.00576
Guseman AJ, Pielak GJ (2017) Cosolute and crowding effects on a side-by-side protein dimer. Biochemistry 56(7):971–976
Hara M (2010) The multifunctionality of dehydrins: an overview. Plant Signal Behav 5(5):503–508
Harada R, Sugita Y, Feig M (2012) Protein crowding affects hydration structure and dynamics. J Am Chem Soc 134(10):4842–4849. https://doi.org/10.1021/ja211115q
Hatters DM, Minton AP, Howlett GJ (2002) Macromolecular crowding accelerates amyloid formation by human apolipoprotein C-II. J Biol Chem 277(10):7824–7830. https://doi.org/10.1074/jbc.M110429200
Hong J, Gierasch LM (2010) Macromolecular crowding remodels the energy landscape of a protein by favoring a more compact unfolded state. J Am Chem Soc 132(30):10445–10452
Hunger J, Ottosson N, Mazur K, Bonn M, Bakker HJ (2015) Water-mediated interactions between trimethylamine-N-oxide and urea. Phys Chem Chem Phys 17(1):298–306
Jia J, Peng X, Qi W, Su R, He Z (2017) Effects of macromolecular crowding on alkaline phosphatase unfolding, conformation and stability. Int J Biol Macromol 101:373–382
Kamlet MJ, Taft R (1976) The solvatochromic comparison method. I. The. Beta.-scale of solvent hydrogen-bond acceptor (HBA) basicities. J Am Chem Soc 98 (2):377-383
Kamlet MJ, Abboud JL, Taft R (1977) The solvatochromic comparison method. 6. The. Pi.* scale of solvent polarities. J Am Chem Soc 99 (18):6027-6038
Kamlet MJ, Abboud JLM, Abraham MH, Taft R (1983) Linear solvation energy relationships. 23. A comprehensive collection of the solvatochromic parameters,. pi.*,. alpha., and. beta., and some methods for simplifying the generalized solvatochromic equation. J Org Chem 48(17):2877–2887
Kilburn D, Roh JH, Guo L, Briber RM, Woodson SA (2010) Molecular crowding stabilizes folded RNA structure by the excluded volume effect. J Am Chem Soc 132(25):8690–8696
Kim IW, Jang MD, Ryu YK, Cho EH, Lee YK, Park JH (2002) Dipolarity, hydrogen-bond basicity and hydrogen-bond acidity of aqueous poly(ethylene glycol) solutions. Anal Sci 18(12):1357–1360. https://doi.org/10.2116/analsci.18.1357
King JT, Arthur EJ, Brooks CL 3rd, Kubarych KJ (2014) Crowding induced collective hydration of biological macromolecules over extended distances. J Am Chem Soc 136(1):188–194. https://doi.org/10.1021/ja407858c
Knowles DB, LaCroix AS, Deines NF, Shkel I, Record MT Jr (2011) Separation of preferential interaction and excluded volume effects on DNA duplex and hairpin stability. Proc Natl Acad Sci USA 108(31):12699–12704. https://doi.org/10.1073/pnas.1103382108

Kulkarni AM, Chatterjee AP, Schweizer KS, Zukoski CF (2000) Effects of polyethylene glycol on protein interactions. J Chem Phys 113(21):9863–9873
Kuznetsova IM, Turoverov KK, Uversky VN (2014) What macromolecular crowding can do to a protein. Int J Mol Sci 15(12):23090–23140. https://doi.org/10.3390/ijms151223090
Kuznetsova IM, Zaslavsky BY, Breydo L, Turoverov KK, Uversky VN (2015) Beyond the excluded volume effects: mechanistic complexity of the crowded milieu. Molecules 20(1):1377–1409. https://doi.org/10.3390/molecules20011377
Li X, Mehler EL (2006) Simulation of molecular crowding effects on an Alzheimer's beta-amyloid peptide. Cell Biochem Biophys 46(2):123–141. https://doi.org/10.1385/cbb:46:2:123
Lin Z-M, Li S (2010) Effect of mixed crowding on refolding of human muscle creatine kinase. Protein Pept Lett 17(11):1426–1435
Ma B, Xie J, Wei L, Li W (2013a) Macromolecular crowding modulates the kinetics and morphology of amyloid self-assembly by beta-lactoglobulin. Int J Biol Macromol 53:82–87. https://doi.org/10.1016/j.ijbiomac.2012.11.008
Ma Q, Hu JY, Chen J, Liang Y (2013b) The role of crowded physiological environments in prion and prion-like protein aggregation. Int J Mol Sci 14(11):21339–21352. https://doi.org/10.3390/ijms141121339
Madeira P, Teixeira JA, Macedo EA, Mikheeva LM, Zaslavsky BY (2008a) Correlations between distribution coefficients of various biomolecules in different polymer/polymer aqueous two-phase systems. Fluid Phase Equil 267:150–157. https://doi.org/10.1016/j.fluid.2008.03.003
Madeira PP, Teixeira JA, Macedo EA, Mikheeva LM, Zaslavsky BY (2008b) "On the Collander equation": protein partitioning in polymer/polymer aqueous two-phase systems. J Chromatogr A 1190(1–2):39–43
Madeira PP, Reis CA, Rodrigues AE, Mikheeva LM, Zaslavsky BY (2010) Solvent properties governing solute partitioning in polymer/polymer aqueous two-phase systems: nonionic compounds. J Phys Chem B 114(1):457–462. https://doi.org/10.1021/jp907346s
Madeira PP, Reis CA, Rodrigues AE, Mikheeva LM, Chait A, Zaslavsky BY (2011) Solvent properties governing protein partitioning in polymer/polymer aqueous two-phase systems. J Chromatogr A 1218(10):1379–1384. https://doi.org/10.1016/j.chroma.2011.01.020
Madeira PP, Bessa A, Alvares-Ribeiro L, Aires-Barros MR, Reis CA, Rodrigues AE, Zaslavsky BY (2012) Salt effects on solvent features of coexisting phases in aqueous polymer/polymer two-phase systems. J Chromatogr A 1229:38–47. https://doi.org/10.1016/j.chroma.2012.01.029
Madeira PP, Bessa A, Teixeira MA, Álvares-Ribeiro L, Aires-Barros MR, Rodrigues AE, Zaslavsky BY (2013) Study of organic compounds–water interactions by partition in aqueous two-phase systems. J Chromatogr A 1322:97–104
Madeira PP, Bessa A, Alvares-Ribeiro L, Raquel Aires-Barros M, Rodrigues AE, Uversky VN, Zaslavsky BY (2014) Amino acid/water interactions study: a new amino acid scale. J Biomol Struct Dyn 32(6):959–968. https://doi.org/10.1080/07391102.2013.800994
Malik A, Kundu J, Mukherjee SK, Chowdhury PK (2012) Myoglobin unfolding in crowding and confinement. J Phys Chem B 116(43):12895–12904. https://doi.org/10.1021/jp306873v
Martin I, Celaya G, Alfonso C, Moro F, Rivas G, Muga A (2014) Crowding activates ClpB and enhances its association with DnaK for efficient protein aggregate reactivation. Biophys J 106(9):2017–2027. https://doi.org/10.1016/j.bpj.2014.03.042
Miklos AC, Sarkar M, Wang Y, Pielak GJ (2011) Protein crowding tunes protein stability. J Am Chem Soc 133(18):7116–7120. https://doi.org/10.1021/ja200067p
Milles S, Huy Bui K, Koehler C, Eltsov M, Beck M, Lemke EA (2013) Facilitated aggregation of FG nucleoporins under molecular crowding conditions. EMBO Rep 14(2):178–183. https://doi.org/10.1038/embor.2012.204
Minton AP (1993) Macromolecular crowding and molecular recognition. J Mol Recognit 6(4):211–214. https://doi.org/10.1002/jmr.300060410
Minton AP (2000a) Implications of macromolecular crowding for protein assembly. Curr Opin Struct Biol 10(1):34–39. https://doi.org/10.1016/s0959-440x(99)00045-7

Minton AP (2000b) Protein folding: thickening the broth. Curr Biol 10(3):R97–R99. https://doi.org/10.1016/s0960-9822(00)00301-8
Minton AP (2005) Models for excluded volume interaction between an unfolded protein and rigid macromolecular cosolutes: macromolecular crowding and protein stability revisited. Biophys J 88(2):971–985. https://doi.org/10.1529/biophysj.104.050351
Moody ML, Willauer HD, Griffin ST, Huddleston JG, Rogers RD (2005) Solvent property characterization of poly(ethylene glycol)/dextran aqueous biphasic systems using the free energy of transfer of a methylene group and a linear solvation energy relationship. Ind Eng Chem Res 44:3749–3760. https://doi.org/10.1021/ie049491c
Morar AS, Olteanu A, Young GB, Pielak GJ (2001) Solvent-induced collapse of alpha-synuclein and acid-denatured cytochrome c. Protein Sci 10(11):2195–2199. https://doi.org/10.1110/ps.24301
Munishkina LA, Cooper EM, Uversky VN, Fink AL (2004) The effect of macromolecular crowding on protein aggregation and amyloid fibril formation. J Mol Recognit 17(5):456–464. https://doi.org/10.1002/jmr.699
Munishkina LA, Ahmad A, Fink AL, Uversky VN (2008a) Guiding protein aggregation with macromolecular crowding. Biochemistry 47(34):8993–9006. https://doi.org/10.1021/bi8008399
Munishkina LA, Fink AL, Uversky VN (2008b) Concerted action of metals and macromolecular crowding on the fibrillation of alpha-synuclein. Protein Pept Lett 15(10):1079–1085. https://doi.org/10.2174/092986608786071102
Murzin AG, Brenner SE, Hubbard T, Chothia C (1995) SCOP: a structural classification of proteins database for the investigation of sequences and structures. J Mol Biol 247(4):536–540. https://doi.org/10.1006/jmbi.1995.0159
Nakano S-i, Sugimoto N (2017) Model studies of the effects of intracellular crowding on nucleic acid interactions. Mol BioSyst 13(1):32–41
Nakano S, Miyoshi D, Sugimoto N (2014) Effects of molecular crowding on the structures, interactions, and functions of nucleic acids. Chem Rev 114(5):2733–2758. https://doi.org/10.1021/cr400113m
Ohto T, Hunger J, Backus EH, Mizukami W, Bonn M, Nagata Y (2017) Trimethylamine-N-oxide: its hydration structure, surface activity, and biological function, viewed by vibrational spectroscopy and molecular dynamics simulations. Phys Chem Chem Phys 19(10):6909–6920
Patel CN, Noble SM, Weatherly GT, Tripathy A, Winzor DJ, Pielak GJ (2002) Effects of molecular crowding by saccharides on α-chymotrypsin dimerization. Protein Sci 11(5):997–1003
Phillip Y, Schreiber G (2013) Formation of protein complexes in crowded environments from in vitro to in vivo. FEBS Lett 587 (8):1046–1052. doi:https://doi.org/10.1016/j.febslet.2013.01.007
Pico G, Bassani G, Farruggia B, Nerli B (2007) Calorimetric investigation of the protein-flexible chain polymer interactions and its relationship with protein partition in aqueous two-phase systems. Int J Biol Macromol 40(3):268–275. https://doi.org/10.1016/j.ijbiomac.2006.08.008
Politi R, Harries D (2010) Enthalpically driven peptide stabilization by protective osmolytes. Chem Commun (Camb) 46(35):6449–6451. https://doi.org/10.1039/c0cc01763a
Politi R, Sapir L, Harries D (2009) The impact of polyols on water structure in solution: a computational study. Chem A Eur J 113(26):7548–7555
Ralston GB (2005) Effects of "crowding" in protein solutions. J Chem Educ 67:857–860. https://doi.org/10.1021/ed067p857
Rezus Y, Bakker H (2009) Destabilization of the hydrogen-bond structure of water by the osmolyte trimethylamine N-oxide. J Phys Chem B 113(13):4038–4044
Rivera E, Straub J, Thirumalai D (2009) Sequence and crowding effects in the aggregation of a 10-residue fragment derived from islet amyloid polypeptide. Biophys J 96(11):4552–4560. https://doi.org/10.1016/j.bpj.2009.03.039
Sahle CJ, Schroer MA, Juurinen I, Niskanen J (2016) Influence of TMAO and urea on the structure of water studied by inelastic X-ray scattering. Phys Chem Chem Phys 18(24):16518–16526

Sarkar M, Smith AE, Pielak GJ (2013) Impact of reconstituted cytosol on protein stability. Proc Natl Acad Sci USA 110(48):19342–19347. https://doi.org/10.1073/pnas.1312678110

Sarma R, Paul S (2013) Exploring the molecular mechanism of trimethylamine-N-oxide's ability to counteract the protein denaturing effects of urea. J Phys Chem B 117(18):5691–5704

Schummel PH, Haag A, Kremer W, Kalbitzer HR, Winter R (2016) Cosolvent and crowding effects on the temperature and pressure dependent conformational dynamics and stability of globular actin. J Phys Chem B 120(27):6575–6586

Schummel PH, Gao M, Winter R (2017) Modulation of the polymerization kinetics of α/β-tubulin by Osmolytes and macromolecular crowding. ChemPhysChem 18(2):189–197

Senske M, Törk L, Born B, Havenith M, Herrmann C, Ebbinghaus S (2014) Protein stabilization by macromolecular crowding through enthalpy rather than entropy. J Am Chem Soc 136(25):9036–9041

Shahid S, Hasan I, Ahmad F, Hassan MI, Islam A (2019) Carbohydrate-based macromolecular crowding-induced stabilization of proteins: towards understanding the significance of the size of the Crowder. Biomolecules 9(9). https://doi.org/10.3390/biom9090477

Sharp KA (2015) Analysis of the size dependence of macromolecular crowding shows that smaller is better. Proc Natl Acad Sci 112(26):7990–7995

Shtilerman MD, Ding TT, Lansbury PT (2002) Molecular crowding accelerates fibrillization of α-synuclein: could an increase in the cytoplasmic protein concentration induce Parkinson's disease? Biochemistry 41(12):3855–3860

Silvério SC, Rodriguez O, Teixeira JA, Macedo EA (2010) Solute partitioning in polymer-salt ATPS: the Collander equation. Fluid Phase Equilib 296:173–177. https://doi.org/10.1016/j.fluid.2010.04.009

Stepanenko OV, Povarova OI, Sulatskaya AI, Ferreira LA, Zaslavsky BY, Kuznetsova IM, Turoverov KK, Uversky VN (2016) Protein unfolding in crowded milieu: what crowding can do to a protein undergoing unfolding? J Biomol Struct Dyn 34(10):2155–2170. https://doi.org/10.1080/07391102.2015.1109554

Sukenik S, Sapir L, Gilman-Politi R, Harries D (2013) Diversity in the mechanisms of cosolute action on biomolecular processes. Faraday Discuss 160:225–237; discussion 311–227. doi:https://doi.org/10.1039/c2fd20101a

Taft R, Kamlet MJ (1976) The solvatochromic comparison method. 2. The. Alpha.-scale of solvent hydrogen-bond donor (HBD) acidities. J Am Chem Soc 98 (10):2886–2894

Tokuriki N, Kinjo M, Negi S, Hoshino M, Goto Y, Urabe I, Yomo T (2004) Protein folding by the effects of macromolecular crowding. Protein Sci 13(1):125–133. https://doi.org/10.1110/ps.03288104

Ueda Y-m, Zouzumi Y-k, Maruyama A, Nakano S-i, Sugimoto N, Miyoshi D (2016) Effects of trimethylamine N-oxide and urea on DNA duplex and G-quadruplex. Sci Technol Adv Mater 17(1):753–759

Uversky VN (1993) Use of fast protein size-exclusion liquid chromatography to study the unfolding of proteins which denature through the molten globule. Biochemistry 32(48):13288–13298. https://doi.org/10.1021/bi00211a042

Uversky VN, Elisa MC, Bower KS, Li J, Fink AL (2002) Accelerated alpha-synuclein fibrillation in crowded milieu. FEBS Lett 515(1–3):99–103. https://doi.org/10.1016/s0014-5793(02)02446-8

Verma PK, Kundu A, Ha J-H, Cho M (2016) Water dynamics in cytoplasm-like crowded environment correlates with the conformational transition of the macromolecular crowder. J Am Chem Soc 138(49):16081–16088

Wang Y, Sarkar M, Smith AE, Krois AS, Pielak GJ (2012) Macromolecular crowding and protein stability. J Am Chem Soc 134(40):16614–16618. https://doi.org/10.1021/ja305300m

Yamin G, Munishkina LA, Karymov MA, Lyubchenko YL, Uversky VN, Fink AL (2005) Forcing nonamyloidogenic beta-synuclein to fibrillate. Biochemistry 44(25):9096–9107. https://doi.org/10.1021/bi048778a

Zaslavsky BY (1994) Aqueous two-phase partitioning: physical chemistry and bioanalytical applications. Marcel Dekker, New York

Zaslavsky BY, Bagirov T, Borovskaya A, Gulaeva N, Miheeva L, Mahmudov A, Rodnikova M (1989) Structure of water as a key factor of phase separation in aqueous mixtures of two nonionic polymers. Polymer 30(11):2104–2111

Zaslavsky BY, Miheeva LM, Masimov EA, Djafarov SF, Reichardt C (1990) Solvent polarity of aqueous polymer solutions as measured by the solvatochromic technique. J Chem Soc Faraday Trans 86(3):519–524

Zaslavsky A, Madeira P, Breydo L, Uversky VN, Chait A, Zaslavsky B (2013) High throughput characterization of structural differences between closely related proteins in solution. Biochim Biophys Acta 1834 2:583–592. https://doi.org/10.1016/j.bbapap.2012.11.004

Zhang DL, Wu LJ, Chen J, Liang Y (2012) Effects of macromolecular crowding on the structural stability of human alpha-lactalbumin. Acta Biochim Biophys Sin Shanghai 44(8):703–711. https://doi.org/10.1093/abbs/gms052

Zhou HX (2004) Loops, linkages, rings, catenanes, cages, and crowders: entropy-based strategies for stabilizing proteins. Acc Chem Res 37(2):123–130. https://doi.org/10.1021/ar0302282

Zhou BR, Liang Y, Du F, Zhou Z, Chen J (2004) Mixed macromolecular crowding accelerates the oxidative refolding of reduced, denatured lysozyme: implications for protein folding in intracellular environments. J Biol Chem 279(53):55109–55116. https://doi.org/10.1074/jbc.M409086200

Zhou HX, Rivas G, Minton AP (2008) Macromolecular crowding and confinement: biochemical, biophysical, and potential physiological consequences. Annu Rev Biophys 37:375–397. https://doi.org/10.1146/annurev.biophys.37.032807.125817

Zhou Z, Fan JB, Zhu HL, Shewmaker F, Yan X, Chen X, Chen J, Xiao GF, Guo L, Liang Y (2009) Crowded cell-like environment accelerates the nucleation step of amyloidogenic protein misfolding. J Biol Chem 284(44):30148–30158. https://doi.org/10.1074/jbc.M109.002832

Zimmerman SB, Minton AP (1993) Macromolecular crowding: biochemical, biophysical, and physiological consequences. Annu Rev Biophys Biomol Struct 22:27–65. https://doi.org/10.1146/annurev.bb.22.060193.000331

Chapter 5
Effects of Macromolecular Crowding on the Structure and Dynamics of Biological Membranes

Semen V. Nesterov

Abstract This chapter reviews and analyzes the effects of crowding on lipid membranes. Presented data shows that due to steric effects crowding on one side of the bilayer causes membrane bending toward that side. The curvature of the membrane increases with increasing crowding, with disordered proteins being more effective in generating deformations. The effect of crowding on lateral diffusion in the vicinity of the membrane is discussed, highlighting the difference between crowding and immobilization by fixed scaffolds. The effect of membrane crowding by integral proteins is also analyzed and it is shown that it leads to lipid deformations near proteins, promotes the formation of lipid microdomains and protein clusters. The role of the crowding agent effects on the physicochemical properties of solutions is reviewed and the evidences are given for the different effects of uniformly distributed macromolecules and of proteins within the membrane-associated condensate, formed by liquid phase separation. In summary, the review and analyzes show that the role of various crowding effects on membranes has been significantly underestimated and a reliable understanding of biological membranes cannot be built without considering the effects of crowding.

Keywords Lipid microdomain · Lipid raft · Membrane structure · Membrane dynamics · Membrane crowding · Membrane curvature · Protein clusters · Liquid-liquid phase separation

Introduction

The macromolecular crowding (hereinafter briefly called crowding) topic is usually raised in the context of the volume phase of solutions containing high-molecular-weight polymers, which are considered as crowding agents that make some volume

S. V. Nesterov (✉)
NRC "Kurchatov Institute", Moscow, Russia
e-mail: semen.v.nesterov@phystech.edu

V. N. Uversky (ed.), *(Macro)Molecular Crowding*, Subcellular Biochemistry 109,
https://doi.org/10.1007/978-3-032-03370-3_5

excluded for the solvent diffusion. When the concentration of macromolecules increases, in addition to expelling part of the volume, the viscosity and the number of interactions between molecules increases, making diffusion more complicated (Alfano et al. 2024). At the same time, in real biological systems one of the key types of macromolecules is lipids. Lipid membranes are always present delineating the boundaries of bulk phases and playing other important roles such as transport, sorting and organizing, signal transduction, formation of electric and chemical potential differences. In this connection, when considering real biological systems, one cannot ignore the interactions of the bulk phase with membranes. In relation to membranes, the phenomenon of crowding is also applicable, although it requires a special approach in consideration.

In principle, membranes are lipid bilayers that separate from water to maximize the entropy of the lipid-water system (due to so called hydrophobic interactions). Lipid bilayers have a liquid-crystalline structure. Hydrocarbon chains of the membrane lipids have liquid properties in lateral directions (Jewell 2011), while their mobility toward the aqueous phase (perpendicular to the membrane) is limited by a high energy penalty. In the terms of macromolecular crowding, the situation is unique for membranes because, depending on the point of view, lipids can be considered both as macromolecules (crowding agent) and as a solvent.

In the first capacity, as crowding agents, lipids act in relation to water-soluble compounds, for which the hydrophobic region of the membrane forms inaccessible volume. Furthermore, vesicles as a whole are poorly permeable to water molecules and ions and also create steric constraints for other large molecules or supramolecular formations. Therefore, lipid vesicles can be also treated as excluded volume.

In a second capacity, as a solvent, lipids act for integral proteins. Large hydrophobic domains of membrane proteins in that case act as crowding agents. However, some authors prefer to consider lipids as crowding agents, which extensively interact with embedded proteins via hydrophobic and electrostatic interactions (Löwe et al. 2020). This approach may indeed help avoiding confusion. However, considering membrane lipids as a two-dimensional fluid is more intuitive and convenient in many cases. Anyway, terminology in no way changes the nature of the phenomena involved.

This chapter will discuss two main sides of the macromolecular crowding phenomenon on the membrane and on it interfaces. First case implies a high protein/lipid ratio in the membrane. The second considers at the membrane surface large amounts of polymers, whichever are membrane-bound or not. Consideration of lipid membranous structures as crowding agents for bulk phase or as a confinement for internal volume will be left beyond the scope of this chapter.

Although the topic of crowding effects on membranes is much less represented than the topic of crowding in solutions, there are at least several hundreds of works considering it in abstract or title (according PubMed). In the field of research approaches, as in most biological questions, three main directions can be distinguished: *in vitro* on model systems, *in silico* using various modeling approaches (Kutti Kandy and Radhakrishnan 2022), and *in situ* or even *in vivo*. Most experimental approaches both *in vitro* and *in vivo* use fluorescent techniques (Guigas and

Weiss 2016), which should be interpreted with great caution due to possible artifacts, such as interaction of the probe with the crowding agent (Silonov et al. 2024). Modern genetically encoded fluorescence resonance energy transfer (FRET)-based probes allows researchers to assess crowding effects in cells both in cytoplasm and on the membrane boundary (Löwe et al. 2023). The aim of this chapter is not to list in detail all the progress on this topic, but to highlight the key features and physicochemical effects resulting from crowding effects on biological membranes. For a more detailed review, including experimental techniques, see elsewhere (Guigas and Weiss 2016; Löwe et al. 2020).

Steric Effects of Crowding Induce Membrane Curvature

Crowding has now been shown to generate spontaneous vesicle formation and tubular morphologies on cell- and model membranes (Ruhoff et al. 2022). Furthermore, steric pressure from protein crowding is a structure-independent fission mechanism (Snead et al. 2017; Snead and Stachowiak 2018). Steric pressure from macromolecules also can assist vesicle fusion process (Terasawa et al. 2012). Disordered proteins/domains have larger hydrodynamic radii than ordered proteins/domains of the same length, and that is why generate more steric pressure and enhance membrane bending (Busch et al. 2015). However, not only proteins are used by cell for membrane reshaping. For example, density of glycocalyx (which is also known as the pericellular matrix or cell coat, and represents a layer of glycoproteins and glycolipids surrounding the cell membranes of bacteria, epithelial cells, and a large variety of animal cells (Martínez-Palomo 1970)) determines membrane morphology (Shurer et al. 2019). Interestingly glycocalyx is rather disordered, which makes it in some physical features similar to the intrinsic disordered proteins. Glycocalyx has a special ability to undergo a phase transition between "mashroom" and "brush" packaging (Kutti Kandy and Radhakrishnan 2022), changing the area and volume occupied by one molecule near the membrane. In some way, this resembles the transition of a protein from a disordered to a globular state. The crowding effects of membrane-bound polymers on membrane morphology are shown in Fig. 5.1. From a physical point of view, volume exclusion due to crowding can be simplistically viewed as a decrease in small molecules concentration, and hence a decrease in the pressure of the moving particles on the membrane. Therefore, the membrane is displaced in the direction of high concentration of macromolecules as if there were a region of reduced hydrostatic pressure in this area.

The ability of the polymer to curve the membrane has important biological implications. For example, the changes in pearls-on-a-string morphology in axons changes action potential conduction velocity (Griswold et al. 2025). Membrane bended regions participate in protein sorting, trafficking and signaling, because they can be sensed independently of lipid or protein composition by special protein sensors (Bhatia et al. 2010).

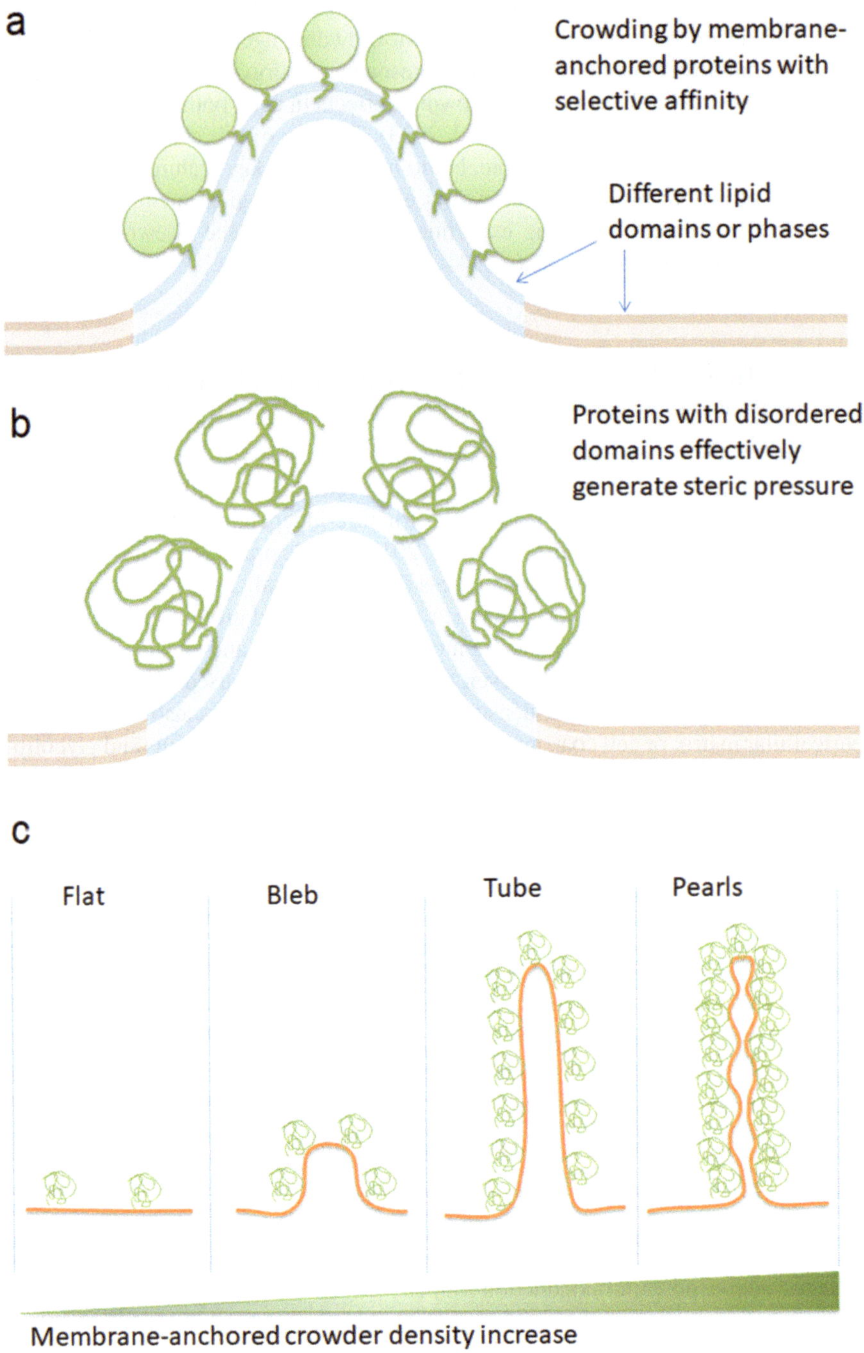

Fig. 5.1 Effects of crowding on membrane. (**a**) Stimulation of membrane phase separation into domains due to interaction with membrane-anchored crowder. The membrane curves toward high protein density. (**b**) The presence of disordered domains increases the membrane bending efficiency of proteins (compared to globular proteins of similar molecular weight). (**c**) Membrane transformation under crowding agent density increase

It should be also noted that transmembrane helixes can induce formation of the membrane domains (Domański et al. 2012), and helix insertion into the membrane also contribute into the membrane curvature formation (Ruhoff et al. 2022). This effect is not crowding per se, because it is based on the lipid-protein interactions, but the crowding with proteins having such lipid-interacting domains is often needed to observe pronounced effects. However, proteins without lipid-interacting domains can also induce membrane bending (reviewed in (Kirchhausen 2012)). Special effects on the membrane morphology beyond the simple crowding may also have biomolecular condensates formed by intrinsically disordered proteins and RNA by liquid-liquid phase separation (LLPS) (Lipowsky 2023).

Effect of Crowding on Lateral Diffusion Within Membrane

Investigation of membrane protein mobility in model lipid membranes showed different effects of increasing membrane protein concentration depending on the phase state of the lipids (Peters and Cherry 1982). In the liquid disordered phase, crowding decreases the diffusion rate, while in the liquid ordered phase, on the contrary, it increases the diffusion coefficient. The overall effect smooth out the large change in the diffusion coefficient that occurs under phase transition without crowding (Peters and Cherry 1982). Crowding also affect the diffusion of membrane-adhered particles (Liu and Beltramo 2023). For example, lateral diffusion due to crowding is strongly impaired in the post-synaptic membrane and changes in macromolecular crowding may regulate synaptic plasticity (Li et al. 2016; Kokolaki et al. 2020). It is important to note that creating a higher crowding in some region in the cell is a non-trivial task. In postsynaptic density, this is solved by LLPS (Zeng et al. 2018). This emphasizes the close connection between crowding and LLPS phenomena in real biological systems.

While crowding in the bulk phase reduces the diffusion rate, according to modeling, there is a depletion zone of 10 to 15 Å near the membrane surface, in which proteins of the bulk phase almost do not enter (Nawrocki et al. 2019). This can probably be caused by polarization in the near-membrane solvent layer. Because of this, diffusion at the intermembrane boundary under crowding conditions may be much faster for a number of compounds that enter there unimpeded than in other regions.

It is worth noting that steric constraints on movement along membranes may not be exclusive to crowding effects. For example, membrane contact with the cytoskeleton restricts the mobility of proteins, while the formation of bleb outgrowths devoid of contact with the cytoskeleton returns mobility to membrane proteins (Webb et al. 1981). The study of obstructed diffusion with polymer-tethered lipids acting as immobile obstacles disclosed that increase of obstacles concentration induce percolation transition to anomalous two-component diffusion mode (Deverall et al. 2005). Therefore, one must distinguish typical crowding (when crowder-agents are mobile) from steric constraints set by almost immobile (on the time scale of diffusion of molecules) structures such as the cytoskeleton, intercellular and inter-organelle contacts, large aggregates, and vesicles. The effects of practically immobile structures should be treated as confinement but not crowding.

Although crowding slows diffusion, *in vivo* effect may be the opposite. The bleb formation may be caused by crowding (Fig. 5.1). As noted in the previous section, the blebs and other similar structures can break the contact of membrane with cytoskeleton. Therefore, despite the intuitive idea that crowding slows down diffusion due to steric constraints, in real biological membranes, the effect of crowding may be reversal due to the release of some membrane parts from the contacts with immobilized structures.

Phase Separation, Aggregation and Clustering of Membrane-Associated Proteins

The presence of lipid membranes, especially charged ones, can influence the distribution of proteins on the surface, thereby affecting the assembly of different complexes or aggregation. There some examples demonstrating that interactions of proteins with lipids of different chemical structure are selective. Aβ can interact with anionic membrane lipids via an electrostatic anchor (Bokvist et al. 2004). It is known that α-synuclein preferentially interacts with the liquid disordered phase rather than the ordered one (Galvagnion et al. 2016). Such selectivity naturally leads to the anisotropic distribution of proteins according to their affinity for the corresponding lipids, especially if membrane rafts are formed (Lingwood et al. 2009). So, rafts can attract proteins and increase crowding around them. Such crowding near the membrane can lead to oligomerization or aggregation of proteins, or to the formation of liquid condensates (Nesterov et al. 2021; Wang et al. 2023). The predominance of one or the other process will depend on the propensity for phase separation of proteins crowded together on the membrane and the presence of disordered domains. It is important to note that crowding does not necessarily lead to aggregation. Crowding can even stabilize the disordered monomeric form of the protein and thus prevent aggregation (McNulty et al. 2006). The elevated degree of disorder and susceptibility to LLPS exhibited by numerous membrane-binding proteins (Nesterov et al. 2025) may be attributable to the necessity of safeguarding against undesirable aggregation. In addition to condensates, polymerization of specific proteins can occur on the membrane. Crowding in bulk phase can stimulate polymerization of cytoskeleton proteins altering membrane curvature and properties (Garenne et al. 2020).

Integral proteins influence the phase separation processes in the membrane not less than near-membrane crowding agents. Crowding of the membrane by transmembrane helixes promotes phase separation with lipid domains formation (Domański et al. 2012). The thickness and shape of the hydrophobic part of the protein play a key role in the interaction with the membrane. When the hydrophobic region of the integral protein differs in thickness from non-deformed lipid membrane hydrophobic zone this is referred to as hydrophobic mismatch. Molecular simulations show that due to hydrophobic mismatching and differences in geometrical shapes, initially randomly distributed proteins in the membrane may eventually form oligomers (Schmidt et al. 2008; Morozova et al. 2012). Mismatch leads to membrane deformation, with lipids experiencing both stretching and bending (Andersen and Koeppe 2007) (Fig. 5.2). In the context of crowding, it is important that the energy of bending sharply rises when the average distance between proteins drops below 10 nm (Schlame 2021). This fact explains why crowding by proteins in the membrane leads to clustering. Total area size, where lipids were bended is smaller for clustered than for randomly distributed proteins. Lipids may reduce the

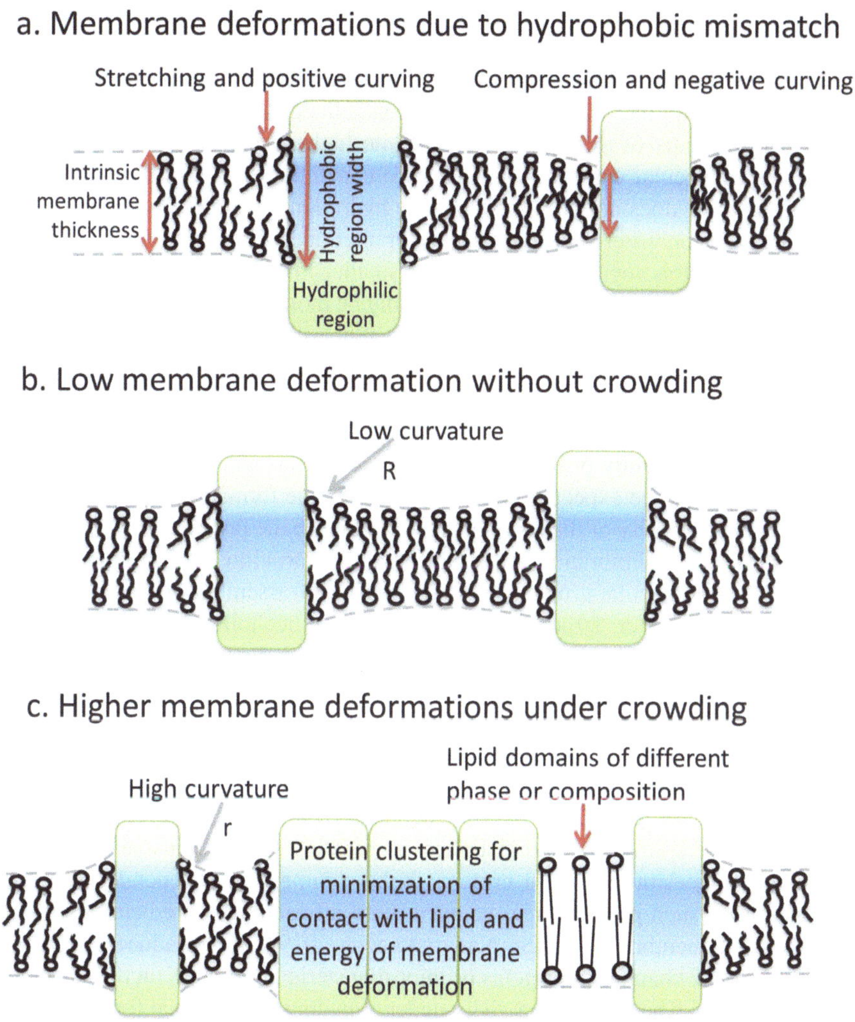

Fig. 5.2 Lipid membrane deformations due to protein insertions. (**a**) Hydrophobic mismatch induces stretching or compression of membrane near integral proteins. (**b**) The strain due to hydrophobic mismatch relaxes away from the protein and causes a slight curvature of the membrane surface. (**c**) Under crowding conditions, membrane deformation lacks room to relax. To reduce the strain energy, protein clusters may form, phase separation of lipids may occur, or lipids may be sorted based on the shape or thickness of their hydrophobic zone

energy penalty of crowding by adjusting their length and their intrinsic curvature as it occurs in the mitochondria, where the membrane would otherwise rupture (Schlame 2021).

Physicochemical Changes of the Solvent Under Crowding Affects the Membrane

Crowding agents induce changes in solvent properties of aqueous media in crowded environment (Ferreira et al. 2016). One manifestation of the special action of crowding on the solvent other than volume exclusion is the difference between the action of crowding agents of different weight. While high concentration of low-weight molecules (molecular crowding) is known to constrain water mobility, macromolecular crowding does not alter the bulk-like hydration dynamics and has a modest crowding effect on water at the crowder-water interface (Verma et al. 2018; Semmeq et al. 2024). In this regard, it appears that smaller molecules crowd better (Sung et al. 2021) although this effect would probably be more accurately attributed to the peculiarities of solvation. Indeed, high molecular weight crowding agents have a different effect on processes such as hydrogel (Hassan et al. 2015) and condensate/coacervate (Biswas et al. 2023) formation. As for the protein stability, small crowders increase it mainly via the enthalpy of denaturation while larger crowders increase protein stability primarily through entropy (Das and Sen 2019). Existence of a difference can be expected also for the membrane formation and organization processes. The polarity of the solvent should influence the intensity of hydrophobic interactions. Extramembranous crowding leads to variations in the lipid packing density. Macromolecular crowding also influence self-assembly of fatty acids (Sun et al. 2023). However, structural changes in phosphocholine-based vesicles are more pronounced with higher molecular weight crowding agents (Quinn et al. 2022). This is generally consistent with the fact that membrane self-organization processes depend more on entropy than enthalpy. In any case, there is currently insufficient experimental data to distinguish properly between the different aspects of crowding effects on the membrane.

Physicochemical changes in the solvent due to crowding can affect not only folding but also the conformation and activity of proteins. For proteins near the membrane, the same estimates as for proteins in the volume will be generally valid, but in some cases such proteins may be even more affected. The predicted depletion zone near the membrane surface (Nawrocki et al. 2019) should induce asymmetry causing specific changes in proteins or imposing steric constraints on the substrate/inhibitor access. The change in activity under crowding is confirmed for lipase (Sun et al. 2023) and membrane-bound alkaline phosphatase (Deshwal and Maiti 2021). In the last case crowding even change the behavior of inhibitors. Moreover, such influence of crowding is not present when the enzyme is attached to a polystyrene microbeads instead of a liposomes (Deshwal and Maiti 2021). This supports the hypothesis that the membrane can modulate the effects of crowding agents. Unfortunately, more details about such effects are not yet available.

In the previous section, it was pointed out that crowding affects the distribution of proteins on and in the membrane. Proteins and lipids are interconnected. Not only lipid domains influence the distribution of proteins, but also proteins can affect distribution of lipids. High crowding at the membrane interface, especially if crowding agents have affinity for certain type of lipids or membrane-embedded proteins, can affect domain formation in the membrane. Non-specific interactions with long polymers suppress domain phase separation in model membranes (Imam et al. 2016). Membrane bound small (such as GFP) and large particles (nanolipoproteins) in giant unilamellar vesicles also induce lipid mixing (Zeno et al. 2016). Thus, steric pressure due to crowding appears to disrupt the lipid domains and to promote membrane homogeneity.

However, it is worth considering that effect of crowding agents depends on their chemical nature and physical properties. For example, the interaction with the membrane of condensates enriched with intrinsically disordered proteins, on the contrary, stimulates the formation of domains in the membrane (Nesterov et al. 2025). Protein condensates couples with ordered lipid domains (Wang et al. 2023). This may be due to the different interactions of different crowding agents with water and lipid heads. A sufficiently inert and uniformly distributed crowding agent, as was mentioned, induces the disruption of lipid domains due to steric pressure. At the same time, an inhomogeneously distributed crowding agent capable to affect significantly the solvent properties (which manifests itself as LLPS) may conversely promote the formation of lipid domains in the contact region (Fig. 5.3).

Thus, crowding can affect biological membrane systems through changing of solution physicochemical properties (alters diffusion, viscosity, phase behavior of lipids) and through direct effects on membrane-anchored enzymes folding and activity. The effects often depends on crowding agent, so, one should be careful when analyzing data about effects of crowding.

Conclusions

Macromolecular crowding is unique biophysical parameter affecting biological membranes in several ways through different mechanisms. First, crowding on one of the membrane surfaces by large polymers induce bending of the membrane, which is well explained by the excluded volume effect. Second, crowding affects the lateral diffusion, which is due to the imposition of steric constraints on the motion of molecules in a crowded medium. Third, the integral membrane proteins distort the membrane more as crowding increases. That is why membrane proteins under conditions of high density form clusters based on the thickness and shape of the hydrophobic zone. In general, macromolecular crowding is an unseen driving force which creates conditions for the formation and dispersion of microdomains in the membrane. It is imperative to comprehend the extent to which crowding at the membrane and on its interface modifies membrane properties to formulate realistic conceptualizations of membranes and to effectively translate model experiment data to biological contexts.

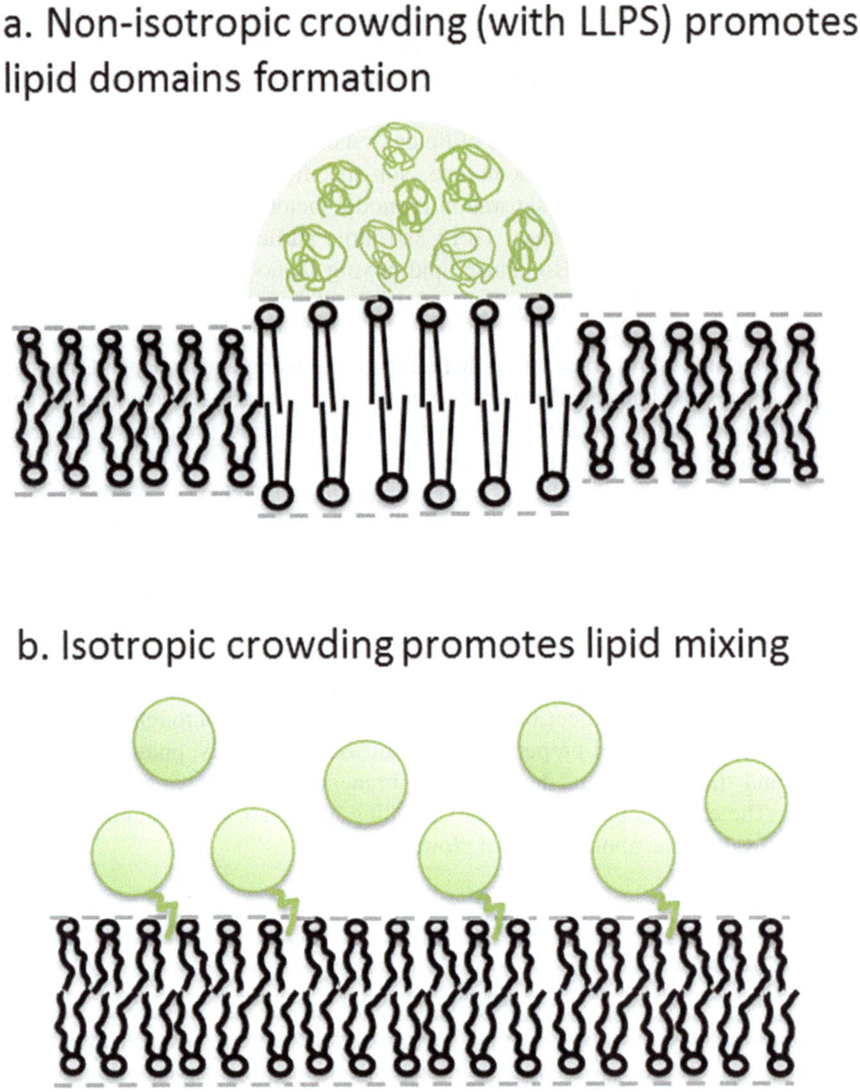

Fig. 5.3 Different effects of crowding on the membrane depending on the properties of the crowding agent. (**a**) Localized (into condensate) crowding by disordered polymers prone to LLPS. Condensate promotes lipid demixing with formation of liquid ordered phase domain. (**b**) Crowding by nearly isotropically dispersed macromolecules with or without affinity to the membrane promotes lipid mixing

Acknowledgements The work was carried out within the state assignment of NRC "Kurchatov institute".

References

Alfano C, Fichou Y, Huber K et al (2024) Molecular crowding: the history and development of a scientific paradigm. Chem Rev 124:3186–3219. https://doi.org/10.1021/acs.chemrev.3c00615

Andersen OS, Koeppe RE (2007) Bilayer thickness and membrane protein function: an energetic perspective. Annu Rev Biophys Biomol Struct 36:107–130. https://doi.org/10.1146/annurev.biophys.36.040306.132643

Bhatia VK, Hatzakis NS, Stamou D (2010) A unifying mechanism accounts for sensing of membrane curvature by BAR domains, amphipathic helices and membrane-anchored proteins. Semin Cell Dev Biol 21:381–390. https://doi.org/10.1016/j.semcdb.2009.12.004

Biswas S, Hecht AL, Noble SA et al (2023) Understanding the impacts of molecular and macromolecular crowding agents on protein–polymer complex coacervates. Biomacromolecules 24:4771–4782. https://doi.org/10.1021/acs.biomac.3c00545

Bokvist M, Lindström F, Watts A, Gröbner G (2004) Two types of Alzheimer's beta-amyloid (1-40) peptide membrane interactions: aggregation preventing transmembrane anchoring versus accelerated surface fibril formation. J Mol Biol 335:1039–1049. https://doi.org/10.1016/j.jmb.2003.11.046

Busch DJ, Houser JR, Hayden CC et al (2015) Intrinsically disordered proteins drive membrane curvature. Nat Commun 6:7875. https://doi.org/10.1038/ncomms8875

Das N, Sen P (2019) Size-dependent macromolecular crowding effect on the thermodynamics of protein unfolding revealed at the single molecular level. Int J Biol Macromol 141:843–854. https://doi.org/10.1016/j.ijbiomac.2019.09.029

Deshwal A, Maiti S (2021) Macromolecular crowding effect on the activity of liposome-bound alkaline phosphatase: a paradoxical inhibitory action. Langmuir 37:7273–7284. https://doi.org/10.1021/acs.langmuir.1c01177

Deverall MA, Gindl E, Sinner E-K et al (2005) Membrane lateral mobility obstructed by polymer-tethered lipids studied at the single molecule level. Biophys J 88:1875–1886. https://doi.org/10.1529/biophysj.104.050559

Domański J, Marrink SJ, Schäfer LV (2012) Transmembrane helices can induce domain formation in crowded model membranes. Biochim Biophys Acta 1818:984–994. https://doi.org/10.1016/j.bbamem.2011.08.021

Ferreira LA, Madeira PP, Breydo L et al (2016) Role of solvent properties of aqueous media in macromolecular crowding effects. J Biomol Struct Dyn 34:92–103. https://doi.org/10.1080/07391102.2015.1011235

Galvagnion C, Brown JWP, Ouberai MM et al (2016) Chemical properties of lipids strongly affect the kinetics of the membrane-induced aggregation of α-synuclein. Proc Natl Acad Sci USA 113:7065–7070. https://doi.org/10.1073/pnas.1601899113

Garenne D, Libchaber A, Noireaux V (2020) Membrane molecular crowding enhances MreB polymerization to shape synthetic cells from spheres to rods. Proc Natl Acad Sci 117:1902–1909. https://doi.org/10.1073/pnas.1914656117

Griswold JM, Bonilla-Quintana M, Pepper R et al (2025) Membrane mechanics dictate axonal pearls-on-a-string morphology and function. Nat Neurosci 28:49–61. https://doi.org/10.1038/s41593-024-01813-1

Guigas G, Weiss M (2016) Effects of protein crowding on membrane systems. Biochim Biophys Acta BBA Biomembr 1858:2441–2450. https://doi.org/10.1016/j.bbamem.2015.12.021

Hassan MM, Martin AD, Thordarson P (2015) Macromolecular crowding and hydrophobic effects on Fmoc-diphenylalanine hydrogel formation in PEG : water mixtures. J Mater Chem B 3:9269–9276. https://doi.org/10.1039/C5TB02139A

Imam ZI, Kenyon LE, Carrillo A et al (2016) Steric pressure among membrane-bound polymers opposes lipid phase separation. Langmuir ACS J Surf Colloids 32:3774–3784. https://doi.org/10.1021/acs.langmuir.6b00170

Jewell SA (2011) Living systems and liquid crystals. Liq Cryst 38:1699–1714. https://doi.org/10.1080/02678292.2011.603846

Kirchhausen T (2012) Bending membranes. Nat Cell Biol 14:906–908. https://doi.org/10.1038/ncb2570

Kokolaki ML, Fauquier A, Renner M (2020) Molecular crowding and diffusion-capture in synapses. iScience 23:101382. https://doi.org/10.1016/j.isci.2020.101382

Kutti Kandy S, Radhakrishnan R (2022) Crowding-induced membrane remodeling: interplay of membrane tension, polymer density, architecture. Biophys J 121:3674–3683. https://doi.org/10.1016/j.bpj.2022.05.031

Li TP, Song Y, MacGillavry HD et al (2016) Protein crowding within the postsynaptic density can impede the escape of membrane proteins. J Neurosci 36:4276–4295. https://doi.org/10.1523/JNEUROSCI.3154-15.2016

Lingwood D, Kaiser H-J, Levental I, Simons K (2009) Lipid rafts as functional heterogeneity in cell membranes. Biochem Soc Trans 37:955–960. https://doi.org/10.1042/BST0370955

Lipowsky R (2023) Remodeling of biomembranes and vesicles by adhesion of condensate droplets. Membranes 13:223. https://doi.org/10.3390/membranes13020223

Liu P, Beltramo PJ (2023) Effects of crowding on the diffusivity of membrane adhered particles. Soft Matter 19:7708–7716. https://doi.org/10.1039/d3sm01269g

Löwe M, Kalacheva M, Boersma AJ, Kedrov A (2020) The more the merrier: effects of macromolecular crowding on the structure and dynamics of biological membranes. FEBS J 287:5039–5067. https://doi.org/10.1111/febs.15429

Löwe M, Hänsch S, Hachani E et al (2023) Probing macromolecular crowding at the lipid membrane interface with genetically-encoded sensors. Protein Sci 32:e4797. https://doi.org/10.1002/pro.4797

Martínez-Palomo A (1970) The surface coats of animal cells1. In: Bourne GH, Danielli JF, Jeon KW (eds) International review of cytology. Academic, pp 29–75

McNulty BC, Young GB, Pielak GJ (2006) Macromolecular crowding in the Escherichia coli periplasm maintains alpha-synuclein disorder. J Mol Biol 355:893–897. https://doi.org/10.1016/j.jmb.2005.11.033

Morozova D, Weiss M, Guigas G (2012) Shape as a determinant of membrane protein cluster formation. Soft Matter 8:11905–11910. https://doi.org/10.1039/C2SM26720A

Nawrocki G, Im W, Sugita Y, Feig M (2019) Clustering and dynamics of crowded proteins near membranes and their influence on membrane bending. Proc Natl Acad Sci USA 116:24562–24567. https://doi.org/10.1073/pnas.1910771116

Nesterov SV, Ilyinsky NS, Uversky VN (2021) Liquid-liquid phase separation as a common organizing principle of intracellular space and biomembranes providing dynamic adaptive responses Biochim Biophys Acta BBA Mol Cell Res 1868:article number 119102. https://doi.org/10.1016/j.bbamcr.2021.119102

Nesterov SV, Ilyinsky NS, Fonin AV, Uversky VN (2025) Signal transduction by phase separation—unnoticed revolution in molecular biology. J Mol Recognit 38:e70003. https://doi.org/10.1002/jmr.70003

Peters R, Cherry RJ (1982) Lateral and rotational diffusion of bacteriorhodopsin in lipid bilayers: experimental test of the Saffman-Delbrück equations. Proc Natl Acad Sci USA 79:4317–4321. https://doi.org/10.1073/pnas.79.14.4317

Quinn SD, Dresser L, Graham S et al (2022) Crowding-induced morphological changes in synthetic lipid vesicles determined using smFRET. Front Bioeng Biotechnol 10. https://doi.org/10.3389/fbioe.2022.958026

Ruhoff VT, Moreno-Pescador G, Pezeshkian W, Bendix PM (2022) Strength in numbers: effect of protein crowding on the shape of cell membranes. Biochem Soc Trans 50:1257–1267. https://doi.org/10.1042/BST20210883

Schlame M (2021) Protein crowding in the inner mitochondrial membrane. Biochim Biophys Acta BBA Bioenerg 1862:148305. https://doi.org/10.1016/j.bbabio.2020.148305

Schmidt U, Guigas G, Weiss M (2008) Cluster formation of transmembrane proteins due to hydrophobic mismatching. Phys Rev Lett 101:128104. https://doi.org/10.1103/PhysRevLett.101.128104

Semmeq A, Del Galdo S, Chiarini M et al (2024) Macromolecular *vs* molecular crowding in aqueous solutions: a comparative study of PEG400 and ethylene glycol. J Mol Liq 394:123713. https://doi.org/10.1016/j.molliq.2023.123713
Shurer CR, Kuo JC, Monét Roberts L et al (2019) Physical principles of membrane shape regulation by the Glycocalyx. Cell 177:1757–1770.e21. https://doi.org/10.1016/j.cell.2019.04.017
Silonov SA, Kuklin AI, Nesterov SV et al (2024) Pitfalls of using ANS dye under molecular crowding conditions. Int J Mol Sci 25:13600. https://doi.org/10.3390/ijms252413600
Snead WT, Stachowiak JC (2018) Structure *versus* stochasticity—the role of molecular crowding and intrinsic disorder in membrane fission. J Mol Biol 430:2293–2308. https://doi.org/10.1016/j.jmb.2018.03.024
Snead WT, Hayden CC, Gadok AK et al (2017) Membrane fission by protein crowding. Proc Natl Acad Sci 114. https://doi.org/10.1073/pnas.1616199114
Sun Y-L, Ge B-Q, Li M-Z et al (2023) The effect of macromolecular crowding degree on the self-assembly of fatty acid and lipid hydrolysis. Npj Sci Food 7:39. https://doi.org/10.1038/s41538-023-00213-2
Sung H-L, Sengupta A, Nesbitt D (2021) Smaller molecules crowd better: Crowder size dependence revealed by single-molecule FRET studies and depletion force modeling analysis. J Chem Phys 154:155101. https://doi.org/10.1063/5.0045492
Terasawa H, Nishimura K, Suzuki H et al (2012) Coupling of the fusion and budding of giant phospholipid vesicles containing macromolecules. Proc Natl Acad Sci USA 109:5942–5947. https://doi.org/10.1073/pnas.1120327109
Verma PK, Kundu A, Cho M (2018) How molecular crowding differs from macromolecular crowding: a femtosecond mid-infrared pump–probe study. J Phys Chem Lett 9:6584–6592. https://doi.org/10.1021/acs.jpclett.8b03153
Wang H-Y, Chan SH, Dey S et al (2023) Coupling of protein condensates to ordered lipid domains determines functional membrane organization. Sci Adv 9:eadf6205. https://doi.org/10.1126/sciadv.adf6205
Webb WW, Barak LS, Tank DW, Wu ES (1981) Molecular mobility on the cell surface. Biochem Soc Symp:191–205
Zeng M, Chen X, Guan D et al (2018) Reconstituted postsynaptic density as a molecular platform for understanding synapse formation and plasticity. Cell 174:1172–1187.e16. https://doi.org/10.1016/j.cell.2018.06.047
Zeno WF, Rystov A, Sasaki DY et al (2016) Crowding-induced mixing behavior of lipid bilayers: examination of mixing energy, phase, packing geometry, and reversibility. Langmuir 32:4688–4697. https://doi.org/10.1021/acs.langmuir.6b00831

Chapter 6
Effects of Molecular Crowding on the Structure, Folding, and Stability of DNA

Victor M. Golyshev, Nadezhda Yu Biziukova, Olga A. Tarasova, and Alexander A. Lomzov

Abstract This chapter comprehensively reviews the effects of molecular crowding on the structure, folding, and thermodynamic stability of various DNA structures. While traditional studies often utilize dilute solutions, cellular environments are characterized by high macromolecular concentrations, ions, and metabolites, which significantly alter DNA behavior. The chapter discusses experimental methods, highlighting their strengths and limitations in studying DNA under crowded conditions. Key structural transformations—such as B-A and B-Z DNA conformational transitions, triplex formation, G-quadruplex (G4) topologies, and i-motif stabilization—are examined in the context of cosolute-induced effects. The chapter emphasizes the challenges of replicating *in vivo* conditions *in vitro* and underscores the need for innovative approaches to study nucleic acids in physiologically relevant environments. These insights advance our understanding of DNA behavior in cellular contexts, with implications for biotechnology, drug design, and gene regulation.

Keywords Molecular crowding · DNA structure · Thermodynamics · Experimental methods · DNA folding · DNA stability · B-DNA · A-DNA · Z-DNA · DNA triplex · G-quadruplex · I-motif

---

V. M. Golyshev (✉) · A. A. Lomzov (✉)
Laboratory of Structural Biology, Institute of Chemical Biology & Fundamental Medicine SB RAS, Novosibirsk, Russia
e-mail: lomzov@niboch.nsc.ru

N. Y. Biziukova · O. A. Tarasova
Department of Bioinformatics, Laboratory of Big Data Analysis for Digital Pharmacology, Institute of Biomedical Chemistry, Moscow, Russia

V. N. Uversky (ed.), *(Macro)Molecular Crowding*, Subcellular Biochemistry 109,
https://doi.org/10.1007/978-3-032-03370-3_6

Introduction

The majority of experimental studies on DNA are performed in dilute solutions, which consist of an analyzed object, dilute buffer, and salts. In many cases, real media *in vivo* drastically differ from experimental ones, and real objects are located in crowding conditions. Often, the term "crowding condition" refers to an environment with a substantial proportion of volume occupied by various macromolecules and cell organelles, or cosolvents, which mimic such conditions *in vivo*. Nonetheless, crowding could mean another situation: confined space, where DNA molecules are located very close to each other. For example, this situation exists on the surface of biochips, on glass support for the synthesis of oligonucleotides, or on nanoparticles for DNA delivery into cells (DNA immobilized on solid surfaces).

The subject of this chapter is DNA in molecular crowding. Nevertheless, this may mean many research subjects: from oligonucleotides and their complexes, synthetic polynucleotides, and plasmid DNA to chromosomes and the whole genome in the nucleus. The influence of molecular crowding itself on DNA is very complicated and may vary: e.g., the effect on structure, dynamics, kinetics, solvation, and hydration. Investigation of oligonucleotides in a crowding condition or in cell-mimicking solutions is an excellent model system that allows studying *in vitro* many aspects of crowding seen *in vivo*. Nonetheless, the trends observed for short DNA fragments are not always in agreement with trends for long DNA strands. Investigation of DNA *in vivo* is a much more difficult task.

Macromolecular crowding means a high concentration of biomacromolecules present in the cell such as nucleic acids, proteins, and polysaccharides. Typical concentration of macromolecules in a prokaryotic cell is in the range 20–40% by weight (Zimmerman and Trach 1991). The majority of its intracellular macromolecules are proteins (>50% of dry mass). The proportion of all types of RNA such as rRNA, mRNA, and tRNA is ~20%, and that of DNA is ~3–5% (van den Berg et al. 2017). In eukaryotic cells, the proportion of components is rather different. For example, in mammalian cells, chemical composition calculated in terms of dry mass is as follows: DNA, 1%; RNA, 4%; proteins, 60%; and lipids, 13% (Feijó Delgado et al. 2013). Besides, in the intracellular medium, there are many other molecules such as metabolites (with the total concentration of more than 200 mM) and ions. For *Escherichia coli* grown on glucose, it has been shown that amino acids constitute 50% of all metabolites, and almost 90% of them is negatively charged glutamate; there is 15% of triphosphates, 9% of glutathione, and 15% of central carbon intermediates (Bennett et al. 2009). In the mammalian cell, the main ions concentrations are 100 mM potassium (K^+), 5–100 mM chloride (Cl^-), 10 mM sodium (Na^+), 10 mM bound magnesium (Mg^{2+}), and 0.5 mM free Mg^{2+}. Concentrations of ions can change by more than an order of magnitude depending on the cell type and physiological and environmental conditions. Many metabolites and macromolecules of the cell are negatively charged such as organic phosphates and acids. In the intercellular fluid, the main positive ions are ~135 mM Na^+ and 5 mM K^+, and negative ones are ~100 mM Cl^- and ~25 mM HCO_3^- (Rollin et al. 2023).

There are a number of reviews in which information is collected about the impact of molecular crowding on different properties of DNA. Professor Sugimoto's group has studied and summarized data about the influence of crowding on the structure, interactions, and functions of DNA (Nakano et al. 2014; Nakano and Sugimoto 2017) and about methods for prediction of structures of nucleic acids under such conditions (Takahashi and Sugimoto 2020). Prof. Singh has reviewed the structure and dynamics of double-stranded DNA in cell-like conditions (Singh et al. 2022). Furthermore, a recent short review on the topic was published by D. Collette et al. (2023).

In this chapter, we compiled and summarized information about the crowding effect on the structure and thermodynamics of DNA molecules having various secondary and tertiary structures.

Methods for the DNA Analysis

We analyzed publications to compile methods used for studying DNA under molecular crowding conditions using text mining (Fig. 6.1). Text mining was used to identify names of chemical compounds, proteins, genes, diseases and methods that may be associated with molecular crowding from abstracts of scientific publications relevant to the study of this phenomenon. Machine learning methods based on the Conditional Random Fields (Bundschus et al. 2008) and Hunflair (Sänger et al. 2024) algorithms were used to recognize chemical compounds, diseases, proteins and genes as named entities. A special dictionary was developed to extract the names of the methods used to study molecular crowding and the effects associated with it. Relationships and associations between recognized named entities were

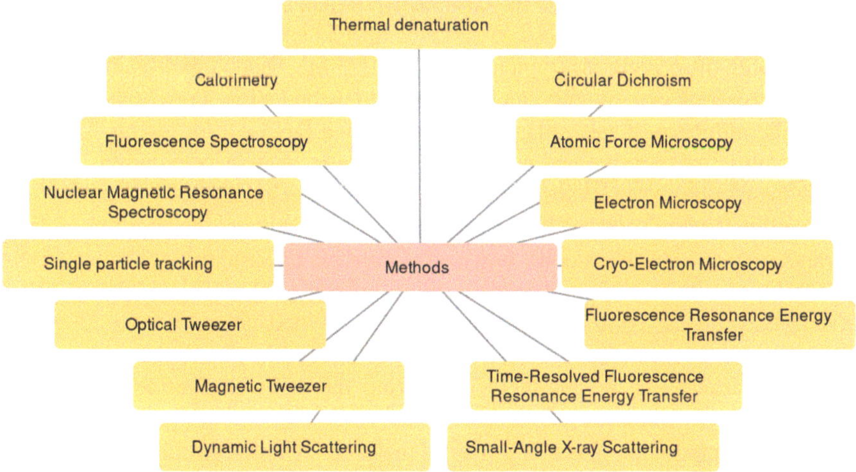

Fig. 6.1 Methods for studying DNA in crowding conditions, identified through text mining

identified using the set of pattern phrases and based on co-occurrences of entities. The entities and the relationships between them were visualized using Cytoscape (Shannon et al. 2003).

The most widely employed techniques for structural and thermal stability analysis are briefly discussed below, along with an analysis of their advantages and disadvantages.

Methods for DNA Structural Analysis

There are a number of methods for DNA secondary and tertiary structure analysis under crowding conditions. The most widely used are circular dichroism (CD) spectroscopy, nondenaturing polyacrylamide gel electrophoresis (PAGE), nuclear magnetic resonance (NMR) and electron paramagnetic resonance (EPR) spectroscopy, fluorescence (Förster) resonance energy transfer (FRET), small-angle X-ray scattering (SAXS), cryo–electron microscopy (cryo-EM), and atomic force microscopy (AFM).

Among the methods of structural analyses, **CD spectroscopy** is one of the simplest to use. It is based on the phenomenon of a difference in absorption between left- and right-handed circularly polarized light in molecules with a chiral center(s) (Berova et al. 2000). Reliable determination of overall DNA conformation can be performed by a comparison of sample CD spectra with spectra of reference samples (Bishop and Chaires 2002; Kypr et al. 2009). Most of studies on DNA conformation involve a UV spectrum range (190–330 nm) where nucleic acids absorb the light. Different helix forms (A-, B-, or Z-form) and G-quadruplexes (G4s) have unique shapes of CD spectra (Bishop and Chaires 2002). In the majority of studies on nucleic-acid structures, CD spectroscopy is the first choice owing to many advantages of the method (Preisler et al. 1995; Rajendran et al. 2010; Pandey et al. 2023). First, the concentration of DNA in samples for CD spectroscopy is small in comparison with other methods (typically 10^{-7} to 10^{-5} M). Second, buffer conditions in CD spectroscopy experiments can be varied widely and allow a researcher to investigate the influence of salts, pH, or cosolvents on a DNA conformation. Third, the equipment is inexpensive and accessible. Moreover, sample preparation is simple while the recording of CD spectra is quite fast (Kypr et al. 2009).

Nondenaturing (native) PAGE of DNA is PAGE without the addition of denaturants such as urea (Harwood). It gives information on structural features of DNA or DNA complexes. Electrophoretic mobility of negatively charged DNA is sensitive to its structure, conformation, and flexibility (Woodson and Koculi 2009). For instance, some researchers have observed a difference in electrophoretic mobility between parallel and antiparallel G4s having the same nucleotide composition [$G_3(TTAG_3)_3$] (Vorlickova 2005).

NMR spectroscopy is one of powerful methods for investigation of nucleic-acid structure, dynamics, and interactions (Plavec 2022). It allows researchers to analyze processes on timescales from picoseconds to hours. An interaction with biomolecules, cosolutes, ligands, ions, and other objects can be studied in detail. NMR enables a scientist to determine the structure of DNA complexes in solution with different solution environments. An atomic nucleus possessing a nonzero magnetic moment (^{1}H, ^{13}C, ^{15}N, or ^{31}P) in strong magnetic fields can absorb electromagnetic radiation at resonance frequencies. A one-dimensional (1D) ^{1}H NMR spectrum allows researchers to observe the imino protons of nucleobases that are not exchanged with a solution, and therefore nucleobases forming hydrogen bonds can be detected. Structure determination is usually based on the nuclear Overhauser effect (NOE) and spin–spin coupling constants between coupled nuclei. 2D and 3D NMR experiments could be used for measurement of distances between protons (nuclear Overhauser effect spectroscopy; NOESY) and of torsion angles (COSY). Such techniques provide information on local geometry of DNA. The dynamics of DNA could be studied via measurement of spin–spin and spin–lattice relaxation times by NOE. A number of other techniques have been developed for the analysis of nucleic-acid structure and dynamics (Plavec 2022).

NMR spectroscopy requires sophisticated equipment, high concentration of DNA (typically in the millimolar range and a volume of 0.5 ml); for 2D experiments, time of spectra collection is long, and buffering conditions are limited. Moreover, NMR data analysis is challenging. All these difficulties are compensated by detailed information on the structure and dynamics of DNA in solution under given conditions.

Electron paramagnetic resonance **(EPR) spectroscopy** is based on the absorption of electromagnetic radiation with resonance frequency of an unpaired electron in a magnetic field. Double electron–electron resonance (DEER) or pulsed electron–electron double resonance (PELDOR) spectroscopy is a pulsed EPR method for determination of a distance distribution between two spin labels. In this technique, microwave radiation is applied as a short pulse sequence for selection of a certain dipole–dipole interaction (weak dipolar coupling) between two spin labels (paramagnetic centers) in the range of distances of 20 to 120 Å (Krumkacheva et al. 2019). In this case, spin-labeled DNA should be used. Spin labels can be introduced into a DNA strand during automatic solid-state synthesis (with a modified nucleotide having an attached spin label or with phosphoramidite building blocks) or can be covalently attached by postsynthetic protocols (for example, at a desired position in a internucleotide PS site) (Shevelev et al. 2014; Chubarov et al. 2024). DEER experiments are mostly performed in a frozen solution, except for specific cases such as spin labels with long T_2 relaxation times (Shevelev et al. 2014). Typical experimental conditions are as follows: volume of the samples 10–80 μl at a concentration of 0.01 to 1 mM, and duration of the measurements is several hours. One of advantages of the DEER method is the small size of the most popular spin labels. A disadvantage of DEER spectroscopy for structural studies about an influence of molecular crowding is the necessity of glycerol or ethylene glycol at high concentration in stock buffers for EPR. In addition, this approach requires covalent attachment of spin labels to nucleic acids.

FRET is a phenomenon of energy transfer between donor and acceptor chromophores through nonradiative dipole–dipole coupling. FRET has been employed to study conformational changes of DNA after addition of crowders. Disadvantages of the approach is the possibility of a fluorescence change with variation of hydrophobicity and necessity of covalent attachment of fluorophores.

SAXS is a technique by which a size and shape distribution of a biomolecule in solution can be quantified. It is a low-resolution structural method (Hutin et al. 2023). In general, a monochromatic beam of X-rays passes through a sample solution and a tiny portion of rays scatter on biomolecules. The scattering radiation is detected at small angles (0.1–10°) (Rambo and Tainer 2010). Typical volume of the sample is ~15–250 μl, and the concentration of the DNA should be in range 0.1–1.0 mg/mL; at synchrotron facilities, collection time of an appropriate SAXS profile [I(q)] should be microseconds to tens of minutes (Burke and Butcher 2012), whereas on a bench top source, this time is hours. Resolution of SAXS is ~20–30 Å and nucleic acids with length of tens to hundreds of nucleotides are suitable for this analysis. The buffer for nucleic-acid samples requires at least 100 mM monovalent ions. Phosphate buffer is not appropriate due to a reduction in sample contrast (Burke and Butcher 2012). Postprocessing of obtained data requires data on DNA secondary structure and data on molecular modeling for generating a set of conformations. One of problems is obtaining realistic ensembles during computational modeling and another one is the presence of a mixture of conformations in solution.

Cryo-EM is electron microscopy performed on samples maintained at cryogenic temperatures, where an electron beam passes through a sample, and scattering on the analyzed objects results in different densities registered by detectors. An analysis of an electron density difference between two adjacent areas (contrast)—which could be amplitude or phase—is used for molecules geometry determination. In case of biological samples, phase contrast is suitable because low molecular weights of atoms in samples (C, N, O, P, or S) are not appropriate for obtaining amplitude contrast. The resolution of modern cryo-EM analyses is approximately 3 Å, and minimal molecular weight of a sample is ~65 kDa. Disadvantages of the method are measurements at cryogenic temperatures and difficulties with data analysis (Renaud et al. 2018).

AFM is a technique involving the scanning of topography of a sample surface. In AFM, a cantilever measures the force between the tip of a probe and molecules on a substrate. For AFM analyses of DNA, they usually use noncontact tapping mode, where the cantilever oscillates at resonance frequency and the phase and amplitude of the oscillation depend on the distance between surface atoms and the cantilever tip. There are two modes of analysis of samples: in air or in an aqueous solution. In the air mode, DNA is placed on mica or highly oriented pyrolytic graphite (HOPG), whose surface can be modified to enhance nucleic-acid affinity for the surface. In aqueous solutions, DNA structure could be determined in different buffers including crowding conditions. Today, this method is quite fast and simple for such an analysis, but obtaining high-resolution images is not a trivial task (Pastré et al. 2007; Benedetti et al. 2015).

All the methods described above with their advantages and disadvantages are summarized in Table 6.1.

Table 6.1 Methods of the investigation of effects of molecular crowders on the structure of nucleic acids

Method	Advantages	Disadvantages
NMR	High resolution; provides dynamic and structural information in solution	Limited to small nucleic acids (<40 kDa); requires isotopic labeling; low sensitivity for large systems
CD	Sensitive to secondary-structure changes; fast and easy to perform	Low resolution; limited to global structural changes; cannot provide atomic-level details
FRET	Measuring structural dynamics across at least 12 orders of magnitude in terms of the timescale and mapping of conformational and functional heterogeneities. Can monitor real-time dynamics and folding of nucleic acids	Requires labeling with donor and acceptor fluorophores, which may perturb native structure. Limited to studying labeled regions, not the entire molecule
EPR(DEER)	High sensitivity to distance changes; useful for studying conformational dynamics and for capture of transient intermediates	Requires spin labeling; limited to specific sites; complicated sample preparation
SAXS	Works in solution; no size limitation; determines overall shape and size of DNA	Low resolution; limited to global structural features; requires high sample purity
Cryo-EM	High resolution for large complexes; minimal sample preparation. Structure of unstained DNA in a frozen state at nanometer resolution could be obtained	Expensive; requires specialized equipment; challenging for small nucleic acids
AFM	High spatial resolution; works in a liquid or air; no labeling is required; analysis of DNA conformational changes and DNA compaction induced by a crowding condition for large DNA molecules	Limited to surface features; may cause sample deformation; low throughput
PAGE	Simple and cost-effective; good for analyzing conformational changes and complexes	Low resolution; qualitative; limited to size and charge differences

Methods for the Investigation of the Impact of Molecular Crowding on Thermal Stability of Nucleic Acids

Experimental methods for evaluation of the influence of molecular crowding on thermal stability of nucleic acids could be categorized into indirect spectroscopic methods and direct experimental techniques. In the indirect techniques, obtained data are model dependent, whereas in the direct measurements, there is direct determination of the heat of a reaction or heat capacity changes.

There are several techniques that allow for measurement thermodynamic properties of nucleic acids denaturation/renaturation. By the indirect measurements, researchers commonly analyze a dependence of some property on temperature. The most popular technique is **UV melting**, where the absorption of ultraviolet light (by nucleic acids at thermodynamic equilibrium) at a given wavelength is registered at different temperatures. Thermodynamic parameters ($\Delta H°$, $\Delta S°$, and $\Delta G°_{37}$) of denaturation/renaturation can be calculated via fitting of the experimental data to a theoretical curve (Marky and Breslauer 1987; Mergny and Lacroix 2009; Schroeder and Turner 2009). For the analysis of obtained melting curves, a few assumptions should be made. The first one is the applicability of the two-state model without any intermediate or additional states. The second assumption is that $\Delta C°_p$ is zero. In addition, at every temperature, nucleic acids should be at thermodynamic equilibrium. The appropriate temperature range is 5–95 °C. The technique requires dry air or nitrogen gas flow through the cell chamber for prevention water condensation on the cell walls in case of humidity >20% at ambient temperature (for optical measurement at low temperatures) and mineral oil on top of the nucleic-acid solution in the cuvette at high temperature to prevent evaporation (Marky and Breslauer 1987). In case of DNA hairpins, duplexes, and triplexes, wavelength 260 and 280 nm is typically used. For G4s and i-motifs, characteristic wavelength is 295 nm, whereas 320 or 330 nm is used as a "baseline." In some cases, denaturation transitions (for example, from a triplex to duplex or from hairpins to single-stranded states) cannot be observed in UV-melting experiments (Davis et al. 1998). Recently, a new assay of stability of i-motifs was invented: 4D UV melting and annealing (4DUVMA). In this approach, UV spectra are recorded at different pH levels (5–8) and temperatures (Cheng et al. 2021).

Alternatively, CD spectroscopy, NMR chemical shifts (for example, 1H), or fluorescence spectroscopy) could be employed for indirect measurements of thermodynamics.

CD spectroscopy allows analyzing nucleic acids' secondary structure depending on temperature. The shape and magnitude of CD spectra enables investigators to simply determine formation of a complex and its secondary structure: duplexes in A-, B-, or Z-form; triplexes; quadruplexes; or i-motifs. This method is very simple as a UV-melting analysis but gives more information about secondary structure and structure rearrangements associated with temperature changes or addition of various compounds, for example cosolvents.

NMR techniques are useful tools for detailed investigation of temperature-induced structural changes in nucleic acids. For example, proton 1D NMR spectra at different temperatures (in the range 5–80 °C) help to determine the type of nucleic-acid structure via analysis of chemical shifts or chemical exchange of imino protons. Examination of the changes in chemical shifts of imino protons with temperature variation allows determining thermodynamic parameters ($\Delta H°$ and $\Delta S°$) of the helix-to-coil transition (Lomzov et al. 2016). In addition to the whole complex denaturation, kinetics and dynamics of base pair opening and closing can be determined by this experimental technique. More sophisticated techniques such as 2D NOESY NMR experiments enable researchers to determine spatial structure of nucleic acids' complexes (Endoh et al. 2023). In these studies, a large quantity of a sample is needed (typically 600 µl of a sample with a concentration in the mM range).

Changes in DNA structure may be monitored by **FRET**. In this method, a fluorophore and quencher are attached to DNA that are close to each other in a folded state (Decian et al. 2007; Luo et al. 2021). With the heating of the sample, the proportion of the folded state should decrease and the intensity of the fluorescence should increase because of a greater distance between the fluorophore and quencher (You et al. 2011). The main advantages of such techniques are a wide range of concentration which could be used for the assay (a nM to mM range) and a huge number of samples in a single experiment. For these types of experiments, real-time PCR machines could be used. Nonetheless, a FRET-based method requires modification of oligonucleotides with a fluorophore and quencher, they also have an impact on thermodynamic properties of model systems.

There are two main direct assays of nucleic-acid thermodynamics: isothermal titration calorimetry (ITC) and differential scanning calorimetry (DSC).

ITC is a technique that allows to directly measure the heat of a reaction (Bastos et al. 2023). After injection of a small amount of a titrant from a syringe into the cell with the titrate, there is a release or consumption of heat (an exothermic or endothermic reaction). Interpretation of ITC data requires a thermodynamic model of interaction between the titrant and titrate. Titration at constant temperature helps to obtain reaction enthalpy [$\Delta H(T)$], the binding constant [$K_b(T)$], and the stoichiometry of the reaction (the number of the titrant molecules bound to the titrate, n). Gibbs free energy $\Delta G(T)$ changes and an entropy change [$\Delta S = (\Delta H - \Delta G)/T$] can be calculated by means of $K_b(T)$. Analysis of reaction heat at different temperatures enables one to determine heat capacity changes of binding ($\Delta C_p = \Delta\Delta H/\Delta T$). Advantages of ITC are direct observation of reaction enthalpy and of the binding constant at a desired temperature (below melting temperature) without labeling of DNA samples. Disadvantages of ITC are a large sample quantity required for the analysis and much lower accessibility of ITC equipment in comparison to instruments for UV-melting analysis (Srivastava and Yadav 2019).

Another direct technique of DNA thermal-stability investigation is **DSC** (Gill et al. 2010). In a DSC experiment, measurements of a sample physical properties are performed with changes in temperature with time, i.e., heat capacity changes are determined during DNA complex disruption/formation at different temperatures. In this method, measurements of heat capacity are used for determination of enthalpy and entropy changes during denaturation/renaturation transition without assumptions about a thermodynamic scheme of the transition. As a result, complicated unfolding transitions with intermediate states could be examined by DSC in detail. For example, an unfolding pathway of a telomeric G4 at different concentration of PEG and K^+ and Na^+ can be identified (Bončina et al. 2016). The minimal quantity of the sample is 0.001 to 1 mM with volume over 300 μl.

The main methods for research on DNA thermodynamics described above with their advantages and disadvantages are summarized in Table 6.2.

In this section, we described the main methods used for structural and thermodynamic analyses of nucleic acids for an analysis of molecular crowding effects.

Table 6.2 Methods for investigation of effects of molecular crowding on hybridization properties of nucleic acids

Method	Advantages	Disadvantages	References
UV melting	Simple method, low quantity of the sample	Indirect method	Schroeder and Turner (2009)
FRET-based melting	Low quantity of the sample	Indirect method; necessity of sample labeling	Dhakal et al. (2021)
CD melt	Simple method, low quantity of the sample, sensitive to secondary structure changes during hybridization	Indirect method	Davis et al. (1998)
NMR	Direct analysis of atomic interactions	Large quantity of the sample, complicated equipment and techniques, expensive equipment	Lomzov et al. (2016)
DSC	Direct measurements of thermodynamic parameters	Model-dependent results. Large quantity of the sample	Olsen et al. (2016)
ITC	Direct measurements of thermodynamic parameters	Large quantity of the sample	Miyoshi et al. (2004) and (Goobes et al. (2003)

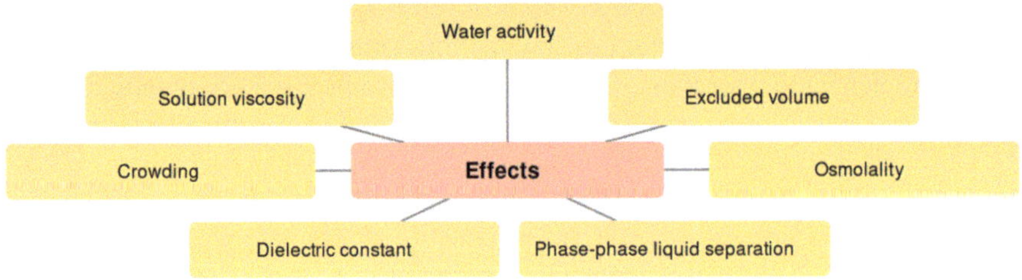

Fig. 6.2 Effects of molecular crowding observed for DNA, identified through text mining

Effect of Molecular Crowding on DNA

The molecular crowding effect is complex and associated with a range of phenomena that arise when cosolutes are added to a DNA solution. Using text mining methods, we analyzed publications to identify effects linked to molecular crowding (Fig. 6.2).

Our analysis revealed that it is often challenging to isolate a single effect caused by the addition of cosolutes. Therefore, to determine the contribution of specific effects, researchers vary the selection of cosolutes and their concentration. Among these, the compounds most widely used for studying molecular crowding phenomena are shown in Fig. 6.3.

Below, we provide a detailed overview of how various cosolutes influence the structure and thermodynamic properties of different structured DNA forms.

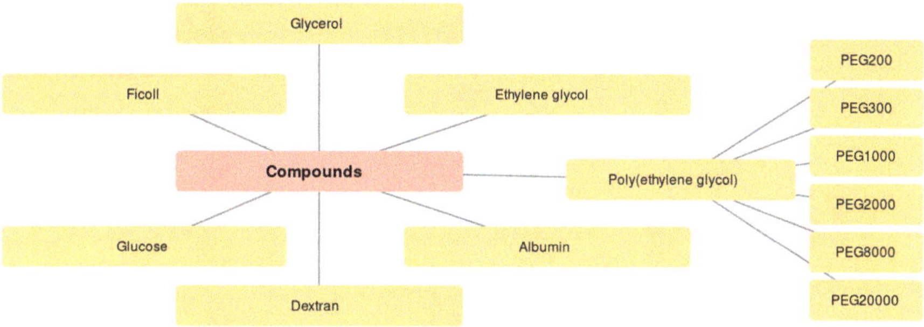

Fig. 6.3 Compounds commonly employed to investigate molecular crowding effects, identified through text mining

DNA Structural Changes Under Crowding Conditions

The effect of molecular crowing on secondary and tertiary structure of DNA strongly depends on its type. Here we collected and analyzed data on double and triple helixes, G4s, and i-motifs as the most widely studied DNA structural elements.

DNA Double Helix

The most widely known structure of DNA is the B-form of a double helix (canonical duplex structure, Fig. 6.4), which is formed by two antiparallel strands owing to stacking interactions and hydrogen bonds between complementary base pairs (Watson–Crick base pairing). DNA has an ability to adopt a variety of structural conformations, depending on environmental conditions. Double-stranded DNA also can be in the A-form under low-humidity conditions (Charney and Chen 1987) or in the Z-form at high ionic strength (Phromsiri et al. 2017) (Fig. 6.4).

The cellular environment depends on the cell type, cell compartment, and stage of the cell cycle. In general, however, it is a solution that contains proteins and nucleic acids, metabolites, and different types of salts. Physicochemical properties of DNA under such conditions differ significantly from those under typical *in vitro* conditions.

The canonical B-form of DNA is a double-stranded right-handed helix formed by two antiparallel complementary strands (Watson–Crick base pairing). The DNA double-helix repeat is ~10–10.5 base pairs per turn, with a helical rise of ~3.36 Å; base pairs are perpendicular to the helical orientation, with C2′-*endo* sugar puckering and the *anti* conformation of all the N-glycoside bonds. Double-stranded DNA has major and minor grooves. Right-handed double-stranded DNA can adopt a variety of helical conformations due to structural polymorphism. They could be categorized into two families: A-DNA and B-DNA.

A-form DNA has a deep narrow major groove and a shallow minor groove, with 11 base pairs per turn and an average rise of 2.56 Å; the average diameter of the

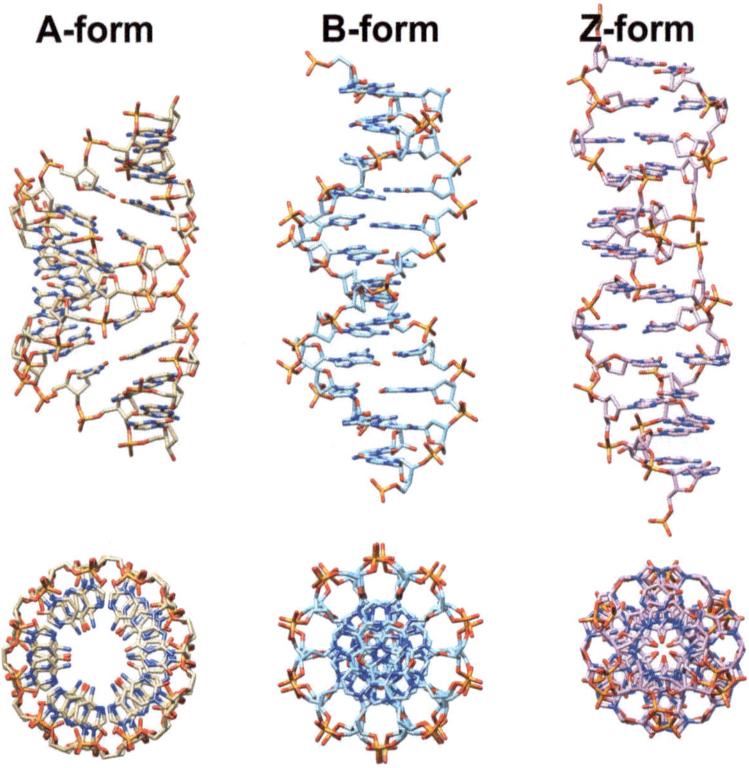

Fig. 6.4 Side and top view of double-stranded DNA in A-, B-, and Z-DNA conformations

helix is 2.3 nm, with the C3′-*endo* conformation of the sugar. A-DNA is less hydrated in comparison with B-DNA. Base pairs are not situated in the center of the helical axis; they are displaced toward the minor groove and are inclined (Ussery 2002). A transition from B-form to A-form DNA is sequence specific, cooperative, and reversible. A reduction in water activity shifts B-form to A-form helix structure. For instance, addition of 80–92% of ethanol leads to aggregation of DNA and A-form formation in monovalent salts (Piškur and Rupprecht 1995; Kulkarni and Mukherjee 2017).

The Z-form of DNA is favored for complexes with a high GC content at low humidity and high ionic strength. Z-DNA is a left-handed double helix with Watson–Crick base pairing where dinucleotide repetitive units $(PuPy)_n$ result in a zig-zag sugar-phosphate backbone (Matos-Rodrigues et al. 2023). A transition from B- to Z-DNA has a high energy barrier, but molecular crowders can lower it (Preisler et al. 1995). Z-DNA has a smaller diameter (18 Å) (as compared to B-DNA and A-DNA), a narrow deep minor groove, with 12 bp per turn and a 3.7 Å distance rise in the base step. In Z-DNA, all cytosines have the *anti* conformation and guanines the *sin* conformation. Sugar puckering in strands is periodic (C2′-*endo* for pyrimidines and C3′-*endo* for purines).

The human genome is estimated to contain approximately 100,000 copies of potential Z-DNA–forming sequences. Although they are found at various positions within the genome, they are seen at a much higher frequency around transcription start sites (Hamada and Kakunaga 1982). Negative supercoiling stabilizes Z-DNA, and the extent of formation varies with sequence; there is an alternating purine/pyrimidine sequence in such DNA when it is circular, and specific binding protein can stabilize the Z-form (Herbert and Rich 1996).

In diluted conditions, poly-d(GC) molecules have the following transition (B-to-Z) concentrations: ~2.5 M for NaCl, 0.66 M for $MgCl_2$, or 60% of ethanol. The enthalpy of such a transition is close to zero (Spingler 2012). Methylated cytosine in poly-d(Gm^5C) reduces mid-transition concentration of ions: it is 700 mM for NaCl, or 0.6 mM for $MgCl_2$ and 50 mM for NaCl, or 20% for ethanol with 50 mM NaCl (Spingler 2012). Addition of 40% of PEG 200 reduces mid-transition concentration of NaCl to ~0.6 M (Phromsiri et al. 2017). Neutral cosolutes such as glycerol, ethylene glycol, 1,3-propanediol (1,3-PDO), 2-methoxyethanol, and 1,2-dimethoxyethane also shift a B–Z equilibrium to the Z-form. It has been shown by CD spectroscopy that $d(GC)_3$ in diluted solutions and under crowded conditions (PEG 200) has a conformation similar to the Z-form because PEG stabilizes the Z-form of DNA without perturbation of structure (Nakano et al. 2008).

DNA Triple Helix

A triplex is a complex of DNA strands where a DNA duplex base pair interacts via Hoogsteen base pairing (purine nucleobase in a Watson–Crick base pair) with a third strand that is located near the major groove of the duplex (Fig. 6.5). Each strand should consist of homopurines or homopyrimidines (Frank-Kamenetskii and Mirkin 1995). There are a number of possible types of triple helixes. An intermolecular DNA parallel triplex is composed of C^+-G:C or T-A:T triads, which require an acidic environment (Petraccone et al. 2004), and is antiparallel, where the third strand consists of purine bases and requires G-G:C or T-A:T with reverse-Hoogsteen hydrogen bonds (Boehm et al. 2018). Formation of the A-A:T triplet is also possible (Mills et al. 2002). Intramolecular DNA triplexes formed by one DNA strand that folds back on itself, to form an intramolecular assembly, are also known as H-DNA. It is stabilized by acidic conditions and by supercoiled structure. The mirror repeat polypurine–polypyrimidine sequence and base motifs are the same as in the intermolecular triplexes with a third pyrimidine strand. Moreover, an intramolecular triplex that is composed of bound pyrimidine–purine–purine units in DNA stretches and of a nonmirror repeat is defined as *H-DNA (Kohwi and Panchenko 1993; Dalla Pozza et al. 2022). Triple helices are less stable than duplex structures owing to electrostatic repulsion of the third strand. In general, formation of triplex structures requires an acidic environment. The formation of C^+-G:C triads requires

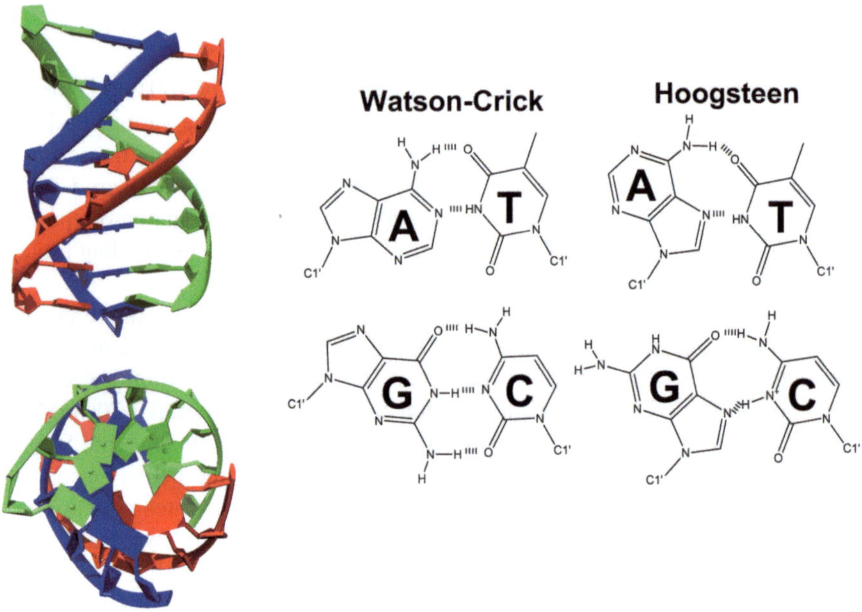

Fig. 6.5 Side and top view of a DNA triplex (PDB ID: 1D3X) (left). Watson-Crick and Hoogsteen base paring (right)

cytosine protonation that occurs under buffering conditions with pH < 5. As a result, triplex complexes with these triads will be destabilized with increasing pH.

There are also G-triplexes formed by G-rich sequences with three G tracts. The strength of G triplex stabilizing effects of ions is reported to be $Ca^{2+} > K^{+} > Mg^{2+} > Na^{+}$. Ca^{2+} promotes the formation of parallel G-triplexes under either dilute or molecular-crowding conditions. Melting temperatures of G-triplexes consisting of two G-triads are lower than physiological temperature, even under molecular-crowding conditions, but G-triplexes containing three G-triads have much higher thermal stability, especially in a molecular-crowding condition (Jiang et al. 2015).

G-Quadruplexes (G4s)

One of the best-studied high-order structures of DNA is the G4. It is formed by guanine-rich nucleic-acid sequences and is composed of two or more stacks of planar G quartets, where nucleobases interact via Hoogsteen pairing, and the bonds are stabilized by monovalent cations such as K^{+} or Na^{+} (Yang 2019). There are many types of quadruplexes formed by one, two, or four strands (Fig. 6.6).

Most relevant in the cell is intramolecular G4 topology, which formed by sequence pattern $(G_{>2}N_{1-7})_3G_{>2}$, and therefore this G4 consists of G-tetrads and loops. These DNA sequences may adopt a variety of G4 conformations (Rhodes and

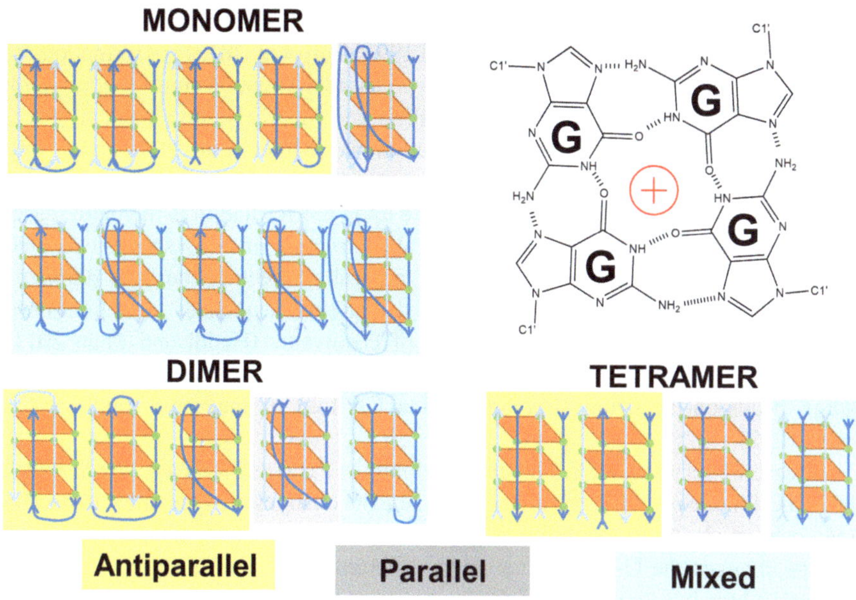

Fig. 6.6 Diversity of G4 DNA topologies and G-quartet structure

Lipps 2015). In case of intramolecular G4s, strands can be oriented parallel, antiparallel (chair), or in a combination (basket or hybrid). There are three types of loops: propeller (double-strand reversal), lateral (edgewise), and diagonal. Diagonal loops link noninteracting bases in the same tetrad, lateral loops link guanidines that form a Hoogsteen pair, and propeller loops link guanidines sharing the same groove but in different tetrads (Webba da Silva 2007). Orientation of the *N*-glycoside bond angle can be *anti* or *syn*. In case of the parallel orientation of strands, intramolecular G4 tetrads are in the *anti* glycosidic conformation and all groves are of medium width. In case of the antiparallel orientation, G4 is in the chair conformation and G-tetrads have *syn-anti-syn-anti* glycosidic angles, widths of grooves alternate between wide and narrow (Yang 2019). In case of the mixed orientation of strands, there are two possible types: the hybrid (3 + 1) type, which is formed by three codirectional G-tracts and one antidirectional G-tract (↑↑↑↓), and the basket(2 + 2) type, which is formed by a pair of parallel strands in different directions. For such types of G4s, there are two medium-width grooves, one narrow, and one wide groove (Chalikian and Macgregor 2023).

Prediction of DNA G4 structure/topology is rather difficult and strongly depends on solution environments. Just a few examples are presented below.

Human telomeric repeat TTAGGG (HT G4) can form different intramolecular G4s. It has been shown *in vitro* in a diluted solution that 4-repeat human telomeric oligonucleotides could arise in K^+ solutions: a propeller-type parallel-stranded G4 (Parkinson et al. 2002), an antiparallel and 2-hybrid (3 + 1) G4 (Dai et al. 2007; Phan et al. 2007), or a basket-type G4 with only two G-tetrad layers (Lim et al.

2009); in the presence of sodium ions in solution, it could form a hybrid form (Wang et al. 2014) or antiparallel (Wang and Patel 1993; Lim et al. 2013) conformation. All of these sequences assume the propeller-type parallel-stranded conformation in the presence of K^+ and molecular crowders PEG and Ficoll (Heddi and Phan 2011). Low-molecular-weight cosolvents like acetonitrile (Miller et al. 2010) and ethanol (Smirnov and Shafer 2007) at a volume proportion of ~40% and KCl stabilize the parallel form of the HT G4. A water release upon G4 formation is the main reason for such a shift in the equilibrium to the parallel structure. In NaCl solutions, molecular crowders do not shift the equilibrium to parallel structures but also stabilize a G4. Furthermore, it has been demonstrated that such sequences form high-order structures at even higher concentrations of cosolvents (Heddi and Phan 2011). Moreover, it has been shown that long (n > 1) telomeric DNA sequences $(T_2AG_3)_{4n}$ adopt beads-on-the-string structure with a G4 fold of the same topology (Yu et al. 2012).

In case of very similar *Tetrahymena* telomeric TTGGGG repeats, which form a hybrid G4 in the presence of Na^+, addition of PEG changes the type of G4 to parallel (Miyoshi et al. 2005). Addition of cosolutes with different numbers of hydroxyl groups changes the G4 conformation in different ways. For example, 40 wt% of glycerol (three hydroxyl groups) does not change topology, but 40 wt% of 1,3-PDO and 1-propanol or 1,2-dimethoxyethane shifts the conformational equilibrium to a dimeric parallel G4. In a potassium-containing buffer, *Tetrahymena* telomeric repeats exist in a parallel-type G4 with monomeric and dimeric forms. Addition of 40% of above-mentioned cosolutes shifts the equilibrium to the only dimeric form of the parallel G4 (Tateishi-Karimata et al. 2020).

G4-forming sequences have been found in many promoter regions of genes and may serve as transcriptional regulators (Qin and Hurley 2008). In the majority of such sequences that could form intramolecular G4s, motif G_3NG_3 N_{long} G_3NG_3 is present and folded into a parallel G4 in K^+ solutions (Yang and Okamoto 2010). Under molecular-crowding conditions (40% PEG 200), there is no change in the parallel type of G4s within promoter sequences, but formation of G4 aggregates is induced (Gao et al. 2023).

In general, G4 structures are more stable in a molecular-crowding condition than in diluted solutions. Existence of such structures in a cellular environment was shown recently (Lam et al. 2013; Lee et al. 2021).

I-Motifs

The intercalated motif or i-motif is formed by two parallel duplexes intercalated in the antiparallel orientation, which is stabilized by hemiprotonated cytosine–cytosine base pairing (CC^+) (Gehring et al. 1993) (Fig. 6.7). The hemiprotonated cytosine–cytosine base pair CC^+ is bonded by three hydrogen bonds, where N3 of one of the pair of cytosines is protonated. There are two types of i-motif topologies, 3′E and 5′E, where terminal outer base pair $C:C^+$ is formed at the 3′ or 5′ end of a strand (Fig. 6.7) (Phan et al. 2000). Base pairs are perpendicular to the main axis of the

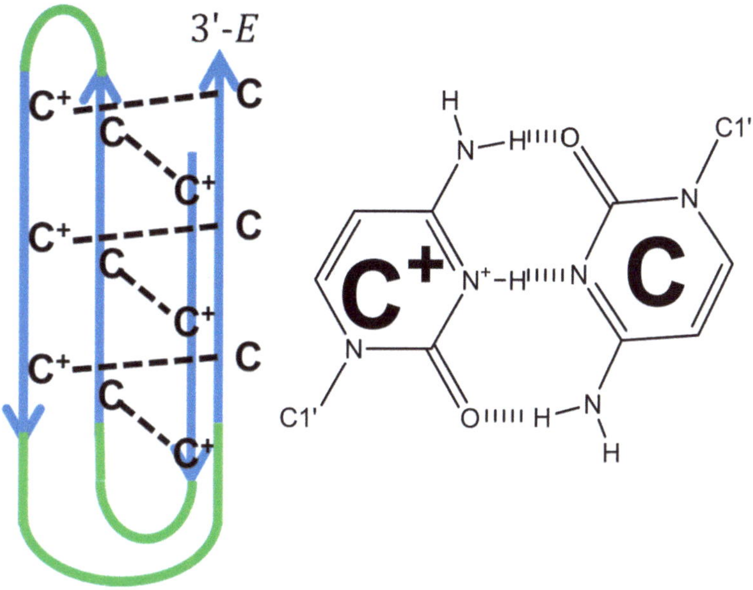

Fig. 6.7 Schematic representation of the i-motif (left) and of CC+ base pairing (right)

i-motif, and the distance between base pairs in one duplex is 6.2 Å (3.1 Å between consecutive base pairs). Such a short distance between negatively charged phosphate groups in comparison with duplexes and quadruplexes leads to destabilization of this type of secondary structure. In an i-motif, one can distinguish two wide grooves and two narrow grooves. The necessity of a protonated cytosine for the formation of an i-motif results in a strong dependence of its stability on solution pH. On the other hand, strong electrostatic repulsion of the phosphates is associated with a high dependence on ionic strength of the solution. Recently, it was demonstrated that an increase of the concentration of potassium ions can lead to stabilization (in MES or bis-Tris buffer) or destabilization (phosphate, citrate, and sodium cacodylate buffer) of an i-motif (Gao and Hou 2021).

I-motifs can be inter- or intramolecular. To date, most of 3D structures determined by X-ray analysis of i-motifs have been intermolecular, and only a few are intramolecular, for example, an insulin-linked polymorphic region (PDB ID: 8AYG) (Guneri et al. 2024), but in NMR spectroscopy, both types are well represented. The presence of such highly ordered structures *in vivo* in human cells has been detected using an antibody; the majority of such structures get folded during the interphase of mitosis (Zeraati et al. 2018; Tang et al. 2020). Stability of an i-motif from the human genome has been researched in live mammalian cells by NMR; it has been proven that under in-cell conditions at 37 °C, the i-motif exists (Dzatko et al. 2018; Víšková et al. 2024).

Changing the concentration of organic solvents in a DNA solution affects many solvent properties at once, making it hard to identify the impact of each property.

Table 6.3 Properties of cosolvents mainly used for the mimicking of molecular-crowding conditions in studies on nucleic acids

Cosolute with water	Type	Feature
DMSO	Polar aprotic solvent	Decreases water activity and dielectric constant
DMF	Polar aprotic solvent	Decreases dielectric constants with small changes in water activity
Ethanol	Polar, protic, neutral, highly water-soluble	Decreases dielectric constants with small changes in water activity
Ethylene glycol	Neutral, highly water-soluble	Decreases dielectric constants and water activity
Sucrose	Neutral, polyol, highly water-soluble	Decreases dielectric constants and water activity
Glycerol	Neutral, polyol, highly water-soluble	Decreases dielectric constants and water activity
PEG 200	Neutral, highly water-soluble	Decreases dielectric constants and water activity
PEG 8000	Neutral, moderately water-soluble	Increases excluded-volume effect and dielectric constants
Ficoll (M_w 400,000	Neutral, polyol, highly water-soluble	Increases excluded-volume effect
Dextran (M_w 10,000)	Neutral, polyol, highly water-soluble	Increases excluded-volume effect

Another option is to methodically adjust a specific solvent property by means of various types of organic solvents having different characteristics. Collected data on properties of widely used solvents are presented in Table 6.3.

Presented data indicate complexity of different factors affecting DNA secondary and tertiary structures. The transition between various DNA forms under crowding conditions is thermodynamically driven. Therefore, next we analyze these transitions through the prism of thermodynamic effects.

Thermal Stability Changes Under Crowding Conditions

Molecular crowders change thermodynamic stability of nucleic acids. Thermal stability of nucleic-acid structures depends on environmental conditions, such as the type and concentration of ions solution pH, and the cosolvent type and concentration. Complex formation by nucleic acids is governed by many factors. First of all, it is hydrogen bond formation via Watson–Crick or Hoogsteen base pairing and stacking interactions between nucleobases. Second, it is solvation, which is important both for base staking and for backbone–solvent interaction. Third, electrostatic interactions also play a role. Negatively charged phosphate groups and nucleobases actively interact with ions via the shielding electrostatic repulsion mechanism, there is formation of salt bridges or ion coordination.

Molecular crowding, from the perspective of the thermodynamic equilibrium between different DNA conformations, shifts this equilibrium owing to effects of various factors. Among these, the following key factors can be highlighted: direct interactions with cosolvent molecules, changes in diffusion, excluded-volume effects, increased osmotic pressure, reduced water activity, and an altered dielectric constant. The water molecules that engage in DNA hydration in a crowding condition show reduced mobility, which leads to higher osmotic pressure and a decline of water activity relative to that of pure water (Conti Nibali and Havenith 2014). The diffusion coefficient of synthetic DNA molecules in the cytoplasm is 5–100 times less than that in water, depending on the size of the molecule; in nucleoplasm, DNA fragments are nearly immobile, in comparison with a DNA molecule of the same size in the cytoplasm. The dielectric constant in the cytoplasm and in nucleus of eukaryotes is ~15 and ~ 60 (Sasmal et al. 2013), respectively, whereas for water, it is ~80.

DNA Duplex Thermal Stability

Thermal stability of a B-form DNA duplex differs considerably between a diluted solution and molecular-crowding conditions and significantly depends on the GC content, length of the duplex, and types and concentrations of cosolvents and ions (Nakano et al. 2014; Nakano and Sugimoto 2017; Takahashi and Sugimoto 2020; Singh et al. 2022; Collette et al. 2023).

The main effect of neutral cosolvents on thermal stability of a short synthetic DNA duplex is changes of water activity. Duplex formation or dissociation at thermodynamic equilibrium occurs in accordance with the two-state model. Duplex formation by two single complementary stands typically proceeds with the uptake or release of water, cations, and/or cosolute. It could be described by the following thermodynamic scheme:

$$A + B \rightleftarrows AB + \Delta n_w H_2O + \Delta n_{cs} CS + \Delta n_{M^+} M^+, \quad (6.1)$$

where A and B denote complementary strands; AB is double-stranded DNA; Δn_w, Δn_{cs}, and $\Delta n_M{}^+$ are numbers of water (w), cosolute (cs), and cation molecules (M^+) released/taken up upon the duplex formation. The observed equilibrium constant (K_{obs}) for the duplex formation is given by

$$K_0 = K_{obs} a_w^{\Delta n_w} a_{CS}^{\Delta n_{cs}} a_{M^+}^{\Delta n_{M^+}}, \quad (6.2)$$

where K_0 and K_{obs} are true and observed equilibrium constants, a_w, a_{CS}, and a_{M^+} are activities of water, cosolute, and cations. Estimation of the number of take up/released molecules could be done by changing activity of one of the solution's components (water, cation, or cosolvent) without changing others. Osmotic-stress experiments, a method that induces water activity changes via the addition of cosolutes that do not interact with nucleic acids, can show the number of taken up/released water molecules upon complex formation (Spink and Chaires 1999; Spink et al. 2007):

$$\frac{\partial \ln K_{obs}}{\partial \ln a_w} = \left(\Delta n_w + \Delta n_{cs} \left(\frac{\partial \ln a_{cs}}{\partial \ln a_w} \right) + \Delta n_{M^+} \left(\frac{\partial \ln a_{M^+}}{\partial \ln a_w} \right) \right). \tag{6.3}$$

In experiments designed to evaluate water binding, ionic strength is almost constant (the dilution of the buffer with a cosolute may change ion concentrations a few percent, but the derivative term is small in magnitude). Cosolvents used in these experiments almost do not interact with DNA directly. In this case, Eq. 6.3 is reduced to

$$\Delta n_w = \left(\frac{\partial \ln K_{obs}}{\partial \ln a_w} \right). \tag{6.4}$$

Nonetheless, effects of changes in thermal stability of a nucleic-acid complex with the addition of a cosolute and a reduction in water activity can be associated with alterations of the dielectric constant of the medium. This phenomenon should also be taken into account (Yildirim et al. 2014).

Many research groups have investigated the effect of osmotic stress on DNA duplex stability using short duplexes. Variation of duplex length, GC content, and sequence have been analyzed (Nakano et al. 2004; Ghosh et al. 2020a). An excellent review by prof. Sugimoto summarizes a lot of these studies (Nakano et al. 2014). In general, for short DNA duplexes, it has been shown that just a few additional water molecules per base pair are taken up during 9–15 bp duplex formation in experiments with PEG 200. A larger number of water molecules (3–8) was noted when ethylene glycol or glycerol was applied (Nakano et al. 2014).

An extended nearest-neighbor model has been developed for predicting DNA duplex stability in media with different cosolutes that mimic molecular crowding in the cell (Ghosh et al. 2019, 2020b). This model allows to predict thermodynamic changes in enthalpy, entropy, and Gibbs free energy upon duplex formation under any crowded conditions.

The nearest-neighbor model implies the possibility of physicochemical-parameter calculation based on the additivity of effects of dinucleotide pairs in a duplex and a few additional impacts (terminal base pair, initiation, and symmetry term).

It has been demonstrated that destabilization with a change in water activity is independent from the ions' concentration (Ghosh et al. 2020b) and changes in Gibbs free energy linearly correlate with the change in water activity under crowded conditions ($\Delta\Delta G^\circ_{37} \sim \Delta a_w$). The coefficient between them depends on the nearest-neighbor type and chemical structure of a cosolute: the main impact is exerted by the number and position of hydroxyl groups present in the chemical compound (Nakano et al. 2012). Excluded-volume effects of high-molecular-weight cosolutes,

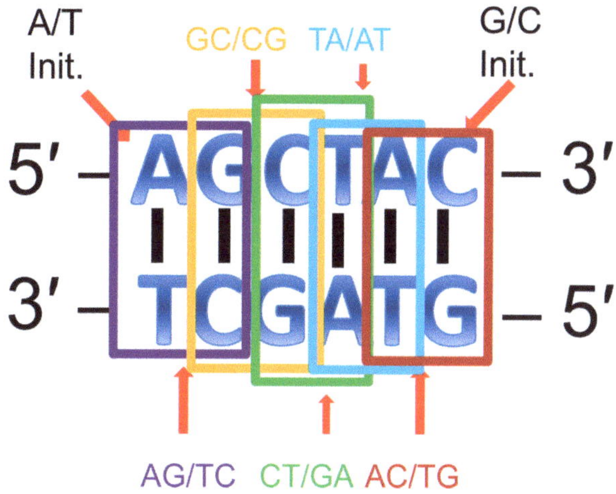

Fig. 6.8 The nearest-neighbor model for prediction DNA thermodynamics

such as PEG (M_W > 4000 Da), dextran, and Ficoll, enhance thermal stability of duplexes (Koumoto et al. 2008; Harve et al. 2009; Singh et al. 2022) Fig. (6.8).

A general method for predicting the stability of short DNA duplexes with different cosolutes is based on the relation between duplex stability and water activity of the cosolute solution. The effects on the crowding are additive, and thermodynamic parameters of hybridization and can be calculated as follows:

$$\Delta G_{NN} = \Delta G_{NN(\text{in diluted solution})} + \Delta G_{NN(\text{crowding})}. \tag{6.5}$$

Parameter $\Delta G_{NN\ (crowding)}$ can be calculated due to a linear function of the change in water activity (Δa_w) in the presence of a crowding agent:

$$\Delta G_{NN(\text{crowding})} = m_{cs} \times \Delta a_w, \tag{6.6}$$

where m_{cs} is a coefficient that depends on the structure of the crowding agent. It has been found that m_{cs} values for cosolutes may be expressed as:

$$m_{cs} = m_{PEG} \times \left(S_{CS} / S_{PEG}\right), \tag{6.7}$$

where S_{CS} and S_{PEG} are slopes for the cosolute of interest and PEG, respectively, in the $\Delta\Delta G°_{37}$ vs. Δa_w plot and could be determined experimentally. Values of the nearest-neighbor parameters have been determined for a solution consisting of 100 mM NaCl and 40% PEG 200 (Table 6.4), as prefactors (m_{cs}) for different cosolutes for recalculating resistance to other conditions (Table 6.5) (Ghosh et al. 2020a). Thermal stability of a DNA duplex in 20 wt% 1,3-PDO can be computed as shown in Fig. 6.9.

Polymeric duplexes [for example, poly(dAdT)·poly(dTdA) or poly(dGdC)·poly(dCdG)] have been studied in osmotic-stress experiments by DSC with solutions containing ethylene glycol, glycerol, sucrose, urea, betaine, PEG

Table 6.4 Nearest-neighbor parameters for DNA/DNA duplex formation in 1 M NaCl without cosolvents (SantaLucia and Hicks 2004) and in the presence of 100 mM NaCl and 40 wt% of PEG 200 (Ghosh et al. 2020a)

	1 M NaCl in the absence of cosolute			100 mM NaCl in the presence of 40 wt% of PEG 200		
Nearest-neighbor set	ΔH°, Kcal/mol	ΔS°, cal/mol/K	$\Delta G°_{37}$, Kcal/mol	ΔH°, Kcal/mol	ΔS°, cal/mol/K	$\Delta G°_{37}$, Kcal/mol
5′-AA-3′/3′-TT-5’	−7.9	−22.2	−1.01	−6.5	−19.2	−0.55
AT/TA	−7.2	−20.4	−0.87	−9.4	−29.4	−0.28
TA/AT	−7.2	−21.3	−0.59	−4.3	−13.3	−0.18
CA/GT	−8.5	−22.7	−1.46	−13.1	−38.8	−1.07
GT/CA	−8.4	−22.4	−1.45	−9.2	−26.8	−0.89
CT/GA	−7.8	−22.1	−0.95	−3.4	−7.9	−0.95
GA/CT	−8.2	−22.2	−1.31	−4.9	−13	−0.87
CG/GC	−10.6	−27.2	−2.16	−6.4	−16.1	−1.41
GC/CG	−9.8	−24.4	−2.23	−4.2	−9.3	−1.32
GG/CC	−8	−19.9	−1.83	−4	−8.9	−1.24
Initiation G/C pair	0.1	−2.8	0.97	−10.1	−35.1	0.79
Initiation A/T pair	2.3	4.1	1.03	−2.9	−12.7	1.04
Symmetry correction	0	−1.4	0.43	0	−1.4	0.43

Table 6.5 Nearest-neighbor parameters for 100 mM NaCl and 40% of PEG 200 with prefactors (m_{cs}) for different cosolutes

Nearest-neighbor set	$\Delta G°_{37\ (without\ cosolute)}$, Kcal/mol	$\Delta G°_{37\ (40\%\ wt\ PEG200)}$, Kcal/mol	$m_{PEG/1,2\ DME}$, Kcal/mol	$m_{EG/GOL}$, Kcal/mol	$m_{1,3PDO/\ 2\text{-}ME}$, Kcal/mol
5′-AA-3′/3′-TT-5’	−0.65	0.10	2.00	0.70	1.30
AT/TA	−0.60	0.32	6.40	2.20	4.20
TA/AT	−0.36	0.20	4.00	1.40	2.60
CA/GT	−1.23	0.23	4.60	1.60	3.00
GT/CA	−1.20	0.31	6.20	2.20	4.10
CT/GA	−1.11	0.20	4.00	1.40	2.60
GA/CT	−0.93	0.06	1.20	0.40	0.80
CG/GC	−1.85	0.47	9.40	3.30	6.20
GC/CG	−2.05	0.72	14.40	5.00	9.50
GG/CC	−1.69	0.44	8.80	3.00	5.80
Initiation G/C pair	0.98	−0.22	−4.4	−1.5	−2.9
Initiation A/T pair	1.03	−0.03	−0.6	−0.2	−0.4

200, or PEG 1450 at increasing osmolality (Spink et al. 2007). Melting temperatures of such polynucleotides decrease with the increase of a cosolute's concentration, with the only exception being PEG 1450. In this case, excluded-volume effects dominate over hydrational changes. In the case of poly(dAdT), the addition of betaine does not lead to any changes in thermal stability. In the case of poly(dA)·poly(dT)

↓ ↓
5′ A – G – C – T – C 3′
3′ T – C – G – A – G 5′
↑ ↑

$\Delta G°_{37,NN} = \Delta G°_{37,NN}$ (100 mM NaCl) + $\Delta G°_{37,NN}$ (20 wt% 1,3 PDO)

$\Delta G°_{37,NN}$ (20 wt% 1,3 PDO) $= m_{1,3\ PDO} \cdot \Delta a_w$

$\Delta G°_{37}$ (AG/CT) = -1.11 + (2.6 · 0.044) = -1.00 kcal/mol

$\Delta G°_{37}$ (GC/CG) = -2.05 + (9.5 · 0.044) = -1.63 kcal/mol

$\Delta G°_{37}$ (TC/GA) = -0.93 + (0.8 · 0.044) = -0.89 kcal/mol

$\Delta G°_{37}$ (Int A/T) = 1.03 + (-0.4 · 0.044) = 1.01 kcal/mol

$\Delta G°_{37}$ (Int G/C) = 0.98 + (-2.9 · 0.044) = 0.85 kcal/mol

$\Delta G°_{37}$ (Pred.) = $\Delta G°_{37}$ (Int A/T) + 2 · $\Delta G°_{37}$(AG/CT) + $\Delta G°_{37}$(GC/CG) + $\Delta G°_{37}$(TC/GA) + $\Delta G°_{37}$ (Int G/C) = 1.01 + 2 · (-1.00) + (-1.63) + (-0.89) + 0.85 = -2.66 kcal/mol

Fig. 6.9 Calculation of Gibbs free energy of duplex formation for d(AGCTAC)/d(GTAGCT) in 100 mM NaCl and in 20 wt% 1,3-PDO with improved NN parameters (Table 6.2). The arrows show NN base pairs. Δa_w of the solution was taken from (Ghosh et al. 2020a)

or *E. coli* DNA, researchers have demonstrated a release of water molecules upon duplex formation in the presence of low-molecular-weight cosolutes, (Spink and Chaires 1999).

Crowded conditions in the cell or nucleolus may be simulated using PEG 200, whereas those in the nucleus resemble more a Ficoll 70–like condition (Takahashi et al. 2019). Aside from cosolvents, proteins *in vitro* can also have the effect of molecular crowding. Proteins in *E. coli* constitute a major fraction of the cellweight (~55% of total dry weight) (van den Berg et al. 2017). As a consequence, the influence of proteins on DNA duplex thermal stability should be taken into account. In general, the presence of such proteins as BSA, myoglobin, and α-lactalbumin does not change melting temperature of a complementary DNA duplex, whereas lysozyme diminishes it. In contrast, all proteins stabilize long internal and bulge loop structures. Effects of basic proteins on the stability of fully matched duplexes are dissimilar to those of basic amino-acid-rich polypeptides and polyamines (Morimoto et al. 2022).

Molecular crowding takes place not only in noncharged media. This phenomenon could involve many charged molecules. It has been shown that high polyanion concentration increases thermal stability of a DNA duplex in comparison with other similar ionic-strength conditions even with nonionic cosolvents and cations (Karimata et al. 2007). Polyethylene sodium phosphate (PEP–Na) increases thermal stability not only of DNA duplexes but also of triplexes and G4s (Moriyama et al. 2015). Histone proteins, protamine, and poly(L-lysine) are positively charged molecules, which stabilize DNA duplexes. Putrescine and spermidine at concentrations of several tens of millimoles/L increase melting temperature of DNA duplexes, but 1,10-decanediamine, the tetrapentylammonium ion, and the benzyltributylammonium ion decrease it (Yamaguchi et al. 2022).

B–Z Transition

It is well known that DNA in the Z-form is less hydrated than B-DNA. The transition from the B- to Z-form of a short duplex in a diluted solution requires 4 M NaCl, but the presence of either 50 wt% of PEG 200 or 20 wt% of PEG 8000 lowers this requirement to 2 M NaCl. Trivalent cobalt hexamine at 200 μM also induces such a transition. Cosolutes with a smaller weight proportion of the hydroxyl group in the molecule give greater stabilization of the Z-form (Nakano et al. 2008). For a methylated polymer, this transition also occurs at much lower concentrations of a transition inducer; e.g., for poly-d(Gm^{5}C), a crowder shifts thermodynamic equilibrium from the B- to Z-form (Preisler et al. 1995). Osmotic-stress experiments have revealed that seven water molecules are released per base pair upon Z-DNA formation with PEG 200 as a cosolute (Nakano et al. 2008).

Other structural forms of DNA—triplexes, G4s, and cruciforms—manifest substantial stabilization in the presence of molecular-crowding conditions, in contrast to the canonical duplex configuration (Takahashi and Sugimoto 2020).

Three-Way Junctions (TWJs)

Branched nucleic-acid structures are where three or more double-stranded regions meet in one place. A TWJ in DNA arises during DNA replication and during recombination involving phages. In general, addition of molecular PEG 200 to a solution destabilizes a TWJ (Muhuri et al. 2009; Miyoshi et al. 2009a). On the other hand, a detailed analysis has revealed that under such conditions, the number of water molecules taken up by the whole TWJ is much smaller than the sum of individual arms (duplexes). Therefore, in a PEG 200 duplex-TWJ equilibrium should be shifted to TWJ structures.

Triplex

Different effects of a cosolvent on triplexes as compared to those on duplexes have been explained by different involvement of water molecules in the formation of a Hoogsteen base pair (a water release reaction) and a Watson–Crick base pair (a water uptake reaction) (Nakano et al. 2014).

In diluted solutions, thermal stability of triplexes goes up with an increase in potassium concentration and intermolecular triplexes have lower thermal stability than intramolecular ones. An increase of low concentrations of potassium from 0 to

0.1 M stabilizes all triplexes significantly. Nucleobase composition, e.g., the C^+-G:C content, has a greater influence on triplex stability under molecular-crowding conditions because cytosine is easily protonated under molecular-crowding conditions (Teng et al. 2020).

Various cosolvents have different effects on the stability of triplexes. Addition of 1,3-PDO, 2-methoxyethanol, 1,2-dimethoxyethane, or PEG 200 results in higher thermal stability (Miyoshi et al. 2009b; Aviñó et al. 2016). By contrast, in the presence of ethanol, acetonitrile, DMSO, glycerol, or ethylene glycol, a decline is observed (Miyoshi et al. 2009b; Aviñó et al. 2016).

The triplex formed by polymeric strands poly(dT)poly(dA)poly(dT) shows a large increase in thermal stability in the presence of PEG 400, 1000, 3400, or 8000. Higher molecular weight of the polymer results in greater stabilization effects. In this context, the excluded-volume effect is the main factor for the stabilization of the triplex (Spink and Chaires 1999).

G4s

G4s' stability undergoes effects opposite to those of a DNA duplex in osmotic-stress experiments. During complex formation, water molecules are released. As a result, in the presence of a cosolvent, thermodynamic equilibrium shifts to a G4. Osmotic pressure experiments have shown rather different numbers of released water molecules for G4s of different topologies (Nakano et al. 2014).

The degree of hydration changes is strongly dependent on folded structure, the number of G-quartets, loop size, metal ions, and cosolutes (Miyoshi et al. 2006; Olsen et al. 2006; Lane et al. 2008; Arora and Maiti 2009a, b; Nakata et al. 2024). Besides, one of major contributors to thermal stability of G4s is loops. It has been observed that at formation of G4 with longer T_n loops releases fewer water molecules than a G4 with shorter loops does, and therefore a G4 with shorter loops is more stable and less hydrated (Arora and Maiti 2009a). Loops contribute to the stability of an antiparallel intramolecular quadruplex because of enthalpy and cause additional uptake of counterions and a greater water release (Olsen et al. 2009). Typical values of Δn_w are in a range of 3–40 for a whole G4 with ethylene glycol, and 60–200 in case of PEG 200. Stability of a G4 decreases linearly with an increase in $\ln a_w$ for all the tested cosolutes (PEG 200, 1,2-dimethoxyethane, 2-methoxyethanol, 1,3-PDO, ethylene glycol, and glycerol), but the slope depends on chemical structure of the cosolute; in general, the number of hydroxyl groups affects the number of water molecules to be released during G4 formation (Miyoshi et al. 2006).

Below a few examples are presented of the effects of cosolvents on G4 thermal stability. For thrombin binding aptamer $d(G_2T_2G_2TGTG_2T_2G_2)$ folded into an antiparallel G4, scientists have found a correlation of thermodynamic properties with the solvent's dielectric properties (Smirnov and Shafer 2007). A decrease in the dielectric constant (in comparison with the water) increases quadruplex stability. There are some exceptions. The presence of formamide diminishes quadruplex stability. Meanwhile, the presence of N-methylformamide (ε = 189) or dimethylformamide (ε = 37) significantly raises melting temperature (Smirnov and Shafer 2007). This observation highlights the importance of electrostatic interactions during quadruplex formation. DMSO can improve the stability of the parallel and antiparallel topologies, especially in the parallel G4 sequence of in promoter of c-KIT gene, but not the hybrid topologies (Yang et al. 2023). In the presence of 20 wt% of PEG 200 and KCl, the antiparallel basket-type G4 conformation has the same thermal stability as that in a diluted solution. The replacement of the salt with NaCl leads to the stabilization of the G4 in the presence of the cosolvent (Fujii et al. 2017).

The presented examples indicate the complexity of molecular crowding effects observed with DNA G4s.

I-Motif

Thermal stability of i-motifs depends on the number of cytosines and their topology (3′E or 5′E). In case of the 5′E topology, the motif is thermodynamically more stable, but the 3′E topology folds much faster. Among all secondary structures of nucleic-acids, i-motifs are the most sensitive to a treatment history: time of storage in a refrigerator, annealing and cooling speed, conditions before dilution, and the temperature at which samples are maintained (Abdelhamid and Waller 2020). Loops greatly affect thermal stability: the longer the loop, the lower is thermal stability and the lower is the pH_T value of the transition (Gurung et al. 2015). A long middle loop destabilizes an i-motif less than the other two long loops do (Wright et al. 2017). An analysis of thermal and pH stability of dC_n strands has revealed that the most stable intramolecular i-motif should contain 4 N—1 nucleotides, where N is 4 to 7 (Fleming et al. 2017). In such an i-motif, the length of each loop is 1 nt, and they have pH_T >7.2 and thermal stability >37 °C at pH 7.0 (Fleming et al. 2017). Thermal stability of an i-motif is not so sensitive to ionic strength of the solution. An increase in monovalent ions' concentration over 100 mM does not cause a significant decrease in thermal stability. In the case of divalent ions at up to 1 mM, a noticeable rize of thermal stability is observed (Saxena et al. 2017).

Ordinary UV or CD experiments cannot be performed on an i-motif because of the nonequilibrium transition during changing temperatures (hysteresis) (Rogers et al. 2021). It has been shown by a CD spectroscopy–based pH titration method that a pH-induced transition manifests hysteresis (Rogers et al. 2018). It has been found that PEG 200, 300, and 8000, 1,2-dimethoxyethane, and 2-methoxyethanol raise the stability of i-motifs, whereas glycerol lowers it, and ethylene glycol, 1,3-PDO, and Ficoll 70 do not change melting temperature much (Rajendran et al. 2010; Bhavsar-Jog et al. 2014; Cheng et al. 2021). Such osmotic-stress experiments point to stabilization of i-motifs in conjunction with the release of water molecules during complex formation. The observed stabilization by PEG could be explained by an increase in pK_a of cytosine with such a cosolute. PEGs and Ficoll 70 significantly decrease thermal hysteresis, owing to accelerating folding kinetics (Cheng et al. 2021).

Conclusion

In this chapter, we collected, analyzed, and summarized data on the structure and stability of DNA complexes under different crowding conditions and methods for their analysis. The variety of already discovered effects of cosolvents on DNA indicates the complexity of molecular crowding *in vivo*. The existing data suggest that *in vitro* experiments only weakly mimic the process in a live cell. The current trend is the development of methods for studying interactions of nucleic acids in live cells (Chen et al. 2021).

Acknowledgements This work was supported by the Russian state-funded project for ICBFM SB RAS (grant number 123021600208-7) for A.A.L. in part of the Methods for analysis and the Structure; Russian Science Foundation (grant number 23-74-01116) for V.M.G. in part of Thermodynamics; and Program of Fundamental Scientific Research in the Russian Federation for the Long-Term Period (2021-2030) (No. 124050800018-9) for O.A.T. and N.Yu.B. in part of text mining analysis.

References

Abdelhamid MAS, Waller ZAE (2020) Tricky topology: persistence of folded human Telomeric i-motif DNA at ambient temperature and neutral pH. Front Chem 8:1–6. https://doi.org/10.3389/fchem.2020.00040

Arora A, Maiti S (2009a) Stability and molecular recognition of Quadruplexes with different loop length in the absence and presence of molecular crowding agents. J Phys Chem B 113:8784–8792. https://doi.org/10.1021/jp809486g

Arora A, Maiti S (2009b) Differential biophysical behavior of human telomeric RNA and DNA quadruplex. J Phys Chem B 113:10515–10520. https://doi.org/10.1021/JP810638N

Aviñó A, Mazzini S, Gargallo R, Eritja R (2016) The effect of small Cosolutes that mimic molecular crowding conditions on the stability of triplexes involving duplex DNA. Int J Mol Sci 17:211. https://doi.org/10.3390/ijms17020211

Bastos M, Abian O, Johnson CM et al (2023) Isothermal titration calorimetry. Nat Rev Methods Prim 3:17. https://doi.org/10.1038/s43586-023-00199-x

Benedetti F, Japaridze A, Dorier J et al (2015) Effects of physiological self-crowding of DNA on shape and biological properties of DNA molecules with various levels of supercoiling. Nucleic Acids Res 43:2390–2399. https://doi.org/10.1093/nar/gkv055

Bennett BD, Kimball EH, Gao M et al (2009) Absolute metabolite concentrations and implied enzyme active site occupancy in Escherichia coli. Nat Chem Biol 5:593–599. https://doi.org/10.1038/nchembio.186

Berova N, Nakanishi K, Woody RW (2000) Circular dichroism: principles and applications, second edi. Wiley

Bhavsar-Jog YP, Van Dornshuld E, Brooks TA et al (2014) Epigenetic modification, dehydration, and molecular crowding effects on the thermodynamics of i-motif structure formation from C-Rich DNA. Biochemistry 53:1586–1594. https://doi.org/10.1021/bi401523b

Bishop GR, Chaires JB (2002) Characterization of DNA structures by circular dichroism. Curr Protoc Nucleic Acid Chem 11:7.11.1–7.11.8. https://doi.org/10.1002/0471142700.nc0711s11

Boehm BJ, Whidborne C, Button AL et al (2018) DNA triplex structure, thermodynamics, and destabilisation: insight from molecular simulations. Phys Chem Chem Phys 20:14013–14023. https://doi.org/10.1039/C8CP02385A

Bončina M, Vesnaver G, Chaires JB, Lah J (2016) Unraveling the thermodynamics of the folding and interconversion of human telomere G-Quadruplexes. Angew Chemie Int Ed 55:10340–10344. https://doi.org/10.1002/anie.201605350

Bundschus M, Dejori M, Stetter M et al (2008) Extraction of semantic biomedical relations from text using conditional random fields. BMC Bioinformatics 9:207. https://doi.org/10.1186/1471-2105-9-207

Burke JE, Butcher SE (2012) Nucleic acid structure characterization by small angle X-ray scattering (SAXS). Curr Protoc Nucleic Acid Chem 51. https://doi.org/10.1002/0471142700.nc0718s51

Chalikian TV, Macgregor RB (2023) The effect of pressure on the conformational stability of DNA. Springer Nature Singapore

Charney E, Chen HH (1987) Structure of A-DNA in solution. Proc Natl Acad Sci USA 84:1546–1549. https://doi.org/10.1073/pnas.84.6.1546

Chen Y-I, Sripati MP, Nguyen TD et al (2021) Recent developments in the characterization of nucleic acid hybridization kinetics. Curr Opin Biomed Eng 19. https://doi.org/10.1016/j.cobme.2021.100305

Cheng M, Chen J, Ju H et al (2021) Drivers of i-DNA formation in a variety of environments revealed by four-dimensional UV melting and annealing. J Am Chem Soc 143:7792–7807. https://doi.org/10.1021/jacs.1c02209

Chubarov AS, Endeward B, Kanarskaya MA, et al (2024) Pulsed dipolar EPR for self-limited complexes of oligonucleotides studies. Biomol 14:887. https://doi.org/10.3390/BIOM14080887

Collette D, Dunlap D, Finzi L (2023) Macromolecular crowding and DNA: bridging the gap between in vitro and in vivo. Int J Mol Sci 24:10–12. https://doi.org/10.3390/ijms242417502

Conti Nibali V, Havenith M (2014) New insights into the role of water in biological function: studying solvated biomolecules using terahertz absorption spectroscopy in conjunction with molecular dynamics simulations. J Am Chem Soc 136:12800–12807. https://doi.org/10.1021/ja504441h

Dai J, Carver M, Punchihewa C et al (2007) Structure of the Hybrid-2 type intramolecular human telomeric G-quadruplex in K+ solution: insights into structure polymorphism of the human telomeric sequence. Nucleic Acids Res 35:4927–4940. https://doi.org/10.1093/nar/gkm522

Dalla Pozza M, Abdullrahman A, Cardin CJ et al (2022) Three's a crowd - stabilisation, structure, and applications of DNA triplexes. Chem Sci 13:10193–10215. https://doi.org/10.1039/d2sc01793h

Davis TM, McFail-Isom L, Keane E, Williams LD (1998) Melting of a DNA hairpin without hyperchromism. Biochemistry 37:6975–6978. https://doi.org/10.1021/bi980047l

Decian A, Guittat L, Kaiser M et al (2007) Fluorescence-based melting assays for studying quadruplex ligands. Methods 42:183–195. https://doi.org/10.1016/j.ymeth.2006.10.004

Dhakal S, Megalathan A, Wijesinghe KM, Ranson L (2021) Single-molecule analysis of nanocircle-embedded i-motifs under crowding. J Phys Chem B 125:2193–2201. https://doi.org/10.1021/acs.jpcb.0c09640

Dzatko S, Krafcikova M, Hänsel-Hertsch R et al (2018) Evaluation of the stability of DNA i-motifs in the nuclei of living mammalian cells. Angew Chemie Int Ed 57:2165–2169. https://doi.org/10.1002/anie.201712284

Endoh T, Tateishi-karimata H, Sugimoto N (2023) Handbook of chemical biology of nucleic acids

Feijó Delgado F, Cermak N, Hecht VC et al (2013) Intracellular water exchange for measuring the dry mass, water mass and changes in chemical composition of living cells. PLoS One 8:e67590. https://doi.org/10.1371/journal.pone.0067590

Fleming AM, Ding Y, Rogers RA et al (2017) 4 n −1 is a "sweet spot" in DNA i-motif folding of 2′-deoxycytidine homopolymers. J Am Chem Soc 139:4682–4689. https://doi.org/10.1021/jacs.6b10117

Frank-Kamenetskii MD, Mirkin SM (1995) Triplex DNA structures. Annu Rev Biochem 64:65–95. https://doi.org/10.1146/annurev.bi.64.070195.000433

Fujii T, Podbevšek P, Plavec J, Sugimoto N (2017) Effects of metal ions and cosolutes on G-quadruplex topology. J Inorg Biochem 166:190–198. https://doi.org/10.1016/j.jinorgbio.2016.09.001

Gao B, Hou X-MM (2021) Opposite effects of potassium ions on the thermal stability of i-motif DNA in different buffer systems. ACS Omega 6:8976–8985

Gao C, Deng J, Anwar N et al (2023) Molecular crowding promotes the aggregation of parallel structured G-quadruplexes. Int J Biol Macromol 240:124442. https://doi.org/10.1016/j.ijbiomac.2023.124442

Gehring K, Leroy J-L, Guéron M (1993) A tetrameric DNA structure with protonated cytosine-cytosine base pairs. Nature 363:561–565. https://doi.org/10.1038/363561a0

Ghosh S, Takahashi S, Endoh T et al (2019) Validation of the nearest-neighbor model for Watson-crick self-complementary DNA duplexes in molecular crowding condition. Nucleic Acids Res 47:3284–3294. https://doi.org/10.1093/nar/gkz071

Ghosh S, Takahashi S, Ohyama T et al (2020a) Nearest-neighbor parameters for predicting DNA duplex stability in diverse molecular crowding conditions. Proc Natl Acad Sci USA 117:14194–14201. https://doi.org/10.1073/pnas.1920886117

Ghosh S, Takahashi S, Ohyama T et al (2020b) Nearest-neighbor parameters for predicting DNA duplex stability in diverse molecular crowding conditions. Proc Natl Acad Sci 117:14194–14201. https://doi.org/10.1073/pnas.1920886117

Gill P, Moghadam TT, Ranjbar B (2010) Differential scanning calorimetry techniques: applications in biology and nanoscience. J Biomol Tech 21:167

Goobes R, Kahana N, Cohen O, Minsky A (2003) Metabolic buffering exerted by macromolecular crowding on DNA–DNA interactions: origin and physiological significance. Biochemistry 42:2431–2440. https://doi.org/10.1021/bi026775x

Guneri D, Alexandrou E, El Omari K et al (2024) Structural insights into i-motif DNA structures in sequences from the insulin-linked polymorphic region. Nat Commun 15:29–39. https://doi.org/10.1038/s41467-024-50553-0

Gurung SP, Schwarz C, Hall JP et al (2015) The importance of loop length on the stability of i-motif structures. Chem Commun 51:5630–5632. https://doi.org/10.1039/C4CC07279K

Hamada H, Kakunaga T (1982) Potential Z-DNA forming sequences are highly dispersed in the human genome. Nature 298:396–398. https://doi.org/10.1038/298396a0

Harve KS, Lareu R, Rajagopalan R, Raghunath M (2009) Understanding how the crowded interior of cells stabilizes DNA/DNA and DNA/RNA hybrids-in silico predictions and in vitro evidence. Nucleic Acids Res 38:172–181. https://doi.org/10.1093/nar/gkp884

Harwood AJ Native polyacrylamide gel electrophoresis. In: Basic DNA and RNA protocols. Humana Press, New Jersey, pp 93–96

Heddi B, Phan AT (2011) Structure of human telomeric DNA in crowded solution. J Am Chem Soc 133:9824–9833. https://doi.org/10.1021/ja200786q

Herbert A, Rich A (1996) The biology of left-handed Z-DNA. J Biol Chem 271:11595–11598. https://doi.org/10.1074/jbc.271.20.11595

Hutin S, Guillotin A, Zubieta C, Tully MD (2023) Structural characterization of protein–DNA complexes using small angle X-ray scattering (SAXS) with contrast variation, pp 163–194

Jiang HX, Cui Y, Zhao T et al (2015) Divalent cations and molecular crowding buffers stabilize G-triplex at physiologically relevant temperatures. Sci Rep 5. https://doi.org/10.1038/srep09255

Karimata H, Nakano SI, Sugimoto N (2007) Effects of polyethylene glycol on DNA duplex stability at different NaCl concentrations. Bull Chem Soc Jpn 80:1987–1994. https://doi.org/10.1246/bcsj.80.1987

Kohwi Y, Panchenko Y (1993) Transcription-dependent recombination induced by triple-helix formation. Genes Dev 7:1766–1778. https://doi.org/10.1101/gad.7.9.1766

Koumoto K, Ochiai H, Sugimoto N (2008) Hydration is an important factor to regulate thermodynamic stability of a DNA duplex under molecular crowding conditions. Chem Lett 37:864–865. https://doi.org/10.1246/cl.2008.864

Krumkacheva OA, Shevelev GY, Lomzov AA et al (2019) DNA complexes with human apurinic/apyrimidinic endonuclease 1: structural insights revealed by pulsed dipolar EPR with orthogonal spin labeling. Nucleic Acids Res 47:7767–7780. https://doi.org/10.1093/nar/gkz620

Kulkarni M, Mukherjee A (2017) Understanding B-DNA to A-DNA transition in the right-handed DNA helix: perspective from a local to global transition. Prog Biophys Mol Biol 128:63–73. https://doi.org/10.1016/j.pbiomolbio.2017.05.009

Kypr J, Kejnovska I, Renciuk D, Vorlickova M (2009) Circular dichroism and conformational polymorphism of DNA. Nucleic Acids Res 37:1713–1725. https://doi.org/10.1093/nar/gkp026

Lam EYN, Beraldi D, Tannahill D, Balasubramanian S (2013) G-quadruplex structures are stable and detectable in human genomic DNA. Nat Commun 4:1796. https://doi.org/10.1038/ncomms2792

Lane AN, Chaires JB, Gray RD, Trent JO (2008) Stability and kinetics of G-quadruplex structures. Nucleic Acids Res 36:5482–5515. https://doi.org/10.1093/NAR/GKN517

Lee WTC, Yin Y, Morten MJ et al (2021) Single-molecule imaging reveals replication fork coupled formation of G-quadruplex structures hinders local replication stress signaling. Nat Commun 12:2525. https://doi.org/10.1038/s41467-021-22830-9

Lim KW, Amrane S, Bouaziz S et al (2009) Structure of the human telomere in K + solution: a stable basket-type G-Quadruplex with only two G-tetrad layers. J Am Chem Soc 131:4301–4309. https://doi.org/10.1021/ja807503g

Lim KW, Ng VCM, Martín-Pintado N et al (2013) Structure of the human telomere in Na+ solution: an antiparallel (2+2) G-quadruplex scaffold reveals additional diversity. Nucleic Acids Res 41:10556–10562. https://doi.org/10.1093/nar/gkt771

Lomzov AA, Sviridov EA, Shernuykov AV et al (2016) Study of a DNA duplex by nuclear magnetic resonance and molecular dynamics simulations. Validation of pulsed dipolar electron paramagnetic resonance distance measurements using Triarylmethyl-based spin labels. J Phys Chem B 120:5125–5133. https://doi.org/10.1021/acs.jpcb.6b03193

Luo Y, Granzhan A, Verga D, Mergny J (2021) FRET-MC: a fluorescence melting competition assay for studying G4 structures in vitro. Biopolymers 112. https://doi.org/10.1002/bip.23415

Marky LA, Breslauer KJ (1987) Calculating thermodynamic data for transitions of any molecularity from equilibrium melting curves. Biopolymers 26:1601–1620. https://doi.org/10.1002/bip.360260911

Matos-Rodrigues G, Hisey JA, Nussenzweig A, Mirkin SM (2023) Detection of alternative DNA structures and its implications for human disease. Mol Cell 83:3622–3641. https://doi.org/10.1016/j.molcel.2023.08.018

Mergny JL, Lacroix L (2009) UV melting of G-Quadruplexes. Curr Protoc nucleic acid Chem Chapter 17. https://doi.org/10.1002/0471142700.NC1701S37

Miller MC, Buscaglia R, Chaires JB et al (2010) Hydration is a major determinant of the G-quadruplex stability and conformation of the human telomere 3′ sequence of d(AG 3(TTAG3)3). J Am Chem Soc 132:17105–17107. https://doi.org/10.1021/ja105259m

Mills M, Arimondo PB, Lacroix L et al (2002) Chemical modification of the third Strand: differential effects on purine and pyrimidine triple helix formation. Biochemistry 41:357–366. https://doi.org/10.1021/bi011122m

Miyoshi D, Matsumura S, Nakano SI, Sugimoto N (2004) Duplex dissociation of telomere DNAs induced by molecular crowding. J Am Chem Soc 126:165–169. https://doi.org/10.1021/ja036721q

Miyoshi D, Karimata H, Sugimoto N (2005) Drastic effect of a single base difference between human and Tetrahymena telomere sequences on their structures under molecular crowding conditions. Angew Chemie Int Ed 44:3740–3744. https://doi.org/10.1002/anie.200462667

Miyoshi D, Karimata H, Sugimoto N (2006) Hydration regulates thermodynamics of G-quadruplex formation under molecular crowding conditions. J Am Chem Soc 128:7957–7963. https://doi.org/10.1021/ja061267m

Miyoshi D, Muhuri S, Mimura K, Sugimoto N (2009a) DNA junction structure stabilized by molecular crowding conditions. Nucleic Acids Symp Ser (Oxf) 53:59–60. https://doi.org/10.1093/nass/nrp030

Miyoshi D, Nakamura K, Tateishi-Karimata H et al (2009b) Hydration of Watson crick base pairs and dehydration of hoogsteen base pairs inducing structural polymorphism under molecular crowding conditions. J Am Chem Soc 131:3522–3531. https://doi.org/10.1021/ja805972a
Morimoto R, Horita M, Yamaguchi D et al (2022) Evaluation of weak interactions of proteins and organic cations with DNA duplex structures. Biophys J 121:2873–2881. https://doi.org/10.1016/j.bpj.2022.07.003
Moriyama R, Iwasaki Y, Miyoshi D (2015) Stabilization of DNA structures with poly(ethylene sodium phosphate). J Phys Chem B 119:11969–11977. https://doi.org/10.1021/acs.jpcb.5b03787
Muhuri S, Mimura K, Miyoshi D, Sugimoto N (2009) Stabilization of three-way junctions of DNA under molecular crowding conditions. J Am Chem Soc 131:9268–9280. https://doi.org/10.1021/ja900744e
Nakano S, Sugimoto N (2017) Model studies of the effects of intracellular crowding on nucleic acid interactions. Mol BioSyst 13:32–41. https://doi.org/10.1039/c6mb00654j
Nakano SI, Karimata H, Ohmichi T et al (2004) The effect of molecular crowding with nucleotide length and cosolute structure on DNA duplex stability. J Am Chem Soc 126:14330–14331. https://doi.org/10.1021/ja0463029
Nakano SI, Wu L, Oka H et al (2008) Conformation and the sodium ion condensation on DNA and RNA structures in the presence of a neutral cosolute as a mimic of the intracellular media. Mol BioSyst 4:579–588. https://doi.org/10.1039/b718806d
Nakano S, Yamaguchi D, Tateishi-Karimata H et al (2012) Hydration changes upon DNA folding studied by osmotic stress experiments. Biophys J 102:2808–2817. https://doi.org/10.1016/j.bpj.2012.05.019
Nakano SI, Miyoshi D, Sugimoto N (2014) Effects of molecular crowding on the structures, interactions, and functions of nucleic acids. Chem Rev 114:2733–2758. https://doi.org/10.1021/cr400113m
Nakata M, Kosaka N, Kawauchi K, Miyoshi D (2024) Quantitative effects of the loop region on topology, thermodynamics, and cation binding of DNA G-quadruplexes. ACS Omega 9:35028–35036. https://doi.org/10.1021/ACSOMEGA.4C05008
Olsen CM, Gmeiner WH, Marky LA (2006) Unfolding of G-quadruplexes: energetic, and ion and water contributions of G-quartet stacking. J Phys Chem B 110:6962–6969. https://doi.org/10.1021/JP0574697
Olsen CM, Lee H-T, Marky LA (2009) Unfolding thermodynamics of intramolecular G-Quadruplexes: base sequence contributions of the loops. J Phys Chem B 113:2587–2595. https://doi.org/10.1021/jp806853n
Olsen CM, Shikiya R, Ganugula R et al (2016) Application of differential scanning calorimetry to measure the differential binding of ions, water and protons in the unfolding of DNA molecules. Biochim Biophys Acta Gen Subj 1860:990–998. https://doi.org/10.1016/j.bbagen.2015.10.002
Pandey A, Roy S, Srivatsan SG (2023) Probing the competition between duplex, G-Quadruplex and i-motif structures of the oncogenic c-Myc DNA promoter region. Chem An Asian J 18:32–35. https://doi.org/10.1002/asia.202300510
Parkinson GN, Lee MPH, Neidle S (2002) Crystal structure of parallel quadruplexes from human telomeric DNA. Nature 417:876–880. https://doi.org/10.1038/nature755
Pastré D, Hamon L, Mechulam A et al (2007) Atomic force microscopy imaging of DNA under macromolecular crowding conditions. Biomacromolecules 8:3712–3717. https://doi.org/10.1021/bm700856u
Petraccone L, Erra E, Mattia CA et al (2004) Linkage of proton binding to the thermal dissociation of triple helix complex. Biophys Chem 110:73–81. https://doi.org/10.1016/j.bpc.2004.01.004
Phan AT, Guéron M, Leroy J-L (2000) The solution structure and internal motions of a fragment of the cytidine-rich strand of the human telomere 1 1Edited by I. Tinoco J Mol Biol 299:123–144. https://doi.org/10.1006/jmbi.2000.3613

Phan AT, Kuryavyi V, Luu KN, Patel DJ (2007) Structure of two intramolecular G-quadruplexes formed by natural human telomere sequences in K + solution †. Nucleic Acids Res 35:6517–6525. https://doi.org/10.1093/nar/gkm706

Phromsiri P, Gerling RR, Blose JM (2017) The effects of a neutral cosolute on the B to Z transition for DNA duplexes incorporating both CG and CA steps. Nucleosides Nucleotides Nucleic Acids 36:690–703. https://doi.org/10.1080/15257770.2017.1388395

Piškur J, Rupprecht A (1995) Aggregated DNA in ethanol solution. FEBS Lett 375:174–178. https://doi.org/10.1016/0014-5793(95)01206-T

Plavec J (2022) NMR study on nucleic acids. Handb Chem Biol Nucleic Acids 1–44. https://doi.org/10.1007/978-981-16-1313-5_8-1, Springer Nature Singapore

Preisler RS, Chen HH, Colombo MF et al (1995) The B form to Z form transition of poly(dG-m5dC) is sensitive to neutral solutes through an osmotic stress. Biochemistry 34:14400–14407. https://doi.org/10.1021/bi00044a017

Qin Y, Hurley LH (2008) Structures, folding patterns, and functions of intramolecular DNA G-quadruplexes found in eukaryotic promoter regions. Biochimie 90:1149–1171. https://doi.org/10.1016/j.biochi.2008.02.020

Rajendran A, Nakano SI, Sugimoto N (2010) Molecular crowding of the cosolutes induces an intramolecular i-motif structure of triplet repeat DNA oligomers at neutral pH. Chem Commun 46:1299–1301. https://doi.org/10.1039/b922050j

Rambo RP, Tainer JA (2010) Bridging the solution divide: comprehensive structural analyses of dynamic RNA, DNA, and protein assemblies by small-angle X-ray scattering. Curr Opin Struct Biol 20:128–137. https://doi.org/10.1016/j.sbi.2009.12.015

Renaud J-P, Chari A, Ciferri C et al (2018) Cryo-EM in drug discovery: achievements, limitations and prospects. Nat Rev Drug Discov 17:471–492. https://doi.org/10.1038/nrd.2018.77

Rhodes D, Lipps HJ (2015) Survey and summary G-quadruplexes and their regulatory roles in biology. Nucleic Acids Res 43:8627–8637. https://doi.org/10.1093/nar/gkv862

Rogers RA, Fleming AM, Burrows CJ (2018) Unusual isothermal hysteresis in DNA i-motif pH transitions: a study of the RAD17 promoter sequence. Biophys J 114:1804–1815. https://doi.org/10.1016/j.bpj.2018.03.012

Rogers RA, Meyer MR, Stewart KM et al (2021) Hysteresis in poly-2′-deoxycytidine i-motif folding is impacted by the method of analysis as well as loop and stem lengths. Biopolymers 112. https://doi.org/10.1002/bip.23389

Rollin R, Joanny J-F, Sens P (2023) Physical basis of the cell size scaling laws. elife 12:1–43. https://doi.org/10.7554/eLife.82490

Sänger M, Garda S, Wang XD et al (2024) HunFlair2 in a cross-corpus evaluation of biomedical named entity recognition and normalization tools. Bioinformatics 40:btae564. https://doi.org/10.1093/BIOINFORMATICS/BTAE564

SantaLucia J, Hicks D (2004) The thermodynamics of DNA structural motifs. Annu Rev Biophys Biomol Struct 33:415–440. https://doi.org/10.1146/annurev.biophys.32.110601.141800

Sasmal DK, Ghosh S, Das AK, Bhattacharyya K (2013) Solvation dynamics of biological water in a single live cell under a confocal microscope. Langmuir 29:2289–2298. https://doi.org/10.1021/LA3043473/SUPPL_FILE/LA3043473_SI_001.PDF

Saxena S, Joshi S, Shankaraswamy J et al (2017) Magnesium and molecular crowding of the cosolutes stabilize the i-motif structure at physiological pH. Biopolymers 107. https://doi.org/10.1002/bip.23018

Schroeder SJ, Turner DH (2009) Optical melting measurements of nucleic acid thermodynamics. Methods Enzymol 468:371–387. https://doi.org/10.1016/s0076-6879(09)68017-4, Elsevier

Shannon P, Markiel A, Ozier O et al (2003) Cytoscape: a software environment for integrated models of biomolecular interaction networks. Genome Res 13:2498. https://doi.org/10.1101/GR.1239303

Shevelev GY, Krumkacheva OA, Lomzov AA et al (2014) Physiological-temperature distance measurement in nucleic acid using triarylmethyl-based spin labels and pulsed dipolar EPR spectroscopy. J Am Chem Soc 136:9874–9877. https://doi.org/10.1021/ja505122n

Singh A, Maity A, Singh N (2022) Structure and dynamics of dsDNA in cell-like environments. Entropy 24:1587. https://doi.org/10.3390/e24111587
Smirnov IV, Shafer RH (2007) Electrostatics dominate quadruplex stability. Biopolymers 85:91–101. https://doi.org/10.1002/bip.20609
Spingler B (2012) Metal ion-promoted conformational changes of oligonucleotides. Metal ions in life sciences, In, pp 103–118
Spink CH, Chaires JB (1999) Effects of hydration, ion release, and excluded volume on the melting of triplex and duplex DNA. Biochemistry 38:496–508. https://doi.org/10.1021/bi9820154
Spink CH, Garbett N, Chaires JB (2007) Enthalpies of DNA melting in the presence of osmolytes. Biophys Chem 126:176–185. https://doi.org/10.1016/j.bpc.2006.07.013
Srivastava VK, Yadav R (2019) Isothermal titration calorimetry. Elsevier Inc.
Takahashi S, Sugimoto N (2020) Stability prediction of canonical and non-canonical structures of nucleic acids in various molecular environments and cells. Chem Soc Rev 49:8439–8468. https://doi.org/10.1039/D0CS00594K
Takahashi S, Yamamoto J, Kitamura A et al (2019) Characterization of intracellular crowding environments with topology-based DNA Quadruplex sensors. Anal Chem 91:2586–2590. https://doi.org/10.1021/acs.analchem.8b04177
Tang W, Niu K, Yu G et al (2020) In vivo visualization of the i-motif DNA secondary structure in the Bombyx mori testis. Epigenetics Chromatin 13:12. https://doi.org/10.1186/s13072-020-00334-y
Tateishi-Karimata H, Banerjee D, Ohyama T et al (2020) Hydroxyl groups in cosolutes regulate the G-quadruplex topology of telomeric DNA. Biochem Biophys Res Commun 525:177–183. https://doi.org/10.1016/j.bbrc.2020.02.045
Teng Y, Tateishi-Karimata H, Ohyama T, Sugimoto N (2020) Effect of potassium concentration on triplex stability under molecular crowding conditions. Molecules 25:387. https://doi.org/10.3390/molecules25020387
Ussery DW (2002) DNA structure: A-, B- and Z-DNA Helix Families. Encycl Life Sci. https://doi.org/10.1038/npg.els.0003122
van den Berg J, Boersma AJ, Poolman B (2017) Microorganisms maintain crowding homeostasis. Nat Rev Microbiol 15:309–318. https://doi.org/10.1038/nrmicro.2017.17
Víšková P, Ištvánková E, Ryneš J et al (2024) In-cell NMR suggests that DNA i-motif levels are strongly depleted in living human cells. Nat Commun 15:1992. https://doi.org/10.1038/s41467-024-46221-y
Vorlickova M (2005) Guanine tetraplex topology of human telomere DNA is governed by the number of (TTAGGG) repeats. Nucleic Acids Res 33:5851–5860. https://doi.org/10.1093/nar/gki898
Wang Y, Patel DJ (1993) Solution structure of the human telomeric repeat d[AG3(T2AG3)3] G-tetraplex. Structure 1:263–282. https://doi.org/10.1016/0969-2126(93)90015-9
Wang ZF, Li MH, Hsu STD, Chang TC (2014) Structural basis of sodium-potassium exchange of a human telomeric DNA quadruplex without topological conversion. Nucleic Acids Res 42:4723–4733. https://doi.org/10.1093/nar/gku083
Webba da Silva M (2007) Geometric formalism for DNA Quadruplex folding. Chem A Eur J 13:9738–9745. https://doi.org/10.1002/chem.200701255
Woodson SA, Koculi E (2009) Analysis of RNA folding by native polyacrylamide gel electrophoresis. Methods Enzymol 469:189–208. https://doi.org/10.1016/S0076-6879(09)69009-1, Elsevier
Wright EP, Huppert JL, Waller ZAE (2017) Identification of multiple genomic DNA sequences which form i-motif structures at neutral pH. Nucleic Acids Res 45:2951–2959. https://doi.org/10.1093/nar/gkx090
Yamaguchi D, Yoshida M, Nakano S (2022) Evaluation of thermal stability of DNA oligonucleotide structures embedded in hydrogels. DNA 2:302–313. https://doi.org/10.3390/dna2040021
Yang D (2019) G-quadruplex DNA and RNA. Methods in molecular biology, In, pp 1–24

Yang D, Okamoto K (2010) Structural insights into G-Quadruplexes: towards new anticancer drugs. Future Med Chem 2:619–646. https://doi.org/10.4155/fmc.09.172

Yang Y, Xiang J, Sun H et al (2023) New insights into the effect of molecular crowding environment induced by dimethyl sulfoxide on the conformation and stability of G-quadruplex. Chinese Chem Lett 34:107695. https://doi.org/10.1016/j.cclet.2022.07.038

Yildirim A, Sharma M, Varner BM et al (2014) Conformational preferences of DNA in reduced dielectric environments. J Phys Chem B 118:10874–10881. https://doi.org/10.1021/jp505727w

You Y, Tataurov AV, Owczarzy R (2011) Measuring thermodynamic details of DNA hybridization using fluorescence. Biopolymers 95:472–486. https://doi.org/10.1002/bip.21615

Yu H, Gu X, Nakano SI et al (2012) Beads-on-a-string structure of long telomeric dnas under molecular crowding conditions. J Am Chem Soc 134:20060–20069. https://doi.org/10.1021/ja305384c

Zeraati M, Langley DB, Schofield P et al (2018) I-motif DNA structures are formed in the nuclei of human cells. Nat Chem 10:631–637. https://doi.org/10.1038/s41557-018-0046-3

Zimmerman SB, Trach SO (1991) Estimation of macromolecule concentrations and excluded volume effects for the cytoplasm of Escherichia coli. J Mol Biol 222:599–620. https://doi.org/10.1016/0022-2836(91)90499-V

Chapter 7
Effects of Molecular Crowding on the Structure, Folding, Stability, and Catalysis of RNA

Shu-ichi Nakano

Abstract Dynamic and thermodynamic properties of RNA molecules vary across different cellular locations through multiple environmental factors, such as excluded volume interactions of cellular substances, alteration of the physical properties of the surrounding medium, and binding interactions with cellular substances. These environmental factors modulate the energy landscape of RNA transitions, thereby consideration of the effects of the intracellular environment on molecular interactions contributing to RNA folding and RNA–protein complexation is important for understanding RNA structures and their interactions in living cells. This chapter describes the dynamics and interactions of several RNA structures, including functional RNAs such as aptamers and ribozymes, in the environment of molecular crowding, revealed by in-cell, in silico, and *in vitro* studies. Specifically, *in vitro* model studies that have used cell-mimicking solutions comprising cellular substances or artificial molecular crowding agents have enhanced our understanding of RNA interactions and their molecular functions under molecular crowding conditions and increased our ability to predict RNA structures in cells. This chapter also provides an overview of the comparison between RNA and DNA interactions because the effects of the environmental factors on DNA are similar to those on RNA.

Keywords RNA duplex · G-quadruplex · Aptamer · Ribozyme · Thermodynamic stability · Electrostatic interaction · Metal ion binding · Cell-mimicking solution

S.-i. Nakano (✉)
Department of Nanobiochemistry, Faculty of Frontiers of Innovative Research in Science and Technology (FIRST), Konan University, Kobe, Japan
e-mail: shuichi@konan-u.ac.jp

V. N. Uversky (ed.), *(Macro)Molecular Crowding*, Subcellular Biochemistry 109,
https://doi.org/10.1007/978-3-032-03370-3_7

1. Introduction

RNA forms many different ordered structures, similar to DNA. Watson–Crick and other types of base-pairing interactions lead to RNA folding and assembly into unimolecular hairpin and pseudoknot structures, double-stranded duplexes, and four-stranded tetraplexes. These RNA structures are primarily stabilized via hydrogen bonding between nucleobases, base stacking, solvation (hydration), and cation binding, which are predominantly governed by noncovalent electrostatic forces, reflecting enthalpy changes. The formation of ordered RNA structures further influences molecular movement and dynamics, reflecting entropy changes. Experimentally, the Gibbs free energy change for the formation of a single Watson–Crick base pair in simple aqueous electrolyte solutions ranges from −1 to −3 kcal mol^{-1} at room temperature (Bloomfield et al. 2000), allowing even short oligonucleotides to form thermodynamically stable structures useful for biotechnology applications. Progress in oligonucleotide sequence design stems from an enhanced understanding of RNA interaction mechanisms and their thermodynamic properties and also from the development of computational algorithms for predicting RNA structures and their stabilization energy (Sun et al. 2017; Zuber et al. 2022). Compared with DNA, RNA adopts more complex structures and plays broad biological roles, which include regulation of gene expression through RNA structural rearrangements, binding of proteins and ligand molecules, and chemical catalysis.

The intracellular environment is characterized by the condition of high concentrations of substances and spatial confinement, called a molecular crowding condition. Protein content in a cell reaches several tens of weight percent, and metabolic molecules are present at concentrations surpassing millimolar levels (Theillet et al. 2014; Milo et al. 2016). The molecular environment in the cytoplasm is distinct from that in the nucleus, and further diversified across subcellular regions compartmentalized through membranes, condensation, and phase separation. Most human cell types express more than ten thousand different cellular RNAs (Ramskold et al. 2009), which undergo massive changes in the surrounding environment during the intracellular transport (Figure 7.1A). Moreover, the intracellular environment changes during cell development, cell aging, and disease progression. It is remarkable that RNA structures and their interactions are influenced by multiple environmental factors, such as excluded volume interactions of cellular substances, alteration of the physical properties of the surrounding medium, and binding interactions with cellular substances (Fig. 7.1b). Excluded volume interactions and solvent property alterations are pronounced in crowded media where substance concentrations are typically higher than several or several tens of percent. The contribution of binding interactions is determined by molecular concentration and binding affinity. These environmental factors modulate dynamic and thermodynamic properties of RNA across a wide range of energy and time scales. In principle, RNA changes its state to a new equilibrium state in response to a change in its surrounding environment in order to minimize the total energy of the system. The difference

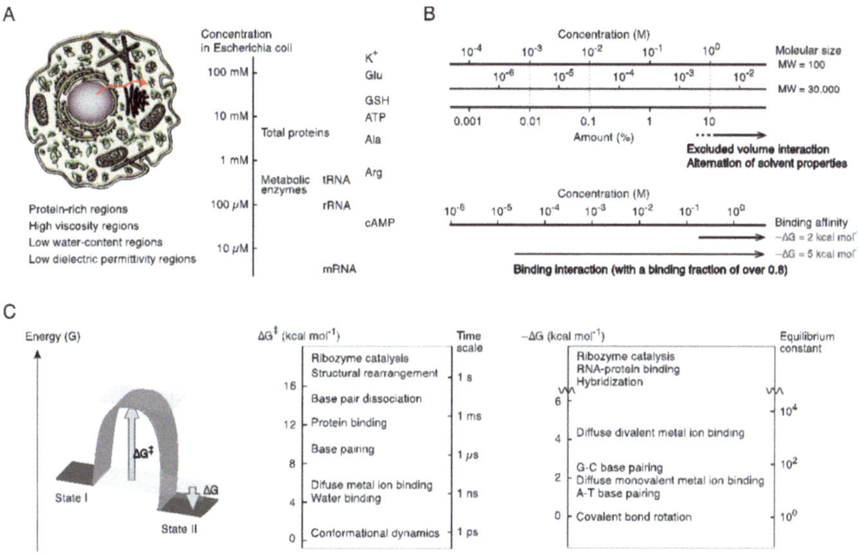

Fig. 7.1 (**a**) Typical characteristics of the cell microenvironment surrounding RNA during the intracellular transport from the nucleus to cytoplasm, and the amounts of cellular substances in *Escherichia coli.* (Milo et al. 2016). (**b**) Environmental factors that influence RNA structures and their interactions, categorized by the molecular weight of cellular substances or the binding affinity between RNA and RNA-binding substance. (**c**) Free energy landscape of an RNA transition between States I and II, representing the free energy barrier ($\Delta G^{\ddagger}$) and the free energy difference (ΔG) that determine the time scale and equilibrium constant of the transition, respectively

in free energy levels between different states (e.g., States I and II representing a two-state transition, depicted in Fig. 7.1c), referred by the the free energy change, determines the equilibrium constant of the transition, and the free energy barrier between the states determines the rate of the transition. The thermodynamic stability of State II (e.g., folded RNA) is determined relative to the energy level of State I (e.g., unfolded RNA). The efficiency of an intramolecular transition is determined based on the free energy change while the efficiency of an intermolecular transition induced by interaction with exogenous substances is determined based on the free energy change and molecular concentration (or activity). The energy level of each RNA state is modulated via physical and chemical interactions with the surrounding environment. Compared with the interactions for protein folding and assembly, electrostatic interactions dominate RNA folding and RNA–protein complexation whereas hydrophobic interactions are less pronounced. Accordingly, consideration of the effects of the intracellular environment on electrostatic interactions of RNA is important for understanding RNA structures and their interactions in cells.

Computational methods, such as molecular modeling and molecular dynamics (MD) simulations, are powerful in analyzing fast dynamic processes of RNA, including thermal fluctuations and transient binding interactions (Šponer et al. 2018). Specifically, supercomputer simulations can capture the fast dynamic processes and transient binding interactions of RNA molecules in cell-mimicking crowded spaces (Yu et al. 2016a; Feig et al. 2017; Trovato and Fumagalli 2017). Experimental methods, such as X-ray crystallography, nuclear magnetic resonance (NMR) spectroscopy, and fluorescence imaging microscopy, are also available for analyzing RNA structures and their interactions in cell and cell-free systems. The subsequent second section briefly describes the intracellular environment and the interactions of cellular RNAs based on the studies that have used in-cell chemical probing, spectroscopy, and microscopy methods. Alongside the observations in a cell, remarkable progress has also been achieved through *in vitro* experiments that have used cell-mimicking solutions comprising cellular substances or artificial molecular crowding agents. The *in vitro* model studies facilitate comparisons of the influences of cellular substances and physical properties of intracellular media (e.g., viscosity, water activity, and dielectric permittivity) on different RNA structures, including functional RNAs such as aptamers and ribozymes. Besides the studies on RNA, studies on DNA are informative because the physical and chemical interactions of DNA are broadly the same as those of RNA. The third section provides an overview of the comparison between RNA and DNA interactions as well as the microscopic mechanisms that influence the conformational dynamics and thermodynamic stability of various RNA and DNA structures. The fourth section introduces *in vitro* studies using cosolute-containing solutions, exploring macroscopic thermodynamic and kinetic properties of RNA that aid in understanding RNA structures and their interactions in the intracellular environment. Based on in-cell, in silico, and *in vitro* investigations, the final section offers insights into RNA interactions and their biological functions in the environment of molecular crowding.

2. Intracellular Environment Surrounding RNA

The intracellular environment varies across subcellular regions, where different types and amounts of macromolecules (e.g., proteins and polysaccharides), small metabolites (e.g., amino acids, sugars, alcohols, polyols, urea, and methylamines), and metal ions (e.g., K^+, Na^+, Mg^{2+}, and Ca^{2+}) are present. These cellular substances vary in size and possess different chemical properties, including polarity, hydrogen-bonding ability, and electric charges. These substances exert highly complex effects on the physical properties of the surrounding medium of RNA. Experimental analyses using fluorescent probes indicate variations in the intracellular medium, such as viscosity, pH, polarity, and temperature (Yang et al. 2023). Intracellular macromolecules create a large excluded volume and reduce the space for movement and conformational fluctuations of RNA. The total amount of free (unbound) water molecules in a cell is reduced owing to the presence of numerous cellular substances that need to be hydrated. In subcellular regions characterized by low water activity, water binding to RNA is less favorable compared with the binding in the medium of pure water and the strength of hydrophobic interactions is decreased. Meanwhile, dielectric permittivity is reduced and varied considerably among subcellular regions (Asami et al. 1976; Parsegian and Rau 1984; Garlid 2000; Wang et al. 2011; Sasmal et al. 2013). In subcellular regions characterized by low dielectric permittivity, the strength of attractive and repulsive electrostatic interactions is enhanced. Intracellular viscosity also exhibits substantial variation, ranging from a value close to the value of water to over a hundred-fold higher (Dix and Verkman 1990; Luby-Phelps et al. 1993; Kuimova et al. 2009; Sasmal et al. 2013). Nonetheless, studies using fluorescently labeled RNA molecules, such as polyadenine and cytoplasmic protein-coding mRNAs, have indicated that RNA diffusion in the nucleus and cytoplasm is not severely restricted (Pederson 1999; Misteli 2008; Wu et al. 2016). However, observations from the fluorescence recovery after the photobleaching assay have shown highly restricted dynamics of mRNAs assembled into ribonucleoprotein complexes (Braga et al. 2007), indicating a pronounced impact of RNA–protein complexation on the diffusion of RNA molecules in cells.

Chemical probing methods have been used for experimental investigation of the structure and interaction of cellular RNAs (Spitale et al. 2015; Bevilacqua et al. 2016). According to several chemical probing analyses listed in Table 7.1, RNA tends to bind to cellular proteins and forms less stable structures in cells than in cell-free uncrowded environments. In-cell spectroscopy and imaging microscopy are employed for RNA transcript analysis and targeted RNA investigations (Giassa et al. 2018; Zhao et al. 2019; Yamaoki et al. 2020), necessitating techniques for maintaining cell viability and metabolic stability during data acquisition as well as those for detecting target RNAs with low background signals. According to several

Table 7.1 Reports for the analysis of RNAs in cells by chemical probing methods

Method[a]	Structures of RNAs in cells	References
DMS probing	Messenger RNAs associated with stress responses in *Arabidopsis* seedling cells exhibit lower levels of organization compared to those *in vitro*	Ding et al. (2014)
DMS probing	Structures of mRNAs in rapidly dividing yeast cells tend to be less organized owing to ATP-dependent unfolding by helicases	Rouskin et al. (2014)
DMS probing	RNA G-quadruplexes in mammalian, yeast, and bacterial cells often remain unfolded possibly due to the action of single-stranded RNA-binding proteins	Guo and Bartel (2016)
SHAPE	Structures of mRNAs of housekeeping genes in *Escherichia coli* resemble those determined in cell-free environments	Mustoe et al. (2018)
SHAPE	Most RNA structures remain stable across different cellular compartments in human and mouse cells, but some RNAs adopt different structures among cellular compartments owing to distinct protein binding	Sun et al. (2019)
RNA-MaP	Structures of mRNAs in human cells are disrupted by binding of cellular proteins	Jarmoskaite et al. (2019)
PAIR-MaP	Many RNA tertiary structures formed in *Escherichia coli* and human cells exhibit alternative and unstable base-pairing interactions not observed *in vitro*	Mustoe et al. (2019) and Spitale and Incarnato (2023)

[a]Abbreviations: DMS (dimethyl sulfide); SHAPE (selective 2′-hydroxyl acylation analyzed by primer extension); RNA-Map (RNA on a massively parallel array); PAIR-MaP (pairing ascertained from interacting RNA strands measured by mutational profiling)

Table 7.2 Reports for the analysis of RNAs in cells by spectroscopy or microscopy

Method	State of RNAs in cells	References
Fluorescence microscopy	Intracellular localization and dynamics of RNA molecules are influenced by RNA-binding proteins involved in gene expression, modification, splicing, transport, and degradation	Urbanek et al. (2014) and Sato et al. (2020)
Fluorescence microscopy	Thermodynamic stability of the hairpin structure of an RNA oligonucleotide exhibits a broad distribution across human cells, with a higher stability in the nucleus than the cytoplasm	Gao et al. (2016) and Yoo and Davis (2022)
NMR	Weak interactions between RNA and protein are abundantly formed in various types of prokaryotic and eukaryotic cells	Majumder et al. (2015)
NMR	Hairpin and G-quadruplex structures of RNA and DNA oligonucleotides formed in human cells exhibit a higher frequency of base-pair opening dynamics compared to those formed *in vitro* owing to interactions with cellular proteins	Dzatko et al. (2018), Yamaoki et al. (2018) and Yamaoki et al. (2022)
Pulsed electron-electron double-resonance	An RNA oligonucleotide duplex in *Xenopus laevis* oocytes adopts more compact helical conformation compared with the conformation observed *in vitro* owing to nonspecific electrostatic interactions with cellular proteins	Collauto et al. (2020)

spectroscopy and microscopy analyses listed in Table 7.2, thermodynamic stability of RNA structures varies across intracellular locations and weak nonspecific interactions between RNA and cellular proteins are common in cells. These studies demonstrate the difficulty in elucidating RNA structures and interactions in living cells based on the knowledge obtained using uncrowded solutions.

3. Similarities and Differences Between RNA and DNA Interactions

The chemical structures of RNA and DNA are almost identical. Both possess the sugar-phosphate backbone and purine and pyrimidine nitrogenous bases with binding sites for water and cations, rendering them soluble in aqueous media containing metal ions. In general, the hydration pattern of biomolecules is determined using their geometric and chemical characteristics (Laage et al. 2017). The intertwined strands of a double-stranded helix form the major and minor grooves, whose width and depth differ between A-form and B-form conformations. These grooves are surrounded by hydration layers comprising an ordered hydrogen-bonded water network, with the 2′-hydroxyl group on the sugar moiety of an RNA duplex (dsRNA) contributing to this interaction network (Egli et al. 1996). Electrostatic binding of cations to RNA is influenced by the charge density distribution on the shape of RNA structures but not directly affected by the 2′-hydroxyl group of RNA (Cheatham and

Kollman 1997; Auffinger and Westhof 2001; Muthukumar 2017; Cruz-Leon and Schwierz 2022). The A-form dsRNA creates a stronger electrostatic field because of the shorter phosphate–phosphate distance compared with the B-form. The electric charge of the phosphate groups generates a diffuse counterion atmosphere along the RNA backbone via Coulomb interactions under rapid exchange of associated metal ions with bulk metal ions owing to weak interactions (Lipfert et al. 2014). The interactions of diffused metal ions with RNA and DNA can be modeled using the polyelectrolyte theory, which posits the condensation of point-charge cations at the surface of a rod-shaped structure, assuming a constant distance between negative charges and ignoring the shape of double-helical grooves and sequence-dependent variation in cation binding (Anderson and Record 1995). According to this theory, the number of cations electrostatically screening the charge of RNA backbone decreases under conditions of increased strength of electrostatic interactions, such as binding of high-valent cations and binding in media with low dielectric permittivity. Single-stranded RNA (ssRNA) and DNA (ssDNA) do not have specific metal ion–binding sites except for phosphate groups. In duplex structures, besides the phosphate groups, nucleobases such as the electronegative N7 and O6 atoms of the guanine base offer weak binding sites for metal ions, modifying water distribution in the grooves (Pasi et al. 2015; Yu and Chen 2018; Kolev et al. 2018). The metal ion binding enables the sugar-phosphate backbones to bring them closer to each other, essential for the formation and stability of base pairs of dsRNA and DNA duplex (dsDNA). Notably, compact tertiary structures cannot be stably formed without divalent metal ions or multivalent organic cations like polyamines (Misra and Draper 1998).

Figure 7.2 shows the free energy landscape of RNA transitions induced by binding or dissociation of exogenous substances, such as water, metal ions, ligands, and

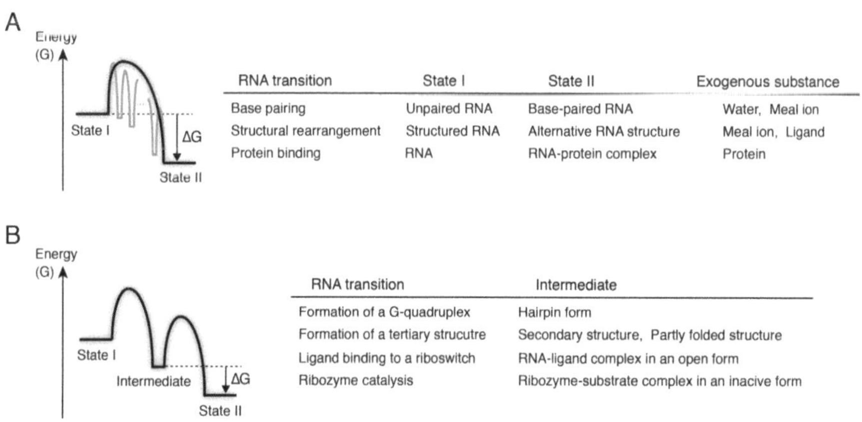

Fig. 7.2 (**a**) Free energy landscape of an RNA transition from States I to II in a highly cooperative or non-cooperative manner (black and gray, respectively), and examples of RNA transitions induced by exogenous substances. (**b**) Free energy landscape of an RNA transition mediated through an intermediate state and examples of such transitions

proteins. For example, the transition of RNA bases from unpaired to paired states involves the formation of hydrogen bonding and stacking of nucleobases and the reorganization of water and metal ion interactions with RNA. Although each interaction energy is not large, the free energy change for base pairing is considerable owing to the simultaneous formation of multiple interactions. Furthermore, formation of an RNA duplex from single strands involves the simultaneous formation of several base pairs in a highly cooperative manner, being energetically coupled to one another described by the energy landscape between two states (Fig. 7.2a). In contrast, formation of complex RNA structures often involves intermediate states (Fig. 7.2b). However, in the cases where multiple transitions occur in a highly cooperative manner, a simplified energy landscape between two states can be applied. Notably, when multiple interactions are simultaneously formed or deformed, influences of the surrounding environment on RNA transitions can be amplified.

Several environmental factors modulating the energy levels of RNA can be considered. RNA molecules in a cell are surrounded by crowder molecules and interfaces of cell organelles and other cellular components. The molecular environments influencing RNA interactions of hydrogen bonding, base stacking, hydration, and cation binding modulate the free energy levels of RNA through entropic and enthalpic factors. Moreover, changes in molecular activity shift the equilibrium of intermolecular reactions. It is plausible that many insights into physical and chemical interactions of DNA are valid for those of RNA as well.

Conformational Dynamics

RNA and DNA strands exhibit dynamic ensembles of diverse conformations with similar free energy levels. In RNA, the presence of the 2′-hydroxyl group restricts the sugar pucker conformation to the C3′-endo position (Fig. 7.3a), rendering it less flexible compared with DNA. An experimental study combining small-angle X-ray scattering (SAXS) and single-molecule fluorescence resonance energy transfer

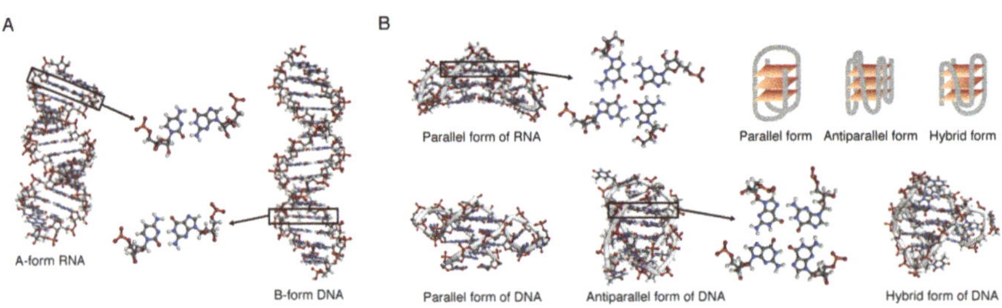

Fig. 7.3 (**a**) Shapes of the RNA and DNA duplexes and their base pairing. (**b**) The RNA G-quadruplex adopting a parallel conformation (PDB ID: 2M18) and DNA G-quadruplexes adopting parallel (PDB ID: 1K8P), antiparallel (PDB ID: 143D), and hybrid (PDB ID: 2K4I) conformations

methods has reported a longer persistence length for an ssRNA (U_{40}) than its ssDNA counterpart (T_{40}) (Chen et al. 2012). Another SAXS study has reported RNA sequence (U_{30} and A_{30})-dependent disordered conformations of ssRNA and indicated that the backbone of ssRNA is stiffer than that of ssDNA, resulted from the salt bridge interactions involving water, metal ions, and the 2′-hydroxyl group (Plumridge et al. 2020). This study also reports more attraction of metal ions (Na^+ and Mg^{2+}) in the counterion atmosphere of ssRNAs than in that of their DNA counterparts (T_{30} and A_{30}, respectively). Similar findings are reported for stem-loop structures comprising single-stranded loops linked to a base-paired stem, known as main secondary structure elements. The loop regions usually exhibit high dynamics, influencing the shape of global structures, stability, and protein binding (Razga et al. 2005; Rau et al. 2012). Computer simulations indicate that the conformational ensembles of an RNA hairpin with a long loop (U_{18}) are more compact than those of a DNA hairpin with a counterpart loop (T_{18}) (Swadling et al. 2018) and that the flexibility of loop nucleotides (5–13 nucleotide lengths) in model RNA and DNA hairpins is affected by metal ion (Na^+ and Mg^{2+}) binding (Tan and Chen 2008). It is not surprising that the flexibility of single-stranded RNA and DNA regions is affected by the strength of base stacking and by the efficiency of metal ion binding that reduces the repulsion between the phosphate groups. Notably, the flexibility of single-stranded regions influences the conformation and stability of RNA tertiary structures (Razga et al. 2005; Lipfert et al. 2014). For example, the single-stranded junction of an RNA pseudoknot derived from the beet western yellow virus is highly flexible, and metal ions (Mg^{2+}) bound at the junction contribute to the stabilization of the loop–single strand interaction of the pseudoknot (Egli et al. 2002, Soto et al. 2007). In the case of an internal loop region of the kink-turn motif derived from 23S rRNA, its kinked conformation is stabilized by a divalent metal ion (Mg^{2+})-dependent conformational change (Goody et al. 2004).

Base-paired double-stranded chains exhibit less flexibility compared with single-stranded chains. The C3′-endo conformation of the sugar in RNA favors the adoption of the A-form as a double-stranded conformation (Fig. 7.3a). Experimental studies employing single-molecule atomic force microscopy report a longer persistence length of dsRNA than dsDNA (Abels et al. 2005; Herrero-Galan et al. 2013). MD simulations indicate differences in the flexibility between RNA and DNA duplexes resulted from distinct hydration states and metal ion binding patterns (Auffinger and Westhof 2000, 2001). The sugar-phosphate backbone of an A-form dsRNA is more rigid than that of a B-form dsDNA with the same sequence; however, the helical conformation of dsRNA is not always less flexible than dsDNA (Pan and Mackerell 2003; Noy et al. 2004; Perez et al. 2004). It has been reported that the stretching flexibility of dsRNA is higher than that of dsDNA (Bao et al. 2017), and the bending flexibility of dsRNA is reduced by multivalent cations bound in the major groove, unlike dsDNA that exhibits increased flexibility (Drozdetski et al. 2016). These reports demonstrate that the effects of metal ion binding on the flexibility of base-paired regions are not the same as those on single-stranded regions and that the correlation between metal ion binding and the mechanical property of RNA chains is complex (Marin-Gonzalez et al. 2021).

In the environment of molecular crowding, crowders may directly influence the conformational dynamics of RNA by displacing water and metal ions bound to RNA. Furthermore, crowders create a significant excluded volume and increase the viscosity of the medium, thereby suppressing the movement and conformational fluctuations of RNA, particularly those of long RNA chains. Considering these mechanisms, it is conceivable that there is no intrinsic difference in the effects of the intracellular environment on the conformational dynamics of RNA and DNA.

Thermodynamic Stability of Secondary Structures

Stability of RNA secondary structures is primarily determined by the strengths of Watson–Crick base pairs and loop interactions among nucleotides in close proximity in the primary sequence. Owing to the differences in inclination, slide, and twist angles of Watson–Crick base pairs between A-form and B-form duplexes, their base stacking arrangements are different. Computer simulation and experimental analysis studies have revealed the complexity of sequence- and geometry-dependent stacking interaction energies (Svozil et al. 2010; Mcdonald et al. 2013). The stacking overlap of base pairs in an A-form duplex is smaller than that in a B-form duplex, and the stacking of uracil paired with adenine lacks the interaction of the methyl group of thymine (Gil et al. 2009; Svozil et al. 2010; Parker et al. 2013; Brown et al. 2015). Nonetheless, both computational and experimental studies have indicated that RNA base pairs usually exhibit higher stability than DNA base pairs (Brown et al. 2015; Zuber et al. 2022). The disparity in the stability between A-form and B-form duplexes is resulted from the differences in their hydration states (Egli et al. 1996, Rozners and Moulder 2004, Liu et al. 2021) and counterion binding (Tan and Chen 2011; Gebala and Herschlag 2019), as well as different stacking arrangements. Similarly, Hoogsteen base pairs formed in A-form dsRNA are energetically more stable than those formed in B-form dsDNA, where the direct influence of the 2′-hydroxyl group of RNA is small (Rangadurai et al. 2018). Formation of loop regions is also stabilized via base stacking interactions, water-mediated interactions, and metal ion binding (Auffinger and Westhof 1997; Shanker and Bandyopadhyay 2017; Šponer et al. 2018), which are shared between RNA and DNA (Kuznetsov et al. 2008; Blose et al. 2009). Additionally, RNA hairpins with a GCUAA-loop or U_{18}-loop exhibit higher stability compared to their DNA counterparts, resulted from hydrogen bonding of the 2′-hydroxyl groups of loop nucleotides (Paladino and Zangi 2013; Swadling et al. 2018). Overall, the stabilizing interactions in loops resemble those in base pairing. This similarity enables us to predict both RNA and DNA secondary structures from their sequences using the nearest-neighbor interaction model. Using experimentally obtained thermodynamic data, the prediction of secondary structures comprising Watson–Crick base pairs, mismatched base pairs, and various loop regions (e.g., hairpin, bulge, and internal loops) in accordance with the principle of free energy minimization is satisfactory (Zuker et al. 1999; Reuter and Mathews 2010; Zuber et al. 2022). Moreover, computer simulations provide predictions of secondary and more complex structures, such as pseudoknots and G-quadruplexes, along with predictions of metal ion binding sites and loop flexibility (Lomzov et al. 2015; Shi et al. 2015; Shi et al. 2018; Jin et al. 2018; Nguyen et al. 2019; Lombardi and Londono-Vallejo 2020; Sato and Kato 2022; Kawaguchi and Kiryu 2023). Recently, machine-learning approaches trained on known RNA structures and thermodynamic data have facilitated the prediction of RNA secondary and tertiary structures (Singh et al. 2019; Sato et al. 2021; Townshend et al. 2021; Zhao et al. 2021).

Regarding the stability of RNA secondary structures in the intracellular environment, we need to pay specific attention to environmental factors capable of modulating the interaction energy of base pairing and loop formation. Crowders influence the interactions of RNA with water and metal ions: polar crowders may displace water molecules hydrating the sugar-phosphate backbone and nucleobases, and charged crowders modify the counterion atmosphere of RNA, specifically, positively charged crowders compete with metal ions for binding to the RNA backbone. Additionally, hydrophobic crowders preferentially bind to hydrophobic sites on RNA. Molecular simulation studies on dsDNAs have demonstrated that hydrophilic or hydrophobic interfaces surrounding the duplex confined in a small space influence the strengths of Watson–Crick hydrogen bonding and base stacking (Elder et al. 2015). Considering the physical properties of the medium is also important because polar and charged crowders and interfaces deplete free water in the medium, leading to reduced water activity and dielectric permittivity, thereby altering the enthalpy and entropy changes of electrostatic binding of water and metal ions to RNA. Furthermore, molecular environments characterized by high viscosity and confined spaces suppress RNA movement and flexibility, consequently changing the entropy of RNA folding and assembly. Effects of these environmental factors on DNA are supposed to be similar to those on RNA; however, disparities in the magnitude of environmental effects can arise owing to the presence of the 2′-hydroxyl group, which influences RNA dynamics, conformation, and interactions with exogenous substances.

Structure and Stability of G-Quadruplex and Tertiary Structures

Specific RNA sequences form highly compact structures mediated by interactions that are not formed in secondary structures comprising Watson–Crick base pairs. Guanine-rich sequences have a tendency to adopt intramolecular and intermolecular four-stranded tetraplexes called G-quadruplexes, consisting of stacked guanine tetrads held together via Hoogsteen base pairs, and the arrangement of base stacking is different between RNA and DNA G-quadruplexes. MD simulation studies on DNA G-quadruplexes have indicated that the stacked guanine tetrad is less flexible than base pairs in a duplex (Castelli et al. 2022) but the loops connecting guanine tracts are relatively flexible (Islam et al. 2018). G-quadruplexes adopt several different folding topologies, classified into parallel, antiparallel, and hybrid conformations, all having similar arrangements of hydrogen-bonded guanine tetrads but different arrangements of loops (Fig. 7.3b). Most RNA G-quadruplexes adopt the parallel conformation, and this preference is enabled by the C3′-endo sugar pucker conformation and hydrogen bonding of the 2′-hydroxyl group (Malgowska et al. 2016; Haase et al. 2018). In contrast, DNA adopts different topologies depending on the length of loops and the conditions of metal ion and molecular crowding (Joachimi et al. 2009; Hatzakis et al. 2010). The mechanism of formation of the parallel conformation is likely similar between RNA and DNA (Havrila et al. 2018); however, the stacking of guanine tetrads in RNA G-quadruplexes is weaker than that in DNA G-quadruplexes owing to the steric hindrance of the 2′-hydroxyl group of RNA (Kankia 2019). Nonetheless, RNA G-quadruplexes usually exhibit higher stability than their DNA counterparts (Joachimi et al. 2009; Arora and Maiti 2009; Guiset Miserachs et al. 2016). The enhanced stability of RNA G-quadruplexes is attributed to the contribution of their 2′-hydroxyl groups to water and metal ion binding (Sacca et al. 2005; Collie et al. 2010; Bugaut and Balasubramanian 2012; Zaccaria and Fonseca Guerra 2018) and the additional stacking of RNA loop nucleotides with guanine tetrads (Kankia 2019). A remarkable feature of G-quadruplexes is their ability to coordinate metal ions to guanine tetrads. Stability of G-quadruplexes largely depends on the size of the metal ion, with particular metal ions like K^+ greatly stabilizing both RNA and DNA structures; however, the effects of different kinds of metal ions are not always consistent between RNA and DNA (Guiset Miserachs et al. 2016).

Many cellular RNAs form tertiary structures via interactions among distal unpaired nucleotides in their secondary structures. Specific structural motifs involving the interactions of the 2′-hydroxyl groups of unpaired nucleotides, such as the A-minor motif and ribose zipper motifs, are found in ribosomal RNAs and ribozymes (Leontis et al. 2002; Xin et al. 2008). During the process of RNA tertiary folding, less folded and extended intermediates are often formed through metal ion-induced collapse, followed by slow formation or rearrangement of RNA interactions (Moghaddam et al. 2009; Woodson 2010). For example, a pseudoknot structure is formed through the folding back of a single-stranded region to the minor groove of the stem of a hairpin loop. Stability of the pseudoknot is determined relative to the energy level of the hairpin state as a partially folded intermediate (Fig. 7.2b). The fold-back interaction of the RNA pseudoknot derived from simian retrovirus type-1 is stabilized via interactions of the ribose zipper motif, base stacking, and specific base-triplet formations (Michiels et al. 2001), resulting in higher stability compared to its DNA counterpart (Reiling-Steffensmeier and Marky 2016). RNAs adopting specific secondary and tertiary structures perform biological functions, including aptamers recognizing target ligand molecules and ribozymes catalyzing substrate chemical reactions. Similarly, several DNA aptamers and DNA enzymes have been discovered through *in vitro* selection methods. These functional RNA and DNA molecules often undergo structural rearrangements between different conformations induced by metal ion, ligand, or substrate binding. The rate and efficiency of the structural rearrangements determine their functions.

The intracellular environment influences the formation of RNA tertiary structures similarly to their secondary structures, making the studies on secondary structures useful to elucidate the influences on tertiary structures. Meanwhile, substantial differences exist in the interaction and folding mechanisms between RNA secondary and tertiary structures. Specifically, folding of long RNA chains to a tertiary structure causes large changes in the solvent-accessible area, resulting in more profound changes in molecular dynamics, occupied volume, and negative charge density compared to folding to secondary structures. Crowders with hydrophobic properties may bind to hydrophobic sites on RNA tertiary structures, such as the surface of guanine tetrad planes and the interior of tightly folded structures. It is conceivable that the environment of molecular crowding has a more pronounced impact on tertiary structure formation than secondary structure formation.

Binding Interactions with Proteins

The average surface–surface distance between proteins in a living cell, calculated based on their cellular abundance, is comparable with the electrostatic screening distance, creating an environment rich in long-range Coulomb interactions of proteins. Therefore, consideration of the interaction mechanism and binding strength between RNA and protein under molecular crowding conditions is important. There are many reports on RNA- and DNA-binding proteins with a wide range of binding affinity (Gerstberger et al. 2014; Jankowsky and Harris 2015), and machine-learning approaches provide predictions for RNA–protein and DNA–protein complexes (Alipanahi et al. 2015; Baek et al. 2024). Coarse-grained simulations comparing RNA and DNA sequence-specific binding proteins report that the binding affinity for ssRNA–protein complexes tends to be lower than that for ssDNA–protein complexes, attributed to the lower flexibility of ssRNA compared to ssDNA (Pal and Levy 2019). RNA sequence-specific binding proteins differentiate target RNA from its DNA counterpart. Proteins containing dsRNA-binding domains bind to target dsRNA over non-target dsDNA, resulted from the interactions with the 2′-hydroxyl group of RNA (Masliah et al. 2013) and recognition of the groove of an A-form duplex (Vukovic et al. 2014). Coulomb interactions between the phosphate group and the side chain of basic amino acid residues (e.g., arginine and lysine) greatly contribute to protein binding to their target hairpins (e.g., U1A, Sac 7d, and p53) (Law et al. 2006; Yuann et al. 2012; Cobb et al. 2013) and G-quadruplexes (e.g., FMRP, CIRBP, zinc finger peptide, and arginine-rich peptide) (Phan et al. 2011; Huang et al. 2018; Meier-Stephenson 2022). More strikingly, Coulomb interactions dominate the binding of RNA sequence-nonspecific binding proteins although the binding is usually weak.

The excluded volume of macromolecules at amounts comparable to those in cells causes increases in the molecular activity by several- to several hundred-fold, evaluated by a hard-sphere molecular interaction model (Minton 1998). The condition of increased molecular activity promotes collision and association between

RNA and protein. MD simulations of a dsRNA in the presence of excess amounts of lysozyme, serving as a protein crowder model, indicate the formation of multiple RNA–lysozyme binding modes via RNA sequence-nonspecific interactions, and in particular, positively charged amino acids, notably arginine residues, bind to the phosphate groups (Collauto et al. 2020). Similarly, simulation studies on a dsDNA in the presence of protein G, which is not a specific DNA-binding protein, have demonstrated interactions between lysine residues and the double-helix backbone via nonspecific Coulomb and hydrogen-bonding interactions (Yildirim et al. 2018). The strength of electrostatic interactions is likely enhanced in media with low dielectric permittivity, which is especially important for the binding of RNA sequence-nonspecific binding proteins to RNA.

4. RNA in Cell-Mimicking Media

Experimental approaches using *in vitro* cell-mimicking solutions offer practical advantages for inferring the dynamics and interactions of RNA structures in the intracellular environment. The approach uses homogeneous aqueous solutions with controlled compositions and quantities of certain cellular substances or artificial molecular crowding agents as cosolutes. The commonly used cosolutes are water-soluble polymers and small organic solvents with various properties, such as molecular size, polarity, and charge. The effect of cosolutes differs depending on how they interact with RNA and thus careful selection of the type and concentration of cosolutes based on the experiment objectives is important. To conduct studies for different RNA structures, it is essential to prepare RNA molecules to adopt desired secondary and tertiary structures. Construction of designated structures using polynucleotides is difficult, and the assessment of structural stability of long RNA chains is not easy owing to their complex folding pathways. Polynucleotides also have a tendency to irreversibly precipitate in the presence of multivalent cations or organic solvents. Alternatively, oligonucleotides are convenient for preparing designated secondary and tertiary structures, including those modified with chemical or fluorescent probes that enable a broad spectrum of experimental studies.

Stability of RNA Structures

RNA in the Environment of Molecular Crowding Created by Neutral Cosolutes

When cosolute binding to RNA is ignored, effects of their excluded volume on RNA can be tested. Solutions containing cosolutes at concentrations of several or several tens of percent create notable excluded volume interactions (Figure 7.1B). Water-soluble neutrally charged polymers, like polyethylene glycol (PEG) and

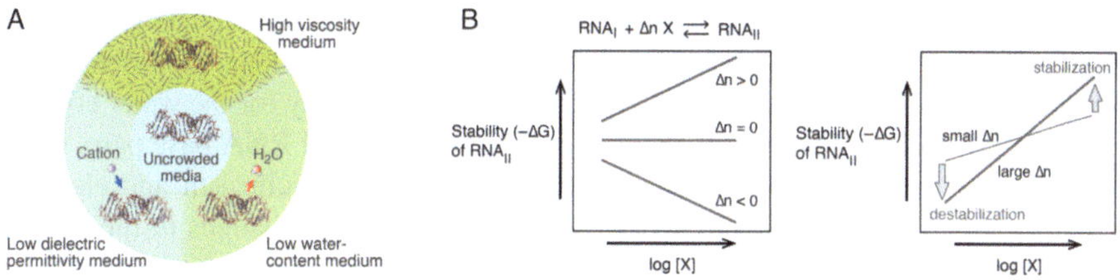

Fig. 7.4 (**a**) Molecular environments of cosolute-containing solutions that may influence the dynamics and interactions of RNA. (**b**) Correlations between the stability of RNA_{II} and the logarithm of X concentration (or activity) for the RNA transition from RNA_I to RNA_{II} induced by an exogenous substance X, depending on the net number Δn of cooperatively bound or released X during the transition

polysaccharides, are widely used for evaluating the excluded volume effect of macromolecules whereas the excluded volume effect of small cosolutes has also been proposed to be considerable (Minton 1998; Sharp 2015). Solutions containing net-neutral cosolutes at concentrations of several tens of percent can be used for evaluating the effects of physical properties of the surrounding medium on RNA (Figure 7.4A) (Nakano and Sugimoto 2016; Ferreira et al. 2017). Spectroscopic analysis and computer simulation studies indicate modification of the hydrogen-bonding network of water owing to the presence of large and small cosolutes, such as PEG, polysaccharides, and small sugars (Verma et al. 2016; You et al. 2021), where the dynamics of water molecules hydrating small cosolutes differ from those hydrating large cosolutes (Verma et al. 2018; Semmeg et al. 2024). Compared with the properties of water, most crowded solutions exhibit lower water activity and dielectric permittivity and polymer cosolute-containing solutions demonstrate higher viscosity (Nakano and Sugimoto 2016). The media with low water activity, low dielectric permittivity, and high viscosity represent the characteristics of the intracellular environment. These physical properties can be controlled using aqueous solutions containing single or several types of cosolutes. However, the addition of cosolutes alters multiple physical properties, making it challenging to distinguish the effect of one property from other properties.

Cosolute-containing solutions are useful for investigating the thermodynamic stability of RNA structures using spectroscopic methods. For example, the stability of an intramolecular hairpin of a 34-mer RNA derived from an RNA thermometer sequence is increased owing to the excluded volume effect of a large PEG (10,000 Da) that shifts the equilibrium toward a compactly folded RNA state (Yoo and Davis 2022). Conversely, short RNA (6-mer – 14-mer) oligonucleotides forming an intermolecular duplex or an intramolecular hairpin undergo destabilization in the crowded solution of a large PEG (8000 Da) (Nakano et al. 2009), indicating no profound stabilizing effect driven by the excluded volume effect of PEG on short oligonucleotides. Rather, this destabilizing effect is resulted from the decreased water content of solution, which is unfavorable to the formation of RNA base-paired structures accompanied by water binding. Osmotic stress assays using small cosolutes (e.g., PEG (200 Da), ethylene glycol, glycerol, acetamide, proline, and trimethylglycine) together with MD simulations have indicated that dsRNA is more hydrated than dsDNA with the same sequence (Rozners and Moulder 2004; Pramanik et al. 2011). Therefore, there seems to be a tendency for the water content in the intracellular environment to influence the formation of RNA base pairs more significantly than DNA base pairs. Studies using circular dichroism and electron paramagnetic resonance spectroscopic methods have shown changes in the helical conformation of single-stranded RNA and DNA oligonucleotides (6-mer–16-mer) in crowded solutions of both large and small cosolutes (e.g., PEGs (8000 and 200 Da), ethylene glycol, glycerol, and 1,2-dimethoxyethane) (Nakano et al. 2008b, 2014). In contrast, these cosolute-containing solutions have no influence on the conformations of duplex and hairpin structures, except for a duplex of an alternating CG sequence that undergoes structural changes to form a left-handed Z-form duplex. There appears to be no noticeable effect of molecular crowding on the hydrogen bonding and base stacking arrangements of most RNA base-paired structures, whereas there is a substantial influence on the conformational ensembles of single strands. Therefore, the energy levels of RNA and DNA in an unfolded state tend to be modulated in the environment of molecular crowding.

The effect of polymer cosolutes on the stability of RNA base pairing may be changed depending on the salt concentration in solution. For example, in the crowded solution of a large PEG (8000 Da), both intermolecular duplexes and intramolecular hairpins of RNA and DNA oligonucleotides undergo destabilization at high salt concentrations (several hundreds of millimolars of NaCl) but undergo stabilization at low salt concentrations (several tens of millimolars of NaCl), ascribed to the altered dependence of metal ion concentrations on the stability of oligonucleotide structures (Karimata et al. 2007; Nakano et al. 2014, 2016) (Fig. 7.4b). This dependence reflects the net number of metal ions cooperatively bound to RNA during the formation of folded structures. Comparative studies using various types of cosolutes (e.g., large and small PEGs, dimethoxyethane, ethanol, and dimethyl sulfoxide) have indicated increased efficiency of screening the charge of RNA backbone by metal ions in media with low dielectric permittivity (Nakano et al. 2014, 2016). This conclusion is supported by a computer simulation study showing a decrease in the net number of metal ions (Na^+ and Mg^{2+}) bound to dsRNA in the presence of globular-shaped crowders resembling PEG (600 Da) (Yu et al. 2016b). Therefore, it seems that intracellular dielectric permittivity influences the stability of base-paired RNA structures, which can be amplified or diminished by the effect of intracellular water activity depending on how water and metal ions contribute to RNA transitions.

In contrast to duplex and hairpin structures, G-quadruplexes are stabilized in crowded solutions of small cosolutes (e.g., PEG (200 Da), ethanol, and acetonitrile), driven by reduced water activity favorable to the formation of stacked guanine tetrads accompanied by water release (Miyoshi et al. 2006; Miller et al. 2010; Verdian Doghaei et al. 2015; Matsumoto et al. 2020). According to the osmotic stress assays using ethylene glycol, RNA G-quadruplexes exhibit lower hydration levels than their DNA counterparts (Arora and Maiti 2009), implying a greater influence of water content in the intracellular environment on RNA G-quadruplexes than DNA G-quadruplexes. There is also a report that dielectric permittivity of solution has a pronounced impact on the stability of DNA G-quadruplexes (Smirnov and Shafer 2007). Other examples include the promotion of docking of two RNA helices in crowded solution of large PEGs (8000 and 1000 Da), driven by the excluded volume effect of PEG (Kilburn et al. 2010; Paudel and Rueda 2014). It is notable that the cooperativity of RNA folding changes in the environment of molecular crowding created by several types of cosolutes (Strulson et al. 2013). Consistent with the effects of cosolute-containing solutions, the condition of spatial confinement that suppresses extended conformations of partially and fully unfolded RNAs stabilizes the RNA pseudoknot derived from the mouse mammary tumor virus (Feng et al. 2021). Experimental studies of DNA structures in confined spaces have also indicated stabilization of DNA G-quadruplexes but destabilization of structures comprising Watson–Crick base pairs, attributed to the decreased water content of solution (Jonchhe et al. 2018, 2020; Yamaguchi et al. 2022). Accordingly, multiple environmental factors influence the stability of RNA tertiary structures, which is more complex than the influence on secondary structures. Overall, the high sensitivity of RNA structural stability to variations in the molecular environment is consistent with the observations in cells that RNA structures exhibit variable stability depending on their intracellular location.

It is important to refer that somewhat ambiguous evaluation of the influences of excluded volume interactions and altered solvent properties arises from preferential binding or exclusion of cosolutes from RNA. Several net-neutral cosolutes, such as urea, trimethylamine *N*-oxide (TMAO), sucrose, glycerol, ethylene glycol, and PEG, at high amounts have been computationally and experimentally shown to bind to RNA and DNA via hydrogen bonding or hydrophobic interactions (Buscaglia et al. 2013; Holmstrom et al.2015; Tateishi-Karimata et al. 2017; Downey et al. 2007; Jaiswal et al. 2018; Trajkovski et al. 2018; Oprzeska-Zingrebe and Smiatek 2018; Jaganade et al. 2019). These direct molecular interactions contribute to the stabilization of RNA tertiary structures, such as the pseudoknot structure of the hairpin ribozyme in solutions containing PEG (8,000 Da) at concentrations of several or several tens of percent (Daher et al. 2018) and the A-minor motif interactions of the group I ribozyme in solutions containing TMAO at concentrations in the range of $1–10^{-1}$ M (Holmstrom et al. 2015). The direct interaction with RNA is more significant for charged cosolutes.

RNA in the Environment with Charged Cosolutes

RNA and DNA have a high potential to bind to positively charged organic molecules. Unlike metal ions, which are regarded as point charges, organic cations cause substantial steric hindrance particularly when their size is rather bulky. Tetraalkylammonium ions are the amphiphilic monovalent cations, and relatively small ones, such as tetramethylammonium ion and tetraethylammonium ion, stabilize RNA and DNA duplexes like metal ions. Conversely, bulky ones, such as tetrabutylammonium ion and tetrapentylammonium ion, at concentrations of $10^{-1}–10^{-2}$ M destabilize rather than stabilize the duplexes, with a weak dependence of the their sequences (Stellwagen et al. 2011; Nakano et al. 2016). Because influences of excluded volume interactions and altered solvent properties are less pronounced at the concentrations examined, direct cation binding is responsible for the duplex destabilization. This destabilization effect of bulky cations is enhanced in crowded solutions of PEG and organic solvents (e.g., large and small PEGs, ethanol, dimethoxyethane, and dimethyl sulfoxide) (Nakano et al. 2016). In contrast, net negatively charged organic anions, such as glutamic acid and glucose-6-phosphate, do not change the stability of dsRNA in the absence of Mg^{2+} but decrease the stability in its presence, indicating the depletion of divalent metal ions available for RNA in media containing organic anions (Nakano et al. 2015).

A protein-rich environment is a more biologically relevant context. Protein mixtures extracted from yeast cells stabilize the RNA pseudoknot structure of the hairpin ribozyme (Daher et al. 2018), and protein mixtures extracted from human serum stabilize DNA G-quadruplexes (Tanabe et al. 2024). Histone proteins and lysine- or arginine-rich polypeptides stabilize the duplex and G-quadruplex structures of DNA, like metal ions. In experiments using model proteins that are not regarded as specific nucleic acid-binding proteins, protein-rich conditions created by globular proteins with net positively charged surfaces (e.g., lysozyme and cytochrome c) decrease the stability of dsDNA (Morimoto et al. 2022), similar to the effects of bulky tetraalkylammonium ions (Nakano et al. 2019a). Conversely, the basic proteins stabilize the duplex and G-quadruplex structures with long loops (Morimoto et al. 2022; Tanabe et al. 2024), again similar to the effects of bulky tetraalkylammonium ions. This stabilization effect on the long loop structures is enhanced in crowded solution of large PEGs and dextrans at concentrations of several tens of percent. Results from the studies using protein solutions are broadly consistent with the observations in cells that RNA–protein complexes are abundant and thermodynamic stability of RNA structures varies across intracellular locations where the distribution of proteins is variable.

Structural Rearrangements of RNA

When the energy levels of distinct RNA folds undergo different changes in the environment of molecular crowding, an RNA structure may be rearranged to another structure to converge to a new equilibrium state. The structural rearrangement is mediated by the exchange of Watson–Crick base-pairing partners and the refolding of secondary structures. The energy barrier for the dissociation of Watson–Crick base pairs is high, resulting in slow rearrangement of RNA structures. In living cells, RNA structural rearrangements during gene transcription and regulation are often driven by protein or ligand binding. Experiments using cosolute-containing solutions at concentrations of several tens of percent have revealed the ability of the environment of molecular crowding to facilitate rearrangement of RNA and DNA

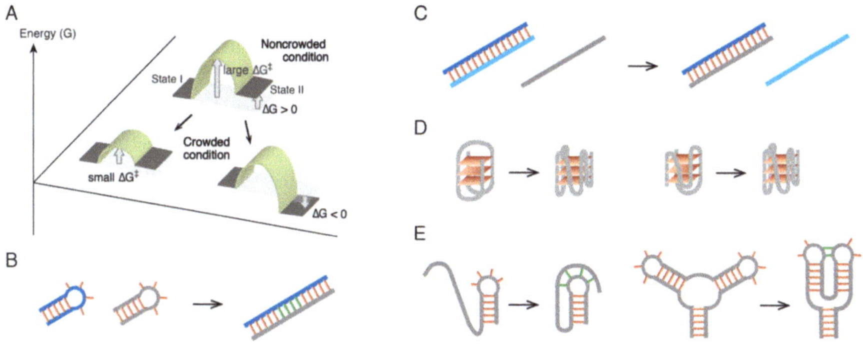

Fig. 7.5 (**a**) Mechanisms for facilitating RNA structural rearrangements from States I to II under molecular crowding conditions. (**b–e**) Structural rearrangements observed for RNA and DNA: (**b**) the association of two complementary hairpins to form an intermolecular duplex, (**c**) strand exchange between a duplex and a single strand, (**d**) changes in the folding topology of G-quadruplexes, and (**e**) formations of tertiary structures via long-range loop–single strand or loop–loop interactions

structures by changing the equilibrium state or the height of the free energy barrier for structural transitions (Fig. 7.5a).

Self-complementary RNA and DNA molecules can form both an intermolecular duplex and an intramolecular hairpin in equilibrium (Fig. 7.5b). The formation of an intermolecular duplex is promoted by increased concentrations of metal ions (e.g., Na^+ and Mg^{2+}), biological polyamines (e.g., putrescine, spermidine, and spermine), and oligolysines (Nakano et al. 2008a). RNA exhibits a higher propensity for the duplex formation compared to DNA with the same sequence, regardless of the structure and charge of cations (Nakano et al. 2007; Nakano et al. 2012). Compared with when using uncrowded solutions, self-complementary oligonucleotides in crowded solutions of neutral cosolutes (large and small PEGs, alcohols, amide compounds, and aprotic compounds) form an intermolecular duplex with a lower amount of cations, mostly independent of their sequence. This is ascribed to the promotion of cation-induced dimerization of RNA and DNA hairpins in media with low dielectric permittivity.

Another example of structural rearrangements is the strand exchange, wherein a base-paired strand exchanges with a single strand (Fig. 7.5c). Studies using RNA and DNA oligonucleotides have indicated that the exchange of Watson–Crick base-pairing partners of a target duplex or hairpin is accelerated in crowded solutions of PEG and organic solvents (e.g., large and small PEGs, diethylene glycol dimethyl ether, dimethoxyethane, methanol, and ethanol) (Dave and Liu 2010; Feng et al. 2015; Zhang et al. 2015; Norden et al. 2021; Wong and Liu 2021). This acceleration is attributed to hydrophobic interactions that facilitate the dissociation of target base pairs. The strand exchange is also accelerated in solutions containing cationic polymers, which bring a target duplex and a single strand into close proximity (Kim et al. 2002). Elucidating the mechanisms of accelerating the exchange of base-pairing partners in the presence of crowders and polycations is important for understanding RNA structural rearrangements occurred in the intracellular environment.

Thermodynamically stable folding topology of G-quadruplexes changes depending on the solution components (Fig. 7.5d). Changing metal ion species from Na^+ to K^+ in solution induces rearrangement of the G-quadruplex of DNA derived from the human telomeric sequence, from an antiparallel to a hybrid conformation (Gray et al. 2009). Similar to the alteration in metal ion condition, the presence of cosolutes (e.g., small PEGs, acetonitrile, and ethanol) changes the folding topology of G-quadruplexes from a hybrid to a parallel conformation. This transition is attributed to the decreased water content of solution and preferential binding of cosolutes to the parallel conformation (Miller et al. 2010; Heddi and Phan 2011; Yu et al. 2012; Buscaglia et al. 2013; Yadav et al. 2022). MD simulations have shown distinct hydration states for different G-quadruplex topologies (Li et al. 2021), and the human telomeric DNA G-quadruplex that forms a hybrid conformation in uncrowded solutions forms a parallel conformation in the crowded solution of ethanol attributed to the decreased water content (Verdian Doghaei et al. 2015). Consequently, the folding topology of G-quadruplexes may be changed by variations in the intracellular environment.

The environment of molecular crowding also has a pronounced impact on the structural rearrangement of riboswitches in response to ligand binding (Fig. 7.5e). The adenine riboswitch, comprising two hairpins and a three-way junction, undergoes long-range loop–loop interactions induced by ligand binding. The ligand binding is stabilized in the presence of large PEGs (12,000 and 35,000 Da) but destabilized in the presence of smaller PEGs, indicating the complexity of physical and chemical interactions of PEG with the riboswitch (Tyrrell et al. 2015). The ligand-induced formation of the pseudoknot structure of the *S*-adenosylmethionine riboswitch from an extended conformation state is promoted in the crowded solution of TMAO, attributed to the excluded volume effect of the crowder (Mcphie et al. 2016). Similarly, enhancement of ligand binding to the lysine riboswitch in the crowded solution of PEG (3000 Da) is attributed to the excluded volume effect of PEG (Sung and Nesbitt 2022). In the study of the flavin mononucleotide riboswitch, ligand binding is enhanced in the crowded solution of a small PEG (200 Da) through the promotion of dehydration upon ligand binding in the medium with low water activity (Rode et al. 2018). These findings indicate that riboswitches in the intracellular environment may exhibit increased sensitivity to their ligands through various mechanisms.

Catalysis of Ribozymes

Ribozymes catalyze chemical reactions on their substrates. The ribozyme catalysis proceeds through several steps: substrate binding, formation of the catalytic core, and catalytic process, followed by the release of reaction products. The catalytic core is formed by rearrangement from the catalytically inactive to the active conformation, often aided by additional metal ions playing crucial roles in the catalytic process and formation of a catalytic core. The catalytic activity of ribozymes is greatly affected by thermodynamic stabilities of the catalytically active conformation and ribozyme–substrate and ribozyme–product complexes, all of which are influenced by the environment of molecular crowding. In crowded solutions that stabilize the active conformation of ribozymes, their catalytic activity remains high even at low concentrations of metal ions. The environment of molecular crowding created by PEGs (1000 and 8000 Da) accelerates the cleavage of substrate RNAs catalyzed by several types of RNA-cleaving ribozymes, including the group I ribozyme consisting of the A-minor motif, hairpin ribozyme comprising two stem-loop domains, human delta virus-like ribozyme comprising two pseudoknot stems, and twister ribozyme comprising two pseudoknot stems (Strulson et al. 2013; Desai et al. 2014; Paudel and Rueda 2014; Panja et al. 2017). These enhanced catalytic activities are attributed to the excluded volume effect of PEG and their weak nonspecific interactions with RNA (Strulson et al. 2013, Desai et al. 2014, Paudel and Rueda 2014, Panja et al. 2017). Moreover, the environment of molecular crowding created by small cosolutes (e.g., PEGs (200 and 400 Da), ethanol, and dimethoxyethane) enhances the catalytic activity of the hairpin and hammerhead ribozymes, attributed to the decreased water content (Herve et al. 2006; Nakano et al. 2014). Prebiotically relevant small molecules, such as ethylene glycol, ribose, and amino acids, have been shown to enhance RNA ligation activity of ligase ribozymes (Dasgupta et al. 2023). Additionally, the catalysis of ribozymes is facilitated by compartmentalization in small vesicles that stabilize compactly folded structures (Strulson et al. 2012; Peng et al. 2022). In contrast, crowded solutions that disrupt the catalytically active conformation suppress the catalysis; however, solutions that moderately decrease base-pair stability can promote the refolding of incorrectly folded RNA molecules to the active conformation and also increase the catalytic turnover of the hammerhead ribozyme (Nakano et al. 2009, 2019b, 2021). In the presence of organic anions, such as acidic amino acids, fatty acids, adenosine phosphate compounds, tRNA, and oligonucleotides, the catalytic activity of the hammerhead ribozyme is suppressed owing to the depletion of free divalent metal ions (Mg^{2+}) in the medium, but this suppression is efficiently recovered in the crowded solution of a large PEG (8000 Da) (Nakano et al. 2015). It is likely that the environment of molecular crowding provides conditions conducive to modulate the catalytic activity of many ribozymes.

5. Conclusion and Future Directions

Recent computational and experimental studies have shown considerable contributions of the intracellular environment to RNA interactions. Compared with proteins, RNAs exhibit higher sensitivity to variations in the molecular environment, which may play a role in regulating biological functions of cellular RNAs. An intriguing aspect is that the effects of molecular crowding on RNA varies depending on interactions of RNA structures. More striking characteristics of the environment of molecular crowding is that the threshold amounts of metal ions and proteins for the binding to RNA are reduced, facilitating the formation and rearrangement of RNA structures. It is noted that RNA refolding and base-pair exchange in response to a change in its surrounding environment occur at much slower rates than crowder binding and the reorganization of water and metal ions bound to RNA. Further examination is warranted to understand the effects of molecular crowding on the kinetics of RNA transitions occurring across different time scales.

The cell-mimicking solutions comprising cellular substances or artificial molecular crowding agents oversimplify the intracellular environment. However, quantitative data obtained using *in vitro* model solutions have increased our ability to predict RNA structures in cells. For example, methods based on the nearest-neighbor interaction model, widely used for predictions in uncrowded solutions, is applicable to predictions in crowded media if the excluded volume effect is negligible, which seems to be valid for short oligonucleotides. In a model crowded solution comprising a small PEG (200 Da), the Gibbs free energy change for the formation of a single Watson–Crick base pair is reduced by 0.3–1 kcal mol^{-1} compared with the values obtained using uncrowded solutions (Adams and Znosko 2019; Ghosh et al. 2020; Ghosh et al. 2023). Other prediction parameters, determined using a solution containing 80% metabolite mixtures of *Escherichia coli*, provide ~0.7 kcal mol^{-1} destabilization compared with the base-pair stability in uncrowded solutions, which improve the prediction accuracy of RNA base pairing formed in the bacterial cell (Sieg et al. 2023). These experimentally obtained data are further useful for machine-learning training, enabling the prediction of cellular RNA structures (Delli Ponti and Tartaglia 2022). These and future *in vitro* studies, together with MD simulations and in-cell measurements, will provide guidelines for understanding RNA structures and their interactions in cells and expand the rational design of oligonucleotides useful for biotechnology.

References

Abels JA, Moreno-Herrero F, Van Der Heijden T, Dekker C, Dekker NH (2005) Single-molecule measurements of the persistence length of double-stranded RNA. Biophys J 88:2737–2744

Adams MS, Znosko BM (2019) Thermodynamic characterization and nearest neighbor parameters for RNA duplexes under molecular crowding conditions. Nucleic Acids Res 47:3658–3666

Alipanahi B, Delong A, Weirauch MT, Frey BJ (2015) Predicting the sequence specificities of DNA- and RNA-binding proteins by deep learning. Nat Biotechnol 33:831–838

Anderson CF, Record MT Jr (1995) Salt-nucleic acid interactions. Annu Rev Phys Chem 46:657–700

Arora A, Maiti S (2009) Differential biophysical behavior of human telomeric RNA and DNA quadruplex. J Phys Chem B 113:10515–10520

Asami K, Hanai T, Koizumi N (1976) Dielectric properties of yeast cells. J Membr Biol 28:169–180

Auffinger P, Westhof E (1997) RNA hydration: three nanoseconds of multiple molecular dynamics simulations of the solvated tRNA(asp) anticodon hairpin. J Mol Biol 269:326–341

Auffinger P, Westhof E (2000) Water and ion binding around RNA and DNA (C,G) oligomers. J Mol Biol 300:1113–1131

Auffinger P, Westhof E (2001) Water and ion binding around r(UpA)$_{12}$ and d(TpA)$_{12}$ oligomers - comparison with RNA and DNA (CpG)$_{12}$ duplexes. J Mol Biol 305:1057–1072

Baek M, Mchugh R, Anishchenko I, Jiang H, Baker D, Dimaio F (2024) Accurate prediction of protein-nucleic acid complexes using RoseTTAFoldNA. Nat Methods 21:117–121

Bao L, Zhang X, Shi YZ, Wu YY, Tan ZJ (2017) Understanding the relative flexibility of RNA and DNA duplexes: stretching and twist-stretch coupling. Biophys J 112:1094–1104

Bevilacqua PC, Ritchey LE, Su Z, Assmann SM (2016) Genome-wide analysis of RNA secondary structure. Annu Rev Genet 50:235–266

Bloomfield VA, Crothers DM, Tinoco I Jr (2000) Nucleic acids: structures, properties and functions. University Science Books Press

Blose JM, Lloyd KP, Bevilacqua PC (2009) Portability of the GN(R)A hairpin loop motif between RNA and DNA. Biochemistry 48:8787–8794

Braga J, Mcnally JG, Carmo-Fonseca M (2007) A reaction-diffusion model to study RNA motion by quantitative fluorescence recovery after photobleaching. Biophys J 92:2694–2703

Brown RF, Andrews CT, Elcock AH (2015) Stacking free energies of all DNA and RNA nucleoside pairs and dinucleoside-monophosphates computed using recently revised AMBER parameters and compared with experiment. J Chem Theory Comput 11:2315–2328

Bugaut A, Balasubramanian S (2012) 5'-UTR RNA G-quadruplexes: translation regulation and targeting. Nucleic Acids Res 40:4727–4741

Buscaglia R, Miller MC, Dean WL, Gray RD, Lane AN, Trent JO, Chaires JB (2013) Polyethylene glycol binding alters human telomere G-quadruplex structure by conformational selection. Nucleic Acids Res 41:7934–7946

Castelli M, Doria F, Freccero M, Colombo G, Moroni E (2022) Studying the dynamics of a complex G-quadruplex system: insights into the comparison of MD and NMR data. J Chem Theory Comput 18:4515–4528

Cheatham TE, Kollman PA (1997) Molecular dynamics simulations highlight the structural differences among DNA:DNA, RNA:RNA, and DNA:RNA hybrid duplexes. J Am Chem Soc 119:4805–4825

Chen H, Meisburger SP, Pabit SA, Sutton JL, Webb WW, Pollack L (2012) Ionic strength-dependent persistence lengths of single-stranded RNA and DNA. Proc Natl Acad Sci USA 109:799–804

Cobb AM, Jackson BR, Kim E, Bond PL, Bowater RP (2013) Sequence-specific and DNA structure-dependent interactions of *Escherichia coli* Muts and human p53 with DNA. Anal Biochem 442:51–61

Collauto A, Von Bulow S, Gophane DB, Saha S, Stelzl LS, Hummer G, Sigurdsson ST, Prisner TF (2020) Compaction of RNA duplexes in the cell. Angew Chem Int Ed Engl 59:23025–23029

Collie GW, Haider SM, Neidle S, Parkinson GN (2010) A crystallographic and modelling study of a human telomeric RNA (TERRA) quadruplex. Nucleic Acids Res 38:5569–5580

Cruz-Leon S, Schwierz N (2022) RNA captures more cations than DNA: insights from molecular dynamics simulations. J Phys Chem B 126:8646–8654

Daher M, Widom JR, Tay W, Walter NG (2018) Soft interactions with model crowders and non-canonical interactions with cellular proteins stabilize RNA folding. J Mol Biol 430:509–523

Dasgupta S, Zhang S, Szostak JW (2023) Molecular crowding facilitates ribozyme-catalyzed RNA assembly. Acs Cent Sci 9:1670–1678

Dave N, Liu J (2010) Fast molecular beacon hybridization in organic solvents with improved target specificity. J Phys Chem B 114:15694–15699

Delli Ponti, R. & Tartaglia, G. G. 2022. Predicting RNA secondary structure using *in vitro* and *in vivo* data. Methods Mol Biol, 2404, 43–52, Springer

Desai R, Kilburn D, Lee HT, Woodson SA (2014) Increased ribozyme activity in crowded solutions. J Biol Chem 289:2972–2977

Ding Y, Tang Y, Kwok CK, Zhang Y, Bevilacqua PC, Assmann SM (2014) *In vivo* genome-wide profiling of RNA secondary structure reveals novel regulatory features. Nature 505:696–700

Dix JA, Verkman AS (1990) Mapping of fluorescence anisotropy in living cells by ratio imaging. Application to cytoplasmic viscosity. Biophys J 57:231–240

Downey CD, Crisman RL, Randolph TW, Pardi A (2007) Influence of hydrostatic pressure and cosolutes on RNA tertiary structure. J Am Chem Soc 129:9290–9291

Drozdetski AV, Tolokh IS, Pollack L, Baker N, Onufriev AV (2016) Opposing effects of multivalent ions on the flexibility of DNA and RNA. Phys Rev Lett 117:028101

Dzatko S, Krafcikova M, Hansel-Hertsch R, Fessl T, Fiala R, Loja T, Krafcik D, Mergny JL, Foldynova-Trantirkova S, Trantirek L (2018) Evaluation of the stability of DNA i-motifs in the nuclei of living mammalian cells. Angew Chem Int Ed Engl 57:2165–2169

Egli M, Portmann S, Usman N (1996) RNA hydration: a detailed look. Biochemistry 35:8489–8494

Egli M, Minasov G, Su L, Rich A (2002) Metal ions and flexibility in a viral RNA pseudoknot at atomic resolution. Proc Natl Acad Sci USA 99:4302–4307

Elder RM, Pfaendtner J, Jayaraman A (2015) Effect of hydrophobic and hydrophilic surfaces on the stability of double-stranded DNA. Biomacromolecules 16:1862–1869

Feig M, Yu I, Wang PH, Nawrocki G, Sugita Y (2017) Crowding in cellular environments at an atomistic level from computer simulations. J Phys Chem B 121:8009–8025

Feng B, Westerlund F, Norden B (2015) Evidence for hydrophobic catalysis of DNA strand exchange. Chem Commun 51:7390–7392

Feng C, Tan YL, Cheng YX, Shi YZ, Tan ZJ (2021) Salt-dependent RNA pseudoknot stability: effect of spatial confinement. Front Mol Biosci 8:666369

Ferreira LA, Uversky VN, Zaslavsky BY (2017) Role of solvent properties of water in crowding effects induced by macromolecular agents and osmolytes. Mol BioSyst 13:2551–2563

Gao M, Gnutt D, Orban A, Appel B, Righetti F, Winter R, Narberhaus F, Muller S, Ebbinghaus S (2016) RNA hairpin folding in the crowded cell. Angew Chem Int Ed Engl 55:3224–3228

Garlid KD (2000) The state of water in biological systems. Int Rev Cytol 192:281–302

Gebala M, Herschlag D (2019) Quantitative studies of an RNA duplex electrostatics by ion counting. Biophys J 117:1116–1124

Gerstberger S, Hafner M, Tuschl T (2014) A census of human RNA-binding proteins. Nat Rev Genet 15:829–845

Ghosh S, Takahashi S, Ohyama T, Endoh T, Tateishi-Karimata H, Sugimoto N (2020) Nearest-neighbor parameters for predicting DNA duplex stability in diverse molecular crowding conditions. Proc Natl Acad Sci USA 117:14194–14201

Ghosh S, Takahashi S, Banerjee D, Ohyama T, Endoh T, Tateishi-Karimata H, Sugimoto N (2023) Nearest-neighbor parameters for the prediction of RNA duplex stability in diverse *in vitro* and cellular-like crowding conditions. Nucleic Acids Res 51:4101–4111

Giassa IC, Rynes J, Fessl T, Foldynova-Trantirkova S, Trantirek L (2018) Advances in the cellular structural biology of nucleic acids. FEBS Lett 592:1997–2011

Gil A, Branchadell V, Bertran J, Oliva A (2009) An analysis of the different behavior of DNA and RNA through the study of the mutual relationship between stacking and hydrogen bonding. J Phys Chem B 113:4907–4914

Goody TA, Melcher SE, Norman DG, Lilley DM (2004) The kink-turn motif in RNA is dimorphic, and metal ion-dependent. RNA 10:254–264

Gray RD, Li J, Chaires JB (2009) Energetics and kinetics of a conformational switch in G-quadruplex DNA. J Phys Chem B 113:2676–2683

Guiset Miserachs H, Donghi D, Borner R, Johannsen S, Sigel RK (2016) Distinct differences in metal ion specificity of RNA and DNA G-quadruplexes. J Biol Inorg Chem 21:975–986

Guo JU, Bartel DP (2016) RNA G-quadruplexes are globally unfolded in eukaryotic cells and depleted in bacteria. Science 353:aaf5371

Haase L, Dickerhoff J, Weisz K (2018) DNA-RNA hybrid quadruplexes reveal interactions that favor RNA parallel topologies. Chemistry 24:15365–15371

Hatzakis E, Okamoto K, Yang D (2010) Thermodynamic stability and folding kinetics of the major G-quadruplex and its loop isomers formed in the nuclease hypersensitive element in the human c-Myc promoter: effect of loops and flanking segments on the stability of parallel-stranded intramolecular G-quadruplexes. Biochemistry 49:9152–9160

Havrila M, Stadlbauer P, Kuhrova P, Banas P, Mergny JL, Otyepka M, Sponer J (2018) Structural dynamics of propeller loop: towards folding of RNA G-quadruplex. Nucleic Acids Res 46:8754–8771

Heddi B, Phan AT (2011) Structure of human telomeric DNA in crowded solution. J Am Chem Soc 133:9824–9833

Herrero-Galan E, Fuentes-Perez ME, Carrasco C, Valpuesta JM, Carrascosa JL, Moreno-Herrero F, Arias-Gonzalez JR (2013) Mechanical identities of RNA and DNA double helices unveiled at the single-molecule level. J Am Chem Soc 135:122–131

Herve G, Tobe S, Heams T, Vergne J, Maurel MC (2006) Hydrostatic and osmotic pressure study of the hairpin ribozyme. Biochim Biophys Acta 1764:573–577

Huang ZL, Dai J, Luo WH, Wang XG, Tan JH, Chen SB, Huang ZS (2018) Identification of G-quadruplex-binding protein from the exploration of RGG motif/G-quadruplex interactions. J Am Chem Soc 140:17945–17955

Islam B, Stadlbauer P, Krepl M, Havrila M, Haider S, Sponer J (2018) Structural dynamics of lateral and diagonal loops of human telomeric G-quadruplexes in extended MD simulations. J Chem Theory Comput 14:5011–5026

Jaganade T, Chattopadhyay A, Pazhayam NM, Priyakumar UD (2019) Energetic, structural and dynamic properties of nucleobase-urea interactions that aid in urea assisted RNA unfolding. Sci Rep 9:8805

Jaiswal AK, Srivastava R, Pandey P, Bandyopadhyay P (2018) Microscopic picture of water-ethylene glycol interaction near a model DNA by computer simulation: concentration dependence, structure, and localized thermodynamics. PLoS One 13:e0206359

Jankowsky E, Harris ME (2015) Specificity and nonspecificity in RNA-protein interactions. Nat Rev Mol Cell Biol 16:533–544

Jarmoskaite I, Denny SK, Vaidyanathan PP, Becker WR, Andreasson JOL, Layton CJ, Kappel K, Shivashankar V, Sreenivasan R, Das R, Greenleaf WJ, Herschlag D (2019) A quantitative and predictive model for RNA binding by human Pumilio proteins. Mol Cell 74:E18–981.e18

Jin L, Shi YZ, Feng CJ, Tan YL, Tan ZJ (2018) Modeling structure, stability, and flexibility of double-stranded RNAs in salt solutions. Biophys J 115:1403–1416

Joachimi A, Benz A, Hartig JS (2009) A comparison of DNA and RNA quadruplex structures and stabilities. Bioorg Med Chem 17:6811–6815

Jonchhe S, Pandey S, Emura T, Hidaka K, Hossain MA, Shrestha P, Sugiyama H, Endo M, Mao H (2018) Decreased water activity in nanoconfinement contributes to the folding of G-quadruplex and i-motif structures. Proc Natl Acad Sci USA 115:9539–9544

Jonchhe S, Pandey S, Karna D, Pokhrel P, Cui Y, Mishra S, Sugiyama H, Endo M, Mao H (2020) Duplex DNA is weakened in nanoconfinement. J Am Chem Soc 142:10042–10049

Kankia B (2019) Stability factors of the parallel quadruplexes: DNA versus RNA. J Phys Chem B 123:1060–1067

Karimata H, Nakano S, Sugimoto N (2007) Effects of polyethylene glycol on DNA duplex stability at different NaCl concentrations. Bull Chem Soc Jpn 80:1987–1994

Kawaguchi, R. K. & Kiryu, H. 2023. Genome-wide RNA secondary structure prediction. Methods Mol Biol, 2586, 35–48, Springer

Kilburn D, Roh JH, Guo L, Briber RM, Woodson SA (2010) Molecular crowding stabilizes folded RNA structure by the excluded volume effect. J Am Chem Soc 132:8690–8696

Kim WJ, Akaike T, Maruyama A (2002) DNA strand exchange stimulated by spontaneous complex formation with cationic comb-type copolymer. J Am Chem Soc 124:12676–12677

Kolev SK, Petkov PS, Rangelov MA, Trifonov DV, Milenov TI, Vayssilov GN (2018) Interaction of Na+, K>+, Mg2+ and Ca2+ counter cations with RNA. Metallomics 10:659–678

Kuimova MK, Botchway SW, Parker AW, Balaz M, Collins HA, Anderson HL, Suhling K, Ogilby PR (2009) Imaging intracellular viscosity of a single cell during photoinduced cell death. Nat Chem 1:69–73

Kuznetsov SV, Ren CC, Woodson SA, Ansari A (2008) Loop dependence of the stability and dynamics of nucleic acid hairpins. Nucleic Acids Res 36:1098–1112

Laage D, Elsaesser T, Hynes JT (2017) Water dynamics in the hydration shells of biomolecules. Chem Rev 117:10694–10725

Law MJ, Rice AJ, Lin P, Laird-Offringa IA (2006) The role of RNA structure in the interaction of U1A protein with U1 hairpin II RNA. RNA 12:1168–1178

Leontis NB, Stombaugh J, Westhof E (2002) The non-Watson-Crick base pairs and their associated isostericity matrices. Nucleic Acids Res 30:3497–3531

Li K, Yatsunyk L, Neidle S (2021) Water spines and networks in G-quadruplex structures. Nucleic Acids Res 49:519–528

Lipfert J, Doniach S, Das R, Herschlag D (2014) Understanding nucleic acid-ion interactions. Annu Rev Biochem 83:813–841

Liu T, Yu T, Zhang S, Wang Y, Zhang W (2021) Thermodynamic and kinetic properties of a single base pair in A-DNA and B-DNA. Phys Rev E 103:042409

Lombardi EP, Londono-Vallejo A (2020) A guide to computational methods for G-quadruplex prediction. Nucleic Acids Res 48:1-15

Lomzov AA, Vorobjev YN, Pyshnyi DV (2015) Evaluation of the Gibbs free energy changes and melting temperatures of DNA/DNA duplexes using hybridization enthalpy calculated by molecular dynamics simulation. J Phys Chem B 119:15221–15234

Luby-Phelps K, Mujumdar S, Mujumdar RB, Ernst LA, Galbraith W, Waggoner AS (1993) A novel fluorescence ratiometric method confirms the low solvent viscosity of the cytoplasm. Biophys J 65:236–242

Majumder S, Xue J, Demott CM, Reverdatto S, Burz DS, Shekhtman A (2015) Probing protein quinary interactions by in-cell nuclear magnetic resonance spectroscopy. Biochemistry 54:2727–2738

Malgowska M, Czajczynska K, Gudanis D, Tworak A, Gdaniec Z (2016) Overview of the RNA G-quadruplex structures. Acta Biochim Pol 63:609–621

Marin-Gonzalez A, Vilhena JG, Perez R, Moreno-Herrero F (2021) A molecular view of DNA flexibility. Q Rev Biophys 54:1–20

Masliah G, Barraud P, Allain FH (2013) RNA recognition by double-stranded RNA binding domains: a matter of shape and sequence. Cell Mol Life Sci 70:1875–1895

Matsumoto S, Tateishi-Karimata H, Takahashi S, Ohyama T, Sugimoto N (2020) Effect of molecular crowding on the stability of RNA G-quadruplexes with various numbers of quartets and lengths of loops. Biochemistry 59:2640–2649

Mcdonald AR, Denning EJ, Mackerell AD Jr (2013) Impact of geometry optimization on base-base stacking interaction energies in the canonical A- and B-forms of DNA. J Phys Chem A 117:1560–1568

Mcphie P, Brown P, Chen B, Dayie TK, Minton AP (2016) Modulation of conformational equilibria in the *S*-adenosylmethionine (SAM) II riboswitch by SAM, Mg2+, and trimethylamine *N*-oxide. Biochemistry 55:5010–5020

Meier-Stephenson V (2022) G4-quadruplex-binding proteins: review and insights into selectivity. Biophys Rev 14:635–654

Michiels PJ, Versleijen AA, Verlaan PW, Pleij CW, Hilbers CW, Heus HA (2001) Solution structure of the pseudoknot of SRV-1 RNA, involved in ribosomal frameshifting. J Mol Biol 310:1109–1123

Miller MC, Buscaglia R, Chaires JB, Lane AN, Trent JO (2010) Hydration is a major determinant of the G-quadruplex stability and conformation of the human telomere 3' sequence of $d(AG_3(TTAG_3)_3)$. J Am Chem Soc 132:17105–17107

Milo R, Phillips R, Orme N (2016) Cell biology by the numbers. Garland Science

Minton, A. P. 1998. Molecular crowding: analysis of effects of high concentrations of inert cosolutes on biochemical equilibria and rates in terms of volume exclusion. Methods Enzymol, 295, 127–149, Elsevier

Misra VK, Draper DE (1998) On the role of magnesium ions in RNA stability. Biopolymers 48:113–135

Misteli T (2008) Physiological importance of RNA and protein mobility in the cell nucleus. Histochem Cell Biol 129:5–11
Miyoshi D, Karimata H, Sugimoto N (2006) Hydration regulates thermodynamics of G-quadruplex formation under molecular crowding conditions. J Am Chem Soc 128:7957–7963
Moghaddam S, Caliskan G, Chauhan S, Hyeon C, Briber RM, Thirumalai D, Woodson SA (2009) Metal ion dependence of cooperative collapse transitions in RNA. J Mol Biol 393:753–764
Morimoto R, Horita M, Yamaguchi D, Nakai H, Nakano SI (2022) Evaluation of weak interactions of proteins and organic cations with DNA duplex structures. Biophys J 121:2873–2881
Mustoe AM, Busan S, Rice GM, Hajdin CE, Peterson BK, Ruda VM, Kubica N, Nutiu R, Baryza JL, Weeks KM (2018) Pervasive regulatory functions of mRNA structure revealed by high-resolution shape probing. Cell 173:E18–195.e18
Mustoe AM, Lama NN, Irving PS, Olson SW, Weeks KM (2019) RNA base-pairing complexity in living cells visualized by correlated chemical probing. Proc Natl Acad Sci USA 116:24574–24582
Muthukumar M (2017) 50th anniversary perspective: a perspective on polyelectrolyte solutions. Macromolecules 50:9528–9560
Nakano S, Sugimoto N (2016) The structural stability and catalytic activity of DNA and RNA oligonucleotides in the presence of organic solvents. Biophys Rev 8:11–23
Nakano S, Kirihata T, Fujii S, Sakai H, Kuwahara M, Sawai H, Sugimoto N (2007) Influence of cationic molecules on the hairpin to duplex equilibria of self-complementary DNA and RNA oligonucleotides. Nucleic Acids Res 35:486–494
Nakano S, Kirihata T, Sugimoto N (2008a) Capture of cationic ligands bound diffusely to base pairs during the DNA refolding. Chem Commun:700–702
Nakano S, Wu L, Oka H, Karimata HT, Kirihata T, Fujii S, Sakai H, Kuwahara M, Sawai H, Sugimoto N (2008b) Conformation and the sodium ion condensation on DNA and RNA structures in the presence of a neutral cosolute as a mimic of the intracellular media. Mol BioSyst 4:579–588
Nakano S, Karimata HT, Kitagawa Y, Sugimoto N (2009) Facilitation of RNA enzyme activity in the molecular crowding media of cosolutes. J Am Chem Soc 131:16881–16888
Nakano S, Hirayama H, Miyoshi D, Sugimoto N (2012) Dimerization of nucleic acid hairpins in the conditions caused by neutral cosolutes. J Phys Chem B 116:7406–7415
Nakano S, Kitagawa Y, Miyoshi D, Sugimoto N (2014) Hammerhead ribozyme activity and oligonucleotide duplex stability in mixed solutions of water and organic compounds. FEBS Open Bio 4:643–650
Nakano S, Kitagawa Y, Miyoshi D, Sugimoto N (2015) Effects of background anionic compounds on the activity of the hammerhead ribozyme in Mg^{2+}-unsaturated solutions. J Biol Inorg Chem 20:1049–1058
Nakano S, Tanino Y, Hirayama H, Sugimoto N (2016) Thermal stability of RNA structures with bulky cations in mixed aqueous solutions. Biophys J 111:1350–1360
Nakano S, Ayusawa T, Tanino Y, Sugimoto N (2019a) Stabilization of DNA loop structures by large cations. J Phys Chem B 123:7687–7694
Nakano S, Yamashita H, Tanabe K, Sugimoto N (2019b) Bulky cations greatly increase the turnover of a native hammerhead ribozyme. RSC Adv 9:35820–35824
Nakano S, Yamashita H, Sugimoto N (2021) Enhancement of the catalytic activity of hammerhead ribozymes by organic cations. ChemBioChem 22:2721–2728
Nguyen HT, Hori N, Thirumalai D (2019) Theory and simulations for RNA folding in mixtures of monovalent and divalent cations. Proc Natl Acad Sci USA 116:21022–21030
Norden B, Brown T, Feng B (2021) Mismatch detection in homologous strand exchange amplified by hydrophobic effects. Biopolymers 112:e23426
Noy A, Perez A, Lankas F, Javier Luque F, Orozco M (2004) Relative flexibility of DNA and RNA: a molecular dynamics study. J Mol Biol 343:627–638
Oprzeska-Zingrebe EA, Smiatek J (2018) Preferential binding of urea to single-stranded DNA structures: a molecular dynamics study. Biophys J 114:1551–1562

Pal A, Levy Y (2019) Structure, stability and specificity of the binding of ssDNA and ssRNA with proteins. PLoS Comput Biol 15:e1006768
Paladino A, Zangi R (2013) Propensities for loop structures of RNA & DNA backbones. Biophys Chem 180-181:110–118
Pan Y, Mackerell AD (2003) Altered structural fluctuations in duplex RNA versus DNA: a conformational switch involving base pair opening. Nucleic Acids Res 31:7131–7140
Panja S, Hua B, Zegarra D, Ha T, Woodson SA (2017) Metals induce transient folding and activation of the twister ribozyme. Nat Chem Biol 13:1109–1114
Parker TM, Hohenstein EG, Parrish RM, Hud NV, Sherrill CD (2013) Quantum-mechanical analysis of the energetic contributions to stacking in nucleic acids versus rise, twist, and slide. J Am Chem Soc 135:1306–1316
Parsegian VA, Rau DC (1984) Water near intracellular surfaces. J Cell Biol 99:196–200
Pasi M, Maddocks JH, Lavery R (2015) Analyzing ion distributions around DNA: sequence-dependence of potassium ion distributions from microsecond molecular dynamics. Nucleic Acids Res 43:2412–2423
Paudel BP, Rueda D (2014) Molecular crowding accelerates ribozyme docking and catalysis. J Am Chem Soc 136:16700–16703
Pederson T (1999) Movement and localization of RNA in the cell nucleus. FASEB J 13(Suppl 2):S238–S242
Peng H, Lelievre A, Landenfeld K, Muller S, Chen IA (2022) Vesicle encapsulation stabilizes intermolecular association and structure formation of functional RNA and DNA. Curr Biol 32:e6
Perez A, Noy A, Lankas F, Luque FJ, Orozco M (2004) The relative flexibility of B-DNA and A-RNA duplexes: database analysis. Nucleic Acids Res 32:6144–6151
Phan AT, Kuryavyi V, Darnell JC, Serganov A, Majumdar A, Ilin S, Raslin T, Polonskaia A, Chen C, Clain D, Darnell RB, Patel DJ (2011) Structure-function studies of FMRP RGG peptide recognition of an RNA duplex-quadruplex junction. Nat Struct Mol Biol 18:796–804
Plumridge A, Andresen K, Pollack L (2020) Visualizing disordered single-stranded RNA: connecting sequence, structure, and electrostatics. J Am Chem Soc 142:109–119
Pramanik S, Nagatoishi S, Saxena S, Bhattacharyya J, Sugimoto N (2011) Conformational flexibility influences degree of hydration of nucleic acid hybrids. J Phys Chem B 115:13862–13872
Ramskold D, Wang ET, Burge CB, Sandberg R (2009) An abundance of ubiquitously expressed genes revealed by tissue transcriptome sequence data. PLoS Comput Biol 5:e1000598
Rangadurai A, Zhou H, Merriman DK, Meiser N, Liu B, Shi H, Szymanski ES, Al-Hashimi HM (2018) Why are Hoogsteen base pairs energetically disfavored in A-RNA compared to B-DNA? Nucleic Acids Res 46:11099–11114
Rau M, Stump WT, Hall KB (2012) Intrinsic flexibility of snRNA hairpin loops facilitates protein binding. RNA 18:1984–1995
Razga F, Koca J, Sponer J, Leontis NB (2005) Hinge-like motions in RNA kink-turns: the role of the second A-minor motif and nominally unpaired bases. Biophys J 88:3466–3485
Reiling-Steffensmeier C, Marky LA (2016) A comparative investigation of the stability of DNA and RNA pseudoknots. Biophys J 110:363a–364a
Reuter JS, Mathews DH (2010) RNAstructure: software for RNA secondary structure prediction and analysis. BMC Bioinform 11:129
Rode AB, Endoh T, Sugimoto N (2018) Crowding shifts the FMN recognition mechanism of riboswitch aptamer from conformational selection to induced fit. Angew Chem Int Ed Engl 57:6868–6872
Rouskin S, Zubradt M, Washietl S, Kellis M, Weissman JS (2014) Genome-wide probing of RNA structure reveals active unfolding of mRNA structures *in vivo*. Nature 505:701–705
Rozners E, Moulder J (2004) Hydration of short DNA, RNA and 2'-OMe oligonucleotides determined by osmotic stressing. Nucleic Acids Res 32:248–254
Sacca B, Lacroix L, Mergny JL (2005) The effect of chemical modifications on the thermal stability of different G-quadruplex-forming oligonucleotides. Nucleic Acids Res 33:1182–1192

Sasmal DK, Ghosh S, Das AK, Bhattacharyya K (2013) Solvation dynamics of biological water in a single live cell under a confocal microscope. Langmuir 29:2289–2298

Sato K, Kato Y (2022) Prediction of RNA secondary structure including pseudoknots for long sequences. Brief Bioinform 23:bbab395

Sato H, Das S, Singer RH, Vera M (2020) Imaging of DNA and RNA in living eukaryotic cells to reveal spatiotemporal dynamics of gene expression. Annu Rev Biochem 89:159–187

Sato K, Akiyama M, Sakakibara Y (2021) RNA secondary structure prediction using deep learning with thermodynamic integration. Nat Commun 12:941

Semmeg A, Del Galdo S, Chiarini M, Daidone I, Casieri C (2024) Macromolecular vs molecular crowding in aqueous solutions: a comparative study of PEG400 and ethylene glycol. J Mol Liquid 394:123713

Shanker S, Bandyopadhyay P (2017) How Mg2+ ion and water network affect the stability and structure of non-Watson-Crick base pairs in *E. coli* loop E of 5S rRNA: a molecular dynamics and reference interaction site model (RISM) study. J Biomol Struct Dyn 35:2103–2122

Sharp KA (2015) Analysis of the size dependence of macromolecular crowding shows that smaller is better. Proc Natl Acad Sci USA 112:7990–7995

Shi YZ, Jin L, Wang FH, Zhu XL, Tan ZJ (2015) Predicting 3D structure, flexibility, and stability of RNA hairpins in monovalent and divalent ion solutions. Biophys J 109:2654–2665

Shi YZ, Jin L, Feng CJ, Tan YL, Tan ZJ (2018) Predicting 3D structure and stability of RNA pseudoknots in monovalent and divalent ion solutions. PLoS Comput Biol 14:e1006222

Sieg JP, Jolley EA, Huot MJ, Babitzke P, Bevilacqua PC (2023) *In vivo*-like nearest neighbor parameters improve prediction of fractional RNA base-pairing in cells. Nucleic Acids Res 51:11298–11317

Singh J, Hanson J, Paliwal K, Zhou Y (2019) RNA secondary structure prediction using an ensemble of two-dimensional deep neural networks and transfer learning. Nat Commun 10:5407

Smirnov IV, Shafer RH (2007) Electrostatics dominate quadruplex stability. Biopolymers 85:91–101

Soto AM, Misra V, Draper DE (2007) Tertiary structure of an RNA pseudoknot is stabilized by "diffuse" Mg2+ ions. Biochemistry 46:2973–2983

Spitale RC, Incarnato D (2023) Probing the dynamic RNA structurome and its functions. Nat Rev Genet 24:178–196

Spitale RC, Flynn RA, Zhang QC, Crisalli P, Lee B, Jung JW, Kuchelmeister HY, Batista PJ, Torre EA, Kool ET, Chang HY (2015) Structural imprints *in vivo* decode RNA regulatory mechanisms. Nature 519:486–490

Šponer J, Bussi G, Krepl M, Banas P, Bottaro S, Cunha RA, Gil-Ley A, Pinamonti G, Poblete S, Jurecka P, Walter NG, Otyepka M (2018) RNA structural dynamics as captured by molecular simulations: a comprehensive overview. Chem Rev 118:4177–4338

Stellwagen E, Muse JM, Stellwagen NC (2011) Monovalent cation size and DNA conformational stability. Biochemistry 50:3084–3094

Strulson CA, Molden RC, Keating CD, Bevilacqua PC (2012) RNA catalysis through compartmentalization. Nat Chem 4:941–946

Strulson CA, Yennawar NH, Rambo RP, Bevilacqua PC (2013) Molecular crowding favors reactivity of a human ribozyme under physiological ionic conditions. Biochemistry 52:8187–8197

Sun LZ, Zhang D, Chen SJ (2017) Theory and modeling of RNA structure and interactions with metal ions and small molecules. Annu Rev Biophys 46:227–246

Sun L, Fazal FM, Li P, Broughton JP, Lee B, Tang L, Huang W, Kool ET, Chang HY, Zhang QC (2019) RNA structure maps across mammalian cellular compartments. Nat Struct Mol Biol 26:322–330

Sung HL, Nesbitt DJ (2022) Synergism in the molecular crowding of ligand-induced riboswitch folding: kinetic/thermodynamic insights from single-molecule spectroscopy. J Phys Chem B 126:6419–6427

Svozil D, Hobza P, Sponer J (2010) Comparison of intrinsic stacking energies of ten unique dinucleotide steps in A-RNA and B-DNA duplexes. Can we determine correct order of stability by quantum-chemical calculations? J Phys Chem B 114:1191–1203

Swadling JB, Ishii K, Tahara T, Kitao A (2018) Origins of biological function in DNA and RNA hairpin loop motifs from replica exchange molecular dynamics simulation. Phys Chem Chem Phys 20:2990–3001
Tan ZJ, Chen SJ (2008) Salt dependence of nucleic acid hairpin stability. Biophys J 95:738–752
Tan ZJ, Chen SJ (2011) Importance of diffuse metal ion binding to RNA. Met Ions Life Sci 9:101–124
Tanabe K, Miyazaki K, Umeno H, Takemoto M, Nakano S (2024) Basic protein- and peptide-induced stabilization of long-loop DNA G-quadruplexes. Biochimie 219:110–117
Tateishi-Karimata H, Ohyama T, Muraoka T, Podbevsek P, Wawro AM, Tanaka S, Nakano SI, Kinbara K, Plavec J, Sugimoto N (2017) Newly characterized interaction stabilizes DNA structure: oligoethylene glycols stabilize G-quadruplexes CH- interactions. Nucleic Acids Res 45:7021–7030
Theillet FX, Binolfi A, Frembgen-Kesner T, Hingorani K, Sarkar M, Kyne C, Li C, Crowley PB, Gierasch L, Pielak GJ, Elcock AH, Gershenson A, Selenko P (2014) Physicochemical properties of cells and their effects on intrinsically disordered proteins (IDPs). Chem Rev 114:6661–6714
Townshend RJL, Eismann S, Watkins AM, Rangan R, Karelina M, Das R, Dror RO (2021) Geometric deep learning of RNA structure. Science 373:1047–1051
Trajkovski M, Endoh T, Tateishi-Karimata H, Ohyama T, Tanaka S, Plavec J, Sugimoto N (2018) Pursuing origins of (poly)ethylene glycol-induced G-quadruplex structural modulations. Nucleic Acids Res 46:4301–4315
Trovato F, Fumagalli G (2017) Molecular simulations of cellular processes. Biophys Rev 9:941–958
Tyrrell J, Weeks KM, Pielak GJ (2015) Challenge of mimicking the influences of the cellular environment on RNA structure by PEG-induced macromolecular crowding. Biochemistry 54:6447–6453
Urbanek MO, Galka-Marciniak P, Olejniczak M, Krzyzosiak WJ (2014) RNA imaging in living cells - methods and applications. RNA Biol 11:1083–1095
Verdian Doghaei A, Housaindokht MR, Bozorgmehr MR (2015) Molecular crowding effects on conformation and stability of G-quadruplex DNA structure: insights from molecular dynamics simulation. J Theor Biol 364:103–112
Verma PK, Kundu A, Ha JH, Cho M (2016) Water dynamics in cytoplasm-like crowded environment correlates with the conformational transition of the macromolecular crowder. J Am Chem Soc 138:16081–16088
Verma PK, Kundu A, Cho M (2018) How molecular crowding differs from macromolecular crowding: a femtosecond mid-infrared pump-probe study. J Phys Chem Lett 9:6584–6592
Vukovic L, Koh HR, Myong S, Schulten K (2014) Substrate recognition and specificity of double-stranded RNA binding proteins. Biochemistry 53:3457–3466
Wang W, Foley K, Shan X, Wang S, Eaton S, Nagaraj VJ, Wiktor P, Patel U, Tao N (2011) Single cells and intracellular processes studied by a plasmonic-based electrochemical impedance microscopy. Nat Chem 3:249–255
Wong KL, Liu J (2021) Factors and methods to modulate DNA hybridization kinetics. Biotechnol J 16:e2000338
Woodson SA (2010) Compact intermediates in RNA folding. Annu Rev Biophys 39:61–77
Wu B, Eliscovich C, Yoon YJ, Singer RH (2016) Translation dynamics of single mRNAs in live cells and neurons. Science 352:1430–1435
Xin Y, Laing C, Leontis NB, Schlick T (2008) Annotation of tertiary interactions in RNA structures reveals variations and correlations. RNA 14:2465–2477
Yadav K, Sardana D, Shweta H, Clovis NS, Sen S (2022) Molecular picture of the effect of cosolvent crowding on ligand binding and dispersed solvation dynamics in G-quadruplex DNA. J Phys Chem B 126:1668–1681
Yamaguchi D, Yoshida M, Nakano S (2022) Evaluation of thermal stability of DNA oligonucleotide structures embedded in hydrogels. DNA 2:302–313

Yamaoki Y, Kiyoishi A, Miyake M, Kano F, Murata M, Nagata T, Katahira M (2018) The first successful observation of in-cell NMR signals of DNA and RNA in living human cells. Phys Chem Chem Phys 20:2982–2985

Yamaoki Y, Nagata T, Sakamoto T, Katahira M (2020) Recent progress of in-cell NMR of nucleic acids in living human cells. Biophys Rev 12:411–417

Yamaoki Y, Nagata T, Kondo K, Sakamoto T, Takami S, Katahira M (2022) Shedding light on the base-pair opening dynamics of nucleic acids in living human cells. Nat Commun 13:7143

Yang R, Zhu T, Xu J, Zhao Y, Kuang Y, Sun M, Chen Y, He W, Wang Z, Jiang T, Zhang H, Wei M (2023) Organic fluorescent probes for monitoring micro-environments in living cells and tissues. Molecules 28:3455

Yildirim A, Brenner N, Sutherland R, Feig M (2018) Role of protein interactions in stabilizing canonical DNA features in simulations of DNA in crowded environments. BMC Biophys 11:8

Yoo H, Davis CM (2022) An *in vitro* cytomimetic of in-cell RNA folding. ChemBioChem 23:e202200406

You X, Shirley JC, Lee E, Baiz CR (2021) Short- and long-range crowding effects on water's hydrogen bond networks. Cell Rep Phys Sci 2:100419

Yu T, Chen SJ (2018) Hexahydrated Mg2+ binding and outer-shell dehydration on RNA surface. Biophys J 114:1274–1284

Yu H, Gu X, Nakano S, Miyoshi D, Sugimoto N (2012) Beads-on-a-string structure of long telomeric DNAs under molecular crowding conditions. J Am Chem Soc 134:20060–20069

Yu I, Mori T, Ando T, Harada R, Jung J, Sugita Y, Feig M (2016a) Biomolecular interactions modulate macromolecular structure and dynamics in atomistic model of a bacterial cytoplasm. eLife 5:e19274

Yu T, Zhu Y, He Z, Chen SJ (2016b) Predicting molecular crowding effects in ion-RNA interactions. J Phys Chem B 120:8837–5544

Yuann JM, Tseng WH, Lin HY, Hou MH (2012) The effects of loop size on Sac7d-hairpin DNA interactions. Biochim Biophys Acta 1824:1009–1015

Zaccaria F, Fonseca Guerra C (2018) RNA versus DNA G-quadruplex: the origin of increased stability. Chemistry 24:16315–16322

Zhang T, Shang CL, Duan RX, Hakeem A, Zhang ZY, Lou XD, Xia F (2015) Polar organic solvents accelerate the rate of DNA strand replacement reaction. Analyst 140:2023–2028

Zhao J, Qian X, Yeung PY, Zhang QC, Kwok CK (2019) Mapping *in vivo* RNA structures and interactions. Trends Biochem Sci 44:555–556

Zhao Q, Zhao Z, Fan X, Yuan Z, Mao Q, Yao Y (2021) Review of machine learning methods for RNA secondary structure prediction. PLoS Comput Biol 17:e1009291

Zuber J, Schroeder SJ, Sun H, Turner DH, Mathews DH (2022) Nearest neighbor rules for RNA helix folding thermodynamics: improved end effects. Nucleic Acids Res 50:5251–5262

Zuker M, Mathews DH, Turner DH (1999) Algorithms and thermodynamics for RNA secondary structure prediction: a practical guide. Springer

Chapter 8
Understanding the Effect of Macromolecular Crowding on Protein Misfolding and Aggregation

Aabgeena Naeem and Gufran Ahmed Siddiqui

Abstract Biological molecules evolved and function within cellular environments that are extremely congested by the presence of other macromolecules. A characteristic of the intracellular environment is that it contains the high concentration of macromolecules, the consequences of crowding inside cells. Interior cell spaces are termed crowed rather than concentrated, because no single macromolecular species is present at high concentration; instead, a diverse array of macromolecules collectively occupies a substantial fraction of the cell's volume. In cellular interior, 20–30% volume is occupied by different macromolecules whose global content is approximated to be in a range of 80-400 mg/ml. Crowding is not only applicable to the intracellular environment but also to the extracellular environment of tissues, such as cartilage and blood plasma that contain 80 g/l of protein, a concentration that is sufficiently high to cause substantial macromolecular crowding effects. The effect of macromolecular crowding becomes more prominent during plasmolysis, when there is decrease in hydration of the cells; i.e., loss of water from cells. Ageing cells, e.g. brain and liver cells, lose water content resulting in the increase level of their crowdedness. In this chapter, we propose that with advancing age, the pathogenesis of protein disorder increases as a result of an increase in the effective concentration of amyloidogenic proteins.

Keywords Protein misfolding · Protein aggregation · Macromolecular crowding · Protein function · Protein-protein interactions

---

A. Naeem (✉) · G. A. Siddiqui
Department of Biochemistry, Faculty of Life Sciences, Aligarh Muslim University, Aligarh, Uttar Pradesh, India
e-mail: anaeem.bc@amu.ac.in

V. N. Uversky (ed.), *(Macro)Molecular Crowding*, Subcellular Biochemistry 109,
https://doi.org/10.1007/978-3-032-03370-3_8

Introduction

Proteins, being the complex biological macromolecules, are required for the proper functioning of a cell. Ordered protein needs to be folded into specific native 3D conformation to perform its intended task (LaPelusa and Kaushik 2020). The process of protein folding is highly controlled and regulated with the help of a complex cellular machinery that includes molecular chaperons among other players. However, at some instances, an ordered protein fails to attain or maintain its correct native structure due to the specific environmental or cellular factors, such as mutations, changes in pH and temperature, or interactions with other molecules. This results in accumulation of the misfolded non-functional protein conformers (Balch et al. 2008; Soto 2003).

Protein Misfolding and Aggregation

For a protein to function properly it should be folded to their native conformations in less than a few seconds. As the protein folds, its free energy decreases and conformational stability increases. This ensures the proper functioning of the cell. Cells eliminate misfiled proteins through a sophisticated quality control system which includes numerous intracellular proteases, most notably the proteasome and molecular chaperones that assist in proper protein folding. Together, these components maintain protein homeostasis by identifying, refolding, or degrading misfolded proteins. Any failure in the protein quality control system results in protein misfolding and aggregation. Aggregation of proteins is enhanced with the hydrophobic core of the protein is exposed to the outer environment and forms hydrophobic interactions between the protein molecules. Aggregated protein molecules have low net charge with high hydrophobicity and β-sheet structure. Misfolded protein confomers also called toxic conformers have a tendency to form aggregates. These aggregates can assume different forms ranging from small oligomers to larger unbranched fibrillar structures known as amyloid fibrils (Louros et al. 2023). Protein fibrillation is a complex multi-step process consisting of several distinct phases, such as nucleation, growth/elongation, and saturation. The aggregates formed by the misfolded proteins can have detrimental effects on the cell. They can disrupt normal cellular processes, interfere with the function of organelles, and form toxic entities that damage cells. Accumulation of misfolded proteins cause diseases called amyloid diseases or amyloidoses, or proteinopathies (protein conformational disorders or protein misfolding diseases). Often, aggregated proteins are implicated in numerous neurodegenerative diseases, such as Alzheimer's, Parkinson's, Huntington's disease, and type 2 diabetes mellitus, where specific proteins misfold and aggregate leading to the development of the characteristic disease pathologies (Nechushtai et al. 2023; Jucker and Walker 2023).

Macromolecular Crowding: A Primer

In a biological context, macromolecular crowding refers to the dense and complex environment found naturally inside a cellular compartment due to the high concentration of various biological macromolecules, such as proteins, nucleic acids, polysaccharides, and others. These biomolecules occupy space and create an environment, where molecular interactions occur within the reduced free volume (Siddiqui and Naeem 2023). This crowding effect has many consequences:

- Macromolecular crowding can influence the association and dissociation rates of molecular interactions, potentially impacting the specificity and efficiency of biological processes.
- It can also affect the functionality and dynamics of biomolecules, altering their folding pathways, rates of diffusion and stability.

Key Aspects of Macromolecular Crowding

- **Cellular compartmentalization:** Various cellular compartments shows varying degrees of crowding due to the specific macromolecules concentrated in particular organelles. This leads to the localized variations in crowding effects, influencing diverse cellular functions (Sarkar et al. 2013; Sadarangani et al. 2023).
- **Exclude volume effect:** When one type of macromolecule occupies space within a confined environment, it creates an exclude volume, restricting the available space for other biomolecules. This effect influences molecular diffusion, reaction rates and molecular interactions (Nayar 2023).
- **High molecular density:** Cells contain a multitude of macromolecules, leading to the space with increased density and amplified molecular collisions. This dense environment alters the physical and chemical properties of the cellular environment (Rickard et al. 2022; Stewart et al. 2023).
- **Influence on molecular interactions:** Macromolecular crowding significantly affects various molecular interactions within cells, including protein-protein binding, enzyme-substrate interactions, and nucleic acid folding. It can modify reaction kinetics and thermodynamics, impacting the equilibrium of biochemical processes (Kim et al. 2010; Ross et al. 2020).

Types of Macromolecules Contributing to Crowding

Macromolecular crowding in cellular milieu arises from the high concentration and diverse presence of various large biomolecules. Here are the primary types of macromolecules that contribute to the crowding phenomena:

Nucleic acids: DNA, RNA and associated proteins form nucleosomes and chromatin structures, contributing to molecular density in nucleus.

Proteins: Enzymes, structural proteins, receptor proteins and other functional proteins contribute significantly to the crowded environment within cellular compartments.

Other metabolites: Other cellular constituents such as polysaccharides, lipids, metabolites and various cellular constituents like metal ions, cofactors and water contribute to the crowded cellular environment.

Relevance in Cellular Environments

Macromolecular crowding plays a vital role in regulating cellular processes, such as protein folding, enzyme kinetics, gene expression, and signal transduction. The crowded environment can modulate the efficiency and specificity of these processes. Crowding alters the conformational dynamics of biomolecules. Proteins may experience altered folding pathways or stability in crowded environments, influencing their functional states and aggregation propensity (Wang and Lu 2018; Samiotakis and Cheung 2011; Sakota et al. 2015).

Understanding macromolecular crowding is crucial for comprehending cellular functionality and disease mechanisms. It has implications for cellular homeostasis, organelle function, and the development of disease related to the protein misfolding-based aggregation and dysfunction. Considering crowding effects is essential in biotechnological applications and drug design as well. It aids in optimizing enzymatic reactions, protein engineering, and developing therapeutics that function effectively within crowded cellular environments. Crowding can be mimicked *in vitro* by adding high concentrations of inert synthetic or natural macromolecules, termed crowding agents. Experimental and computational studies have verified the large effects of crowding on the thermodynamics and kinetics of many biological processes, including protein folding, protein-ligand binding, and aggregation, as well as misfolding. Experimental crowding studies have involved large complex proteins (i.e., multidomain and/or disulfide containing proteins) and often extreme solvent conditions, such as acidic pH or high temperature, whereas computational simulations have focused on very small proteins or peptides.

Significance of Studying the Impact of Macromolecular Crowding on Protein Behavior

Cellular Realism

Naturally, cells are inherently crowded with different biomolecules thereby representing specific environment, in which the protein has to do its physiological work. Understanding how crowded environment affects protein behavior and function provides insights into the realistic conditions mimicking cellular environment, under which biological processes occur.

Protein Structure and Function

Macromolecular crowding may affect protein folding kinetics, stability, and conformational changes. Studying these impacts help in understanding how the crowded environment alters the protein structure and thereby influencing its function.

Protein-Protein Interactions

Altered interactions: Crowding influences protein-protein interactions, affecting binding affinities, association rates, and complex formations. Understanding these alterations provides insights into signaling pathways, enzymatic reactions and regulatory processes (Phillip and Schreiber 2013).

Biological Processes

Impact on enzymatic reactions: Crowding alters enzyme kinetics and substrate accessibility, influencing biochemical reactions crucial for cellular function (Ostrowska et al. 2023; Bulthuis et al. 2023). Gene expression regulation: crowding can influence gene expression by affecting transcription factor binding and RNA polymerase activity, thereby modulating cellular processes (Leake 2018).

Disease Relevance

Protein misfolding and aggregation are implicated in various diseases. Understanding how crowding influences these processes provides insights into disease mechanism, especially in neurodegenerative disorders, such as Alzheimer's and Parkinson's diseases.

Therapeutic Development

Considering crowding effects in drug design and development is critical. Therapeutics designed to function in crowded cellular environments are more likely to be effective and have improved pharmacological properties.

Biotechnological Applications

Understanding crowding is crucial for biotechnology and protein engineering, as it aids in optimizing enzyme reactions, protein engineering, and the development of bio-based materials, enhancing their efficiency and stability.

Computational Modelling and Simulations

Studying crowding is also important for modeling complex systems, as itprovides essential data for computational models and simulations, aiding in the prediction and understanding of cellular processes and protein behavior.

Role of Exclude Volume Effect in Macromolecular Crowding

The effects of macromolecular crowding on protein structure, function, and dynamics are predicted on the basis of the excluded volume effect (Zhou et al. 2008). Excluded volume is considered to be the most dominant factor inside the cells making folded proteins more stable due to unfavorable entropic factors than the loosely or unfolded polypeptide chain. Excluded volume origin is being entropic in nature, a complete disorder state. Low concentration of crowders have major perturbations to the protein structure and are unlikely to take place while minor perturbations are not readily visible (Zhou et al. 2008; Zimmerman and Minton 1993). Protein solvation; i.e., increase in water content provides a unique signature for capturing local dynamics. This allows one to decouple hard-sphere interactions from enthalpic soft sphere ones. Fast fluctuations play an important role in determining the functional characteristics and dynamics of enzymes and proteins. Such motions are prone to be modulated even when the cellular interior crowding conditions are quite calm and relaxed. By the time the excluded volume effects come into the picture in the physiological milieu, modulations of functionally important protein or enzyme motions that need lower activation energy have already resulted due to the enthalpic (soft) interactions (Kuznetsova et al. 2015). In cell, the cellular interior environment is highly crowded because of the presence of considerable amounts of soluble and insoluble macromolecules, such as nucleic acids, proteins, and carbohydrates, which makes the environment congested. This means that a significant fraction of the intracellular space is not available to other macromolecular species that could be pathogenic including bacteria and viruses.

Importance of Understanding Macromolecular Crowding on Drug Design and Therapeutic Interventions

The impact of macromolecular crowding on drug transport forms a critical area of investigation in the development of advanced drug delivery systems. Crowded environments, akin to those found within intracellular environment of biological cells, influence the movement and distribution of therapeutic molecules. In interior of cell environments, the presence of a high concentration of different macromolecules alters the physical behavior of drugs and their carriers thus affecting the diffusion rates and transport mechanism (Dey et al. 2022). Firstly, the mechanism of drug diffusion in crowded media is affected due to alterations in solvent quality and the available free volume. In crowded environment, molecules experience restricted diffusion, which is a deviation from their behavior in dilute solutions. This is due to the hindrance caused in the presence of other macromolecules, which leads to an anomalous diffusion process that is neither purely Brownian nor entirely immobilized. Experimental procedures employed to study these phenomena include single-particle tracking fluorescence recovery after photobleaching, and nuclear magnetic resonance spectroscopy. These techniques provide information about the mobility of molecules within crowded systems of the cell interior and help to quantify the extent of the crowding effect.

Secondly, advanced computational models and simulations are being used to predict and visualize drug transport dynamics (Lin et al. 2020). Computational models include variables such as the size, shape, and charge of both the crowding agents and the drug molecules for modeling drug transport in crowded conditions. These computational models are useful in analyzing the molecular interactions in crowded environments, affecting diffusion coefficients of drugs and macromolecules in the intracellular environment and transport kinetics.

Consequence of Protein Misfolding in Cell Function and Disease

The etiologies of neurodegenerative diseases may be varied; however, a common pathological denominator is the formation of abnormal protein conformers and the occurrence of pathogenic proteinaceous deposits. A major step towards the understanding of mechanisms of neurodegenerative disease, implicated in neuronal degeneration is the identification of genes, which are responsible for familial variants of neurodegenerative diseases. Approaches based on these disease-associated genes concluded the two faces of protein misfolding in neurodegeneration: a gain of toxic function and a loss of physiological function, which can also occur in combination.

Aberrant proteins as the result of production errors in the genes, inherited or acquired amino acid substitutions, especially due to oxidative modification, will not fold correctly and be trapped in misfolded conformations. Due to amino acid substitutions many inherited disorders exhibit loss-of-function pathogenesis because the aberrant or misfolded protein is eliminated by one of the protein quality control systems, examples are cystic fibrosis, sickle cell anemia and phenylketonuria. However, not all aberrant proteins are eliminated and the misfolded protein may accumulate and form toxic oligomers and/or aggregated inclusions. Soluble oligomeric species are associated with early pathological alterations, and oligomeric assemblies of diverse disease-associated proteins share common structural features. Here, in this case the loss of function may be accompanied by a gain-of-function pathogenesis. This will in turn determines the pathological and clinical features of diseases e.g. Parkinson and Huntington diseases (Terry 2020). Two faces of protein aggregation and misfolding contribute to the pathogenic mechanisms of Alzheimer's disease, Parkinson's disease, frontotemporal lobar degeneration and prion diseases [28]. The use of chemical or pharmacological chaperones with specific effects on the misfolded protein can decrease the amounts of accumulated and aggregated proteins. Gain in function positive examples are enzyme enhancement in a number of lysosomal disorders.

Misfolding and Aggregation of Intrinsically Disordered Proteins

Most of the proteins in the human body are well-defined and consist of stable three-dimensional structure that is required for their physiological function. There is another class of proteins that is natively unstructured and are referred to as intrinsically disordered proteins (Uversky 2002). IDPs can quickly change their conformations and participate in a variety of biological processes. Because IDPs contain multiple charged side chains and few hydrophobic residues. However, still IDPs are not found in extended states but exist as compact states due to hydrogen bonds and salt bridges (Uversky 2002). IDPs are highly regulated in the cell, and aberrant regulation results in aggregation of IDPs. An example of IDPs is human tau protein, which exemplifies the dynamics and characteristics of an IDP. It is physiologically relevant in the Alzheimer's disease and other degenerative amyloid disorders. Characterizing the conformational changes and aggregation propensities of IDPs aggregating fragment of tau and critical interactions with the surrounding environment conclude these interactions are involved in folding and fibril formation of amyloidal proteins. IDPs form highly stable complexes when involved in signaling interactions (Uversky 2002). They are in constant "bound–unbound" transitions and acts as dynamic and sensitive "on–off" switches. The ability of IDPs to return to their highly flexible native conformations after the completion of a particular process, and their tendency to gain different conformations depending on the environmental conditions, are the unique physiological properties of IDPs. This allows them to exert different functions in different cellular contests according to a specific conformational state (Uversky and Dunker 2010).

Overview of Protein Aggregation in Neurodegenerative Disease

Neurodegenerative diseases such as Alzheimer's disease, Parkinson's disease, Huntington's disease, amyotrophic lateral sclerosis and prion diseases have common cellular and molecular mechanisms including protein aggregation, oligomeric species and inclusion body formation. The aggregates usually containing misfolded protein consist of fibers with a beta-sheet conformation and are termed amyloid. There is partial but no perfect common characteristics among the cells in which aberrant abnormal proteins are deposited and these cells undergo degeneration. The inclusions, amyloids, and visible protein aggregates signify an end stage of a molecular cascade of several steps (Ross et al. 2020). The earlier steps such as formation of oligomeric species in the cascade may be more directly tied to pathogenesis than the inclusions themselves. In prion disorder, genetic variants assist in illumination of the pathogenesis of the more common sporadic forms.

References

Balch WE, Morimoto RI, Dillin A, Kelly JW (2008) Adapting proteostasis for disease intervention. Science 319(5865):916–919. https://doi.org/10.1126/science.1141448

Bulthuis EP, Dieteren CEJ, Bergmans J, Berkhout J, Wagenaars JA, van de Westerlo EMA, Podhumljak E, Hink MA, Hesp LFB, Rosa HS, Malik AN, Lindert MK, Willems P, Gardeniers H, den Otter WK, Adjobo-Hermans MJW, Koopman WJH (2023) Stress-dependent macromolecular crowding in the mitochondrial matrix. EMBO J 42(7):e108533. https://doi.org/10.15252/embj.2021108533

Dey D, Nunes-Alves A, Wade RC, Schreiber G (2022) Diffusion of small molecule drugs is affected by surface interactions and crowder proteins. Iscience 25(10):105088

Jucker M, Walker LC (2023) Alzheimer's disease: from immunotherapy to immunoprevention. Cell 186(20):4260–4270

Kim YC, Best RB, Mittal J (2010) Macromolecular crowding effects on protein-protein binding affinity and specificity. J Chem Phys 133(20):205101. https://doi.org/10.1063/1.3516589

Kuznetsova IM, Zaslavsky BY, Breydo L, Turoverov KK, Uversky VN (2015) Beyond the excluded volume effects: mechanistic complexity of the crowded milieu. Molecules 20(1):1377–1409. https://doi.org/10.3390/molecules20011377

LaPelusa A, Kaushik R (2020) Physiology, proteins. StatPearls Publishing, Treasure Island
Leake MC (2018) Transcription factors in eukaryotic cells can functionally regulate gene expression by acting in oligomeric assemblies formed from an intrinsically disordered protein phase transition enabled by molecular crowding. Transcription 9(5):298–306
Lin X, Li X, Lin X (2020) A review on applications of computational methods in drug screening and design. Molecules 25(6). https://doi.org/10.3390/molecules25061375
Louros N, Schymkowitz J, Rousseau F (2023) Mechanisms and pathology of protein misfolding and aggregation. Nat Rev Mol Cell Biol 24(12):912–933. https://doi.org/10.1038/s41580-023-00647-2
Nayar D (2023) Molecular crowders can induce collapse in hydrophilic polymers via soft attractive interactions. J Phys Chem B 127(28):6265–6276
Nechushtai L, Frenkel D, Pinkas-Kramarski R (2023) Autophagy in Parkinson's Disease. Biomolecules 13(10):1435
Ostrowska N, Feig M, Trylska J (2023) Varying molecular interactions explain aspects of crowder-dependent enzyme function of a viral protease. PLoS Comput Biol 19(4):e1011054. https://doi.org/10.1371/journal.pcbi.1011054
Phillip Y, Schreiber G (2013) Formation of protein complexes in crowded environments--from in vitro to in vivo. FEBS Lett 587 (8):1046–1052. doi:https://doi.org/10.1016/j.febslet.2013.01.007
Rickard MM, Luo H, De Lio A, Gruebele M, Pogorelov TV (2022) Impact of the cellular environment on adenosine triphosphate conformations. J Phys Chem Lett 13(42):9809–9814
Ross ML, Kunkel J, Long S, Asuri P (2020) Combined effects of confinement and macromolecular crowding on protein stability. Int J Mol Sci 21(22). https://doi.org/10.3390/ijms21228516
Sadarangani V, Kalia A, Kausar T, Murarka P, Sau AK (2023) Effect of the macromolecular crowding agents on the structure and function of human arginase-I, a therapeutically important enzyme. J Phys Chem B 127(41):8749–8761. https://doi.org/10.1021/acs.jpcb.3c02940
Sakota K, Tabata D, Sekiya H (2015) Macromolecular crowding modifies the impact of specific Hofmeister ions on the coil-globule transition of PNIPAM. J Phys Chem B 119(32):10334–10340. https://doi.org/10.1021/acs.jpcb.5b01255
Samiotakis A, Cheung MS (2011) Folding dynamics of Trp-cage in the presence of chemical interference and macromolecular crowding. J Chem Phys 135(17):175101. https://doi.org/10.1063/1.3656691
Sarkar M, Smith AE, Pielak GJ (2013) Impact of reconstituted cytosol on protein stability. Proc Natl Acad Sci USA 110(48):19342–19347. https://doi.org/10.1073/pnas.1312678110
Siddiqui GA, Naeem A (2023) Connecting the dots: macromolecular crowding and protein aggregation. J Fluoresc 33(1):1–11. https://doi.org/10.1007/s10895-022-03082-2
Soto C (2003) Unfolding the role of protein misfolding in neurodegenerative diseases. Nat Rev Neurosci 4(1):49–60. https://doi.org/10.1038/nrn1007
Stewart CJ, Olgenblum GI, Propst A, Harries D, Pielak GJ (2023) Resolving the enthalpy of protein stabilization by macromolecular crowding. Protein Sci 32(3):e4573. https://doi.org/10.1002/pro.4573
Terry C (2020) Insights from nature: a review of natural compounds that target protein misfolding in vivo. Curr Res Biotechnol 2:131–144
Uversky VN (2002) Natively unfolded proteins: a point where biology waits for physics. Protein Sci 11(4):739–756. https://doi.org/10.1110/ps.4210102
Uversky VN, Dunker AK (2010) Understanding protein non-folding. Biochim Biophys Acta 1804(6):1231–1264. https://doi.org/10.1016/j.bbapap.2010.01.017
Wang Z, Lu HP (2018) Single-molecule spectroscopy study of crowding-induced protein spontaneous denature and crowding-perturbed unfolding-folding conformational fluctuation dynamics. J Phys Chem B 122(26):6724–6732. https://doi.org/10.1021/acs.jpcb.8b03119

Zhou HX, Rivas G, Minton AP (2008) Macromolecular crowding and confinement: biochemical, biophysical, and potential physiological consequences. Annu Rev Biophys 37:375–397. https://doi.org/10.1146/annurev.biophys.37.032807.125817

Zimmerman SB, Minton AP (1993) Macromolecular crowding: biochemical, biophysical, and physiological consequences. Annu Rev Biophys Biomol Struct 22:27–65. https://doi.org/10.1146/annurev.bb.22.060193.000331

Chapter 9
The Role of Macromolecular Crowding in Cytoskeletal Organization

Judith Herzfeld

Abstract The cytoskeleton is not only responsible for cell morphology and rheology, but is also involved in such functions as cell motility, intracellular transport and cell division. Key to all these roles is the suitable deployment of its labile parts, a process that occurs in a highly crowded environment. Here we consider the effects of crowding on the assembly and organization of cytoskeletal filaments, and how the cell exploits or evades these effects, as needed. Due to entropic trade-offs among different degrees of freedom, long protein filaments crowded by globular proteins spontaneously form segregated bundles. This behavior can be recruited to specific locations by proteins that nucleate filaments and can be adjusted by proteins that cross-link parallel filaments in a fashion that stabilizes polarity, spacing and register within the bundle. Alternatively, spontaneous bundling can be prevented by limiting filament growth with capping and severing proteins or by frustrating filament alignment with oblique cross-links or the formation of branches. Thus, building upon the effects of macromolecular crowding, a modest library of regulatory proteins is able to achieve versatile results.

Keywords Macromolecular crowding · Excluded volume · Entropically-driven order · Cytoskeleton · Actin · Microtubules · Intermediate filaments · Cell morphology · Cell rheology · Sickle cell hemoglobin

J. Herzfeld (✉)
Department of Chemistry, Brandeis University, Waltham, MA, USA
e-mail: herzfeld@brandeis.edu

V. N. Uversky (ed.), *(Macro)Molecular Crowding*, Subcellular Biochemistry 109,
https://doi.org/10.1007/978-3-032-03370-3_9

Introduction

Most of the effects of macromolecular crowding are of a quantitative nature. These include changes in rates of diffusion and folding, and in equilibrium constants for conformations and associations. Here we will focus instead on dramatic qualitative effects of crowding. These are phenomena that owe to "entropically-driven order", the order that occurs in systems with two or more distinguishable degrees of freedom, under conditions where the entropy lost in restricting some is exceeded by the entropy gained in others. This crowding-induced order impacts filaments that form the scaffolds of cells, tracks of intracellular transport, and machinery of cell motility and division.

The filaments are formed by the open-ended, reversible association of otherwise ordinary proteins. Broadly speaking, there are two classes of filaments, polar and nonpolar. Both microtubules and actin filaments are polar and therefore support directional activities, including as hosts for molecular motors and as props for leading edges in cell motility and cell division. On the other hand, intermediate filaments (so named because they were first identified by widths of about 10 nm, intermediate between the 6 nm of F-actin and the 25 nm of microtubules (Saez and Gonzalez-Granado 2022)) are nonpolar and suited to balancing activities, including cohesion and stretch sensing.

As befits their distinct roles, polar and nonpolar filaments also differ mechanically and in surface presentation.

— Polar filaments are composed of globular molecules (actin or tubulin) that are fitted together in a compact, repeating pattern, much as a skilled mason might build a stone structure. The resulting sturdiness makes these filaments capable of sustaining compressive loads as needed, for example, to ratchet out cell extensions. At the same time, the regularity of the surface provides reliable stepping stones for molecular motors.

— On the other hand, the fundamental unit of an intermediate filament is an antiparallel dimer with a coiled-coil core that is flanked by dangling N- and C-termini. In forming filaments, the coiled-coils associate side-to-side and end-to-end, with the shorter N-termini involved in assembly and the longer C-termini protruding to create a bottle brush-like structure. By virtue of the coiled-coils running along their lengths, intermediate filaments are relatively floppy, much like long springs. In flexing and stretching, these cores provide elasticity. At the same time, the intrinsically unstructured sidearms at the surface are capable of interacting with a variety of partners in a stretch-sensitive fashion.

Polar and nonpolar filaments also change length in different ways.

- Polar filaments generally change length by monomer addition and subtraction (under the influence of triphosphates).
- On the other hand, intermediate filaments are also observed to shorten by breaking and lengthen by joining of fragments (under the influence of phosphorylation and dephosphorylation of the N-termini).

What the cytoskeletal filaments all have in common are highly elongated shapes. In uncrowded conditions, elongated particles are spontaneously dispersed and randomly oriented. However, under crowded conditions elongated particles are spontaneously aligned and bundled. Without understanding this default arrangement, we cannot understand the mechanisms by which cells manipulate the cytoskeleton. The investment is enormous, ranging from the great variety of binding proteins that stabilize, cross-link, cap and sever the filaments, to the kinases and phosphatases that modulate the activities of the various players.

Here we will explore the effects of crowding in steps that serially add competing degrees of freedom. We begin with monodisperse rod-shaped particles, focusing on trade-offs between their translational and orientational distributions. Next, we consider bidisperse particles, with the possibility of demixing. The complexity then increases with consideration of the broad length distributions that result from reversible self-assembly. To that we add the coexisting majority of intracellular proteins that do not participate in self-assembly. Finally, we consider mixtures of self-assembled filaments with different physical properties.

The insights from these systems leads to a picture of the default condition of cytoskeletal proteins in crowded solutions, one with filaments of each type collected in segregated bundles. The question then becomes how cells modulate this spontaneous behavior to carry out different functions. Therefore, we end the chapter with a review of the different types of tools that cells use to choreograph cytoskeletal elements.

Alignment of Monodisperse Rods

Onsager was the first to connect particle shape to the symmetry-breaking behavior of colloid solutions with increasing volume fraction (Onsager 1949). He noted that the space that one hard sphere of radius r excludes to the center of another is a sphere of radius $2r$, which has a volume 8 times that of the sphere. Similarly, the space that one hard spherocylinder of radius r and cylinder length l excludes to the center of another that is parallel, is a spherocylinder of radius $2r$ and cylinder length $2\ l$, which has a volume 8 times that of the spherocylinder. On the other hand, the space excluded to the center of an orthogonal spherocylinder is a square spheroplatelet of thickness $4r$ and area l^2, which has an additional volume of $4rl^2$. Comparing the two, we see that, for l large compared to r, the volume excluded to the orthogonal spherocylinder is ~(l/r) times the volume excluded to a parallel spherocylinder. At intermediate orientations, the excess excluded volume due to non-alignment scales smoothly with $\sin\theta$, where θ is the angle between the spherocylinder axes. Thus, depending on the aspect ratio (l/r), there comes a concentration where unobstructed volume to be gained by alignment becomes significant compared to the unoccupied volume.

This phenomenon is illustrated by the game of "pick-up sticks" shown in Fig. 9.1. As disordered as they are, the sticks on the left actually have very little freedom to

Fig. 9.1 Entropically-driven order among rods is illustrated by difficult (left) and easy (right) starting configurations for a game of "pick-up sticks". The object of the game is to remove one stick at a time without perturbing any other stick. (A player's turn ends when s/he perturbs another stick.) This is clearly much more difficult for randomly oriented sticks (left) than for aligned sticks (right)

rotate or translate due to mutual obstruction. While the aligned sticks on the right have still less rotational freedom, they have considerably more translational freedom. Thus, at a certain level of jamming, the loss in rotational freedom upon alignment is less than the gain in translational freedom.

In quantifying this effect without the benefit of any computers, Onsager had to truncate his calculation of the partition function at the two-body term. This is an adequate approximation at low concentrations and sufficed to show an alignment transition for rods with large aspect ratios. Subsequent lattice models were able to take multibody interactions into account for rods restricted to three mutually orthogonal axes (DiMarzio 1961; Flory 1956; Herzfeld 1982). Later, the full range of orientations was modeled by scaled particle theory (Cotter and Martire 1970).

However, for the present purposes, it suffices to do a rough calculation of the concentration at which N rods of length l and radius r align spontaneously in a volume V. In this context, the most relevant measure of concentration is the occupied volume fraction $\varphi = N(\pi r^2 l)/V = c\ (\pi\ r^2 l)$, where c = number concentration = N/V. Tumbling rods of length l begin to interfere with each other when the average interparticle distance $(V/N)^{1/3} = (1/c)^3 < l$, i.e., when $c > l^{-3}$ or $\varphi > \pi(r/l)^2$. As expected, the dependence on aspect ratio, (r/l), is strong and obstruction sets in at very low volume occupancy φ when the aspect ratio (l/r) large. Even in suspensions of TMV, with an aspect ratio of only (300 nm / 18 nm) =17, birefringence is observed at volume occupancies $\varphi > 1$–2%, consistent with $\pi\ (r/l)^2 = 0.011$.

Technically, excluded volume only references hard repulsions between particles. However, since biological macromolecules typically carry net charge, there are also soft repulsions with an extent that depends on the degree of screening by ions in the medium. This entails some quantitative adjustment, but has little impact on entropically-driven order. Onsager suggested that the effect of soft repulsions could be well-approximated by a small increase in the lengths and diameters of the rods (Onsager 1949). The result would be two countervailing effects: a modest decrease in the aspect ratio which would reduce the benefits of alignment and a modest increase in the volume fraction which would enhance the drive toward alignment.

Demixing of Different Particles

Adding a second type of particle introduces new degrees of freedom with new opportunities for entropy trade-offs. Other things being equal, entropy favors mixing and demixing usually owes to interaction energies that are more favorable between like particles than between unlike particles (as in mixtures of polar and non-polar molecules). However, excluded volumes lead to new trade-offs in entropy that can also mitigate against mixing. Setting aside particle shape for the moment, let's consider spheres of two sizes. At low concentrations, they will be mixed. As the concentration increases, the average spacing between the fully dispersed large spheres decreases until it is too small accommodate the smaller spheres. Under these circumstances demixing occurs because the entropy lost by the large spheres in tighter packing is less than the entropy gained by the small spheres in accessing previously excluded space. Looking at it another way, for a given volume occupancy, there are more small particles than large ones. Therefore, the smaller ones compete for space more effectively because, with more coordinates, they gain more entropy per unit of available space. The result is that the smaller spheres tend to drive the larger ones together.

This phenomenon also occurs if the larger particles are elongated. As illustrated in Fig. 9.2, interspersing rods and spheres severely constrains both and, under

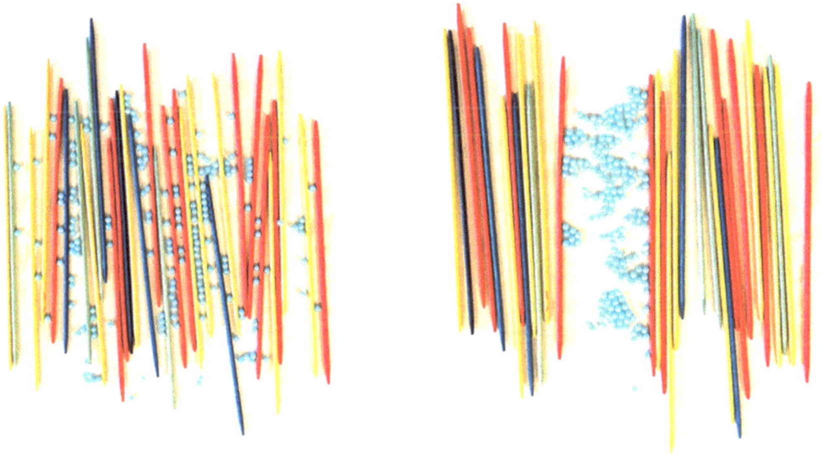

Fig. 9.2 Entropically-driven order in a mixture of rods and spheres

crowded conditions, the entropy is greater if they separate. Moreover, as the spheres compete for space more effectively, they tend to drive rods closer together.

Non-specific demixing is a long-known feature of colloidal systems. In the 1950's, Asakura and Oosawa analyzed the apparent attraction between the larger particles in terms of an effective short-range "depletion force" (Asakura and Oosawa 1954, 1958). Their idea was that, to the extent that smaller particles are excluded from space between larger ones, the osmotic pressure due to the smaller ones is less between the larger ones than elsewhere and that will drive the larger ones together. This construct is useful for estimating the magnitude of the effect. However, nomenclature that invokes a force of attraction suggests that energy is involved when the effect is actually entirely entropic (as is osmotic pressure in general) (Oosawa 2021). The construct also has the disadvantage of modeling the large and small particles differently which is awkward and not useful in systems with still greater polydispersity, such as the cytoskeletal systems of interest here. Lattice models and scaled particle theory provide more rigorous accounts.

Polydispersity via Open-Ended Reversible Self-Assembly

Protein filaments generally self-assemble in two stages, a nucleation stage followed by a growth stage. Nucleation is a relatively tentative process because only a subset of contacts is available to stabilize monomer addition. However, once a full "nucleus" is formed, each new monomer can form a full complement of contacts and the probability of each further monomer addition is the same. In dilute solution, the result at equilibrium is an exponential distribution of filament lengths with a decay length that depends only on the monomer concentration and monomer association constant. However, under crowded conditions, obstruction also limits filament growth and aligned filaments will obstruct each other less than unaligned filaments. As a result, while the distribution of lengths remains exponential in any given orientation, the decay length will be longer for filaments oriented closer to a preferred direction that develops spontaneously during self-assembly (Herzfeld 1988; Herzfeld and Briehl 1981).

These features have been observed in vitro for microtubules (Hitt et al. 1990). They are also well-illustrated in vivo by the behavior of hemoglobin in sickle cell disease. Hemoglobin is normally an exceptionally soluble protein, remaining unaggregated at the very high concentrations present inside erythrocytes. Even the sickle cell mutant is highly soluble when oxygenated. However, deoxygenated sickle cell hemoglobin forms thick filaments. Under these conditions, the normally flexible, biconcave cell becomes stiff and distorted in a manner that depends on the rate of deoxygenation (Kaul and Xue 1991). When deoxygenation is slow, nucleation events are few and far between. The result is a large domain of aligned filaments that distorts the erythrocyte into the eponymous sickle shape. On the other hand, when deoxygenation is fast, nucleation occurs quickly, at many locations and the result is a spiculated cell, with numerous small domains of aligned filaments distorting different portions of the cell membrane in different directions.

Alignment is relatively easy to observe for filaments as stiff as microtubules and sickle cell hemoglobin filaments. Semi-flexible filaments behave in rod-like fashion over contour lengths up to their persistence lengths and aligned domains can develop to that extent. Whereas microtubules have a persistence length > 1 mm, persistence lengths for F-actin and intermediate filaments are only 10 μm and 1 μm, respectively (Patteson et al. 2021). The corresponding aspect ratios of 1667 and about 100, for F-actin and intermediate filaments respectively, suffice for crowding-induced alignment, but the spatial extent of the aligned domains will be smaller and more difficult to observe than those for microtubules.

Bundling and Sorting of Self-Assembled Filaments

In ordinary cells, filaments comprise only a fraction of the protein and crowding is due primarily to the abundance of protein that is globular. As discussed above, the latter will crowd the larger filaments into aligned bundles (Madden and Herzfeld 1993). Such crowding-induced bundling has been observed *in vitro* for both actin filaments (Goverman et al. 1996; Cuneo et al. 1992; Suzuki et al. 1989) and microtubules (Guo et al. 2008; Needleman et al. 2004).

However, different types of filaments coexist *in vivo*. This presents further opportunities for trade-offs that increase entropy. As for spheres of different diameters, filaments of different diameters will tend to segregate. In addition, flexible filaments have more ways of using space than stiffer ones. Both of these features contribute to the prediction of spontaneous segregation of microtubules from F-actin and intermediate filaments, as microtubules are both thicker and stiffer than either of the other filaments (Kulp and Herzfeld 1995). The same phenomenon has been exploited to separate rigid DNA duplexes from flexible single DNA strands (Zanchetta et al. 2008).

On the other hand, sorting of intermediate filaments vs. F-actin is less obvious because the former.

- while generally more flexible, are a bit thicker than the latter,
- have variable dimensions due to elasticity of the core and disordered folding of the sidearms, and.
- comprise an enormous class of proteins, coded by >70 genes that are differentially expressed in cell-type specific fashion (Saez and Gonzalez-Granado 2022).

To summarize, whereas protein filaments are isotropically dispersed in dilute solution, they form tight, often segregated bundles under crowded conditions. This behavior occurs spontaneously, without the help of any accessory proteins. However, the resulting bundles will be rather crude. Comprising filaments that are randomly parallel or antiparallel, they will have no overall polarity even when the filaments themselves are polar. In addition, even where neighboring filaments are pointed in the same direction, they are unlikely to be in register. Moreover, the spacing between neighboring filaments will be irregular. To serve cellular functions, it may be necessary for cells to tune bundles for polarity, registration, and spacing. Cells may also need to control the locations of bundles and whether they occur at all.

Control Points

Cells are constantly remodeling the cytoskeleton and are heavily invested in strategies for both exploiting and evading the effects of crowding. Comprehensive reviews of identified players are available for actin filaments (Pollard 2016), microtubules (Peng and Nakamura 2023), and intermediate filaments (Coulombe et al. 2000). In addition, a variety of cross-linkers and molecular motors effect coupling between all three classes of filaments (Coulombe et al. 2000; Huber et al. 2015).

Opportunities for choreographing filaments occur in nucleation, in growth and disassembly, and in either tuning bundles or avoiding them. Let's consider these in turn, specifically for the polar filaments which have been more thoroughly studied to-date than the non-polar, intermediate filaments.

— Since homogeneous nucleation is difficult, especially when most monomers are tied up in filaments, a cell can easily control where new filaments form by providing heterogeneous nucleation sites in suitable numbers and at suitable locations. This process is particularly prominent in mitosis.

- Once filaments form, there are several ways to control length, which as we have seen is an important determinant of spontaneous alignment and demixing. First the concentrations of triphosphates influence the rates of monomer association relative to dissociation, and thus the rate of net growth (or disassembly). However, triphosphates have so many roles in the cell that their concentrations need to answer to many needs. On the other hand, capping proteins are available specifically to obstruct growth by binding selectively to filament ends. Large concentrations of capping proteins can keep filaments too short to spontaneously align and demix (Madden and Herzfeld 1994). Filament length is also controlled by proteins capable of severing long filaments, counteracting spontaneous annealing of short ones.
- Even when filaments are long, it is possible to prevent them from bundling by cross-linking them at oblique angles. The cross-linked network forms a gel, avoiding the stiffer frame generated by filament bundles. The formation of branched filaments also frustrates bundling.
- Other filament binding proteins cross-link filaments in parallel configurations, biasing the growth of neighboring filaments in the same direction. In addition, acting as spacers, these binding proteins can control the lateral distance between filaments. Furthermore, given specific binding sites, these cross-linkers can stabilize registration between neighboring filaments.

Conclusion

To understand how cells regulate cytoskeletal organization, it is necessary to understand how the filament forming components would behave absent such regulation. In particular, it is important to understand that, due to crowding by all the other proteins in the cell, long filaments will tend to form segregated bundles. The cell therefore needs to control nucleation events, control filament growth, and either regularize the structures of the filament bundles or frustrate their formation by oblique cross-linking or branching. Proteins have been identified for each of these roles, often with activities modulated by kinases.

We might ask what difference all this makes. Suppose, counterfactually, that filaments are spontaneously dispersed and randomly oriented, so that the default condition of the cytoplasm is a soft gel and the bundles required for scaffolding, motility and intracellular transport form only under the influence of cross-linking. Under this scenario, bundle formation would be sluggish. But, perhaps just as important, because the cross-linking would need to be stronger, disassembly and redeployment of monomers would also be sluggish. In other words, the entropically-driven order owed to macroscopic crowding makes an important contribution to the responsiveness of the cytoskeleton to the needs of the cell.

Compliance with Ethical Standards

The author declares no conflict of interest.

References

Asakura S, Oosawa F (1954) On interaction between two bodies immersed in a solution of macromolecules. J Chem Phys 22:1255–1256. https://doi.org/10.1063/1.1740347
Asakura S, Oosawa F (1958) Interaction between particles suspended in solutions of macromolecules. J Polym Sci 33:183 192. https://doi.org/10.1002/pol.1958.1203312618
Cotter M, Martire D (1970) Statistical mechanics of rodlike particles. III. A fluid of rigid spherocylinders with restricted orientational freedom. J Chem Phys 53:4500–4511
Coulombe PA, Bousquet O, Ma L, Yamada S, Wirtz D (2000) The 'ins' and 'outs' of intermediate filament organization. Trends Cell Biol 10:420–428. https://doi.org/10.1016/S0962-8924(00)01828-6
Cuneo P, Magri E, Verzola A, Grazi E (1992) 'Macromolecular crowding' is a primary factor in the organization of the cytoskeleton. Biochem J 281:507–512. https://doi.org/10.1042/bj2810507
DiMarzio E (1961) Statistics of orientation effects in linear polymer molecules. J Chem Phys 35:658–669
Flory PJ (1956) Phase equilibria in solutions of rod-like particles. Proc Roy Soc London Ser A 234:73–89

Goverman J, Schick LA, Newman J (1996) The bundling of actin with polyethylene glycol 8000 in the presence and absence of gelsolin. Biophys J 71:1485–1492. https://doi.org/10.1016/S0006-3495(96)79349-9
Guo Y, Liu Y, Oldenbourg R, Tang JX, James M, Valles J (2008) Effects of osmotic force and torque on microtubule bundling and pattern formation. Phys Rev E 78:041910. https://doi.org/10.1103/PhysRevE.78.041910
Herzfeld J (1982) Lattice statistics for highly heterogeneous populations of rigid rectangular particles. J Chem Phys 76:4185–4190
Herzfeld J (1988) Liquid crystalline order in self-assembling systems: orientation dependence of the particle size distribution. J Chem Phys 88:2776–2779. https://doi.org/10.1063/1.454009
Herzfeld J, Briehl RW (1981) Length distributions and the alignment transition of polymers formed by linear reversible polymerization. Macromolecules 14:1209–1214. https://doi.org/10.1021/ma50006a014
Hitt AL, Cross AR, Robley C, Williams J (1990) Microtubule solutions display nematic liquid crystal structure. J Biol Chem 265(3):1639–1647. https://doi.org/10.1016/S0021-9258(19)40064-1
Huber F, Boire A, López MP, Koenderink GH (2015) Cytoskeletal crosstalk: when three different personalities team up. Curr Opinion Cell Biol 32:39–47. https://doi.org/10.1016/j.ceb.2014.10.005
Kaul DK, Xue H (1991) Rate of deoxygenation and rheologic behavior of blood in sickle cell anemia. Blood 77:1353–1361. https://doi.org/10.1182/blood.V77.6.1353.1353
Kulp DT, Herzfeld J (1995) Crowding-induced organization of cytoskeletal elements: III. Spontaneous bundling and sorting of self-assembled filaments with different flexibilities. Biophys Chem 57:93–102
Madden TL, Herzfeld J (1993) Crowding-induced organization of cytoskeletal elements: I. Spontaneous demixing of cytosolic proteins and model filaments to form filament bundles. Biophys J 65:1147–1154
Madden TL, Herzfeld J (1994) Crowding-induced organization of cytoskeletal elements: II. Dissolution of spontaneously formed filament bundles by capping proteins. J Cell Biol 126(1):169–174. https://doi.org/10.1083/jcb.126.1.169
Needleman DJ, Ojeda-Lopez MA, Raviv U, Ewert K, Jones JB, Miller HP, Wilson L, Safinya CR (2004) Synchrotron X-ray diffraction study of microtubules buckling and bundling under osmotic stress: a probe of interprotofilament interactions. Phys Rev Lett 93:198104. https://doi.org/10.1103/PhysRevLett.93.198104
Onsager L (1949) The effects of shape on the interaction of colloidal particles. Ann N Y Acad Sci 51:627–659
Oosawa F (2021) The history of the birth of the Asakura–Oosawa theory. J Chem Phys 155:084104. https://doi.org/10.1063/5.0049350
Patteson AE, Carroll RJ, Iwamoto DV, Janmey PA (2021) The vimentin cytoskeleton: when polymer physics meets cell biology. Phys Biol 18:011001. https://doi.org/10.1088/1478-3975/abbcc2
Peng N, Nakamura F (2023) Microtubule-associated proteins and enzymes modifying tubulin. Cytoskeleton 80:60–76. https://doi.org/10.1002/cm.21748
Pollard TD (2016) Actin and actin-binding proteins. Cold Spring Harb Perspect Biol 8:a018226. https://doi.org/10.1101/cshperspect.a018226
Saez A, Gonzalez-Granado JM (2022) Recent advances in intermediate filaments—volume 1. Int J Mol Sci 23:5308. https://doi.org/10.3390/ijms23105308
Suzuki A, Yamazaki M, Ito T (1989) Osmoelastic coupling in biological structures: formation of parallel bundles of actin filaments in a crystalline-like structure caused by osmotic stress. Biochemistry 28:6513–6518. https://doi.org/10.1021/bi00441a052
Zanchetta G, Nakata M, Buscaglia M, Clark NA (2008) Phase separation and liquid crystallization of complementary sequences in mixtures of nanoDNA oligomers. Proc Natl Acad Sci USA 105:1111–1117. https://doi.org/10.1073/pnas.0711319105

Chapter 10
Macromolecular Crowding in Cytoplasm and Cellular Organelles

Muhamed Adilović, Vladimir N. Uversky, and Jasmin Šutković

Abstract A high concentration of diverse macromolecules causes a densely packed environment, influencing both function and structural change of the cell organelles, cytoplasm, and molecules within. The concept of molecular crowding was for the first time introduced in the 70 s but was only accepted at the end of the 90 s. Crowding demonstrates how the high concentration of macromolecules, such as proteins, nucleic acids, and polysaccharides, influences the behavior and interactions of these molecules. This systematic review explored recent literature that impacted the understanding of molecular crowding, emphasizing the effects of molecular crowding on cytoplasm, nucleus, endoplasmic reticulum, Golgi apparatus, and chloroplasts. Numerous scientific literatures have confirmed that crowding impacts practically all aspects of these organelles' structure and functions, including gene expression regulation, protein folding, enzyme activity, and the organization of cellular structures like the cytoskeleton and chromatin, and consequently clearly affects the regulation of cell metabolic activities. Moderate macromolecular crowding contributes to the normal functioning of these cellular processes, however excessive crowding triggers various responses, disrupts the normal functioning of the cell, and can result in various diseases, ranging from neurodegenerative disorders to diabetes. Different therapeutic options are being explored to help mitigate the negative effects of excessive macromolecular crowding and prevent disease development.

Keywords Macromolecular crowding · Homeostasis · Enzyme activity · Metabolism · Disorders · Therapeutic approaches

---

M. Adilović (✉) · J. Šutković (✉)
Department of Genetics and Bioengineering, Faculty of Engineering and Natural Sciences, International University of Sarajevo, Sarajevo, Bosnia and Herzegovina
e-mail: madilovic@ius.edu.ba; jsutkovic@ius.edu.ba

V. N. Uversky
Department of Molecular Medicine and USF Health Byrd Alzheimer's Research Institute, Morsani College of Medicine, University of South Florida, Tampa, FL, USA

V. N. Uversky (ed.), *(Macro)Molecular Crowding*, Subcellular Biochemistry 109,
https://doi.org/10.1007/978-3-032-03370-3_10

Introduction

Macromolecular crowding is a phenomenon that focuses on enzymatic reactions in cellular conditions, where high concentrations of macromolecules affect the dynamics and functionality of all proteins in the cell (Alfano et al. 2024). Cytoplasm and body fluids experience crowding and restore homeostasis. Molecular crowding affects the chemical rate constant and balance of all cellular interactions and highly controls interactive forces (attractive, repulsive, and non-specific molecular interactive forces). Further, cytosolic crowding influences intracellular water mobility, and macromolecular complex formation (Subramanya and Boyd-Shiwarski 2024).

The concept of molecular crowding was first discussed in the 1970s, by scientists like A. P. Minton, who recognized that the dense packing of macromolecules (such as proteins, nucleic acids, and carbohydrates) inside cells can significantly affect biochemical processes (Minton and Wilf 1981; Minton 2000). Later, in the 80 s it became more popular but became accepted only by the end of the 90 s (Monterroso et al. 2024).

Most of the cellular functions, including cell cycle, regulation of cell size, cell differentiation, and embryonic development, are associated with macromolecular crowding. The cellular density, with the protein and enzyme interaction within the nucleus and cytoplasm, depends on the scale and time of interaction, including non-specific biomolecules present in the cytoplasm (Tabaka et al. 2014; Rollin et al. 2023).

The mobility and interactions between biomolecules are affected by molecular crowding, causing oscillations in inter- and intramolecular reactions, and influencing the stereoscopic, functional, and reactive molecular states (Blanco et al. 2021; Grassmann et al. 2024).

For a detailed study of biological macromolecules under controlled conditions biophysical studies are usually done in dilute solutions with low concentrations of the molecules involved, just enough buffer and salts to keep everything at pH, etc. Nevertheless, there are many factors that differentiate the *in vivo* environment. The pronounced feature is the concentration of diverse macromolecules. In *E. coli*, it is estimated that there are on the order of 300–400 g/l or macromolecules in the intracellular fluid and make up to ~30–40% of cell volume (Zimmerman and Trach 1991). Allen Milon proposed that values are those of the highest known concentration in prokaryotic cells for macromolecules (Minton 1980). Later, Model et al. demonstrated and presented for 15 different organisms the estimated amounts of macromolecules, stating that in *E. coli* the fluid with macromolecules remains at pegged values of 300–400 g/L; whereas in rat kidneys the values are lower, with 9 g/L (Illmer et al. 1999; Model et al. 2021). Further, Illmer et al. demonstrated that the blood plasma contains 80 g L^{-1} of proteins (Illmer et al. 1999). This observation confirmed that the intracellular environment is crowded, emphasizing the role of cell volume on protein function and stability. The initial theory stated that macromolecules are fully crowding the cell volume, representing prokaryotic cells as "bags full of macromolecules," in which cell volume is taken up largely by other large molecules rather than accessible space for proteins (McGuffee and Elcock 2010). This theory was misleading the scientific community, giving the false impression that the entire volume of the cell is filled with macromolecules, resulting in justified criticism by other scientists (Srere 2000) and James Clegg (1984).

Today, it's clear that macromolecular concentrations are extremely high due to cell confinement. Cells are homogeneously compressed (confined) structures, where the cell movement is restricted to spatial movement of cells or their interior within a limited space (Alfano et al. 2024). To summarize, molecular crowding together with confinement are the key determinants of the activity of the proteins in the cell, controlling cellular functions and processes such as signaling, metabolism, and gene expression (Zhou et al. 2008).

Molecular crowding affects different organelles in the cell. For example, in the nucleus, all the chemical processes, and biochemical and molecular reactions depend on ionic strength, pH, and reactant concentration. Therefore, macromolecular crowding (with all the large biomolecules such as chromatin and proteins) influences the behavior and interactions, leading to compact conformations and reduced diffusion, affecting all nuclear functions such as gene transcription, RNA splicing, and DNA replication by maintaining specific assemblies of macromolecular complexes, see Fig. 10.1 as an approximate illustration of nucleus crowding (Richter et al. 2008; Shim et al. 2020; Takahashi et al. 2020b; Endoh et al. 2023).

The effects of crowding on the structure of chromatin, with its DNA in telomeres, has been shown to cause the structural transition from antiparallel to parallel form (Sugimoto 2014). Crowding effects are also demonstrated on RNAs and ribonucleoprotein structures, where the folding and stability of these molecules are significantly enhanced by crowding (Strulson et al. 2014; Dupuis et al. 2014).

Macromolecular crowding facilitates liquid-liquid phase separation (LLPS), which is at the basis of the formation of membranelles organelles (MLOs) within the cell (Vweza et al. 2021). For example, LLPS is one of the main steps that drive the formation of heterochromatin, a repressive DNA domain (Larson et al. 2017). Further, LLPS plays major roles in nuclear architecture, DNA repair, and genome stability (Chiolo et al. 2011). In addition, the environment itself in the organelles is crowded, dictating specialized conditions. Therefore, a clear understanding of the

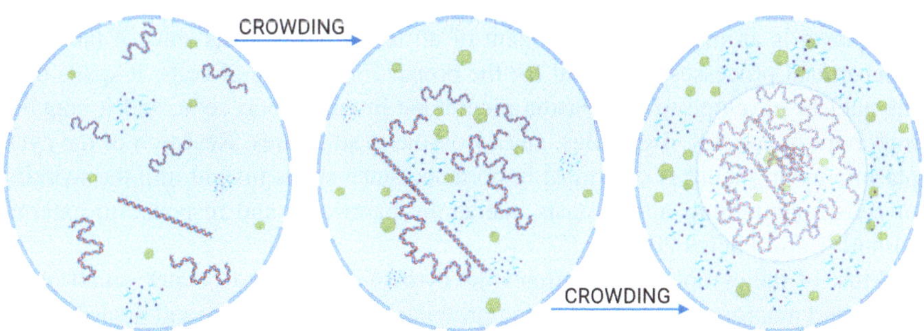

Fig. 10.1 Nucleus crowding illustration (Figure generated with BioRender)

contribution of macromolecular crowding to LLPS formation and functionality is essential (Collette et al. 2023).

Crowding has been reported to strongly influence the chloroplast functioning. Due to the complicated mixture of proteins involved in photosynthesis and other protein complexes, embedded between stacked and unstacked regions, macromolecular crowding strongly modulates all the interactions and controls the membrane surface charge (Wietrzynski et al. 2024).

In 2007, Haferkamp and Kirchhoff demonstrated the significance of molecular protein crowding on Photosystem II (PSII) function in the chloroplast, which catalyzes the key reactions of photosynthesis. They have concluded that the addition of lipids leads to a dilution of the protein density (increasing the crowding effect) causes a separation of PSII and LHCII (main light-harvesting complex), a decreased excitonic coupling (Haferkamp and Kirchhoff 2008).

If compared to other organelles the membrane of endoplasmic reticulum (ER) is differently affected by crowding. ET is comparatively thin and loosely set in comparison to the plasma membrane because sphingolipids and sterol molecules are more embedded into the plasma membrane (Löwe et al. 2020). The ER is structurally and functionally highly dynamic cellular organelles; it changes its shape in response to the cellular indications, stage of the cell cycle, stage of the developmental process, and type of cell. Here, macromolecular crowding is shown to intervene in protein denaturation and disrupt the movement of cargo proteins (Schwarz and Blower 2016).

With this review we aim to further elaborate and explain the macromolecular crowding within the cytoplasm and cellular organelles, such as the nucleus, chloroplasts, and endoplasmic reticulum, and demonstrate how crowding significantly affects biochemical reactions, influencing protein folding, enzyme activity, and central dogma.

Macromolecular Crowding in Cytoplasm

Overview of the Cytoplasmic Environment

Cytoplasm is an important component of all living active cells since it facilitates biochemical processes important for the proper functioning of cells. It spans from the nuclear envelope to the plasma membrane in eukaryotic cells, and it contains important molecules, organelles, and cytoskeletal structures. Research of the cytoplasmic environment is of utmost importance since sheds insight into the working of cells, maintaining homeostasis, metabolic pathways, and response to external stimuli (Luby-Phelps 2000).

Most of the cytoplasm constitutes the cytosol consisting of water, small molecules, and macromolecules, including proteins, polysaccharides, and nucleic acids. Distribution and crowding of macromolecules contribute to the viscosity of the cytoplasm, and it has an effect on other processes, including enzyme activity, protein folding, molecular transport, diffusion rates, and other molecular interactions (Minton 2001; Ellis 2001).

Proteins

Proteins constitute the largest portion of macromolecules in the cytoplasm and greatly contribute to macromolecular crowding. They are present in different forms, including soluble enzymes and multiprotein complexes like ribosomes or proteasomes. They perform vital functions including enzymatic reactions and intracellular signaling, and they interact with other cellular components, contributing to the proper functioning (Zhou et al. 2008).

Additionally, different types of protein macromolecules: microfilaments, microtubules, and intermediary filaments, make up the cytoskeleton, maintaining cellular shape, providing mechanical support, enabling cellular motility, contributing to the spatial distribution of organelles, and facilitating intracellular transport. Different motor proteins interact with the cytoskeleton and help with organelle trafficking (Fletcher and Mullins 2010).

The crowding nature of the cytoplasm is important for proper functioning. Apart from affecting reaction kinetics and promoting molecular interactions, it promotes proper protein folding and stabilization of native states, as well as oligomerization (Minton 2006). This is accomplished through different means, including ideal ionic strength maintained through specific ion concentrations, as well as a near-neutral pH of 7.12 However, different conditions, including stress, can promote excessive protein aggregation, having negative consequences on proper cellular metabolism and contributing to diseases including neurodegeneration (Zhou 2004; Siddiqui and Naeem 2023).

Nucleic Acids

The second major macromolecular component of the cytoplasm is nucleic acids. DNA, while generally present in the nucleus, occupies cytoplasmic space during cellular divisions. RNA molecules are the ones mainly present in the cytoplasm, having a major contribution to protein synthesis. Messenger RNAs serve as templates, while transfer RNAs and ribosomal RNAs participate in their translation into polypeptides which will be folded into proteins. Specific crowding of nucleic acids in the cytoplasm can affect and regulate their accessibility and function (Ellis 2001).

Crowding can also affect the organization of cellular components. They can form membrane-less organelles like stress granules through phase separation. These in turn compartmentalize biochemical reactions and regulate molecular availability (Patel et al. 2017).

Carbohydrates

Carbohydrates do not generally constitute a major percentage of the macromolecular component of the cytoplasm; however, there are specialized cells where carbohydrates can be found in a larger proportion. Specifically, carbohydrates are present as glycogen in energy-storing cells like liver hepatocytes, as well as cells that spend a lot of energy like various types of muscle cells. These types of macromolecules occupy a significant percentage of cytoplasmic space, but also interact with other molecules, contributing to different biochemical reactions and having an effect on intracellular diffusion (Neoh et al. 2024).

Other Biomacromolecules

Lipids

Lipids account for ~2% of the total cellular mass. Similar to proteins, lipid concentration varies with cell types and changes with a physiological state of cells (Oh et al. 2022). The fact that different lipids are required for different functions is further reflected in the significant variability of the composition of lipids of different organelles, cell types, and tissues (Saghatelian et al. 2006; Klose et al. 2013), where, e.g., mammalian cells are expressing tens of thousands of different lipid species and use hundreds of proteins to synthesize, metabolize, and transport them (Muro et al. 2014). In fact, besides being building blocks of cellular membranes, lipids, which are not polymeric molecules, act as energy storage and play many other important physiological roles (Muro et al. 2014). Since the macromolecular crowding phenomenon is not limited to the cytoplasm and can be observed in the densely protein-packed, nonhomogeneous cellular membranes and at the membrane interfaces, lipids are known to play a crucial role in cellular crowding by affecting the fluidity and structure of cell membranes (Löwe et al. 2023). In its turn, the membrane fluidity and structure can modulate the movement and interactions of proteins within the membrane, impacting various cellular processes, such as signaling and transport. In essence, the composition and arrangement of lipids within the membrane can directly modulate how crowded the membrane environment feels for other molecules like proteins, affecting their function and dynamics (Frechin et al. 2015; Jeon et al. 2016; Löwe et al. 2023).

Polyamines

Polyamines are small polycationic compounds with a positive charge, which are abundantly found in all living cells, with three primary types of polyamines present in mammalian cells being putrescine (Put, [$NH_2(CH_2)_4NH_2$]), spermidine (Spd, $NH_2(CH_2)_4NH(CH_2)_3NH_2$), and spermine (Spm, [$NH_2(CH_2)_3NH(CH_2)_4NH(CH_2)_3NH_2$]) (Thomas and Thomas 2001; Wallace et al. 2003; Xuan et al. 2023). Although the intracellular polyamine concentrations are generally within the micromolar to low millimolar range, due to their high positive charge, they are compartmentalized and/or bound to nucleic acids and other negatively charged molecules, such as proteins and phospholipids (Igarashi and Kashiwagi 2000; Bachrach 2005). Therefore, all polyamines are strongly involved in cell growth and viability in mammalian cells, being responsible for regulation functions and physical state of molecules they are interacting with (Igarashi and Kashiwagi 2019). For example, they can induce and stabilize liquid crystalline DNA (Saminathan 2002), accelerate the aggregation of proteins (Goers et al. 2003) or promote liquid-liquid phase separation of disordered proteins (Percival et al. 2024).

Variability of Crowding Across Cell Types

While macromolecular crowding is a phenomenon observed in all living cells, its magnitude as well as consequences differ between cell types, tissues, and organisms. Crowding is therefore, although a crucial factor in cellular physiology, also highly variable due to various reasons, including cells' metabolic activity, size, specialization, as well as environmental conditions (Guigas et al. 2007). For example, hepatocytes are metabolically more active, thus experiencing higher levels of both molecular crowding in forms of glycogen storage, as well as the crowding of organelles like mitochondria that make a lot of precursors for macromolecules (Ellis 2001; Spinelli and Haigis 2018). On the other hand, certain dormant/resting cells, like those of the immune system, generally exhibit less molecular crowding in their cytoplasm (Costa et al. 2022).

Moreover, certain specialized cells exhibit unique molecular crowding. For example, neurons are polarized cells containing axons and dendrites—specific compartments where molecular crowding can differ on a local level. These in turn have an effect on cellular transport, secretion, and signaling (Wnek 2016; Berth and Lloyd 2023).

On a larger scale, the differences in the cellular physiology, structure, and organization between prokaryotes and eukaryotes also result in the differences in macromolecular crowding. Prokaryotes are smaller and lack distinct compartments, resulting in a more uniform cytoplasm, while eukaryotic cells, being more complex and having compartmentalization in the form of organelles, can result in observable crowding differences, depending on the cells' microenvironment (Zhou et al. 2008). Finally, environmental conditions can also influence macromolecular crowding, sometimes to a great extent. The larger the increase in the environmental change, the larger the macromolecular crowding patterns can be, which is specifically related to extremophiles, including acidophiles, thermophiles, and halophiles, which must be able to survive under extreme pH conditions, temperatures, or salinity. These alter their cytoplasm and have an impact on macromolecular crowding (Tehei and Zaccai 2005; Cowan et al. 2024).

Ways to Study Crowding in Cytoplasm

The study of cytoplasm and its complex components is done through advanced and innovative imaging techniques including fluorescence correlation spectroscopy, nuclear magnetic resonance, and cryo-electron tomography (Deniz et al. 2008; Rivas and Minton 2016; Weber et al. 2019). When it comes to macromolecular crowding, it is specific since it cannot be easily and accurately mimicked during *in vitro* experiments. This is why there are specific approaches, for instance using crowding sensitive probes (Gnutt and Ebbinghaus 2016), or using different computational modeling techniques to simulate macromolecular behavior in a cytoplasmic environment is also of great importance (Rivas and Minton 2022).

Impact of Macromolecular Crowding on Cellular Functions

Macromolecular crowding has a range of effects on different cellular functions, enzyme activity and kinetics, molecular binding, metabolic pathways, nucleic acid dynamics, and signal transduction (Mourão et al. 2014; Mittal et al. 2015; Fuentes-Lemus et al. 2021). It increases the effective concentration of molecules and their collision frequency through the process known as the excluded volume effect, which in turn modulates their interactions and the formation of molecular complexes, stabilizes protein-protein interactions, and helps with the assembly of large macromolecular complexes including ribosomes (Kuznetsova et al. 2015). The summary of the different effects of macromolecular crowding on cellular functions is shown in Table 10.1.

Enzyme Activity and Metabolic Pathways

Enzymatic activity can be affected in different ways through the modulation of kinetics, thermodynamics, substrate availability, and metabolic regulation. It can enhance binding affinity, including that of a substrate, through the excluded volume effect—promoting interactions by limiting the available diffusion space, resulting in the increase of effective reactant concentration. Molecular crowding can also stabilize macromolecular structures, enzyme conformations, and intermediates, by lowering the free energy required for a reaction. This influences metabolic flux and can, in turn, increase catalytic activity, promote enzyme-substrate binding, and protein-protein interactions, as well as the formation of multi-protein complexes—metabolons—which have important roles in various metabolic pathways (Kuznetsova et al. 2014). Molecular crowding can also promote energy-efficient pathways through the regulation of reaction thermodynamics, as well as regulate metabolite and cofactor diffusion rates, by confining reactions through microdomains (Silverstein and Slade 2019).

Table 10.1 Impact of macromolecular crowding on different cellular functions

Biological function	Mode of impact of macromolecular crowding
Enzyme activity	Can promote enzyme activity, enhance binding affinity, stabilize enzyme structure, and lower free energy required for a reaction. Excessive crowding can decrease enzyme efficiency
Metabolic pathways	Can promote energy-efficient pathways, regulate metabolite and cofactor diffusion rates, and force cells to adapt to stress
Protein folding and stability	Can accelerate protein folding, stabilize the native state, and reduce misfolding and aggregation. Excessive crowding promotes aggregation and stabilization of misfolded proteins or intermediates
Signal transduction	Helps facilitate signaling through the modulation of the concentration of signaling molecules and the formation of protein complexes. Excessive crowding can slow down the diffusion of molecules involved in a signaling pathway and interfere with signal transduction
Cellular organization	Various macromolecules, molecular complexes, and cellular structures are formed through macromolecular crowding
Cellular homeostasis and adaptation	Important in facilitating various cellular processes responsible for homeostasis, as well as for the cell's adaptation to stress. Excessive crowding triggers different metabolic and signaling pathways, allowing cells to further adapt to changing conditions

However, it can also have the opposite effect—decreasing the rate of reactions by limiting the movement of molecules through steric hindrance and increasing viscosity (Ellis 2001). Specific multi-step complex reactions can further benefit from macromolecular crowding through the increase of efficiency by regulating transient species more effectively. This is particularly useful during different scenarios of cellular stress, caused by environmental changes. Cells are able to respond to changes in energy demands by modulating anabolic and catabolic pathways (Patel et al. 2017).

Protein Folding and Stability

Protein folding can also be affected by macromolecular crowding in different ways. Under ideal conditions, it can promote the structural integrity of proteins and proper folding of polypeptides through the excluded volume effect and decrease of entropy, which promotes tightly folded structures and reduces the chance of misfolding and aggregation (Christiansen et al. 2013; Ådén and Wittung-Stafshede 2014). This is especially noticeable in small and medium-sized proteins, allowing them to quickly fold into their native states. Folding rates are accelerated through the increase of folding intermediates, which further helps in proper protein folding (Zhou et al. 2008), as shown in different studies (Stagg et al. 2007; Gulotta et al. 2024). Once folded, native protein structures can be further stabilized through molecular crowding by the reduction of surface area available to solvent. This can protect proteins from denaturation by minimizing the energy needed for the maintenance of a native conformation (Wang et al. 2012).

However, protein aggregation can be promoted by macromolecular crowding under stress (for example oxidative stress or heat shock). This can result in cytotoxic effects contributing to pathological effects which are relevant for diseases like Parkinson's or Alzheimer's. Additionally, crowding can also have an effect on partially folded or unfolded proteins, further destabilizing them, and contributing to amyloidosis disease (Patel et al. 2017). One of the ways folding intermediates are stabilized, preventing unwanted macromolecular crowding through protein aggregation, is through the usage of molecular chaperons, including Hsp70 and Hsp90, as well as through the liquid-liquid phase separation mechanism (Oakes and Papa 2015; Julius et al. 2019).

Signal Transduction

Proper cellular signaling requires specific localization and precise interaction of molecules and proteins involved in signaling pathways. Molecular crowding facilitates this through the increase of local concentration of signaling molecules, as well as the formation of protein complexes necessary for signal transduction (Rohwer et al. 1998). A general signal transduction pathway is composed of a transmembrane receptor and downstream propagation of a signal through scaffolding proteins. These are helped through macromolecular crowding, including receptor clustering and the formation of scaffolding complexes, like mitogen-activated protein kinase signaling (Lodish 2021).

Molecular crowding can also modulate the dynamics of signaling molecules. It can enhance signal specificity and reduce crosstalk between different signaling pathways by restricting diffusion and creating local microdomains, as shown in different studies (Xu et al. 2017; Nussinov et al. 2021; Tun et al. 2022). However, under specific conditions, too much regulation and restriction in the form of excessive crowding can have a negative effect and interfere with signal transmission by slowing down the diffusion of signaling molecules and secondary messengers (Minton 2001; Nussinov et al. 2021).

Cellular Organization

Membrane-less organelles are formed through the liquid-liquid phase separation in the cytoplasm, influenced by macromolecular crowding. They represent the crowding of different macromolecules in a local region of the cytoplasm, forming ribosomes, cytoskeleton, or stress granules, critical for proper cellular functioning and response to external stimuli. They participate in various processes, including signal transduction, translation, and protein modification (Li et al. 2021, 2022).

Cellular Homeostasis and Adaptation

Macromolecular crowding in the cytoplasm has an important role in cellular homeostasis as well as adaptive responses to the environment, through the effect on processes already covered in this chapter: biochemical reactions and enzymatic activity, protein folding, molecular interactions, formation of multi-protein complexes, intracellular transport, as well as structural organization (Minton 2001; Zhou et al. 2008).

Additionally, cellular processes responsible for the adaptation to stress also rely on macromolecular crowding. For example, osmotic stress triggers the synthesis of osmolytes maintaining the intracellular pressure and adjusting concentrations of different macromolecules to avoid their aggregation (Speer et al. 2022; Meneses-Reyes et al. 2024). Moreover, exposure to prolonged stress can result in alternative signaling pathways which can trigger different cellular processes including apoptosis (Rana et al. 2020).

Disease and Therapeutic Implications

Neurodegenerative Diseases

Macromolecular crowding overall has an important role in various neurodegenerative diseases, including Alzheimer's disease, Huntington's disease, Parkinson's disease, amyloidosis, and amyotrophic lateral sclerosis, listed in Table 10.2, together with other diseases caused due to crowding in endoplasmic reticulum, section

Table 10.2 Overview of diseases related to protein misfolding and aggregation

Disease	Mechanism
Cystic fibrosis	Mutated CFTR cannot fold properly; UPR pathway is triggered
Diabetes	Increased insulin production, excessive macromolecular crowding, chronic UPR and apoptosis of β-cells
Alzheimer's disease	Accumulation of different proteins, including amyloid-beta peptides and tau protein
Huntington's disease	Mutated huntingtin is more prone to cleavage, misfolding, and aggregation; it affects various processes in the cell, including chaperons
Parkinson's disease	Formation of Lewy bodies in the cytoplasm due to the aggregation of α-synuclein protein. Aggregation affects normal cellular functions and contributes to nerve cell degradation
Amyotrophic lateral sclerosis	Formation of Bunina8 bodies in the cytoplasm due to the aggregation of different proteins, including TDP-43, SOD1, or FUS. They affect normal cellular functioning, can spread further, and result in neuron deterioration
Cancer	Not a lot of information is present on cancer related to macromolecular crowding, however, research shows that it can be important in specific cancers, affecting the localization of key proteins and the further spread of cancer cells

"Macromolecular crowding in endoplasmic reticulum" of this chapter. The pathology of these conditions revolves around macromolecular crowding, impacting folding, aggregation, clearance, organelle dynamics, and intracellular transport. Under normal conditions, crowding helps proteins achieve their native structure and stabilizes it. However, it can also have similar effects on misfolded proteins or partially folded intermediates, including α-synuclein, amyloid-beta (Munishkina et al. 2004; Ma et al. 2012), huntingtin (Groover et al. 2021), and tau proteins (Wu et al. 2016), increasing the chance of their aggregation. These aggregates can affect cellular functions, become toxic to cells, and result in neuronal death (Mittal et al. 2015; Takahashi et al. 2020a; Cicardi et al. 2021). Intrinsically disordered proteins, including huntingtin and tau, are more prone to aggregation induced by the crowding effect, and have been shown to have an increased rate of crowding under specific conditions (Uversky 2009; Cino et al. 2012).

Additionally, stress granules can be affected by macromolecular crowding. They are cytoplasmic granules that isolate mRNAs and RNA-binding proteins during stress when functioning normally. However, problems in macromolecular crowding can result in abnormal stress granules which can worsen cellular dysfunction under stress, and are linked to amyotrophic lateral sclerosis and dementia (Bounedjah et al. 2012).

Crowding can accelerate the nucleation stage of aggregation, which can result in rapid amyloid fibrils formation. Aggregation hotspots can also form, especially in cytoplasmic regions that exhibit impaired proteostasis (Minton 2001; Werner et al. 2021). Moreover, neurodegenerative diseases often have a malfunction in protein quality control systems. In turn, different pathways important for the clearance of misfolded proteins and protein aggregates, including proteasomes and autophagy pathways, become overwhelmed, resulting in further cellular crowding (Zhou et al. 2008; Amzallag and Hornstein 2022).

Therapeutics

Macromolecular crowding, while extremely useful in certain circumstances, can lead to problems and protein aggregation disorders. Understanding and analysis of these pathways can give us insights into potential therapeutic strategies. Currently, there are different approaches to developing therapeutics for disorders caused by macromolecular crowding in cytoplasm.

Osmolytes are small molecules involved in the regulation of a cell's osmotic pressure. They do not directly affect protein function, but they can contribute to the stabilization of proteins' structures and reduce aggregation by preventing and decreasing intermediate protein conformations that can contribute to disorders. These compounds have been found useful in the treatment of diseases like Alzheimer's and Huntington's (Kushwah et al. 2020; Aravindan et al. 2020; Khan et al. 2023).

Proteostasis—the process responsible for the homeostasis of proteins and their quality control—is another approach that has potential benefits for aggregation-related diseases. Molecules involved in the proper folding and clearing of misfolded proteins, including chaperons and proteasome activators, can also counter the negative effects of macromolecular crowding. Targeting these through specific drugs, and focusing on autophagy-inducing pathways as well, could prove beneficial (Amzallag and Hornstein 2022).

Other approaches to the design of the treatment include targeting biochemical pathways dysregulated in crowding, including those responsible for stress granule formation, through stress granule modulators, which could help with amyotrophic lateral sclerosis (Portz et al. 2021; Yang et al. 2023). Additionally, using nanotechnology holds potential as well, and nanoparticles that could modulate crowding, or deliver anti-aggregation agents to aggregation hotspots could be helpful as well (Asefy et al. 2021).

Macromolecular Crowding in Nucleus

The nucleus, as a core structural unit of the cell, is a place where the most fundamental biochemical processes take place, such as DNA replication, transcription, splicing, and ribosome biogenesis. In total, about 2 meters of DNA is found in the nucleus, encapsulated by the nuclear envelope, and nucleolus, and crosslinked with a complex network of proteins and membrane (Rowat et al. 2008). In its spherical nature, the nucleolus primarily functions as a factory for the assembling of the cell's ribosomes before transported into the cytoplasm, ribosomal RNA (rRNA) transcription, rRNA processing (Dubois and Boisvert 2016). The long genomic DNA molecule is compacted and stratified within chromosomes. Chromosomes are made of chromatin structure, a mixture composed of DNA molecules and histone proteins (Luijsterburg et al. 2008). The chromatin three-dimensional structure undergoes dynamic remodeling during cell differentiation, playing a critical role in gene silencing and strongly influencing gene expression (García Fernández et al. 2023).

The chromatin is organized in 3 basic levels. In the lowest level DNA is wrapped around nucleosomes like a beads-on-a-string structure (Miyoshi and Sugimoto 2008). The nucleosome core particle (NCP) is composed of a central heterotetramer formed by 8 histones molecules (H2A, H2B, H3 and H4), in a left-handed 1.7 superhelical turn and a long dsDNA molecule (147 bp) wrapped in left-handed superhelical turns, as presented in Fig. 10.2 (Koch et al. 2013; Li and Zhu 2015). The 3D structure of the NCP has been determined by X-ray crystallography method, at the 1.9–2.8 Å resolution (Li and Zhu 2015).

In addition, chromosome packaging is organized by linker histones (H1 and H5), which bind to the DNA linker regions (DNA entry site) and exit to the NCP, organizing the nucleosome structure into a condensed 30-nm diameter helical chromatin fiber, representing a second level of DNA compaction (Ramakrishnan 1997; Li and Zhu 2015). This level represents a euchromatin structure, which is highly dynamic and unfolds into an approximate 11 nm fiber (Morrison and Thakur 2021). At the highest level of chromatin organization, DNA is packaged into the metaphase

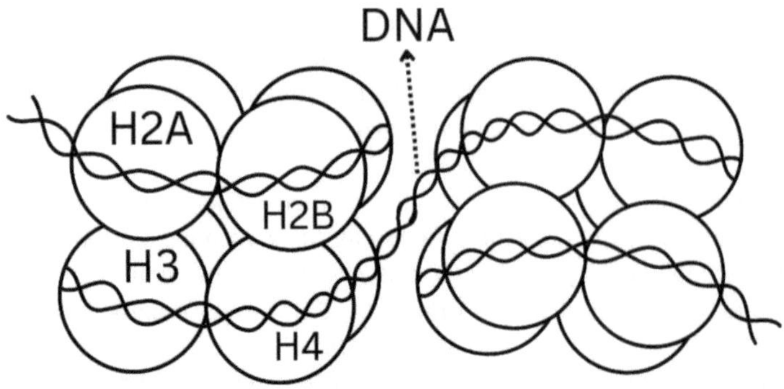

Fig. 10.2 Histone octamer with associated dsDNA (Figure generated with BioRender)

chromosome, during Mitosis and Meiosis (Miyoshi and Sugimoto 2008; Corless and Gilbert 2016). Due to the dynamic behaviors of all organelles in the cell, the chromatin structure changes during the cell cycle progress. Therefore, the crowding of its complex fluids affects the entire range of macromolecules contained in the nucleus (Ma et al. 2015).

So far, it's clear that the molecular conditions in living cells are different from those *in vitro*, in which purified biomolecules exist in a homogeneous and dilute solution. *In vivo* experimental evidence showed that osmotic pressure within the cell is a crucial factor in molecular crowding of the nucleus. One of the studies demonstrated the significant influence of the osmotic hypertonic effects on nuclear ultrastructure, where the supplementation of sucrose, sorbitol, or NaCl to the cell growth medium shifted the nuclear organization (Richter et al. 2007).

Further, the thermodynamics of DNA and RNA duplexes with the nucleus are deeply affected by an uptake of water molecules. In addition, in the last decade, several studies have evaluated the interesting behavior of nucleic acids in cellular environments using various co-solutes. The DNA/RNA duplex stability is largely affected by co-solute concentrations (Marky and Kupke 2000). For example, Spink and Chaires demonstrated that *E. coli* DNA duplexes are destabilized by small co-solutes, whereas the duplexes are stabilized by large co-solutes (Spink and Chaires 1999).

In laboratory conditions, Polyethylene glycol (PEG) is a very common and typical molecular crowding agent, shown to cause conformation change to DNA. Several studies demonstrated that the concentration of PEG affects and induce the B – Z transition of DNA duplexes (Preisler et al. 1995; Kültz 2000; Li et al. 2008). Similar effects are demonstrated by other solutes, such as 1,2-Dimethoxyethane and 1-propanol. These molecules regulate the topological change in DNA G-quadruplex, where the telomeric repeat from hybrid to parallel topology is observed (Tateishi-Karimata et al. 2020).

A computational biology analysis from 2016 concluded that macromolecular crowding restricts the movement of the transcription factors weakens the transcriptional bursting and causes a more homogenous cell population (Golkaram et al.

2016). The gene expression regulation and control significantly depend on the structures and stabilities of nucleic acids, and it is a fact that their function and stability have been optimized to function in a crowding molecular environment (Endoh et al. 2023).

It is shown that molecular crowding affects the polymerization of DNA with unnatural nucleobases and sugars. A study demonstrated that in uncrowded conditions, pyrimidine dNTPs had high incorporation efficiency but low accuracy, while purine dNTPs had the opposite trend. However, under crowded conditions, the DNA polymerization efficiency stayed largely unchanged (Takahashi et al. 2020b).

Macromolecular Crowding in Chloroplast

Chloroplast is an essential organelle in plants, mainly involved in the production of energy through photosynthesis and oxygen-release processes (Song et al. 2021). During this process the chloroplasts coordinate the production of active compounds such as amino acids, nucleotides, lipids, phytohormones, vitamins, and secondary metabolites (Bobik and Burch-Smith 2015). Chloroplast is made of stroma, thylakoids, lamella, intermembrane space and inner membrane. A stack made of thylakoids suspended in the stroma is called granum, where light reactions of photosynthesis occur (Hameed et al. 2021).

Macromolecular crowding in chloroplasts is a key factor in optimizing photosynthetic efficiency and ensuring the proper functioning of the photosynthetic machinery under varying environmental conditions (Kirchhoff et al. 2008). Two macromolecular complexes, Photosystem II (PSII) with Light-Harvesting Complex II (LHCII) are responsible for the photosynthesis process located in grana thylakoids (Kirchhoff 2019). PSII is a large protein complex that controls the light-dependent reactions of photosynthesis using light energy by splitting water molecules into oxygen, protons, and electrons. LHCII, on the other hand, is a multidomain protein that captures light energy and transfers it to the PSII reaction center (Pan et al. 2020). Additionally, Photosystem I (PSI) with Light-Harvesting Complex and the ATPase enzyme are found preferentially in unstacked regions (stromal lamellae) (Johnson 2011).

The complicated mixture of PSII and other protein complexes in chloroplast are driven by close interactions between stacked and unstacked regions, but mainly modulated by macromolecular crowding effects and membrane surface charge (Wietrzynski et al. 2024). In addition, the photosynthetic membranes are the densest of all bio-membranes (Kirchhoff 2008). It is presented that grana thylakoid membranes impose unique structural limitations on the movement of molecules between the stacked and unstacked regions. For example, the diffusion of protein complexes between these regions depends crucially on the relationship between the width of the stromal gap in grana stacks and the extent of protein protrusion into the stroma (Kirchhoff 2014).

Scientists have used the percolation theory to calculate the contact between molecular cluster in biological membranes, determines whether long-distance diffusion of biomolecules is possible or not (Deniz 2022). The actual protein number for grana and whole thylakoids is derived from several studies, mainly using atomic force microscopy and other biochemical data (Kirchhoff et al. 2002; Ostermeier et al. 2024; Rolo et al. 2024).

The stromal gap in grana is only about 3.5 nm, in this way excluding PSI and the ATPase as a structural compartment from stacked grana because their stromal protrusions are larger (Amunts and Nelson 2009; Kirchhoff et al. 2011). The width of thylakoid lumen is also a critical parameter for protein diffusion, and in several species it was determined to be around 9 nm (Perez-Boerema et al. 2024). The majority of the total membrane area is occupied by grana stacks, having the majority of the total pigment content (Albertsson 2001). However, there is an observable differences in proportion of grana relative to the stroma thylakoids on the light condition (Anderson et al. 2012).

Further, the protein density in stroma lamellae membranes is about 1/3 lower than in grana membranes leading to the conclusion that the difference between stacked and unstacked thylakoid membranes is not high (Kirchhoff 2014). Several studies presented that more than 1300 proteins are organized within the chloroplast sub-compartments (two membranes, stroma, thylakoid membranes, thylakoid lumen, plastoglobuli, and nucleoids), confirming the strong crowding effect in chloroplast (Sun et al. 2009; Ferro et al. 2010).

Macromolecular crowding has a physiological advantage in photosynthetic membranes, it allows protein and protein bound pigments to interact in higher rates, increasing the likelihood of absorption of sunlight, which is a molecularly diluted energy source (Kirchhoff 2014). For example, over 20 years ago Kirchhoff et al., concluded that the chlorophyll concentration in thylakoid membranes was about 0.3 M (Kirchhoff et al. 2002). Further, in 2014 the same author claimed that for an efficient intermolecular exciton energy transfer between LHCII complexes and PSII, a high protein density in grana membranes is favored (Kirchhoff 2014), confirming molecular crowding positive effects.

Macromolecular Crowding in Endoplasmic Reticulum

Overview of Endoplasmic Reticulum Structure and Function

The endoplasmic reticulum (ER) is an important organelle in eukaryotic cells, involved in protein and lipid synthesis, calcium storage, and cellular signaling. Although ER is a single organelle composed of a continuous membrane system, it has two separate regions—rough and smooth ER. Rough ER has ribosomes and is therefore involved in the synthesis of proteins, while smooth ER is involved in lipid metabolism. Nascent polypeptides are folded in the lumen of ER into functioning proteins, where they are also post-translationally modified (Voeltz et al. 2002; Ellgaard and Helenius 2003). ER is functionally connected to other organelles, including the Golgi apparatus, mitochondria, and endosomes, participating with them in metabolite and molecule exchange, as well as intracellular communication (Rowland and Voeltz 2012).

Unlike cytoplasm, ER's environment is slightly oxidizing, which favors the formation of disulfide bonds. This condition also influences the methods used to study macromolecular crowding in ER. Molecular proximity and dynamics are used to gain insights into the level of crowding using Förster Resonance Energy Transfer (FRET) and fluorescence correlation spectroscopy (Lippincott-Schwartz et al. 2000). Due to the oxidizing environment, specially designed FRET sensors are used, allowing for precise measurements of concentration and spatial distributions of macromolecules. These types of assays indicate that a high degree of macromolecular crowding is present in ER and might at certain times be even higher than in cytoplasm. It can vary based on different factors, including protein synthesis rates and stress. Stress specifically increases the presence of chaperons and accumulation of misfolded proteins (Lecinski et al. 2021).

Main Macromolecules in the ER

Proteins are the main macromolecules found in the lumen of ER, and we can classify them further into chaperons, enzymes, and structural components.

Chaperone proteins comprise the most abundant macromolecular component in the ER. They are involved in the proper folding of proteins and the prevention of their aggregation. The central chaperon is binding immunoglobulin protein—a member of the heat shock protein 70 family. It can bind to nascent proteins and stabilize them until they reach native confirmation, decreasing the chance of misfolding (Schwarz and Blower 2016). Other chaperone types, including calnexin and calreticulin, are involved in the folding of glycoproteins through the recognition of glycan structures (Ellgaard and Ruddock 2005). Under stress and specific conditions, misfolded proteins might accumulate, which can trigger unfolded protein response—a pathway responsible for either restoring homeostasis or inducing apoptosis (Hetz 2012).

Enzymes in the ER lumen help with post-translational modification which are the main steps in the protein maturation journey. Protein disulfide isomerases are an example of such enzymes, helping form and rearrange disulfide bonds which stabilize proteins' tertiary and quaternary structure (Anelli and Sitia 2008). Peptidyl-prolyl isomerases enhance the folding process through the isomerization of proline residues, while glycosyltransferases help in the attachment of sugar molecules to proteins, ensuring structural stability (Alberts 2022).

Impact of Crowding on Protein Folding

Protein Folding

Since ER is the main organelle for the proper folding of proteins, processes there can influence macromolecular crowding and be connected to the problems related to cellular homeostasis, protein aggregation, and disease. Normal operation helps proteins fold properly, while misfolded proteins are targeted for degradation through the ER-associated degradation (ERAD) pathway (Krshnan et al. 2022).

ER-Associated Degradation Pathway

The ERAD pathway identifies misfolded proteins and retrotranslocates them into the cytoplasm to be degraded through the ubiquitin-proteasome mechanism, which is done in three stages. First molecular chaperons, including BiP and calnexin, recognize misfolded proteins; and normal levels of macromolecular crowding can help since they increase the chance of molecular interactions and consequently the efficiency of detection of misfolded proteins. They are then retrotranslocated to the cytoplasm through the coordinated work between proteins found in ER lumen and cytoplasm, including Sec61 translocons and ubiquitin ligases. Finally, misfolded proteins are ubiquitinated and degraded in the cytoplasm (Smith et al. 2011; Krshnan et al. 2022).

However, excessive crowding, especially during stress, can promote the accumulation of misfolded proteins due to an increase in aggregation risk and molecular collisions, as well as due to conditions such as oxidative stress, which can overwhelm the quality control mechanisms found in ER. Moreover, abnormal crowding can result in partially folded proteins having their hydrophobic regions expressed for prolonged periods of time, increasing the chance for non-desirable interactions and aggregation, further burdening the folding mechanisms in ER. This can trigger unfolded protein response (UPR) and, under continued stress, can contribute to diseases including neurodegeneration and diabetes (Hetz 2012; Rivas and Minton 2022).

Unfolded Protein Response

UPR is activated by proteins bound in the ER membrane, including inositol-requiring enzyme 1 (IRE1), protein kinase R-like ER kinase (PERK), and activating transcription factor 6 (ATF6). It adapts the ER's folding and stabilization capacities in several ways, shown in Table 10.3. It can enhance the folding capacities by upregulating the transcription of genes for chaperons and folding enzymes through ATF6 and IRE1. UPR can reduce the load of protein synthesis by decreasing translation rates through PERK. It can also improve the degradation of misfolded proteins by enhancing ERAD pathway through IRE1. And it can modulate the expansion of ER

Table 10.3 Unfolded protein response main mechanisms

Mechanism	Mode of action
Enhanced folding capacity	Upregulation of chaperons and folding enzymes
Reduced protein load	Decrease of translation rates
Degradation of misfolded proteins	Enhancing ERAD pathway
ER expansion	Increase of ER volume to decrease crowding

to provide more space and reduce crowding that way (Oakes and Papa 2015; Hetz et al. 2020).

Macromolecular crowding in the ER can also affect the functionality of chaperons, apart from increasing their demand. Increased macromolecular concentrations can change the diffusion rates of chaperones and their targets, which impacts folding kinetics (Gorensek-Benitez et al. 2017).

Protein Transport

ER is the place where proteins are sorted for proper transport and secretion, through the vesicles coated with COPII protein. Crowding can affect these processes by impacting protein-protein interaction kinetics, and vesicle formation efficiency, which can further have an impact on secretion rate of large and aggregation-prone proteins (Zhou et al. 2008; Das and Sen 2022).

Disease and Therapeutic Implications

Stress-induced macromolecular crowding in ER is responsible for different types of diseases. Moreover, chaperons have an important role in maintaining proteostasis, especially in crowded environments, while protein aggregation and stress are observed in different disorders, especially due to chaperone dysfunction (Oakes and Papa 2015). Brief overview of the diseases is listed in Table 10.2.

Cystic Fibrosis

In cystic fibrosis, different mutations, the most common one being ΔF508—the deletion of Phenylalanine at 508th position in cystic fibrosis transmembrane protein (CFTR), result in improper folding of CFTR. Under normal conditions, misfolded CFTR is targeted for degradation via ERAD. However, due to the disease, it accumulates and increases stress on ER. Moreover, continued stress and macromolecular crowding can overwhelm ER pathways, including the UPR, which can further contribute to inflammation and pathogenesis, as well as trigger apoptosis (Bartoszewski et al. 2011).

Diabetes

Diabetes is another disease implicated in problems with macromolecular crowding in ER. In type 2 diabetes, macromolecular crowding in the ER of pancreatic β-cells is observed and can contribute to their dysfunction. Large amounts of insulin are produced, which in turn requires an increase in protein folding, providing additional burden on cells. Other problems including hyperglycemia and chronic metabolic stress can further increase the demand for insulin, contributing to the crowded environment (Schwarz and Blower 2016). This can result in different problems: slowing down of the folding and transport of insulin, excessive protein crowding, as well as chronic UPR activation which can lead to the apoptosis of β-cells, further decreasing insulin production and worsening hyperglycemia (Hetz 2012).

Neurodegenerative Disorders

Different neurodegenerative disorders are caused and can be exacerbated due to protein misfolding and aggregation, the majority of which are mentioned in 2.3.1 of this chapter. For instance, Alzheimer's disease is characterized by the aggregation of amyloid-beta peptides and tau protein (Gallardo and Holtzman 2019). Huntington's disease is caused by mutations in huntingtin, resulting in various changes in cells, including the misfolding and crowding of proteins in both cytoplasm and ER. This further impairs the function of chaperons and proteasomes, preventing the clearance of toxic aggregates (Ellgaard and Ruddock 2005). Parkinson's disease is not understood well. It is characterized by the deterioration of nerve cells, and the cause involves the aggregation of α-synuclein—a protein involved in the regulation of synaptic trafficking and neurotransmitter release. Their excessive crowding forms Lewy bodies in the cytoplasm, displacing other cellular structures and affecting function (Morris et al. 2024). Amyotrophic lateral sclerosis is another disease that is poorly understood, with protein aggregates forming in the cytoplasm in the form of Bunina bodies, affecting cellular functions, and resulting in the deterioration of primarily motor neurons. The aggregation of proteins can spread like prions, and different mutations have been shown to have an effect on the type of proteins comprising the aggregates, including TDP-43, SOD1, or FUS (Meyer 2021; Wang and Zeng 2024).

Cancer

There isn't a lot of research on macromolecular crowding and cancer. Moreover, cancer comprises a variety of diseases that have different causes and different manifestations (de Visser and Joyce 2023). Studying crowding would require a specific approach for each individual cancer type. One study shows that macromolecular crowding is important in the specific localization of a key protein involved in cancer—SIRT1 (Sun and Fang 2016). It is a class III histone deacetylase that could act both as a tumor suppressor and a tumor promoter, depending on the cancer type and signaling pathways it is involved in (Lin and Fang 2013). Study on ovarian cancer and extracellular macromolecular crowding shows different effects, ranging impeding the aggregation of cancer cells, affecting intracellular cytoskeleton and nucleus, to promoting their spread to other cells and tissues (Bascetin et al. 2021).

Therapeutic Approaches

Different options are available for the mitigation and treatment of excessive macromolecular crowding in the ER. Molecules enhancing the efficiency of chaperons and proteasomes, or those reducing the influx of proteins into ER by decreasing translation through integrated stress response inhibitor may decrease the effects of crowding (Hetz and Papa 2018). Activation of alternative degradation pathways, including autophagy, is also seen as an option to decrease the burden of macromolecular crowding on ERAD pathway (Ruggiano et al. 2014).

Increasing the transcription of endogenous chaperons, as well as introducing chemical chaperons are also one of the potential therapeutic strategies, which would help with crowded environment and restore the function of ER. Restoring calcium levels through SERCA activators is important for proper functioning of chaperons in ER and can help with excessive macromolecular crowding. Moreover, reduction of oxidative stress through antioxidants can also decrease aggregation and improve protein stability (Bhandary et al. 2012). Additionally, small molecules that can stabilize misfolded proteins could contribute to the efficiency of the ERAD pathway, decrease chronic ER stress, and help with the diseases characterized by prolonged aggregation of proteins (Wolff et al. 2014).

Regarding cancer, the macromolecular crowding approach has proven useful specifically for the study of cancer development. One study has shown that using polyethylene glycol modulates collagen matrix architecture and can strongly influence breast cancer behavior (Ranamukhaarachchi et al. 2019). Another study shows the potential of using macromolecular crowding for the development of *in vitro* tumor models by enhancing cell-derived matrices. They have demonstrated better mimicking of tumor microenvironment, making the model more accurate for cancer research (Shologu et al. 2022).

Macromolecular Crowding in Golgi Apparatus

The Golgi apparatus or Golgi complex (also known as Golgi body) is an integral part of the cellular endomembrane system, which besides Golgi apparatus includes the ER and lysosomes. Although these endomembrane system components constitute a network of membranes, they are not continuous with one another. Structurally, Golgi apparatus, being a key membrane-bound organelle in the secretory pathway, represents the densely packed cisternae that are layered on top of each other to form stacks and therefore is typically depicted as a stack of pancake- or pita-shaped disks which require sophisticated arrangement of various proteins and membrane faces (Li et al. 2019). In mammals, the stacks are further connected by tubular membranes into an elongated, twisted, but continuous structure named the Golgi ribbon (Wei and Seemann 2017). "The Golgi apparatus is morphologically distinct from the ER because of its juxtanuclear position, and at the ultrastructural level, appears as a ribbon-like structure of laterally interconnected stacks of flattened cisternae, each having a network of extensive fenestrated membranes associated both at their *cis*- and *trans*-face" (Smirle et al. 2013).

The Golgi apparatus is a complex organelle responsible for sorting, modification, and packaging of proteins, fats, and sugars for transport within a cell. Therefore, the Golgi apparatus is often defined as the Grand Central Station of all membrane traffic, acting as part factory, part shipping station, and part switchboard. This organelle plays a number of crucial roles in various essential cellular functions, such as cell-cell communication, control of cell polarity, DNA damage response, immune signaling, mitosis, and shaping the cell surface, with these and other functions being critically dependent on the integrity of Golgi apparatus and protein homeostasis (Buzuk and Hellerschmied 2023). In fact, sorting at the Golgi apparatus is protein folding state-dependent, with unfolded proteins being efficiently segregated from folded proteins in the Golgi (Hellerschmied et al. 2019). Therefore, it is not surprising that the Golgi apparatus is characterized by the multifaceted role in neuronal plasticity, development, neurogenesis, and Alzheimer's disease (Toader et al. 2023).

It was emphasized that the membranes of the Golgi apparatus contain from 30,000 to 40,000 individual membrane proteins per μm^2 (Quinn et al. 1984), clearly representing a highly crowded milieu. The Golgi apparatus can become further crowded during several cellular processes, such as mitosis and apoptosis, cell migration, and differentiation, as well as can originate from high protein synthesis rate leading to the Golgi overload with cargo (Gomez-Navarro et al. 2020), or due to the stress conditions disrupting normal Golgi function and resulting in protein accumulation and crowding (Kim et al. 2023), or because of the impaired retrograde transport, where disruption of the recycling of proteins back from the Golgi to the ER can contribute to crowding (Hellerschmied et al. 2019; Arab et al. 2024). During these events, the Golgi apparatus undergoes significant structural changes, often being fragmented into smaller vesicles, which can lead to a build-up of materials within the organelle as it attempts to process and package molecules needed for the cell's changing conditions (Wei and Seemann 2017). The enhanced crowding of the Golgi apparatus is linked to a number of issues, such as (Hellerschmied et al. 2019; Gomez-Navarro et al. 2020):

- Protein misfolding, as unfolded proteins can accumulate in the Golgi apparatus, which can lead to proteinopathies;
- Golgi fragmentation caused, e.g., viral infections and leading to a rapid increase in viral trafficking and release;
- Malfunction leading to improper protein modification, which can affect protein function and stability, or to failure to accurately sort proteins to their correct destinations, potentially leading to misfolded or incorrectly targeted proteins.

Conclusions

Macromolecular crowding is an important process, affecting the efficiency and specificity of biochemical processes and molecular interactions in all parts of cells. Cytoplasm contains a large number of macromolecules, ranging from nucleic acids and proteins including the cytoskeleton, to their complexes, including ribosomes. Moreover, different cell types have different levels of macromolecular crowding as well as different molecules present, including muscle and liver cells which exhibit higher concentrations of carbohydrates in the form of glycogen. Crowding in the cytoplasm has a wide range of effects on the formation and storage of these macromolecules, cellular shape and integrity, as well as metabolic processes. Nucleus has a high macromolecular concentration and crowding influences chromatin's structural organization, nucleic acid stability, and regulation of gene expression.

Crowding of macromolecules in chloroplasts is essential for efficient photosynthesis, allowing for tight packing of proteins, pigments, and membranes required for optimal solar energy capture, conversion, utilization, and transfer. Study of crowding in such specific environments can help us improve crop productivity and energy capture and conversion in artificial systems.

Endoplasmic reticulum is home to different processes, including protein and lipid synthesis, as well as calcium storage. Proper folding of proteins and activity of chaperons is sensitive to the level of macromolecular crowding and disruptions, including excessive crowding, can trigger a wide range of responses, from the decrease of protein synthesis, to increase of folding efficiency, to protein aggregation.

Various diseases can be caused or worsened by macromolecular crowding, including neurogenerative disorders, diabetes, cystic fibrosis, diabetes, and cancer. Overall, the effects of macromolecular crowding are crucial for understanding normal cellular function as well as its disruption and disease progression. Further study and deeper understanding of crowding and its effects is essential for the advancement of biological research and therapeutic strategies aiming to combat crowding-related problems.

Compliance with Ethical Standards Authors declare no conflict of interest.

Funding Source Not applicable.

References

Ådén J, Wittung-Stafshede P (2014) Folding of an unfolded protein by macromolecular crowding in vitro. Biochemistry 53:2271–2277. https://doi.org/10.1021/bi500222g

Alberts B (2022) Molecular biology of the cell, 7th edn. W. W. Norton & Company, New York

Albertsson P-Å (2001) A quantitative model of the domain structure of the photosynthetic membrane. Trends Plant Sci 6:349–354. https://doi.org/10.1016/S1360-1385(01)02021-0

Alfano C, Fichou Y, Huber K et al (2024) Molecular crowding: the history and development of a scientific paradigm. Chem Rev 124:3186–3219. https://doi.org/10.1021/acs.chemrev.3c00615

Amunts A, Nelson N (2009) Plant photosystem I design in the light of evolution. Structure 17:637–650. https://doi.org/10.1016/j.str.2009.03.006

Amzallag E, Hornstein E (2022) Crosstalk between biomolecular condensates and Proteostasis. Cells 11:2415. https://doi.org/10.3390/cells11152415

Anderson JM, Horton P, Kim E-H, Chow WS (2012) Towards elucidation of dynamic structural changes of plant thylakoid architecture. Philos Trans R Soc B 367:3515–3524. https://doi.org/10.1098/rstb.2012.0373

Anelli T, Sitia R (2008) Protein quality control in the early secretory pathway. EMBO J 27:315–327. https://doi.org/10.1038/sj.emboj.7601974

Arab M, Chen T, Lowe M (2024) Mechanisms governing vesicle traffic at the Golgi apparatus. Curr Opin Cell Biol 88:102365. https://doi.org/10.1016/j.ceb.2024.102365

Aravindan S, Chen S, Choudhry H et al (2020) Osmolytes dynamically regulate mutant huntingtin aggregation and CREB function in Huntington's disease cell models. Sci Rep 10:15511. https://doi.org/10.1038/s41598-020-72613-3

Asefy Z, Hoseinnejhad S, Ceferov Z (2021) Nanoparticles approaches in neurodegenerative diseases diagnosis and treatment. Neurol Sci 42:2653–2660. https://doi.org/10.1007/s10072-021-05234-x

Bachrach U (2005) Naturally occurring polyamines: interaction with macromolecules. CPPS 6:559–566. https://doi.org/10.2174/138920305774933240

Bartoszewski R, Rab A, Fu L, et al (2011) CFTR expression regulation by the unfolded protein response. Methods Enzymol 491:3–24. https://doi.org/10.1016/B978-0-12-385928-0.00001-8, Elsevier

Bascetin R, Laurent-Issartel C, Blanc-Fournier C et al (2021) A biomimetic model of 3D fluid extracellular macromolecular crowding microenvironment fine-tunes ovarian cancer cells dissemination phenotype. Biomaterials 269:120610. https://doi.org/10.1016/j.biomaterials.2020.120610

Berth SH, Lloyd TE (2023) Disruption of axonal transport in neurodegeneration. J Clin Invest 133:e168554. https://doi.org/10.1172/JCI168554

Bhandary B, Marahatta A, Kim H-R, Chae H-J (2012) An involvement of oxidative stress in endoplasmic reticulum stress and its associated diseases. Int J Mol Sci 14:434–456. https://doi.org/10.3390/ijms14010434

Blanco PM, Madurga S, Garcés JL et al (2021) Influence of macromolecular crowding on the charge regulation of intrinsically disordered proteins. Soft Matter 17:655–669. https://doi.org/10.1039/D0SM01475C

Bobik K, Burch-Smith TM (2015) Chloroplast signaling within, between and beyond cells. Front Plant Sci 6. https://doi.org/10.3389/fpls.2015.00781

Bounedjah O, Hamon L, Savarin P et al (2012) Macromolecular crowding regulates assembly of mRNA stress granules after osmotic stress: new role for compatible osmolytes. J Biol Chem 287:2446–2458. https://doi.org/10.1074/jbc.M111.292748

Buzuk L, Hellerschmied D (2023) Ubiquitin-mediated degradation at the Golgi apparatus. Front Mol Biosci 10:1197921. https://doi.org/10.3389/fmolb.2023.1197921

Chiolo I, Minoda A, Colmenares SU et al (2011) Double-Strand breaks in heterochromatin move outside of a dynamic HP1a domain to complete Recombinational repair. Cell 144:732–744. https://doi.org/10.1016/j.cell.2011.02.012

Christiansen A, Wang Q, Cheung MS, Wittung-Stafshede P (2013) Effects of macromolecular crowding agents on protein folding in vitro and in silico. Biophys Rev 5:137–145. https://doi.org/10.1007/s12551-013-0108-0

Cicardi ME, Marrone L, Azzouz M, Trotti D (2021) Proteostatic imbalance and protein spreading in amyotrophic lateral sclerosis. EMBO J 40:e106389. https://doi.org/10.15252/embj.2020106389

Cino EA, Karttunen M, Choy W-Y (2012) Effects of molecular crowding on the dynamics of intrinsically disordered proteins. PLoS One 7:e49876. https://doi.org/10.1371/journal.pone.0049876

Clegg JS (1984) Properties and metabolism of the aqueous cytoplasm and its boundaries. Am J Phys Regul Integr Comp Phys 246:R133–R151. https://doi.org/10.1152/ajpregu.1984.246.2.R133

Collette D, Dunlap D, Finzi L (2023) Macromolecular crowding and DNA: bridging the gap between in vitro and in vivo. IJMS 24:17502. https://doi.org/10.3390/ijms242417502

Corless S, Gilbert N (2016) Effects of DNA supercoiling on chromatin architecture. Biophys Rev 8:245–258. https://doi.org/10.1007/s12551-016-0210-1

Costa E, Machado M, Pintado M, Silva S (2022) Biological macromolecules as immunomodulators. In: Biological macromolecules. Elsevier, pp 273–287

Cowan DA, Albers SV, Antranikian G et al (2024) Extremophiles in a changing world. Extremophiles 28:26. https://doi.org/10.1007/s00792-024-01341-7

Das N, Sen P (2022) Macromolecular crowding: how shape and interaction affect the structure, function, conformational dynamics and relative domain movement of a multi-domain protein. Phys Chem Chem Phys 24:14242–14256. https://doi.org/10.1039/D1CP04842B

de Visser KE, Joyce JA (2023) The evolving tumor microenvironment: from cancer initiation to metastatic outgrowth. Cancer Cell 41:374–403. https://doi.org/10.1016/j.ccell.2023.02.016

Deniz AA (2022) Percolation physics and density transition frameworks converge in biomolecular condensation. Proc Natl Acad Sci USA 119:e2210177119. https://doi.org/10.1073/pnas.2210177119

Deniz AA, Mukhopadhyay S, Lemke EA (2008) Single-molecule biophysics: at the interface of biology, physics and chemistry. J R Soc Interface 5:15–45. https://doi.org/10.1098/rsif.2007.1021

Dubois M-L, Boisvert F-M (2016) The nucleolus: structure and function. In: Bazett-Jones DP, Dellaire G (eds) The functional nucleus. Springer, Cham, pp 29–49

Dupuis NF, Holmstrom ED, Nesbitt DJ (2014) Molecular-crowding effects on single-molecule RNA folding/unfolding thermodynamics and kinetics. Proc Natl Acad Sci USA 111:8464–8469. https://doi.org/10.1073/pnas.1316039111

Ellgaard L, Helenius A (2003) Quality control in the endoplasmic reticulum. Nat Rev Mol Cell Biol 4:181–191. https://doi.org/10.1038/nrm1052

Ellgaard L, Ruddock LW (2005) The human protein disulphide isomerase family: substrate interactions and functional properties. EMBO Rep 6:28–32. https://doi.org/10.1038/sj.embor.7400311

Ellis RJ (2001) Macromolecular crowding: obvious but underappreciated. Trends Biochem Sci 26:597–604. https://doi.org/10.1016/s0968-0004(01)01938-7

Endoh T, Tateishi-Karimata H, Sugimoto N (2023) Effects of molecular crowding on structures and functions of nucleic acids. In: Sugimoto N (ed) Handbook of chemical biology of nucleic acids. Springer Nature Singapore, Singapore, pp 1127–1171

Ferro M, Brugière S, Salvi D et al (2010) AT_CHLORO, a comprehensive chloroplast proteome database with Subplastidial localization and curated information on envelope proteins. Mol Cell Proteomics 9:1063–1084. https://doi.org/10.1074/mcp.M900325-MCP200

Fletcher DA, Mullins RD (2010) Cell mechanics and the cytoskeleton. Nature 463:485–492. https://doi.org/10.1038/nature08908

Frechin M, Stoeger T, Daetwyler S et al (2015) Cell-intrinsic adaptation of lipid composition to local crowding drives social behaviour. Nature 523:88–91. https://doi.org/10.1038/nature14429

Fuentes-Lemus E, Reyes JS, Gamon LF et al (2021) Effect of macromolecular crowding on protein oxidation: consequences on the rate, extent and oxidation pathways. Redox Biol 48:102202. https://doi.org/10.1016/j.redox.2021.102202

Gallardo G, Holtzman DM (2019) Amyloid-β and tau at the crossroads of Alzheimer's disease. Adv Exp Med Biol 1184:187–203. https://doi.org/10.1007/978-981-32-9358-8_16, Springer Singapore

García Fernández F, Huet S, Miné-Hattab J (2023) Multi-scale imaging of the dynamic Organization of Chromatin. IJMS 24:15975. https://doi.org/10.3390/ijms242115975

Gnutt D, Ebbinghaus S (2016) The macromolecular crowding effect—from in vitro into the cell. Biol Chem 397:37–44. https://doi.org/10.1515/hsz-2015-0161

Goers J, Uversky VN, Fink AL (2003) Polycation-induced oligomerization and accelerated fibrillation of human α-synuclein in vitro. Protein Sci 12:702–707. https://doi.org/10.1110/ps.0230903

Golkaram M, Hellander S, Drawert B, Petzold LR (2016) Macromolecular crowding regulates the gene expression profile by limiting diffusion. PLoS Comput Biol 12:e1005122. https://doi.org/10.1371/journal.pcbi.1005122

Gomez-Navarro N, Melero A, Li X-H et al (2020) Cargo crowding contributes to sorting stringency in COPII vesicles. J Cell Biol 219:e201806038. https://doi.org/10.1083/jcb.201806038

Gorensek-Benitez AH, Smith AE, Stadmiller SS et al (2017) Cosolutes, crowding, and protein folding kinetics. J Phys Chem B 121:6527–6537. https://doi.org/10.1021/acs.jpcb.7b03786

Grassmann G, Miotto M, Desantis F, et al (2024) Computational approaches to predict protein–protein interactions in crowded cellular environments. Chem Rev 124:3932–3977. https://doi.org/10.1021/acs.chemrev.3c00550

Groover SE, Adegbuyiro A, Fan CK et al (2021) Macromolecular crowding in solution alters huntingtin interaction and aggregation at interfaces. Colloids Surf B Biointerfaces 206:111969. https://doi.org/10.1016/j.colsurfb.2021.111969

Guigas G, Kalla C, Weiss M (2007) The degree of macromolecular crowding in the cytoplasm and nucleoplasm of mammalian cells is conserved. FEBS Lett 581:5094–5098. https://doi.org/10.1016/j.febslet.2007.09.054

Gulotta A, Bucciarelli S, Roosen-Runge F et al (2024) Testing mixing rules for structural and dynamical quantities in multi-component crowded protein solutions. APL Bioeng 8:026116. https://doi.org/10.1063/5.0204201

Haferkamp S, Kirchhoff H (2008) Significance of molecular crowding in grana membranes of higher plants for light harvesting by photosystem II. Photosynth Res 95:129–134. https://doi.org/10.1007/s11120-007-9253-2

Hameed A, Ahmed MZ, Hussain T et al (2021) Effects of salinity stress on chloroplast structure and function. Cells 10:2023. https://doi.org/10.3390/cells10082023

Hellerschmied D, Serebrenik YV, Shao L et al (2019) Protein folding state-dependent sorting at the Golgi apparatus. MBoC 30:2296–2308. https://doi.org/10.1091/mbc.E19-01-0069

Hetz C (2012) The unfolded protein response: controlling cell fate decisions under ER stress and beyond. Nat Rev Mol Cell Biol 13:89–102. https://doi.org/10.1038/nrm3270

Hetz C, Papa FR (2018) The unfolded protein response and cell fate control. Mol Cell 69:169–181. https://doi.org/10.1016/j.molcel.2017.06.017

Hetz C, Zhang K, Kaufman RJ (2020) Mechanisms, regulation and functions of the unfolded protein response. Nat Rev Mol Cell Biol 21:421–438. https://doi.org/10.1038/s41580-020-0250-z

Igarashi K, Kashiwagi K (2000) Polyamines: mysterious modulators of cellular functions. Biochem Biophys Res Commun 271:559–564. https://doi.org/10.1006/bbrc.2000.2601

Igarashi K, Kashiwagi K (2019) The functional role of polyamines in eukaryotic cells. Int J Biochem Cell Biol 107:104–115. https://doi.org/10.1016/j.biocel.2018.12.012

Illmer P, Erlebach C, Schinner F (1999) A practicable and accurate method to differentiate between intra- and extracellular water of microbial cells. FEMS Microbiol Lett 178:135–139. https://doi.org/10.1111/j.1574-6968.1999.tb13769.x

Jeon J-H, Javanainen M, Martinez-Seara H et al (2016) Protein crowding in lipid bilayers gives rise to non-Gaussian anomalous lateral diffusion of phospholipids and proteins. Phys Rev X 6:021006. https://doi.org/10.1103/PhysRevX.6.021006

Johnson GN (2011) Physiology of PSI cyclic electron transport in higher plants Biochimica et Biophysica Acta (BBA) Bioenerg 1807:384–389. https://doi.org/10.1016/j.bbabio.2010.11.009

Julius K, Weine J, Gao M, et al (2019) Impact of macromolecular crowding and compression on protein–protein interactions and liquid–liquid phase separation phenomena. Macromolecules 52:1772–1784. https://doi.org/10.1021/acs.macromol.8b02476

Khan S, Siraj S, Shahid M et al (2023) Osmolytes: wonder molecules to combat protein misfolding against stress conditions. Int J Biol Macromol 234:123662. https://doi.org/10.1016/j.ijbiomac.2023.123662

Kim WK, Choi W, Deshar B et al (2023) Golgi stress response: new insights into the pathogenesis and therapeutic targets of human diseases. Mol Cells 46:191–199. https://doi.org/10.14348/molcells.2023.2152

Kirchhoff H (2008) Molecular crowding and order in photosynthetic membranes. Trends Plant Sci 13:201–207. https://doi.org/10.1016/j.tplants.2008.03.001

Kirchhoff H (2014) Diffusion of molecules and macromolecules in thylakoid membranes. Biochimica et Biophysica Acta (BBA) Bioenerg 1837:495–502. https://doi.org/10.1016/j.bbabio.2013.11.003

Kirchhoff H (2019) Chloroplast ultrastructure in plants. New Phytol 223:565–574. https://doi.org/10.1111/nph.15730

Kirchhoff H, Mukherjee U, Galla H-J (2002) Molecular architecture of the thylakoid membrane: lipid diffusion space for Plastoquinone. Biochemistry 41:4872–4882. https://doi.org/10.1021/bi011650y

Kirchhoff H, Haferkamp S, Allen JF et al (2008) Protein diffusion and macromolecular crowding in thylakoid membranes. Plant Physiol 146:1571–1578. https://doi.org/10.1104/pp.107.115170

Kirchhoff H, Hall C, Wood M et al (2011) Dynamic control of protein diffusion within the granal thylakoid lumen. Proc Natl Acad Sci USA 108:20248–20253. https://doi.org/10.1073/pnas.1104141109

Klose C, Surma MA, Simons K (2013) Organellar lipidomics—background and perspectives. Curr Opin Cell Biol 25:406–413. https://doi.org/10.1016/j.ceb.2013.03.005

Koch MW, Metz LM, Kovalchuk O (2013) Epigenetics and miRNAs in the diagnosis and treatment of multiple sclerosis. Trends Mol Med 19:23–30. https://doi.org/10.1016/j.molmed.2012.10.008

Krshnan L, van de Weijer ML, Carvalho P (2022) Endoplasmic reticulum-associated protein degradation. Cold Spring Harb Perspect Biol 14:a041247. https://doi.org/10.1101/cshperspect.a041247

Kültz D (2000) Osmotic regulation of DNA activity and the cell cycle. In: Cell and molecular response to stress. Elsevier, pp 157–179

Kushwah N, Jain V, Yadav D (2020) Osmolytes: a possible therapeutic molecule for ameliorating the neurodegeneration caused by protein Misfolding and aggregation. Biomolecules 10:132. https://doi.org/10.3390/biom10010132

Kuznetsova I, Turoverov K, Uversky V (2014) What macromolecular crowding can do to a protein. IJMS 15:23090–23140. https://doi.org/10.3390/ijms151223090

Kuznetsova IM, Zaslavsky BY, Breydo L et al (2015) Beyond the excluded volume effects: mechanistic complexity of the crowded milieu. Molecules 20:1377–1409. https://doi.org/10.3390/molecules20011377

Larson AG, Elnatan D, Keenen MM et al (2017) Liquid droplet formation by HP1α suggests a role for phase separation in heterochromatin. Nature 547:236–240. https://doi.org/10.1038/nature22822

Lecinski S, Shepherd JW, Frame L, et al (2021) Investigating molecular crowding during cell division and hyperosmotic stress in budding yeast with FRET. Curr Top Membr 88:75–118. https://doi.org/10.1016/bs.ctm.2021.09.001, Elsevier

Li G, Zhu P (2015) Structure and organization of chromatin fiber in the nucleus. FEBS Lett 589:2893–2904. https://doi.org/10.1016/j.febslet.2015.04.023

Li Z, Wu J, Wang Z-G (2008) Osmotic pressure and packaging structure of caged DNA. Biophys J 94:737–746. https://doi.org/10.1529/biophysj.107.112508

Li J, Ahat E, Wang Y (2019) Golgi structure and function in health, stress, and diseases. Results Probl Cell Differ 67:441–485. https://doi.org/10.1007/978-3-030-23173-6_19, Springer International Publishing

Li W, Jiang C, Zhang E (2021) Advances in the phase separation-organized membraneless organelles in cells: a narrative review. Transl Cancer Res 10:4929–4946. https://doi.org/10.21037/tcr-21-1111

Li C, Zhang X, Dong M, Han X (2022) Progress on crowding effect in cell-like structures. Membranes (Basel) 12:593. https://doi.org/10.3390/membranes12060593

Lin Z, Fang D (2013) The roles of SIRT1 in cancer. Genes Cancer 4:97–104. https://doi.org/10.1177/1947601912475079

Lippincott-Schwartz J, Roberts TH, Hirschberg K (2000) Secretory protein trafficking and organelle dynamics in living cells. Annu Rev Cell Dev Biol 16:557–589. https://doi.org/10.1146/annurev.cellbio.16.1.557

Lodish HF (2021) Molecular cell biology, 9th edn. Macmillan Learning, Austin

Löwe M, Kalacheva M, Boersma AJ, Kedrov A (2020) The more the merrier: effects of macromolecular crowding on the structure and dynamics of biological membranes. FEBS J 287:5039–5067. https://doi.org/10.1111/febs.15429

Löwe M, Hänsch S, Hachani E et al (2023) Probing macromolecular crowding at the lipid membrane interface with genetically-encoded sensors. Protein Sci 32:e4797. https://doi.org/10.1002/pro.4797

Luby-Phelps K (2000) Cytoarchitecture and physical properties of cytoplasm: volume, viscosity, diffusion, intracellular surface area. Int Rev Cytol 192:189–221. https://doi.org/10.1016/s0074-7696(08)60527-6, Elsevier

Luijsterburg MS, White MF, Van Driel R, Dame RT (2008) The major architects of chromatin: architectural proteins in bacteria, archaea and eukaryotes. Crit Rev Biochem Mol Biol 43:393–418. https://doi.org/10.1080/10409230802528488

Ma Q, Fan J-B, Zhou Z et al (2012) The contrasting effect of macromolecular crowding on amyloid fibril formation. PLoS One 7:e36288. https://doi.org/10.1371/journal.pone.0036288

Ma Y, Kanakousaki K, Buttitta L (2015) How the cell cycle impacts chromatin architecture and influences cell fate. Front Genet 6. https://doi.org/10.3389/fgene.2015.00019

Marky LA, Kupke DW (2000) Enthalpy-entropy compensations in nucleic acids: contribution of electrostriction and structural hydration. In: Methods in enzymology. Elsevier, pp 419–441

McGuffee SR, Elcock AH (2010) Diffusion, Crowding & Protein Stability in a dynamic molecular Model of the bacterial cytoplasm. PLoS Comput Biol 6:e1000694. https://doi.org/10.1371/journal.pcbi.1000694

Meneses-Reyes GI, Rodriguez-Bustos DL, Cuevas-Velazquez CL (2024) Macromolecular crowding sensing during osmotic stress in plants. Trends Biochem Sci 49:480–493. https://doi.org/10.1016/j.tibs.2024.02.002

Meyer T (2021) Amyotrophic lateral sclerosis (ALS) – diagnosis, course of disease and treatment options. Dtsch Med Wochenschr 146:1613–1618. https://doi.org/10.1055/a-1562-7882

Minton AP (1980) Excluded volume as a determinant of protein structure and stability. Biophys J 32:77–79. https://doi.org/10.1016/S0006-3495(80)84917-4

Minton AP (2000) Implications of macromolecular crowding for protein assembly. Curr Opin Struct Biol 10:34–39. https://doi.org/10.1016/S0959-440X(99)00045-7

Minton AP (2001) The influence of macromolecular crowding and macromolecular confinement on biochemical reactions in physiological media. J Biol Chem 276:10577–10580. https://doi.org/10.1074/jbc.R100005200

Minton AP (2006) How can biochemical reactions within cells differ from those in test tubes? J Cell Sci 119:2863–2869. https://doi.org/10.1242/jcs.03063

Minton AP, Wilf J (1981) Effect of macromolecular crowding upon the structure and function of an enzyme: glyceraldehyde-3-phosphate dehydrogenase. Biochemistry 20:4821–4826. https://doi.org/10.1021/bi00520a003

Mittal S, Chowhan RK, Singh LR (2015) Macromolecular crowding: macromolecules friend or foe. Biochim Biophys Acta Gen Subj 1850:1822–1831. https://doi.org/10.1016/j.bbagen.2015.05.002

Miyoshi D, Sugimoto N (2008) Molecular crowding effects on structure and stability of DNA. Biochimie 90:1040–1051. https://doi.org/10.1016/j.biochi.2008.02.009

Model MA, Hollembeak JE, Kurokawa M (2021) Macromolecular crowding: a hidden link between cell volume and everything Else. Cell Physiol Biochem 55:25–40. https://doi.org/10.33594/000000319

Monterroso B, Margolin W, Boersma AJ et al (2024) Macromolecular crowding, phase separation, and homeostasis in the orchestration of bacterial cellular functions. Chem Rev 124:1899–1949. https://doi.org/10.1021/acs.chemrev.3c00622

Morris HR, Spillantini MG, Sue CM, Williams-Gray CH (2024) The pathogenesis of Parkinson's disease. Lancet 403:293–304. https://doi.org/10.1016/S0140-6736(23)01478-2

Morrison O, Thakur J (2021) Molecular complexes at euchromatin, heterochromatin and centromeric chromatin. IJMS 22:6922. https://doi.org/10.3390/ijms22136922

Mourão MA, Hakim JB, Schnell S (2014) Connecting the dots: the effects of macromolecular crowding on cell physiology. Biophys J 107:2761–2766. https://doi.org/10.1016/j.bpj.2014.10.051

Munishkina LA, Cooper EM, Uversky VN, Fink AL (2004) The effect of macromolecular crowding on protein aggregation and amyloid fibril formation. J Mol Recogn 17:456–464. https://doi.org/10.1002/jmr.699

Muro E, Atilla-Gokcumen GE, Eggert US (2014) Lipids in cell biology: how can we understand them better? MBoC 25:1819–1823. https://doi.org/10.1091/mbc.e13-09-0516

Neoh GKS, Tan X, Chen S et al (2024) Glycogen metabolism and structure: a review. Carbohydr Polym 346:122631. https://doi.org/10.1016/j.carbpol.2024.122631

Nussinov R, Tsai C-J, Jang H (2021) Signaling in the crowded cell. Curr Opin Struct Biol 71:43–50. https://doi.org/10.1016/j.sbi.2021.05.009

Oakes SA, Papa FR (2015) The role of endoplasmic reticulum stress in human pathology. Annu Rev Pathol 10:173–194. https://doi.org/10.1146/annurev-pathol-012513-104649

Oh S, Lee C, Yang W et al (2022) Protein and lipid mass concentration measurement in tissues by stimulated Raman scattering microscopy. Proc Natl Acad Sci USA 119:e2117938119. https://doi.org/10.1073/pnas.2117938119

Ostermeier M, Garibay-Hernández A, Holzer VJC et al (2024) Structure, biogenesis, and evolution of thylakoid membranes. Plant Cell 36:4014–4035. https://doi.org/10.1093/plcell/koae102

Pan X, Cao P, Su X et al (2020) Structural analysis and comparison of light-harvesting complexes I and II. Biochimica et Biophysica Acta (BBA) Bioenerget 1861:148038. https://doi.org/10.1016/j.bbabio.2019.06.010

Patel A, Malinovska L, Saha S et al (2017) ATP as a biological hydrotrope. Science 356:753–756. https://doi.org/10.1126/science.aaf6846

Percival M, Pantoja CF, Cima-Omori M-S, et al (2024) Polyamines promote disordered protein phase separation

Perez-Boerema A, Engel BD, Wietrzynski W (2024) Evolution of thylakoid structural diversity. Annu Rev Cell Dev Biol 40:169–193. https://doi.org/10.1146/annurev-cellbio-120823-022747

Portz B, Lee BL, Shorter J (2021) FUS and TDP-43 phases in health and disease. Trends Biochem Sci 46:550–563. https://doi.org/10.1016/j.tibs.2020.12.005

Preisler RS, Chen HH, Colombo MF et al (1995) The B form to Z form transition of poly(dG-m5dC) is sensitive to neutral solutes through an osmotic stress. Biochemistry 34:14400–14407. https://doi.org/10.1021/bi00044a017

Quinn P, Griffiths G, Warren G (1984) Density of newly synthesized plasma membrane proteins in intracellular membranes II. Biochemical studies. J Cell Biol 98:2142–2147. https://doi.org/10.1083/jcb.98.6.2142

Ramakrishnan V (1997) Histone structure and the organization of the nucleosome. Annu Rev Biophys Biomol Struct 26:83–112. https://doi.org/10.1146/annurev.biophys.26.1.83

Rana PS, Kurokawa M, Model MA (2020) Evidence for macromolecular crowding as a direct apoptotic stimulus. J Cell Sci 133:jcs243931. https://doi.org/10.1242/jcs.243931

Ranamukhaarachchi SK, Modi RN, Han A et al (2019) Macromolecular crowding tunes 3D collagen architecture and cell morphogenesis. Biomater Sci 7:618–633. https://doi.org/10.1039/c8bm01188e

Richter K, Nessling M, Lichter P (2007) Experimental evidence for the influence of molecular crowding on nuclear architecture. J Cell Sci 120:1673–1680. https://doi.org/10.1242/jcs.03440

Richter K, Nessling M, Lichter P (2008) Macromolecular crowding and its potential impact on nuclear function Biochimica et Biophysica Acta (BBA) Mol Cell Res 1783:2100–2107. https://doi.org/10.1016/j.bbamcr.2008.07.017
Rivas G, Minton AP (2016) Macromolecular crowding in vitro, in vivo, and in between. Trends Biochem Sci 41:970–981. https://doi.org/10.1016/j.tibs.2016.08.013
Rivas G, Minton AP (2022) Influence of nonspecific interactions on protein associations: implications for biochemistry in vivo. Annu Rev Biochem 91:321–351. https://doi.org/10.1146/annurev-biochem-040320-104151
Rohwer JM, Postma PW, Kholodenko BN, Westerhoff HV (1998) Implications of macromolecular crowding for signal transduction and metabolite channeling. Proc Natl Acad Sci USA 95:10547–10552. https://doi.org/10.1073/pnas.95.18.10547
Rollin R, Joanny J-F, Sens P (2023) Physical basis of the cell size scaling laws. eLife 12:e82490. https://doi.org/10.7554/eLife.82490
Rolo D, Schöttler MA, Sandoval-Ibáñez O, Bock R (2024) Structure, function, and assembly of PSI in thylakoid membranes of vascular plants. Plant Cell 36:4080–4108. https://doi.org/10.1093/plcell/koae169
Rowat AC, Lammerding J, Herrmann H, Aebi U (2008) Towards an integrated understanding of the structure and mechanics of the cell nucleus. BioEssays 30:226–236. https://doi.org/10.1002/bies.20720
Rowland AA, Voeltz GK (2012) Endoplasmic reticulum-mitochondria contacts: function of the junction. Nat Rev Mol Cell Biol 13:607–625. https://doi.org/10.1038/nrm3440
Ruggiano A, Foresti O, Carvalho P (2014) Quality control: ER-associated degradation: protein quality control and beyond. J Cell Biol 204:869–879. https://doi.org/10.1083/jcb.201312042
Saghatelian A, McKinney MK, Bandell M et al (2006) A FAAH-regulated class of *N*- acyl Taurines that activates TRP ion channels. Biochemistry 45:9007–9015. https://doi.org/10.1021/bi0608008
Saminathan M (2002) Polyamine structural effects on the induction and stabilization of liquid crystalline DNA: potential applications to DNA packaging, gene therapy and polyamine therapeutics. Nucleic Acids Res 30:3722–3731. https://doi.org/10.1093/nar/gkf503
Schwarz DS, Blower MD (2016) The endoplasmic reticulum: structure, function and response to cellular signaling. Cell Mol Life Sci 73:79–94. https://doi.org/10.1007/s00018-015-2052-6
Shim AR, Nap RJ, Huang K et al (2020) Dynamic crowding regulates transcription. Biophys J 118:2117–2129. https://doi.org/10.1016/j.bpj.2019.11.007
Shologu N, Gurdal M, Szegezdi E et al (2022) Macromolecular crowding in the development of a three-dimensional organotypic human breast cancer model. Biomaterials 287:121642. https://doi.org/10.1016/j.biomaterials.2022.121642
Siddiqui GA, Naeem A (2023) Connecting the dots: macromolecular crowding and protein aggregation. J Fluoresc 33:1–11. https://doi.org/10.1007/s10895-022-03082-2
Silverstein TP, Slade K (2019) Effects of macromolecular crowding on biochemical systems. J Chem Educ 96:2476–2487. https://doi.org/10.1021/acs.jchemed.9b00399
Smirle J, Au CE, Jain M et al (2013) Cell biology of the endoplasmic reticulum and the Golgi apparatus through proteomics. Cold Spring Harb Perspect Biol 5:a015073–a015073. https://doi.org/10.1101/cshperspect.a015073
Smith MH, Ploegh HL, Weissman JS (2011) Road to ruin: targeting proteins for degradation in the endoplasmic reticulum. Science 334:1086–1090. https://doi.org/10.1126/science.1209235
Song Y, Feng L, Alyafei MAM et al (2021) Function of chloroplasts in plant stress responses. IJMS 22:13464. https://doi.org/10.3390/ijms222413464
Speer SL, Stewart CJ, Sapir L et al (2022) Macromolecular crowding is more than hard-Core repulsions. Annu Rev Biophys 51:267–300. https://doi.org/10.1146/annurev-biophys-091321-071829
Spinelli JB, Haigis MC (2018) The multifaceted contributions of mitochondria to cellular metabolism. Nat Cell Biol 20:745–754. https://doi.org/10.1038/s41556-018-0124-1
Spink CH, Chaires JB (1999) Effects of hydration, ion release, and excluded volume on the melting of triplex and duplex DNA. Biochemistry 38:496–508. https://doi.org/10.1021/bi9820154

Srere PA (2000) Macromolecular interactions: tracing the roots. Trends Biochem Sci 25:150–153. https://doi.org/10.1016/S0968-0004(00)01550-4

Stagg L, Zhang S-Q, Cheung MS, Wittung-Stafshede P (2007) Molecular crowding enhances native structure and stability of α/β protein flavodoxin. Proc Natl Acad Sci USA 104:18976–18981. https://doi.org/10.1073/pnas.0705127104

Strulson CA, Boyer JA, Whitman EE, Bevilacqua PC (2014) Molecular crowders and cosolutes promote folding cooperativity of RNA under physiological ionic conditions. RNA 20:331–347. https://doi.org/10.1261/rna.042747.113

Subramanya AR, Boyd-Shiwarski CR (2024) Molecular crowding: physiologic sensing and control. Annu Rev Physiol 86:429–452. https://doi.org/10.1146/annurev-physiol-042222-025920

Sugimoto N (2014) Noncanonical structures and their thermodynamics of DNA and RNA under molecular crowding. In: International review of cell and molecular biology. Elsevier, pp 205–273

Sun L, Fang J (2016) Macromolecular crowding effect is critical for maintaining SIRT1's nuclear localization in cancer cells. Cell Cycle 15:2647–2655. https://doi.org/10.1080/15384101.2016.1211214

Sun Q, Zybailov B, Majeran W et al (2009) PPDB, the plant proteomics database at Cornell. Nucleic Acids Res 37:D969–D974. https://doi.org/10.1093/nar/gkn654

Tabaka M, Kalwarczyk T, Szymanski J et al (2014) The effect of macromolecular crowding on mobility of biomolecules, association kinetics, and gene expression in living cells. Front Phys 2. https://doi.org/10.3389/fphy.2014.00054

Takahashi A, Nagao C, Murakami K et al (2020a) Effects of molecular crowding environment on the acquisition of toxic properties of wild-type SOD1. Biochim Biophys Acta Gen Subj 1864:129401. https://doi.org/10.1016/j.bbagen.2019.07.010

Takahashi S, Herdwijn P, Sugimoto N (2020b) Effect of molecular crowding on DNA polymerase reactions along unnatural DNA templates. Molecules 25:4120. https://doi.org/10.3390/molecules25184120

Tateishi-Karimata H, Banerjee D, Ohyama T et al (2020) Hydroxyl groups in cosolutes regulate the G-quadruplex topology of telomeric DNA. Biochem Biophys Res Commun 525:177–183. https://doi.org/10.1016/j.bbrc.2020.02.045

Tehei M, Zaccai G (2005) Adaptation to extreme environments: macromolecular dynamics in complex systems. Biochim Biophys Acta Gen Subj 1724:404–410. https://doi.org/10.1016/j.bbagen.2005.05.007

Thomas T, Thomas TJ (2001) Polyamines in cell growth and cell death: molecular mechanisms and therapeutic applications. Cell Mol Life Sci CMLS 58:244–258. https://doi.org/10.1007/PL00000852

Toader C, Eva L, Covache-Busuioc R-A et al (2023) Unraveling the multifaceted role of the Golgi apparatus: insights into neuronal plasticity, development, neurogenesis, Alzheimer's disease, and SARS-CoV-2 interactions. Brain Sci 13:1363. https://doi.org/10.3390/brainsci13101363

Tun MT, Yang S, Forti FL et al (2022) Macromolecular crowding amplifies allosteric regulation of T-cell protein tyrosine phosphatase. J Biol Chem 298:102655. https://doi.org/10.1016/j.jbc.2022.102655

Uversky VN (2009) Intrinsically disordered proteins and their environment: effects of strong denaturants, temperature, pH, counter ions, membranes, binding partners, osmolytes, and macromolecular crowding. Protein J 28:305–325. https://doi.org/10.1007/s10930-009-9201-4

Voeltz GK, Rolls MM, Rapoport TA (2002) Structural organization of the endoplasmic reticulum. EMBO Rep 3:944–950. https://doi.org/10.1093/embo-reports/kvf202

Vweza A-O, Song C-G, Chong K-T (2021) Liquid–liquid phase separation in the presence of macromolecular crowding and state-dependent kinetics. IJMS 22:6675. https://doi.org/10.3390/ijms22136675

Wallace HM, Fraser AV, Hughes A (2003) A perspective of polyamine metabolism. Biochem J 376:1–14. https://doi.org/10.1042/bj20031327

Wang H, Zeng R (2024) Aberrant protein aggregation in amyotrophic lateral sclerosis. J Neurol 271:4826–4851. https://doi.org/10.1007/s00415-024-12485-z

Wang Y, Sarkar M, Smith AE et al (2012) Macromolecular crowding and protein stability. J Am Chem Soc 134:16614–16618. https://doi.org/10.1021/ja305300m

Weber MS, Wojtynek M, Medalia O (2019) Cellular and structural studies of eukaryotic cells by cryo-electron tomography. Cells 8:57. https://doi.org/10.3390/cells8010057

Wei J-H, Seemann J (2017) Golgi ribbon disassembly during mitosis, differentiation and disease progression. Curr Opin Cell Biol 47:43–51. https://doi.org/10.1016/j.ceb.2017.03.008

Werner TER, Horvath I, Wittung-Stafshede P (2021) Response to crowded conditions reveals compact nucleus for amyloid formation of folded protein. QRB Discov 2:e2. https://doi.org/10.1017/qrd.2020.17

Wietrzynski W, Lamm L, Wood WHJ, et al (2024) Molecular architecture of thylakoid membranes within intact spinach chloroplasts

Wnek GE (2016) Perspective: do macromolecules play a role in the mechanisms of nerve stimulation and nervous transmission? J Polym Sci B Polym Phys 54:7–14. https://doi.org/10.1002/polb.23898

Wolff S, Weissman JS, Dillin A (2014) Differential scales of protein quality control. Cell 157:52–64. https://doi.org/10.1016/j.cell.2014.03.007

Wu Y, Teng N, Li S (2016) Effects of macromolecular crowding and osmolyte on human tau fibrillation. Int J Biol Macromol 90:27–36. https://doi.org/10.1016/j.ijbiomac.2015.11.091

Xu G, Zhao J, Cheng K et al (2017) The effects of macromolecular crowding on Calmodulin structure and function. Chemistry 23:6736–6740. https://doi.org/10.1002/chem.201700367

Xuan M, Gu X, Li J et al (2023) Polyamines: their significance for maintaining health and contributing to diseases. Cell Commun Signal 21:348. https://doi.org/10.1186/s12964-023-01373-0

Yang C, Wang Z, Kang Y et al (2023) Stress granule homeostasis is modulated by TRIM21-mediated ubiquitination of G3BP1 and autophagy-dependent elimination of stress granules. Autophagy 19:1934–1951. https://doi.org/10.1080/15548627.2022.2164427

Zhou H (2004) Protein folding and binding in confined spaces and in crowded solutions. J Mol Recog 17:368–375. https://doi.org/10.1002/jmr.711

Zhou H-X, Rivas G, Minton AP (2008) Macromolecular crowding and confinement: biochemical, biophysical, and potential physiological consequences. Annu Rev Biophys 37:375–397. https://doi.org/10.1146/annurev.biophys.37.032807.125817

Zimmerman SB, Trach SO (1991) Estimation of macromolecule concentrations and excluded volume effects for the cytoplasm of Escherichia coli. J Mol Biol 222:599–620. https://doi.org/10.1016/0022-2836(91)90499-V

Chapter 11
Macromolecular Crowding in Mitochondria

Semen V. Nesterov and Vladimir N. Uversky

Abstract The chapter reviews the various effects of crowding on mitochondrial structure and function. Data illustrate that, as a rule, the concentration of macromolecules in the mitochondrial matrix and inner membrane is at least as high as in other parts of the cytoplasm, while the intermembrane space is sparse. The effect of crowding on mitochondrial shape, the role of disordered protein domains, and the role of changes in crowding during mitochondrial swelling are discussed. Because of the excluded volume in the matrix, real changes in ion and metabolite concentrations under swelling are substantially higher than believed, and the inner membrane is highly curved at least partly due to crowding of the matrix. The high concentration of integral proteins in the mitochondrial inner membrane leads to enrichment with non-bilayer lipid cardiolipin to compensate for their induced membrane deformations. Also crowding may be one of the stimuli for the formation of enzyme supercomplexes. All reviewed data suggest that the structure of mitochondria is adapted exactly to the conditions of high crowding, and crowding itself is one of the key factors in the regulation of mitochondrial structure and function, the role of which is significantly underestimated in the scientific literature.

Keywords Mitochondrial structure · Mitochondrial function · Mitochondrial swelling · Mitochondrial shape · Intermembrane space · Matrix · Enzyme supercomplexes · Intrinsically disordered proteins

S. V. Nesterov (✉)
NRC "Kurchatov Institute", Moscow, Russia
e-mail: semen.v.nesterov@phystech.edu

V. N. Uversky
Department of Molecular Medicine and USF Health Byrd Alzheimer's Research Institute, Morsani College of Medicine, University of South Florida, Tampa, FL, USA

V. N. Uversky (ed.), *(Macro)Molecular Crowding*, Subcellular Biochemistry 109,
https://doi.org/10.1007/978-3-032-03370-3_11

Introduction

Mitochondria are key organelles for eukaryotic cells, realizing a significant part of metabolic reactions, producing through oxidative phosphorylation (OxPhos) most of ATP under aerobic conditions, and initiating the most important signaling processes, including initiation of response to hypoxia, immune response, and apoptosis. Moreover, mitochondria have a certain degree of autonomy, are surrounded by an outer membrane, have their own circular DNA and ribosomes. According to the dominant theory, mitochondria are endosymbionts that evolved from a bacterium engulfed several hundred million years ago by a eukaryotic ancestral cell. It is for bacteria that the role of crowding in their cytoplasm is well known and high molecular crowding is characteristic of bacteria (Van Den Berg et al. 2017). In this connection, it seems highly probable that in mitochondria crowding should also play significant role in the processes of internal organization and functioning of enzyme systems. In spite of this, there is rather little literature on mitochondria regarding the phenomenon of crowding or displaced volume. PubMed database finds only several dozen articles till 2024 in the abstract or title mentioning "excluded volume" or "crowding" along with "mitochondria", while there are several hundred along with "bacteria" and about two thousand along with "cell". Yet even fewer articles actually focus specifically on studying the effects of crowding on mitochondrial structure and function, rather than simply mentioning mitochondria in a different context along with the phenomenon. This chapter analyzes how crowding effects can affect mitochondrial processes.

Effects of Crowding on Mitochondrial Shape

As is widely known the term "mitochondria" comes from the Greek μίτος, mitos, "thread", and χονδρίον, chondrion, "granule". As the name implies, the mitochondria observed in the cells are mostly elongated in shape. At the same time, most electron microscopy images of isolated mitochondria, which are often used by scientists to assess ultrastructure, show these organelles close to a spheroid (Hackenbrock 1966; Krasinskaia et al. 1989; Mannella 2006). One reason for this discrepancy between *in vivo* and *in vitro* conditions is that in vivo mitochondria form a mitochondrial network (Naguro et al. 2012; Zamponi et al. 2018), which may be observed as threads. However, in addition to the fission of the mitochondrial network into small pieces, there is another reason why individual mitochondria often do not retain an ellipsoid shape after isolation. This is because standard isotonic media for incubating mitochondria are based on low-molecular nonmetabolizable sugars and/or membrane low permeable salts. In opposite, in the cytosol mitochondria are surrounded by a large number of proteins and other polymers, are in contact with the cytoskeleton and other organelles. In muscle mitochondria are even sandwiched between myofibrils. Under natural intracellular environment mitochondria are subjected not only to steric effects but also to molecular crowding. It has long been shown that the shape of isolated mitochondria depends on the presence of high molecular weight compounds in the incubation medium, such as Ficoll and Dextran (Laties and Treffry 1969). The addition of high molecular weight crowding agents to the medium induces mitochondrial elongation independently of metabolic conditions (Laties and Treffry 1969).

This raises the question of whether the *in vitro* results on mitochondrial structure are at all consistent with *in vivo* reality. Naturally, this issue is not new. The available consensus suggests that the change in structure during mitochondrial rounding hardly changes the volume of the mitochondrial matrix and does not impair the primary function of the mitochondrion, ATP synthesis. The increased amount of data show a strong correlation of mitochondrial ultrastructure abnormalities with functional abnormalities (Glancy et al. 2020; Gasanoff et al. 2021; Jenkins et al. 2024). This calls for a more acute approach to the study of mitochondrial structure in vitro and to move towards non-invasive methods of investigation directly in the cytoplasm.

High Protein Density in Mitochondria

Mitochondria are very dense organelles with a high concentration of protein in both membranes and matrix. This is clearly seen in cryoelectron tomography data, where there are no contrasting agents and proteins are more dense (deeper gray color) than in the aqueous phase. There is definitely a difference in the amount of proteins in the mitochondrial matrix in different species and even in different tissues. Presumably, species more dependent on oxidative phosphorylation have denser mitochondria. This assumption is generally supported by the fact that in mitochondria of unicellular algae (Waltz et al. 2024; Dietrich et al. 2024) protein concentration in the matrix visually observed on tomography is markedly less than in rat or aerobic yeast mitochondria (Nesterov et al. 2024b). But more research is needed.

At least in mammals, mitochondria are crowded more than the rest of the cytoplasm or at least approximately at the same extent (Fig. 11.1). This suggests that it is reasonable to consider the real volume of fluid in the matrix with a correction for the excluded volume, rather than as the entire volume bounded by the inner membrane. The gaps between large protein complexes and supercomplexes (metabolons) of the mitochondrial matrix are in most cases no larger than a few nanometers. Such an environment can be compared to a sponge full of small capillaries. However, even large proteins in vivo retain some mobility rather than forming an interlocked skeleton, as happens when the sample is chemically fixed.

Naturally, protein density and hence the degree of crowding can vary from organism to organism, from tissue to tissue, or due to external factors. Stress-dependent macromolecular crowding changes were shown in the mitochondrial matrix (Bulthuis et al. 2023). Interestingly, the cytoplasm of bacteria can be glass-like and it fluidized when metabolic activity occurs (Parry et al. 2014). Thus, changes in the

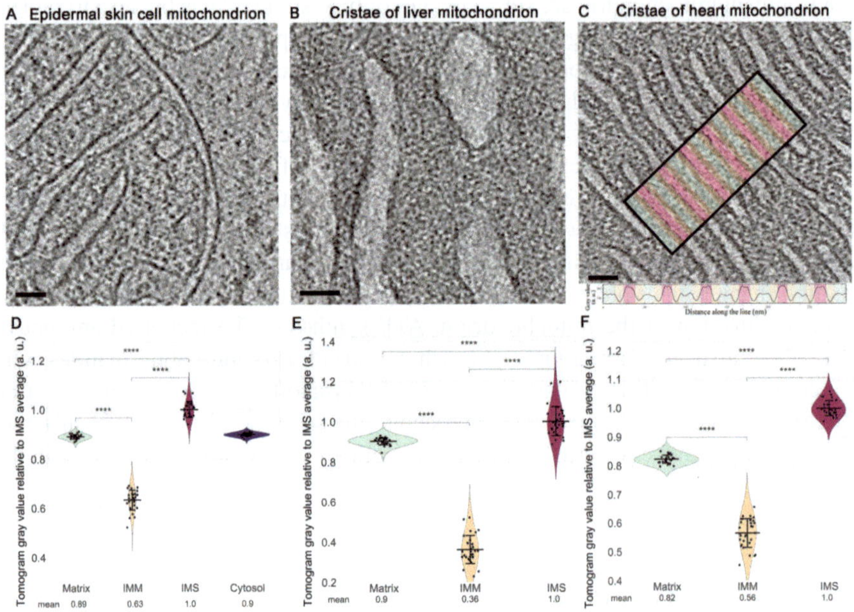

Fig. 11.1 Analysis of cryogenic electron tomography slices. The figures show cristae with lighter colored IMS in them and darker matrix around. (**a**) Cultured human keratinocyte mitochondria in native cytosol environment. (**b**) Rat liver mitochondria. (**c**) Rat heart mitochondrial interior. Intensity averaging was performed on the selected and highlighted area and within 30 nm on neighboring tomographic slices, the results of which are shown in the insert under the panel (**d–f**). Analysis of electron transparency of different regions of mitochondria of keratinocyte, rat liver and heart respectively. There are very few proteins in the IMS and it is most physicochemically close to the true liquid aqueous phase. Scale bars are 50 nm. Figure reproduced from (Nesterov et al. 2024a) with permission

physicochemical properties of the cytoplasm are also known for bacteria. This indirectly supports the idea that there are similarities between the mechanisms of regulation of the bacterial cytoplasm and the mitochondrial matrix. This topic requires further research.

Intrinsic Disorder in Mitochondria

In the context of macromolecular crowding, not only the density of proteins but also their packing is important. Disordered polypeptide chains have a much larger radius and affect the environment differently than similar weight globular proteins. Therefore, this section will focus on disordered domains in mitochondrial proteins.

Earlier, it was established that mitochondria represents one of the most ordered compartments of human cell (Zhao et al. 2021). In fact, analysis of the proteomes of

24 subcellular compartments revealed that mitochondrion (GO:0005739) and mitochondrial membrane (GO:0031966) are characterized by the average protein-level disorder content of 0.070 and 0.059, respectively, and the proteomes of these compartments contain 25.9% and 20.5% proteins with long disordered regions (LDRs) defined as the sequence segments with at least 30 consecutive disordered amino acids (Zhao et al. 2021). In comparison with other 22 subcellular compartments, mitochondrion and mitochondrial membrane are located at the eighteenth and twentieth positions, respectively, based on their average protein-level disorder contents. As a reference, the average protein-level disorder content of proteins in the most "disordered" compartment, nucleus, is 0.171, and, in whole human proteome, it is 0.128. Furthermore, 57.4% and 44.2% of nuclear and human proteins contain putative LDRs (Zhao et al. 2021). In the least disordered cellular compartment, peroxisome, the average protein-level disorder content is 0.035, and 18.9% of proteins contain putative LDRs (Zhao et al. 2021). The idea of low overall disorder content in the mitochondrial proteome is further illustrated by Fig. 11.2 showing

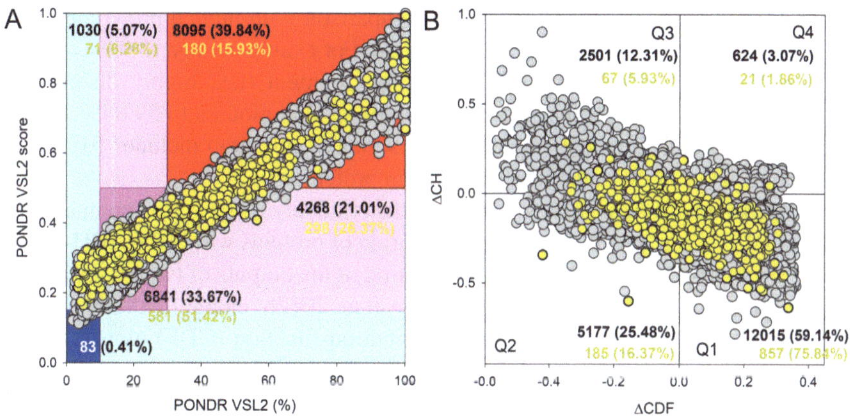

Fig. 11.2 Evaluation of the global disorder status of 20,317 human proteins (gray circles) and 1130 human mitochondrial proteins (small yellow circles) (mitochondrial protein IDs for this study were taken from (Morgenstern et al. 2021)). (**a**) PONDR® VSL2 score *vs*. PONDR® VSL2 (%) plot. Here, each point corresponds to a query protein, coordinates of which are evaluated from the corresponding PONDR® VSL2 data as its average disorder score (ADS) and percent of the predicted intrinsically disordered residues (PPIDR). Color blocks are used to visualize proteins based on the accepted classification, with red, pink/light pink, and blue/light blue regions containing highly disordered, moderately disordered, and ordered proteins, respectively. Dark blue or pink regions correspond to the regions, where PPIDR agrees with ADS, whereas areas in which only one of these criteria applies are shown by light blue or light pink. (**b**) ΔCH-ΔCDF plot, where coordinates for a protein are calculated as the average distance of its CDF curve from the CDF boundary (X axis) and its distance from the CH boundary. Protein classification is based on the quadrant, where it is located: Q1, protein predicted to be ordered by both predictors. Q2, protein predicted to be ordered to by CH-plot and disordered by CDF. Q3, protein predicted to be disordered by both predictors. Q4, protein predicted to be disordered by CH-plot and ordered by CDF. Data for these plots were generated using the Rapid Intrinsic Disorder Analysis Online (RIDAO) platform (Dayhoff and Uversky 2022)

comparison of the overall disorder predisposition of mitochondrial proteins with global disorder status of human proteome. In these analyses, two established approaches were utilized, classification of proteins based on the outputs of the per-residue disorder predictor PONDR® VSL2, and the ΔCH-ΔCDF plot. The PONDR® VSL2-based classification groups proteins according to their predicted intrinsically disordered residues (PPIDR) values (Rajagopalan et al. 2011), where proteins with PPIDR <10% are considered as ordered or mostly ordered; proteins with 10% ≤PPIDR <30% are considered as moderately disordered; whereas proteins with the PPIDR ≥30% are considered as highly disordered. Since in the outputs of the per-residue disorder predictors, there is no direct correlation between the average disorder score (ADS, calculated for each protein as a protein length-normalized sum of all the per-residue disorder scores) and PPIDR values estimated for a query protein (e.g., a protein whose ADS is ranging from 0.5 to 1.0 is predicted as 100% disordered based on the PPIDR criteria, whereas a protein that has any ADS <0.5 is predicted to have the PPIDR of 0%), proteins can also be classified based on their corresponding ADS values, being annotated as highly ordered (ADS <0.15), moderately disordered or flexible (ADS between 0.15 and 0.5) and highly disordered (ADS ≥0.5). Figure 11.2a shows that in comparison with human proteome, mitochondrial proteins are noticeably more ordered. For example, only 15.93% of mitochondrial proteins are located within the red segment (i.e., highly disordered), whereas for human proteome, the corresponding value mounts to 39.84%. On the contrary, dark pink segment (moderately disordered proteins) includes 51.42% of mitochondrial proteins and 33.67% of human proteome.

Figure 11.2b provides further support of the relatively low disorder content in the mitochondrial proteome by showing distribution of proteins within the ΔCH-ΔCDF plot. This approach is based on the combination of the outputs of two binary predictors (Mohan et al. 2008; Sun et al. 2011; Xue et al. 2012b) charge hydropathy (CH) (Uversky et al. 2000) and cumulative distribution function (CDF) plots (Oldfield et al. 2005), which enables rapid discrimination between flavors of disorder (Huang et al. 2012, 2014). To this end, for each query protein, ΔCH, the vertical distance of the corresponding point in CH plot from the boundary, is calculated, whereas ΔCDF is computed as the average distance between the order-disorder boundary and the CDF curve. Then, ΔCH is plotted against ΔCDF resulting in a ΔCH-ΔCDF plot. Proteins in the bottom-right quadrant (Q1) are predicted to be ordered by both CH and CDF, the ones in the bottom-left quadrant (Q2) are predicted to be ordered by CH and disordered by CDF (molten globular or hybrid proteins containing significant levels of order and disorder), the proteins in the top-left quadrant (Q3) are predicted to be disordered by both CH and CDF, the ones in the top-right quadrant (Q4) are predicted to be disordered by CH and ordered by CDF. Figure 11.2b shows that although about 40% of the human proteome are predicted to have one or another disorder flavor (being spread between quadrants Q2, Q3, and Q4), almost 76% of the mitochondrial proteins are predicted is ordered (i.e., located within the Q1 quadrant).

The relative low levels of intrinsic disorder in mitochondria can be explained by its bacterial origin (John and Whatley 1975; Cavalier-Smith 1987; Andersson et al.

2003; Cavalier-Smith 2006). In fact, it is well known that the eukaryotic proteomes are significantly more disordered than the prokaryotic proteomes (Ward et al. 2004; Uversky and Dunker 2010; Xue et al. 2012a; Peng et al. 2015). It is also possible that the low disorder content of the mitochondrial proteome can be attributed to the high level of crowding within these organelles and their limited volume. In fact, studies have shown that normal mitochondria have an average volume of around 0.24 ± 0.15 μm³ (or (2.4 ± 1.5) × 10^{11} Å³) (Shami et al. 2021). By other estimates, in synapses, the mitochondrial volume is on average 0.11 ± 0.08 μm³ (Honma et al. 2001). Matrix contains vast majority of mitochondrial proteins and occupies 63.16 ± 3.28% of mitochondrial volume (Gerencser et al. 2012) (i.e., ranges from ~0.7 × 10^{11} to ~1.5 × 10^{11} Å³). According to (Kaasik et al. 2007), the absolute values of neuronal mitochondria volume measured by deconvolution confocal microscopy are in the range of 0.04–0.08 μm³. Back-of-the-envelope calculations show that if all the mitochondrial proteins would be completely disordered (i.e., behaved as native coils or native pre-molten globules) they would not be able to be fitted within the mitochondria (see Fig. 11.3), as their total hydrodynamic volume is exceeding

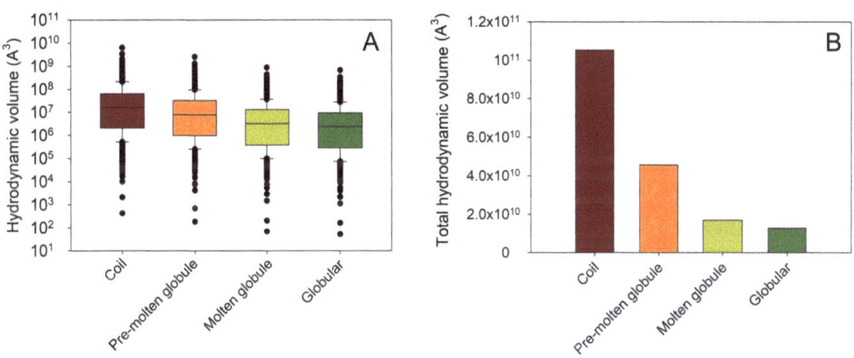

Fig. 11.3 Back-of-the-envelope calculations of the hydrodynamic volumes occupied by the mitochondrial proteins in different conformational states. (**a**) Distributions of the hydrodynamic volumes of individual proteins in coil-like (coil), pre-molten globule-like (PMG), molten globule-like (MG), and globular forms. The corresponding values were calculated using the reported length and copy-number values of the human mitochondrial proteins (Morgenstern et al. 2021) and known length-volume correlations for the polypeptides in the different structural forms, since it is known that the hydrodynamic volume occupied by a given polypeptide of a given length is dependent on its conformational state (Uversky 1993, 2002a, b; Tcherkasskaya and Uversky 2003):

$$\log\left(V_h^{Globular}\right)=(2.197\pm0.037)+(1.072\pm0.015)\log(N),\log\left(V_h^{MG}\right)=(2.46\pm0.13)+(1.020\pm0.053)\log(N),$$

$$\log\left(V_h^{PMG}\right)=(2.33\pm0.12)+(1.234\pm0.047)\log(N),$$

$$\log\left(V_h^{Coil}\right)=(1.997\pm0.078)+(1.498\pm0.035)\log(N).$$

(b) Total hydrodynamic volumes taken by the human mitochondrial proteins in different conformational **states** or at least comparable to the aforementioned estimates of the mitochondrial matrix volume.

One might ask a question on how macromolecular crowding might affect the behavior of intrinsically disordered proteins (IDPs) and intrinsically disordered regions (IDRs). Although based on the excluded-volume effects present in crowded environments, IDPs/IDRs were expected to fold in the presence of high concentrations of crowding agents or in the crowded cellular milieu, accumulated evidence indicated that, based on their response to the presence of crowders IDPs/IDRs can be grouped into three major categories, foldable, non-foldable, and unfoldable (for a systematic review, see (Fonin et al. 2018)). This response of IDPs/IDRs to crowding is defined by the active nature of crowded cellular fluids, components of which can be engaged in direct physical interactions and soft interactions with target proteins causing their (local) folding or unfolding. As a result, macromolecular crowding can differently affect structures of different IDPs (Fonin et al. 2018). Furthermore, mitochondria are known to contain mitochondrial stress granules (MitoSGs) (Sun et al. 2023), a form of membrane-less organelles (MLOs) originating as a result of liquid-liquid phase separation (LLPS) of specific IDPs. These MLOs are self-crowded and are characterized by locally increased concentrations of IDPs responsible for their biogenesis. Such MLO-resident IDPs/IDRs, being capable of phase separation, are typically minimally foldable or even non-foldable at all (at least within the physiologically safe time-frame of normal PMLO existence) (Fonin et al. 2018). Brief FuzDrop-based analysis (Hardenberg et al. 2020) revealed that less than half (47.7%) of the 172 most disordered mitochondrial proteins with PPIDR values exceeding 50% were predicted to have high droplet-promoting potential. This is an interesting observation suggesting that although high disorder content is crucial for LLPS, not all highly disordered proteins are capable of spontaneous liquid-liquid phase separation. Detailed analysis of these correlations is outside the scopes of this chapter and will be conducted elsewhere.

Mitochondrial Subcompartments Can Change Their Volume and Protein Density

Mitochondria have a unique inner membrane structure with many folds capable of ultrastructural changes. When considering mitochondria, it is more correct to speak of two subcompartments, the matrix and the intermembrane space. In this case, the whole mitochondrion has a limitation on the maximum volume given by the outer membrane. It can be approximately considered that without synthesis of new lipids and proteins in the processes of mitochondria interaction (fission, fusion, and change of shape) the observed surface of the outer membrane should be preserved. Since a sphere has the highest volume-to-area ratio for equal surface area, the sphere shape should be characteristic of swollen mitochondria. However, in most processes of mitochondrial plasticity, only the inner membrane is involved, while the outer membrane acts only as a restrictor. This is due to the fact that the outer membrane

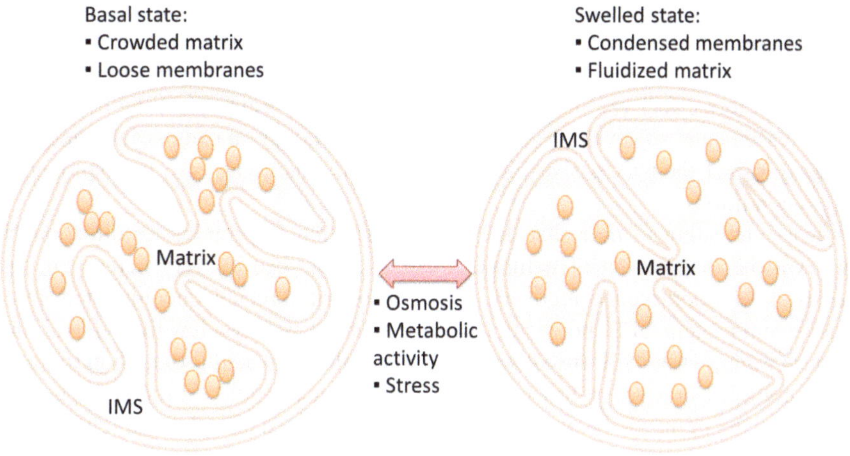

Fig. 11.4 The simplistic scheme of mitochondrial reversible swelling. The figure shows slices of mitochondria with near-circle outer and folded inner membrane. The ovals represent matrix proteins. *IMS* intermembrane space

is permeable for most ions, so it does not participate in osmotic volume regulation. Figure 11.4 shows the simplified illustration of how volume regulation caused by osmotic changes due to metabolic or stress factors may influence the concentration of proteins and thus the crowding in the matrix. Such osmotic low-amplitude swelling which does not disrupt the membranes is well known in mitochondria (Halestrap 1989).

Crowding Amplifies Matrix Liquid Phase Volume Changes

The fact of high crowding raises the importance of osmotic regulation in mitochondria. Even increasing the water fraction by tens of percent under such constricted conditions can change the available volume up to an order of magnitude. Consider as a hypothetical example a densely packed compressed mitochondrion with observed matrix volume of 0.3 μm^3, 0.25 μm^3 of which is occupied by ribosomes, enzyme complexes, DNA and RNA biomolecular condensates. In this case, a 66% increase of the matrix volume (to 0.5 μm^3) due to osmotic water influx will lead to an increase in the solvent volume from 0.05 to 0.25 μm^3, which is as much as 500%. Given the lack of information on what percentage of the volume can be considered as excluded in mitochondria, a real gain in volume regulation cannot be deduced. However, the given example clearly shows the presence of such amplification.

Amplification of free volume change may be used even for mitochondrial volume regulation system itself. Increase in volume cause decrease in magnesium ion concentration and activates K^+/H^+ antiporters, which decrease matrix volume while increased magnesium concentration does not inhibit channels again (the Mg^{2+} brake hypothesis is reviewed in (Garlid and Paucek 2003)). The existence of a gain between the apparent increase in matrix volume and the increase in effective solute volume may make the regulatory action of calcium and magnesium ions in mitochondria more effective. Such an effect can work both ways. Swelling in response to increased entry of divalent cations into the mitochondrion can partially and temporarily offset their excessive biological effects, which can also be exploited in stress conditions.

It is interesting to note that due to the clustering of proteins in mitochondria in vivo, the distribution of natural crowding agents can be very heterogeneous. Analysis of fluorescence recovery of mitochondria-directed unconjugated green fluorescent protein GFP showed that its diffusion rate is only a few times lower than in pure water, while GFP fused with subunits of the fatty acid β-oxidation multienzyme complex shows very low mobility (Partikian et al. 1998). Authors suggest that the clustering of matrix enzymes in membrane-associated complexes might serve to establish a relatively uncrowded aqueous space in which solutes can freely diffuse.

Water Polarization Accompanies Crowding

It should be noted that in compressed mitochondrial cristae, even in the intermembrane space (IMS), in which there are much less proteins than in matrix, one can hardly speak of the usual liquid phase of water. The lumen of mitochondrial disk-shaped cristae can be less than 20 nm (Plokhikh et al. 2024; Waltz et al. 2024). If we consider the size of a water molecule to be about 2.7 A, a simple calculation would place less than one hundred water molecules between the membranes in lamellar region of cristae. At such a small distance from the charged membrane, all water dipoles will be polarized (Cheng et al. 2003). Despite the above, the volume of solvent the IMS is close to that observed on electron microscopy because of the low concentration of proteins there. In the first approximation displaced volume can be neglected. Obviously, not only the membrane surface but also the charged surfaces of proteins can cause water polarization. Therefore, in the crowded matrix, even at some distance from the membrane, water is not fully liquid as it is in the bulk. In the presence of a membrane potential, the water in narrow cristae will be even more polarized. Thus, simple physical estimates suggest that it is not at all correct to speak of a liquid phase of water in mitochondria even in IMS with low extent of crowding. This topic is raised in the context of crowding for the reason that there is no clear boundary defining the difference between bulk water, which is considered as a free solvent, and protein- or membrane-bound water, whose mobility is limited. In this connection, it is difficult to say unambiguously which volume around proteins and membranes should be considered as excluded. It is also debatable whether the magnitude of the membrane potential affects the amount of excluded volume near membrane.

This also leads to the conclusion that all estimates of the mitochondrial transmembrane potential made on the basis of the Nernt or other equations applicable to dilute electrolyte solutions should be treated with caution. In principle, this applies to the rest of the cell cytoplasm as well. Fortunately, both protons and potential-dependent dyes largely accumulate in perimembrane zone (Lee et al. 2024), so volume changes for them are not as important in practice as one might expect. Due to this fact, the methods available in the literature for qualitative or semi-quantitative estimation of membrane potential are nearly independent of the solvent volume in the matrix and IMS. This is partly why the problem of crowding in mitochondria has managed almost to be ignored for so long. The impact of crowding is simply outside of the mitochondrial parameters traditionally evaluated by recognized methods.

Impact of Crowding on Membrane Architecture

The crowding affects not only matrix but also inner membrane proteins (Schlame 2021). Low lipid-to-protein ratios in inner mitochondrial membrane may cause anomalous surface diffusion (Jeon et al. 2016), protein clustering (Guigas and Weiss 2016), and membrane shape instability (Chen et al. 2016). Crowding strains the hydrophobic coupling between proteins and lipids, while association of OxPhos proteins to supercomplexes and membrane enrichment with specific lipids, such as cardiolipin, compensate this effects (Schlame 2021; Xu et al. 2021). The unusual lipid composition of mitochondrial and chloroplast bioenergetic membranes enriched with non-bilayer lipids (Garab et al. 2022) may be, among other things, an adaptation to abnormally high concentrations of membrane proteins.

Another interesting feature is the presence of asymmetry in the distribution of domains on different sides of the membrane in many proteins of the OxPhos system. This leads, accordingly, to a different excluded volume near the outer and inner membrane surfaces. This will be particularly noticeable for complexes I, III and ATP synthases, which have huge subunits protruding into the matrix. This asymmetry is one of the driving forces for the curvature of the membrane surface due to steric effects. On the side where there is more protein, the membrane will be convex, and where there is less, correspondingly concave (Stachowiak et al. 2012). This corresponds well with the real experimental evidence of cristae membrane folds with the greatest curvature near ATP synthases (Blum et al. 2019), which are very asymmetrical protein complexes in relation to the membrane. However, the crowding effect in this case is rather auxiliary, since the membrane domains of ATP synthase dimers also have a membrane-bending structure.

There are some evidences that steric effects, including protein size, shape and crowding may solely drive the formation of membrane curvature and coexisting domains of different fluidity in the membrane (Frese et al. 2008). In turn, domain formation in the mitochondrial membrane may be important for protein sorting and functional cluster formation. So crowding can potentially influence the formation of

OxPhos system clusters, which were discovered in mitochondria (Nesterov et al. 2021). It should be noted that crowding can also affect membrane fusion processes (Snead et al. 2017), which are very important for mitochondrial dynamics (Zamponi et al. 2018).

Mitochondria Evolved to Function Under Crowded Conditions

Several mitochondrial enzymes were studied under different crowding agents and concentrations. On malate dehydrogenase *in vitro* it was founded that crowding slowed malate oxidation significantly less than reverse process of oxaloacetate reduction (Poggi and Slade 2015). By doing so, crowding stimulates the enzyme to work in the right direction of the Krebs cycle. The experiments with isolated citrate synthase showed that crowding has a stronger effect on its activity above the critical percolation concentration after which polymer chains of the crowding agent overlaps forming a mesh (Wilcox et al. 2020). Below this concentration, enzyme activity depend more on the chemical nature of the crowding agent and is nearly independent on crowder concentration or medium viscosity (Wilcox et al. 2020). As was mention in the previous subsection, the inner mitochondrial membrane although adapts to protein crowding through enrichment by non-bilayer lipids such as cardiolipin (Schlame 2021; Xu et al. 2021). In sum these suggest that mitochondrial enzymes and membranes have evolved to function optimally under the crowded constraints and in a viscous environment.

Conclusions

The topic of crowding in mitochondria is surprisingly underrepresented in the scientific literature, despite the fact that the high concentration of protein in them has long been known and the problem of crowding has long been raised and discussed. The issue is complex. First, crowding on the outside of the mitochondrion affects its shape and interaction with other parts of the cytoplasm. Second, the internal crowding strongly affects mitochondrial matrix and inner membrane enzymes. The analysis showed that the proteins and even the lipid composition of the mitochondrial inner membrane are adapted to high crowding conditions. The largest crowding effects are observed under stress conditions: when the matrix volume changes due to swelling, the real available solvent volume increases significantly more. As a result, both the concentration of metabolites and ions and their diffusion can change significantly, and conditions for rearrangements of the supramolecular structure of proteins, which tend to cluster in mitochondria under crowding conditions, can be created. The topic of crowding in mitochondria requires more experimental investigation using modern biophysics and molecular biology techniques.

References

Andersson GE, Karlberg O, Canbäck B, Kurland CG (2003) On the origin of mitochondria: a genomics perspective. Philos Trans R Soc Lond Ser B Biol Sci 358:165–179. https://doi.org/10.1098/rstb.2002.1193

Blum TB, Hahn A, Meier T et al (2019) Dimers of mitochondrial ATP synthase induce membrane curvature and self-assemble into rows. Proc Natl Acad Sci USA 116:4250–4255. https://doi.org/10.1073/pnas.1816556116

Bulthuis EP, Dieteren CEJ, Bergmans J et al (2023) Stress-dependent macromolecular crowding in the mitochondrial matrix. EMBO J 42:e108533. https://doi.org/10.15252/embj.2021108533

Cavalier-Smith T (1987) The simultaneous symbiotic origin of mitochondria, chloroplasts, and microbodies. Ann N Y Acad Sci 503:55–71. https://doi.org/10.1111/j.1749-6632.1987.tb40597.x

Cavalier-Smith T (2006) Origin of mitochondria by intracellular enslavement of a photosynthetic purple bacterium. Proc R Soc B Biol Sci 273:1943–1952. https://doi.org/10.1098/rspb.2006.3531

Chen Z, Atefi E, Baumgart T (2016) Membrane shape instability induced by protein crowding. Biophys J 111:1823–1826. https://doi.org/10.1016/j.bpj.2016.09.039

Cheng J-X, Pautot S, Weitz DA, Xie XS (2003) Ordering of water molecules between phospholipid bilayers visualized by coherent anti-stokes Raman scattering microscopy. Proc Natl Acad Sci 100:9826–9830. https://doi.org/10.1073/pnas.1732202100

Dayhoff GW, Uversky VN (2022) Rapid prediction and analysis of protein intrinsic disorder. Protein Sci Publ Prot Soc 31:e4496. https://doi.org/10.1002/pro.4496

Dietrich L, Agip A-NA, Kunz C et al (2024) In situ structure and rotary states of mitochondrial ATP synthase in whole *Polytomella* cells. Science 385:1086–1090. https://doi.org/10.1126/science.adp4640

Fonin AV, Darling AL, Kuznetsova IM et al (2018) Intrinsically disordered proteins in crowded milieu: when chaos prevails within the cellular gumbo. Cell Mol Life Sci 75:3907–3929. https://doi.org/10.1007/s00018-018-2894-9

Frese RN, Pàmies JC, Olsen JD et al (2008) Protein shape and crowding drive domain formation and curvature in biological membranes. Biophys J 94:640–647. https://doi.org/10.1529/biophysj.107.116913

Garab G, Yaguzhinsky LS, Dlouhý O et al (2022) Structural and functional roles of non-bilayer lipid phases of chloroplast thylakoid membranes and mitochondrial inner membranes. Prog Lipid Res 86:101163. https://doi.org/10.1016/j.plipres.2022.101163

Garlid KD, Paucek P (2003) Mitochondrial potassium transport: the K+ cycle. Biochim Biophys Acta BBA Bioenerg 1606:23–41. https://doi.org/10.1016/S0005-2728(03)00108-7

Gasanoff ES, Yaguzhinsky LS, Garab G (2021) Cardiolipin, non-bilayer structures and mitochondrial bioenergetics: relevance to cardiovascular disease. Cells 10:1721. https://doi.org/10.3390/cells10071721

Gerencser AA, Chinopoulos C, Birket MJ et al (2012) Quantitative measurement of mitochondrial membrane potential in cultured cells: calcium-induced de- and hyperpolarization of neuronal mitochondria. J Physiol 590:2845–2871. https://doi.org/10.1113/jphysiol.2012.228387

Glancy B, Kim Y, Katti P, Willingham TB (2020) The functional impact of mitochondrial structure across subcellular scales. Front Physiol 11. https://doi.org/10.3389/fphys.2020.541040

Guigas G, Weiss M (2016) Effects of protein crowding on membrane systems. Biochim Biophys Acta BBA Biomembr 1858:2441–2450. https://doi.org/10.1016/j.bbamem.2015.12.021

Hackenbrock CR (1966) Ultrastructural bases for metabolically linked mechanical activity in mitochondria. I. Reversible ultrastructural changes with change in metabolic steady state in isolated liver mitochondria. J Cell Biol 30:269–297. https://doi.org/10.1083/jcb.30.2.269

Halestrap AP (1989) The regulation of the matrix volume of mammalian mitochondria in vivo and in vitro and its role in the control of mitochondrial metabolism. Biochim Biophys Acta 973:355–382

Hardenberg M, Horvath A, Ambrus V et al (2020) Widespread occurrence of the droplet state of proteins in the human proteome. Proc Natl Acad Sci USA 117:33254–33262. https://doi.org/10.1073/pnas.2007670117

Honma S, Moritani M, Zhang L-F et al (2001) Quantitative ultrastructure of synapses on functionally identified primary afferent neurons in the cat trigeminal mesencephalic nucleus. Exp Brain Res 137:150–162. https://doi.org/10.1007/s002210000632

Huang F, Oldfield C, Meng J et al (2012) Subclassifying disordered proteins by the CH-CDF plot method. Pac Symp Biocomput Pac Symp Biocomput:128–139

Huang F, Oldfield CJ, Xue B et al (2014) Improving protein order-disorder classification using charge-hydropathy plots. BMC Bioinform 15:S4. https://doi.org/10.1186/1471-2105-15-S17-S4

Jenkins BC, Neikirk K, Katti P et al (2024) Mitochondria in disease: changes in shapes and dynamics. Trends Biochem Sci 49:346–360. https://doi.org/10.1016/j.tibs.2024.01.011

Jeon J-H, Javanainen M, Martinez-Seara H et al (2016) Protein crowding in lipid bilayers gives rise to non-Gaussian anomalous lateral diffusion of phospholipids and proteins. Phys Rev X 6:021006. https://doi.org/10.1103/PhysRevX.6.021006

John P, Whatley FR (1975) Paracoccus denitrificans and the evolutionary origin of the mitochondrion. Nature 254:495–498. https://doi.org/10.1038/254495a0

Kaasik A, Safiulina D, Zharkovsky A, Veksler V (2007) Regulation of mitochondrial matrix volume. Am J Physiol-Cell Physiol 292:C157–C163. https://doi.org/10.1152/ajpcell.00272.2006

Krasinskaia IP, Litvinov IS, Zakharov SD et al (1989) Two qualitatively different structuro-functional states of mitochondria. Biokhimii͡a Mosc Russ 54:1550–1556

Laties GG, Treffry T (1969) Reversible changes in conformation of mitochondria of constant volume. Tissue Cell 1:575–592. https://doi.org/10.1016/S0040-8166(69)80032-7

Lee C, Wallace DC, Burke PJ (2024) Super-resolution imaging of voltages in the interior of individual, vital mitochondria. ACS Nano 18:1345–1356. https://doi.org/10.1021/acsnano.3c02768

Mannella CA (2006) Structure and dynamics of the mitochondrial inner membrane cristae. Biochim Biophys Acta BBA Mol Cell Res 1763:542–548. https://doi.org/10.1016/j.bbamcr.2006.04.006

Mohan A, Sullivan WJ Jr, Radivojac P et al (2008) Intrinsic disorder in pathogenic and non-pathogenic microbes: discovering and analyzing the unfoldomes of early-branching eukaryotes. Mol BioSyst 4:328. https://doi.org/10.1039/b719168e

Morgenstern M, Peikert CD, Lübbert P et al (2021) Quantitative high-confidence human mitochondrial proteome and its dynamics in cellular context. Cell Metab 33:2464–2483.e18. https://doi.org/10.1016/j.cmet.2021.11.001

Naguro T, Nakane H, Inaga S (2012) Mitochondrial continuous intracellular network-structures visualized with high-resolution field-emission scanning electron microscopy. In: Schatten H (ed) Scanning electron microscopy for the life sciences. Cambridge University Press, Cambridge, pp 50–70

Nesterov S, Chesnokov Y, Kamyshinsky R et al (2021) Ordered clusters of the complete oxidative phosphorylation system in cardiac mitochondria. Int J Mol Sci 22:1462. https://doi.org/10.3390/ijms22031462

Nesterov SV, Ilyinsky NS, Plokhikh KS et al (2024a) Order wrapped in chaos: on the roles of intrinsically disordered proteins and RNAs in the arrangement of the mitochondrial enzymatic machines. Int J Biol Macromol 267:131455. https://doi.org/10.1016/j.ijbiomac.2024.131455

Nesterov SV, Plokhikh KS, Chesnokov YM et al (2024b) Safari with an electron gun: visualization of protein and membrane interactions in mitochondria in natural environment. Biochem Mosc 89:257–268. https://doi.org/10.1134/S0006297924020068

Oldfield CJ, Cheng Y, Cortese MS et al (2005) Comparing and combining predictors of mostly disordered proteins. Biochemistry 44:1989–2000. https://doi.org/10.1021/bi047993o

Parry BR, Surovtsev IV, Cabeen MT et al (2014) The bacterial cytoplasm has glass-like properties and is fluidized by metabolic activity. Cell 156:183–194. https://doi.org/10.1016/j.cell.2013.11.028

Partikian A, Ölveczky B, Swaminathan R et al (1998) Rapid diffusion of green fluorescent protein in the mitochondrial matrix. J Cell Biol 140:821–829. https://doi.org/10.1083/jcb.140.4.821

Peng Z, Yan J, Fan X et al (2015) Exceptionally abundant exceptions: comprehensive characterization of intrinsic disorder in all domains of life. Cell Mol Life Sci 72:137–151. https://doi.org/10.1007/s00018-014-1661-9

Plokhikh KS, Nesterov SV, Chesnokov YM et al (2024) Association of 2-oxoacid dehydrogenase complexes with respirasomes in mitochondria. FEBS J 291:132–141. https://doi.org/10.1111/febs.16965

Poggi CG, Slade KM (2015) Macromolecular crowding and the steady-state kinetics of malate dehydrogenase. Biochemistry 54:260–267. https://doi.org/10.1021/bi5011255

Rajagopalan K, Mooney SM, Parekh N et al (2011) A majority of the cancer/testis antigens are intrinsically disordered proteins. J Cell Biochem 112:3256–3267. https://doi.org/10.1002/jcb.23252

Schlame M (2021) Protein crowding in the inner mitochondrial membrane. Biochim Biophys Acta BBA Bioenerg 1862:148305. https://doi.org/10.1016/j.bbabio.2020.148305

Shami GJ, Cheng D, Verhaegh P et al (2021) Three-dimensional ultrastructure of giant mitochondria in human non-alcoholic fatty liver disease. Sci Rep 11:3319. https://doi.org/10.1038/s41598-021-82884-z

Snead WT, Hayden CC, Gadok AK et al (2017) Membrane fission by protein crowding. Proc Natl Acad Sci 114. https://doi.org/10.1073/pnas.1616199114

Stachowiak JC, Schmid EM, Ryan CJ et al (2012) Membrane bending by protein–protein crowding. Nat Cell Biol 14:944–949. https://doi.org/10.1038/ncb2561

Sun X, Xue B, Jones WT et al (2011) A functionally required unfoldome from the plant kingdom: intrinsically disordered N-terminal domains of GRAS proteins are involved in molecular recognition during plant development. Plant Mol Biol 77:205–223. https://doi.org/10.1007/s11103-011-9803-z

Sun C-L, Van Gilst M, Crowder CM (2023) Hypoxia-induced mitochondrial stress granules. Cell Death Dis 14:448. https://doi.org/10.1038/s41419-023-05988-6

Tcherkasskaya O, Uversky V (2003) Polymeric aspects of protein folding: a brief overview. Protein Pept Lett 10:239–245. https://doi.org/10.2174/0929866033478988

Uversky VN (1993) Use of fast protein size-exclusion liquid chromatography to study the unfolding of proteins which denature through the molten globule. Biochemistry 32:13288–13298. https://doi.org/10.1021/bi00211a042

Uversky VN (2002a) Natively unfolded proteins: a point where biology waits for physics. Protein Sci 11:739–756. https://doi.org/10.1110/ps.4210102

Uversky VN (2002b) What does it mean to be natively unfolded? Eur J Biochem 269:2–12. https://doi.org/10.1046/j.0014-2956.2001.02649.x

Uversky VN, Dunker AK (2010) Understanding protein non-folding. Biochim Biophys Acta BBA Prot Proteom 1804:1231–1264. https://doi.org/10.1016/j.bbapap.2010.01.017

Uversky VN, Gillespie JR, Fink AL (2000) Why are "natively unfolded" proteins unstructured under physiologic conditions? Proteins 41:415–427. https://doi.org/10.1002/1097-0134(20001115)41:3<415::aid-prot130>3.0.co;2-7

Van Den Berg J, Boersma AJ, Poolman B (2017) Microorganisms maintain crowding homeostasis. Nat Rev Microbiol 15:309–318. https://doi.org/10.1038/nrmicro.2017.17

Waltz F, Righetto RD, Kelley R et al (2024) In-cell architecture of the mitochondrial respiratory chain

Ward JJ, Sodhi JS, McGuffin LJ et al (2004) Prediction and functional analysis of native disorder in proteins from the three kingdoms of life. J Mol Biol 337:635–645. https://doi.org/10.1016/j.jmb.2004.02.002

Wilcox XE, Ariola A, Jackson JR, Slade KM (2020) Overlap concentration and the effect of macromolecular crowding on citrate synthase activity. Biochemistry 59:1737–1746. https://doi.org/10.1021/acs.biochem.0c00073

Xu Y, Erdjument-Bromage H, Phoon CKL et al (2021) Cardiolipin remodeling enables protein crowding in the inner mitochondrial membrane. EMBO J 40:e108428. https://doi.org/10.15252/embj.2021108428

Xue B, Dunker AK, Uversky VN (2012a) Orderly order in protein intrinsic disorder distribution: disorder in 3500 proteomes from viruses and the three domains of life. J Biomol Struct Dyn 30:137–149. https://doi.org/10.1080/07391102.2012.675145

Xue B, Oldfield CJ, Van Y-Y et al (2012b) Protein intrinsic disorder and induced pluripotent stem cells. Mol BioSyst 8:134–150. https://doi.org/10.1039/C1MB05163F

Zamponi N, Zamponi E, Cannas SA et al (2018) Mitochondrial network complexity emerges from fission/fusion dynamics. Sci Rep 8:363. https://doi.org/10.1038/s41598-017-18351-5

Zhao B, Katuwawala A, Uversky VN, Kurgan L (2021) IDPology of the living cell: intrinsic disorder in the subcellular compartments of the human cell. Cell Mol Life Sci CMLS 78:2371–2385. https://doi.org/10.1007/s00018-020-03654-0

Chapter 12
Molecular Crowing in Nuclear Pore

Masahiro Kumeta

Abstract Nuclear pores serve as the sole gates mediating nucleocytoplasmic molecular communication. They constantly accept heavy molecular traffic at a rate of ~1000 molecules per second, selected from a vast number of molecules randomly approaching the pores. The central channel of the pores are highly crowded with an intrinsically disordered region of pore-forming subunits and this channel functions as a selective permeability barrier. Recently, the phase separation properties of the hydrophobic subunits of pores have been reported, together with the flexible amphiphilic nature of the transporting molecules. These findings suggest that phase separation is a fundamental mechanism of action in nuclear pores. In this chapter, the entire nucleocytoplasmic transport system and composition of the nuclear pore complex are reviewed, followed by a detailed review of recent studies focusing on the characteristic features of both nuclear pores and transporting molecules. Finally, intrinsic and extrinsic factors that adaptively affect the function of the molecular crowding barrier are introduced.

Keywords Nuclear pore complex · NPC · Nucleoporin · Nup · Nuclear transport · Karyopherin · Hydrogel · FG-Nup

M. Kumeta (✉)
Graduate School of Biostudies, Kyoto University, Kyoto, Japan

Center for Living Systems Information Science (CeLiSIS), Kyoto University, Kyoto, Japan
e-mail: kumeta@lif.kyoto-u.ac.jp

V. N. Uversky (ed.), *(Macro)Molecular Crowding*, Subcellular Biochemistry 109,
https://doi.org/10.1007/978-3-032-03370-3_12

Overview of the Nucleocytoplasmic Molecular Transport System

In contrast to prokaryotes, eukaryotic cells insulate their genetic material from the cytoplasm and organize gene activity in a multi-step manner in the nucleus. Therefore, the proper control of nucleocytoplasmic molecular communication is essential for all eukaryotic cells to ensure both basal and adaptive cellular activities. The nucleus is the largest and most characteristic endomembrane compartment in eukaryotic cells and is enclosed by a double lipid membrane with many accessory proteins, collectively called the nuclear envelope (NE). The nuclear pore complex (NPC) is a giant macromolecular complex that forms a pore penetrating the NE and functions as the sole gate mediating molecular traffic between the cytoplasm and nucleoplasm (reviewed in (Cowburn and Rout 2023, Dultz et al. 2022, Fernandez-Martinez and Rout 2021)).

Two major modes of molecular passage through the NPC have been proposed: passive diffusion and facilitated transport (Fig. 12.1).

1. Passive diffusion

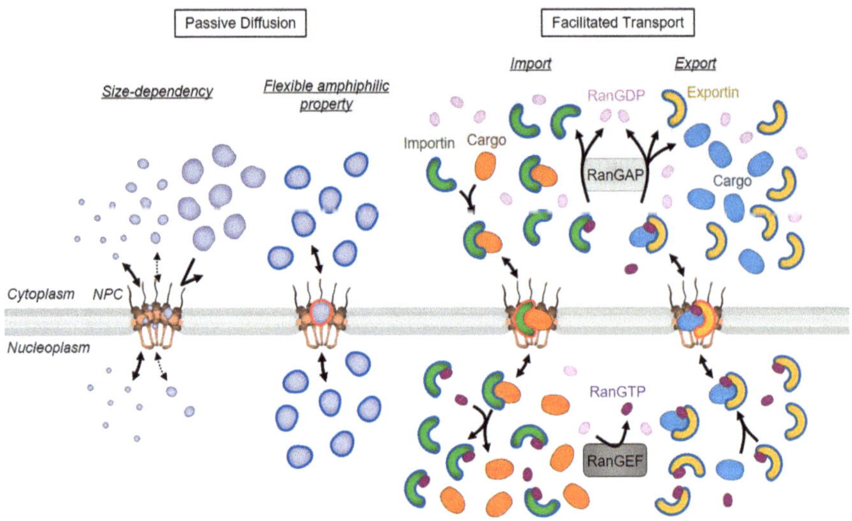

Fig. 12.1 Overview of the nucleocytoplasmic molecular transport system. Selective permeability for passive diffusion is dependent on the size or flexible amphiphilic property of the passing molecule. Amphiphilic molecules transiently expose hydrophobic moieties to the molecular surface (described in red) to adapt to the hydrophobic crowding environment in the nuclear pore. Facilitated transport is mediated by karyopherins; importin (green) for nuclear import and exportin (yellow) for export. Catch-and-release of the cargo is regulated by the small GTPase Ran; GTP-bound Ran (purple) binds to karyopherin and facilitate dissociation or association with the cargo for import (orange) and export (blue), respectively. Cytoplasmic RanGAP (gray) and nuclear chromatin-bound RanGEF (black) maintain high levels of RanGTP in the nucleus and RanGDP (pink) in the cytoplasm, resulting in the directed transport of respective cargos

Small molecules diffusively pass through the NPC (Keminer and Peters 1999). This idea was supported by the observation that a single green fluorescent protein (GFP) expressed in cells or applied for *in vitro* nuclear transport assay reached the nucleus, whereas a three tandem GFP molecule was excluded, demonstrating that molecular mass is a crucial factor determining passage (Kumeta et al. 2012; Sakakida et al. 2005; Glodowski et al. 2002; Adam et al. 1990). The nature of the NPC barrier function that accepts the passage of small molecules has been explained by size-dependent filtration with a threshold cutoff size of 30–60 kDa; however, recent studies have suggested the existence of a soft barrier property that intensifies gradually with increasing molecular mass (Timney et al. 2016). In addition to molecular mass, the amphiphilic property and structural flexibility of the molecule also affect passages (Kumeta et al. 2018; Yoshimura et al. 2014). These factors are related to the adaptability of molecules to the environment within the NPC. Therefore, passive diffusion is determined by the molecular properties required to efficiently enter and stably exist in the barrier region of the NPC.

2. Facilitated transport

   A set of molecules, collectively called nuclear transport receptors (NTRs), mediate the molecular passage. They bind to the target cargo molecule directly or through adaptor molecules to form a transport complex, pass through the NPC, and release the cargo. Karyopherin is one of the largest families of NTRs that contains importins and exportins and plays a major role in transporting cargo with signal peptides, such as the nuclear localization signal (NLS) and nuclear export signal (NES) (reviewed in (Wing et al. 2022, Kimura and Imamoto 2014, Mosammaparast and Pemberton 2004)). The small GTPase Ran plays a regulatory role in cargo binding and release. In the case of importin, GTP-bound Ran binds to the transport complex and attenuates the interaction between importin and the cargo, whereas GDP-bound Ran lacks affinity for the complex. Cytoplasmic localization of Ran GTPase-activating protein (RanGAP) and nuclear localization of Ran guanine nucleotide exchange factor (RanGEF) establish the dominant localization of GTP-bound Ran in the nucleus, ensure the binding of the cargo in the cytoplasm, and release it in the nucleus. This catch-and-release mechanism allows importin to import its cargo, although its passage through NPC is bidirectional (Lolodi et al. 2016). Ran acts in opposition to the exportin cargo, establishing the nuclear export pathway.

Structural Organization of the Nuclear Pore Complex

NPC is one of the largest macromolecular complexes in eukaryotic cells, with an outer diameter of 100–130 nm and a height of 50–80 nm (Akey et al. 2022; Schuller et al. 2021; Allegretti et al. 2020). The inner diameter of the NPC ranges from 40 to 70 nm depending on the species and cell types, forming a molecular transport channel large enough to pass ribosomal subunits (approximately 25 nm in diameter) or artificial capsid-like particles (up to 36 nm) (Ruland et al. 2021; Paci et al. 2020). Typical mammalian cells contain 2000–5000 NPC per nucleus, mediating a mass flow of ~100 MDa and a rate of ~1000 selective molecular passages per NPC per second (Ribbeck and Gorlich 2001; Maul and Deaven 1977). In addition to the large central pore, the NPC structure is characterized by an eight-fold rotational symmetry and contains cytoplasmic filaments and nuclear basket structures. The structure and composition of NPC varies considerably in different species, cell types, and cell states (reviewed in (Fernandez-Martinez and Rout 2021, Field and Rout 2019). Owing to their large and complicated composition, NPCs can be easily constructed as custom-made devices by selecting slightly different building blocks to fulfill different cellular needs for its function (Raices and D'angelo 2012).

The structure of the NPC was first reported in 1949, using electron microscopy to observe the nuclear membrane of *Xenopus* oocyte (Callan and Tomlin 1950; Callan et al. 1949). Since then, intensive structural studies using electron tomography, X-ray crystallography, nuclear magnetic resonance spectroscopy, and cryo-electron microscopy have been performed to elucidate the fine structure of NPC at the submolecular level (reviewed in (Cristi et al. 2023; Dultz et al. 2022; Hampoelz et al. 2019; Kim et al. 2018)).

The NPC is composed of multiple copies of ~30 different proteins called nucleoporins (Nups), an assembly of more than 500 proteins in yeast and ~ 1000 proteins in mammalian cells, with a total molecular mass of approximately 66 MDa in yeast and 125 MDa in vertebrates (Rout and Blobel 1993; Reichelt et al. 1990). Nups are assembled into several ring-like substructures (Fig. 12.2). The transmembrane ring contains Nups with transmembrane domains including Gp210, Ndc1, and Pom121, and functions to anchor the entire NPC structure to the nuclear membrane. The cytoplasmic and nucleoplasmic outer rings are almost identical and symmetrically

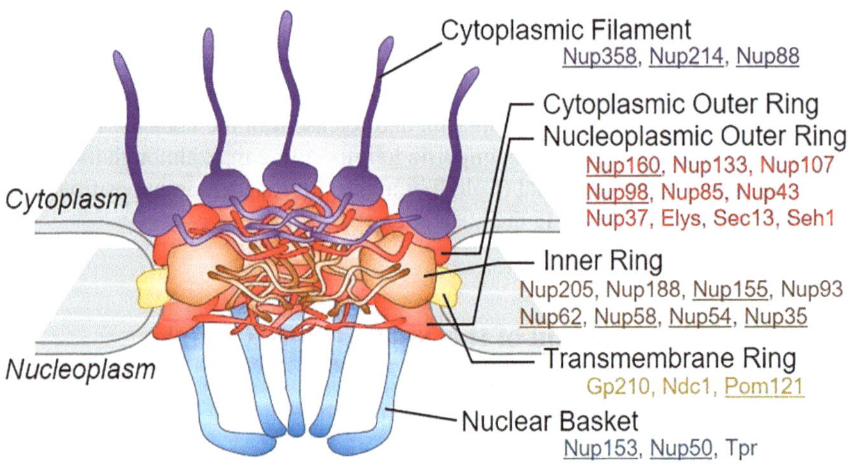

Fig. 12.2 Structure and subunit composition of the NPC. Illustration of the major ring-like substructures in NPC. Human subunits composing each substructure is listed. FG-Nups are underlined

located on both sides of the pore. A Y-shaped rigid sub-complex made of six to nine Nups, including Nup160, Nup133, Nup107, Nup96, Nup85, Nup37, Sec13, and Seh1 in case of human, is arranged to form a scaffold for the outer rings, which is essential for NPC formation (Kelley et al. 2015; Walther et al. 2003; Lutzmann et al. 2002). The cytoplasmic filament consists of an elongated Nup358 tail and a central Nup214-Nup88 sub-complex, and serves as an mRNA export platform (Fernandez-Martinez et al. 2016; Bernad et al. 2006). The structure of the nuclear basket is not fully understood because of the intrinsic flexibility of the major subunit Tpr. The function of the nuclear basket is to exclude heterochromatin formation to maintain an open chromatin structure around the nucleoplasmic side of the NPC (Iglesias et al. 2020; Wong et al. 2015). The inner ring forms the central channel of the NPC. It consists of a membrane-bound layer (Nup205, Nup188, and Nup155), a connector layer (Nup93 and Nup35), and an inner layer (Nup62, Nup58, and Nup54), and it organizes the central transporter region of the NPC. Although all Nups exhibited uniformly slow turnover rates, they are constantly exchanged to ensure maintenance and damage repair in the NPC (Hakhverdyan et al. 2021).

Organization of the Molecular Crowding Barrier in the Nuclear Pore

Among the ~30 Nups, approximately half contain characteristic phenylalanine-glycine (FG) repeated motifs and play a central role in forming molecular crowding barrier in the nuclear pore. These Nups, called FG-Nups, are enriched in intrinsically disordered regions that contain ~50 hydrophobic FG-repeat units separated by hydrophilic spacers (Denning et al. 2003). The total amount of FG-repeat regions, including the FG, FxFG, and GLFG motifs, accounts for ~12% of the total mass of NPC, and their local concentration is estimated to be approximately 100 mg/ml in NPC (Rout et al. 2000; Bayliss et al. 1999). Under this highly crowded environment, many of the FG-Nups spontaneously assemble and undergo phase separation (reviewed in (Zilman 2018)). *In vitro* formation of an elastic hydrogel of Nsp1p (the yeast homologue of mammalian Nup62) was first reported when its FG-repeat domain was purified and dissolved in aqueous solution at a concentration of 26 mg/ml (Frey et al. 2006). Substitution of phenylalanine with serine abolished hydrogel formation, demonstrating that hydrophobic interactions between phenylalanines are essential for phase separation (Frey et al. 2006). This physicochemical property is widely shared among FG-Nups, as revealed by both experimental and simulation analyses (Fig. 12.3) (Dekker et al. 2023; Schmidt and Gorlich 2015; Labokha et al. 2012). Electron microscopy has revealed several different types of amyloid-like fibril structures formed in the hydrogel by FG-FG interactions, which were unstable and reversible nature (Ibanez De Opakua et al. 2022). Taken all together, the FG-repeat regions rapidly interchange between liquid-like transient interactions and

Fig. 12.3 Hydrogel made by FG-Nup. FG-repeat region of Nup98 forms a hydrogel-like solid structure. Solid and dashed lines of the grid sheet indicate 10 and 5 mm width, respectively

amyloid-like cohesive interactions, forming a hydrophobic molecular crowding barrier in NPC.

The FG-Nup hydrogel is highly permeable to NTR molecules and mimic the selective permeability of NPC. A saturated FG-Nup hydrogel or phase-separated FG particles accept a diffusive influx of NTRs >1000 times faster than a similarly sized macromolecule, and also mediate the facilitated transport of NTR-mediated cargos (Schmidt and Gorlich 2015; Frey and Gorlich 2007). Mobility measurements by fluorescence recovery after photobleaching (FRAP) revealed that the influx of NTRs is as fast as the diffusion of the surrounding buffer into the FG particles, although FG-repeat molecules, the constituents of the FG particles, are less dynamic (Schmidt and Gorlich 2015). In addition, NTRs affect hydrogel properties by preventing amyloid formation (Milles et al. 2013). Considering that ~100 NTR molecules are consistently stacked within the pore, the nature of the NPC barrier is controlled not only by FG-Nups, but also by NTRs transiently residing within the pore (Lowe et al. 2015).

The status of the molecular crowding barrier in NPC has been examined *in vivo* using several approaches. In-cell measurements of a probed FG-Nup using highly time-resolved fluorescence microscopy revealed that the FG domain of Nup98 adopts an extended conformation inside the NPC and that it fluctuates extensively to form a dynamic barrier across the central channel (Yu et al. 2023). The distribution and crowding states of each FG-Nups varied in the different regions of the pore. FG-Nups located in the central channel tend to contain the GLFG motif and form many short-lived FG-FG interactions to constitute a highly dynamic network, whereas peripheral FG-Nups are enriched in FxFG and FG motifs and likely form an entropic brush (Dekker et al. 2023; Huang et al. 2020; Patel et al. 2007). *In vivo* direct measurement of crowding states within the NPC using a molecular crowding-sensitive FRET probe revealed stronger peripheral crowding than that in the inner region (Konishi et al. 2017). Because peripheral regions need to select passing cargo from a vast number of randomly approaching molecules, such localized organization of the crowding barrier may contribute to establishing rapid and efficient decision-making at the pore.

Physicochemical Features of the Cargo Molecules

The molecular passage through a single NPC is estimated to be ~1000 molecules per second (Ribbeck and Gorlich 2001; Maul and Deaven 1977). Molecules pass through the NPC either by size-dependent passive diffusion or by facilitated transport with the help of NTRs. From an integrative view of the two mechanisms utilizing the same channel, the question arises as to how the NTRs efficiently pass through the NPC despite their large size. Structural studies focusing on karyopherins, one of the largest families of NTRs including importins and exportins, have revealed their characteristic features that are optimized for transport.

The karyopherin family consists of 14–20 members depending on species and include a major import mediator importin-β and a major export mediator CRM1 (reviewed in (Wing et al. 2022, Kimura and Imamoto 2014, Mosammaparast and Pemberton 2004)). They are typically composed of ~20 repeats of HEAT motif, each compromising of two α-helices connected with a short linker region (Chook and Blobel 1999; Cingolani et al. 1999). These α-helices are characteristic in its amphiphilic property; locating hydrophobic amino acids in one side and hydrophilic amino acids in another side, that makes two consecutive helices facing with each other spontaneously by hydrophobic interaction (Fig. 12.4a) (Yoshimura et al. 2014). As the entire molecule is composed of tandem repeats of this helical structure, karyopherins have a solenoid structure and show high structural flexibility

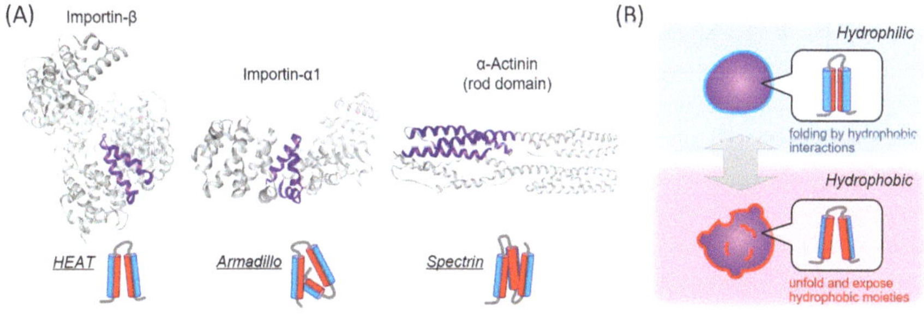

Fig. 12.4 Structural property of the NPC-permeable amphiphilic molecules. (**a**) 3D structures of importin-β (PDB: 1QGK), importin-α1 (6BVT) and a part of the rod domain of dimerized α-actinin (1HCI). Single unit of amphiphilic HEAT repeats, armadillo repeats, and spectrin repeats are shown in purple. Schematic illustration describes the conformation of amphiphilic α-helices with hydrophilic face in blue and hydrophobic face in red. (**b**) Flexible structural change of the amphiphilic molecules in hydrophilic (blue) and hydrophobic (red) solvent

(Forwood et al. 2010). This molecular feature enables importin-β to undergo dynamic conformational change in a hydrophobic solvent, which attenuates interaction between amphiphilic α-helices and expose more hydrophobic moieties to the surface of the molecule (Fig. 12.4b) (Yoshimura et al. 2014). Such a conformational change occurs rapidly in the order of microseconds, which is short enough considering the typical molecular passage duration through the NPC to be several milliseconds long (Yoshimura et al. 2014; Yang et al. 2004).

Based on structural studies, NTRs are expected to pass through the molecular crowding barrier in the NPC according to the following steps: (1) Approach the NPC channel either diffusively or by interaction with Nups; (2) Upon entry into the peripheral crowding barrier, NTRs undergo structural changes to adapt to the hydrophobic environment by exposing more hydrophobic amino acids to the surface of the molecule; (3) Diffusion remained in the pore; (4) Upon reaching the peripheral region of the barrier and taking a hydrophilic solvent, the NTRs reconvert their structure to adapt to the hydrophilic environment and exit (Fig. 12.5).

HEAT repeat is also found in other group of proteins including protein phosphatases (PP2A and PP4A), condensins (CAP-D2 and CAP-G), and cohesins (SA2 and PDS5) (reviewed in (Yoshimura and Hirano 2016)). Despite their large molecular sizes, many of these proteins pass through the NPC by themselves or even with their interacting partners, similar to NTRs (Yoshimura et al. 2014). Considering their

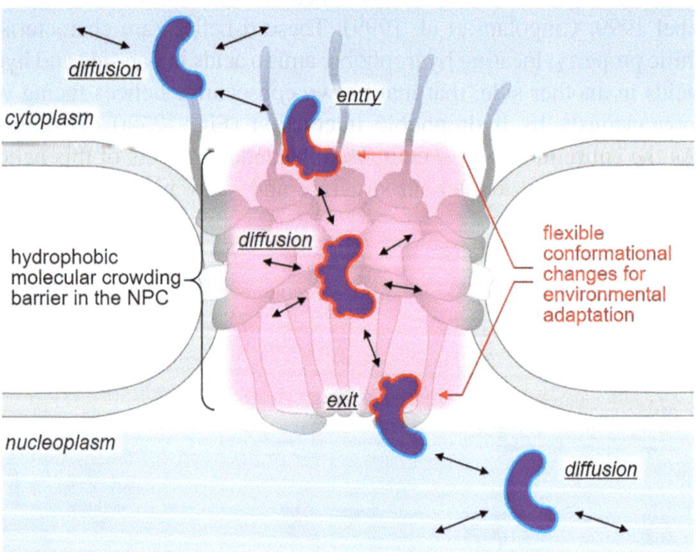

Fig. 12.5 Mechanism of passage through the molecular crowding barrier in the NPC. Molecules diffusively approach to the NPC. Upon entry, the molecules change their conformation to adapt to the hydrophobic crowding environment and diffuse in the central pore. When reaching to the peripheral region again, they are folded back to the original structure to adapt to the hydrophilic environment and released. All the steps are reversible. Hydrophilic and hydrophobic areas are described in blue and red, respectively

roles in large protein complexes and condensed chromatin, HEAT repeats are likely to play a shared role in protein function in crowded environments (Yoshimura and Hirano 2016). Molecules composed of other amphiphilic structural motifs share similar features. Importin-α, which is a mediator of the interaction between importin-β and NLS-containing cargo, is composed of a tandem array of armadillo repeat formed by hydrophobic interaction of three amphiphilic α-helices (Fig. 12.4a) (Huber et al. 1997). The armadillo repeat is also found in β-catenin which transmits Wnt signal to the nucleus and pass through the NPC by directly interacting with Nups (Yokoya et al. 1999; Fagotto et al. 1998). The spectrin repeat is also made of three amphiphilic α-helices facing with each other to form a bundle structure (Fig. 12.4a) (Yan et al. 1993). Similar to the HEAT repeat, the spectrin repeat undergoes a flexible conformational change in a hydrophobic environment and passes through the NPC by itself (Kumeta et al. 2012). Although most spectrin repeat-containing proteins function as rigid cytoskeletal components, many of these cytoplasmic proteins can potentially pass through the NPC and may have another role in the nucleus (Kumeta et al. 2011). Several isoforms of α-actinins and spectrins function in the nucleus as well (Kumeta et al. 2010; Sridharan et al. 2003). These studies, focusing on nuclear transport of non-NTR proteins, have revealed that adaptability to a hydrophobic crowding environment is at least one of the key features which determines passage through the NPC. HEAT repeats and other similar motifs are suitable molecular units that pass through the crowding barrier owing to their amphiphilic and highly flexible properties.

Adaptive Control of the Crowding Barrier in the Nuclear Pore

Changes in the properties of the molecular crowding barrier in NPC occur due to intrinsic and extrinsic factors. Phosphorylation and glycosylation are major intrinsic factors regulating the structure and function of the NPC, whereas extrinsic factors, such as environmental stress and viral infection, can also affect the selective permeability of the barrier.

Several kinases target Nups. For example, extracellular signal-regulated kinase (ERK) phosphorylates Nup214, Nup153, and Nup60, and Rho-associated coiled-coil-containing protein kinase (ROCK) targets Nup62, both of which result in the attenuation of nuclear transport (Hazawa et al. 2018; Kosako and Imamoto 2010; Kosako et al. 2009). Molecular dynamics simulations have revealed a partial loss of the phosphorylated FG-Nup from the central channel, as phosphorylation confers a hydrophilic moiety and attenuates hydrophobic interactions among them (Mishra et al. 2019). Phosphorylation regulates the mitotic disassembly and reassembly of NPC as well (reviewed in (Otsuka and Ellenberg 2018, Champion et al. 2016)). Cell cycle-dependent kinase 1 (CDK1), polo-like kinase 1 (PLK1), and NIMA-related kinases (NEKs) are responsible for the mitotic breakdown of NPC by introducing stepwise hyperphosphorylation of Nups including Nup98 (Linder et al. 2017; Laurell et al. 2011). Both disassembly and reassembly are well organized in the mitotic time course in order to accurately set up this huge complex (Otsuka et al. 2023; Otsuka et al. 2017). Subcellular *O*-glycosylation is highly concentrated in NPC components and this affects the stability of a group of Nups, including Nup358, Nup214, Nup153, Nup93, Nup62, and Sec13 (Zhu et al. 2016; Li and Kohler 2014). *O*-glycosylation is involved in the interaction among Nups, and the loss of this modification results in a defect in the selective permeability of NPC (Zhu et al. 2016; Mizuguchi-Hata et al. 2013). Because both phosphorylation and glycosylation are known to directly affect phase separation events, these modifications serve as effective ways to adaptively control the molecular crowding barrier and regulate nuclear transport activity (Yamazaki et al. 2022; Nosella et al. 2021).

Several extrinsic factors directly affect molecular crowding barrier in NPC. Oxidative stress affects nuclear transport by acting on both NPC components and transport factors (Kodiha et al. 2008; Miyamoto et al. 2004; Stochaj et al. 2000). Multiple Nups including Nup358, Nup155, Nup153, and Nup62 form disulfide bonds in response to oxidative stress and suppress nuclear transport (Yoshimura et al. 2013). Further studies revealed separate roles of Nups in sensing oxidative stress; a group of them (Nup214, Nup153, Nup98, Nup62, and Pom121) increases disulfide bond formation under the oxidative environment to increase molecular crowding, whereas an another group (Nup88, Nup54, and Nup88) decreases disulfide bonds in the reduced condition to decrease crowding degree in pores (Zhang et al. 2020). The redox environment directly affects the frequency of nuclear transport events and also the retention time of cargo in NPC, suggesting a role for NPC in this environmental stress response (Zhang et al. 2020). Infection with various viruses alters NPC to control host cell activity, which is suitable for viral replication (reviewed in (De Jesus-Gonzalez et al. 2021, Shen et al. 2021)). Proteases from several positive-strand RNA viruses target FG-Nups, including Nup153, Nup98, and Nup62, and alter the nuclear transport activity by processing those Nups (De Jesus-Gonzalez et al. 2020; Walker et al. 2013). Although quantitative details of the effect of viral proteases on the molecular crowding states of NPC have not yet been fully characterized, the molecular crowding barrier in NPC serves as a crucial machinery in host-virus interactions, as it is one of the major targets of virus hijacking cell activities.

Conclusions and Perspectives

Eukaryotic cells employ a molecular crowding barrier to control nucleocytoplasmic molecular communication, which is essential for their survival. This system is beneficial to the cells in the following ways:

1. Autonomous and stable maintenance of the barrier
   NPC is comprised of several hundred proteins. Although there are several rigid molecular units in the scaffold, the major channel region is composed of molecular crowding; the assembly of intrinsically disordered regions of FG-Nups con-

densed by multivalent interactions among them. When compared to the molecular machineries made by 'lock-and-key' type tight bindings, this assembly mechanism allows autonomous and dynamic maintenance of the barrier function in the pore. In addition, having many components are beneficial for the system to stably express its function; for example, the accidental loss of one molecule has little effect on the entire function of the NPC.

2. Spontaneous and low-cost operation of the selection system
The NPC does not adopt an apparently 'closed' conformation similar to typical membrane-bound channels; thus, it continues to select passing molecules from a large number of randomly approaching molecules throughout its lifetime. Small or flexible amphiphilic molecules adapt well to the crowded environment in the pores, enabling diffusive entry into the barrier region. Molecules that could not adapt to the pore environment are entropically rejected. All the processes involve inner and intermolecular dynamics and operate spontaneously. This is significant from an energy perspective. Because of the constant heavy traffic through the NPC, if one ATP molecule is consumed per molecular passage, it requires approximately 3×10^6 ATP per second, which accounts for ~30% of the total ATP production in cells (Flamholz et al. 2014). By utilizing spontaneous interactions between the barrier and the passing molecule in this system, cells save metabolic energy and achieve the desired molecular distribution at a low cost.

The structural and functional mechanisms of the NPC and transport system serve as good examples of how cells utilize molecular crowding in controlling biological systems. The molecular crowding/phase separation barrier is a recently emerging model that is expected to achieve an integrated understanding of classically proposed passive and facilitated transport events. Further studies focusing on the basal and adaptive regulation of molecular crowding phenomena will uncover the fundamental mechanisms of NPC and the entire orchestration of eukaryotic nuclear maintenance strategies.

Acknowledgement I thank Dr. Yoshimura SH (Kyoto University, Japan) and Dr. Konishi HA (The Rockefeller University, USA) for providing Fig. 12.3 image.

Conflicts of Interest The authors have no conflicts of interest to declare that are relevant to the content of this chapter.

References

Adam SA, Marr RS, Gerace L (1990) Nuclear protein import in Permeabilized mammalian cells requires soluble cytoplasmic factors. J Cell Biol 111:807–816

Akey CW, Singh D, Ouch C, Echeverria I, Nudelman I, Varberg JM, Yu Z, Fang F, Shi Y, Wang J, Salzberg D, Song K, Xu C, Gumbart JC, Suslov S, Unruh J, Jaspersen SL, Chait BT, Sali A, Fernandez-Martinez J, Ludtke SJ, Villa E, Rout MP (2022) Comprehensive structure and functional adaptations of the yeast nuclear pore complex. Cell 185(361–378):E25–378.e25

Allegretti M, Zimmerli CE, Rantos V, Wilfling F, Ronchi P, Fung HKH, Lee CW, Hagen W, Turonova B, Karius K, Bormel M, Zhang X, Muller CW, Schwab Y, Mahamid J, Pfander B, Kosinski J, Beck M (2020) In-cell architecture of the nuclear pore and snapshots of its turnover. Nature 586:796–800

Bayliss R, Ribbeck K, Akin D, Kent HM, Feldherr CM, Gorlich D, Stewart M (1999) Interaction between Ntf2 and Xfxfg-containing nucleoporins is required to mediate nuclear import of Rangdp. J Mol Biol 293:579–593

Bernad R, Engelsma D, Sanderson H, Pickersgill H, Fornerod M (2006) Nup214-Nup88 nucleoporin subcomplex is required for Crm1-mediated 60 S Preribosomal nuclear export. J Biol Chem 281:19378–19386

Callan HG, Tomlin SG (1950) Experimental studies on amphibian oocyte nuclei. I. Investigation of the structure of the nuclear membrane by means of the electron microscope. Proc R Soc Lond B Biol Sci 137:367–378

Callan HG, Randall JT, Tomlin SG (1949) An electron microscope study of the nuclear membrane. Nature 163:280

Champion L, Linder MI, Kutay U (2016) Cellular reorganization during mitotic entry. Trends Cell Biol 27:26–41

Chook YM, Blobel G (1999) Structure of the nuclear transport complex Karyopherin-Beta2-ran X Gppnhp. Nature 399:230–237

Cingolani G, Petosa C, Weis K, Muller CW (1999) Structure of importin-Beta bound to the Ibb domain of importin-alpha. Nature 399:221–229

Cowburn D, Rout M (2023) Improving the hole picture: towards a consensus on the mechanism of nuclear transport. Biochem Soc Trans 51:871–886

Cristi AC, Rapuri S, Coyne AN (2023) Nuclear pore complex and nucleocytoplasmic transport disruption in neurodegeneration. FEBS Lett 597:2546–2566

De Jesus-Gonzalez LA, Cervantes-Salazar M, Reyes-Ruiz JM, Osuna-Ramos JF, Farfan-Morales CN, Palacios-Rapalo SN, Perez-Olais JH, Cordero-Rivera CD, Hurtado-Monzon AM, Ruiz-Jimenez F, Gutierrez-Escolano AL, Del Angel RM (2020) The nuclear pore complex: a target for Ns3 protease of dengue and Zika viruses. Viruses 12

De Jesus-Gonzalez LA, Palacios-Rapalo S, Reyes-Ruiz JM, Osuna-Ramos JF, Cordero-Rivera CD, Farfan-Morales CN, Gutierrez-Escolano AL, Del Angel RM (2021) The nuclear pore complex is a key target of viral proteases to promote viral replication. Viruses 13

Dekker M, Van Der Giessen E, Onck PR (2023) Phase separation of intrinsically disordered Fg-Nups is driven by highly dynamic Fg motifs. Proc Natl Acad Sci USA 120:E2221804120

Denning DP, Patel SS, Uversky V, Fink AL, Rexach M (2003) Disorder in the nuclear pore complex: the Fg repeat regions of nucleoporins are natively unfolded. Proc Natl Acad Sci USA 100:2450–2455

Dultz E, Wojtynek M, Medalia O, Onischenko E (2022) The nuclear pore complex: birth, life, and death of a cellular behemoth. Cells 11

Fagotto F, Gluck U, Gumbiner BM (1998) Nuclear localization signal-independent and importin/Karyopherin-independent nuclear import of Beta-catenin. Curr Biol 8:181–190

Fernandez-Martinez J, Rout MP (2021) One ring to rule them all? Structural and functional diversity in the nuclear pore complex. Trends Biochem Sci 46:595–607

Fernandez-Martinez J, Kim SJ, Shi Y, Upla P, Pellarin R, Gagnon M, Chemmama IE, Wang J, Nudelman I, Zhang W, Williams R, Rice WJ, Stokes DL, Zenklusen D, Chait BT, Sali A, Rout MP (2016) Structure and function of the nuclear pore complex cytoplasmic Mrna export platform. Cell 167(1215–1228):E25–1228.e25

Field MC, Rout MP (2019) Pore timing: the evolutionary origins of the nucleus and nuclear pore complex. F1000res 8

Flamholz A, Phillips R, Milo R (2014) The quantified cell. Mol Biol Cell 25:3497–3500

Forwood JK, Lange A, Zachariae U, Marfori M, Preast C, Grubmuller H, Stewart M, Corbett AH, Kobe B (2010) Quantitative structural analysis of importin-Beta flexibility: paradigm for solenoid protein structures. Structure 18:1171–1183

Frey S, Gorlich D (2007) A saturated Fg-repeat hydrogel can reproduce the permeability properties of nuclear pore complexes. Cell 130:512–523
Frey S, Richter RP, Gorlich D (2006) Fg-rich repeats of nuclear pore proteins form a three-dimensional meshwork with hydrogel-like properties. Science 314:815–817
Glodowski DR, Petersen JM, Dahlberg JE (2002) Complex nuclear localization signals in the matrix protein of vesicular stomatitis virus. J Biol Chem 277:46864–46870
Hakhverdyan Z, Molloy KR, Keegan S, Herricks T, Lepore DM, Munson M, Subbotin RI, Fenyo D, Aitchison JD, Fernandez-Martinez J, Chait BT, Rout MP (2021) Dissecting the structural dynamics of the nuclear pore complex. Mol Cell 81(153–165):E7–165.e7
Hampoelz B, Andres-Pons A, Kastritis P, Beck M (2019) Structure and assembly of the nuclear pore complex. Annu Rev Biophys 48:515–536
Hazawa M, Lin DC, Kobayashi A, Jiang YY, Xu L, Dewi FRP, Mohamed MS, Hartono N, Meguro-Horike M, Horike SI, Koeffler HP, Wong RW (2018) Rock-dependent phosphorylation of Nup62 regulates P63 nuclear transport and squamous cell carcinoma proliferation. EMBO Rep 19:73–88
Huang K, Tagliazucchi M, Park SH, Rabin Y, Szleifer I (2020) Nanocompartmentalization of the nuclear pore lumen. Biophys J 118:219–231
Huber AH, Nelson WJ, Weis WI (1997) Three-dimensional structure of the armadillo repeat region of Beta-catenin. Cell 90:871–882
Ibanez De Opakua A, Geraets JA, Frieg B, Dienemann C, Savastano A, Rankovic M, Cima-Omori MS, Schroder GF, Zweckstetter M (2022) Molecular interactions of Fg nucleoporin repeats at high resolution. Nat Chem 14:1278–1285
Iglesias N, Paulo JA, Tatarakis A, Wang X, Edwards AL, Bhanu NV, Garcia BA, Haas W, Gygi SP, Moazed D (2020) Native chromatin proteomics reveals a role for specific nucleoporins in heterochromatin organization and maintenance. Mol Cell 77(51–66):E8–66.e8
Kelley K, Knockenhauer KE, Kabachinski G, Schwartz TU (2015) Atomic structure of the Y complex of the nuclear pore. Nat Struct Mol Biol 22:425–431
Keminer O, Peters R (1999) Permeability of single nuclear pores. Biophys J 77:217–228
Kim SJ, Fernandez-Martinez J, Nudelman I, Shi Y, Zhang W, Raveh B, Herricks T, Slaughter BD, Hogan JA, Upla P, Chemmama IE, Pellarin R, Echeverria I, Shivaraju M, Chaudhury AS, Wang J, Williams R, Unruh JR, Greenberg CH, Jacobs EY, Yu Z, De La Cruz MJ, Mironska R, Stokes DL, Aitchison JD, Jarrold MF, Gerton JL, Ludtke SJ, Akey CW, Chait BT, Sali A, Rout MP (2018) Integrative structure and functional anatomy of a nuclear pore complex. Nature 555:475–482
Kimura M, Imamoto N (2014) Biological significance of the importin-Beta family-dependent nucleocytoplasmic transport pathways. Traffic 15:727–748
Kodiha M, Tran D, Qian C, Morogan A, Presley JF, Brown CM, Stochaj U (2008) Oxidative stress Mislocalizes and retains transport factor importin-alpha and nucleoporins Nup153 and Nup88 in nuclei where they generate high molecular mass complexes. Biochim Biophys Acta 1783:405–418
Konishi HA, Asai S, Watanabe TM, Yoshimura SH (2017) In vivo analysis of protein crowding within the nuclear pore complex in interphase and mitosis. Sci Rep 7:5709
Kosako H, Imamoto N (2010) Phosphorylation of nucleoporins: signal transduction-mediated regulation of their interaction with nuclear transport receptors. Nucleus 1:309–313
Kosako H, Yamaguchi N, Aranami C, Ushiyama M, Kose S, Imamoto N, Taniguchi H, Nishida E, Hattori S (2009) Phosphoproteomics reveals new Erk map kinase targets and links Erk to nucleoporin-mediated nuclear transport. Nat Struct Mol Biol 16:1026–1035
Kumeta M, Yoshimura SH, Harata M, Takeyasu K (2010) Molecular mechanisms underlying nucleocytoplasmic shuttling of Actinin-4. J Cell Sci 123:1020–1030
Kumeta M, Yoshimura SH, Hejna J, Takeyasu K (2011) Nucleocytoplasmic shuttling of cytoskeletal proteins: molecular mechanism and biological significance. Int J Cell Biol 2012:494902
Kumeta M, Yamaguchi H, Yoshimura SH, Takeyasu K (2012) Karyopherin-independent spontaneous transport of amphiphilic proteins through the nuclear pore. J Cell Sci 125:4979–4984

Kumeta M, Konishi HA, Zhang W, Sakagami S, Yoshimura SH (2018) Prolines in the alpha-helix confer the structural flexibility and functional integrity of importin-Beta. J Cell Sci 131
Labokha AA, Gradmann S, Frey S, Hulsmann BB, Urlaub H, Baldus M, Gorlich D (2012) Systematic analysis of barrier-forming Fg hydrogels from Xenopus nuclear pore complexes. EMBO J 32:204–218
Laurell E, Beck K, Krupina K, Theerthagiri G, Bodenmiller B, Horvath P, Aebersold R, Antonin W, Kutay U (2011) Phosphorylation of Nup98 by multiple kinases is crucial for Npc disassembly during mitotic entry. Cell 144:539–550
Li B, Kohler JJ (2014) Glycosylation of the nuclear pore. Traffic 15:347–361
Linder MI, Kohler M, Boersema P, Weberruss M, Wandke C, Marino J, Ashiono C, Picotti P, Antonin W, Kutay U (2017) Mitotic disassembly of nuclear pore complexes involves Cdk1- and Plk1-mediated phosphorylation of key interconnecting nucleoporins. Dev Cell 43(141–156):E7–156.e7
Lolodi O, Yamazaki H, Otsuka S, Kumeta M, Yoshimura SH (2016) Dissecting in vivo steady-state dynamics of Karyopherin-dependent nuclear transport. Mol Biol Cell 27:167–176
Lowe AR, Tang JH, Yassif J, Graf M, Huang WY, Groves JT, Weis K, Liphardt JT (2015) Importin-Beta modulates the permeability of the nuclear pore complex in a ran-dependent manner. Elife 4
Lutzmann M, Kunze R, Buerer A, Aebi U, Hurt E (2002) Modular self-assembly of a Y-shaped multiprotein complex from seven nucleoporins. EMBO J 21:387–397
Maul GG, Deaven L (1977) Quantitative determination of nuclear pore complexes in cycling cells with differing Dna content. J Cell Biol 73:748–760
Milles S, Huy Bui K, Koehler C, Eltsov M, Beck M, Lemke EA (2013) Facilitated aggregation of Fg nucleoporins under molecular crowding conditions. EMBO Rep 14:178–183
Mishra A, Sipma W, Veenhoff LM, Van Der Giessen E, Onck PR (2019) The effect of Fg-Nup phosphorylation on Npc selectivity: a one-bead-per-amino-acid molecular dynamics study. Int J Mol Sci 20
Miyamoto Y, Saiwaki T, Yamashita J, Yasuda Y, Kotera I, Shibata S, Shigeta M, Hiraoka Y, Haraguchi T, Yoneda Y (2004) Cellular stresses induce the nuclear accumulation of importin alpha and cause a conventional nuclear import block. J Cell Biol 165:617–623
Mizuguchi-Hata C, Ogawa Y, Oka M, Yoneda Y (2013) Quantitative regulation of nuclear pore complex proteins by O-Glcnacylation. Biochim Biophys Acta 1833:2682–2689
Mosammaparast N, Pemberton LF (2004) Karyopherins: from nuclear-transport mediators to nuclear-function regulators. Trends Cell Biol 14:547–556
Nosella ML, Tereshchenko M, Pritisanac I, Chong PA, Toretsky JA, Lee HO, Forman-Kay JD (2021) O-linked-N-Acetylglucosaminylation of the Rna-binding protein Ews N-terminal low complexity region reduces phase separation and enhances condensate dynamics. J Am Chem Soc 143:11520–11534
Otsuka S, Ellenberg J (2018) Mechanisms of nuclear pore complex assembly - two different ways of building one molecular machine. FEBS Lett 592:475–488
Otsuka S, Steyer AM, Schorb M, Heriche JK, Hossain MJ, Sethi S, Kueblbeck M, Schwab Y, Beck M, Ellenberg J (2017) Postmitotic nuclear pore assembly proceeds by radial dilation of small membrane openings. Nat Struct Mol Biol 25:21–28
Otsuka S, Tempkin JOB, Zhang W, Politi AZ, Rybina A, Hossain MJ, Kueblbeck M, Callegari A, Koch B, Morero NR, Sali A, Ellenberg J (2023) A quantitative map of nuclear pore assembly reveals two distinct mechanisms. Nature 613:575–581
Paci G, Zheng T, Caria J, Zilman A, Lemke EA (2020) Molecular determinants of large cargo transport into the nucleus. elife 9
Patel SS, Belmont BJ, Sante JM, Rexach MF (2007) Natively unfolded nucleoporins gate protein diffusion across the nuclear pore complex. Cell 129:83–96
Raices M, D'angelo MA (2012) Nuclear pore complex composition: a new regulator of tissue-specific and developmental functions. Nat Rev Mol Cell Biol 13:687–699

Reichelt R, Holzenburg A, Buhle EL Jr, Jarnik M, Engel A, Aebi U (1990) Correlation between structure and mass distribution of the nuclear pore complex and of distinct pore complex components. J Cell Biol 110:883–894

Ribbeck K, Gorlich D (2001) Kinetic analysis of translocation through nuclear pore complexes. EMBO J 20:1320–1330

Rout MP, Blobel G (1993) Isolation of the yeast nuclear pore complex. J Cell Biol 123:771–783

Rout MP, Aitchison JD, Suprapto A, Hjertaas K, Zhao Y, Chait BT (2000) The yeast nuclear pore complex: composition, architecture, and transport mechanism. J Cell Biol 148:635–651

Ruland JA, Kruger AM, Dorner K, Bhatia R, Wirths S, Poetes D, Kutay U, Siebrasse JP, Kubitscheck U (2021) Nuclear export of the pre-60s ribosomal subunit through single nuclear pores observed in real time. Nat Commun 12:6211

Sakakida Y, Miyamoto Y, Nagoshi E, Akashi M, Nakamura TJ, Mamine T, Kasahara M, Minami Y, Yoneda Y, Takumi T (2005) Importin alpha/Beta mediates nuclear transport of a mammalian circadian clock component, Mcry2, together with Mper2, through a bipartite nuclear localization signal. J Biol Chem 280:13272–13278

Schmidt HB, Gorlich D (2015) Nup98 Fg domains from diverse species spontaneously phase-separate into particles with nuclear pore-like Permselectivity. elife 4

Schuller AP, Wojtynek M, Mankus D, Tatli M, Kronenberg-Tenga R, Regmi SG, Dip PV, Lytton-Jean AKR, Brignole EJ, Dasso M, Weis K, Medalia O, Schwartz TU (2021) The cellular environment shapes the nuclear pore complex architecture. Nature 598:667–671

Shen Q, Wang YE, Palazzo AF (2021) Crosstalk between nucleocytoplasmic trafficking and the innate immune response to viral infection. J Biol Chem 297:100856

Sridharan D, Brown M, Lambert WC, Mcmahon LW, Lambert MW (2003) Nonerythroid Alphaii Spectrin is required for recruitment of Fanca and Xpf to nuclear foci induced by Dna Interstrand cross-links. J Cell Sci 116:823–835

Stochaj U, Rassadi R, Chiu J (2000) Stress-mediated inhibition of the classical nuclear protein import pathway and nuclear accumulation of the small Gtpase Gsp1p. FASEB J 14:2130–2132

Timney BL, Raveh B, Mironska R, Trivedi JM, Kim SJ, Russel D, Wente SR, Sali A, Rout MP (2016) Simple rules for passive diffusion through the nuclear pore complex. J Cell Biol 215:57–76

Walker EJ, Younessi P, Fulcher AJ, Mccuaig R, Thomas BJ, Bardin PG, Jans DA, Ghildyal R (2013) Rhinovirus 3c protease facilitates specific nucleoporin cleavage and Mislocalisation of nuclear proteins in infected host cells. PLoS One 8:E71316

Walther TC, Alves A, Pickersgill H, Loiodice I, Hetzer M, Galy V, Hulsmann BB, Kocher T, Wilm M, Allen T, Mattaj IW, Doye V (2003) The conserved Nup107-160 complex is critical for nuclear pore complex assembly. Cell 113:195–206

Wing CE, Fung HYJ, Chook YM (2022) Karyopherin-mediated nucleocytoplasmic transport. Nat Rev Mol Cell Biol 23:307–328

Wong RW, Mamede JI, Hope TJ (2015) Impact of nucleoporin-mediated chromatin localization and nuclear architecture on Hiv integration site selection. J Virol 89:9702–9705

Yamazaki H, Takagi M, Kosako H, Hirano T, Yoshimura SH (2022) Cell cycle-specific phase separation regulated by protein charge Blockiness. Nat Cell Biol 24:625–632

Yan Y, Winograd E, Viel A, Cronin T, Harrison SC, Branton D (1993) Crystal structure of the repetitive segments of Spectrin. Science 262:2027–2030

Yang W, Gelles J, Musser SM (2004) Imaging of single-molecule translocation through nuclear pore complexes. Proc Natl Acad Sci USA 101:12887–12892

Yokoya F, Imamoto N, Tachibana T, Yoneda Y (1999) Beta-catenin can be transported into the nucleus in a ran-unassisted manner. Mol Biol Cell 10:1119–1131

Yoshimura SH, Hirano T (2016) Heat repeats - versatile arrays of amphiphilic helices working in crowded environments? J Cell Sci 129:3963–3970

Yoshimura SH, Otsuka S, Kumeta M, Taga M, Takeyasu K (2013) Intermolecular disulfide bonds between nucleoporins regulate Karyopherin-dependent nuclear transport. J Cell Sci 126:3141–3150

Yoshimura SH, Kumeta M, Takeyasu K (2014) Structural mechanism of nuclear transport mediated by importin Beta and flexible amphiphilic proteins. Structure 22:1699–1710

Yu M, Heidari M, Mikhaleva S, Tan PS, Mingu S, Ruan H, Reinkemeier CD, Obarska-Kosinska A, Siggel M, Beck M, Hummer G, Lemke EA (2023) Visualizing the disordered nuclear transport machinery in situ. Nature 617:162–169

Zhang W, Watanabe R, Konishi HA, Fujiwara T, Yoshimura SH, Kumeta M (2020) Redox-sensitive Cysteines confer proximal control of the molecular crowding barrier in the nuclear pore. Cell Rep 33:108484

Zhu Y, Liu TW, Madden Z, Yuzwa SA, Murray K, Cecioni S, Zachara N, Vocadlo DJ (2016) Post-translational O-Glcnacylation is essential for nuclear pore integrity and maintenance of the pore selectivity filter. J Mol Cell Biol 8:2–16

Zilman A (2018) Aggregation, phase separation and spatial morphologies of the assemblies of Fg nucleoporins. J Mol Biol 430:4730–4740

Chapter 13
Catalytic Droplets: Enzyme Containing Microcompartments

Munishwar Nath Gupta and Vladimir N. Uversky

Abstract Bacterial microcompartments (BMCs) and biomolecular condensates are two different forms of protein compartmentalization with different aims and functional advantages, representing specific designs of sequestering enzymes or segments of metabolic pathways. BMCs represent special proteinaceous organelles that are entirely composed of protein and are typically characterized by polyhedral shapes. By encapsulating and organizing metabolic enzymes with their substrates and cofactors BMCs act as specialized compartments within bacterial cells that promote and enhance specific biochemical pathways. They also serve important protective functions shielding vulnerable enzymes within a defined microenvironment and sequestering toxic or volatile intermediates. On the other hand, biomolecular condensates (also known as membrane-less organelles, MLOs) are dynamic, cell size–dependent, cytoplasmic and nucleoplasmic entities that typically contain both RNA and protein. They have unique morphologies, specific distribution patterns, are characterized by specific set of resident proteins, but their structural integrity is not supported by encapsulation in the membrane. Instead, their biogenesis is driven by liquid-liquid phase separation, and their structure is entirely controlled and mediated by the protein–protein, protein–RNA, and/or protein–DNA interactions. MLOs represent a different liquid state of cytoplasm or nucleoplasm (or mitochondrial matrix or chloroplastic stroma), whose major biophysical properties are rather similar to those of the rest of the intracellular fluid. Often, MLOs emerge in response to some specific environmental cues, being exploited by cells to respond in real time in a smart stimuli-responsive manner. BMCs are more permanent entities with selective transport through the protein shell. In that way and in many respects, they are closer to intracellular membrane-bounded organelles of eukaryotes than to

M. N. Gupta
Department of Biochemical Engineering and Biotechnology, Indian Institute of Technology, New Delhi, India

V. N. Uversky (✉)
Department of Molecular Medicine and USF Health Byrd Alzheimer's Research Institute, Morsani College of Medicine, University of South Florida, Tampa, FL, USA
e-mail: vuversky@usf.edu

V. N. Uversky (ed.), *(Macro)Molecular Crowding*, Subcellular Biochemistry 109,
https://doi.org/10.1007/978-3-032-03370-3_13

MLOs. This chapter discusses diverse functions of BMCs and considers the ways by which they contribute to metabolic innovation in bacteria. Some functional roles of MLOs are considered as well.

Keywords Compartmentalization · Bacterial microcompartments · Multi-protein complexes · Membrane-less organelles · Endosymbiotic hypothesis · Biomolecular condensates · Organelle · Inclusion bodies · RuBisCO · Carboxysomes · Carbon dioxide fixation

Introduction

Enzymes catalyze (almost) all chemical reactions in a cell and extracellular parts in a living organism. Hierarchically speaking, one generally talks of cellular organelles as the next level of organization. Organization in biology, as Albert Imre Szent-Györgyi (1893–1986) used to say (Gyorgyi 1960), is expressed by 2 + 2 exceeding 4 and is one of the major tricks biology uses to create cell, tissues, organs, organism, and social communities. This chapter is about a sort of sublevel of organization for many enzymes/ proteins in bacterial microcompartments. The large multi-protein complexes and membrane-less organelles (also known as biomolecular condensates) are also covered. While cellular organelles of eukaryotic cells have been part of textbooks for a long time, microcompartments and biomolecular condensates are not yet an integral part of cellular biology (Alberts et al. 1994; Uversky 2017a; Fonin et al. 2022; Antifeeva et al. 2022; Nesterov et al. 2021; Uversky 2021; Uversky and Finkelstein 2019; Darling et al. 2019b; Turoverov et al. 2019; Uversky 2017b). In this chapter, the words protein and enzyme are used interchangeably. We recognize that this is an oversimplification, as proteins can do much more than simply serve as highly efficient biological catalysts. Furthermore, the moonlighting phenomenon has shown that nature does not sharply distinguish between the two (enzymes and non-enzymes). A non-catalytic protein can behave as an enzyme and *vice versa*. So, rigorously speaking, it is best to rely upon the context in which either of the term is used (Gupta and Uversky 2023). Since the major focus of the chapter is actually on enzymes, which are proteins, both terms will be used here.

The issue of compartmentalization of proteins has obvious importance in the context of molecular crowding. A protein freely floating around has more space to share with other biological molecules. Whereas, put in any kind of enclosure, the shared space becomes limited and introduces crowding. This, of course, has several biological advantages. In this chapter, we discuss two designs for the "protein containers", namely bacterial microcompartments and biomolecular condensates.

Importance of Compartmentalization

It is curious how we humans often go back and forth in search of higher productivity/efficiency and better regulation. More than a decade back, rest of the world emulated Japan and switched over to the cubicle-less offices. It was believed that it will lead to less hierarchical environments and much better communication. Sadly, it did not work out quite that way, and, in general, the disillusioned world is shifting back to cubicles. In the case of prokaryotes, organisms and their cells have exclusively these cubicle-less (i.e. membrane-less) compartments. As the complexity increased in case of eukaryotes, number of biomacromolecules (especially proteins, —a human body (considering the adult male anatomical model as a reference) contains at least 1.4×10^{10} km of DNA, assuming that there are on average, 7×10^{12} nucleated cells in a body of an adult male (Hatton et al. 2023), and each of nucleated cells contains ~2 m of DNA (Alberts et al. 1994). Note that this summed up length of DNA in one person is almost 100-fold larger than an average distance of 1.5×10^8 km from the Earth to the Sun, is definitely of astronomic proportions, making an interesting and rather unexpected link between microcosm and universe As the number of proteins became much larger; many of these organized with specific functions systems (which were membrane-less) got enclosed with membranes. It also led to semi-autonomy like a federal system where many of the aspects are self-governed by states. However, eukaryotic cells still have enormous number of very different (by content and by the content-related function) membrane-less organelles.

As an English geneticist and the 2001 Nobel Prize in Physiology or Medicine winner Paul Nurse (2021) wrote in his book (Nurse 2021), the compartmentalization by cell wall/membranes has a thermodynamic implication; "Within their insulating membranes, cells can establish and cultivate the order they need to operate, whilst at the same time creating disorder in their local surroundings outside the cell. That way life does not contravene the second law of thermodynamics." The difference between the less complex prokaryotes and more sophisticated eukaryotes is the way the "federal structure" in the cell operates.

As Gideon Schreiber and Amy E. Keating (2011) pointed out, not all molecular encounters are biologically relevant (Schreiber and Keating 2011). Each time, we bump into a co-worker at the office cafeteria, corridor, or water fountain does not necessarily leads to any meaningful outcome. However, putting people in a group/committee is expected to lead to a meaningful interaction. That is what organization achieves. Compartmentalization involves similar philosophy.

When the streets are crowded, and we move slowly in a car/cab, this does affect our mood and functionality. The crowded intracellular conditions are actually akin to more like a crowded market. The compartmentalization leads to some sort of seclusion, cutting off many noisy (biologically irrelevant) interactions, increasing proximity and effective concentration. As has been pointed out by Teilum et al., effective concentration is a parameter, which influences biological specificity of the protein (Teilum et al. 2021). We now discuss the different kinds of "compartmentalization" in prokaryotes and eukaryotes.

Membrane-Encapsulated Compartments in Eukaryotes and Endosymbiotic Hypothesis

A eukaryotic cell has many classical organelles, all having an outer membrane. The word eukaryote actually means "true nucleus" (from the ancient Greek ευ (eu), "good, true," and κάρυον (karion), kernel, core, or nucleus). The nucleus is a large organelle that contains chromosomes. Within the nucleus is nucleolus, which contains some genes for rRNA. The double-layered membrane of the nucleus has perinuclear space and the transport of molecules between nucleus and cytoplasm is via nuclear pores, which are lined inside with proteins. The very important point to be noted here is that all kinds of cellular compartments are never cut-off totally from the rest of the cell. The exchange of materials is however regulated.

It is relevant to recall the endosymbiotic hypothesis, which proposes that at some early point in evolution, eukaryotic cells entered into a symbiotic relationship with a bacteria and acquired ability to carry out oxidative phosphorylation. These bacteria later evolved into mitochondia. This was before eukaryotic cells evolved differently into plant and animal cells. In eukaryotic cells which had symbiosis with ancestors of cyanobacteria-like organisms, the bacteria evolved into chloroplasts. Both chloroplasts and mitochondria share common features (e.g., they contain their own genomes). Also, there are quite a few similarities in proteins and lipids of these organelle and prokaryotic cells. Christopher Snyder and George B. Stefano (2015) provided more details about the hypothesis and other functions of these organelles (Snyder and Stefano 2015). In those early days of evolution, the bacteria harbored by the eukaryotic cells escaped destruction by their lysosomes—the other cellular organelle of the eukaryotic cells.

There are three important distinctions between cellular organelles of eukaryotic (once again, those which are embedded in the membranes and are considered as classic distinctions of eukaryotic cells) and bacterial microcompartments. Firstly, the latter are of much smaller sizes and have similar shapes (mostly spherical). They are liquid droplets containing proteins (and nucleic acids) and do not have the bilayered lipid membranes surrounding them. Thirdly, the role of many of them is primarily to help the organism to survive in the environments, which have little to offer in terms of nutrients. While cellular organelles of eukaryotes have dedicated functions, which are more general in nature, bacterial microcompartments are more specific in the sense that each kind deals with utilization of a specific class of compounds though with a very similar strategy.

We should remind here once again that in addition to the "classic" membrane-bound organelles, eukaryotic cells contain numerous membrane-less organelles (MLOs) or biomolecular condensates (BCs), biogenesis of which is driven by the liquid-liquid phase separation (LLPS), and which, therefore, resemble bacterial microcompartments. MLOs/BCs are dynamic, cell size–dependent, cytoplasmic and nucleoplasmic entities that typically contain both RNA and protein and are therefore known by their generic names, such as ribonucleoprotein (RNP) granules/bodies, or RNP droplets (Brangwynne 2013; Uversky 2017b). They exhibit liquid-like behavior, such as wetting, dripping, and relaxation to spherical structures upon fusion (Brangwynne et al. 2009; Brangwynne et al. 2011; Feric and Brangwynne 2013; Wippich et al. 2013). They are optically resolvable as spherical micron-sized droplets (Nott et al. 2015). They have unique morphologies, specific distribution patterns, and, at the molecular level their structure, is determined by a specific set of resident proteins (Uversky 2017b). The structural integrity of MLOs is not supported by encapsulation in the membrane, with their biogenesis being entirely controlled and mediated by protein–protein, protein–RNA, and/or protein–DNA interactions (Dundr and Misteli 2010). The lack of membrane ensures the direct contact of the MLO components with the surrounding cytoplasm or nucleoplasm and defines the highly dynamic nature of these organelles (Phair and Misteli 2000; Pederson 2001). Biophysical characterization of several MLOs revealed that they are just slightly denser than the rest of the cytoplasm or nucleoplasm (Handwerger et al. 2005; Updike et al. 2011), possess high level of internal dynamics, and, therefore, can be classified as liquid-droplet phases of the nucleoplasm/cytoplasm (Brangwynne et al. 2009; Brangwynne et al. 2011; Li et al. 2012; Aggarwal et al. 2013; Feric and Brangwynne 2013; Wippich et al. 2013). In other words, MLOs/BCs represent a different liquid state of cytoplasm or nucleoplasm (or mitochondrial matrix or chloroplastic stroma), whose major biophysical properties are rather similar to those of the rest of the intracellular fluid (Brangwynne 2013). Importantly, it was also shown that structurally, MLOs can include layers, which might have different biophysical properties and functions (Hirose et al. 2023). MLOs can exist in the form of 1D, 2D and 3D assemblages, where 1D structures are found on the chromatin and cytoskeleton, 2D MLOs originates from the phase separation in/on the membranes, and the 3D MLOs are formed via the LLPS in the bulk volume of the nucleoplasm/cytoplasm/matrix/stroma (Nesterov et al. 2021). Importantly, the corresponding 1D, 2D and 3D MLOs can influence each other, where 1D and 2D structures can nucleate formation of 3D condensates (Nesterov et al. 2021). It was also hypothesized that LLPS, being triggered by environmental change, can propagate within the cell interior moving along the cytoskeleton or as an autowave. Such quick propagation of a new phase utilizes a low amount of energy and can adjust cell signaling and metabolic systems to new demands (Nesterov et al. 2021). These considerations suggested an intriguing possibility, where "the interconnected phase separation phenomena in different dimensions represent a previously unexplored system of intracellular communication and regulation which cannot be ignored when considering both physiological and pathological cell processes" (Nesterov et al. 2021).

It is likely that there are hundreds of different MLOs, which are very diverse (Shin and Brangwynne 2017; Hnisz et al. 2017; Banani et al. 2017; Alberti 2017; Saha et al. 2016; Courchaine et al. 2016; Chong and Forman-Kay 2016; Alberti and Hyman 2016; Hyman et al. 2014; Uversky 2017b; Zaslavsky and Uversky 2018), present in all domains of life, and have numerous important roles in the organization of various intracellular processes (Hyman et al. 2014; Toretsky and Wright 2014; Forman-Kay et al. 2018; Shin and Brangwynne 2017; Alberti and Dormann 2019; Alberti et al. 2019; Gomes and Shorter 2019). Therefore, MLOs/BCs are many

(Uversky 2021; Antifeeva et al. 2022; Hyman et al. 2014; Toretsky and Wright 2014; Forman-Kay et al. 2018; Shin and Brangwynne 2017; Alberti and Dormann 2019; Alberti et al. 2019; Gomes and Shorter 2019; Hnisz et al. 2017; Banani et al. 2017; Alberti 2017; Saha et al. 2016; Courchaine et al. 2016; Chong and Forman-Kay 2016; Alberti and Hyman 2016; Uversky 2017a, b; Zaslavsky and Uversky 2018; Darling and Uversky 2023). A very incomplete list of the most known MLOs is given below. In cytoplasm, one can find stress granules (Wippich et al. 2013), processing bodies or P-bodies (Decker et al. 2007), germline P-granules (germ cell granules or nuage) (Brangwynne et al. 2009; Chuma et al. 2009), neuronal RNA granules (Kiebler and Bassell 2006), and centrosomes (Decker et al. 2011), whereas among nuclear MLOs are nucleoli (Shav-Tal et al. 2005), nuclear speckles or inter-chromatin granule clusters (Lamond and Spector 2003), nuclear stress bodies (nSBs) (Biamonti and Vourc'h 2010; Biamonti 2004), histone locus bodies (HLBs) (Nizami et al. 2010), Cajal bodies (CBs; (Strzelecka et al. 2010)), promyelocytic leukaemia nuclear bodies (PML nuclear bodies) or nuclear dots (PODs) (Maul et al. 2000), paraspeckles (Fox et al. 2002), perinucleolar compartment (PNC) (Huang 2000), the Sam68 nuclear body (SNB) (Huang 2000), chromatin (Li et al. 2007), cleavage bodies (Li et al. 2006), nuclear gems (Gemini of coiled of Cajal bodies) (Matera and Frey 1998; Gubitz et al. 2004), nuclear pores (Grossman et al. 2012), Oct1/PTF/ transcription (OPT) domains (Harrigan et al. 2011), and PcG bodies (polycomb bodies, subnuclear organelles containing polycomb group proteins) (Pirrotta and Li 2012).

Bacterial Microcompartments

Just to recall the more well-known facts, bacterial cytoplasm is not featureless and monophasic (Heritage et al. 1996). (Important note: this statement is also applicable to the eukaryotic cytoplasm, nucleoplasm, matrix, and stroma). Apart from ribosomes engaged in protein synthesis, bacterial cytoplasm invariably has inorganic or organic inclusion bodies. The latter are exemplified by those formed by glycogen or β-hydroxybuterate, which store energy for the microorganisms. Volutin or mono-chromatin granules are common in cornebacteria and are made up of polyphosphates. The bacteria, who convert hydrogen sulfide to sulfur have sulfur granules. These are, of course, not to be confused with protein inclusion bodies formed frequently when proteins are overexpressed in the recombinant bacterial hosts (Mukherjee and Nath Gupta 2016). Besides, magnetotactic bacteria, which occur in aquatic environments have, magnetosomes containing crystalline mineral surrounded by a lipid membrane (Lefevre and Bazylinski 2013). Furthermore, the membrane invaginations also leads to the formation of vesicles, tubules, lamelle, or mesosomes (Arechaga 2013).

Bacterial microcompartments (BMCs) are unique microsized organelles, which have been discovered relatively recently (they were first visualized in the late 1950s and 1960s inside cyanobacteria as electron dense, polyhedrally shaped bodies, reminiscent of viruses or phage capsids (Drews and Niklowitz 1956; Gantt and Conti 1969; Tsai and Yeates 2011)) and which are composed entirely of protein. In fact, these inclusion bodies have polyhedral shape (e.g., α-carboxysome BMCs are nearly icosahedral in shape, where 20 roughly flat triangular facets come together at 30 edges and 12 pentagonal vertices (Schmid et al. 2006; Iancu et al. 2007)) and are bound by a unilamellar protein shell (Codd and Marsden 1984; Shively and English 1991). Therefore, while the classic eukaryotic cellular organelles, such as nucleus, lysosomes, mitochondria, Golgi, etc. have lipid biomembranes, many BMCs have a protein shell enclosing the enzymes catalyzing segments of metabolic activity (Kerfeld et al. 2005). This segregation is believed to achieve the following advantages for the organisms:

- Any toxic metabolites remain confined to the microcompartment;
- The biological specificity operates by many promiscuous substrates being not available around the enzyme/metabolic segments. This confinement prevents "side reactions" or unwanted reactions by enzymes on the both sides of the divide;
- Just like in the case of multi-enzyme complexes, the co-localization in a confined space increases the metabolic flux for the confined enzyme activities.

The BMCs mostly confine catabolic activities and are known as metabolosomes. One well-recognized example of a BMC is carboxysomes found in all cynobacteria and chemotrophic bacteria (Cannon et al. 2001). This BMC consists of carbonic anhydrase and ribulose 1,5-bisphosphate carboxylase/oxygenase (RuBisCO) and fixes carbon dioxide. The well-known examples of BMCs, are those which catabolize propanediol, ethanolamine, fucose, rhamnose, and are expressed when these compounds are available. Catalytic BMCs occur in all phyla of bacteria, which underlines their biochemical significance. BMC enable bacteria to grow in niche, where such compounds act as the carbon source. BMC typically use a core strategy for catabolism employing signature enzymes. CoA act as a source of phosphorus and substrate level phosphorylation is used to produce ATP. The genes for the enzymes, shell proteins and transporters (which bring the signature substrates inside the BMC) are all parts of a superloci (Kerfeld et al. 2005).

The building block of the shell protein is Pfam00936 domain, which forms hexamers of different architecture to produce three proteins called BMC-H (most abundant), BMC-T, and BMC- P. At the center of this proteinacous shell is a pore for transport of the substrates (Kerfeld et al. 2005).

The early insights into the 3D structures of shell protein came from studying their crystal structures in the case of different carboxysomes. These studies also provided insights into how structural variations can lead to minor differences in the architectures of the shells. It also provided clues about how the selective transport of molecules across the shells may take place (Kerfeld et al. 2005). MD simulation with shell protein BMC-H showed that chloride ions blocked the permeation through positively charged pores. This explained why BMC showed decreased transport at high salt concentration (Trettel et al. 2023).

An early comprehensive review by Kerfeld et al. (2010) is a good source of information about BMCs (Kerfeld et al. 2010). It also highlights how the early insights came from carboxysomes. Their modular nature and strong probability of horizontal transfer of the genes among various bacteria are also discussed (Kerfeld et al. 2010). Although the core enzyme assembly can be followed by shell enclosing it or both steps can happen simultaneously, the step-wise assembly is believed to be more common (Kerfeld et al. 2018). Another article from the same group [Kerfeld et al. 2015] emphasized that BMCs should be viewed as "genetic, metabolic, structural, and evolutionary modules" for the bacteria, which enable them to utilize the resources available in the niche environments (Kerfeld and Erbilgin 2015). The focus on the modular nature naturally leads to harnessing engineered BMC for various possible applications (Kerfeld and Erbilgin 2015).

The more recent review by Henning Kirst and Cheryl A. Kerfeld is an updated introductory overview on BMC (Kirst and Kerfeld 2019). By 2019, BMC were identified in 23 out of 29 established bacterial phyla. BMCs represent an elegant design, in which diverse catabolic pathways are reduced to a common strategy (Kirst and Kerfeld 2019). As aforementioned, the substrate specific enzyme (called signature enzyme) is supported by dehydrogenases of aldehyde and alcohol. The volatile toxic aldehyde intermediates can damage proteins and DNA and hence its containment (and further metabolism) prevents cellular damage or even death and loss of carbon because the aldehyde can escape because of its volatile nature (Kirst and Kerfeld 2019). The authors mentioned that one advantage of containment is that the cross-talk with other metabolic pathways, which can compete for the same substrate and hence deplete its concentration, is avoided (Kirst and Kerfeld 2019). This glosses over the fact that the amount of substrate available to the BMC will anyway be dictated by the operation of the competing metabolic pathways. It is, however, fair to say that the substrate molecules upon entry into the BMC are cut-off from the competing pathways and are committed to the degradative pathway of the BMC (Kirst and Kerfeld 2019).

Another interesting feature, which has emerged by that time is that many BMCs have gated pores, in which the closed or open conformations could be regulated by ligand binding (Kirst and Kerfeld 2019). In fact, substantial work has been carried out to manipulate selective entry of the substates by modifying protein shells. The shells have also been impregnated with redox chemistry to bring in electron flow through shells (Kirst and Kerfeld 2019). The review also describes early impressive efforts towards applications of BMCs and their design for multiple required applications. This includes designed *E. coli*, in which BMC locus from other organisms has been transplanted for degradation of propanediol (Kirst and Kerfeld 2019).

An unusual BMC has been reported in *Actinobacteria*, which has pathways for the degradation of both 2-aminophenol and catechol (Doron et al. 2023). The first two enzymes for aminophenol degradation pathway are 2-aminophenol dioxygenase and a semi-aldehyde dehydrogenase with pyruvate and acetyl CoA as the final end products. The genes for the shell proteins were also present in the gene cluster (Doron et al. 2023). The authors surmised that the BMC actually degrades some precursor molecules of the aminophenol and catechol and the compartmentalization is to segregate the toxic metabolic intermediates (Doron et al. 2023).

Ferlez et al. (2016) focused on a fairly widespread type of BMC, which use glycyl radical enzyme (Ferlez et al. 2019). This is unlike the other BMC types, which exploit B12-dependent radical chemistry. These BMCs have singular specificities, which include degradative pathways for choline, 1,2-propanediol, L-fuculose phosphate, and L-rhamnulose phosphate (Ferlez et al. 2019). Many of these substrates arise in gut from degradation of glycan and lipids. Hence, it is not surprising that this type of BMCs is common in the gut microbiome. Now, gut microbiome is considered as a new organ with the ever increasing recognition of the importance in human health. These BMCs include an activating enzyme and an external source of electrons to produce the glycyl radical (Ferlez et al. 2019). The review covers whatever we know about the reducing proteins present in these BMCs (Ferlez et al. 2019).

Sutter et al. (2021) presented an important study that contains the updated database for the BMCs, which now are known to occur in 45 bacterial phyla (Sutter et al. 2021). 7000 BMC loci corresponds to 68 types/subtypes of BMCs (Sutter et al. 2021). Overall, six types of shell proteins have been described. The general picture of icosohedral protein shells with pentagons at 12 vertices and hexagons forming the facets continues to be correct, as is the range of 40–200 nm for the sizes of BMCs (Sutter et al. 2021). The authors with the availability of the extensive data reflect that different ecological niches require different metabolic capabilities, which, in turn, defines the functional properties of BMCs of the bacteria which inhabit these niche (Sutter et al. 2021).

The distribution of BMCs in human body has been investigated by using bioinformatics tools (Asija et al. 2021). The authors pointed out that BMCs "could be used to prebiotically alter microbiomes or treat dysbiosis". For that and similar applications, the data in this survey could be a good starting point (Asija et al. 2021).

In general, it is believed that intrinsically disordered regions (IDRs) play an important role in the self-assembly of BMC. Kumar et al. (2023) have used bioinformatics tools to look at the distribution of such regions in the enzymes and the shell proteins of BMC associated with the degradation of propanediol (Kumar et al. 2023). It is believed that the IDRs present in the shell proteins also help in their self-assembly by providing them with the desired flexibility (Kumar et al. 2023).

In this context, it is interesting to mention a recent report by Poudyal et al. (2023), which negates the over-riding importance of IDRs in liquid-liquid phase separation (LLPS) which also is involved in the self-assembly of BMCs (Poudyal et al. 2023). Working with diverse kinds of proteins, these researchers could achieve LLPS by using macromolecular crowding agents. The authors conclude that "Our results confirm that both folded and disordered proteins can be driven to form dynamic, reversible liquid-like condensates in a concentration-dependent manner. The driving forces and kinetics for assembly vary from protein to protein, and the observed saturation concentrations scale directly with the apparent intermolecular binding constant (K_D)" (Poudyal et al. 2023) As has been pointed out, there are no ideal macromolecular crowders, which would be capable to best simulate the *in vivo* situations *in vitro*. PEG, used by these authors brings in the extraneous factors in play (Zhou et al. 2008). An important design would be to limit the study to compare concentration (in the range, in which the proteins are likely to occur) dependence of both structured and intrinsically disordered proteins to establish whether intrinsic disorder does facilitate LLPS. The presented extensive data indicates that IDPs/IDRs do play a positive role in LLPS (Uversky et al. 2015). That structured proteins also self-assemble and are part of BMC is anyway a known information.

The importance of disorder for LLPS is supported by multiple observations. This was experimentally demonstrated for several MLOs (such as nuages (Nott et al. 2015), nucleolus (Mitrea et al. 2016), P-granules (Elbaum-Garfinkle et al. 2015), and RNA granules (Lin et al. 2015) to name a few), computationally/bioinformatically shown for several nuclear MLOs (Meng et al. 2015) and for 4796 human proteins from 20 MLOs (Darling et al. 2018), hypothesized for several "assemblages" (Toretsky and Wright 2014; Csizmok et al. 2016), and generalized for all MLOs and complex biological coacervates (Uversky et al. 2015; Antifeeva et al. 2022; Kulkarni et al. 2022; Nesterov et al. 2021; Uversky 2021; Turoverov et al. 2019; Fonin et al. 2018; Uversky 2017a). Fluidity of MLOs originates from the multivalent interactions between intrinsicallu disordered proteins (IDPs) or hybrid proteins containing ordered domains and IDRs that are not accompanied by noticeable alteration in the structure of proteins undergoing LLPSs (Uversky 2017a, b; Darling et al. 2018). Therefore, MLOs represent a special form of disorder-based protein complexes (Uversky 2017b, a; Uversky et al. 2015; Nott et al. 2015), and can be considered as illustrations of the disorder-based emergent behavior of proteins (Turoverov et al. 2019; Uversky 2013, 2019). The lack of noticeable structural changes in IDPs forming MLOs is supported by the NMR analysis of several MLOs or liquid droplets formed by the Alzheimer-related protein tau (Ambadipudi et al. 2017; Ambadipudi et al. 2019), elastin-like polypeptides (ELPs) (Reichheld et al. 2017), low-complexity domain of the RNA-binding protein fused in sarcoma (FUS) (Murthy et al. 2019), heterogeneous nuclear ribonucleoprotein A2 (hnRNPA2) (Ryan et al. 2018), Parkinson's disease-related α-synuclein (Takamuku et al. 2022), p53 (Usluer et al. 2021), SERPINE1 mRNA binding protein 1 (SERBP1) (Baudin et al. 2021), TAR DNA-binding protein 43 (TDP-43) (Li et al. 2018), Probable ATP-dependent RNA helicase DDX4, Synaptic functional regulator FMR1, and Ubiquilin-2 (UBQLN2) (Murthy and Fawzi 2020) to name a few.

The shells of BMC are being developed for designing nanocages and nanoreactors for novel applications. The binding capacity of a recombinant GRM2-type BMC shell for iron and cobalt could be increased by functionalizing with short peptides rich in cysteine residues (Kalnins et al. 2023). Beal et al. (2023) have mentioned the usefulness of BMC shells as drug carriers and display scaffolds for polyvalent antigens (Beal et al. 2023). These researchers used copper click chemistry to attach a spytag on a protein for display on the surface of a BMC shell modified suitably with the spyCatcher. This is potentially a powerful strategy for vaccine development (Beal et al. 2023). Jiang et al. (2023) have used a protein shell of a carboxysome from a chemoautotroph and sequestered a [Ni-Fe]-hydrogenase from *E. coli* in it (Jiang et al. 2023). The synthetic BMC produced hydrogen under both aerobic and anaerobic conditions. This is an interesting bio-inspired design of a robust biocatalyst for production of green fuel (Jiang et al. 2023). Rose et al. (2023) have provided a good discussion on applications of BMC in design of novel biomaterials (Rose et al. 2023). Their review also covers the BMC structures and how their design can be adapted for creating scaffolds, catalysts and drug delivery vehicles. They also point out how pathogenic bacteria exploit their BMC (Rose et al. 2023).

RuBisCO, Carboxysomes, and Carbon Dioxide Fixation

Ribulose 1,5-bisphosphate carboxylase/oxygenase (RuBisCO) is the most abundant enzyme on earth, is an important part of the dark reactions of the photosynthesis which fixes carbon dioxide, and is considered by many as one of the most essential factors for all life on earth. RuBisCO constitutes up to 50% of the leaf proteins in plants. It converts the 5 carbon containing bisphosphate molecule into 2 molecules of 3-phosphoglycerate by fixing 1 molecule of carbon dioxide. It has low catalytic efficiency with k_{cat} of about 3 s^{-1} only. In photorespiration, RuBsCO actually acts as an oxygenase and converts the bisphosphate to form 1 molecule of the phosphoglycerate together with 1 molecule of 2-phosphoglycolate with consumption of 1 molecule of oxygen (Voet et al. 2008). The phosphoglycolate is a toxic compound, and hence carboxysomes prevent formation of this toxic metabolite by ensuring that high concentration of carbon dioxide by selective entry lowers the oxygenase activity. Therefore, RuBsCO is a promiscuous enzyme (Gupta et al. 2020).

In the initial structural work, RuBsCO was found to have 8 large subunits and 8 small subunits forming a massive heterohexadecamer (~540,000 Da) (Andersson and Backlund 2008). In this L_8S_8 structure, an α/β-barrel domain has the active site. RuBisCO is active when a lysine residue inside of its active site is carbamylated and bind a Mg^{2+} ion. Later work has shown that this corresponds to the isoform I structure. Recently, Vladimir Camel and Gaston Zolla (2021) have discussed the structural dynamics and catalytic properties of the four isoforms of RuBisCO (Camel and Zolla 2021).

For RuBisCO to be active, its active site lysine residue must be carbamylated at the epsilon-amino group by a nonsubstrate carbon dioxide (a proton is lost from the amino group and the change in composition is NHCO), the reaction is catalyzed by RuBisCO activase and requires ATP (Andersson and Backlund 2008). This modification of lysine is required for both carboxylase and oxygenase activity (Voet et al. 2008). RuBisCO is a rather inefficient enzyme and is rapidly self-inactivated under the physiological conditions, because its active site can undergo spontaneous decarbamylation (loosing CO_2) and strongly binds a 5-carbon sugar, ribulose-1,5-bisphosphate (RuBP), or any of the sugar inhibitor analogs such us 3-ketoarabinitol-1,5-bisphosphate (KABP), D-xylulose-1,5-bisphosphate (XuBP), D-glycero-2,3-pentodiulose-1,5-bisphosphate (PDBP), 2-carboxytetritol-1,4-bisphosphate (CTBP), and 2-carboxy-D-arabinitol 1-phosphate (CA1P). Therefore, the rate of CO_2 fixation and the rate of photosynthetic electron transport depend on the RuBisCO regulation (Perchorowicz et al. 1981). As a result, RuBisCO is a highly regulated machine used to control carbon fixation reaction in response of environmental fluctuations (Parry et al. 2008). This control is managed by the RuBisCO activase (RCA), which is an enzyme responsible for the RuBisCO activation by uncoupling RuBisCO and RuBP interaction or releasing the bound RuBP or its analog from RuBisCO (Salvucci et al. 1985). It also allows the rapid formation of the critical carbamate in the active site of RuBisCO by binding the activator CO_2 (Portis and Parry 2007). It is believed that the presence of RCAs defines the RuBisCO ability to function at its maximal capacity in sub-optimal CO_2 concentration that would normally not permit carbamylation in vivo (Portis et al. 1986). That is why carboxysomes contain, besides carbonic anhydrase and RuBisCO, activase enzyme as well.

It has been estimated that as much as 90% conversion of carbon into biomass on earth happens because of RuBisCO (Erb and Zarzycki 2018). RuBisCO evolved before there was oxygen on the earth. Later, in order to become somewhat more specific towards carbon dioxide, it evolved by lowering its catalytic efficiency. The promiscuous activity could not be totally abolished as it is difficult for enzymes to distinguish similar molecules (Gupta et al. 2020). It is debatable whether RuBisCO evolved from the RuBisCO like proteins (RLP) or vice versa. RLPs are involved mostly in S metabolism but their catalytic mechanisms shows considerable similarity with RuBisCO. Considering RuBisCO catalyzes a more complex reaction than most of RLP, it is likely that it evolved from an ancestor, which was a non-carbon dioxide fixing catalyst (Erb and Zarzycki 2018). Further down the line, carboxysomes evolved to compartmentalize the carbon dioxide fixing biocatalytic activity.

Yeats et al. (2008) have described the discovery of carboxysomes, as well as early studies on their isolation, purification, and characterization (Yeates et al. 2008). The early BMCs were discovered in *E. coli* and *S. enterica*. This review captures what was already known about shell proteins and the assembly of the core enzymes, and it is also worth reading as it discusses early speculations about various aspects of BMC (Yeates et al. 2008).

Ni et al. (2023) have used cryo-electron tomography to look at RuBisCO organization inside the two different types of α-carboxysomes from a marine organism. RuBisCO has three concentric layers forming "inter-twinning spirals" (Ni et al. 2023). These authors also have investigated the role of the intrinsically disordered linker protein CsoS2 in the organization of the enzyme in the carboxysomes (Ni et al. 2023). They showed that at the shell formation, the intrinsically disordered CsoS2 C-terminus undergoes disorder-to-order transition to gain a well-structured organization that acts as a universal "molecular thread" stitching through multiple shell protein interfaces (Ni et al. 2023).

The recent article by Basalla et al. (2023) summarized our current understanding of the two-protein system involved in the maintenance of carboxysome distribution (McdAB), which (a) ensures correct spacing of carboxysomes in the bacteria and (b) is involved in LLPS to form biocondensates of carboxysomes (Basalla et al. 2023). McdA, which is a ParA/MinD ATPase (i.e., a member of a protein family involved in the positioning of diverse cellular structures in bacteria) interacts with McdB (which is associated with carboxysomes) to ensure equal spacing of carboxysomes. McdA is an ATPase which is known to be involved in similar role for other cellular organelles. McdB has specificity with respect to particular carboxysome and is capable of LLPS. The authors have looked closely at this adaptor protein from a cyanobacteria (Basalla et al. 2023). McdB connects the ATPase to its cargo. It has an IDR at the N-terminal region, a middle coiled-coil domain, which dimerizes to aid condensation and a C-terminal domain, which trimerises the McdB dimers to ensure polyvalency required to further facilitate condensation. This finer detail should be helpful in designing synthetic carboxysomes for different applications (Basalla et al. 2023).

Recently, Chen et al. (2023) have successfully generated morphologically correct carboxysomes in tobacco chloroplasts by transforming nine carboxysome genetic components derived from a proteobacterium and showed that the engineered carboxysomes could fix carbon dioxide (Chen et al. 2023). The authors concluded that they provided "proof-of-concept for a route to engineering fully functional CO_2-fixing modules and entire CO_2-concentrating mechanisms into chloroplasts to improve crop photosynthesis and productivity" (Chen et al. 2023).

Biomolecular Condensates

Nucleolus was the first membrane-less organelle (MLO) seen in 1774, when Abbé Gasparo Ferdinando Felice Fontana (1730–1805) described a cellular compartment that becomes visible under the microscope during the interphase of the eukaryotic cell division (reviewed in (Pederson 2011)). However, it took more than 60 years to provide the first properly documented accounts of this MLO, which was introduced by Rudolf Wagner (1805–1864) in 1835, as the *macula germinativa*, an important formation in the ovum of several mammalian species (Wagner 1835) and independently, described by Gabriel Gustav Valentin (1810–1883) in 1837, as "a kind of second nucleus within the nucleus" (Valentin 1837). In 1898, Thomas Harrison Montgomery Jr. (1873–1912) published a 318-page-long monograph containing 346 hand-drawn color figures of nuclei and nucleoli from different species (Montgomery 1898). Later, many such membrane-less organelles were identified in eukaryotes. In a recent overview of the field, it was emphasized that one can find almost 100 different MLOs in bacteria and eukaryotes (Darling and Uversky 2023). These include nuclear speckles and Cajal bodies in nucleus and stress granules, processing bodies, and centriole in cytoplasm. MLOs were eventually found implicated in many biological processes, such as stress responses and control of gene expression and signal transduction (Boeynaems et al. 2018; Darling and Uversky 2023; Darling et al. 2018; Nesterov et al. 2021; Uversky 2017b). Their defective functioning was found to lead to diseases including age-related disorders, such as amylotrophic lateral sclerosis (Darling and Shorter 2021; Darling et al. 2019b; Darling et al. 2019a; Uversky 2017c). Liquid-liquid phase separation (LLPS) as the basis of the formation of these MLOs was established first in the case of P granules (Brangwynne et al. 2009). LLPS occurs when supersaturated solution spontaneously separates into a denser phase (MLO) and the dilute phase which merges with the nucleoplasm or cytoplasm (Uversky and Finkelstein 2019). MLOs are liquid-like inasmuch as they tend to assume spherical shape, show deformability and exchange materials with the surroundings in a dynamic manner. Yet, they contain molecules (mostly proteins and nucleic acids), which are bound by physical cross-links unlike simple (what we call generally) liquids, which have molecules associated by van der Waals forces. While MLOs are generally described as liquid-like, it has been pointed out that "in reality, an organelle can be a liquid, some form of solid, a liquid–gel, a solid–gel, a crystalline–solid, a semi-crystalline–solid, or liquid–crystalline depending on the extent of spatial ordering and the directional preferences for spatial ordering" (Boeynaems et al. 2018). Overall, parameters, such as molecular concentration and extent of cross-links etc. determine the physical state of the MLO.

The MLOs are also known as biomoleular condensates (BCs), and their formation obviously requires multivalency and high conformational flexibility to create liquid-like networks. SLiMs as short sequences identified in IDPs/IDRs are believed to play important role, both as linkers of structured domains as well as for facilitating assembly formation by fly-casting. Oligomers of proteins also can create multivalent species. Not all supersaturated situations lead to phase separation, so there is some switching mechanisms. One such mechanism could be PTM which alters the protein-protein interactions (Boeynaems et al. 2018).

In recent years, many BCs have been found in cytoplasm and on the membranes as well (Nesterov et al. 2021). They all share similar assembly process but of course differ in composition, location, and function (Banani et al. 2017; Nesterov et al. 2021). Their composition is not constant and exchange with the surroundings can take place within minutes or even seconds.

A recent article by Alberty et al. (2019) proposed guidelines for characterizing a process as based upon LLPS in both *in vitro* and in cells (Alberti et al. 2019). These authors pointed out that LLPS should be viewed as a stimuli-responsive process, in which several parameters play a role. It was also highlighted that the low complexity sequence regions (LCRs) as a subset of IDRs are often involved in LLPS of which prion or prion-like proteins are prominent example(Alberti et al. 2019).

Lyon et al. (2021) have pointed out that LLPS is not the only mechanism by which BC formation takes place (Lyon et al. 2021). One other important mechanism is when molecules bind to existing entities like RNA aggregates or genomic DNA. In such cases, multiple factors, such as availability and number of binding sites on these structures, affinity of the molecule towards these sites, cooperativity and molecular concentration come into play (Lyon et al. 2021).

Apart from accelerating (by mass action) or suppressing (by excluding potential substrates from the condensed phase) reaction rates, BC formation can affect structural dynamics of proteins or their oligomerization and thus modulate catalysis (Lyon et al. 2021). An interesting example cited by Lyons et al. (2021) is that of RuBisCO (Lyon et al. 2021). Photosynthetic organisms have multivalent ligands, which bind to octamers of the enzyme to push it into a condensate. This again favors availability of carbon dioxide over oxygen and also improves the otherwise poor rate of carbon dioxide fixation by the enzyme (Lyon et al. 2021). The review also cites numerous examples of the variety of biological processes and the mechanisms by which BC formation affects them. Such mechanisms include exclusion of factors, regulating folding state of macromolecules, vectorial organization, sensing and switching, and establishing mesoscale architecture (Lyon et al. 2021).

Lin et al. (2023) have recently looked into how compositionally distinct BC containing a protein and different RNA molecules can be formed (Lin et al. 2023). They find that if all RNA molecules are available simultaneously, there is a well-mixed BC. If some RNA is available first, it gets “dynamically arrested” and addition of another RNA will lead to a different composition (Lin et al. 2023). Therefore, which and how many RNA are simultaneously available in adequate concentration dictate the composition of the protein- and RNA-containing BC (Lin et al. 2023).

Vijayaraghavan Rangachari (2023) has recently written an excellent account of the timeline of the various developments related to MLO and their evolutionary aspects (Rangachari 2023). As it was pointed out, the idea of the importance of LLPS in biology was first given in 1929 by Hendrik Gerard Bungenberg de Jong (1893–1977) and Hugo Rudolph Kruyt (1882–1959) (Bungenberg de Jong and Kruyt 1929) and also suggested by Alexander Oparin in 1938 as a model according to which the formation of liquid coacervates due to phase separations was a necessary element in the formation of primitive life (Oparin 1938). It is important to emphasize here that in reality, Soviet Biochemist Alexander Ivanovich Oparin (1894–1980) proposed his model much earlier—his original work was published in Russian in 1924, (Oparin 1924) whereas the work cited by Rangachari is the English translation of that classic piece published 14 years later (note that this book was translated into about 50 languages!).

The evidence of LLPS leading to the formation of MLO in cells was described 1899 by Edmund Beecher Wilson (1856–1939), who was the first to point that the protoplasm of the eggs of several species of sea stars and sea urchins he studied, although being different visually, represents an emulsion consisting of a solid substance and droplets of various sizes and, apparently, of various chemical nature, and that these droplets are liquid, being able to coalesce (Wilson 1899; Antifeeva et al. 2022). This important observation was made based on a fortunate combination of circumstances: the microspheres in the protoplasm of the oocytes of one of the sea star species he worked on were natively pigmented, and following destruction of the egg, he was able to observe how granules having the same color merge when colliding with each other, forming a larger droplet (coalescing) (Wilson 1899; Antifeeva et al. 2022). More than 100 years later, a similar approach was used by Clifford Brangwynn and Anthony Hyman to confirm the liquid nature of *C. elegans* P-granules (Brangwynne et al. 2009). LLPS is a protein and RNA concentration-dependent process that arises from transient, multivalent, and non-stoichiometric weak interactions, which include ionic bonds, hydrohen bonding, hydrophobic interactions, π-π, and cation-π interactions. Denser (as compared to their surrounding medium) liquid-like phase characterizes these BCs. As their formation and dissolution does not require any ATP, it is an economical strategy to create a spatiotemporal niche. Absence of membranes enables facile reversibility.

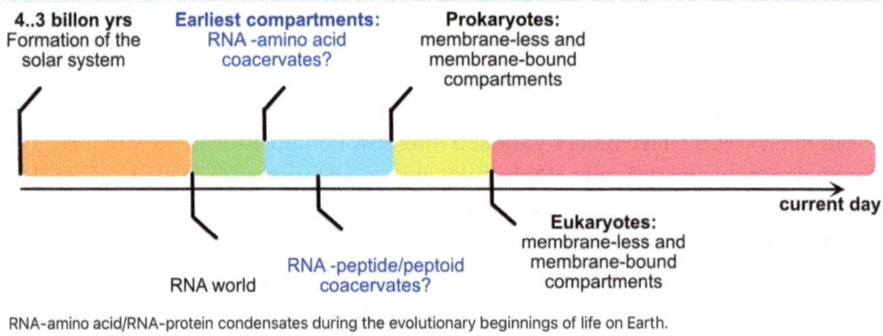

Fig. 13.1 Possible sequence of events leading to the formation of MLO before membrane enclosed organelles. The amino acids, peptides and later proteins would have interacted with RNA to form MLOs. (Adapted from Rangachari 2023)

Rangachari (2023) suggested that in the RNA world, these dense droplets of RNA may have formed to achieve higher rates and selectivity. This may have been followed by RNA coacrevating with other molecules to form reaction hubs of the prebiotic world (Fig. 13.1) (Rangachari 2023).

Recently, Nunzio et al. (2023] have discussed how the biological condensates are involved in the replication cycles of RNA viruses with focus on the example of HIV-1 (see Fig. 13.2) (Di Nunzio et al. 2023). These and other authors pointed out that different viruses utilize LLPS to invade the host cell and that one can distinguish LLPS associated with viral replication and trafficking of viral components, from the LLPS-mediated interference of viruses with host cell functions (Di Nunzio et al. 2023; Bianchi et al. 2023; Brocca et al. 2020).

MLOs or biomolecular condensates involve associations of proteins but can often also involve nucleic acids or even other biopolymers. When electrostatic interactions alone are involved, these are also referred to as coacervates. Nucleoli, nuclear speckles, P-bodies are examples of large MLOs, whereas signaling granules, DNA damage foci and receptor clusters exemplify smaller ones (Ramsak et al. 2023). The molecular mechanisms of the interactions involved in the formation of BCs are not always well understood. Coiled-coil association, frequently observed in BCs, is one of the well-characterized protein-protein interactions, in which the intrinsically disordered components come together to form dimeric helical structures. Ramsak et al. (2023) have used engineered *de novo* BCs in the mammalian cells as models to gain understanding of the finer details of parameters which influences BC formation (Ramsak et al. 2023). As expected, polyvalency was shown to be critical for the BC biogenesis. The general model has been that "stickers" (interaction-prone regions on polypeptides) associate and are separated by linkers (regions which don't actively participate in binding). In this work, it was found that the nature of linkers (in terms of their length and flexibility) is also crucial (Ramsak et al. 2023). The network formation requires mismatch between the nature of linkers on the associating polypeptides (Ramsak et al. 2023). An interesting feature of this work was that through the modular design of the phase-separating proteins, the

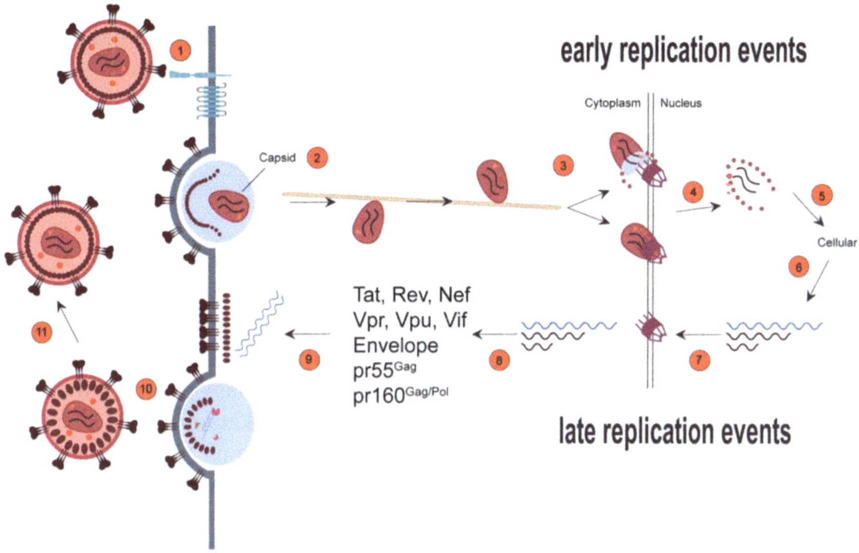

Fig. 13.2 Evolving and potential roles for biomolecular condensates at various steps of the retroviral replication cycle. 1, engagement with host cell receptors and viral entry; 2, membrane fusion and intracellular trafficking of the viral capsid; 3, nuclear entry; 4, uncoating and completion of reverse transcription; 5, integration into host DNA; 6, transcription, intranuclear trafficking of the viral ribonucleoprotein; 7, nucleocytoplasmic mRNA export through the nuclear pore with completely spliced mRNAs exported first then the singly-spliced and unspliced mRNAs; 8, mRNA engagement with the host translation machinery to generate early (Tat, Rev. and Nef) and late (pr55Gag, pr160Gag/Pol, Vpr, Vpu and Vif) viral gene products, as indicated, and genomic RNA selection; 9, genomic RNA and viral protein trafficking to virus assembly sites; 10, virus budding; and 11, virus maturation mediated by the viral protease substrates, pr55Gag and pr160Gag/Pol. Fig reproduced from Nunzio et al. (2023) under CC 4.0 Attribution license

authors were able to "achieve multiple coexisting condensates, chemical regulation of LLPS, condensate fusion, formation from either one or two polypeptide components or LLPS regulation by a third polypeptide chain" (Ramsak et al. 2023).

Recently, Leshem et al. (2023) have investigated the formation of BCs by a library of designer minimalistic peptides (Baruch Leshem et al. 2023). The authors showed that the LLPS propensity, dynamics, and encapsulation efficiency of compartments formed by the members of a library of LLPS-promoting peptide building blocks composed of various assembly domains, can be tuned by changes to the peptide composition A notable finding was that peptides containing both arginine and aromatic amino acids facilitated the condensation of the peptides. The cation-π, π-π, and electrostatic interactions were involved in LLPS in case of these peptides (Baruch Leshem et al. 2023).

Chilkoti's group have tried to identify rules for designing programmable synthetic condensates (Dai et al. 2023a; Dai et al. 2023b). As a prelude to that, the review is also a good account of physicochemical principles governing formation of BCs by LLPS (Dai et al. 2023b). Earlier, the same group had investigated self-assembly of IDPs of low complexity (Simon et al. 2017) and BC formation by synthetic IDPs (Dzuricky et al. 2020). These studies indicated that designing of proteins and peptides capable of the controllable LLPS is an achievable goal and, therefore, has very strong promises in biotechnology.

Conclusions and Future Perspectives

Bacterial microcompartments and biomolecular condensates indicate two different designs (with different aims and functional advantages) of sequestering enzymes or segments of metabolic pathways. BMCs are more permanent entities with selective transport through the protein shell. In that way and in many respects, these are closer to intracellular membrane-bounded organelles of eukaryotes. It may be interesting to ponder over the comparative virtues and limitations of a simple protein shell (of BMC) and lipid containing biomembranes of cells and their intracellular organelles.

One striking feature is the kind of shapes one can achieve by using proteins alone. Similarity of icosohedral shells found in numerous BMC and finite ways viral coats are organized indicate the limitations of working with ordered proteins alone. The phospholipids of biomembranes, which are excellent amphiphiles (surfactants) lend considerable versatility to the design in many ways. They bring that stickiness, fashion nonpolar pores, and channels and combine with integral and peripheral proteins to expand innovation space in a highly versatile manner. The importance of intrinsic disorder of membrane proteins in influencing membrane curvatures, for example, reflects that more complex functional needs required something beyond just proteins to create enclosures in the eukaryotic cells. Nevertheless, it is astonishing how much is achieved by BMCs in prokaryotes.

With new tools becoming available, we will learn more about structure-function relationships at the organelle-level by investigating BMC from different organisms. The increasing amount of work appearing in the current literature on synthetic or engineered BMC reflects their immense potential as modular vehicles for drug delivery and targeting. BMC fit in well with design principles of synthetic biology. So, we can anticipate lots of applications of BMC in the synthetic biology domain.

Biomolecular condensates, on the other hand are transients and often exploit simple physical chemistry of phase separation being exploited for cells to respond in real time in a smart stimuli-responsive manner. They show a smarter way by which cells achieve 'crowded environments' without creating any enclosures. They are part of a grand strategy by which cells exploit both structure and intrinsic disorder in a seamless fashion. Again, as design principles, they have immense potential for diverse applications as we gain experience with regulating their formation and dissolution. Perhaps, we will see a rich interface developing with smart biomaterials and biomolecular condensates. Learning to regulate and control selective release from these entities will be a prelude to that.

References

Aggarwal S, Snaidero N, Pahler G, Frey S, Sanchez P, Zweckstetter M, Janshoff A, Schneider A, Weil MT, Schaap IA, Gorlich D, Simons M (2013) Myelin membrane assembly is driven by a phase transition of myelin basic proteins into a cohesive protein meshwork. PLoS Biol 11(6):e1001577. https://doi.org/10.1371/journal.pbio.1001577

Alberti S (2017) The wisdom of crowds: regulating cell function through condensed states of living matter. J Cell Sci 130(17):2789–2796. https://doi.org/10.1242/jcs.200295

Alberti S, Dormann D (2019) Liquid-Liquid Phase Separation in Disease. Annu Rev Genet 53. https://doi.org/10.1146/annurev-genet-112618-043527

Alberti S, Hyman AA (2016) Are aberrant phase transitions a driver of cellular aging? BioEssays 38(10):959–968. https://doi.org/10.1002/bies.201600042

Alberti S, Gladfelter A, Mittag T (2019) Considerations and Challenges in Studying Liquid-Liquid Phase Separation and Biomolecular Condensates. Cell 176(3):419–434. https://doi.org/10.1016/j.cell.2018.12.035

Alberts B, Bray D, Raff M, Roberts K (1994) Molecular biology of the cell. Garland Science, New York/London

Ambadipudi S, Biernat J, Riedel D, Mandelkow E, Zweckstetter M (2017) Liquid-liquid phase separation of the microtubule-binding repeats of the Alzheimer-related protein Tau. Nat Commun 8(1):275. https://doi.org/10.1038/s41467-017-00480-0

Ambadipudi S, Reddy JG, Biernat J, Mandelkow E, Zweckstetter M (2019) Residue-specific identification of phase separation hot spots of Alzheimer's-related protein tau. Chem Sci 10(26):6503–6507. https://doi.org/10.1039/c9sc00531e

Andersson I, Backlund A (2008) Structure and function of Rubisco. Plant Physiol Biochem 46(3):275–291. https://doi.org/10.1016/j.plaphy.2008.01.001

Antifeeva IA, Fonin AV, Fefilova AS, Stepanenko OV, Povarova OI, Silonov SA, Kuznetsova IM, Uversky VN, Turoverov KK (2022) Liquid-liquid phase separation as an organizing principle of intracellular space: overview of the evolution of the cell compartmentalization concept. Cell Mol Life Sci 79(5):251. https://doi.org/10.1007/s00018-022-04276-4

Arechaga I (2013) Membrane invaginations in bacteria and mitochondria: common features and evolutionary scenarios. J Mol Microbiol Biotechnol 23(1–2):13–23. https://doi.org/10.1159/000346515

Asija K, Sutter M, Kerfeld CA (2021) A survey of bacterial microcompartment distribution in the human microbiome. Front Microbiol 12:669024. https://doi.org/10.3389/fmicb.2021.669024

Banani SF, Lee HO, Hyman AA, Rosen MK (2017) Biomolecular condensates: organizers of cellular biochemistry. Nat Rev Mol Cell Biol 18(5):285–298. https://doi.org/10.1038/nrm.2017.7

Baruch Leshem A, Sloan-Dennison S, Massarano T, Ben-David S, Graham D, Faulds K, Gottlieb HE, Chill JH, Lampel A (2023) Biomolecular condensates formed by designer minimalistic peptides. Nat Commun 14(1):421. https://doi.org/10.1038/s41467-023-36060-8

Basalla JL, Mak CA, Byrne JA, Ghalmi M, Hoang Y, Vecchiarelli AG (2023) Dissecting the phase separation and oligomerization activities of the carboxysome positioning protein McdB. elife 12. https://doi.org/10.7554/eLife.81362

Baudin A, Moreno-Romero AK, Xu X, Selig EE, Penalva LOF, Libich DS (2021) Structural characterization of the RNA-binding protein SERBP1 reveals intrinsic disorder and atypical RNA binding modes. Front Mol Biosci 8:744707. https://doi.org/10.3389/fmolb.2021.744707

Beal DM, Liang M, Brown I, Budge JD, Burrows ER, Howland K, Lee P, Martin S, Morrell A, Nemoto-Smith E, Roobol J, Stanley M, Smales CM, Warren MJ (2023) Modification of bacterial microcompartments with target biomolecules via post-translational SpyTagging. Mater Adv 4(14):2963–2970. https://doi.org/10.1039/d3ma00071k

Biamonti G (2004) Nuclear stress bodies: a heterochromatin affair? Nat Rev Mol Cell Biol 5(6):493–498. https://doi.org/10.1038/nrm1405

Biamonti G, Vourc'h C (2010) Nuclear stress bodies. Cold Spring Harb Perspect Biol 2(6):a000695. https://doi.org/10.1101/cshperspect.a000695

Bianchi G, Brocca S, Longhi S, Uversky VN (2023) Liaisons dangereuses: Intrinsic Disorder in Cellular Proteins Recruited to Viral Infection-Related Biocondensates. Int J Mol Sci 24(3). https://doi.org/10.3390/ijms24032151

Boeynaems S, Alberti S, Fawzi NL, Mittag T, Polymenidou M, Rousseau F, Schymkowitz J, Shorter J, Wolozin B, Van Den Bosch L, Tompa P, Fuxreiter M (2018) Protein Phase Separation: A New Phase in Cell Biology. Trends Cell Biol 28(6):420–435. https://doi.org/10.1016/j.tcb.2018.02.004

Brangwynne CP (2013) Phase transitions and size scaling of membrane-less organelles. J Cell Biol 203(6):875–881. https://doi.org/10.1083/jcb.201308087

Brangwynne CP, Eckmann CR, Courson DS, Rybarska A, Hoege C, Gharakhani J, Julicher F, Hyman AA (2009) Germline P granules are liquid droplets that localize by controlled dissolution/condensation. Science 324(5935):1729–1732. https://doi.org/10.1126/science.1172046

Brangwynne CP, Mitchison TJ, Hyman AA (2011) Active liquid-like behavior of nucleoli determines their size and shape in Xenopus laevis oocytes. Proc Natl Acad Sci USA 108(11):4334–4339. https://doi.org/10.1073/pnas.1017150108

Brocca S, Grandori R, Longhi S, Uversky V (2020) Liquid-Liquid Phase Separation by Intrinsically Disordered Protein Regions of Viruses: Roles in Viral Life Cycle and Control of Virus-Host Interactions. Int J Mol Sci 21(23). https://doi.org/10.3390/ijms21239045

Bungenberg de Jong H, Kruyt H (1929) Coacervation (partial miscibility in colloid systems). In: Proc K Ned Akad Wet, pp 849–856

Camel V, Zolla G (2021) An Insight of RuBisCO Evolution through a Multilevel Approach. Biomolecules 11(12). https://doi.org/10.3390/biom11121761

Cannon GC, Bradburne CE, Aldrich HC, Baker SH, Heinhorst S, Shively JM (2001) Microcompartments in prokaryotes: carboxysomes and related polyhedra. Appl Environ Microbiol 67(12):5351–5361. https://doi.org/10.1128/AEM.67.12.5351-5361.2001

Chen T, Hojka M, Davey P, Sun Y, Dykes GF, Zhou F, Lawson T, Nixon PJ, Lin Y, Liu LN (2023) Engineering alpha-carboxysomes into plant chloroplasts to support autotrophic photosynthesis. Nat Commun 14(1):2118. https://doi.org/10.1038/s41467-023-37490-0

Chong PA, Forman-Kay JD (2016) Liquid-liquid phase separation in cellular signaling systems. Curr Opin Struct Biol 41:180–186. https://doi.org/10.1016/j.sbi.2016.08.001

Chuma S, Hosokawa M, Tanaka T, Nakatsuji N (2009) Ultrastructural characterization of spermatogenesis and its evolutionary conservation in the germline: germinal granules in mammals. Mol Cell Endocrinol 306(1–2):17–23. https://doi.org/10.1016/j.mce.2008.11.009

Codd G, Marsden W (1984) The carboxysomes (polyhedral bodies) of autotrophic prokarygtes. Biol Rev 59(3):389–422

Courchaine EM, Lu A, Neugebauer KM (2016) Droplet organelles? EMBO J 35(15):1603–1612. https://doi.org/10.15252/embj.201593517

Csizmok V, Follis AV, Kriwacki RW, Forman-Kay JD (2016) Dynamic protein interaction networks and new structural paradigms in signaling. Chem Rev 116. https://doi.org/10.1021/acs.chemrev.5b00548

Dai Y, Farag M, Lee D, Zeng X, Kim K, Son HI, Guo X, Su J, Peterson N, Mohammed J, Ney M, Shapiro DM, Pappu RV, Chilkoti A, You L (2023a) Programmable synthetic biomolecular condensates for cellular control. Nat Chem Biol 19(4):518–528. https://doi.org/10.1038/s41589-022-01252-8

Dai Y, You L, Chilkoti A (2023b) Engineering synthetic biomolecular condensates. Nat Rev Bioeng 1:1–15. https://doi.org/10.1038/s44222-023-00052-6

Darling AL, Shorter J (2021) Combating deleterious phase transitions in neurodegenerative disease. Biochim Biophys Acta, Mol Cell Res 1868 5:118984. https://doi.org/10.1016/j.bbamcr.2021.118984

Darling AL, Uversky VN (2023) Known types of membrane-less organelles and biomolecular condensates. In: Droplets of life. Elsevier, pp 271–335

Darling AL, Liu Y, Oldfield CJ, Uversky VN (2018) Intrinsically disordered proteome of human membrane-less organelles. Proteomics 18(5–6):e1700193. https://doi.org/10.1002/pmic.201700193

Darling AL, Breydo L, Rivas EG, Gebru NT, Zheng D, Baker JD, Blair LJ, Dickey CA, Koren J 3rd, Uversky VN (2019a) Repeated repeat problems: Combinatorial effect of C9orf72-derived dipeptide repeat proteins. Int J Biol Macromol 127:136–145. https://doi.org/10.1016/j.ijbiomac.2019.01.035

Darling AL, Zaslavsky BY, Uversky VN (2019b) Intrinsic Disorder-Based Emergence in Cellular Biology: Physiological and Pathological Liquid-Liquid Phase Transitions in Cells. Polymers (Basel) 11(6). https://doi.org/10.3390/polym11060990

Decker CJ, Teixeira D, Parker R (2007) Edc3p and a glutamine/asparagine-rich domain of Lsm4p function in processing body assembly in Saccharomyces cerevisiae. J Cell Biol 179(3):437–449. https://doi.org/10.1083/jcb.200704147

Decker M, Jaensch S, Pozniakovsky A, Zinke A, O'Connell KF, Zachariae W, Myers E, Hyman AA (2011) Limiting amounts of centrosome material set centrosome size in C. elegans embryos. Curr Biol CB 21(15):1259–1267. https://doi.org/10.1016/j.cub.2011.06.002

Di Nunzio F, Uversky VN, Mouland AJ (2023) Biomolecular condensates: insights into early and late steps of the HIV-1 replication cycle. Retrovirology 20(1):4. https://doi.org/10.1186/s12977-023-00619-6

Doron L, Sutter M, Kerfeld CA (2023) Characterization of a novel aromatic substrate-processing microcompartment in Actinobacteria. MBio 14(4):e0121623. https://doi.org/10.1128/mbio.01216-23

Drews G, Niklowitz W (1956) Cytology of Cyanophycea. II. Centroplasm and granular inclusions of Phormidium uncinatum. Arch Mikrobiol 24 (2):147–162

Dundr M, Misteli T (2010) Biogenesis of nuclear bodies. Cold Spring Harb Perspect Biol 2(12):a000711. https://doi.org/10.1101/cshperspect.a000711

Dzuricky M, Rogers BA, Shahid A, Cremer PS, Chilkoti A (2020) De novo engineering of intracellular condensates using artificial disordered proteins. Nat Chem 12(9):814–825. https://doi.org/10.1038/s41557-020-0511-7

Elbaum-Garfinkle S, Kim Y, Szczepaniak K, Chen CC, Eckmann CR, Myong S, Brangwynne CP (2015) The disordered P granule protein LAF-1 drives phase separation into droplets with tunable viscosity and dynamics. Proc Natl Acad Sci USA 112(23):7189–7194. https://doi.org/10.1073/pnas.1504822112

Erb TJ, Zarzycki J (2018) A short history of RubisCO: the rise and fall (?) of Nature's predominant CO(2) fixing enzyme. Curr Opin Biotechnol 49:100–107. https://doi.org/10.1016/j.copbio.2017.07.017

Feric M, Brangwynne CP (2013) A nuclear F-actin scaffold stabilizes ribonucleoprotein droplets against gravity in large cells. Nat Cell Biol 15(10):1253–1259. https://doi.org/10.1038/ncb2830

Ferlez B, Sutter M, Kerfeld CA (2019) Glycyl radical enzyme-associated microcompartments: redox-replete bacterial organelles. mBio 10(1). https://doi.org/10.1128/mBio.02327-18

Fonin AV, Darling AL, Kuznetsova IM, Turoverov KK, Uversky VN (2018) Intrinsically disordered proteins in crowded milieu: when chaos prevails within the cellular gumbo. Cell Mol Life Sci 75(21):3907–3929. https://doi.org/10.1007/s00018-018-2894-9

Fonin AV, Antifeeva IA, Kuznetsova IM, Turoverov KK, Zaslavsky BY, Kulkarni P, Uversky VN (2022) Biological soft matter: intrinsically disordered proteins in liquid-liquid phase separation and biomolecular condensates. Essays Biochem 66(7):831–847. https://doi.org/10.1042/EBC20220052

Forman-Kay JD, Kriwacki RW, Seydoux G (2018) Phase Separation in Biology and Disease. J Mol Biol 430(23):4603–4606. https://doi.org/10.1016/j.jmb.2018.09.006

Fox AH, Lam YW, Leung AK, Lyon CE, Andersen J, Mann M, Lamond AI (2002) Paraspeckles: a novel nuclear domain. Current biology : CB 12(1):13–25

Gantt E, Conti SF (1969) Ultrastructure of blue-green algae. J Bacteriol 97(3):1486–1493. https://doi.org/10.1128/jb.97.3.1486-1493.1969

Gomes E, Shorter J (2019) The molecular language of membraneless organelles. J Biol Chem 294(18):7115–7127. https://doi.org/10.1074/jbc.TM118.001192

Grossman E, Medalia O, Zwerger M (2012) Functional architecture of the nuclear pore complex. Annu Rev Biophys 41:557–584. https://doi.org/10.1146/annurev-biophys-050511-102328

Gubitz AK, Feng W, Dreyfuss G (2004) The SMN complex. Exp Cell Res 296(1):51–56. https://doi.org/10.1016/j.yexcr.2004.03.022. S0014482704001338 [pii]

Gupta MN, Uversky VN (2023) Moonlighting enzymes: when cellular context defines specificity. Cell Mol Life Sci 80(5):130. https://doi.org/10.1007/s00018-023-04781-0

Gupta MN, Alam A, Hasnain SE (2020) Protein promiscuity in drug discovery, drug-repurposing and antibiotic resistance. Biochimie 175:50–57. https://doi.org/10.1016/j.biochi.2020.05.004

Gyorgyi AS (1960) Introduction to a submolecular biology. Academic

Handwerger KE, Cordero JA, Gall JG (2005) Cajal bodies, nucleoli, and speckles in the Xenopus oocyte nucleus have a low-density, sponge-like structure. Mol Biol Cell 16(1):202–211. https://doi.org/10.1091/mbc.E04-08-0742

Harrigan JA, Belotserkovskaya R, Coates J, Dimitrova DS, Polo SE, Bradshaw CR, Fraser P, Jackson SP (2011) Replication stress induces 53BP1-containing OPT domains in G1 cells. J Cell Biol 193(1):97–108. jcb.201011083 [pii]. https://doi.org/10.1083/jcb.201011083

Hatton IA, Galbraith ED, Merleau NSC, Miettinen TP, Smith BM, Shander JA (2023) The human cell count and size distribution. Proc Natl Acad Sci USA 120(39):e2303077120. https://doi.org/10.1073/pnas.2303077120

Heritage J, Evans EGV, Killington R (1996) Introductory microbiology, vol 3. Cambridge University Press

Hirose T, Ninomiya K, Nakagawa S, Yamazaki T (2023) A guide to membraneless organelles and their various roles in gene regulation. Nat Rev Mol Cell Biol 24(4):288–304. https://doi.org/10.1038/s41580-022-00558-8

Hnisz D, Shrinivas K, Young RA, Chakraborty AK, Sharp PA (2017) A Phase Separation Model for Transcriptional Control. Cell 169(1):13–23. https://doi.org/10.1016/j.cell.2017.02.007

Huang S (2000) Review: perinucleolar structures. J Struct Biol 129(2–3):233–240. https://doi.org/10.1006/jsbi.2000.4247

Hyman AA, Weber CA, Julicher F (2014) Liquid-liquid phase separation in biology. Annu Rev Cell Dev Biol 30:39–58. https://doi.org/10.1146/annurev-cellbio-100913-013325

Iancu CV, Ding HJ, Morris DM, Dias DP, Gonzales AD, Martino A, Jensen GJ (2007) The structure of isolated Synechococcus strain WH8102 carboxysomes as revealed by electron cryotomography. J Mol Biol 372(3):764–773. https://doi.org/10.1016/j.jmb.2007.06.059

Jiang Q, Li T, Yang J, Aitchison CM, Huang J, Chen Y, Huang F, Wang Q, Cooper AI, Liu LN (2023) Synthetic engineering of a new biocatalyst encapsulating [NiFe]-hydrogenases for enhanced hydrogen production. J Mater Chem B 11(12):2684–2692. https://doi.org/10.1039/d2tb02781j

Kalnins G, Bertins M, Viksna A, Tars K (2023) Functionalization of bacterial microcompartment shell interior with cysteine containing peptides enhances the iron and cobalt loading capacity. Biometals 37. https://doi.org/10.1007/s10534-023-00538-1

Kerfeld CA, Erbilgin O (2015) Bacterial microcompartments and the modular construction of microbial metabolism. Trends Microbiol 23(1):22–34. https://doi.org/10.1016/j.tim.2014.10.003

Kerfeld CA, Sawaya MR, Tanaka S, Nguyen CV, Phillips M, Beeby M, Yeates TO (2005) Protein structures forming the shell of primitive bacterial organelles. Science 309(5736):936–938. https://doi.org/10.1126/science.1113397

Kerfeld CA, Heinhorst S, Cannon GC (2010) Bacterial microcompartments. Ann Rev Microbiol 64:391–408. https://doi.org/10.1146/annurev.micro.112408.134211

Kerfeld CA, Aussignargues C, Zarzycki J, Cai F, Sutter M (2018) Bacterial microcompartments. Nat Rev Microbiol 16(5):277–290. https://doi.org/10.1038/nrmicro.2018.10

Kiebler MA, Bassell GJ (2006) Neuronal RNA granules: movers and makers. Neuron 51(6):685–690. https://doi.org/10.1016/j.neuron.2006.08.021

Kirst H, Kerfeld CA (2019) Bacterial microcompartments: catalysis-enhancing metabolic modules for next generation metabolic and biomedical engineering. BMC Biol 17(1):79. https://doi.org/10.1186/s12915-019-0691-z

Kulkarni P, Bhattacharya S, Achuthan S, Behal A, Jolly MK, Kotnala S, Mohanty A, Rangarajan G, Salgia R, Uversky V (2022) Intrinsically disordered proteins: critical components of the wetware. Chem Rev 122(6):6614–6633. https://doi.org/10.1021/acs.chemrev.1c00848

Kumar G, Hazra JP, Sinha S (2023) Disordered regions endow structural flexibility to shell proteins and function towards shell-enzyme interactions in 1,2-propanediol utilization microcompartment. J Biomol Struct Dyn 41(18):8891–8901. https://doi.org/10.1080/07391102.2022.2138552

Lamond AI, Spector DL (2003) Nuclear speckles: a model for nuclear organelles. Nat Rev Mol Cell Biol 4(8):605–612. https://doi.org/10.1038/nrm1172

Lefevre CT, Bazylinski DA (2013) Ecology, diversity, and evolution of magnetotactic bacteria. Microbiol Mol Biol Rev 77(3):497–526. https://doi.org/10.1128/MMBR.00021-13

Li L, Roy K, Katyal S, Sun X, Bleoo S, Godbout R (2006) Dynamic nature of cleavage bodies and their spatial relationship to DDX1 bodies, Cajal bodies, and gems. Mol Biol Cell 17(3):1126–1140. E05-08-0768 [pii]. https://doi.org/10.1091/mbc.E05-08-0768

Li B, Carey M, Workman JL (2007) The role of chromatin during transcription. Cell 128(4):707–719. https://doi.org/10.1016/j.cell.2007.01.015

Li P, Banjade S, Cheng HC, Kim S, Chen B, Guo L, Llaguno M, Hollingsworth JV, King DS, Banani SF, Russo PS, Jiang QX, Nixon BT, Rosen MK (2012) Phase transitions in the assembly of multivalent signalling proteins. Nature 483(7389):336–340. https://doi.org/10.1038/nature10879

Li HR, Chiang WC, Chou PC, Wang WJ, Huang JR (2018) TAR DNA-binding protein 43 (TDP-43) liquid-liquid phase separation is mediated by just a few aromatic residues. J Biol Chem 293(16):6090–6098. https://doi.org/10.1074/jbc.AC117.001037

Lin Y, Protter DS, Rosen MK, Parker R (2015) Formation and maturation of phase-separated liquid droplets by RNA-binding proteins. Mol Cell 60(2):208–219. https://doi.org/10.1016/j.molcel.2015.08.018

Lin AZ, Ruff KM, Dar F, Jalihal A, King MR, Lalmansingh JM, Posey AE, Erkamp NA, Seim I, Gladfelter AS, Pappu RV (2023) Dynamical control enables the formation of demixed biomolecular condensates. Nat Commun 14(1):7678. https://doi.org/10.1038/s41467-023-43489-4

Lyon AS, Peeples WB, Rosen MK (2021) A framework for understanding the functions of biomolecular condensates across scales. Nat Rev Mol Cell Biol 22(3):215–235. https://doi.org/10.1038/s41580-020-00303-z

Matera AG, Frey MR (1998) Coiled bodies and gems: Janus or gemini? Am J Hum Genet 63(2):317–321. https://doi.org/10.1086/301992

Maul GG, Negorev D, Bell P, Ishov AM (2000) Review: properties and assembly mechanisms of ND10, PML bodies, or PODs. J Struct Biol 129(2–3):278–287. https://doi.org/10.1006/jsbi.2000.4239

Meng F, Na I, Kurgan L, Uversky VN (2015) Compartmentalization and functionality of nuclear disorder: intrinsic disorder and protein-protein interactions in intra-nuclear compartments. Int J Mol Sci 17(1). https://doi.org/10.3390/ijms17010024

Mitrea DM, Cika JA, Guy CS, Ban D, Banerjee PR, Stanley CB, Nourse A, Deniz AA, Kriwacki RW (2016) Nucleophosmin integrates within the nucleolus via multi-modal interactions with proteins displaying R-rich linear motifs and rRNA. elife 5. https://doi.org/10.7554/eLife.13571

Montgomery TH (1898) Comparative cytological studies with especial regard to the morphology of the nucleolus. Ginn & Company, Boston

Mukherjee J, Nath Gupta M (2016) Paradigm shifts in our views on inclusion bodies. Curr Biochem Eng 3(1):47–55

Murthy AC, Fawzi NL (2020) The (un)structural biology of biomolecular liquid-liquid phase separation using NMR spectroscopy. J Biol Chem 295(8):2375–2384. https://doi.org/10.1074/jbc.REV119.009847

Murthy AC, Dignon GL, Kan Y, Zerze GH, Parekh SH, Mittal J, Fawzi NL (2019) Molecular interactions underlying liquid-liquid phase separation of the FUS low-complexity domain. Nat Struct Mol Biol 26(7):637–648. https://doi.org/10.1038/s41594-019-0250-x

Nesterov SV, Ilyinsky NS, Uversky VN (2021) Liquid-liquid phase separation as a common organizing principle of intracellular space and biomembranes providing dynamic adaptive responses. Biochim Biophys Acta, Mol Cell Res 1868 11:119102. https://doi.org/10.1016/j.bbamcr.2021.119102

Ni T, Jiang Q, Ng PC, Shen J, Dou H, Zhu Y, Radecke J, Dykes GF, Huang F, Liu LN, Zhang P (2023) Intrinsically disordered CsoS2 acts as a general molecular thread for alpha-carboxysome shell assembly. Nat Commun 14(1):5512. https://doi.org/10.1038/s41467-023-41211-y

Nizami Z, Deryusheva S, Gall JG (2010) The Cajal body and histone locus body. Cold Spring Harb Perspect Biol 2(7):a000653. cshperspect.a000653 [pii]. https://doi.org/10.1101/cshperspect.a000653

Nott TJ, Petsalaki E, Farber P, Jervis D, Fussner E, Plochowietz A, Craggs TD, Bazett-Jones DP, Pawson T, Forman-Kay JD, Baldwin AJ (2015) Phase transition of a disordered nuage protein generates environmentally responsive membraneless organelles. Mol Cell 57(5):936–947. https://doi.org/10.1016/j.molcel.2015.01.013

Nurse P (2021) What is life?: Five great ideas in biology. WW Norton & Company, New York

Oparin AI (1924) The origine of Life. Moscow Worker, Moscow

Oparin AI (1938) The origin of life. Macmillan, New York

Parry MA, Keys AJ, Madgwick PJ, Carmo-Silva AE, Andralojc PJ (2008) Rubisco regulation: a role for inhibitors. J Exp Bot 59(7):1569–1580. https://doi.org/10.1093/jxb/ern084

Pederson T (2001) Protein mobility within the nucleus--what are the right moves? Cell 104 (5):635–638

Pederson T (2011) The nucleoulus. Cold Spring Harb Perspect Biol 3(3):a000638. https://doi.org/10.1101/cshperspect.a000638

Perchorowicz JT, Raynes DA, Jensen RG (1981) Light limitation of photosynthesis and activation of ribulose bisphosphate carboxylase in wheat seedlings. Proc Natl Acad Sci USA 78(5):2985–2989. https://doi.org/10.1073/pnas.78.5.2985

Phair RD, Misteli T (2000) High mobility of proteins in the mammalian cell nucleus. Nature 404(6778):604–609. https://doi.org/10.1038/35007077

Pirrotta V, Li HB (2012) A view of nuclear Polycomb bodies. Curr Opin Genet Dev 22(2):101–109. https://doi.org/10.1016/j.gde.2011.11.004

Portis AR Jr, Parry MA (2007) Discoveries in Rubisco (Ribulose 1,5-bisphosphate carboxylase/oxygenase): a historical perspective. Photosynth Res 94(1):121–143. https://doi.org/10.1007/s11120-007-9225-6

Portis AR, Salvucci ME, Ogren WL (1986) Activation of Ribulosebisphosphate Carboxylase/Oxygenase at Physiological CO(2) and Ribulosebisphosphate Concentrations by Rubisco Activase. Plant Physiol 82(4):967–971. https://doi.org/10.1104/pp.82.4.967

Poudyal M, Patel K, Gadhe L, Sawner AS, Kadu P, Datta D, Mukherjee S, Ray S, Navalkar A, Maiti S, Chatterjee D, Devi J, Bera R, Gahlot N, Joseph J, Padinhateeri R, Maji SK (2023) Intermolecular interactions underlie protein/peptide phase separation irrespective of sequence and structure at crowded milieu. Nat Commun 14(1):6199. https://doi.org/10.1038/s41467-023-41864-9

Ramsak M, Ramirez DA, Hough LE, Shirts MR, Vidmar S, Elersic Filipic K, Anderluh G, Jerala R (2023) Programmable de novo designed coiled coil-mediated phase separation in mammalian cells. Nat Commun 14(1):7973. https://doi.org/10.1038/s41467-023-43742-w

Rangachari V (2023) Biomolecular condensates - extant relics or evolving microcompartments? Commun Biol 6(1):656. https://doi.org/10.1038/s42003-023-04963-3

Reichheld SE, Muiznieks LD, Keeley FW, Sharpe S (2017) Direct observation of structure and dynamics during phase separation of an elastomeric protein. Proc Natl Acad Sci USA 114(22):E4408–E4415. https://doi.org/10.1073/pnas.1701877114

Rose SM, Radhakrishnan A, Sinha S (2023) Innate and engineered attributes of bacterial microcompartments for applications in bio-materials science. J Mater Chem B 11(22):4842–4854. https://doi.org/10.1039/d3tb00098b

Ryan VH, Dignon GL, Zerze GH, Chabata CV, Silva R, Conicella AE, Amaya J, Burke KA, Mittal J, Fawzi NL (2018) Mechanistic view of hnRNPA2 low-complexity domain structure, interactions, and phase separation altered by mutation and arginine methylation. Mol Cell 69(3):465–479. e467. https://doi.org/10.1016/j.molcel.2017.12.022

Saha S, Weber CA, Nousch M, Adame-Arana O, Hoege C, Hein MY, Osborne-Nishimura E, Mahamid J, Jahnel M, Jawerth L, Pozniakovski A, Eckmann CR, Julicher F, Hyman AA (2016) Polar positioning of phase-separated liquid compartments in cells regulated by an mRNA competition mechanism. Cell 166(6):1572–1584. e1516. https://doi.org/10.1016/j.cell.2016.08.006

Salvucci ME, Portis AR Jr, Ogren WL (1985) A soluble chloroplast protein catalyzes ribulosebisphosphate carboxylase/oxygenase activation in vivo. Photosynth Res 7(2):193–201. https://doi.org/10.1007/BF00037012

Schmid MF, Paredes AM, Khant HA, Soyer F, Aldrich HC, Chiu W, Shively JM (2006) Structure of Halothiobacillus neapolitanus carboxysomes by cryo-electron tomography. J Mol Biol 364(3):526–535. https://doi.org/10.1016/j.jmb.2006.09.024

Schreiber G, Keating AE (2011) Protein binding specificity versus promiscuity. Curr Opin Struct Biol 21(1):50–61. https://doi.org/10.1016/j.sbi.2010.10.002

Shav-Tal Y, Blechman J, Darzacq X, Montagna C, Dye BT, Patton JG, Singer RH, Zipori D (2005) Dynamic sorting of nuclear components into distinct nucleolar caps during transcriptional inhibition. Mol Biol Cell 16(5):2395–2413. https://doi.org/10.1091/mbc.E04-11-0992

Shin Y, Brangwynne CP (2017) Liquid phase condensation in cell physiology and disease. Science 357(6357). https://doi.org/10.1126/science.aaf4382

Shively J, English R (1991) The carboxysome, a prokaryotic organelle: a mini-review. Can J Bot 69(5):957–962

Simon JR, Carroll NJ, Rubinstein M, Chilkoti A, Lopez GP (2017) Programming molecular self-assembly of intrinsically disordered proteins containing sequences of low complexity. Nat Chem 9(6):509–515. https://doi.org/10.1038/nchem.2715

Snyder C, Stefano GB (2015) Mitochondria and chloroplasts shared in animal and plant tissues: significance of communication. Med Sci Monit 21:1507–1511. https://doi.org/10.12659/MSM.894481

Strzelecka M, Trowitzsch S, Weber G, Luhrmann R, Oates AC, Neugebauer KM (2010) Coilin-dependent snRNP assembly is essential for zebrafish embryogenesis. Nat Struct Mol Biol 17(4):403–409. https://doi.org/10.1038/nsmb.1783

Sutter M, Melnicki MR, Schulz F, Woyke T, Kerfeld CA (2021) A catalog of the diversity and ubiquity of bacterial microcompartments. Nat Commun 12(1):3809. https://doi.org/10.1038/s41467-021-24126-4

Takamuku M, Sugishita T, Tamaki H, Dong L, So M, Fujiwara T, Matsuki Y (2022) Evolution of alpha-synuclein conformation ensemble toward amyloid fibril via liquid-liquid phase separation (LLPS) as investigated by dynamic nuclear polarization-enhanced solid-state MAS NMR. Neurochem Int 157:105345. https://doi.org/10.1016/j.neuint.2022.105345

Teilum K, Olsen JG, Kragelund BB (2021) On the specificity of protein-protein interactions in the context of disorder. Biochem J 478(11):2035–2050. https://doi.org/10.1042/BCJ20200828

Toretsky JA, Wright PE (2014) Assemblages: functional units formed by cellular phase separation. J Cell Biol 206(5):579–588. https://doi.org/10.1083/jcb.201404124

Trettel DS, Neale C, Zhao M, Gnanakaran S, Gonzalez-Esquer CR (2023) Monatomic ions influence substrate permeation across bacterial microcompartment shells. Sci Rep 13(1):15738. https://doi.org/10.1038/s41598-023-42688-9

Tsai SJ, Yeates TO (2011) Bacterial microcompartments: insights into the structure, mechanism, and engineering applications. Prog Mol Biol Transl Sci 103:1–20

Turoverov KK, Kuznetsova IM, Fonin AV, Darling AL, Zaslavsky BY, Uversky VN (2019) Stochasticity of biological soft matter: emerging concepts in intrinsically disordered proteins and biological phase separation. Trends Biochem Sci 44(8):716–728. https://doi.org/10.1016/j.tibs.2019.03.005

Updike DL, Hachey SJ, Kreher J, Strome S (2011) P granules extend the nuclear pore complex environment in the C. elegans germ line. J Cell Biol 192 (6):939-948. https://doi.org/10.1083/jcb.201010104

Usluer S, Spreitzer E, Bourgeois B, Madl T (2021) p53 Transactivation Domain Mediates Binding and Phase Separation with Poly-PR/GR. Int J Mol Sci 22(21). https://doi.org/10.3390/ijms222111431

Uversky VN (2013) Unusual biophysics of intrinsically disordered proteins. Biochim Biophys Acta 1834(5):932–951. https://doi.org/10.1016/j.bbapap.2012.12.008

Uversky VN (2017a) Intrinsically disordered proteins in overcrowded milieu: Membrane-less organelles, phase separation, and intrinsic disorder. Curr Opin Struct Biol 44:18–30. https://doi.org/10.1016/j.sbi.2016.10.015

Uversky VN (2017b) Protein intrinsic disorder-based liquid-liquid phase transitions in biological systems: Complex coacervates and membrane-less organelles. Adv Colloid Interf Sci 239:97–114. https://doi.org/10.1016/j.cis.2016.05.012

Uversky VN (2017c) The roles of intrinsic disorder-based liquid-liquid phase transitions in the "Dr. Jekyll-Mr. Hyde" behavior of proteins involved in amyotrophic lateral sclerosis and frontotemporal lobar degeneration. Autophagy 13 (12):2115–2162. https://doi.org/10.1080/15548627.2017.1384889

Uversky VN (2019) Intrinsically disordered proteins and their "mysterious" (meta)physics. Front Phys 7:Article 10 (18 pages). https://doi.org/10.3389/fphy.2019.00010

Uversky VN (2021) Recent developments in the field of intrinsically disordered proteins: intrinsic disorder-based emergence in cellular biology in light of the physiological and pathological liquid-liquid phase transitions. Annu Rev Biophys 50:135–156. https://doi.org/10.1146/annurev-biophys-062920-063704

Uversky VN, Finkelstein AV (2019) Life in phases: intra- and inter- molecular phase transitions in protein solutions. Biomolecules 9(12). https://doi.org/10.3390/biom9120842

Uversky VN, Kuznetsova IM, Turoverov KK, Zaslavsky B (2015) Intrinsically disordered proteins as crucial constituents of cellular aqueous two phase systems and coacervates. FEBS Lett 589(1):15–22. https://doi.org/10.1016/j.febslet.2014.11.028

Valentin G (1837) Repertorium für anatomie und physiologie

Voet D, Voet JG, Pratt CW (2008) Principles of biochemistry, vol 4. Wiley, New York

Wagner R (1835) Einige bemerkungen und fragen über das keimbläschen (vesicular germinativa). Müller's Archiv Anat Physiol Wissenschaft Med 268:373–377

Wilson EB (1899) The structure of protoplasm. Science 10(237):33–45. https://doi.org/10.1126/science.10.237.33

Wippich F, Bodenmiller B, Trajkovska MG, Wanka S, Aebersold R, Pelkmans L (2013) Dual specificity kinase DYRK3 couples stress granule condensation/dissolution to mTORC1 signaling. Cell 152(4):791–805. https://doi.org/10.1016/j.cell.2013.01.033

Yeates TO, Kerfeld CA, Heinhorst S, Cannon GC, Shively JM (2008) Protein-based organelles in bacteria: carboxysomes and related microcompartments. Nat Rev Microbiol 6(9):681–691. https://doi.org/10.1038/nrmicro1913

Zaslavsky BY, Uversky VN (2018) In Aqua Veritas: The Indispensable yet Mostly Ignored Role of Water in Phase Separation and Membrane-less Organelles. Biochemistry 57(17):2437–2451. https://doi.org/10.1021/acs.biochem.7b01215

Zhou HX, Rivas G, Minton AP (2008) Macromolecular crowding and confinement: biochemical, biophysical, and potential physiological consequences. Annu Rev Biophys 37:375–397. https://doi.org/10.1146/annurev.biophys.37.032807.125817

Chapter 14
Macromolecular Crowding in Cell Stress and Death

Michael A. Model

Abstract To study macromolecular crowding (MC) in living cells, one needs a method to measure it. Several existing approaches to quantify MC address slightly different aspects of crowding. If we define MC through protein concentration, it can be measured by quantitative phase imaging coupled with volume determination; both can be realized on a standard bright-field microscope. Osmotic cell theory can help identify the essential factors that control MC. Nevertheless, there are still many gaps in our understanding of MC regulation and, in particular, of the interrelationship between MC and cell stress or damage. Experiments show that MC is subject to homeostatic control and returns to its resting values following various disturbances. Severe cell damage causes an accumulation of water and a decrease in MC; however, based on limited data, water accumulation is restricted to one area of the cell (necrotic bleb), while the rest of the cell remains at normal density. Similar heterogeneous water distribution is observed in vacuolated mammalian cells. Intermediate degrees of stress tend to produce dehydration and an increase in MC. Apoptotic shrinkage is one common example of stress-induced dehydration, but the effect may be more general. A hypothesis on its mechanism is proposed.

Keywords Macromolecular crowding · Dehydration · Intracellular water · Apoptotic volume decrease · Necrotic volume increase · Quantitative phase contrast · Transmission-through-dye · Cell volume

High density of intracellular proteins and other macromolecules is one of the basic facts of life. Moreover, cells seem to take great care to maintain it within narrow limits. But what makes macromolecular crowding (MC) so important for the well-being of the cell? The question has two sides: (1) how does cell death (or, more accurately, cell injury or stress that may lead to death) affect the crowding? (2) how do changes in the crowding affect cell death or survival? It seems that the present

M. A. Model (✉)
Department of Biological Science, Kent State University, Kent, OH, USA
e-mail: mmodel@kent.edu

V. N. Uversky (ed.), *(Macro)Molecular Crowding*, Subcellular Biochemistry 109,
https://doi.org/10.1007/978-3-032-03370-3_14

state of knowledge is too limited to attempt a comprehensive review. Therefore, the present chapter can be viewed rather as an assembly of various bits of information, hypotheses, and observations.

How Real Is Macromolecular Crowding?

An introduction to the principles of macromolecular crowding can be found in numerous reviews (e.g., Zhou et al. 2008; Rivas and Minton 2016; Kuznetsova et al. 2014; Burg 2000) and will not be repeated here. We can start by defining macromolecular crowding as it will be understood in this chapter: the mass/volume concentration of macromolecules or "proteins" (we will be using the word "protein" as a placeholder for all organic macromolecules):

$$MC = \frac{m_p}{V} \tag{14.1}$$

The magnitude of MC is usually close to 0.2 g/ml (Model 2021), although many authors, for reasons not always clear, quote much higher values. Equation (14.1) is not the only possible definition of MC, and section "Detection and quantification of macromolecular crowding" will discuss it in more detail.

An important assumption implicit in Eq. 14.1 is that proteins are distributed more or less uniformly. In a remarkable review published in 1984, James Clegg presented evidence against such a notion and argued that the cytosol consists rather of protein clusters separated by watery lacunas (Clegg 1984). If this is true, then the average protein concentration would be physically meaningless as a measure of MC because cell compression or swelling should mostly affect soluble, but not immobilized, proteins. Later results have not sufficiently clarified the issue. It is now widely accepted that many enzymes form at least transient complexes with each other (Zhang and Fernie 2021; Vonck and Schäfer 2009) or with the cytoskeleton (Hudder et al. 2003). Experiments on diffusion show that a significant fraction of cellular proteins (about a half, according to some results) is immobilized (Luby-Phelps 1999) and therefore may be more immune to crowding effects. This can be kept in mind as a possible caveat.

Macromolecular Crowding, Hydration, and Cell Volume: The Unavoidable Triangle

Although active research in MC in live cells picked up relatively recently, some of its aspects have been studied de facto for several decades, only under different names: cell water content and cell volume.

Table 14.1 The approximate relationships between MC (defined as dry mass concentration in g/ml) and some other frequently used quantities. The conversions are based on the buoyant density of 1.4 g/ml (Model and Petruccelli 2018); however, the density of organisms with high RNA content may be larger

Quantity	Relationship to MC	Value at MC 0.2
Buoyant density	1 + 0.28MC	1.056
Water volume/Cell volume	1–0.7MC	0.86
Water mass/Cell mass	$\frac{1-0.7MC}{1+0.28MC}$	0.814
Water mass/Protein mass	$\frac{1-0.7MC}{0.98MC}$	4.4
Refractive index dn/dc	$n = n_o + (dn/dc) \cdot MC$	1.373 (at $\lambda = 0.5$ μm)
Proteins[a] DNA[b] Sucrose[b] NaCl[c]	0.185 ml/g 0.18 ml/g 0.145 ml/g = 0.05 M^{-1} 0.01 M^{-1}	

[a]Theisen et al. (2000)
[b]Tumolo et al. (2004)
[c]Schmidt et al. (2014)

MC and cell hydration. Since proteins (or dry mass, which in mammalian cells is mostly due to proteins) occupy a significant volume in an otherwise watery cell, MC is related to water content; thus follows a simple relationship between MC and hydration (Table 14.1). However, the role of intracellular water is not limited to the maintenance of MC.

All proteins contain bound water molecules. Some of them, enclosed within tight pockets, seem to be practically excluded from the exchange with bulk water. The existence of such osmotically inactive water is supported by numerous lines of evidence (Hinke 1970; Garlid 1999; Fullerton et al. 2006; Zelenina and Brismar 2000; De Bruijne and Van Steveninck 1970; Hempling et al. 1977; Dupre and Hempling 1978; Du et al. 1994; Cameron et al. 2006; Cameron et al. 2011; Bagatolli et al. 2021; Kreuzer et al. 2012; Urbonaite et al. 2015), including extensive research done by food scientists (Zielbauer et al. 2016). More superficially located water can be significantly slower than free water but not slow enough to be counted as osmotically inactive. In a crowded environment, much of intracellular water is located in sufficient vicinity of proteins to be affected by them; the experimental numbers vary widely (Mulkern et al. 1989; Luby-Phelps 1999; Cameron et al. 2011; Cameron and Fullerton 2014), which may reflect not only the difficulties in studying intracellular water but variability in conditions. Our recent measurement using TIE/TTD (see below) showed that osmotically inactive water in HeLa cells constitutes about 40% of protein mass, which is similar to the published values (Model 2014).

The amount of bound water depends on protein conformation. Normal protein activity, such as oxygenation of hemoglobin, is accompanied by binding or unbinding of multiple water molecules (Ball 2008). An increase in exposed protein surface area promotes the association of loosely bound water molecules (Vorob'ev, 2003). In contrast, tightly bound water is lost during protein denaturation or oligomerization (Fullerton et al., 2006; Hishida et al., 2023; Cameron et al., 2006); the familiar dehydration of boiled meat illustrates this latter effect (You et al., 2025). It is perhaps not surprising then that the state of intracellular water can be labile as well: it has been reported, for example, that the amount of osmotically inactive water changes significantly during the cell cycle (Dupre and Hempling 1978).

MC and cell volume. MC is inversely proportional to cell volume, especially in nonproliferating cells, and cell volume regulation has been studied in much detail. The major fact about cell volume is that many cells tend to restore it after an abrupt exposure to a different osmolarity; osmotic shrinkage in a hypertonic solution is often followed by full or partial volume restoration, known as the regulatory volume increase (RVI), and osmotic swelling in a hypotonic solution is followed by the regulatory volume decrease (RVD). These changes in volume are naturally matched by the corresponding changes in MC. But there are also important differences between cell volume regulation and MC regulation; in particular, the word “nonproliferating” is important.

Volume regulation experiments are often carried out on cultured cells that grow and divide, and therefore, cell volume measurements must be limited to a fraction of an hour; as to what happens afterwards, that has traditionally been a subject of a different discipline, cell size regulation, which appears to be little related to MC. Leaving aside a major question of how the *average* cell sizes are maintained in a population, we only point out that prolonged culturing of cells at different osmolarities affects their average sizes in a manner opposite to the immediate change: more concentrated solutions produce larger cells, and vice versa (Ozturk and Palsson 1991; Zhang et al. 2010; Liu et al. 2019).

Clearly, cell volume can hardly be a target of homeostatic control over a long time, but the intrinsic parameter of MC can. MC slowly returns to its original value of around 0.2 g/ml within a wide range of external osmolarities, even in cells that do not exhibit rapid regulatory responses. This has been observed in HeLa and 3T3 cells (Mudrak et al. 2018; Hollembeak and Model 2021), mouse embryonic fibroblasts (Unpublished), and bacteria (Liu et al. 2019). The exposure to a cationic ionophore gramicidin causes at least an hour-long swelling (which is fully anticipated), but this swelling turns into compression and an increase in MC by 24 h (Hollembeak and Model 2021).

Whether or not MC regulation should be viewed as a more generalized form of volume regulation depends on the mechanism. Acute membrane or cytoskeleton stretch during swelling can directly activate ion channels responsible for RVD (Jentsch 2016) independently of MC. In other cases, MC appears to act as the primary sensor of swelling or shrinkage (see below), and then cell volume takes a secondary role.

Detection and Quantification of Macromolecular Crowding

Macromolecular crowding and, in particular, its nonspecific volume exclusion effects manifest themselves in more than one way: they slow down diffusion, compress proteins into more compact shapes, and promote aggregation and formation of phase inclusions (Rivas and Minton 2016). Accordingly, MC can be detected and quantified using various approaches.

Quantification by diffusion. One common method to quantify MC is based on the mobility of a fluorescent tracer. Diffusion in the cytoplasm is a complex process, which depends not only on the concentration of dissolved proteins (through viscosity) but also on the cytoskeleton, the presence of inclusions, and a large-scale, metabolism-dependent agitation of the cytoplasm (Luby-Phelps 1999; Parry et al. 2014; Munder et al. 2016). Thus, it has been found that crowding, as quantified by dry weight per volume, is only weakly correlated to the diffusion of GFP (Konopka et al. 2009; Cayley et al. 1991). Therefore, the interpretation of diffusion measurements in terms of crowding may imply a broader definition of crowding, including the "crowding" of the cytoplasm by ribosomes (Marini et al. 2020; Delarue et al. 2018) or by other molecular aggregates (Munder et al. 2016). Likewise, rotational diffusion of EGFP probed by fluorescence anisotropy was only loosely related to changes in cell volume (Biswas et al. 2022).

Quantification by FRET. The other group of experimental approaches targets the other essential MC effect: the transition of proteins from loose to more compact molecular shapes. A number of FRET probes have been synthesized, whose fluorescence reflects conformational changes (e.g., Gnutt et al. 2015; Boersma et al. 2015; Miyagi et al. 2021; Lecinski et al. 2021). Their downside is that consistent quantification of MC by FRET is difficult and may depend on the probe and the cell type. Occasionally, FRET probes produced paradoxical results, such as an apparent increase in crowding despite a large increase in cell volume (Pittas et al. 2023).

Quantification by buoyant density. Direct measurements of intracellular water (or the reciprocal quantity of dry mass) combined with an independent estimate of the average cell volume is an old approach that is still used occasionally (Joyner et al. 2016). Buoyant cell density measurements have been traditionally used on cell populations, but nowadays can be extended to single cells using microfluidic techniques (Son et al. 2015). Since proteins are heavier than water, buoyant density can be converted to mass/volume concentration of proteins. One caveat of this approach is that the density of nucleic acids is higher than that of proteins (Neurohr and Amon 2020), and therefore, organisms with an abundance of RNA may appear having deceptively high MC.

Quantification by refractive index. Optical methods to measure MC in live cells are based on the fact that dissolved organic matter increases the refractive index n over that of the solvent n_0 according to

$$\Delta n = n - n_0 \approx 0.185 \cdot MC(g/ml) \tag{14.2}$$

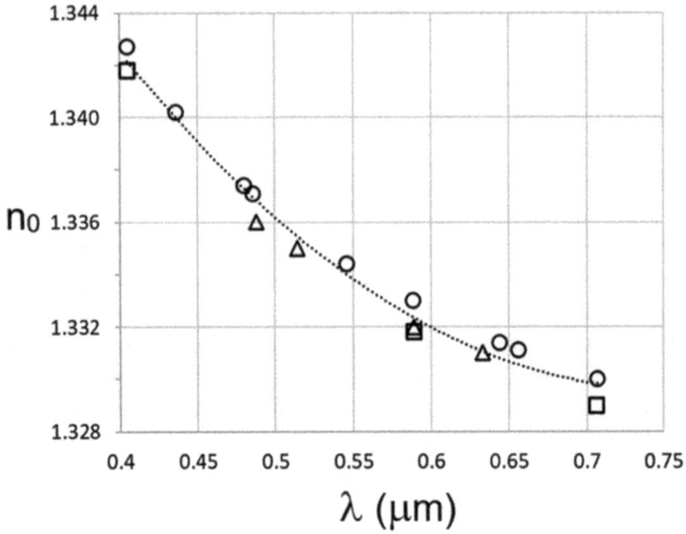

Fig. 14.1 Data for the refractive index of water at 25-30 °C and 1 atm compiled from three sources: Daimon and Masumura 2007 (circles), Harvey et al. 1998 (triangles), and Thormählen et al. 1985 (squares). The overall best-fit quadratic curve has the form $n_0(\lambda) = 0.1053\lambda^2 – 0.158 \lambda + 1.3888$

where 0.185 is the average value applicable to proteins, amino acids, and nucleic acids (Theisen et al. 2000). Therefore, for a typical cell with 0.2 g/ml dry mass, Δn can be expected to be on the order of $37 \cdot 10^{-3}$. Inorganic salts add little to refractive index, so that 0.15 M NaCl would increase it only by $1.5 \cdot 10^{-3}$ (Porus et al. 2011; Schmidt et al. 2014); moreover, since intracellular and extracellular concentrations of salts are usually comparable, the difference between the cell and external solution would be even less affected. On the other hand, sucrose or mannitol, which are often used to create hypertonicity, may appreciably increase the refractive index: a 0.15 M extracellular sugar will raise Δn by $7.5 \cdot 10^{-3}$. The other factor to consider is that the refractive index of water is a strong function of wavelength (Fig. 14.1), so the expected difference of ~$37 \cdot 10^{-3}$ due to MC is comparable to the $10 \cdot 10^{-3}$ difference in n_0 between 400 nm and 600 nm. Moreover, some Internet sources, such as http://www.philiplaven.com/p20.html, give numbers quite different from the data used in Fig. 14.1, and care is needed in interpreting the absolute values of n_0.

With all precautions taken, Eq. (2) can be used to evaluate MC, at least in mammalian cells. The reason for the latter qualification is that most of organic mass of mammalian cells is indeed associated with macromolecules (Model and Petruccelli 2018), but marine invertebrates synthesize large amounts of small organic osmolytes; their contribution to refractive index is similar to that of macromolecules, but they do not generate significant volume exclusion effects. In those cells, FRET-based techniques may be more appropriate. Table 14.1 gives the approximate relationships between MC and other related parameters.

The optical method we use in our lab is based on a combination of transmission-through-dye (TTD) microscopy to measure cell volume and a variant of quantitative phase microscopy, known as the transport-of-intensity equation (TIE) imaging, for protein mass (Zuo et al. 2020; Wittkopp et al. 2020). The combined TTD-TIE method works as follows. A nontoxic extracellular dye Acid Blue 9 is added to the sample medium; since live cells exclude the dye, they appear brighter than the background, when imaged in transmission at the wavelength of maximum dye absorption (630 nm). That makes it possible to convert image contrast to cell volume. After that, the illumination is switched to 485 nm, where the dye is transparent, and two bright-field images separated by a small vertical distance are collected. They are processed by the TIE equation, which relates the spatial distribution of intensity to the optical path; the knowledge of both the optical path and cell volume gives the refractive index (Clements et al. 2025).

Controlling MC in the experiment. In addition to the question of measurement, another methodological issue must be recognized. In order to study biological effects of crowding, one should be able to control it in a specific manner. But how can it be done? External osmolarity provides the easiest way to change the crowding, at least temporarily, but it is not specific enough. Osmotic swelling or shrinkage also affect intracellular ion concentrations, as well as the tension and organization of the membrane and the cytoskeleton; each of these factors may act as a primary responder that sets in motion secondary processes (Lüscher et al. 2020). For example, hypotonic shock followed by RVD results in chloride depletion, which causes subsequent vacuolation (Zook et al. 2024). In fact, the multitude of osmolarity-induced effects may have been the reason for the skeptical attitude to MC that prevailed for many years. And since MC cannot be inhibited or genetically modified, the best approach to studying its specific roles is to eliminate the alternatives (Summers et al. 1997; Rana et al. 2020).

The Determinants of MC

If we accept the definition of MC as the mass/volume concentration of proteins, we can ask what cellular processes are responsible for controlling it. Since in the absence of protein degradation or synthesis, MC is directly related to cell volume, we can start by reviewing some results from the theory of cell volume regulation. The main premises of that theory are as follows (Armstrong 2003; Fraser and Huang 2007; Kay 2017):

- CV is approximated by the volume occupied by intracellular water.
- The amount of water is determined by the requirement of isoosmolarity with the external solution (the small turgor in mammalian cells can be neglected). All osmotically active substances are assumed to behave in an ideal manner, i.e., with the activity coefficient of one. The latter is hardly correct, but at the present

state of knowledge, it is the only way to obtain results without introducing hard-to-verify parameters.

- Intracellular osmolarity is created by inorganic ions (mainly by K^+, Na^+, and Cl^-), whose transport is regulated by various channels and transporters, and by membrane-impermeant organic osmolytes, the majority of which are small molecules, such as amino acids, sugars, urea, and others (Häussinger 1996).
- The relative amounts of ions and organic osmolytes are additionally constrained by the requirement of electroneutrality (or near-neutrality, if one considers the exceedingly small imbalance responsible for the membrane potential). The large excess of inorganic cations over anions means that the missing negative charge must be provided by organic anions - proteins and especially nucleic acids (Model et al. 2023). Nevertheless, it is customary to combine all the membrane-impermeant organic compounds into a single pool and characterize them by the total molar amount N and the average charge per molecule z.
- The starting point for computing ionic and osmotic behavior can be a set of two equations, expressing the conditions of electroneutrality and isoosmolarity with the external medium (with osmolarity E):

$$\begin{aligned} \left[Na^+\right]+\left[K^+\right]+\left[Cl^-\right]+\left[N^{z-}\right]&=E \\ \left[Na^+\right]+\left[K^+\right]+z\left[N^{z-}\right]-\left[Cl^-\right]&=0 \end{aligned} \quad (14.3)$$

To that can be added expressions for ion fluxes, the effect of the Na^+-K^+ pump, and the membrane potential.

Below are some well-established results of the cell volume theory that are relevant to MC.

- Disruption of membrane integrity causes cell swelling and, therefore, a decrease in MC.
- Inhibition of the Na^+-K^+ pump is also expected to produce swelling and a decrease in MC. In reality, this process may be developing slowly in some cell types (Orlov et al. 2013) and, according to our observations, the treatment of cells with the Na^+-K^+ pump inhibitor ouabain can even cause dehydration.
- As the cell grows and synthesizes new protein, while keeping the same protein composition and the same set of transporters, MC remains constant.
- By solving Eqs. (14.3), cell volume at a steady state can be expressed through intracellular chloride concentrations:

$$V=\frac{(1-z)N^{z-}}{E-2\left[Cl^-\right]} \quad (14.4)$$

where N^{z-} without square brackets denotes the total molar amount of organic osmolytes. Note that the ratio N^{z-}/V is not equivalent to the physical MC because the MC effects are determined by the mass/volume concentration of large molecules, while

N^{z-}/V primarily reflects the contribution of small organic molecules to osmolarity. The sign of z is usually negative because the isoelectric point of nucleic acids and of most proteins is in the acidic range. Therefore, intracellular acidification will reduce the quantity (1—z) and, if $[Cl^-]$ remains constant, may result in shrinkage and an increased MC (Fraser and Huang 2007).

- It also follows from Eq. (14.4) that chloride is a major determinant of V; the reason for its special role is that, within the limits of the given model, it is the only membrane-permeant anion. By contrast, potassium loss (down its normally outward electrochemical gradient) may be cancelled by sodium gain (down its normally inward electrochemical gradient).
- An additional factor to consider is the effect of variability in N. To get a better insight into it, we can divide the entire organic pool into two groups: large molecules ("proteins") and small molecules ("amino acids"). We can approximate their roles by stating that "amino acids", due to their numbers, contribute mostly to osmolarity, while "proteins", due to their mass, contribute mostly to electric charge. This leads to the equation:

$$MC = \frac{N_p \mu_p \left(E - [I]\right)}{N_{aa}} \tag{14.5}$$

Here μ_p is the average molecular weight of proteins, [I] is the intracellular osmolarity due to inorganic ions, and N with subscripts stand for the numbers of corresponding molecules (Model et al. 2021). Suppose that some proteins have been hydrolyzed into amino acids, which remain inside the cell. That would reduce the numerator, increase the denominator, and thus reduce the MC; although this increase may be partly compensated by the removal of some inorganic ions (to maintain osmolarity), the expression in the parentheses cannot exceed E, while N_{aa} can increase significantly. Thus, protein degradation and accumulation of amino acids is expected to cause water accumulation, and this process is believed to be responsible for the formation of large intracellular vacuoles (Hu et al. 2022). Conversely, anabolic reactions and protein synthesis are expected to increase MC. This conclusion, however, is based on an unrealistic assumption that the cell is insulated from the environment. As shown by Häussinger and others, inhibitors of proteolysis produce cell swelling through an increased uptake of amino acids and accumulation of ions (Häussinger 1996). Although cell swelling also promotes protein synthesis (or rather, shrinkage inhibits it), changes in the amount of protein may not be able to compensate changes in the amount of water. It is not clear if MC has been measured under these conditions.

To conclude this section, it must be noted that the standard model, while proved correct in many of its predictions, is likely incomplete, as it assumes an ideal behavior of cellular proteins, ions, and water, and ignores complex interactions between them, including the gel-like properties of the cytosol. Osmotic swelling and shrinkage in the absence of an intact membrane (Fels et al. 2009; Platonova et al. 2014) clearly does not fit into the existing theory. This inherent osmosensitivity of the cytosol would be expected to modify cell responses mediated by membrane channels and transporters, but clear understanding of these complex interactions has yet to be achieved.

MC Homeostasis in a Healthy Cell

Despite a large variability in cell sizes, MC levels are very similar in most cell types and are close to 0.2 g/ml (Model 2021; Model et al. 2021), which is equivalent to 15% by volume (Table 14.1). The known exceptions are red blood cells with MC ~ 0.33 g/ml and, according to one report, chondrocytes with variable but generally low density (Cooper et al. 2013). The numbers for bacteria vary among authors (Oldewurtel et al. 2021; Bakken and Olsen 1983; https://bionumbers.hms.harvard.edu/bionumber.aspx?id=108263) and may depend on the bacteria type or their physiological state, but our measurements on E. coli produced values that are still close to 0.2 g/ml (Model et al. 2024).

Not only is MC consistent across cell types but it also appears to be under strict homeostatic control within individual cells. As already mentioned, its preservation during growth and protein synthesis occurs naturally , but cells also tend to return to their resting densities following various kinds of perturbations (Summers et al. 1997; Hollembeak and Model 2021; Mudrak et al. 2018). When perturbation comes in the form of anisotonicity, MC restoration can be related to cell volume regulation through RVD and RVI. These classical responses are mediated by ion channels (Okada et al. 2001; Pasantes-Morales 2016) and often bring cells to their initial volumes within minutes or tens of minutes of a step change in osmolarity. But how does the cell know its volume? The nature of volume sensor has fascinated physiologists for many years (Minton et al. 1992; Burg 2000; Mongin and Orlov 2001; Hoffmann et al. 2009; Orlov et al. 2018; Schliess et al. 2007). MC seems well suited for this role (Rivas and Minton 2016; Van Den Berg et al. 2017; Neurohr and Amon 2020; Holt and Delarue 2023), in particular due to a highly nonlinear relationship between the effective and nominal concentrations of crowders, ensuring high sensitivity to small deviations. The first experimental evidence of a link between MC and volume regulation has been provided by two NIH scientists, G. Colclasure and J. Parker, who showed that MC controls the channels responsible for volume regulation in erythrocytes (Colclasure and Parker 1991; Colclasure and Parker 1992). Another study from the 1990s demonstrated the effect of MC on calcium currents and volume regulation in perfused barnacle muscle cells (Summers et al. 1997). Further progress had to wait until 2022, when a large group of scientists lead by R. Subramanyan mapped out the sequence of events leading from hypertonic shrinkage to RVI. They showed that an increase in crowding drives the assembly of WNK1, SPAK, and OSR1 kinases into liquid phase condensates. There, they become activated and phosphorylate ion transporters, turning the NKCC active and KC inactive. As a result, extra ions flow into the cell, bringing more water inside and restoring cell volume (Boyd-Shiwarski et al. 2022). Whether the WNK-mediated mechanism of RVI participates in the long-term maintenance of MC is unclear because the condensates disassemble after a few minutes of hypotonic stress (Boyd-Shiwarski et al. 2022; Subramanya and Boyd-Shiwarski 2023).

On a longer time scale, the protein kinase complex mTORC1 appears to be well-positioned to control intracellular MC, as it is the main regulator of synthesis and degradation. Its activation facilitates synthetic reactions (which are supposed to increase MC, according to Eq. (14.5)), and its inhibition facilitates proteolysis (Zhao et al. 2015). Indeed, it has been shown that hypertonicity suppresses mTORC1 (Su and Dai 2017), which suggests a stabilizing negative feedback: Hypertonicity → Loss of water → Increase in MC → Inhibition of mTORC1 → Proteolysis → Decrease in N_p/N_{aa} → Restoration of MC. However, this hypothesis is called into question by experiments in which mTORC1 was inhibited with rapamycin. This treatment only increased the diffusion of large tracer particles by reducing the number of obstacles - ribosomes; at the same time, the diffusion of smaller proteins, which is expected to better reflect the concentration of proteins in solution, was unaffected (Delarue et al. 2018).

Self-regulation of MC would be another attractive hypothesis, if a plausible mechanism could be found. Could the complexity of intracellular water hold a key to this? An interesting mechanism of instantaneous stabilization of water potential by exposed protein surfaces was proposed in a recent extensive study by the Derivery group (Watson et al. 2023). This, however, does not seem to be immediately applicable to MC regulation.

Local MC Homeostasis in Necrotic Death and Vacuolation

Necrotic cell death results from irreparable damage to the cell. Its morphological features include cell swelling, formation of watery membrane protrusions (blebs), disruption of organelles, and ultimately, cell lysis (Yan et al. 2020). Necrosis has been traditionally viewed as a purely passive response to severe insult; however, more recent research has found that, necrosis, like apoptosis, can be subject to some regulation (Barros et al. 2001; Proskuryakov et al. 2003), producing special forms of cell death known as necroptosis and pyroptosis (Linkermann and Green 2014; Walle and Lamkanfi 2016). Of the listed signs of necrosis, swelling is perhaps the most essential and is sometimes referred to as necrotic volume increase (Barros et al. 2001).

Our study of free-radical-induced necrotic death (Rana et al. 2017) adds an additional detail to this picture: water accumulation is only limited to membrane blebs, while the rest of the cell remains at normal density. This appears to violate the condition of osmotic balance between the cytoplasm and a membraneless bleb: in the absence of any barrier, ions are expected to distribute evenly, but what compensates for the lower oncotic pressure in the bleb? At least part of the answer lies in the extensive aggregation of cytoplasmic proteins caused by free radicals, which reduces their effective molar concentration. It is unknown at present if such compartmentalization of water is universal for various necrotic stimuli, but the observed

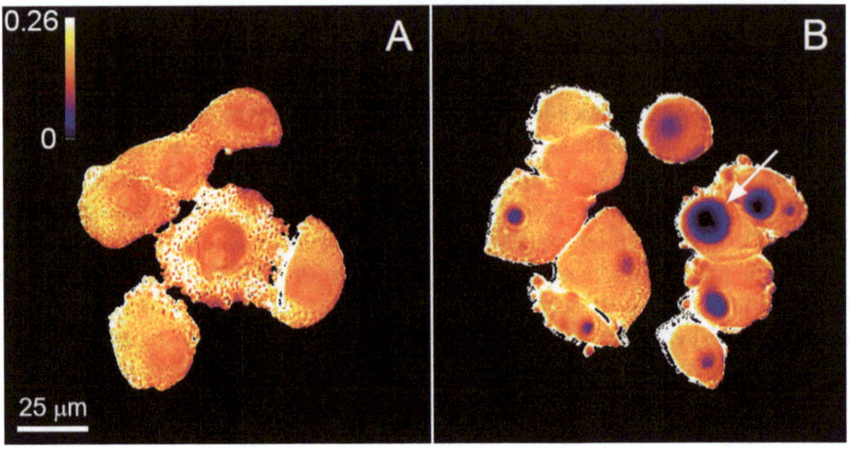

Fig. 14.2 Macromolecular crowding in HeLa cells: untreated (**A**) and exposed to 50% hypotonicity for 24 hours (**B**). Low-protein vacuoles have developed in hypotonically stressed cells; however, the density of the cytoplasm outside the vacuoles (one of them indicated by an arrow) is normal. The color-coded vertical bar shows MC in g/ml. (The figure is modified from Hollembeak and Model 2021)

stability of MC outside the bleb, even under extreme metabolic disruption, is noteworthy.

Experiments in which cells were subjected to severe and prolonged hypotonic stress (at 25–50% of normal osmolarity) provide another example of the remarkable stability of cytoplasmic MC. After the initial volume recovery, cells start developing vacuoles that reach up to several microns in size (Iwasa et al. 2001; the optical path and cell Zook et al, 2024. But, quite interestingly, the parts of the cell outside the vacuoles still have the protein concentration close to the healthy 0.2 g/ml (Fig. 14.2), much like in necrotic blebbing. It seems as if the cytosol actively rejects extra water by segregating it into blebs or vacuoles.

The "insistence" of cells on returning to normal MC makes it difficult to investigate the specific consequences of cytoplasmic dilution. But there are at least two ways to keep cells uniformly and continuously swollen. One is to apply hypotonic solution, while inhibiting volume-regulated anion channels (VRAC) that mediate RVD; the other is to disrupt ion balance with ionophores (Rana et al. 2019). Such cells remain viable for many hours, and can provide a promising model for studying the effects of reduced crowding.

MC in Apoptosis

Swelling upon cell damage, as is observed in necrosis, is a natural consequence of the disruption of active ion extrusion driven by the Na^+-K^+ pump. Therefore, when shrunken dead cells were observed for the first time (Kerr 1971), it became clear that a totally new mechanism was at play.

Since then, hundreds of thousands of papers have been published about apoptosis. Scientists have established that it is a type of genetically programmed cell death characterized by shrinkage and overall abnormal morphology, mitochondrial depolarization, activation of specific apoptotic proteases (caspases), inversion of phosphatidylserine in the plasma membrane, condensation of chromatin, membrane blebbing (without water accumulation, unlike necrotic blebs), and cell fragmentation (Elmore 2007). When apoptotic cells develop in the body (whose number is estimated at 200 billion a day (Jiang et al. 2017)), apoptotic fragments are phagocytosed by other cells without producing inflammation; by contrast, necrosis is proinflammatory. The same adverse stimulus can produce either apoptosis or necrosis, depending on the strength of the stimulus: when intracellular ATP levels are high, apoptosis results, and if not, the cell is destined for necrosis, swelling, and membrane rupture (Leist et al. 1997; Tsujimoto 1997). In other words, when a cell experiences slight damage, it activates the apoptotic pathway; when the resources for apoptosis are lacking, necrosis results.

Of the multiple signs of apoptosis, shrinkage and morphological changes are often the earliest and most reliable ones. Although shrinkage is not sufficiently specific, it clearly separates apoptosis from necrosis when taken in combination with other markers, and it has been given a special name, apoptotic volume decrease (AVD) (Bortner and Cidlowski 2020). Exceptions are known, and in some cases, shrinkage can be preceded by temporary swelling due to increased sodium permeability (Hortelano et al. 2002; Yurinskaya et al. 2005; Kasim et al. 2013; Gibbons et al. 2016; Model et al. 2018), but that does not change the overall picture. AVD is often, but not always, equivalent to an increase in MC: fragmentation (which occurs at a later apoptotic stage) also produces a decrease in cell size without necessarily a change in hydration.

Apoptotic dehydration is believed to result from an increase in membrane permeability for potassium followed by water efflux (Okada et al. 2001) (this hypothesis will be further discussed in section "Hypothesis on the mechanism of dehydration"). Blocking potassium or chloride efflux prevents not only AVD but all other apoptotic reactions, hinting at an essential role of AVD in apoptosis. Conversely, forced shrinkage without any additional stimulus drives cells into apoptosis (Bortner and Cidlowski 1998). This gave rise to the idea that shrinkage causes apoptosis via potassium depletion (Yu 2003); as for macromolecular crowding, it was not viewed as a sound hypothesis twenty years ago.

We revisited the question by controlling intracellular ion composition with ionophores and testing additionally for the cytoskeletal involvement (Rana et al. 2020). The results clearly showed that persistent osmotic compression causes apoptosis regardless of intracellular sodium or potassium concentrations or the state of actin cytoskeleton; the logical conclusion from these experiments is that an increase in MC must act as a direct apoptotic stimulus. At the same time, strong potassium dependence of caspase activation was observed as well, but only in swollen cells with reduced MC. It remains to be determined which protein in the apoptotic signaling chain becomes directly activated by persistent elevation of MC. Several reactions necessary for the activation of proteolysis involve protein binding or oligomerization (Yang et al. 1998; Chang et al. 2003; Rana et al. 2020), which is expected to be enhanced at increased MC. Some of these reactions could possibly link MC to apoptosis.

The other interesting observation concerns the apoptosis signal-regulating kinase 3 (ASK3). It mediates apoptosis induced by oxidative stress or by the activation of apoptotic receptors (Kaji et al. 2010); at the same time, ASK3 responds to osmotic stress. However, its response to osmolarity takes the form opposite to what could be expected: hypotonicity activates ASK3, and hypertonicity inactivates it by directing it into biomolecular condensates (Watanabe et al. 2021). Therefore, this observation is not consistent with the role of ASK3 in hypertonicity-induced apoptosis.

Although apoptosis used to be regarded as a feature of multicellular organisms, it is also well documented in yeast, where it exhibits many similar markers, including shrinkage (Granot et al. 2003; Reiter et al. 2005). As with mammalian cells, higher concentrations of apoptotic agents applied to yeast produce necrotic swelling and rupture (Reiter et al. 2005; Eisenberg et al. 2010).

MC in Cell Stress

The results for necrosis vs. apoptosis suggest a pattern: moderate impacts (such as low doses of a noxious agent) produce shrinkage and an increase in MC, and stronger impacts produce swelling and a decrease in MC (Fig. 14.3). Dehydration at intermediate impacts may mark an early stage of a cell death process, such as apoptosis or autophagy, or may represent an independent response. Here we will try to assess the generality of this trend.

A frequently used method to experimentally produce stress is by energy deprivation; this can be achieved by interfering with ATP synthesis or simply by removing nutrients from the medium. Generally, a loss of ATP that is profound enough to disable ion pumps and dissipate ion gradients is expected to cause water accumulation in any organism lacking strong walls. This has, indeed, been observed in bacteria (Linker 1987) and yeast (Eisenberg et al. 2010; Naito et al. 2005). In terms of crowding (measured with a FRET probe), the uncoupling of mitochondria by protonophore FCCP produced swelling in *E. coli*, which developed as rapidly as within 2 minutes of the addition of the drug (Liu et al. 2019). A publication from the Boersma group (Pittas et al. 2023) reported a paradoxical result on *E. coli* treated with penicillin: they observed an increase in the FRET signal from a crowding sensor despite a several-fold increase in volume. It seems that the volume increase should be a more reliable sign that the extent of damage has exceeded the ability of cells to control their internal milieu. Even more unexpected are the results of Lewis and coauthors (Lewis et al. 2014) that show an increase in the density of dead bacteria. Thus, barring some unusual results with bacteria and heterogeneity of water distribution (Rana et al. 2017), the region *c* of the curve depicted in Fig. 14.3 is the standard sign of necrosis.

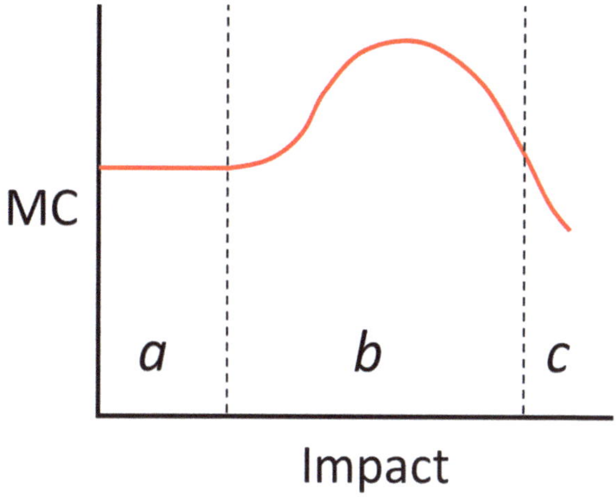

Fig. 14.3 Schematic (and of hypothetical generality) dependence of MC on the strength of a noxious stimulus (which can be a drug dose, for example)

It is much less obvious whether the curve depicted in Fig. 14.3 must rise in the region *b*. When certain gram-positive bacteria experience a lack of nutrients, they reduce themselves to highly dehydrated spores. The low water content of spores (30–50%) is essential for their survival upon exposure to UV, high temperature, complete desiccation, and other harsh conditions (Moeller et al. 2009; Beaman et al. 1982; Setlow 2016; Huang and Hull 2017; de Vries 2006). Not only bacteria, but some protists and fungi can enter dormancy and form dehydrated spores that are resistant to adverse environments.

Sporulation certainly represents an extreme case of dehydration, but there are indications that limited water loss (and hence an increase in MC) may indeed be a common response to stress. A loss of water upon starvation has been observed in E. coli (Shi et al. 2021). Yeast also respond to energy deprivation by shrinkage with preservation of dry mass (i.e., an increase in MC, by our definition) (Joyner et al. 2016). Consistently with this observation, the mobility of intracellular fluorescent tracers was reduced. A reduced diffusion of lipid droplets and fragmented mitochondria in starved yeast were reported by Heimlicher and coauthors (Heimlicher et al. 2019), although the diffusion of proteins was unchanged. In another work (Xie et al. 2023), the mobility of 50–100 nm nanoparticles was increased in yeast following glucose removal, whereas mRNA granules became relatively immobilized. As noted earlier, the mobility of tracers carries information of its own and does not always correlate with protein concentration: an increase in the number of ribosomes in starved yeast (Marini et al. 2020) or the formation of protein precipitates caused by acidification (Munder et al. 2016) are likely to affect translational mobility. In another work, acidification resulted both in an increase in MC and a decrease in the mobility of fluorescent particles (Abuhattum et al. 2018).

As we turn to mammalian cells, a question arises: how to tell apoptosis-unrelated shrinkage from AVD. Indeed, AVD often precedes the appearance of other apoptotic markers (Okada et al. 2001; Bortner and Cidlowski 1998). The question may be partly semantic—whether to regard AVD as an early stage of apoptosis or as a

Table 14.2 The effect of gramicidin on the relative changes in MC in HeLa cells. (Reproduced from Hollembeak and Model 2021)

	MC (Relative)		
Gramicidin concentration (μM)	25 min	1 h	24 h
0.1	0.93	0.945	1.085 ± 0.054 (3)
0.5	0.87	0.99	1.121 ± 0.122 (3)
2.5	0.71	0.60	1.234 ± 0.135 (3)

separate process. But we did not observe any signs of apoptosis in HeLa cells treated with sublethal doses of the Na^+-K^+ ionophore gramicidin. A short-term exposure produced water accumulation and a decrease in MC (as would be expected when sodium permeability is increased), but by 24 hours, swelling gave way to shrinkage (Hollembeak and Model 2021). Moreover, the extent of shrinkage correlated with gramicidin concentration (Table 14.2). In these experiments, the evolution of MC was recorded by TIE/TTD.

The results of long-term (24–48 hours) incubation in a strongly hypotonic medium provided another striking confirmation of this trend (Hollembeak and Model 2021). Prolonged exposure of HeLa cells to extremely dilute media caused significant dehydration and an increase in MC (Fig. 14.4), even though the immediate response to hypotonicity is, of course, the opposite. All the cells that remained attached by the end of incubation did not stain for caspase-3 (a specific apoptotic marker), and a positive caspase-3 staining was only observed in detached cells. Neither dehydration nor apoptosis were observed in hypertonic media. More recently, we have observed significant dehydration in 3T3 cells kept in low concentrations of ouabain, contrary to its expected effect (In preparation).

These very limited data agree with the notion that dehydration and an increase in MC may be a ubiquitous response to stress. On the organelle level, the impairment of HeLa mitochondria with chloramphenicol increases the matrix viscosity, which may also be a sign of an increased MC (Bulthuis et al. 2023). Clearly, much more work is needed to establish firm facts in this barely explored area.

A separate question is whether water loss that presumably occurs at intermediate stages of cell stress brings any benefits to the cell. This seems possible because the molecular chaperone Hsp90, which plays a major role in protein stabilization and homeostasis, is strongly activated by macromolecular crowding in vitro (Halpin et al. 2016). The expression of Hsp27, Hsp40, and HSP70 increased in several organs of the African clawed frog upon water loss (Luu et al. 2018). HSF1, the DNA-binding protein which regulates chaperone expression, requires trimerization for its activation (Fulda et al. 2010), which can be enhanced by crowding. Although its assembly into nuclear stress bodies (foci) is inversely correlated with its activity (Gaglia et al. 2020), the proteotoxic treatments used in that study were different from dehydration. The effect of crowding on heat shock proteins has been reviewed (Roman et al. 2017).

The protective action of MC is suggested by the results of (Maiti et al. 2023) who subjected cells to 40 °C for 4 days and measured the mobility of the green

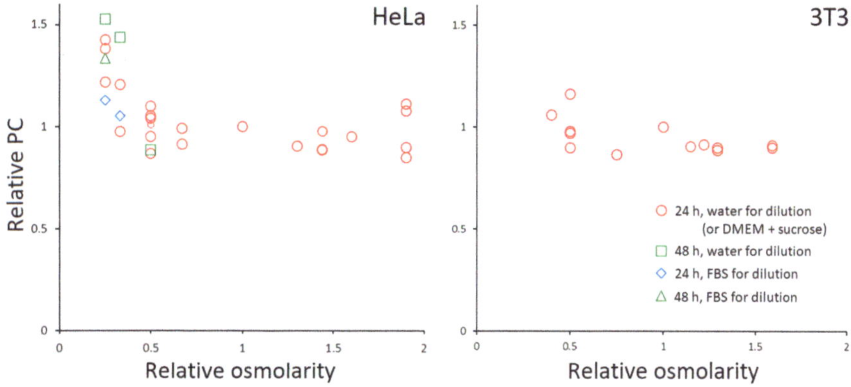

Fig. 14.4 The dependence of MC (denoted here as PC) on the osmolarity of the medium. Each point represents the average for a separate experiment, in which 5–15 cells or cell groups were analyzed. MC was normalized to that in control cells kept in 100% DMEM with 10% FBS. The coefficient of variation for a single experiment was typically between 0.05 and 0.1. (Reproduced from Hollembeak and Model 2021 with permission)

fluorescent protein by fluorescence recovery after photobleaching (FRAP). The mobility was unchanged in wild type cells but increased in cells lacking the heat shock protein Hsp90. The authors conclude that the protective response mediated by Hsp90 is responsible for the maintenance of stable MC during thermal stress; in the absence of Hsp90, the cytoplasm becomes diluted and MC decreases.

MC in Senescence

The amount of intracellular water in various body organs and tissues diminishes with age (Lustyik 1986; Lustyik and Nagy 1988; Lustyik and Nagy 1985; von Zglinicki and Bimmler 1987; von Zglinicki 1988; Steen 1988; Minton 2020; Lee 2024). Incidentally, the well-documented dehydration of aged red blood cells may not belong to the same category, because this effect has a more specific explanation: the accumulation of calcium and acceleration of calcium-dependent potassium efflux (the Gardos effect) (Klei et al. 2020; Lew 2022). No age-related difference in buoyant density was noted in platelets (Savage et al. 1986), but it unexpectedly decreased in senescent bone marrow cells (Xu et al. 2020) and in astrocytes (Unsworth and Fleming 1980). However, in this case, the buoyant density may not be directly related to MC because senescent cells have both vacuoles and smaller nuclei. Water content in senescent human mesenchymal endometrium cells has not been measured directly, but a combination of stable sodium and reduced potassium per gram protein (Shatrova et al. 2022) suggests dehydration.

Although aging is often thought of as a feature of multicellular organisms, a similar process has been extensively studied in yeast and even in bacteria. The overall MC in aged yeast (i.e., the cells that failed to replicate) is not so easy to determine because such cells develop large vacuoles; but local MC has been measured with a FRET crowding sensor. It was found little changed in the cytosol but significantly increased in mitochondria (Mouton et al. 2020). Interestingly, the data on rat myocytes and hepatocytes are in agreement with this result in yeast: mitochondria were the only organelles that became dehydrated in aged animals (von Zglinicki 1988).

Whether MC plays an active role in senescence is unknown: the reduced nucleus/cytoplasm ratio is a factor with a more obvious and immediate impact (Neurohr et al. 2019). But long-term consequences on the organismal level are always harder to study, and it is quite possible that an increase in MC or a decrease in protein hydration compromises some important cellular functions.

Since a decrease in protein solubility is another hallmark of aging in diverse organisms, including *E. coli* (Minton 2020; Lindner and Demarez 2009; Maisonneuve et al. 2008), these effects may be related—potentially through the release of osmotically inactive water, as described in the next section. An interesting observation in this regard is that water loss in organs and tissues from aged individuals is limited to the tightly bound fraction (Kerch 2018).

Hypothesis on the Mechanism of Dehydration

The potassium-mediated mechanism of AVD (section "MC in apoptosis") is not without problems: "The variety of K^+ channels that have been implicated in apoptosis suggests that in principle any type of K^+ channel can support apoptosis, but it is not entirely clear how this can operate as a specific switch driving cells into regulated cell death" (Kunzelmann 2016). Therefore, other possibilities have been considered (Model 2014). One alternative model envisions cytoskeletal compression creating a slightly increased turgor that is being continuously relieved by the outflow of water; potassium and chloride follow water rather than the other way around (Rana and Model 2020). (A similar reverse-osmosis model had been earlier proposed for bacterial spores (Algie 1980)). But this mechanism, too, requires an organization that seems a little too purposeful. It would be tempting to think of a process by which dehydration occurs naturally, for instance, from changes in the structure of multiple proteins. Such changes could be anticipated during stress. When cells are placed in an unfavorable environment, they may experience free radical damage, acidification, a drop in ATP levels, or proteolysis; each of these factors is capable of causing protein unfolding and aggregation (Munder et al. 2016; Mouton et al. 2023). Protein hydration and the amount of cell water would be affected by this because misfolding and aggregation of proteins (though not of all proteins (Zielbauer et al. 2016)) are known to produce water loss. One example is inhibition of proteosomes with bortezomib, which causes both the accumulation of misfolded proteins and pronounced dehydration (Liu et al. 2023). This particular case, however, may look less convincing because bortezomib also causes apoptosis (Zhao et al. 2008; Luo

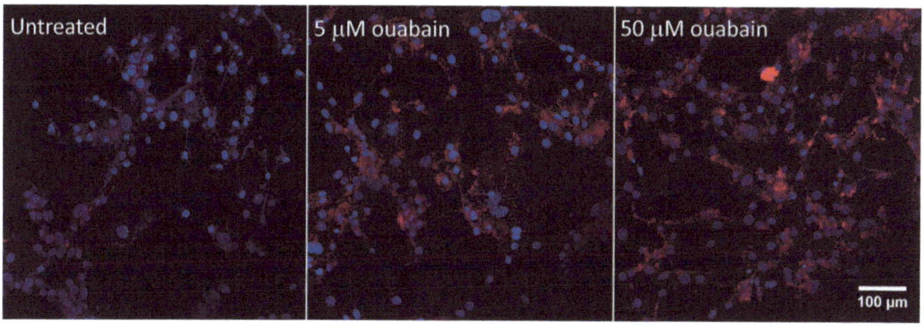

Fig. 14.5 3T3 cells were incubated with indicated concentrations of ouabain and stained with an aggregation probe Proteostat (red) and a DNA probe Hoechst (blue)

et al. 2010), and specific signaling to ion channels cannot be excluded. Perhaps the most unequivocal data come from food research that mostly deals with dead tissues. Water loss from boiled or otherwise thermally treated meat is a well-known phenomenon (even used occasionally in literature and movies), and it results from the disruption of native protein structure (Pang et al. 2021; Oillic et al. 2011; Huff-Lonergan and Lonergan 2005; You et al, 2025. The same mechanism can conceivably operate in living cells.

One difficulty with this hypothesis is the assumption of a large-scale protein damage. The early stages of dehydration often take place in cells that look morphologically almost normal and showing no signs of major dismantling of proteins. However, our recent results with 3T3 cells exposed to as little as 5 μM ouabain suggest that the effect can be quite significant. Despite the fact that such concentrations of ouabain caused little apoptosis, they produced both dehydration and a noticeable increase in staining by the aggregation probe PROTEOSTAT® (Fig. 14.5; the details will be published separately). There has been little investigation of protein damage in other cellular models during stress conditions.

The presumed mechanistic link between protein denaturation/aggregation and dehydration would be the release of tightly bound, osmotically inactive water (see section "Macromolecular crowding, hydration, and cell volume: the unavoidable triangle"). As the mobility and thermodynamic activity of water increases, it makes the cell hypotonic and comes out following the osmotic gradient, leaving behind a dehydrated and shrunken cell. Dehydration due to this effect is limited by the total amount of osmotically inactive water. Our preliminary measurements on HeLa cells are in agreement with this hypothesis: osmotically inactive water constitutes 12% of all water in healthy cells and near zero in amphotericin-treated cells Model 2014

The other possibility (admittedly more far-fetched) would be the formation of a gradually tightening network of crosslinked proteins, providing a reverse-osmosis mechanism similar to the one mentioned earlier (Rana and Model 2020) but not necessarily dependent on active cytoskeletal contraction. Such structures, nicknamed microtrabecular lattice, have been observed in the 1970s by Keith Porter and coworkers (Wolosewick and Porter 1979) but later have fallen out of favor and were ascribed to artefacts of sample preparation (Heuser 2002). Publications of remaining proponents of the microtrabecular theory, such as Hisatake Kondo (2015), seem to make little impact on the prevailing opinion. Nevertheless, the question is worth revisiting because the objects of previous studies were healthy cells, and it remains to be seen if the microstructure of stressed and dehydrated cells is different. There is convincing evidence that cells do possess properties indicative of crosslinking that still await explanation (Fels et al. 2009).

Acknowledgements I am grateful to Mikael Akke (Lund University), Emmanuel Derivery (University of Cambridge), Michael Feig (Michigan State University), Gary Fullerton, Allen Minton (NIH), Arohan Subramanya (University of Pittsburgh), and Priyanka Rana (Case Western University) for advice and discussion. Emily Zook suggested many improvements to the manuscript and has performed the experiments with Proteostat. The work was supported by the Research Council of Kent State University.

References

Abuhattum S, Kim K, Franzmann TM, Eßlinger A, Midtvedt D, Schlüßler R, Möllmert S, Kuan HS, Alberti S, Zaburdaev V, Guck J (2018) Intracellular mass density increase is accompanying but not sufficient for stiffening and growth arrest of yeast cells. Front Phys 6:131

Algie JE (1980) The heat resistance of bacterial spores due to their partial dehydration by reverse osmosis. Curr Microbiol 3:287–290

Armstrong CM (2003) The Na/K pump, Cl ion, and osmotic stabilization of cells. Proc Natl Acad Sci USA 100:6257–6262

Bagatolli LA, Mangiarotti A, Stock RP (2021) Cellular metabolism and colloids: Realistically linking physiology and biological physical chemistry. Prog Biophys Mol Bio 162:79–88

Bakken LR, Olsen RA (1983) Buoyant densities and dry-matter contents of microorganisms: conversion of a measured biovolume into biomass. Appl Environ Microbiol 45:1188–1195

Ball P (2008) Water as an active constituent in cell biology. Chem Rev 108:74–108

Barros LF, Hermosilla T, Castro J (2001) Necrotic volume increase and the early physiology of necrosis. Comp Biochem Physiol A Mol Integr Physiol 130:401–409

Beaman TC, Greenamyre JT, Corner TR, Pankratz HS, Gerhardt P (1982) Bacterial spore heat resistance correlated with water content, wet density, and protoplast/sporoplast volume ratio. J Bacteriol 150:870–877

Biswas P, Ray D, Jana S, Das J, Sinha B, Sinha DK (2022) Probing the role of macromolecular crowding in cell volume regulation using fluorescence anisotropy. BioRxiv:2022–2004

Boersma AJ, Liu B, Poolman B (2015) A sensor for quantification of macromolecular crowding in living cells. Biophys J 108:114a

Bortner CD, Cidlowski JA (1998) A necessary role for cell shrinkage in apoptosis. Biochem Pharmacol 56:1549–1559

Bortner CD, Cidlowski JA (2020) Ions, the movement of water and the apoptotic volume decrease. Front Cell Dev Biol 8:611211

Boyd-Shiwarski CR, Shiwarski DJ, Griffiths SE, Beacham RT, Norrell L, Morrison DE, Wang J, Mann J, Tennant W, Anderson EN, Franks J, Calderon M, Connolly KA, Cheema MU, Weaver CJ, Nkashama LJ, Weckerly CC, Querry KE, Pandey UB, Donnelly CJ, Sun D, Rodan AR, Subramanya AR (2022) WNK kinases sense molecular crowding and rescue cell volume via phase separation. Cell 185:4488–4506

Bulthuis EP, Dieteren CE, Bergmans J, Berkhout J, Wagenaars JA, van de Westerlo EM, Podhumljak E, Hink MA, Hesp LF, Rosa HS, Malik AN (2023) Stress-dependent macromolecular crowding in the mitochondrial matrix. EMBO J 42:e108533

Burg MB (2000) Macromolecular crowding as a cell volume sensor. Cell Physiol Biochem 10:251–256

Cameron I, Fullerton G (2014) Properties and size of multiple non-bulk water fractions on proteins and in cells. Water 4:76–90
Cameron IL, Kanal KM, Fullerton GD (2006) Role of protein conformation and aggregation in pumping water in and out of a cell. Cell Biol Int 30:78–85
Cameron IL, Lanctot AC, Fullerton GD (2011) The molecular stoichiometric hydration model (SHM) as applied to tendon/collagen, globular proteins and cells. Cell Biol Int 35:1205–1215
Cayley S, Lewis BA, Guttman HJ, Record MT Jr (1991) Characterization of the cytoplasm of Escherichia coli K-12 as a function of external osmolarity: implications for protein-DNA interactions in vivo. J Mol Biol 222:281–300
Chang DW, Ditsworth D, Liu H, Srinivasula SM, Alnemri ES, Yang X (2003) Oligomerization is a general mechanism for the activation of apoptosis initiator and inflammatory procaspases. J Biol Chem 278:16466–16469
Clements RJ, Guo R, Petruccelli JC, Model MA. Label-Free quantification of protein density in living cells. Current Protocols. 2025 Apr;5(4):e70130
Clegg JS (1984) Properties and metabolism of the aqueous cytoplasm and its boundaries. Am J Physiol Regul Integr Comp Physiol 246:R133–R151
Colclasure GC, Parker JC (1991) Cytosolic protein concentration is the primary volume signal in dog red cells. J Gen Physiol 98:881–892
Colclasure GC, Parker JC (1992) Cytosolic protein concentration is the primary volume signal for swelling-induced [K-Cl] cotransport in dog red cells. J Gen Physiol 100:1–10
Cooper KL, Oh S, Sung Y, Dasari RR, Kirschner MW, Tabin CJ (2013) Multiple phases of chondrocyte enlargement underlie differences in skeletal proportions. Nature 495:375–378
Daimon M, Masumura A (2007) Measurement of the refractive index of distilled water from the near-infrared region to the ultraviolet region. Appl Opt 46:3811–3820
De Bruijne AW, Van Steveninck J (1970) Apparent nonsolvent water and osmotic behavior of yeast cells. Biochim Biophys Acta Biomembr 196:45–52
de Vries YP (2006) Bacillus cereus spore formation, structure and germination. Dissertation, Wageningen University
Delarue M, Brittingham GP, Pfeffer S, Surovtsev IV, Pinglay S, Kennedy KJ, Schaffer M, Gutierrez JI, Sang D, Poterewicz G, Chung JK (2018) mTORC1 controls phase separation and the biophysical properties of the cytoplasm by tuning crowding. Cell 174:338–349
Du J, Tao J, Kleinhans FW, Mazur P, Critser JK (1994) Water volume and osmotic behaviour of mouse spermatozoa determined by electron paramagnetic resonance. Reprod 101:37–42
Dupre AM, Hempling HG (1978) Osmotic properties of Ehrlich ascites tumor cells during the cell cycle. J Cell Physiol 97:381–395
Eisenberg T, Carmona-Gutierrez D, Büttner S, Tavernaraki N, Madeo F (2010) Necrosis in yeast. Apoptosis 15:257–268
Elmore S (2007) Apoptosis: a review of programmed cell death. Toxicol Pathol 35:495–516
Fels J, Orlov SN, Grygorczyk R (2009) The hydrogel nature of mammalian cytoplasm contributes to osmosensing and extracellular pH sensing. Biophys J 96:4276–4285
Fraser JA, Huang CLH (2007) Quantitative techniques for steady-state calculation and dynamic integrated modelling of membrane potential and intracellular ion concentrations. Prog Biophys Mol Biol 94:336–372
Fulda S, Gorman AM, Hori O, Samali A (2010) Cellular stress responses: cell survival and cell death. Int J Cell Biol 2010:214074
Fullerton GD, Kanal KM, Cameron IL (2006) On the osmotically unresponsive water compartment in cells. Cell Biol Int 30:74–77
Gaglia G, Rashid R, Yapp C, Joshi GN, Li CG, Lindquist SL, Sarosiek KA, Whitesell L, Sorger PK, Santagata S (2020) HSF1 phase transition mediates stress adaptation and cell fate decisions. Nature Cell Biol 22:151–158
Garlid KD (1999) The state of water in biological systems. Int Rev Cytol 192:281–302, Elsevier
Gibbons BA, Kharel P, Robinson LC, Synowicki RA, Model MA (2016) Volume measurements and fluorescent staining indicate an increase in permeability for organic cation transporter substrates during apoptosis. Exp Cell Res 344:112–119

Gnutt D, Gao M, Brylski O, Heyden M, Ebbinghaus S (2015) Excluded-volume effects in living cells. Angew Chem Int Ed 54:2548–2551

Granot D, Levine A, Dor-Hefetz E (2003) Sugar-induced apoptosis in yeast cells. FEMS Yeast Res 4:7–13

Halpin JC, Huang B, Sun M, Street TO (2016) Crowding activates heat shock protein 90. J Biol Chem 291:6447–6455

Harvey AH, Gallagher JS, Sengers JL (1998) Revised formulation for the refractive index of water and steam as a function of wavelength, temperature and density. J Phys Chem Ref Data 27:761–774

Häussinger D (1996) The role of cellular hydration in the regulation of cell function. Biochem J 313:697–710

Heimlicher MB, Bächler M, Liu M, Ibeneche-Nnewihe C, Florin EL, Hoenger A, Brunner D (2019) Reversible solidification of fission yeast cytoplasm after prolonged nutrient starvation. J Cell Sci 132:jcs231688

Hempling HG, Thompson S, Dupre A (1977) Osmotic properties of human lymphocyte. J Cell Physiol 93:293–302

Heuser J (2002) Whatever happened to the 'microtrabecular concept'? Biol Cell 94:561–596

Hinke JAM (1970) Solvent water for electrolytes in the muscle fiber of the giant barnacle. J Gen Physiol 56:521–541

Hishida M, Kaneko A, Yamamura Y, Saito K (2023) Contrasting Changes in Strongly and Weakly Bound Hydration Water of a Protein upon Denaturation. J Phys Chem B 127:6296–6305

Hoffmann EK, Lambert IH, Pedersen SF (2009) Physiology of cell volume regulation in vertebrates. Physiol Rev 89:193–277

Hollembeak JE, Model MA (2021) Stability of intracellular protein concentration under extreme osmotic challenge. Cells 10:3532

Holt LJ, Delarue M (2023) Macromolecular crowding: sensing without a sensor. Curr Opin Cell Biol 85:102269

Hortelano S, Zeini M, Castrillo A, Alvarez AM, Bosca L (2002) Induction of apoptosis by nitric oxide in macrophages is independent of apoptotic volume decrease. Cell Death Differ 9:643–650

Hu M, Zhou N, Cai W, Xu H (2022) Lysosomal solute and water transport. J Cell Biol 221:e202109133

Huang M, Hull CM (2017) Sporulation: how to survive on planet Earth (and beyond). Curr Genet 63:831–838

Hudder A, Nathanson L, Deutscher MP (2003) Organization of mammalian cytoplasm. Mol Cell Biol 23:9318–9326

Huff-Lonergan E, Lonergan SM (2005) Mechanisms of water-holding capacity of meat: The role of postmortem biochemical and structural changes. Meat Sci 71:194–204

Iwasa Y, Hirono C, Sugita M, Takemoto K, Shiba Y (2001) External Cl^--dependent formation of watery vacuoles by long-term hypotonic shock in 3T3-L1 cells. Cell Physiol Biochem 11:311–320, External Cl⁻-Dependent Formation of Watery Vacuoles by Long-Term Hypotonic Shock in 3T3-L1 Cells

Jentsch TJ (2016) VRACs and other ion channels and transporters in the regulation of cell volume and beyond. Nature Rev Mol Cell Biol 17:293–307

Jiang L, Paone S, Caruso S, Atkin-Smith GK, Phan TK, Hulett MD, Poon IK (2017) Determining the contents and cell origins of apoptotic bodies by flow cytometry. Sci Rep 7:14444

Joyner RP, Tang JH, Helenius J, Dultz E, Brune C, Holt LJ, Huet S, Müller DJ, Weis K (2016) A glucose-starvation response regulates the diffusion of macromolecules. elife 5:e09376

Kaji T, Yoshida S, Kawai K, Fuchigami Y, Watanabe W, Kubodera H, Kishimoto T (2010) ASK3, a novel member of the apoptosis signal-regulating kinase family, is essential for stress-induced cell death in HeLa cells. Biochem Biophys Res Commun 395:213–218

Kasim NR, Kuželová K, Holoubek A, Model MA (2013) Live fluorescence and transmission-through-dye microscopic study of actinomycin D-induced apoptosis and apoptotic volume decrease. Apoptosis 18:521–532

Kay AR (2017) How cells can control their size by pumping ions. Front Cell Dev Biol 5:41

Kerch G (2018) Distribution of tightly and loosely bound water in biological macromolecules and age-related diseases. Int J Biol Macromol 118:1310–1318

Kerr JFR (1971) Shrinkage necrosis: a distinct mode of cellular death. J Pathol 105:13–20

Klei TR, Dalimot JJ, Beuger BM, Veldthuis M, Ichou FA, Verkuijlen PJ, Seignette IM, Ligthart PC, Kuijpers TW, van Zwieten R, van Bruggen R (2020) The Gardos effect drives erythrocyte senescence and leads to Lu/BCAM and CD44 adhesion molecule activation. Blood Adv 4:6218–6229

Kondo H (2015) Ultrastructure of cytoplasmic matrix. J Integr Creat Stud 2015:1–9

Konopka MC, Sochacki KA, Bratton BP, Shkel IA, Record MT, Weisshaar JC (2009) Cytoplasmic protein mobility in osmotically stressed Escherichia coli. J Bacteriol 191:231–237

Kreuzer HW, Quaroni L, Podlesak DW, Zlateva T, Bollinger N, McAllister A, Lott MJ, Hegg EL (2012) Detection of metabolic fluxes of O and H atoms into intracellular water in mammalian cells. PLoS One 7:e39685

Kunzelmann K (2016) Ion channels in regulated cell death. Cell Mol Life Sci 73:2387–2403

Kuznetsova IM, Turoverov KK, Uversky VN (2014) What macromolecular crowding can do to a protein. Int J Mol Sci 15:23090–23140

Lecinski S, Shepherd JW, Frame L, Hayton I, MacDonald C, Leake MC (2021) Investigating molecular crowding during cell division and hyperosmotic stress in budding yeast with FRET. Curr Top Membr 88:75–118, Elsevier

Lee LD (2024) Aging changes cell mechanics and dynamics with a backbone of cytoplasmic crowding. Doctoral dissertation, Massachusetts Institute of Technology

Leist M, Single B, Castoldi AF, Kühnle S, Nicotera P (1997) Intracellular adenosine triphosphate (ATP) concentration: a switch in the decision between apoptosis and necrosis. J Exp Med 185:1481–1486

Lew VL (2022) The predictive power of cellular homeostasis models illustrated with a study of the Gardos effect. Paracelsus Proc Exp Med 1:1–13

Lewis CL, Craig CC, Senecal AG (2014) Mass and density measurements of live and dead gram-negative and gram-positive bacterial populations. Appl Environ Microbiol 80:3622–3631

Lindner AB, Demarez A (2009) Protein aggregation as a paradigm of aging. Biochim Biophys Acta Gen Subj 1790:980–996

Linker CS (1987) Mechanisms of volume regulation in Mycoplasma gallisepticum. Dissertation, Harvard University

Linkermann A, Green DR (2014) Necroptosis. N Engl J Med 370:455–465

Liu B, Hasrat Z, Poolman B, Boersma AJ (2019) Decreased effective macromolecular crowding in Escherichia coli adapted to hyperosmotic stress. J Bacteriol 201:e00708–e00718

Liu X, Moscvin M, Oh S, Chen T, Choi W, Evans B, Rowell SM, Nadeem O, Mo CC, Sperling AS, Anderson KC (2023) Characterizing dry mass and volume changes in human multiple myeloma cells upon treatment with proteotoxic and genotoxic drugs. Clin Exp Med 23:3821–3832

Luby-Phelps K (1999) Cytoarchitecture and physical properties of cytoplasm: volume, viscosity, diffusion, intracellular surface area. Int Rev Cytol 192:189–221, Elsevier

Luo P, Lin M, Lin M, Zhu D, Wang Z, Shen J, Yang B, He Q (2010) Bortezomib induces apoptosis in human neuroblastoma CHP126 cells. Int J Pharm Sci 65:213–218

Lüscher BP, Vachel L, Ohana E, Muallem S (2020) Cl^- as a bona fide signaling ion. Am J Physiol Cell Physiol 318:C125–C136

Lustyik G, Nagy I (1985) Alterations of the intracellular water and ion concentrations in brain and liver cells during aging as revealed by energy dispersive X-ray microanalysis of bulk specimens. Scanning Electron Microsc 1:323–337

Lustyik G, Nagy I (1988) Age dependent dehydration of postmitotic cells as measured by X-ray microanalysis of bulk specimens. Scanning Microsc 2:289–299

Lustyik G (1986) Age-dependent alterations of the intracellular water and electrolyte content of heart and muscle cells. Arch Gerontol Geriatr 5:291–296
Luu BE, Wijenayake S, Malik AI, Storey KB (2018) The regulation of heat shock proteins in response to dehydration in Xenopus laevis. Cell Stress Chaperones 23:45–53
Maisonneuve E, Ezraty B, Dukan S (2008) Protein aggregates: an aging factor involved in cell death. J Bacteriol 190:6070–6075
Maiti S, Bhattacharya K, Wider D, Hany D, Panasenko O, Bernasconi L, Hulo N, Picard D (2023) Hsf1 and the molecular chaperone Hsp90 support a "rewiring stress response" leading to an adaptive cell size increase in chronic stress. BioRxiv:2023–2001
Marini G, Nüske E, Leng W, Alberti S, Pigino G (2020) Reorganization of budding yeast cytoplasm upon energy depletion. Mol Biol Cell 31:1232–1245
Minton AP (2020) Water loss in aging erythrocytes provides a clue to a general mechanism of cellular senescence. Biophys J 119:2039–2044
Minton AP, Colclasure GC, Parker JC (1992) Model for the role of macromolecular crowding in regulation of cellular volume. Proc Natl Acad Sci USA 89:10504–10506
Miyagi T, Yamanaka Y, Harada Y, Narumi S, Hayamizu Y, Kuroda M, Kanekura K (2021) An improved macromolecular crowding sensor CRONOS for detection of crowding changes in membrane-less organelles under stressed conditions. Biochem Biophys Res Commun 583:29–34
Model MA, Petruccelli JC (2018) Intracellular macromolecules in cell volume control and methods of their quantification. Curr Top Membr 81:237–289, Elsevier
Model MA (2014) Possible causes of apoptotic volume decrease: an attempt at quantitative review. Am J Physiol Cell Physiol 306:C417–C424
Model MA (2021) Studying cell volume beyond cell volume. Curr Top Membr 88:165–188, Elsevier
Model MA, Guo R, Fasina K, Jin R, Clements RJ, Leff LG (2024) Measurement of protein concentration in bacteria and small organelles under a light transmission microscope. J Mol Recognit 37:e3099
Model MA, Hollembeak JE, Kurokawa M (2021) Macromolecular crowding: a hidden link between cell volume and everything else. Cell Physiol Biochem 55:25–40
Model MA, Mahajan A, Düsterwald KM, Dmitriev AV (2023) Nucleic acids regulate intracellular ions and membrane potential. Paracelsus Proc Exp Med 2:13–29
Model MA, Mudrak NJ, Rana PS, Clements RJ (2018) Staurosporine-induced apoptotic water loss is cell-and attachment-specific. Apoptosis 23:449–455
Clements, Robert J., Ruixin Guo, Jonathan C. Petruccelli, and Michael A. Model. "Label-Free quantification of protein density in living cells." Current Protocols 5, no. 4 (2025): e70130.
Mongin AA, Orlov SN (2001) Mechanisms of cell volume regulation and possible nature of the cell volume sensor. Pathophysiology 8:77–88
Mouton SN, Boersma AJ, Veenhoff LM (2023) A physicochemical perspective on cellular ageing. Trends Biochem Sci 48:949–962
Mouton SN, Thaller DJ, Crane MM, Rempel IL, Terpstra OT, Steen A, Kaeberlein M, Lusk CP, Boersma AJ, Veenhoff LM (2020) A physicochemical perspective of aging from single-cell analysis of pH, macromolecular and organellar crowding in yeast. elife 9:e54707
Mudrak NJ, Rana PS, Model MA (2018) Calibrated brightfield-based imaging for measuring intracellular protein concentration. Cytometry A 93:297–304
Mulkern RV, Bleier AR, Adzamli IK, Spencer RG, Sandor T, Jolesz FA (1989) Two-site exchange revisited: a new method for extracting exchange parameters in biological systems. Biophys J 55:221–232
Munder MC, Midtvedt D, Franzmann T, Nuske E, Otto O, Herbig M, Ulbricht E, Mülle P, Taubenberger A, Maharana S, Malinovska L (2016) A pH-driven transition of the cytoplasm from a fluid-to a solid-like state promotes entry into dormancy. elife 5:e09347
Naito Y, Kaliaperumal V, Hamaguchi HO (2005) Inorganic phosphate dynamics and starvation induced necrosis in yeast cells by Raman microspectroscopy. Spectroscopy 36:837–839

Neurohr GE, Amon A (2020) Relevance and regulation of cell density. Trends Cell Biol 30:213–225
Neurohr GE, Terry RL, Lengefeld J, Bonney M, Brittingham GP, Moretto F, Miettinen TP, Vaites LP, Soares LM, Paulo JA, Harper JW (2019) Excessive cell growth causes cytoplasm dilution and contributes to senescence. Cell 176:1083–1097
Oillic S, Lemoine E, Gros JB, Kondjoyan A (2011) Kinetic analysis of cooking losses from beef and other animal muscles heated in a water bath—Effect of sample dimensions and prior freezing and ageing. Meat Sci 88:338–346
Okada Y, Maeno E, Shimizu T, Dezaki K, Wang J, Morishima S (2001) Receptor-mediated control of regulatory volume decrease (RVD) and apoptotic volume decrease (AVD). J Physiol 532:3–16
Oldewurtel ER, Kitahara Y, van Teeffelen S (2021) Robust surface-to-mass coupling and turgor-dependent cell width determine bacterial dry-mass density. Proc Natl Acad Sci USA 118:e2021416118
Orlov SN, Platonova AA, Hamet P, Grygorczyk R (2013) Cell volume and monovalent ion transporters: their role in cell death machinery triggering and progression. Am J Physiol Cell Physiol 305:C361–C372
Orlov SN, Shiyan A, Boudreault F, Ponomarchuk O, Grygorczyk R (2018) Search for upstream cell volume sensors: the role of plasma membrane and cytoplasmic hydrogel. Curr Topics Membr 81:53–82
Ozturk SS, Palsson BØ (1991) Effect of medium osmolarity on hybridoma growth, metabolism, and antibody production. Biotechnol Bioeng 37:989–993
Pang B, Yu X, Bowker B, Zhang J, Yang Y, Zhuang H (2021) Effect of meat temperature on moisture loss, water properties, and protein profiles of broiler pectoralis major with the woody breast condition. Poult Sci 100:1283–1290
Parry BR, Surovtsev IV, Cabeen MT, O'Hern CS, Dufresne ER, Jacobs-Wagner C (2014) The bacterial cytoplasm has glass-like properties and is fluidized by metabolic activity. Cell 156:183–194
Pasantes-Morales H (2016) Channels and volume changes in the life and death of the cell. Mol Pharmacol 90:358–370
Pittas T, Zuo W, Boersma AJ (2023) Cell wall damage increases macromolecular crowding effects in the Escherichia coli cytoplasm. iScience 26:106367
Platonova A, Boudreault F, Kapilevich LV, Maksimov GV, Ponomarchuk O, Grygorczyk R, Orlov SN (2014) Temperature-induced inactivation of cytoplasmic biogel osmosensing properties is associated with suppression of regulatory volume decrease in A549 cells. J Membr Biol 247:571–579
Porus M, Labbez C, Maroni P, Borkovec M (2011) Adsorption of monovalent and divalent cations on planar water-silica interfaces studied by optical reflectivity and Monte Carlo simulations. J Chem Phys 135:064701
Proskuryakov SY, Konoplyannikov AG, Gabai VL (2003) Necrosis: a specific form of programmed cell death? Exp Cell Res 283:1–16
Rana PS, Gibbons BA, Vereninov AA, Yurinskaya VE, Clements RJ, Model TA, Model MA (2019) Calibration and characterization of intracellular Asante Potassium Green probes, APG-2 and APG-4. Anal Biochem 567:8–13
Rana PS, Model MA (2020) A reverse-osmosis model of apoptotic shrinkage. Front Cell Dev Biol 8:588721
Rana PS, Kurokawa M, Model MA (2020) Evidence for macromolecular crowding as a direct apoptotic stimulus. J Cell Sci 133:243931
Rana PS, Mudrak NJ, Lopez R, Lynn M, Kershner L, Model MA (2017) Phase separation in necrotic cells. Biochem Biophys Res Commun 492:300–303
Reiter J, Herker E, Madeo F, Schmitt MJ (2005) Viral killer toxins induce caspase-mediated apoptosis in yeast. J Cell Biol 168:353–358
Rivas G, Minton AP (2016) Macromolecular crowding in vitro, in vivo, and in between. Trends Biochem Sci 41:970–981

Roman SG, Chebotareva NA, Kurganov BI (2017) Anti-aggregation activity of small heat shock proteins under crowded conditions. Int J Biol Macromol 100:97–103

Savage B, McFadden PR, Hanson SR, Harker LA (1986) The relation of platelet density to platelet age: survival of low-and high-density 111indium-labeled platelets in baboons. Blood 68:386–393

Schliess F, Reinehr R, Häussinger D (2007) Osmosensing and signaling in the regulation of mammalian cell function. FEBS J 274:5799–5803

Schmidt S, Flueckiger J, Wu W, Grist SM, Fard ST, Donzella V, Khumwan P, Thompson ER, Wang Q, Kulik P, Wang X (2014) Improving the performance of silicon photonic rings, disks, and Bragg gratings for use in label-free biosensing. In: Biosensing and nanomedicine VII, vol 9166, 71–108. SPIE

Setlow P (2016) In: Driks A, Eichenberger P (eds) Spore resistance properties. In The bacterial spore: from molecules to systems. ASM Press, pp 201–215

Shatrova A, Burova E, Pugovkina N, Domnina A, Nikolsky N, Marakhova I (2022) Monovalent ions and stress-induced senescence in human mesenchymal endometrial stem/stromal cells. Sci Rep 12:11194

Shi H, Westfall CS, Kao J, Odermatt PD, Anderson SE, Cesar S, Sievert M, Moore J, Gonzalez CG, Zhang L, Elias JE (2021) Starvation induces shrinkage of the bacterial cytoplasm. Proc Natl Acad Sci USA 118:e2104686118

Son S, Kang JH, Oh S, Kirschner MW, Mitchison TJ, Manalis S (2015) Resonant microchannel volume and mass measurements show that suspended cells swell during mitosis. J Cell Biol 211:757–763

Steen B (1988) Body composition and aging. Nutr Rev 46:45–51

Su KH, Dai C (2017) mTORC1 senses stresses: coupling stress to proteostasis. BioEssays 39:1600268

Subramanya AR, Boyd-Shiwarski CR (2023) Molecular crowding: physiologic sensing and control. Annu Rev Physiol 86:429–452

Summers JC, Trais L, Lajvardi R, Hergan D, Buechler R, Chang H, Pena-Rasgado C, Rasgado-Flores H (1997) Role of concentration and size of intracellular macromolecules in cell volume regulation. Am J Physiol Cell Physiol 273:C360–C370

Theisen A, Johann C, Deacon MP, Harding SE (2000) Refractive increment data-book for polymer and biomolecular scientists. Nottingham University Press, Nottingham

Thormählen I, Straub J, Grigull U (1985) Refractive index of water and its dependence on wavelength, temperature, and density. J Phys Chem Ref Data 14:933–945

Tsujimoto Y (1997) Apoptosis and necrosis: intracellular ATP level as a determinant for cell death modes. Cell Death Differ 4:429–434

Tumolo T, Angnes L, Baptista MS (2004) Determination of the refractive index increment (dn/dc) of molecule and macromolecule solutions by surface plasmon resonance. Anal Biochem 333:273–279

Unsworth BR, Fleming LH (1980) Aging in rodent brain: alteration in astrocyte population. Neurobiol Aging 1:169–173

Urbonaite V, De Jongh HHJ, Van Der Linden E, Pouvreau L (2015) Protein aggregates may differ in water entrapment but are comparable in water confinement. J Agric Food Chem 63:8912–8920

Van Den Berg J, Boersma AJ, Poolman B (2017) Microorganisms maintain crowding homeostasis. Nat Rev Microbiol 15:309–318

von Zglinicki T, Bimmler M (1987) Intracellular water and ionic shifts during growth and ageing of rats. Mech Ageing Dev 38:179–187

von Zglinicki T (1988) Intracellular water distribution and aging as examined by X-ray microanalysis. Scanning Microsc 2:1791–1804

Vonck J, Schäfer E (2009) Supramolecular organization of protein complexes in the mitochondrial inner membrane. Biochim Biophys Acta, Mol Cell Res 1793:117–124

Vorob'ev M (2003) Bound water measurements for aqueous protein solutions and food gels. Colloids Surf B Biointerfaces 31:133–140

Walle LV, Lamkanfi M (2016) Pyroptosis. Curr Biol 26:R568–R572
Watanabe K, Morishita K, Zhou X, Shiizaki S, Uchiyama Y, Koike M, Naguro I, Ichijo H (2021) Cells recognize osmotic stress through liquid–liquid phase separation lubricated with poly (ADP-ribose). Nat Commun 12:1353
Watson JL, Seinkmane E, Styles CT, Mihut K, McNally KE, Derivery E (2023) Macromolecular condensation buffers intracellular water potential. Nature 623(7988):842–852
Wittkopp JM, Khoo TC, Carney S, Pisila K, Bahreini SJ, Tubbesing K, Mahajan S, Sharikova A, Petruccelli JC, Khmaladze A (2020) Comparative phase imaging of live cells by digital holographic microscopy and transport of intensity equation methods. Opt Express 28:6123–6133
Wolosewick JJ, Porter KR (1979) Microtrabecular lattice of the cytoplasmic ground substance. Artifact or reality J Cell Biol 82:114–139
Xie Y, Gresham D, Holt LJ (2023) Increased mesoscale diffusivity in response to acute glucose starvation. microPubl biol:10.17912
Xu F, Zhang Q, Wang H (2020) Establishing a density-based method to separate proliferating and senescent cells from bone marrow stromal cells. Aging (Albany NY) 12:15050–15057
Yang, X., Chang, H.Y. and Baltimore, D., 1998. Autoproteolytic activation of pro-caspases by oligomerization. Molecular Cell, 1(2), pp.319-325
Yan GE, Elbadawi M, Efferth T (2020) Multiple cell death modalities and their key features. World Acad Sci J 2:39–48
You, L., Zhang, Y., Ma, Y., Wang, Y., & Wei, Z. (2025). Effect of boling time on the color, water, protein secondary structure, and volatile compunds of beef. Foods, 14(8), 1372.
Yu SP (2003) Regulation and critical role of potassium homeostasis in apoptosis. Prog Neurobiol 70:363–386
Yurinskaya V, Goryachaya T, Guzhova I, Moshkov A, Rozanov Y, Sakuta G, Shirokova A, Shumilina E, Vassilieva I, Lang F, Vereninov A (2005) Potassium and sodium balance in U937 cells during apoptosis with and without cell shrinkage. Cell Physiol Biochem 16:155–162
Zelenina M, Brismar H (2000) Osmotic water permeability measurements using confocal laser scanning microscopy. Eur Biophys J 29:165–171
Zhang X, Garcia IF, Baldi L, Hacker DL, Wurm FM (2010) Hyperosmolarity enhances transient recombinant protein yield in Chinese hamster ovary cells. Biotechnol Lett 32:1587–1592
Zhang Y, Fernie AR (2021) Metabolons, enzyme–enzyme assemblies that mediate substrate channeling, and their roles in plant metabolism. Plant Commun 2:100081
Zhao J, Zhai B, Gygi SP, Goldberg AL (2015) mTOR inhibition activates overall protein degradation by the ubiquitin proteasome system as well as by autophagy. Proc Natl Acad Sci USA 112:15790–15797
Zhao X, Qiu W, Kung J, Zhao X, Peng X, Yegappan M, Yen-Lieberman B, Hsi ED (2008) (2008) Bortezomib induces caspase-dependent apoptosis in Hodgkin lymphoma cell lines and is asso ciated with reduced c-FLIP expression: a gene expression profiling study with implications for potential combination therapies. Leuk Res 32(2):275–285
Zhou HX, Rivas G, Minton AP (2008) Macromolecular crowding and confinement: biochemical, biophysical, and potential physiological consequences. Annu Rev Biophys 37:375–397
Zielbauer BI, Franz J, Viezens B, Vilgis TA (2016) Physical aspects of meat cooking: Time dependent thermal protein denaturation and water loss. Food Biophys 11:34–42
Zook E, Pan YE, Wipplinger A, Kerschbaum HH, Clements RJ, Ritter M, Stauber T, Model MA. Delayed vacuolation in mammalian cells caused by hypotonicity and ion loss. Scientific reports. 2024 Nov 26;14(1):29354
Zuo C, Li J, Sun J, Fan Y, Zhang J, Lu L, Zhang R, Wang B, Huang L, Chen Q (2020) Transport of intensity equation: a tutorial. Opt Lasers Eng 135:106187

Chapter 15
Heterogeneity of Molecular Crowding and Liquid-Liquid Phase Separation

Mitsuki Tsuruta, Sumit Shil, Keiko Kawauchi, and Daisuke Miyoshi

Abstract The inside of a living cell is highly crowded with extremely diverse biomacromolecules, small metabolites and osmolytes. The molecular conditions in cells change dynamically and rapidly depending on the cell cycle and state, organelle, and compartment. Much remains unknown regarding how biomolecular interactions and reactions can proceed in a spatiotemporally specific manner in such crowded, heterogeneous, and dynamic molecular environments. Selective condensation/droplet formation of biomolecules via liquid-liquid phase separation may be critical for interactions and reactions inside cells. In this chapter, we briefly describe the heterogeneity of molecular environments inside cells and the biological roles of liquid-liquid phase separation that allows biomolecular interactions and reactions in such heterogenous molecular environments. Finally, we discuss the mutual relationship between molecular crowding and liquid-liquid phase separation.

Keywords Liquid-liquid phase separation · Molecular crowding · Intrinsic disorder protein · Weak interaction · Neurodegenerative disease

Heterogeneity of Molecular Crowding Inside Living Cells

Life on Earth probably originated approximately 4 billion years ago, when the environment was significantly different from the present. All living organisms have evolved to adapt to their environment to survive. Since the evolution of living organisms occurs through the evolution of biomolecules, the biomolecules in cells have adapted to optimize their structure and function within the cellular environment. Protein denaturation due to changes in temperature and pH show that the structure, stability, and function of biomolecules are highly dependent on the surrounding

M. Tsuruta · S. Shil · K. Kawauchi · D. Miyoshi (✉)
Faculty of Frontiers of Innovative Research in Science and Technology (FIRST), Konan University, Kobe, Hyogo, Japan
e-mail: miyoshi@konan-u.ac.jp

V. N. Uversky (ed.), *(Macro)Molecular Crowding*, Subcellular Biochemistry 109,
https://doi.org/10.1007/978-3-032-03370-3_15

molecular environment. Therefore, one must take the cellular environment into account to study these biomolecular properties. In this section, we introduce briefly but as quantitatively as possible the molecular environment inside an *E. coli* cell as an example.

Molecules Inside Cells

The volume of an *E. coli* cell is approximately 1 μm^3 (Fig. 15.1) and contains approximately 3×10^6 protein molecules (Milo and Phillips 2015). Thus, the protein concentration is about 5 mM. Assuming that an average protein comprises 300 amino acids and has a molecular weight of 30,000, one can estimate that an *E. coli*

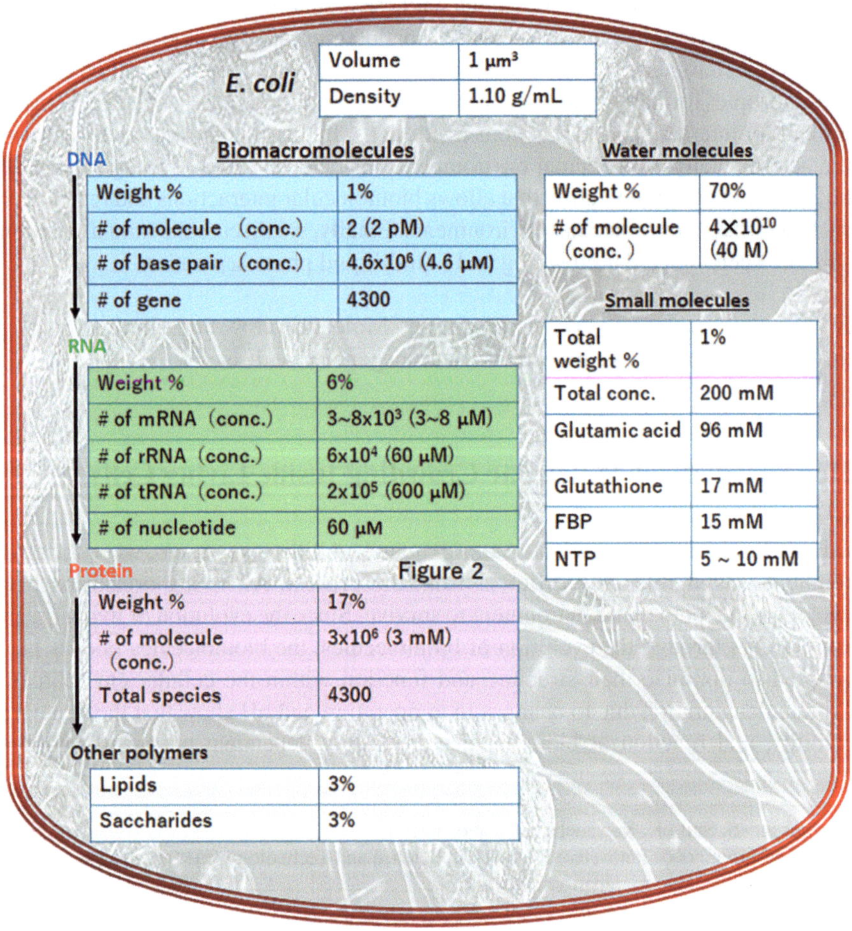

Fig. 15.1 Schematic illustration of the molecular contents of an *E. coli* cell

cell contains approximately 150 g/L of protein. Protein solubility in aqueous solution varies from almost insoluble to hundreds of grams per liter (Kramer et al. 2012) and is dependent on the amino acid composition, the protein's structure, and on the molecular environment, such as pH, the ion species and their concentrations, temperature, and other components of the solution (Riès-Kautt and Ducruix 1997). For example, the solubility of lysozyme is several tens of grams per liter, whereas that of serum albumin is around ten grams per liter (Kramer et al. 2012). These protein solubilities suggest that proteins can supersaturate living cells.

Cells also contain nucleic acids (RNA and DNA), saccharides, lipids, and a wide variety of metabolic products and osmotic regulation molecules. DNA and RNA account for 1% and 6%, respectively, of the total mass of an *E. coli* cell, with lipids accounting for 3% and carbohydrates another 3%. Amino acids are the most abundant metabolites, with glutamic acid (100 mM) being the most abundant. The total concentration of metabolites is 200 mM or higher (Taymaz-Nikerel et al. 2010). Additionally, proline, sucrose, betaine, glycine betaine, trimethyl-N-oxide (TMAO), and urea are present at up to 600 mM as osmotic regulators (Yancey 2005).

Ion concentrations have also been extensively studied and can vary significantly, depending on the measurement method used and the conditions. In *E. coli*, K^+ ranges from 30 to 300 mM, Na^+ is approximately 10 mM, Mg^{2+} ranges from 30 to 100 mM, whereas Ca^{2+} is below 100 nM. The product of concentration and charge within a cell is approximately 190 mM for both cations and anions, but is 150 mM outside the cell. The most abundant intracellular cations are approximately 150 mM K^+ and approximately 10 mM Na^+, but this is reversed outside the cell, with Na^+ being approximately 145 mM and K^+ approximately 5 mM. Interestingly, most anionic components inside a cell are organic compounds such as glutamic acid (~100 mM), fructose 2,6-bisphosphate (~10 mM), nucleotide triphosphate (ATP ~ 10 mM).

An *E. coli* cell can contain up to approximately 400 g/L of biomolecules, which occupy around 40% of the cell volume (Fig. 15.2) (Nakano et al. 2014), giving an occupancy rate similar to the packing density of atoms in a diamond crystal structure (0.34) or the volume ratio of sphere packing (packing fraction: 0.52). In contrast, biochemical laboratory experiments are generally conducted with solutions with biomolecule concentrations in the range of a few μmol, providing extremely dilute environments. The intracellular environment significantly differs from the molecular environment provided by dilute solutions in test tubes, and is closer to that of a solid state. Indeed, *E. coli* cells reportedly exhibit glass-like properties (Parry et al. 2014). Importantly, even though the volumes of human cells are thousands of times that of *E. coli*, the molecular concentrations in both cell types are similar. The components of procaryotic and eukaryotic cells are under molecular crowding conditions caused by highly concentrated biomolecules.

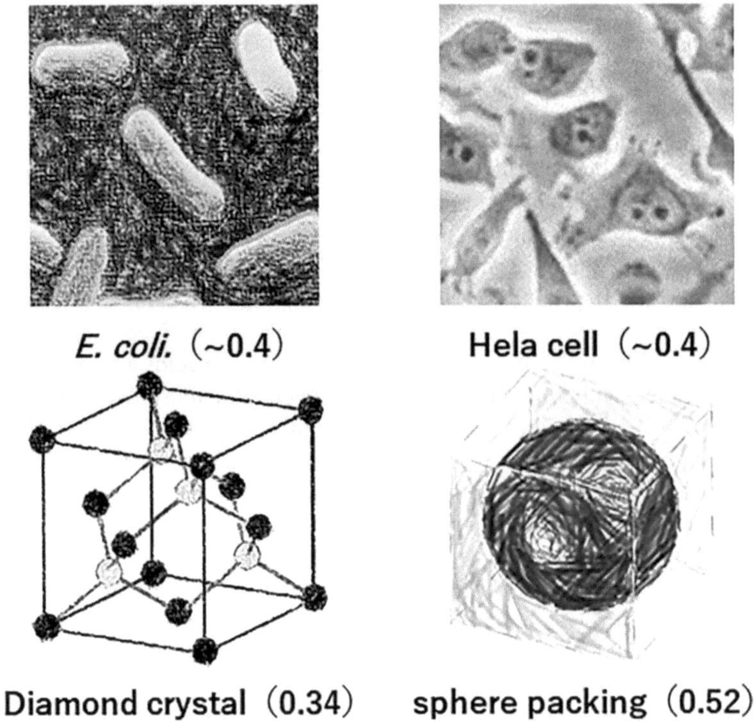

Fig. 15.2 Schematic illustration of an *E. coli* cell (top left), a Hela cell (top right), a diamond crystal (bottom left), and sphere packing (bottom right). Volume occupancy is given in parentheses

Dynamic Alteration of Molecular Environment

The molecular condition in a cell varies dynamically depending on the cell cycle and cellular state, and is different in organelles and other compartments. Cellular volume changes depend on gene expression (RNA and protein), the balance between endocytosis and exocytosis, water and ion transport across membranes, osmotic balance, hydrostatic pressure, and membrane turnover (Cadart et al. 2022). The cell volume increases during interphase by a factor of 2 ~ 3 (Clark and Paluch 2011). During M-phase, the cell volume decreases by around 30% from prophase to metaphase, then returns to the prophase volume (Clark and Paluch 2011). The cell shape changes between being spherical and having a spread morphology (Clark and Paluch 2011).

The cell volume also depends on the cell status. For example, collective invasion by breast cancer is driven by a combination of global cell volume expansion (Chang et al. 2024). These changes in cell volume and shape alter the degree of molecular crowding and make the molecular environment more heterogeneous. Carcinogenic agents decrease the concentration of potassium and increase the concentration of sodium in cancer cells, while anticarcinogenic agents have the opposite effect

(Jansson 1996). Both effects are at least partly due to potassium ion channels playing a significant role in cancer development, including proliferation, invasion, and adaptive drug resistance (Boyle et al. 2022). The ATP concentration and solution pH are known to change depending on the cell status. These reports regarding intracellular environments reveal that cell interiors are heterogeneous and dynamically changing molecular environments.

Liquid-Liquid Phase Separation of Biomolecules

The complex molecular conditions in cells likely promote the binding and reaction of biomolecules in a spatiotemporal manner. The specific binding of one type of protein with one type of substrate cannot be achieved in laboratory experiments under conditions mimicking the presence of excess proteins and compounds found inside cells, especially given that some of these compounds are chemically and structurally similar to the protein and substrate of interest. Moreover, cells maintain homeostasis by fine-tuning of numerous signaling and metabolic pathways and by responding to external stimuli. The failure of even one pathway results in a severe cell status, including disease. Droplets (condensates, membraneless organelles, coacervates) formed via biomolecular liquid-liquid phase separation (LLPS) likely play important roles in intermolecular reactions in complex cellular environments.

Biomolecular LLPS occurs when solutions do not mix uniformly and separate into two or more phases (Fig. 15.3a). Droplets formed via LLPS are similar to oil droplets in water, observed when water and oil are vigorously mixed, such as in salad dressings and in lava lumps. These droplets are unique in that they slowly change shape and move like molten lava (Fig. 15.3b, c). Biomolecular droplets, such as the nucleolus and nuclear bodies in the nucleus, and stress granules and p-bodies in the cytosol, are also formed in cells via LLPS. These biomolecular droplets have attracted attention as "new phases" in living cells, and their biological functions have been elucidated (Brangwynne et al. 2009; Li et al. 2018; Hallegger et al. 2021). Biomolecular droplets can also be formed in test tubes. For example, when the synthetic polymers polyethylene glycol (PEG) and dextran (DEX) are mixed at about 10 weight percent each, LLPS occurs to form separate PEG and DEX phases (Fig. 15.3d). Droplets are also formed when polyglutamic acid and immunoglobulin G or ovalbumin and lysozyme are mixed (Iwashita et al. 2018; Matsuda et al. 2018): interactions between the polypeptide chains separate the protein into a concentrated phase (inside the droplet) and a dilute phase (outside the droplet).

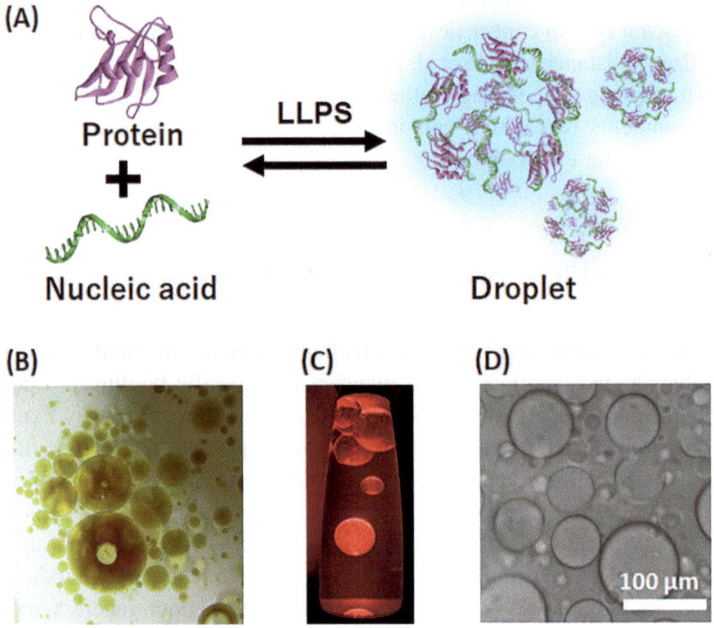

Fig. 15.3 (**a**) Scheme for LLPS of a protein-nucleic acid mixture. (**b**) Oil droplets in water. (**c**) Lava lump. (**d**) Droplets observed in a PEG-DEX mixture

Mechanism of Formation of LLPS Droplets

The interaction of two types of molecules in solution can be represented by a phase diagram, as shown in Fig. 15.4. The mixing ratio for the two molecules (molecular concentration: C) is plotted against an environmental factor such as temperature, pH, salt type or concentration. The boundary between the one-phase region and the two-phase region is depicted in the phase diagram. Inside the boundary, the molecules separate into a dilute phase ($C = C_L$) and a dense phase ($C = C_D$) by forming droplets.

If the concentration of molecules is below the minimum concentration (C_{sat}) required for droplet formation, the molecules will mix homogeneously. This is called condition 1. Conditions 2 and 3 are conditions for which LLPS occurs. For condition 2, the volume fraction occupied by the concentrated phase is lower than that of the dilute phase, while for condition 3, the volume fraction occupied by the concentrated phase is higher than that of the dilute phase. When the molecular concentration in the solution is higher than C_D, as in condition 4, LLPS is not induced, and the solution remains in a single phase.

The parallel line between conditions 2 and 3 is called a tie line. The concentration of molecules in the dilute and dense phases on a tie line remains constant under all condition. However, on a tie line, the volume fraction of dilute and dense phases

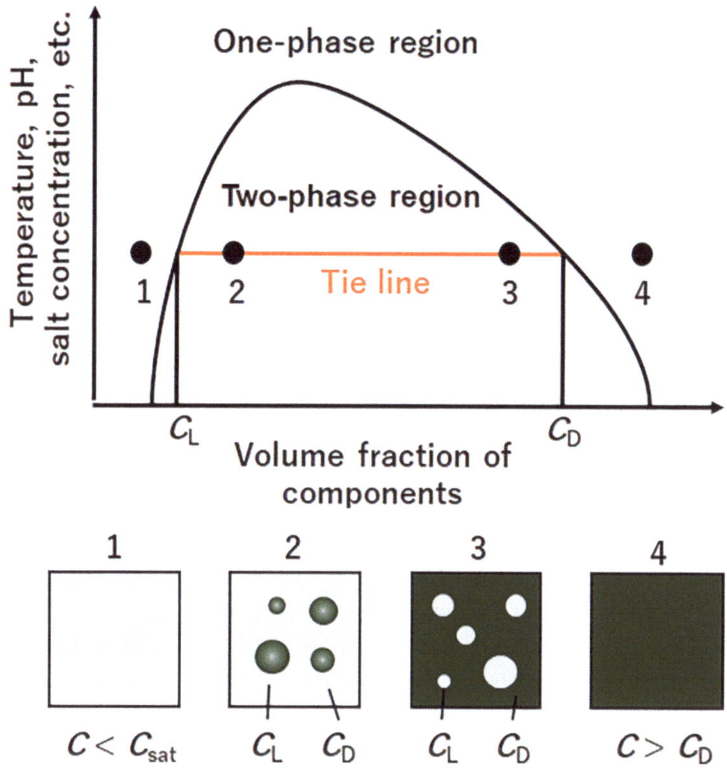

Fig. 15.4 Phase diagram for two types of molecules in solution

varies. A phase diagram can show the extent to which the chemical properties of the molecules affect the formation of droplets.

Properties and Functions of Biomolecular Droplets Formed by LLPS

Droplets have characteristic physical properties that distinguish them from gels and aggregates. First, their formation and dissociation are reversible (Fig. 15.5a). Since these droplets lack a membrane, molecules can move relatively freely between phases (between the inside and the outside the droplet). The desired reaction can be promoted by incorporating and concentrating inside the droplet only the molecules necessary for the reaction, or the desired reaction can be inhibited by incorporating only the molecules that catalyze the reaction (Fig. 15.5b, c). The responsiveness of droplets to external stimul is another important physical property of droplets (Fig. 15.5d). Droplets are also useful for separating and localizing biomolecules (Fig. 15.5E), and they act as a buffer to help maintain the concentration of

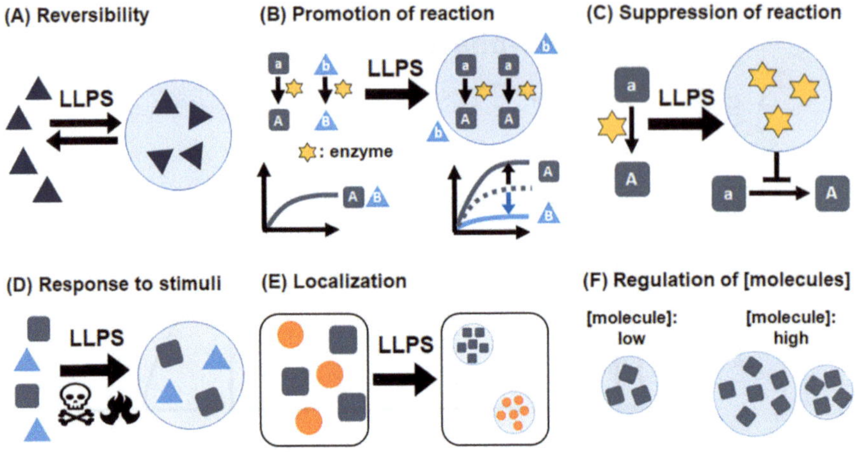

Fig. 15.5 Schematic illustration of the functions and properties of LLPS

molecules outside the droplets constant; this is achieved by altering the number and size of the droplets (Fig. 15.5f). These unique physical properties of droplets allow control of a wide variety of biological reactions in living cells.

Oparin reported the formation of biomolecular droplets in 1924 (Oparin 1924), after observing that droplets are formed upon mixing negatively charged arabicagum and positively charged histones (Fig. 15.6a). He also found that enzymes and substrates are incorporated into the formed droplets (coacervates), and enzymatic reactions are promoted as shown in (Fig. 15.6b). That a coacervate can separate an inner from an exterior environment and can promote chemical reactions using compounds inside the coacervate led Oparin to propose the "chemical evolution theory" as a hypothesis for the origin of life. The chemical evolution theory considers that simple organic compounds such as methane, aldehydes, and ammonia were produced prior to the origin of life. Subsequently, amino acids, nucleobases, and monosaccharides were produced and combined to generate macromolecular compounds such as proteins, nucleic acids, and polysaccharides. The prototypic cell supporting metabolic pathways may have arisen from a partitioned coacervate.

The idea that a coacervate may represent a primitive cell regained the attention of biologists when Brangwynne and Hyman reported in 2009 that *C. elegans* P-granules exhibit fluid droplet-like properties (Brangwynne et al. 2009). In 2011, these authors reported that the nucleolus also exhibits droplet-like properties (Brangwynne et al. 2011). In 2012, Rosen et al. and McKnight et al. independently reported that proteins form droplets in vitro (Li et al. 2012; Kato et al. 2012). These pioneering findings suggest that LLPS is an intrinsic property of biomolecules.

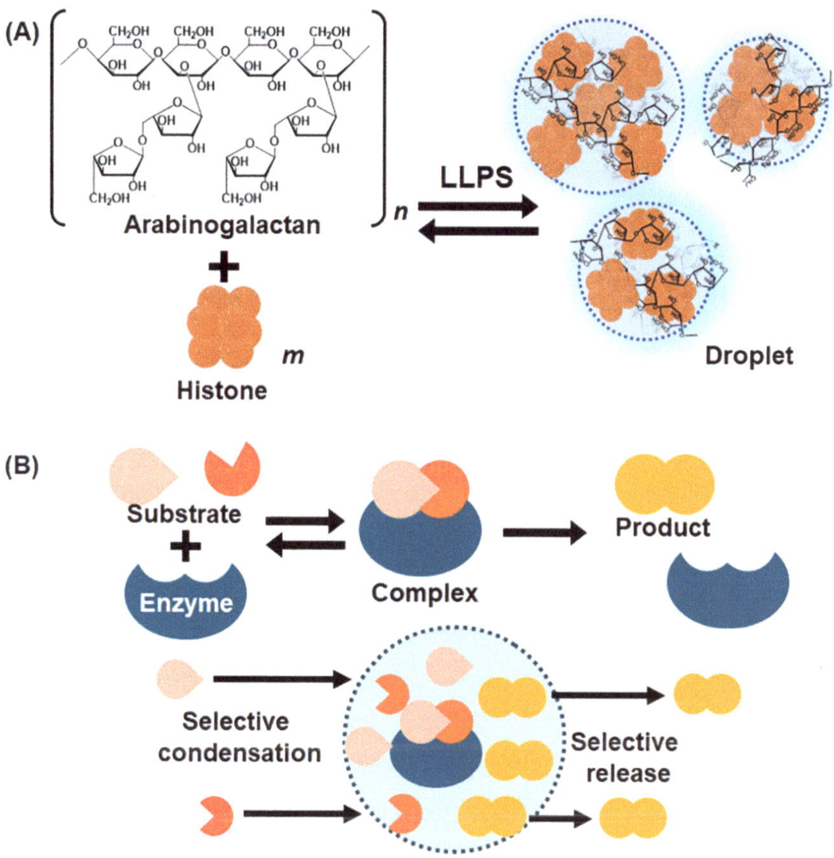

Fig. 15.6 (**a**) LLPS induced by a mixture of arabinogalactan and histone protein. (**b**) Schematic illustration of the selective condensation of a substrate and the selective release of the reaction product from a droplet

Proteins Undergoing LLPS

Many proteins that undergo LLPS contain intrinsically disordered regions (IDRs), such as prion-like domains (PrLDs) and RGG domains (Table 15.1) (Qamar et al. 2018a; Asamitsu et al. 2021; Elbaum-Garfinkle et al. 2015; Molliex et al. 2015; Henninger et al. 2021; Wang et al. 2018; Timothy et al. 2015). IDRs are structureless regions of a polypeptide chain that often contain repeated sequences of several amino acids. Proteins might undergo LLPS through multivalent weak interactions between IDRs. PrLDs containing proteins such as TIA1 in stress granules and the Lsm family proteins in P-bodies are 80% comprised of serine, glutamine, tyrosine, and glycine, and stacking interactions between these side chains facilitate droplet formation (Gilks et al. 2004; Crick et al. 2006; Reijns et al. 2008). The RGG domain

Table 15.1 Proteins with IDRs observed in cellular droplets

Domain	Protein	Droplet	References
Prion-like	TIA1	Stress granule	Qamar et al. (2018a)
	Lsm family	P body	Asamitsu et al. (2021)
RGG	LAF1	P granule	Elbaum-Garfinkle et al. (2015)
	hnRNP family	Stress granule	Molliex et al. (2015)
	TDP-43	Stress granule / Paraspeckle	Henninger et al. (2021)
	FUS	Stress granule / Paraspeckle	Wang et al. (2018)
Phe/Gly	DDX4	Germ granule	Timothy et al. (2015)
Poly-Gly	FMRP	RNP granule	Asamitsu et al. (2021)

contains repeats of arginine and glycine and many positive charges. RGG domains are found in various proteins, including FUS, TDP-43, and the family of hnRNP proteins (Elbaum-Garfinkle et al. 2015; Molliex et al. 2015; Wang et al. 2018; Qamar et al. 2018b). Arginine forms hydrogen bonds, and electrostatic and cation-π interactions with other amino acid residues and with nucleic acids (Chong et al. 2018). DDX4 protein in germ granules forms cation-π and stacking interactions through its Phe/Gly domains (Timothy et al. 2015). A Poly-Gly domain is formed following the mutation of FMRP protein, and may cause droplet formation through hydrophobic interactions (Asamitsu et al. 2021).

Droplets in Cells

Various droplets are formed via the LLPS of biomolecules in cells. Stress granules, P-bodies, and germ granules are observed in the cytoplasm. The nucleus contains nuclear speckles, paraspeckles, and PML bodies (Weber and Brangwynne 2015; Hanazawa et al. 2011). These droplets form at specific locations in the cell, contain specific molecules, and participate in a variety of cellular reactions and processes, such as transcription, RNA splicing, and translation, depending on their function, as shown in Fig. 15.5.

Nucleoli in nematode embryos are formed when FIB1, a major component protein of the nucleolus, reaches concentrations above its solubility threshold (Weber and Brangwynne 2015). The volume of the nucleolus, which is the condensed phase of FIB1, is determined by the degree to which the FIB1 concentration exceeds this threshold. PML bodies and stress granules show similar behavior (Nott et al. 2015). The threshold concentration for LLPS changes, depending on the post-translational modifications of proteins. Mixing unmodified FMRP and CAPRIN1 proteins in a test tube does not generate droplets, whereas when one protein is phosphorylated, droplets are formed (Kim et al. 2019). Droplets in cells can selectively incorporate biomolecules such as proteins and RNAs. PML protein in PML bodies, TIA1 in stress granules, mRNA in P bodies, and NEAT1 long noncoding (lnc)RNA in paraspeckles are essential for the formation of each type of droplet (Decker and Parker

2012; Ishov et al. 1999; Kedersha et al. 1999; Luo et al. 2018). These essential molecules are important for the selective incorporation of biomolecules inside of each droplet. Various biological reactions can be controlled in droplets by the selective incorporation of biomolecules. For example, the incorporation of ribozymes and substrate RNA into a droplet increases the activity of the ribozyme by a factor of over 70 (Poudyal et al. 2019). The polymerization of actin is accelerated when actin-related proteins and N-WASP, a protein that promotes actin polymerization, are incorporated into the droplets (Li et al. 2012). It is also possible to suppress a reaction by separating the substrates required for the reaction from the catalyst. Histone mRNA suppresses its own processing by encapsulating a protein required for processing into a histone protein droplet (Tatomer et al. 2016). The specificity of intermolecular reactions can be improved by incorporating molecules within the droplet, and some signal transductions and metabolic pathways utilize droplets. For example, T cell receptor signaling maintains a phosphorylation-dependent protein complex by incorporating kinases inside droplets that exclude phosphatases (Su et al. 2016).

Bimolecular LLPS in Cell-Like Molecular Environments

Proteins and nucleic acids form droplets through a network of various interactions, such as electrostatic interactions, hydrogen bonds, cation-π interactions, π-π interactions, and hydrophobic interactions (11) (Fig. 15.7). Droplet formation can thus be inhibited and droplets can be broken by increasing the reaction temperature to increase molecular motion, or by adding salts or charged compounds as shown

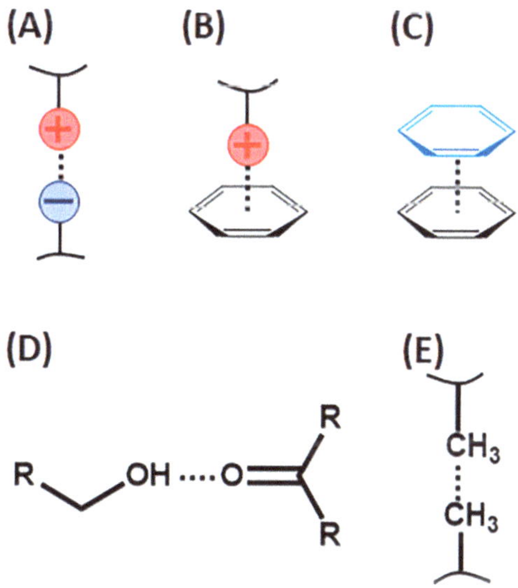

Fig. 15.7 Schematic illustration showing electrostatic (**a**), cation-π, (**b**), stacking (**c**), hydrogen bonding (**d**), and hydrophobic (**e**) interactions

below. The environmental dependence and reversibility of droplets are extremely important for the function of biomolecular LLPS.

Hydrotropes

The addition of a high concentration of monovalent cations such as sodium and potassium or of divalent cations such as magnesium to a droplet destabilizes electrostatic interactions and causes the droplet to disassemble. The addition of negatively charged small molecules or RNA can also lead to degradation of the droplet (Babinchak et al. 2020; Maharana et al. 2018; Decker and Parker 2012). ATP is a negatively charged compound present in cells at concentrations as high as several mM. A new biological function for ATP was reported in 2007: ATP acts as a hydrotrope in cells (Patel et al. 2017). Hydrotropes are relatively low-molecular-weight compounds containing hydrophilic and hydrophobic groups. Unlike surfactants, hydrotropes do not form micelles. Hydrotropes increase the water solubility of relatively hydrophobic molecules, such as proteins and drugs. The 2007 paper reported that heat-induced egg white coagulation was inhibited by the addition of 10 mM ATP, demonstrating that ATP acts as a hydrotrope. ATP has a similar effect on droplets formed by FUS, a protein critical to the onset of amyotrophic lateral sclerosis (ALS). FUS forms droplets in solution in the presence of 1 mM ATP, whereas 8 mM ATP inhibits FUS droplet formation (Patel et al. 2015). The effects of ATP on protein aggregation and droplet formation can be altered by intracellular ATP concentrations. The depletion of ATP might accelerate protein aggregation.

The effect of organic solvents on droplet formation has also been studied. 1,6-Hexanediol leads to the inhibition and dissolution of biomolecular droplets both in vivo and in vitro, whereas the structural isomer 2,5-hexanediol does not cause droplet dissolution. Neither1,6-hexanediol nor 2,5-hexanediol can degrade biomolecular aggregates. Therefore, 1,6-hexanediol but not 2,5-hexanediol shows hydrotropic properties, similar to ATP, against biomolecular LLPS. Utilizing these differences in responsiveness to hexanediols allows us to determine whether a granule is a droplet or an aggregate (Kroschwald et al. 2017). The effect of hexanediols has been observed in many studies, but the underlying molecular mechanism remains unclear.

Small Molecule Crowding

Osmolytes present at concentrations as high as several hundred mM in cells greatly affect the physical properties of proteins, including their ability to undergo LLPS. Osmolytes include amino acids, methylammonium, and sugars that regulate osmotic stress outside the cell. Typical osmolytes such as trimethylamine-N-oxide (TMAO) and betaine stabilize proteins, and TMAO reportedly stabilizes droplets and suppresses amyloid aggregation by TDP-43 protein (Choi et al. 2018). Droplets composed of immunoglobulin G and polyglutamine disintegrate upon the addition of saccharides (Mimura et al. 2019). It remains unclear how osmolytes alter the properties of biological macromolecules and how they affect biomolecular LLPS. Understanding the underlying mechanisms will necessitate detailed systematic studies.

Macromolecular Crowding

Biopolymers inside cells are present at concentrations as high as about 400 mg/mL. The effect of molecular crowding on biomolecules is often investigated by adding about 5 ~ 30 wt% of synthetic polymers (PEG, DEX, Ficoll, etc.) or proteins (such as albumin) to a dilute solution of the biomolecule of interest (Zimmerman and Minton 1993; Nakano et al. 2014). Generally, molecular crowding promotes intermolecular and dehydration reactions by altering hydration and the excluded volume effect (Zimmerman and Minton 1993). These effects promote the formation of droplets of various proteins, such as FUS, TAF 15, and hnRNP A1, in molecularly crowded environments (Patel et al. 2015; Wang et al. 2018). The proteins TDP-43 and G3 BP1 require the addition of several to 10 wt% of molecular crowding agents to form droplets (Yang et al. 2020; Wang et al. 2018). The LLPS ability of biomolecules probably increases, and droplets are more likely to form, in conditions that reproduce a crowded intracellular molecular environment than under the dilute conditions typical of in vitro experiments. Table 15.2 lists compounds used to study the effects of macromolecular crowding on the LLPS of proteins (Guillén-Boixet et al. 2020; Patel et al. 2015; Wang et al. 2018; Maharana et al. 2018; Boeynaems et al. 2017; Ferrolino et al. 2018; Rayman et al. 2018; Wegmann et al. 2018; Ray et al. 2020; Lin et al. 2015; Molliex et al. 2015; Yao et al. 2019; Vega et al. 2019; Yang et al. 2020; Mario Hofweber et al. 2018; Protter et al. 2018). The effects of molecular crowding on biomolecular LLPS have been discussed in recent review articles (Wang et al. 2024; André and Spruijt 2020).

Conclusion: Interrelationship Between Molecular Crowding and LLPS

Multimolecular crowding conditions comprise highly concentrated and/or oversaturated macromolecules, small osmolytes and metabolites, thereby severely restricting the diffusion of molecules, and especially of macromolecules. These conditions make it difficult to imagine how biomolecular reaction cascades, such as intracellular signal transductions, metabolic cycles and gene regulation can proceed in a spatiotemporal manner. Biomolecular LLPS allows biomolecules to participate in

Table 15.2 Crowding reagents that promote protein LLPS

Crowding agent	[Crowding agent]	Protein	References
PEG	1–8%	G3BP1	Guillén-Boixet et al. (2020)
	2–30%	FUS	Patel et al. (2015), Wang et al. (2018), Maharana et al. (2018), and Boeynaems et al. (2017)
	5–15%	NPM1	Ferrolino et al. (2018)
	10%	TIA1	Rayman et al. (2018)
	10%	Tau	Wegmann et al. (2018)
	10%	α-Synuclein	Ray et al. (2020)
	10–20%	hnRNPA1	Lin et al. (2015 and Molliex et al. (2015)
	30%	FMRP	Boeynaems et al. (2017)
Dextran	10%	FUS	Patel et al. (2015), Wang et al. (2018), and Maharana et al. (2018)
		TDP-43	Maharana et al. (2018)
		hnRNPA1	Wang et al. (2018)
		hnRNPA3	Wang et al. (2018)
		EWS	Wang et al. (2018)
		TAF15	Wang et al. (2018)
		FBL	Yao et al. (2019)
		Tau	Vega et al. (2019)
	15%	NPM1	Ferrolino et al. (2018)
Ficoll	5–20%	G3BP1	Yang et al. (2020)
	10%	hnRNPA1	Lin et al. (2015) and Molliex et al. (2015)
	12.5%	Tau	Wegmann et al. (2018)
	15%	FUS	Mario Hofweber et al. (2018)
	15%	NPM1	Ferrolino et al. (2018)
BSA	1.5–10%	hnRNPA1	Protter et al. (2018)
	10%	FUS	Lin et al. (2015)

specific binding and other reactions in multimolecular crowding biosystems and highlight the importance of partitioning molecules inside and outside droplets. Partitioning can be affected by molecular properties such as hydrophobicity, charge, H-bond donors and acceptors, and interactions with molecules inside droplets, as well as molecular factors outside the droplet. Molecular crowding reportedly greatly affects biomolecular LLPS. Crowding reagents are utilized in most protein LLPS studies conducted in test tubes, as listed in Table 15.2. Clearly, molecular crowding is essential for LLPS. The roles played by macromolecular crowding and LLPS in bacterial cellular functions were discussed in a recent review article (Monterroso et al. 2024) in which the authors emphasized the many unsolved questions concerning the structure, function, and regulation of biomolecular LLPS in cells. Elucidating the molecular mechanism underlying the mutual relationship between macromolecular crowding and LLPS is important to a wide range of research fields such as biomolecular physics, the study of cellular robustness and the origin of life, and to the development of molecularly targeted drugs, given that most biomolecules in cells may be phase separated and encapsulated in liquid droplets.

References

André AAM, Spruijt E (2020) Liquid-liquid phase separation in crowded environments. Int J Mol Sci 21(16). https://doi.org/10.3390/ijms21165908

Asamitsu S, Yabuki Y, Ikenoshita S, Kawakubo K, Kawasaki M, Usuki S, Nakayama Y, Adachi K, Kugoh H, Ishii K, Matsuura T, Nanba E, Sugiyama H, Fukunaga K, Shioda N (2021) CGG repeat RNA G-quadruplexes interact with FMRpolyG to cause neuronal dysfunction in fragile X-related tremor/ataxia syndrome. Sci Adv 7:eabd9440

Babinchak WM, Dumm BK, Venus S, Boyko S, Putnam AA, Jankowsky E, Surewicz WK (2020) Small molecules as potent biphasic modulators of protein liquid-liquid phase separation. Nat Commun 11(1):5574. https://doi.org/10.1038/s41467-020-19211-z

Boeynaems S, Bogaert E, Kovacs D, Konijnenberg A, Timmerman E, Volkov A, Guharoy M, De Decker M, Jaspers T, Ryan VH, Janke AM, Baatsen P, Vercruysse T, Kolaitis RM, Daelemans D, Taylor JP, Kedersha N, Anderson P, Impens F, Sobott F, Schymkowitz J, Rousseau F, Fawzi NL, Robberecht W, Van Damme P, Tompa P, Van Den Bosch L (2017) Phase separation of C9orf72 dipeptide repeats perturbs stress granule dynamics. Mol Cell 65(6):1044–1055.e1045. https://doi.org/10.1016/j.molcel.2017.02.013

Boyle Y, Johns TG, Fletcher EV (2022) Potassium ion channels in malignant central nervous system cancers. Cancers (Basel) 14(19). https://doi.org/10.3390/cancers14194767

Brangwynne CP, Eckmann CR, Courson DS, Rybarska A, Hoege C, Gharakhani J, Jülicher F, Hyman AA (2009) Germline P granules are liquid droplets that localize by controlled dissolution/condensation. Science 324(5935):1729–1732. https://doi.org/10.1126/science.1172046

Brangwynne CP, Mitchison TJ, Hyman AA (2011) Active liquid-like behavior of nucleoli determines their size and shape in Xenopus laevis oocytes. Proc Natl Acad Sci USA 108(11):4334–4339. https://doi.org/10.1073/pnas.1017150108

Cadart C, Venkova L, Piel M, Cosentino Lagomarsino M (2022) Volume growth in animal cells is cell cycle dependent and shows additive fluctuations. elife 11. https://doi.org/10.7554/eLife.70816

Chang J, Saraswathibhatla A, Song Z, Varma S, Sanchez C, Alyafei NHK, Indana D, Slyman R, Srivastava S, Liu K, Bassik MC, Marinkovich MP, Hodgson L, Shenoy V, West RB, Chaudhuri O (2024) Cell volume expansion and local contractility drive collective invasion of the basement membrane in breast cancer. Nat Mater 23(5):711–722. https://doi.org/10.1038/s41563-023-01716-9

Choi KJ, Tsoi PS, Moosa MM, Paulucci-Holthauzen A, Liao SJ, Ferreon JC, Ferreon ACM (2018) A chemical chaperone decouples TDP-43 disordered domain phase separation from fibrillation. Biochemistry 57(50):6822–6826. https://doi.org/10.1021/acs.biochem.8b01051

Chong PA, Vernon RM, Forman-Kay JD (2018) RGG/RG motif regions in RNA binding and phase separation. J Mol Biol 430(23):4650–4665. https://doi.org/10.1016/j.jmb.2018.06.014

Clark AG, Paluch E (2011) Mechanics and regulation of cell shape during the cell cycle. Results Probl Cell Differ 53:31–73. https://doi.org/10.1007/978-3-642-19065-0_3

Crick SL, Jayaraman M, Frieden C, Wetzel R, Pappu RV (2006) Fluorescence correlation spectroscopy shows that monomeric polyglutamine molecules form collapsed structures in aqueous solutions. Proc Natl Acad Sci USA 103(45):16764–16769. https://doi.org/10.1073/pnas.0608175103

Decker CJ, Parker R (2012) P-bodies and stress granules: possible roles in the control of translation and mRNA degradation. Cold Spring Harb Perspect Biol 4(9):a012286. https://doi.org/10.1101/cshperspect.a012286

Elbaum-Garfinkle S, Kim Y, Szczepaniak K, Chen CC, Eckmann CR, Myong S, Brangwynne CP (2015) The disordered P granule protein LAF-1 drives phase separation into droplets with tunable viscosity and dynamics. Proc Natl Acad Sci USA 112(23):7189–7194. https://doi.org/10.1073/pnas.1504822112

Ferrolino MC, Mitrea DM, Michael JR, Kriwacki RW (2018) Compositional adaptability in NPM1-SURF6 scaffolding networks enabled by dynamic switching of phase separation mechanisms. Nat Commun 9(1):5064. https://doi.org/10.1038/s41467-018-07530-1

Gilks N, Kedersha N, Ayodele M, Shen L, Stoecklin G, Dember LM, Anderson P (2004) Stress granule assembly is mediated by prion-like aggregation of TIA-1. Mol Biol Cell 15(12):5383–5398. https://doi.org/10.1091/mbc.e04-08-0715

Guillén-Boixet J, Kopach A, Holehouse AS, Wittmann S, Jahnel M, Schlüßler R, Kim K, Trussina I, Wang J, Mateju D, Poser I, Maharana S, Ruer-Gruß M, Richter D, Zhang X, Chang YT, Guck J, Honigmann A, Mahamid J, Hyman AA, Pappu RV, Alberti S, Franzmann TM (2020) RNA-induced conformational switching and clustering of G3BP drive stress granule assembly by condensation. Cell 181(2):346–361.e317. https://doi.org/10.1016/j.cell.2020.03.049

Hallegger M, Chakrabarti AM, Lee FCY, Lee BL, Amalietti AG, Odeh HM, Copley KE, Rubien JD, Portz B, Kuret K, Huppertz I, Rau F, Patani R, Fawzi NL, Shorter J, Luscombe NM, Ule J (2021) TDP-43 condensation properties specify its RNA-binding and regulatory repertoire. Cell 184 (18):4680–4696 e4622. https://doi.org/10.1016/j.cell.2021.07.018

Hanazawa M, Yonetani M, Sugimoto A (2011) PGL proteins self associate and bind RNPs to mediate germ granule assembly in C. Elegans. J Cell Biol 192 (6):929–937. https://doi.org/10.1083/jcb.201010106

Henninger JE, Oksuz O, Shrinivas K, Sagi I, LeRoy G, Zheng MM, Andrews JO, Zamudio AV, Lazaris C, Hannett NM, Lee TI, Sharp PA, Cisse II, Chakraborty AK, Young RA (2021) RNA-mediated feedback control of transcriptional condensates. Cell 184 (1):207–225:e224. https://doi.org/10.1016/j.cell.2020.11.030

Ishov AM, Sotnikov AG, Negorev D, Vladimirova OV, Neff N, Kamitani T, Yeh ET, Strauss JF 3rd, Maul GG (1999) PML is critical for ND10 formation and recruits the PML-interacting protein daxx to this nuclear structure when modified by SUMO-1. J Cell Biol 147(2):221–234. https://doi.org/10.1083/jcb.147.2.221

Iwashita K, Handa A, Shiraki K (2018) Coacervates and coaggregates: liquid-liquid and liquid-solid phase transitions by native and unfolded protein complexes. Int J Biol Macromol 120(Pt A):10–18. https://doi.org/10.1016/j.ijbiomac.2018.08.063

Jansson B (1996) Potassium, sodium, and cancer: a review. J Environ Pathol Toxicol Oncol 15(2–4):65–73

Kato M, Han TW, Xie S, Shi K, Du X, Wu LC, Mirzaei H, Goldsmith EJ, Longgood J, Pei J, Grishin NV, Frantz DE, Schneider JW, Chen S, Li L, Sawaya MR, Eisenberg D, Tycko R, McKnight SL (2012) Cell-free formation of RNA granules: low complexity sequence domains form dynamic fibers within hydrogels. Cell 149(4):753–767. https://doi.org/10.1016/j.cell.2012.04.017

Kedersha NL, Gupta M, Li W, Miller I, Anderson P (1999) RNA-binding proteins TIA-1 and TIAR link the phosphorylation of eIF-2 alpha to the assembly of mammalian stress granules. J Cell Biol 147(7):1431–1442. https://doi.org/10.1083/jcb.147.7.1431

Kim TH, Tsang B, Vernon RM, Sonenberg N, Kay LE, Forman-Kay JD (2019) Phospho-dependent phase separation of FMRP and CAPRIN1 recapitulates regulation of translation and deadenylation. Science 365(6455):825–829. https://doi.org/10.1126/science.aax4240

Kramer RM, Shende VR, Motl N, Pace CN, Scholtz JM (2012) Toward a molecular understanding of protein solubility: increased negative surface charge correlates with increased solubility. Biophys J 102(8):1907–1915. https://doi.org/10.1016/j.bpj.2012.01.060

Kroschwald S, Maharana S, Simon A (2017) Hexanediol: a chemical probe to investigate the material properties of membrane-less compartments. Matters. https://doi.org/10.19185/matters.201702000010

Li P, Banjade S, Cheng HC, Kim S, Chen B, Guo L, Llaguno M, Hollingsworth JV, King DS, Banani SF, Russo PS, Jiang QX, Nixon BT, Rosen MK (2012) Phase transitions in the assembly of multivalent signalling proteins. Nature 483(7389):336–340. https://doi.org/10.1038/nature10879

Li HR, Chiang WC, Chou PC, Wang WJ, Huang JR (2018) TAR DNA-binding protein 43 (TDP-43) liquid-liquid phase separation is mediated by just a few aromatic residues. J Biol Chem 293(16):6090–6098. https://doi.org/10.1074/jbc.AC117.001037

Lin Y, Protter DS, Rosen MK, Parker R (2015) Formation and maturation of phase-separated liquid droplets by RNA-binding proteins. Mol Cell 60(2):208–219. https://doi.org/10.1016/j.molcel.2015.08.018

Luo Y, Na Z, Slavoff SA (2018) P-bodies: composition, properties, and functions. Biochemistry 57(17):2424–2431. https://doi.org/10.1021/acs.biochem.7b01162

Maharana S, Wang J, Papadopoulos DK, Richter D, Pozniakovsky A, Poser I, Bickle M, Rizk S, Guillén-Boixet J, Franzmann TM, Jahnel M, Marrone L, Chang YT, Sterneckert J, Tomancak P, Hyman AA, Alberti S (2018) RNA buffers the phase separation behavior of prion-like RNA binding proteins. Science 360(6391):918–921. https://doi.org/10.1126/science.aar7366

Mario Hofweber SH, Bourgeois B, Spreitzer E, Niedner-Boblenz A, Schifferer M, Ruepp M-D, Simons M, Niessing D, Madl T, Dormann D (2018) Phase separation of FUS is suppressed by its nuclear import receptor and arginine methylation. Cell 173:706–719

Matsuda A, Mimura M, Maruyama T, Kurinomaru T, Shiuhei M, Shiraki K (2018) Liquid droplet of protein-polyelectrolyte complex for high-concentration formulations. J Pharm Sci 107(10):2713–2719. https://doi.org/10.1016/j.xphs.2018.06.021

Milo R, Phillips R (2015) Cell biology by the numbers. Garland Science

Mimura M, Tsumura K, Matsuda A, Akatsuka N, Shiraki K (2019) Effect of additives on liquid droplet of protein-polyelectrolyte complex for high-concentration formulations. J Chem Phys 150(6):064903. https://doi.org/10.1063/1.5063378

Molliex A, Temirov J, Lee J, Coughlin M, Kanagaraj AP, Kim HJ, Mittag T, Taylor JP (2015) Phase separation by low complexity domains promotes stress granule assembly and drives pathological fibrillization. Cell 163(1):123–133. https://doi.org/10.1016/j.cell.2015.09.015

Monterroso B, Margolin W, Boersma AJ, Rivas G, Poolman B, Zorrilla S (2024) Macromolecular crowding, phase separation, and homeostasis in the orchestration of bacterial cellular functions. Chem Rev 124(4):1899–1949. https://doi.org/10.1021/acs.chemrev.3c00622

Nakano S, Miyoshi D, Sugimoto N (2014) Effects of molecular crowding on the structures, interactions, and functions of nucleic acids. Chem Rev 114(5):2733–2758. https://doi.org/10.1021/cr400113m

Nott TJ, Petsalaki E, Farber P, Jervis D, Fussner E, Plochowietz A, Craggs TD, Bazett-Jones DP, Pawson T, Forman Kay JD, Baldwin AJ (2015) Phase transition of a disordered nuage protein generates environmentally responsive membraneless organelles. Mol Cell 57(5):936–947. https://doi.org/10.1016/j.molcel.2015.01.013

Oparin AI (1924) The origin of life. Moskovskii rabochii

Parry BR, Surovtsev IV, Cabeen MT, O'Hern CS, Dufresne ER, Jacobs-Wagner C (2014) The bacterial cytoplasm has glass-like properties and is fluidized by metabolic activity. Cell 156(1–2):183–194. https://doi.org/10.1016/j.cell.2013.11.028

Patel A, Lee HO, Jawerth L, Maharana S, Jahnel M, Hein MY, Stoynov S, Mahamid J, Saha S, Franzmann TM, Pozniakovski A, Poser I, Maghelli N, Royer LA, Weigert M, Myers EW, Grill S, Drechsel D, Hyman AA, Alberti S (2015) A liquid-to-solid phase transition of the ALS protein FUS accelerated by disease mutation. Cell 162(5):1066–1077. https://doi.org/10.1016/j.cell.2015.07.047

Patel A, Malinovska L, Saha S, Wang J, Alberti S, Krishnan Y, Hyman AA (2017) ATP as a biological hydrotrope. Science 356(6339):753–756. https://doi.org/10.1126/science.aaf6846

Poudyal RR, Guth-Metzler RM, Veenis AJ, Frankel EA, Keating CD, Bevilacqua PC (2019) Template-directed RNA polymerization and enhanced ribozyme catalysis inside membraneless compartments formed by coacervates. Nat Commun 10(1):490. https://doi.org/10.1038/s41467-019-08353-4

Protter DSW, Rao BS, Van Treeck B, Lin Y, Mizoue L, Rosen MK, Parker R (2018) Intrinsically disordered regions can contribute promiscuous interactions to RNP granule assembly. Cell Rep 22(6):1401–1412. https://doi.org/10.1016/j.celrep.2018.01.036

Qamar S, Wang G, Randle SJ, Ruggeri FS, Varela JA, Lin JQ, Phillips EC, Miyashita A, Williams D, Strohl F, Meadows W, Ferry R, Dardov VJ, Tartaglia GG, Farrer LA, Kaminski Schierle GS, Kaminski CF, Holt CE, Fraser PE, Schmitt-Ulms G, Klenerman D, Knowles T, Vendruscolo M, St George-Hyslop P (2018a) FUS phase separation is modulated by a molecular chaperone and methylation of arginine cation-pi interactions. Cell 173 (3):720–734 e715. https://doi.org/10.1016/j.cell.2018.03.056

Qamar S, Wang G, Randle SJ, Ruggeri FS, Varela JA, Lin JQ, Phillips EC, Miyashita A, Williams D, Ströhl F, Meadows W, Ferry R, Dardov VJ, Tartaglia GG, Farrer LA, Kaminski Schierle GS, Kaminski CF, Holt CE, Fraser PE, Schmitt-Ulms G, Klenerman D, Knowles T, Vendruscolo M, St George-Hyslop P (2018b) FUS phase separation is modulated by a molecular chaperone and methylation of arginine cation-π interactions. Cell 173(3):720–734.e715. https://doi.org/10.1016/j.cell.2018.03.056

Ray S, Singh N, Kumar R, Patel K, Pandey S, Datta D, Mahato J, Panigrahi R, Navalkar A, Mehra S, Gadhe L, Chatterjee D, Sawner AS, Maiti S, Bhatia S, Gerez JA, Chowdhury A, Kumar A, Padinhateeri R, Riek R, Krishnamoorthy G, Maji SK (2020) Alpha-Synuclein aggregation nucleates through liquid-liquid phase separation. Nat Chem 12(8):705–716. https://doi.org/10.1038/s41557-020-0465-9

Rayman JB, Karl KA, Kandel ER (2018) TIA-1 self-Multimerization, phase separation, and recruitment into stress granules are dynamically regulated by Zn(2). Cell Rep 22(1):59–71. https://doi.org/10.1016/j.celrep.2017.12.036

Reijns MA, Alexander RD, Spiller MP, Beggs JD (2008) A role for Q/N-rich aggregation-prone regions in P-body localization. J Cell Sci 121(Pt 15):2463–2472. https://doi.org/10.1242/jcs.024976

Riès-Kautt M, Ducruix A (1997) [3] inferences drawn from physicochemical studies of crystallogenesis and precrystalline state. Methods Enzymol 276:23–59. doi:https://doi.org/10.1016/s0076-6879(97)76049-x, Elsevier

Su X, Ditlev JA, Hui E, Xing W, Banjade S, Okrut J, King DS, Taunton J, Rosen MK, Vale RD (2016) Phase separation of signaling molecules promotes T cell receptor signal transduction. Science 352(6285):595–599. https://doi.org/10.1126/science.aad9964

Tatomer DC, Terzo E, Curry KP, Salzler H, Sabath I, Zapotoczny G, McKay DJ, Dominski Z, Marzluff WF, Duronio RJ (2016) Concentrating pre-mRNA processing factors in the histone locus body facilitates efficient histone mRNA biogenesis. J Cell Biol 213(5):557–570. https://doi.org/10.1083/jcb.201504043

Taymaz-Nikerel H, Borujeni AE, Verheijen PJ, Heijnen JJ, van Gulik WM (2010) Genome-derived minimal metabolic models for Escherichia coli MG1655 with estimated in vivo respiratory ATP stoichiometry. Biotechnol Bioeng 107(2):369–381. https://doi.org/10.1002/bit.22802

Timothy JN, Evangelia Petsalaki, Patrick Farber DJ, 4 Eden Fussner,1 Anne Plochowietz,5 Timothy D. Craggs,5 David P. Bazett-Jones,3,6 Tony Pawson,1,7 Julie D. Forman-Kay,3,6,* and Andrew J. BaldwinJulie D. Forman-Kay, Andrew J. Baldwin (2015) Phase transition of a disordered Nuage protein generates environmentally responsive Membraneless organelles. Mol Cell 57:936–947. doi:https://doi.org/10.1016/j.molcel.2015.01.013

Vega IE, Umstead A, Kanaan NM (2019) EFhd2 affects tau liquid-liquid phase separation. Front Neurosci 13:845. https://doi.org/10.3389/fnins.2019.00845

Wang J, Choi JM, Holehouse AS, Lee HO, Zhang X, Jahnel M, Maharana S, Lemaitre R, Pozniakovsky A, Drechsel D, Poser I, Pappu RV, Alberti S, Hyman AA (2018) A molecular grammar governing the driving forces for phase separation of prion-like RNA binding proteins. Cell 174(3):688–699.e616. https://doi.org/10.1016/j.cell.2018.06.006

Wang X, Liu J, Mao C, Mao Y (2024) Phase separation-mediated biomolecular condensates and their relationship to tumor. Cell Commun Signal 22(1):143. https://doi.org/10.1186/s12964-024-01518-9

Weber SC, Brangwynne CP (2015) Inverse size scaling of the nucleolus by a concentration-dependent phase transition. Curr Biol 25(5):641–646. https://doi.org/10.1016/j.cub.2015.01.012

Wegmann S, Eftekharzadeh B, Tepper K, Zoltowska KM, Bennett RE, Dujardin S, Laskowski PR, MacKenzie D, Kamath T, Commins C, Vanderburg C, Roe AD, Fan Z, Molliex AM, Hernandez-Vega A, Muller D, Hyman AA, Mandelkow E, Taylor JP, Hyman BT (2018) Tau protein liquid-liquid phase separation can initiate tau aggregation. EMBO J 37(7). https://doi.org/10.15252/embj.201798049

Yancey PH (2005) Organic osmolytes as compatible, metabolic and counteracting cytoprotectants in high osmolarity and other stresses. J Exp Biol 208(Pt 15):2819–2830. https://doi.org/10.1242/jeb.01730

Yang P, Mathieu C, Kolaitis RM, Zhang P, Messing J, Yurtsever U, Yang Z, Wu J, Li Y, Pan Q, Yu J, Martin EW, Mittag T, Kim HJ, Taylor JP (2020) G3BP1 is a tunable switch that triggers phase separation to assemble stress granules. Cell 181(2):325–345.e328. https://doi.org/10.1016/j.cell.2020.03.046

Yao RW, Xu G, Wang Y, Shan L, Luan PF, Wang Y, Wu M, Yang LZ, Xing YH, Yang L, Chen LL (2019) Nascent pre-rRNA sorting via phase separation drives the assembly of dense Fibrillar components in the human nucleolus. Mol Cell 76(5):767–783.e711. https://doi.org/10.1016/j.molcel.2019.08.014

Zimmerman SB, Minton AP (1993) Macromolecular crowding: biochemical, biophysical, and physiological consequences. Annu Rev Biophys Biomol Struct 22:27–65. https://doi.org/10.1146/annurev.bb.22.060193.000331

Chapter 16
Reshuffling Overcrowded Milieu: Stress-Induced Reorganization of the Eukaryotic Membrane-Less Organelles

Anastasia A. Gavrilova, Eva A. Shmidt, Vladimir N. Uversky, and Alexander V. Fonin

Abstract Cells are crowded entities, but the intracellular space represents an inhomogeneously crowded environment, where the concentrations of macromolecules (proteins, nucleic acids, etc.) are not uniformly distributed throughout the cell resulting in regions with different levels of crowding. Liquid-liquid phase separation (LLPS)-driven formation of various membrane-less organelles (MLOs) represents a means for the control, regulation, and redistribution of cellular crowded environment. Because MLOs contain the high concentrations of biological macromolecules (proteins and RNAs), often significantly exceeding those of the surrounding cytoplasm or nucleoplasm, their inside represents an overcrowded milieu. It is well-known that the appearance of the stress-induced MLOs represents a reaction to various types of stresses, enabling the protection of the genetic and protein material during hostile conditions. However, stress can also cause structural, functional, and compositional changes in the MLOs, which are constitutively present in the cells, thereby causing the reshuffling of the overcrowded environment. This chapter describes stress-induced changes in several MLOs (nucleolus, Cajal bodies, paraspeckles, nuclear speckles, NELF Bodies, nucleolar stress bodies, PML-bodies, stress-granules, and P-bodies) found in the eukaryotic cells.

Keywords Membrane-less organelles · Liquid-liquid-phase separation · Overcrowding · Nucleolus · Cajal bodies · Paraspeckles · Nuclear speckles · NELF-bodies · Nucleolar stress bodies · PML-bodies · Stress-granules · P-bodies

---

A. A. Gavrilova · E. A. Shmidt · A. V. Fonin (✉)
Institute of Cytology of the Russian Academy of Sciences, St. Petersburg, Russia
e-mail: alexfonin@incras.ru

V. N. Uversky
Department of Molecular Medicine and USF Health Byrd Alzheimer's Research Institute, Morsani College of Medicine, University of South Florida, Tampa, FL, USA

V. N. Uversky (ed.), *(Macro)Molecular Crowding*, Subcellular Biochemistry 109,
https://doi.org/10.1007/978-3-032-03370-3_16

Introduction

Cells are inhomogeneously crowded possessing regions with different levels of crowding. The most evident illustration of this inhomogeneity is the presence of numerous membrane-less organelles (MLOs) or biomolecular condensates that are formed as a result of the tightly controlled liquid-liquid phase separation (LLPS). The major drivers of LLPS are intrinsically disordered proteins (IDPs) or proteins containing intrinsically disordered regions (IDRs), as structural pliability and the capability of IDPs/IDRs to be involved in weak multivalent interactions define the broad engagement of these proteins in the biological LLPS and MLO biogenesis (Uversky 2015, 2017a, b, 2021; Antifeeva et al. 2022; Darling et al. 2018, 2019; Fonin et al. 2022a; Gupta and Uversky 2024; Kulkarni et al. 2025; Nesterov et al. 2021, 2025; Turoverov et al. 2019; Uversky and Finkelstein 2019; Uversky et al. 2015; Kulkarni et al. 2022). In fact, fluidity of MLOs originates from the multivalent interactions between IDPs or proteins containing IDRs that are not accompanied by noticeable alteration in the structure of proteins undergoing LLPS (Uversky 2017a, b; Darling et al. 2018). Therefore, MLOs represent a special form of disorder-based protein complexes (Uversky 2017a, b; Uversky et al. 2015; Nott et al. 2015) that can be considered as an illustration of the disorder-based emergent behavior of proteins (Darling et al. 2019; Turoverov et al. 2019; Uversky 2013, 2019).

Because of their highly dynamic nature, transformability, and pliability, as well as the capability of free exchange of their content with the environment, MLOs are characterized by the abilities to form rapidly in response to minimal changes in the intracellular environment (temperature, pH, ionic strength, concentration of the target protein, etc.) (Banani et al. 2017) and to undergo reorganization if conditions are changed. In-depth coverage of the MLOs that are either formed *de novo* or reorganized during stress -response in prokaryotic and eukaryotic cells can be found in a recent comprehensive review (Fefilova et al. 2022). In this chapter, we describe stress-induced changes in several MLOs (nucleolus, Cajal bodies, paraspeckles, nuclear speckles, NELF-Bodies, nucleolar stress bodies, PML-bodies, stress-granules, and P-bodies) found in the eukaryotic cells.

Nucleolus

The nucleolus is the largest substructure of the nucleus and the largest membrane-less organelle in a eukaryotic cell. Being the best known and most studied of the membrane-less organelles (MLOs), nucleolus is also one of the subcellular compartments known to the scientific community for the longest time (it was discovered about 250 years ago, being first mentioned by Felice Fontana in 1781 and named "nucleolus" by Valentin in 1839 (Wagner 1835; Valentin 1836), as reviewed in (Pederson 2011)). The main function of the nucleolus, the biogenesis of ribosomal subunits, was determined in the mid-1960s (Yang et al. 2018).

In 1999, following the discovery that the cell cycle regulators phosphatase CDC14 (cell division control protein 14) and E3 ubiquitin-protein ligase MDM2 (mouse double minute 2 homolog) could be transiently localized to the nucleolus (Visintin et al. 1999), the nucleolar sequestration hypothesis was put forward (Bachant and Elledge 1999). According to this hypothesis, the nucleolus can temporarily sequester or trap certain proteins and other molecules in response to cellular stress or other stimuli, with such sequestration possessing regulatory effects on various cellular processes, such as cell cycle progression, gene expression, and stress response. The nucleolus triggers a p53-dependent cellular response to various stimuli and is a sensor of cellular stress (Yang et al. 2018). This cellular reaction is defined by the term "nucleolar stress". Currently, the concept of nucleolar stress is used to describe various distortions in the morphology and function of the nucleolus, which ultimately lead to disturbances in the cell homeostasis. Nucleolar stress is a common phenomenon in neurodegenerative disorders, such as Parkinson's disease (Parlato and Liss 2014), Alzheimer's disease (Arnold et al. 2013), and Huntington's disease (Kim et al. 2021). Although these diseases can have many causes, nucleolar stress may be one of the significant mediators of neuronal degeneration or loss.

Nucleolus is characterized by a complex, multiphase structure, where one can find the following elements: fibrillar center (FC), dense fibrillar component (DFC), and granular component (GC) (Pederson 2011). FC is a place where transcription from ribosomal DNA (rDNA) occurs. Transcriptionally active rDNA genes are located in the FC/DFC border region. rDNA genes form loop structures, which increases the rate of transcription, since this structure allows you to switch from termination to reinitiation of transcription (Denissov et al. 2011). The DFC contains synthesized pre-ribosomal RNA (pre-rRNA), and processing of the 47S pre-rRNA transcript precursor occurs here as well. Pre-rRNA is either co- or post-transcriptionally processed by ribosome assembly factors (AFs) and modified by small nucleolar ribonucleoproteins (snoRNPs) to produce 28S, 18S and 5.8S rRNA. These snoRNP-mediated modifications include 2'-O-methylation and pseudouridine formation (Mangan et al. 2017; Boulon et al. 2010). AFs also facilitate the binding of ribosomal proteins (RPs) and the folding of maturing subunits. Defects in RPs at the organismal level can cause a class of rare human diseases called ribosomopathies. The DFC scaffold protein is likely a fibrillarin 2′-O-methyltransferase forming a cluster network in the DFC surrounding the FC (Yao et al. 2019). Late rRNA processing and ribosomal unit assembly occur in the GC surrounding the FC and DFC. Interacting with RP, 28S, 18S, and 5.8S rRNA form small and large pre-ribosomal subunits, each of which is exported to the cytoplasm, modified and involved in the formation of mature ribosomal subunits 40S and 60S (Frottin et al. 2019; Correll et al. 2024).

Various types of stress cause reorganization of the nucleolus and the formation of stress-specific intranucleolar structures. For example, nucleolar caps form when DNA is damaged by ionizing radiation or RNA polymerase I is inhibited by actinomycin D. Nucleolar caps contain FC and DFC around GC components (Puvion-Dutilleul et al. 1992). Nucleolar functions are reversibly altered when cells are treated with 5,6-dichloro-1-β-D-ribofuranosylbenzimidazole (DRB). DRB reduces the rate but does not inhibit rDNA transcription by RNA polymerase I (POL-I). As a result, the nucleolus reorganizes to a "nucleolar necklace", occurs when the nucleolus loses material and unravels into beaded strands, giving it the appearance of a necklace. Each necklace bead includes FC and DFC (i.e., corresponds to the FC-DFC module) and contains rDNA, rRNA, POL-I, DNA topoisomerase I (TOP1), rRNA 2′-O-methyltransferase fibrillarin (FBL), and UBF (Upstream Binding Factor, which is a nucleolar phosphoprotein that plays a crucial role in the transcription of rRNA genes by POL-I) (Louvet et al. 2005). The formation of such structures under stress has several advantages. First, damaged rDNA exposed on the surface of the nucleolus becomes available for repair. Second, nucleolar caps or necklace beads concentrate high levels of homologous sequences in close proximity to each other, which may facilitate repair by homologous recombination.

In response to heat shock and acidosis, the nucleoli reorganize into reversible amyloid bodies (A-bodies). These formations are interesting, first of all, because, having a structural similarity to pathological amyloids, they perform important biological functions and can reversibly assemble and disassemble. Exposure to stress mediates transcription from the intergenic spacer (IGS) region of rDNA, which is rich in dinucleotide repeats. These transcribed regions are located ~16 kb, ~22 kb, and ~ 28 kb downstream of the rRNA transcription start site. Some proteins are sequestered in the nucleolus, leading to the formation of several "foci". These "foci" are spherical in shape, contain mobile proteins and can undergo fusion, thereby exhibiting properties inherent to dynamic liquid condensates. These lesions then mature into aggregates containing immobilized proteins. Further in line, the capture and immobilization of free proteins occurs. A-body maturation ends once pools of cellular mobile proteins are depleted, resulting in a distinct amyloid structure. Disassembly of A-bodies occurs within 1–2 hours after the stimulus ceases. This process requires the heat shock proteins hsp70 and hsp90.

Proteins recruited to A-bodies are thought to have an ACM (amyloid-converting motif), which has a bipartite structure: an arginine-histidine-rich domain that, due to its positive charge, interacts with RNA, and a motif with a high propensity for amyloid formation. This ensures, on the one hand, interaction with RNA, and on the other, amyloidogenesis. Examples of such proteins are von Hippel-Lindau disease tumor suppressor (VHL), cell division control protein 73 (CDC73), MDM2, DNA polymerase delta catalytic subunit (POLD1), etc. It has been shown that the proteomic composition of A-bodies formed as a result of exposure to heat shock and low pH overlaps by approximately 20% (Marijan et al. 2024). It was demonstrated in (Marijan et al. 2019) that some proteins have temperature-sensitive structures and can mediate specific recruitment to A-bodies. However, previous acidosis is known to prevent sequestration of the heat shock-specific protein FEN1 (Flap endonuclease 1). It is clear that if heat alone was sufficient to cause refolding, the protein would be recruited to A-bodies (Marijan et al. 2019). It is likely that acidotic conditions block elements of the heat shock signaling pathway, such as heat shock-induced transcript expression. Accordingly, protein sequestration in A-bodies is suppressed. The mechanism of specific recruitment is likely based not only on the properties of the protein, but also on the structure of the non-coding RNA (ncRNA) transcribed from IGS.

A study by a group led by Stephen Lee (Bokros et al. 2024) found the accumulation of polyadenylated RNA in A-bodies. It has been shown that stress-induced polyadenylated RNA may be involved in physiological amyloidogenesis. A targeted screening of canonical and non-canonical poly(A) polymerases showed that TENT4b is the main polymerase involved in stress-induced polyadenylation of A-body RNA under temperature stress. Thus, it is likely that amyloidogenesis is not always a passive event, but rather a controlled enzymatic process.

Cajal Bodies

In 1903, Spanish cytologist Santiago Ramón y Cajal (1852–1934) developed a method for identifying cellular structures based on the reduction of silver ions bound to proteins and ribonucleoproteins (Cajal 1903). The new silver staining technique allowed Cajal to characterize various intracellular components, mainly nuclear and cytoskeletal. Along the way, Cajal discovered an important nuclear structure, later defined as Cajal bodies (CB) (Gall 2000). CBs are dynamic nuclear nonmembranous organelles that are functionally and often physically associated with the nucleolus. Depending on the organism, they have a size from 0.2 to 2 microns, whereas their number can vary from 1 to 6 (Love et al. 2017). The number and composition of canonical CBs change throughout the cell cycle, with CB size being greatest at the G1/S phase transition (Boulon et al. 2010). During mitosis, CBs are disassembled and reassembled at the beginning of G1 when transcription resumes (Strzelecka et al. 2010).

The main function of Cajal bodies is considered to be participation in the metabolism of small nuclear ribonucleoproteins (snRNPs). Precursors of the small nuclear RNAs (pre-snRNAs) are modified in the CB before export to the cytoplasm. After returning from the cytoplasm to the nucleus, the main snRNPs (snRNA + spliceosomal Sm proteins) return to the CB, undergoing additional modifications there. Mature snRNPs leave the CB and localize to nuclear speckles. Defective snRNPs are sequestered into the CB by an unknown mechanism. Human telomerase RNA (hTR) and small nucleolar RNA (snoRNA) are also modified in CB. These nuclear structures are transcription-dependent, and their number depends on the cell's need for splicing and ribosome biogenesis (Lafarga et al. 2017). They are especially prominent in neurons, where they are often associated with the nucleolus. Disruption and loss of neuronal CBs are known to be associated with severe neuronal dysfunction in neurological disorders such as motor neuron diseases. A study (Tapia et al. 2012) showed that in spinal muscular atrophy type I (SMA type I), the most common genetic cause of childhood mortality caused by the absence of exons 7 and 8, CB reorganization occurs. Compared with control samples, CBs from patients with CMA type I were smaller in size and intranucleolar in location. CB depletion and reorganization reflect severe dysfunction of pre-mRNA splicing in motor neurons, which is a cause of neuronal degeneration (Tapia et al. 2012).

CBs are complex, highly dynamic structures formed in the nucleoplasm as a result of liquid-liquid phase separation (LLPS) (Dundr 2011). Components of these nuclear bodies can rapidly exchange with the surrounding nucleoplasm, however, the molecular mechanisms of how CBs form and maintain their structural integrity are still not fully understood. In a study (Kaiser et al. 2008), the authors conducted experiments on the immobilization of the structural components of the Cajal body on chromatin in living cells. Using this approach, they demonstrated that any component of the Cajal body can initiate the formation of an entire nuclear body. Therefore, it is likely that CBs are not formed by hierarchical ordered assembly, but rather represent a stochastically self-organizing structure. One of the key CB proteins is the IDR-containing protein coilin. It should be noted that some organisms (e.g., *Drosophila*, *C. elegans*, yeast) lack coilin or its obvious homologs, making the identification of CBs homologues difficult. In budding yeast, a CB analogue called the "nucleolar body" is found within the nucleolus.

CBs overlap compositionally and spatially with the nucleoli (e.g., the proteins fibrillarin, nucleolin, Nopp140 (nucleolar and coiled-body phosphoprotein 1), and nucleolar protein NAP57 (also known as H/ACA ribonucleoprotein complex subunit DKC1) are common to the two organelles) and, as a result, have a close functional relationship. Given the important role of the nucleolus in the stress response, it should not be surprising that CBs also respond to stress (Boulon et al. 2010).

In response to various types of stress, CBs are reorganized and core proteins are dislocated (e.g., coilin (Cioce et al. 2006)) or undergo proteasomal degradation (e.g., FLASH protein also known as CASP8-associated protein 2 (Bongiorno-Borbone et al. 2010)). Cell starvation has been shown to reduce the number and size of CBs (Andrade et al. 1993). Upon UV irradiation, some CB components, such as coilin and snRNP, are redistributed to the nucleus, but others (fibrillarin and Nopp140) remain (Zhang and Lu 2009).

Cold stress of the soybean root meristem cells was shown to reduce the amount of CBs, followed by a recovery in their amount after stress removal. However, this decrease may be caused by difficulty in the formation of CBs or their fusion (Stępiński 2020). CB disassembly may be caused by changes in intermolecular interactions associated with stress-induced post-translational modifications of proteins (e.g., SUMOylation of CB proteins under stress (Navascues et al. 2008)) and/or degradation of CB components through the proteasome pathway. Thus, the participation of the proteasome activator subunit PA28g in the nuclear redistribution of coilin caused by UV-C was clearly demonstrated (Cioce et al. 2006). It has also been shown that coilin does not degrade during stress as its cellular levels remain constant, but rather changes its localization.

Viral infections can lead to different CB responses and reactions. For example, Herpes simplex virus 1 (HSV-1) infection causes the relocation of certain CB proteins (coilin, SMN and fibrillarin) to damaged centrosomes (Morency et al. 2007). Adenoviruses cause reorganization of CB into smaller structures and redistribution of coilin and some other CB components at the periphery of viral replication centers to participate in the processing of virial transcripts (James et al. 2010). Influenza A virus infection causes an increase in the number of CBs and a simultaneous decrease in their size (Fortes et al. 1995). In plants, groundnut rosette virus (GRV) causes the fusion of CBs containing the viral protein ORF3 with the nucleolus (Kim et al. 2007). However, evidence regarding the functional importance of these structural changes has been rather inconsistent. For example, it has been reported that knockdown of coilin in Nicotiana plants can increase the accumulation of barley stripe mosaic virus (BSMV) and tomato golden mosaic virus (TGMV), promoting virus spread. On the other hand, the same study using the same knockdown system reported reduced virus accumulation in turnip vein clearing virus (TVCV) and potato virus Y (PVY), which was also associated with suppression of symptom progression (Love et al. 2017; Shav-Tal et al. 2005).

Overall, available evidence suggests that CBs are highly sensitive to various types of stress. However, the question remains whether the observed structural changes are a consequence of the cell's response to stress or part of its regulation, and if the latter, then the exact mechanisms remain to be elucidated.

Paraspeckles

Paraspeckles are stress-sensitive nuclear bodies whose assembly is controlled by the long non-coding RNA NEAT1 (Nuclear Paraspeckle Assembly Transcript 1). They were discovered in the interchromatin space of the nucleus of mammalian cells (Fox and Lamond 2010; Visa et al. 1993). The number of paraspeckles varies among different cell types. For example, each HeLa cell contains about 10–30 paraspeckles (Wang et al. 2018b). The formation of paraspeckles begins with the transcribing of single-exon RNA from the NEAT1 gene, which is spliced in human cells. This results in a short 3.7-kb (NEAT1_1) and a long 22.7-kb (NEAT1_2) isoform (Sasaki et al. 2009), with the latter playing a dominant role in the formation of paraspeckles (Fox and Lamond 2010). This non-coding RNA probably co-transcriptionally binds to members of the DBHS (Drosophila Behavior Human Splicing) family of proteins SFPQ and NONO, which together form functionally active SFPQ-NONO heterodimers. SFPQ-NONO are structural components of paraspeckles, and interaction with RNA leads to the oligomerization of SFPQ-NONO heterodimers into longer polymer chains along transcripts, which increases the multivalency of the system (Nakagawa et al. 2011). As it matures, additional proteins such as FUS, TDP43, and Rbm14 are recruited into the structure. Currently, more than 40 proteins have been discovered in the structure of paraspeckles. (West et al. 2014). In 2016, a model was proposed according to which paraspeckles are a spheroid consisting of a core and a shell. The core contains the middle hydrophobic portion of the NEAT1_2 transcripts, as well as the SFPQ, NONO, and PSPC1 proteins. The shell is formed by the 5′ and 3′ hydrophilic ends of NEAT1_2 and various proteins, such as TDP43. Paraspeckles were initially described as constitutive structures, however, it was later shown that their number increases under stress. Thus, it is known that the number of paraspeckles increases during infection with viruses, inflammation, aging and cancer (McCluggage and Fox 2021).

The function of paraspeckles is the regulation of genes due to the sequestration of proteins (for example, the structural protein of paraspeckles—SFPQ (splicing factor, proline- and glutamine-rich), regulates specific genes associated with immunity (Sirri et al. 2008)). Also, the amount of mRNA in paraspeckles changes in response to cellular stresses (Govoni et al. 1994) and during the regulation of the circadian rhythm (Rubbi and Milner 2003). NEAT1 knockout mice that lack paraspeckles have been shown to be viable, but nearly half of naturally mated female mice stochastically fail to become pregnant, likely due to impaired corpus luteum function (Nakagawa et al. 2014).

It is known that paraspeckles can have 2 shapes: spherical and elongated, which is not typical for most MLOs. Under normal conditions in mouse corpus luteum cells, most paraspeckles appear as dispersed spheroids, but a small subpopulation of paraspeckles exhibit elongated paraspeckles (West et al. 2014), which can be formed by proteasome inhibition. Stress events such as heat shock and hypoxia induce the formation of spherical paraspeckles (Choudhry et al. 2015; Lellahi et al. 2018), and mitochondrial stress and the concomitant loss of mitochondrial proteins lead to the formation of elongated paraspeckles. Such morphological changes are associated with changes in posttranscriptional processing of NEAT1, favoring the formation of long NEAT1_2 compared to short NEAT1_1. Thus, there is a crosstalk between mitochondria and paraspeckles: changes in NEAT1 transcriptional regulation mediated by mitochondrial stressors lead to changes in paraspeckle morphology (Wang et al. 2018b).

During viral infections, the number of paraspeckles increases. Increased numbers of paraspeckles enhance sequestration of the SFPQ protein, which is a suppressor of several antiviral genes (eg, RIG-I and IL-8). The de repression of the corresponding genes occurs. This mechanism has been observed for Hepatitis D, Influenza, polyI:C infection, and Hantavirus (Imamura et al. 2014; Beeharry et al. 2018; Ma et al. 2017). Furthermore, paraspeckles are involved in the nuclear sequestration of viral mRNA, for example, REV-dependent HIV-1 transcripts (Zhang et al. 2013). Paraspeckles also play an important role in the antibacterial immune response. For example, upregulation of NEAT1 is observed in response to Salmonella infection (Imamura et al. 2018). Overall, these data indicate that paraspeckles, although not essential MLOs, help cells adapt to certain environmental conditions.

Nuclear Speckles

Nuclear speckles (NS) are one of the largest membrane-less nuclear organelles and exist in almost all cells of a wide range of eukaryotes. These are dynamic structures characterized by high concentrations of pre-mRNA processing factors, including snRNP and SR proteins (a family of RNA-binding Serine/Arginine (SR) proteins that play crucial roles in gene expression regulation, particularly in pre-mRNA splicing) (Spector and Lamond 2011). Nuclear speckles also contain phosphatidylinositol phosphates (Sobol et al. 2018) and RNA. Metastasis associated lung adenocarcinoma transcript 1 (MALAT1) is a long non-coding RNA, a single-exon transcript over 7 kb in length, identified in nuclear speckles, but is not essential for NS formation or cell viability (Zhang et al. 2012).

It has been shown that MALAT1 is localized in actively expressed regions of the genome, probably due to interaction with partner proteins that direct this long non-coding RNA to newly synthesized pre-mRNA transcripts (Engreitz et al. 2014; West et al. 2014). NS have been found to be associated with chromosomal regions where high levels of RNA polymerase II transcription activity are observed (Quinodoz et al. 2018; Kim et al. 2020). It is likely that MALAT1 functions as a molecular guide that delivers the splicing machinery located in the NS to sites of active gene transcription (Kopp and Mendell 2018). Moreover, it was experimentally shown that the association of the Hsp70 genes and four genes located near the Hsp70 locus with nuclear speckles leads to a significant increase in their expression after heat shock. The authors hypothesized that this NS-dependent regulation arises from decreased exosomal degradation of the nascent transcript coupled with an increased transcription rate (Kim et al. 2020). Based on these data, the "gene expression amplification" model was proposed, according to which nuclear speckles act as gene expression hubs (Kim et al. 2020). Disruptions in the NS biogenesis affects expression of 1282 different genes (Hu et al. 2019). Therefore, there is no doubt that NSs are important regulators of gene expression.

NSs accumulate various splicing factors and components of the splicing machinery. Alternative splicing of mRNA is significantly affected by stress factors, which is manifested in changes in localization, interactions, expression and chemical modifications of splicing factors and spliceosome components (Dutertre et al. 2011). In addition, cells use changes in alternative splicing patterns to regulate gene expression in response to stress. In response to heat shock, splicing is suppressed for a significant portion of the transcriptome, with the exception of genes involved in stress responses (Biamonti and Caceres 2009). NSs are clearly involved in these regulatory pathways, although the specific mechanisms remain unclear.

Similar morphological changes in the NS are observed under different stress conditions. Increasing and rounding of NS condensates accompanied by a decrease in their total quantity is commonly reported. This was demonstrated by heat shock-induced transcriptional arrest (45 °C for 15 minutes) (Spector et al. 1991), as well as by treatment with transcription inhibitors such as actinomycin D (Melcák et al. 2000; Kim et al. 2019), etoposide-induced genotoxic stress (Raina and Rao 2022), heavy metal stress (Kim et al. 2019), and osmotic stress (Spector et al. 1993). This abnormal morphology has been explained by two processes: (1) migration of proteins back to the NS for storage under stress conditions; (2) fusion of NS particles.

During a viral infection, nuclear speckles are redistributed to the periphery of the nucleus. Nuclear speckles are found adjacent to viral replication centers (VRCs) during infections caused by herpes simplex virus (HSV-1), Kaposi's sarcoma-associated herpes virus (KSHV), human immunodeficiency virus (HIV), adenovirus, and influenza virus (Alkalay et al. 2020; Jiménez-García and Spector 1993;

Mor et al. 2016; Francis et al. 2020). HSV-1 infection causes a redistribution of both nuclear speckles and paraspeckles; this was found to lead to the association of SRSF2 with various paraspeckle factors and influence histone modifications located near viral genes (Li and Bingham 1991). Interestingly, the movement of NS, characterized by directed migration over long distances through the interchromatic space, often ends with the fusion of condensates (Kim et al. 2019). Вероятно, такое слияние является не стохастическим, а контролируемым процессом. Probably, such a merger is not a stochastic, but a controlled process.

NELF-Bodies

In response to stress, cells reprogram regulatory pathways and turn off processes that are not necessary for survival. A key step is the suppression of transcription and translation of housekeeping genes. One of the negative regulators of transcription is the negative elongation factor (NELF) protein. Under normal conditions, NELF activity is blocked by phosphorylation, which is accompanied by activation of pol-II (Kwak and Lis 2013). Under stress conditions, NELF is dephosphorylated and sumoylated, promoting phase separation and the formation of NELF-Bodies at transcriptional housekeeping gene loci. NELF contains intrinsically disordered regions that mediate condensate formation, interaction with chromatin, and enhanced inhibitory potential (Rawat et al. 2021). Thus, NELF-Bodies suppress transcription and help cells under stress (Rawat et al. 2021).

Nucleolar Stress Bodies

Nuclear stress bodies (nSBs) are primate-specific MLOs that arise in response to various types of stress. The size and number of nSBs depend on the type and duration of stress exposure; however, on average, one cell can contain several nSBs measuring 1–2 μm (Chujo and Hirose 2017).

Upon heat shock, cadmium sulfate, and UV exposure, nSBs form around architectural noncoding RNAs, HSATIII arcRNAs, or simply HSATIII, from the pericentromeric HSATIII repeat arrays of chromosomes 9 (locus 9q12). It is known that transcription activation is under the control of heat shock factors 1 and 2 (HSF1 and HSF2) (Alastalo et al. 2003; Akerfelt et al. 2010; Kovács et al. 2022). As a result of heat exposure, HSF1 recruits the histone transferases HAT, TIP60 (also known as KAT5) and p300, which leads to a redistribution of acetylation on histones and triggers the process of chromatin remodeling, recruitment of RNA polymerase II and activation of HSATIII transcription. Dynamic molecular condensates form around HSATIII transcription sites (Col et al. 2017; Gaglia et al. 2020). With increasing duration of stress exposure, the dynamics of the exchange of HSF1 with the nucleoplasm sharply decreases, which indicates the gradual hardening of HSF1-nSB (Gaglia et al. 2020).

Transcriptional activation of HSATIII in stressed cells influences gene expression through massive recruitment of HATs and transcription factors to nSB (Col et al. 2017; Ninomiya and Adachi 2020; Sandqvist et al. 2009). Another important function of these biocondensates is splicing control. nSBs serve dual functions: as "molecular sponges" for YTHDC1 sequestration and as "reaction crucibles" for SRSF phosphorylation. Both mechanisms contribute to the suppression of pre-mRNA splicing during stress recovery (Sandqvist et al. 2009; Ninomiya et al. 2021).

In a study (Goenka et al. 2016), the authors induced knockdown of normal HSATIII expression and detected a significant decrease (~70%) in the survival of HeLa cells after thermal exposure compared to control cells (Goenka et al. 2016). Therefore, although the biological functions of nSB are not fully understood, it is clear that they play an important role in the adaptation of cells to stressful conditions.

PML-Bodies

In the nuclear matrix of most cell lines and tissues, various isoforms of the promyelocytic leukemia protein (PML) form special spherical structures known as nuclear PML-bodies (Daniel et al. 1993; Maul et al. 2000; Stuurman et al. 1990; Corpet et al.; Silonov et al. 2023b; Salomoni and Pandolfi 2002). These entities are MLOs that recruit many different proteins and may be associated with various cellular processes, such as transcription, protein degradation, DNA repair, and senescence (Corpet et al.). In PML bodies, a number of proteins undergo post-translational modifications (especially sumoylation), which lead to their isolation in PML bodies, activation or degradation (Corpet et al.).

Typical PML nuclear bodies are small spheres with a diameter of 0.2–0.5 μm. Based on their morphological characteristics, several subtypes of PML bodies are distinguished, with most of them being characterized by the presence of an electron-dense shell and a less dense inner core region (nucleus) (Sahin et al. 2014; Fonin et al. 2021, 2022b).

PML bodies include specific proteins, such as SUMO (small ubiquitin-related modifier), DAXX (death domain-associated protein), nuclear autoantigen SP100 (Speckled 100 kDa, also known as Nuclear dot-associated Sp100 protein), LYSP100 (lymphoid-restricted homolog of SP100), ISG20 (interferon-stimulated gene 20 kDa protein), PML, BLM (the Bloom syndrome gene product), and nuclear dot protein NDP55 (Silonov et al. 2023a). Other proteins, such as Ubiquitin-homology domain protein PIC1 (also known as SUMO-1 and sentrin), usually associated with nuclear pores, may also be part of PML bodies (Corpet et al.). All of these proteins that make up PML bodies are capable of being reorganized and released from their structure when exposed to stress ies are capable of being reorganized and released from the PML body structure when exposed to stress (Corpet et al.).

Certain stress exposures such as genotoxic stress, aging, UV damage and ionizing radiation lead to localization of endogenous PML and PML isoforms (PML-I, -IV and -V) in the nucleolus (Corpet et al.; Imrichova et al. 2019). As a result of stress, PML accumulates at the border of the nucleolar cap or forms spherical structures containing nucleolar material, collectively these structures are called PML nucleolar associations (PNAs) (Imrichova et al. 2019).

When cells are exposed to the topoisomerase inhibitor doxorubicin, 4 main subtypes of PNAs are formed: PML 'bowls', PML 'funnels', PML 'balloons' and PML nucleolus-derived structures (PML-NDS) (Imrichova et al. 2019) (see Fig. 16.1). They are named according to their 3D visualization obtained using high-resolution microscopy. It is worth noting that each of these morphological forms represents a distinct stage in a continuous dynamic process that is closely related to RNAPI inhibition and reactivation. When cells are exposed to doxorubicin, RNAPI segregation first produces PML cups at the nucleolar border at the site of formation of the nucleolar cap. The cup can then transform into a PML funnel or PML ball that encloses the entire nucleolus. After doxorubicin is washed away, the activity of the

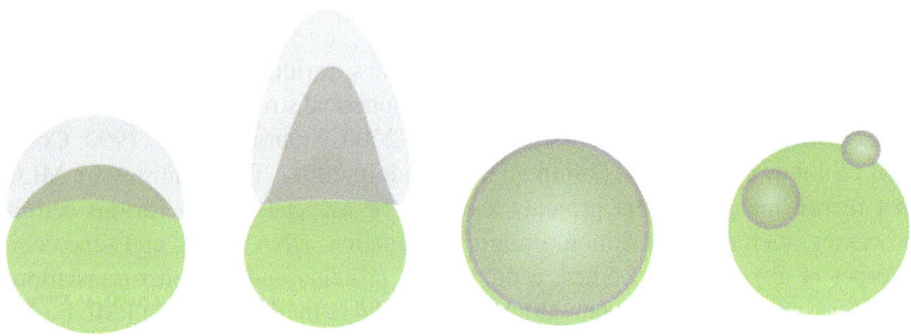

Fig. 16.1 Morphological diversity of PML nucleolar associations (PNAs). Four main subtypes of PNAs are shown: PML 'bowls', PML 'funnels', PML 'balloons', and PML nucleolus-derived structures (PML-NDS)PML 'bowls' are partially open, concave PML structures resembling a bowl or hemisphere. They are formed near damaged regions of rDNA and probably serve as an intermediate stage between standard PML bodies and more complex PML compartments, participating in the assembly and organization of repair factors around damaged rDNAPML 'funnels' (PML funnels) are bodies that expand towards the periphery of the nucleolus and narrow towards the center. They are formed under a high degree of topological stress in rDNA, often after inhibition of topoisomerases or RNA polymerase I. PML funnels are presumably involved in the outflow of damaged rDNA regions from the active zones of the nucleolus, directing them to areas where effective repair is possiblePML 'balloons' (PML balls) - Large spherical structures completely surrounded by a shell of PML protein. They are formed in response to massive rDNA damage, especially when repair pathways such as homologous recombination (HDR) are blocked. Can serve as a repository for damaged rDNA, isolating it from the rest of the nucleolar structure and preventing its involvement in transcriptional processesPML-nucleolus derived structures (PML-NDS) - the final stage, unlike other types of PNAs, move outside the nucleolus, participating in the global response of the cell to DNA damage. Similar structures are involved in long-term isolation of damaged rDNA under severe genotoxic stress

nucleolus and RNAPI is resumed, and PML-NDS are formed in the form of single bodies containing a certain amount of nucleolar components in the PML shell (Imrichova et al. 2019).

PML ‘bowls’ are partially open, concave PML structures resembling a bowl or hemisphere. They are formed near damaged regions of rDNA and probably serve as an intermediate stage between standard PML bodies and more complex PML compartments, participating in the assembly and organization of repair factors around damaged rDNA (Imrichova et al. 2019). PML ‘funnels’ (PML funnels) are bodies that expand towards the periphery of the nucleolus and narrow towards the center. They are formed under high degrees of topological stress in rDNA, often after inhibition of topoisomerases or RNA polymerase I. PML funnels are presumably involved in the outflow of damaged rDNA regions from the active zones of the nucleolus, directing them to areas where effective repair is possible. PML ‘balloons’ are large spherical structures completely surrounded by a shell of PML protein. They are formed in response to massive rDNA damage, especially when repair pathways such as homologous recombination (HDR) are blocked. They can serve as a repository for damaged rDNA, isolating it from the rest of the nucleolar structure and preventing its involvement in transcriptional processes. PML-NDS, which are formed at the final stage, unlike other types of PNAs, move outside the nucleolus, participating in the global response of the cell to DNA damage. Similar structures are involved in long-term isolation of damaged rDNA under severe genotoxic stress (Imrichova et al. 2019).

As cells age, larger PMLs containing a nucleolar component are formed. Such bodies are called senescence-associated nuclear bodies (SANB) and in general they resemble the bodies observed during genotoxic stress (Condemine et al. 2007). A distinctive feature of SANB is the presence of the poly-ubiquitin conjugates observed in most bodies, the localization of which depends on the stage of cell senescence. Therefore, as cells age in the bodies, the content of the nucleolar component and fibrilarin decreases, but the content of poly-ubiquitin increases (Condemine et al. 2007).

SNABs can crosslink with classical nuclear bodies, but their division or budding from the nucleolus has not been observed, which may indicate their relative stability. When proteasomes are inhibited, for example, using MG132, a significant redistribution of PML and its associated proteins: Sp100 and SUMO-1 is observed. Within a few hours after the onset of inhibition, these proteins disappear from PML bodies and begin to accumulate in the nucleolus. At the same time, a change occurs in the subnuclear distribution of proteasomes, which begin to concentrate in euchromatic regions and inside the nucleolus (Mattsson et al. 2001). Presumably, these observations indicate the possible involvement of the nucleolus in the regulation of proteasome-dependent protein degradation. The relocation of PML and its partners to the nucleolus may indicate that these structures play a role in controlling protein stability and post-translational modification. The interaction of PML bodies with the nucleolus under conditions of impaired proteasomal degradation indicates the potential role of the nucleolus as an alternative compartment for the accumulation and, possibly, subsequent utilization of proteins (Mattsson et al. 2001).

PML NBs are hypothesized to be dynamic sensors of cellular stress by distributing associated repair factors to sites of DNA damage (Dellaire and Bazett-Jones 2004; Niwa-Kawakita et al. 2018). Inhibition of rRNA synthesis, not DNA damage, is the main cause of PML nucleolar translocation (Urbancokova et al. 2024). Direct rDNA damage caused by endonuclease I-PpoI provokes PML interactions with nucleoli (Urbancokova et al. 2024). And the most powerful stimuli that form PNA are the simultaneous inhibition of topoisomerase (enzymes involved in maintaining DNA topology) and RNA polymerase I (RNAPI), which is responsible for rDNA transcription. When topoisomerase activity is disrupted, unresolved DNA supercoiling and entanglement occurs, especially in repetitive regions of rDNA, leading to DNA double-strand breaks (DSBs). RNAPI inhibition increases stress by preventing rDNA transcription and causing the formation of nucleolar caps that sequester damaged rDNA.

Defects in the DNA repair pathway, namely homologous recombination, enhance the formation of PML bodies. Non-homologous end joining (NHEJ) does not have a significant effect on PML body formation. Thus, the formation of nucleolar PML associations is triggered by damage to nucleolar rDNA (NORs), resulting from pathological stress and inhibition of rDNA transcription associated with unresolved, HDR-dependent DNA repair processes.

Stress-Granules

Stress granules (SGs) are transiently assembled in eukaryotic cells in response to various types of endogenous and exogenous stress stimuli (Mahboubi and Stochaj 2017; Khong et al. 2017). Halting translation under stressful conditions causes the accumulation of mRNA, translation initiation factors, and RNA-binding proteins in these non-membrane organelles, thereby ensuring the isolation of these molecules from the aggressive external environment (Youn et al. 2019; Protter and Parker 2016). After stress cessation, stress granules are disassembled in a chaperone-dependent manner, releasing sequestered molecules for recycling. At the same time, mRNA encoding proteins of the cell's stress response, for example, heat shock proteins, is not sequestered into stress granules, which allows cells to change their expression program in a stress-dependent manner (Verma et al. 2021). This ensures savings in cellular energy required for the degradation of mRNA and proteins, as well as spent during their subsequent synthesis. Impaired assembly/disassembly of stress granules and their degradation into toxic aggregates of amyloid fibrils are associated with the development of many neurodegenerative diseases, including amyotrophic lateral sclerosis and frontotemporal dementia (Rhine et al. 2022; Alberti and Hyman 2016; Wolozin and Ivanov 2019). SGs are also involved in the processes of oncogenesis, and their formation increases the resistance of cancer cells to therapeutic effects (Mahboubi and Stochaj 2017). Furthermore, SGs are also involved in the replication processes of viruses, in particular SARS-CoV-2 (Poblete-Durán et al. 2016; He et al. 2023).

There are two main pathways for the formation of stress granules, canonical and non-canonical, associated with phosphorylation of the translation initiation factors eIF4 and eIF2 (eukaryotic translation initiation factors 4 and 2), respectively. Such post-translational modifications of these proteins cause dissociation of mRNA and polyribosomes, and, accordingly, translation arrest. Complex coacervation of free mRNA and scaffold internally disordered stress granule proteins (G3BP1/2 (Ras GTPase-activating protein-binding protein), TIA-1 (T-cell-restricted intracellular antigen-1, also known as cytotoxic granule associated RNA binding protein TIA1), and a number of others) under stress conditions ensures the formation of "seeds" of stress granules (Protter and Parker 2016; Yang et al. 2020; Wheeler et al. 2016). Subsequent maturation of stress granules is due to multiple protein-protein and RNA-protein interactions, as a result of which a dynamic layer of stress granule client proteins is formed around the sedentary central "core" of stress granules containing the framework proteins of these non-membrane organelles (Wheeler et al. 2016). Changes in the composition of client proteins due to the liquid-droplet properties of stress granules determine the functional activity of these non-membrane organelles. The constant exchange of the contents of stress granules with the cytoplasm ensures their rapid response in response to changes in external conditions and to the cessation of stress. Disassembly of stress granules after normalization of intracellular environmental conditions allows the retained mRNA to again participate in translation processes. Dissociation of stress granules usually occurs in a chaperone-dependent manner (Moujaber et al. 2017). The breakdown (decomposition) of stress granules can occur via the autophagosomal mechanism, despite the fact that stress granules contain virtually no ubiquitinated proteins (Maharjan et al. 2017). The interaction of the autophagy regulator p62 with the hexanucleotide repeats of C9orf72 contributes to the attraction of the formed complexes to the methylated arginine residues of internally disordered stress granule proteins and the subsequent binding of stress granules to the LC3-II autophagosomal membrane protein *(which is a lapidated and membrane-bound form of the microtubule-associated protein 1 light chain 3 gamma, also known as autophagy-related protein LC3 C)* (Chitiprolu et al. 2018; Maharjan et al. 2017). *Mutations in C9orf72 inhibit this process and promote the transformation of stress granules into amyloid-like fibrils and the development of neurodegenerative diseases (Maharjan* et al. 2017*). Accordingly, the permanent presence of stress granules in the cell cytoplasm can provoke the development of neurodegenerative diseases, which is observed in aging cells* (Lechler et al. 2017).

The transformation of stress granules into aggregates of amyloid fibrils can also be caused by a change in the profile of RNA-protein interactions, in particular, when modifying mRNA (Rhine et al. 2022). RNA modifications can rearrange the network of RNA-protein and RNA-RNA interactions, recruit specific proteins and RNA into the composition of stress granules, regulate the assembly/disassembly of these non-membrane organelles by modulating LLPS RNA and intrinsically disordered proteins, and change the functional activity of stress granules (Linsenmeier et al. 2022; Ranganathan and Shakhnovich 2021; Hojo et al. 2020; Shi et al. 2019a; Su et al. 2021; Campos-Melo et al. 2021; Das et al. 2022). It is known that mRNA molecules localized in stress granules in response to stress are more susceptible to m6A modification (N6-methyladenosine) compared to mRNAs that are unable to be incorporated into these organelles (Sun et al. 2023). At the same time, m6A modification of mRNA molecules has a greater effect on their inclusion in stress granules compared to the length of the molecules of these biopolymers (Ries et al. 2023). N1-methyladenosine (m1A) and N4-acetylcytidine (ac4C) modified RNAs are also capable of inclusion in stress granules (Kudrin et al. 2024; Alriquet et al. 2021). There is known evidence that modification of the internal sites of RNA molecules N7-methylguanosine (m7G) ensures the selective recruitment of such RNAs into the composition of stress granules (Zhao et al. 2023). In mRNA transcripts, m6A modifications are recognized by "reader" proteins belonging to the YTH domain protein family (Shi et al. 2019b). Such proteins, by binding to the modified mRNA, ensure the implementation of its function, mediating the regulation of the expression of downstream genes (Shi et al. 2019b). "Reader" proteins include members of the YTH N6-methyladenosine RNA-binding proteins (YTHDFs) and YTH domain-containing (YTHDCs) families. All proteins of the YTHDF family are capable of LLPS, which under stress conditions is modulated by polymethylated mRNA, and proteins of this family colocalize with the stress granule scaffold protein G3BP1 within these organelles (Fu and Zhuang 2020). At the same time, knockdown of YTHDF1/3 in eukaryotic cells causes a significant decrease in the number of stress granules and a decrease in the content of polymethylated mRNA in these organelles (Fu and Zhuang 2020).

In recent years, the idea that the epitranscriptome, i.e. a set of RNA modifications is one of the most promising markers of biological aging (Wagner and Schosserer 2022; Jiapaer et al. 2022). In particular, the level of m6A modification of RNA molecules, including transcripts of genes involved in the stress response, changes significantly in human cells during the development of neurodegenerative diseases and cellular aging (Bu et al. 2024; McMahon et al. 2021). In this case, changes in the expression of proteins of the YTHDFs family and structural and functional changes in stress granules are often observed (McMillan et al. 2023; Wang et al. 2022). In addition, an abnormal level of m6A modification of RNA molecules in eukaryotic cells was also detected during replicative aging (Zhang and Xia 2023).

P-Bodies

P-bodies are dynamic condensates consisting of translationally inactive mRNAs and proteins involved in mRNA processing, particularly deadenylation of the 3′-polyadenosine [poly(A)] tail of mRNA, 5′ decapping of mRNA, 5′ to 3' mRNA degradation, nonsense-mediated mRNA decay, and microRNA-directed gene silencing (Luo et al. 2018). At the same time, P-bodies are often considered as a depot for storing mRNA (Luo et al. 2018). Furthermore, according (Shan et al. 2023), m6A-methylation of mRNA molecules promotes their dissociation from polysomes and inclusion in P-bodies. The YTHDF2 and YTHDC proteins, in addition to stress granules, are also capable of localizing in P-bodies (Youn et al. 2019).

P-bodies are directly involved in the regulation of a large number of stress-related signaling pathways. Thus, in aging C. elegans cells, P-bodies regulate proteostasis by recruiting the IFE-2 isoform of the transcription initiation factor eIF4E into these organelles, which helps block protein biosynthesis and increases cell lifespan (Rieckher et al. 2018).

Unlike P-bodies that are constantly present in eukaryotic cells, P-bodies in yeast appear in response to stress. Yeast P bodies contain the proteins Dcp1p, Dcp2p, Edc3p, Dhh1p, Pat1p, Lsm1p, Xrn1p, Ccr4p, and Pop2p. Studies of yeast P bodies indicate that recruitment of Dcp1p to P bodies is mediated by Dcp2p. In addition, Pat1p is required for recruitment of the Lsm1-7p complex (Teixeira and Parker 2007). However, in yeast, deletion of any of the genes encoding P-body components does not lead to the dissociation of these MLOs.

The composition of P-bodies depends significantly on the type of stress. It was shown that during the formation of P-bodies in yeast in response to glucose starvation and osmotic shock caused by the introduction of high concentrations of salts into the studied cells, the pool of mRNA sequestered into these non-membrane organelles differs markedly both in nucleotide composition and in the length of transcripts (Wang et al. 2018a).

P-bodies and stress granules are functionally related and have overlapping compositions (Youn et al. 2019; Mokin et al. 2024). Furthermore, it has been shown that the formation of stress granules is associated with the functional activity of P-bodies (Youn et al. 2019).

References

Akerfelt M, Morimoto RI, Sistonen L (2010) Heat shock factors: integrators of cell stress, development and lifespan. Nat Rev Mol Cell Biol 11(8):545–555. https://doi.org/10.1038/nrm2938

Alastalo TP, Hellesuo M, Sandqvist A, Hietakangas V, Kallio M, Sistonen L (2003) Formation of nuclear stress granules involves HSF2 and coincides with the nucleolar localization of Hsp70. J Cell Sci 116(Pt 17):3557–3570. https://doi.org/10.1242/jcs.00671

Alberti S, Hyman AA (2016) Are aberrant phase transitions a driver of cellular aging? BioEssays 38(10):959–968. https://doi.org/10.1002/bies.201600042

Alkalay E, Gam Ze Letova Refael C , Shoval I, Kinor N, Sarid R (2020) The sub-nuclear localization of RNA-binding proteins in KSHV-infected cells. 9(9). https://doi.org/10.3390/cells9091958

Alriquet M, Calloni G, Martínez-Limón A, Delli Ponti R, Hanspach G, Hengesbach M, Tartaglia GG, Vabulas RM (2021) The protective role of m1A during stress-induced granulation. J Mol Cell Biol 12(11):870–880. https://doi.org/10.1093/jmcb/mjaa023

Andrade LE, Tan EM, Chan EK (1993) Immunocytochemical analysis of the coiled body in the cell cycle and during cell proliferation. Proc Natl Acad Sci USA 90(5):1947–1951. https://doi.org/10.1073/pnas.90.5.1947

Antifeeva IA, Fonin AV, Fefilova AS, Stepanenko OV, Povarova OI, Silonov SA, Kuznetsova IM, Uversky VN, Turoverov KK (2022) Liquid-liquid phase separation as an organizing principle

of intracellular space: overview of the evolution of the cell compartmentalization concept. Cell Mol Life Sci 79(5):251. https://doi.org/10.1007/s00018-022-04276-4
Arnold SE, Louneva N, Cao K, Wang LS, Han LY, Wolk DA, Negash S, Leurgans SE, Schneider JA, Buchman AS, Wilson RS, Bennett DA (2013) Cellular, synaptic, and biochemical features of resilient cognition in Alzheimer's disease. Neurobiol Aging 34(1):157–168. https://doi.org/10.1016/j.neurobiolaging.2012.03.004
Bachant JB, Elledge SJ (1999) Mitotic treasures in the nucleolus. Nature 398(6730):757–758. https://doi.org/10.1038/19641
Banani SF, Lee HO, Hyman AA, Rosen MK (2017) Biomolecular condensates: organizers of cellular biochemistry. Nat Rev Mol Cell Biol 18(5):285–298. https://doi.org/10.1038/nrm.2017.7
Beeharry Y, Goodrum G, Imperiale CJ, Pelchat M (2018) The Hepatitis Delta virus accumulation requires paraspeckle components and affects NEAT1 level and PSP1 localization. Sci Rep 8(1):6031. https://doi.org/10.1038/s41598-018-24500-1
Biamonti G, Caceres JF (2009) Cellular stress and RNA splicing. Trends Biochem Sci 34(3):146–153. https://doi.org/10.1016/j.tibs.2008.11.004
Bokros M, Balukoff NC, Grunfeld A, Sebastiao M, Beurel E, Bourgault S, Lee S (2024) RNA tailing machinery drives amyloidogenic phase transition. Proc Natl Acad Sci USA 121(23):e2316734121. https://doi.org/10.1073/pnas.2316734121
Bongiorno-Borbone L, De Cola A, Barcaroli D, Knight RA, Di Ilio C, Melino G, De Laurenzi V (2010) FLASH degradation in response to UV-C results in histone locus bodies disruption and cell-cycle arrest. Oncogene 29(6):802–810. https://doi.org/10.1038/onc.2009.388
Boulon S, Westman BJ, Hutten S, Boisvert FM, Lamond AI (2010) The nucleolus under stress. Mol Cell 40(2):216–227. https://doi.org/10.1016/j.molcel.2010.09.024
Bu FT, Wang HY, Xu C, Song KL, Dai Z, Wang LT, Ying J, Chen J (2024) The role of m6A-associated membraneless organelles in the RNA metabolism processes and human diseases. Theranostics 14(12):4683–4700. https://doi.org/10.7150/thno.99019
Cajal R (1903) Un sencillo metodo de coloracione del reticulo protoplasmieo y sus efectos en los diversos centras nerviosos de vertrebratos y invertebratos. Trabajos del lab de investig biolog de la Univ de Madrid 2:129–129
Campos-Melo D, Hawley ZCE, Droppelmann CA, Strong MJ (2021) The integral role of RNA in stress granule formation and function. Front Cell Dev Biol 9:621779. https://doi.org/10.3389/fcell.2021.621779
Chitiprolu M, Jagow C, Tremblay V, Bondy-Chorney E, Paris G, Savard A, Palidwor G, Barry FA, Zinman L, Keith J, Rogaeva E, Robertson J, Lavallée-Adam M, Woulfe J, Couture JF, Côté J, Gibbings D (2018) A complex of C9ORF72 and p62 uses arginine methylation to eliminate stress granules by autophagy. Nat Commun 9(1):2794. https://doi.org/10.1038/s41467-018-05273-7
Choudhry H, Albukhari A, Morotti M, Haider S, Moralli D, Smythies J, Schödel J, Green CM, Camps C, Buffa F, Ratcliffe P, Ragoussis J, Harris AL, Mole DR (2015) Tumor hypoxia induces nuclear paraspeckle formation through HIF-2α dependent transcriptional activation of NEAT1 leading to cancer cell survival. Oncogene 34(34):4482–4490. https://doi.org/10.1038/onc.2014.378
Chujo T, Hirose T (2017) Nuclear bodies built on architectural long noncoding RNAs: unifying principles of their construction and function. Mol Cells 40(12):889–896. https://doi.org/10.14348/molcells.2017.0263
Cioce M, Boulon S, Matera AG, Lamond AI (2006) UV-induced fragmentation of Cajal bodies. J Cell Biol 175(3):401–413. https://doi.org/10.1083/jcb.200604099
Col E, Hoghoughi N, Dufour S, Penin J, Koskas S, Faure V, Ouzounova M, Hernandez-Vargash H, Reynoird N, Daujat S, Folco E, Vigneron M, Schneider R, Verdel A, Khochbin S, Herceg Z, Caron C, Vourc'h C (2017) Bromodomain factors of BET family are new essential actors of pericentric heterochromatin transcriptional activation in response to heat shock. Sci Rep 7(1):5418. https://doi.org/10.1038/s41598-017-05343-8

Condemine W, Takahashi Y, Le Bras M, de Thé H (2007) A nucleolar targeting signal in PML-I addresses PML to nucleolar caps in stressed or senescent cells. J Cell Sci 120(Pt 18):3219–3227. https://doi.org/10.1242/jcs.007492

Corpet A, Kleijwegt C, Roubille S, Juillard F, Jacquet K, Texier P, Lomonte P PML nuclear bodies and chromatin dynamics: catch me if you can! (1362–4962 (Electronic))

Correll CC, Rudloff U, Schmit JD, Ball DA, Karpova TS, Balzer E, Dundr M (2024) Crossing boundaries of light microscopy resolution discerns novel assemblies in the nucleolus. Histochem Cell Biol 162(1):161–183

Daniel MT, Koken M, Romagné O, Barbey S, Bazarbachi A, Stadler M, Guillemin MC, Degos L, Chomienne C, de Thé H (1993) PML protein expression in hematopoietic and acute promyelocytic leukemia cells. Blood 82(6):1858–1867

Darling AL, Liu Y, Oldfield CJ, Uversky VN (2018) Intrinsically disordered proteome of human membrane-less organelles. Proteomics 18(5–6):e1700193. https://doi.org/10.1002/pmic.201700193

Darling AL, Zaslavsky BY, Uversky VN (2019) Intrinsic disorder-based emergence in cellular biology: physiological and pathological liquid-liquid phase transitions in cells. Polymers (Basel) 11(6). https://doi.org/10.3390/polym11060990

Das S, Santos L, Failla AV, Ignatova Z (2022) mRNAs sequestered in stress granules recover nearly completely for translation. RNA Biol 19(1):877–884. https://doi.org/10.1080/15476286.2022.2094137

Dellaire G, Bazett-Jones DP (2004) PML nuclear bodies: dynamic sensors of DNA damage and cellular stress. BioEssays 26(9):963–977. https://doi.org/10.1002/bies.20089

Denissov S, Lessard F, Mayer C, Stefanovsky V, van Driel M, Grummt I, Moss T, Stunnenberg HG (2011) A model for the topology of active ribosomal RNA genes. EMBO Rep 12(3):231–237. https://doi.org/10.1038/embor.2011.8

Dundr M (2011) Seed and grow: a two-step model for nuclear body biogenesis. J Cell Biol 193(4):605–606. https://doi.org/10.1083/jcb.201104087

Dutertre M, Sanchez G, Barbier J, Corcos L, Auboeuf D (2011) The emerging role of pre-messenger RNA splicing in stress responses: sending alternative messages and silent messengers. RNA Biol 8(5):740–747. https://doi.org/10.4161/rna.8.5.16016

Engreitz JM, Sirokman K, McDonel P, Shishkin AA, Surka C, Russell P, Grossman SR, Chow AY, Guttman M, Lander ES (2014) RNA-RNA interactions enable specific targeting of non-coding RNAs to nascent pre-mRNAs and chromatin sites. Cell 159(1):188–199. https://doi.org/10.1016/j.cell.2014.08.018

Fefilova AS, Antifeeva IA, Gavrilova AA, Turoverov KK, Kuznetsova IM, Fonin AV (2022) Reorganization of cell compartmentalization induced by stress. Biomolecules 12(10). https://doi.org/10.3390/biom12101441

Fonin AV, Antifeeva IA, Kuznetsova IM, Turoverov KK, Zaslavsky BY, Kulkarni P, Uversky VN (2022a) Biological soft matter: intrinsically disordered proteins in liquid-liquid phase separation and biomolecular condensates. Essays Biochem 66(7):831–847. https://doi.org/10.1042/EBC20220052

Fonin AV, Silonov SA, Fefilova AS, Stepanenko OV, Gavrilova AA, Petukhov AV, Romanovich AE, Modina AL, Zueva TS, Nedelyaev EM, Pleskach NM, Kuranova ML, Kuznetsova IM, Uversky VN, Turoverov KK (2022b) New evidence of the importance of weak interactions in the formation of PML-bodies. Int J Mol Sci 23(3). https://doi.org/10.3390/ijms23031613

Fonin AV, Silonov SA, Shpironok OG, Antifeeva IA, Petukhov AV, Romanovich AE, Kuznetsova IM, Uversky VN, Turoverov KK (2021) The role of non-specific interactions in canonical and ALT-associated PML-bodies formation and dynamics. Int J Mol Sci 22(11). https://doi.org/10.3390/ijms22115821

Fortes P, Lamond AI, Ortín J (1995) Influenza virus NS1 protein alters the subnuclear localization of cellular splicing components. J Gen Virol 76(4):1001–1007

Fox AH, Lamond AI (2010) Paraspeckles. Cold Spring Harb Perspect Biol 2(7):a000687. https://doi.org/10.1101/cshperspect.a000687

Francis AC, Marin M, Singh PK, Achuthan V, Prellberg MJ, Palermino-Rowland K, Lan S, Tedbury PR, Sarafianos SG, Engelman AN, Melikyan GB (2020) HIV-1 replication complexes accumulate in nuclear speckles and integrate into speckle-associated genomic domains. Nat Commun 11(1):3505. https://doi.org/10.1038/s41467-020-17256-8

Frottin F, Schueder F, Tiwary S, Gupta R, Körner R, Schlichthaerle T, Cox J, Jungmann R, Hartl F, Hipp M (2019) The nucleolus functions as a phase-separated protein quality control compartment. Science 365(6451):342–347

Fu Y, Zhuang X (2020) M(6)A-binding YTHDF proteins promote stress granule formation. Nat Chem Biol 16(9):955–963. https://doi.org/10.1038/s41589-020-0524-y

Gaglia G, Rashid R, Yapp C, Joshi GN, Li CG, Lindquist SL, Sarosiek KA, Whitesell L, Sorger PK, Santagata S (2020) HSF1 phase transition mediates stress adaptation and cell fate decisions. Nat Cell Biol 22(2):151–158. https://doi.org/10.1038/s41556-019-0458-3

Gall JG (2000) Cajal bodies: the first 100 years. Annu Rev Cell Dev Biol 16:273–300. https://doi.org/10.1146/annurev.cellbio.16.1.273

Goenka A, Sengupta S, Pandey R, Parihar R, Mohanta GC, Mukerji M, Ganesh S (2016) Human satellite-III non-coding RNAs modulate heat-shock-induced transcriptional repression. 129(19):3541–3552

Govoni M, Farabegoli F, Pession A, Novello F (1994) Inhibition of topoisomerase II activity and its effect on nucleolar structure and function. Exp Cell Res 211(1):36–41. https://doi.org/10.1006/excr.1994.1055

Gupta MN, Uversky VN (2024) Protein structure-function continuum model: emerging nexuses between specificity, evolution, and structure. Protein Sci 33(4):e4968. https://doi.org/10.1002/pro.4968

He S, Gou H, Zhou Y, Wu C, Ren X, Wu X, Guan G, Jin B, Huang J, Jin Z, Zhao T (2023) The SARS-CoV-2 nucleocapsid protein suppresses innate immunity by remodeling stress granules to atypical foci. FASEB J 37(12):e23269. https://doi.org/10.1096/fj.202201973RR

Hojo H, Yashiro Y, Noda Y, Ogami K, Yamagishi R, Okada S, Hoshino SI, Suzuki T (2020) The RNA-binding protein QKI-7 recruits the poly(A) polymerase GLD-2 for 3′ adenylation and selective stabilization of microRNA-122. J Biol Chem 295(2):390–402. https://doi.org/10.1074/jbc.RA119.011617

Hu S, Lv P, Yan Z, Wen B (2019) Disruption of nuclear speckles reduces chromatin interactions in active compartments. Epigenetics Chromatin 12(1):43. https://doi.org/10.1186/s13072-019-0289-2

Imamura K, Imamachi N, Akizuki G, Kumakura M, Kawaguchi A, Nagata K, Kato A, Kawaguchi Y, Sato H, Yoneda M, Kai C, Yada T, Suzuki Y, Yamada T, Ozawa T, Kaneki K, Inoue T, Kobayashi M, Kodama T, Wada Y, Sekimizu K, Akimitsu N (2014) Long noncoding RNA NEAT1-dependent SFPQ relocation from promoter region to paraspeckle mediates IL8 expression upon immune stimuli. Mol Cell 53(3):393–406. https://doi.org/10.1016/j.molcel.2014.01.009

Imamura K, Takaya A, Ishida YI, Fukuoka Y, Taya T, Nakaki R, Kakeda M, Imamachi N, Sato A, Yamada T, Onoguchi-Mizutani R, Akizuki G, Tanu T, Tao K, Miyao S, Suzuki Y, Nagahama M, Yamamoto T, Jensen TH, Akimitsu N (2018) Diminished nuclear RNA decay upon Salmonella infection upregulates antibacterial noncoding RNAs 37(13). https://doi.org/10.15252/embj.201797723

Imrichova T, Hubackova S, Kucerova A, Kosla J, Bartek J, Hodny Z, Vasicova P (2019) Dynamic PML protein nucleolar associations with persistent DNA damage lesions in response to nucleolar stress and senescence-inducing stimuli. Aging (Albany NY) 11(17):7206–7235. https://doi.org/10.18632/aging.102248

James NJ, Howell GJ, Walker JH, Blair GE (2010) The role of Cajal bodies in the expression of late phase adenovirus proteins. Virology 399(2):299–311. https://doi.org/10.1016/j.virol.2010.01.013

Jiapaer Z, Su D, Hua L, Lehmann HI, Gokulnath P, Vulugundam G, Song S, Zhang L, Gong Y, Li G (2022) Regulation and roles of RNA modifications in aging-related diseases. Aging Cell 21(7):e13657. https://doi.org/10.1111/acel.13657

Jiménez-García LF, Spector DL (1993) In vivo evidence that transcription and splicing are coordinated by a recruiting mechanism. Cell 73(1):47–59. https://doi.org/10.1016/0092-8674(93)90159-n

Kaiser TE, Intine RV, Dundr M (2008) De novo formation of a subnuclear body. Science 322(5908):1713–1717. https://doi.org/10.1126/science.1165216

Khong A, Matheny T, Jain S, Mitchell SF, Wheeler JR, Parker R (2017) The stress granule transcriptome reveals principles of mRNA accumulation in stress granules. Mol Cell 68 (4):808–820 e805. doi:https://doi.org/10.1016/j.molcel.2017.10.015

Kim C, Yousefian-Jazi A, Choi S-H, Chang I, Lee J, Ryu H (2021) Non-cell autonomous and epigenetic mechanisms of Huntington's disease. Int J Mol Sci 22(22):12499

Kim J, Han KY, Khanna N (2019) Nuclear speckle fusion via long-range directional motion regulates speckle morphology after transcriptional inhibition. 132 (8). 132. https://doi.org/10.1242/jcs.226563

Kim J, Venkata NC, Hernandez Gonzalez GA, Khanna N, Belmont AS (2020) Gene expression amplification by nuclear speckle association. J Cell Biol 219(1). https://doi.org/10.1083/jcb.201904046

Kim SH, Ryabov EV, Kalinina NO, Rakitina DV, Gillespie T, MacFarlane S, Haupt S, Brown JW, Taliansky M (2007) Cajal bodies and the nucleolus are required for a plant virus systemic infection. EMBO J 26(8):2169–2179. https://doi.org/10.1038/sj.emboj.7601674

Kopp F, Mendell JT (2018) Functional classification and experimental dissection of long noncoding RNAs. Cell 172(3):393–407. https://doi.org/10.1016/j.cell.2018.01.011

Kovács D, Kovács M, Ahmed S, Barna J (2022) Functional diversification of heat shock factors 73 (4):427–439. https://doi.org/10.1007/s42977-022-00138-z

Kudrin P, Singh A, Meierhofer D, Kuśnierczyk A, Ørom UAV (2024) N4-acetylcytidine (ac4C) promotes mRNA localization to stress granules. EMBO Rep 25(4):1814–1834. https://doi.org/10.1038/s44319-024-00098-6

Kulkarni P, Bhattacharya S, Achuthan S, Behal A, Jolly MK, Kotnala S, Mohanty A, Rangarajan G, Salgia R, Uversky V (2022) Intrinsically disordered proteins: critical components of the wetware. Chem Rev 122(6):6614–6633. https://doi.org/10.1021/acs.chemrev.1c00848

Kulkarni P, Porter L, Chou TF, Chong S, Chiti F, Schafer JW, Mohanty A, Ramisetty S, Onuchic JN, Tuite M, Uversky VN, Weninger KR, Koonin EV, Orban J, Salgia R (2025) Evolving concepts of the protein universe. iScience 28(3):112012. https://doi.org/10.1016/j.isci.2025.112012

Kwak H, Lis JT (2013) Control of transcriptional elongation. Annu Rev Genet 47:483–508. https://doi.org/10.1146/annurev-genet-110711-155440

Lafarga M, Tapia O, Romero AM, Berciano MT (2017) Cajal bodies in neurons. RNA Biol 14(6):712–725. https://doi.org/10.1080/15476286.2016.1231360

Lechler MC, Crawford ED, Groh N, Widmaier K, Jung R, Kirstein J, Trinidad JC, Burlingame AL, David DC (2017) Reduced insulin/IGF-1 signaling restores the dynamic properties of key stress granule proteins during aging. Cell Rep 18(2):454–467. https://doi.org/10.1016/j.celrep.2016.12.033

Lellahi SM, Rosenlund IA, Hedberg A, Kiær LT, Mikkola I, Knutsen E, Perander M (2018) The long noncoding RNA NEAT1 and nuclear paraspeckles are up-regulated by the transcription factor HSF1 in the heat shock response. J Biol Chem 293(49):18965–18976. https://doi.org/10.1074/jbc.RA118.004473

Li H, Bingham PM (1991) Arginine/serine-rich domains of the su(wa) and tra RNA processing regulators target proteins to a subnuclear compartment implicated in splicing. Cell 67(2):335–342. https://doi.org/10.1016/0092-8674(91)90185-2

Linsenmeier M, Hondele M, Grigolato F, Secchi E, Weis K, Arosio P (2022) Dynamic arrest and aging of biomolecular condensates are modulated by low-complexity domains, RNA and biochemical activity. Nat Commun 13(1):3030. https://doi.org/10.1038/s41467-022-30521-2

Louvet E, Junéra HR, Le Panse S, Hernandez-Verdun D (2005) Dynamics and compartmentation of the nucleolar processing machinery. Exp Cell Res 304(2):457–470. https://doi.org/10.1016/j.yexcr.2004.11.018

Love AJ, Yu C, Petukhova NV, Kalinina NO, Chen J, Taliansky ME (2017) Cajal bodies and their role in plant stress and disease responses. RNA Biol 14(6):779–790. https://doi.org/10.1080/15476286.2016.1243650

Luo Y, Na Z, Slavoff SA (2018) P-bodies: composition, properties, and functions. Biochemistry 57(17):2424–2431. https://doi.org/10.1021/acs.biochem.7b01162

Ma H, Han P, Ye W, Chen H, Zheng X, Cheng L, Zhang L, Yu L, Wu X, Xu Z, Lei Y, Zhang F (2017) The long noncoding RNA NEAT1 exerts Antihantaviral effects by acting as positive feedback for RIG-I signaling. J Virol 91(9). https://doi.org/10.1128/jvi.02250-16

Maharjan N, Künzli C, Buthey K, Saxena S (2017) C9ORF72 regulates stress granule formation and its deficiency impairs stress granule assembly, hypersensitizing cells to stress. Mol Neurobiol 54(4):3062–3077. https://doi.org/10.1007/s12035-016-9850-1

Mahboubi H, Stochaj U (2017) Cytoplasmic stress granules: dynamic modulators of cell signaling and disease. Biochim Biophys Acta Mol basis Dis 1863 4:884–895. https://doi.org/10.1016/j.bbadis.2016.12.022

Mangan H, Gailín M, McStay B (2017) Integrating the genomic architecture of human nucleolar organizer regions with the biophysical properties of nucleoli. FEBS J 284(23):3977–3985. https://doi.org/10.1111/febs.14108

Marijan D, Momchilova EA, Burns D, Chandhok S, Zapf R, Wille H, Potoyan DA, Audas TE (2024) Protein thermal sensing regulates physiological amyloid aggregation. Nat Commun 15(1):1222. https://doi.org/10.1038/s41467-024-45536-0

Marijan D, Tse R, Elliott K, Chandhok S, Luo M, Lacroix E, Audas TE (2019) Stress-specific aggregation of proteins in the amyloid bodies. 593 (22):3162–3172. doi:https://doi.org/10.1002/1873-3468.13597

Mattsson K, Pokrovskaja K, Kiss C, Klein G, Szekely L (2001) Proteins associated with the promyelocytic leukemia gene product (PML)-containing nuclear body move to the nucleolus upon inhibition of proteasome-dependent protein degradation. Proc Natl Acad Sci USA 98(3):1012–1017. https://doi.org/10.1073/pnas.98.3.1012

Maul GG, Negorev D, Bell P, Ishov AM (2000) Review: properties and assembly mechanisms of ND10, PML bodies, or PODs. J Struct Biol 129(2–3):278–287. https://doi.org/10.1006/jsbi.2000.4239

McCluggage F, Fox AH (2021) Paraspeckle nuclear condensates: Global sensors of cell stress? 43 (5):e2000245. doi:https://doi.org/10.1002/bies.202000245

McMahon M, Forester C, Buffenstein R (2021) Aging through an epitranscriptomic lens. Nat Aging 1(4):335–346. https://doi.org/10.1038/s43587-021-00058-y

McMillan M, Gomez N, Hsieh C, Bekier M, Li X, Miguez R, Tank EMH, Barmada SJ (2023) RNA methylation influences TDP43 binding and disease pathogenesis in models of amyotrophic lateral sclerosis and frontotemporal dementia. Mol Cell 83 (2):219–236 e217. doi:https://doi.org/10.1016/j.molcel.2022.12.019

Melcák I, Cermanová S, Jirsová K, Koberna K, Malínský J, Raska I (2000) Nuclear pre-mRNA compartmentalization: trafficking of released transcripts to splicing factor reservoirs. Mol Biol Cell 11(2):497–510. https://doi.org/10.1091/mbc.11.2.497

Mokin YI, Ilyinsky NS, Nesterov SV, Smirnov EY, Sergeeva OS, Romanovich AE, Kuznetsova IM, Turoverov KK, Uversky VN, Fonin AV (2024) Stress-granules, P-bodies, and cell aging: a bioinformatics study. Biochem Biophys Res Commun 694:149404. https://doi.org/10.1016/j.bbrc.2023.149404

Mor A, White A, Zhang K, Thompson M, Esparza M, Muñoz-Moreno R, Koide K, Lynch KW, García-Sastre A, Fontoura BM (2016) Influenza virus mRNA trafficking through host nuclear speckles. Nat Microbiol 1(7):16069. https://doi.org/10.1038/nmicrobiol.2016.69

Morency E, Sabra M, Catez F, Texier P, Lomonte P (2007) A novel cell response triggered by interphase centromere structural instability. J Cell Biol 177(5):757–768. https://doi.org/10.1083/jcb.200612107

Moujaber O, Mahboubi H, Kodiha M, Bouttier M, Bednarz K, Bakshi R, White J, Larose L, Colmegna I, Stochaj U (2017) Dissecting the molecular mechanisms that impair stress granule

formation in aging cells. Biochim Biophys Acta, Mol Cell Res 1864 3:475–486. https://doi.org/10.1016/j.bbamcr.2016.12.008
Nakagawa S, Naganuma T, Shioi G, Hirose T (2011) Paraspeckles are subpopulation-specific nuclear bodies that are not essential in mice. J Cell Biol 193(1):31–39. https://doi.org/10.1083/jcb.201011110
Nakagawa S, Shimada M, Yanaka K, Mito M, Arai T, Takahashi E, Fujita Y, Fujimori T, Standaert L, Marine JC, Hirose T (2014) The lncRNA Neat1 is required for corpus luteum formation and the establishment of pregnancy in a subpopulation of mice. Development 141(23):4618–4627. https://doi.org/10.1242/dev.110544
Navascues J, Bengoechea R, Tapia O, Casafont I, Berciano MT, Lafarga M (2008) SUMO-1 transiently localizes to Cajal bodies in mammalian neurons. J Struct Biol 163(2):137–146. https://doi.org/10.1016/j.jsb.2008.04.013
Nesterov SV, Ilyinsky NS, Fonin AV, Uversky VN (2025) Signal transduction by phase separation-unnoticed revolution in molecular biology. J Mol Recognit 38(2):e70003. https://doi.org/10.1002/jmr.70003
Nesterov SV, Ilyinsky NS, Uversky VN (2021) Liquid-liquid phase separation as a common organizing principle of intracellular space and biomembranes providing dynamic adaptive responses. Biochim Biophys Acta, Mol Cell Res 1868 11:119102. https://doi.org/10.1016/j.bbamcr.2021.119102
Ninomiya K, Adachi S (2020) LncRNA-dependent nuclear stress bodies promote intron retention through SR protein phosphorylation 39 (3):e102729. doi:https://doi.org/10.15252/embj.2019102729
Ninomiya K, Iwakiri J, Aly MK, Sakaguchi Y, Adachi S, Natsume T, Terai G, Asai K, Suzuki T (2021) m(6) A modification of HSATIII lncRNAs regulates temperature-dependent splicing 40 (15):e107976. doi:https://doi.org/10.15252/embj.2021107976
Niwa-Kawakita M, Wu HC, Thé H, Lallemand-Breitenbach V (2018) PML nuclear bodies, membrane-less domains acting as ROS sensors? Semin Cell Dev Biol 80:29–34. https://doi.org/10.1016/j.semcdb.2017.11.007
Nott TJ, Petsalaki E, Farber P, Jervis D, Fussner E, Plochowietz A, Craggs TD, Bazett-Jones DP, Pawson T, Forman-Kay JD, Baldwin AJ (2015) Phase transition of a disordered nuage protein generates environmentally responsive membraneless organelles. Mol Cell 57(5):936–947. https://doi.org/10.1016/j.molcel.2015.01.013
Parlato R, Liss B (2014) How Parkinson's disease meets nucleolar stress. Biochim Biophys Acta (BBA) - Mol Basis Dis 1842(6):791–797. https://doi.org/10.1016/j.bbadis.2013.12.014
Pederson T (2011) The nucleolus. Cold Spring Harb Perspect Biol 3(3). https://doi.org/10.1101/cshperspect.a000638
Poblete-Durán N, Prades-Pérez Y, Vera-Otarola J, Soto-Rifo R, Valiente-Echeverría F (2016) Who regulates whom? An overview of RNA granules and viral infections. Viruses 8(7). https://doi.org/10.3390/v8070180
Protter DSW, Parker R (2016) Principles and properties of stress granules. Trends Cell Biol 26(9):668–679. https://doi.org/10.1016/j.tcb.2016.05.004
Puvion-Dutilleul F, Mazan S, Nicoloso M, Pichard E, Bachellerie JP, Puvion E (1992) Alterations of nucleolar ultrastructure and ribosome biogenesis by actinomycin D. Implications for U3 snRNP function. Eur J Cell Biol 58 (1):149–162
Quinodoz SA, Ollikainen N, Tabak B, Palla A, Schmidt JM, Detmar E, Lai MM, Shishkin AA, Bhat P, Takei Y, Trinh V, Aznauryan E, Russell P, Cheng C, Jovanovic M, Chow A, Cai L, McDonel P, Garber M, Guttman M (2018) Higher-order inter-chromosomal hubs shape 3D genome Organization in the Nucleus. Cell 174(3):744–757.e724. https://doi.org/10.1016/j.cell.2018.05.024
Raina K, Rao BJ (2022) Mammalian nuclear speckles exhibit stable association with chromatin: a biochemical study. Nucleus (Austin, Tex) 13(1):58–73. https://doi.org/10.1080/19491034.2021.2024948

Ranganathan S, Shakhnovich E (2021) Effect of RNA on morphology and dynamics of Membraneless organelles. J Phys Chem B 125(19):5035–5044. https://doi.org/10.1021/acs.jpcb.1c02286

Rawat P, Boehning M, Hummel B, Aprile-Garcia F, Pandit AS, Eisenhardt N, Khavaran A, Niskanen E, Vos SM, Palvimo JJ, Pichler A, Cramer P, Sawarkar R (2021) Stress-induced nuclear condensation of NELF drives transcriptional downregulation. Mol Cell 81(5):1013–1026.e1011. https://doi.org/10.1016/j.molcel.2021.01.016

Rhine K, Al-Azzam N, Yu T, Yeo GW (2022) Aging RNA granule dynamics in neurodegeneration. Front Mol Biosci 9:991641. https://doi.org/10.3389/fmolb.2022.991641

Rieckher M, Markaki M, Princz A, Schumacher B, Tavernarakis N (2018) Maintenance of Proteostasis by P body-mediated regulation of eIF4E availability during aging in Caenorhabditis elegans. Cell Rep 25 (1):199–211 e196. https://doi.org/10.1016/j.celrep.2018.09.009

Ries RJ, Pickering BF, Poh HX, Namkoong S, Jaffrey SR (2023) M(6)a governs length-dependent enrichment of mRNAs in stress granules. Nat Struct Mol Biol 30(10):1525–1535. https://doi.org/10.1038/s41594-023-01089-2

Rubbi CP, Milner J (2003) Disruption of the nucleolus mediates stabilization of p53 in response to DNA damage and other stresses. EMBO J 22(22):6068–6077. https://doi.org/10.1093/emboj/cdg579

Sahin U, de The H, Lallemand-Breitenbach V (2014) PML nuclear bodies: assembly and oxidative stress-sensitive sumoylation. Nucleus (Austin, Tex) 5(6):499–507. https://doi.org/10.4161/19491034.2014.970104

Salomoni P, Pandolfi PP (2002) The role of PML in tumor suppression. Cell 108(2):165–170. https://doi.org/10.1016/s0092-8674(02)00626-8

Sandqvist A, Björk JK, Akerfelt M, Chitikova Z, Grichine A, Vourc'h C, Jolly C, Salminen TA, Nymalm Y, Sistonen L (2009) Heterotrimerization of heat-shock factors 1 and 2 provides a transcriptional switch in response to distinct stimuli. Mol Biol Cell 20(5):1340–1347. https://doi.org/10.1091/mbc.e08-08-0864

Sasaki YT, Ideue T, Sano M, Mituyama T, Hirose T (2009) MENepsilon/beta noncoding RNAs are essential for structural integrity of nuclear paraspeckles. Proc Natl Acad Sci USA 106(8):2525–2530. https://doi.org/10.1073/pnas.0807899106

Shan T, Liu F, Wen M, Chen Z, Li S, Wang Y, Cheng H, Zhou Y (2023) m(6)A modification negatively regulates translation by switching mRNA from polysome to P-body via IGF2BP3. Mol Cell 83 (24):4494–4508 e4496. https://doi.org/10.1016/j.molcel.2023.10.040

Shav-Tal Y, Blechman J, Darzacq X, Montagna C, Dye BT, Patton JG, Singer RH, Zipori D (2005) Dynamic sorting of nuclear components into distinct nucleolar caps during transcriptional inhibition. Mol Biol Cell 16(5):2395–2413. https://doi.org/10.1091/mbc.e04-11-0992

Shi H, Liu B, Nussbaumer F, Rangadurai A, Kreutz C, Al-Hashimi HM (2019a) NMR chemical exchange measurements reveal that N(6)-Methyladenosine slows RNA annealing. J Am Chem Soc 141(51):19988–19993. https://doi.org/10.1021/jacs.9b10939

Shi H, Wei J, He C (2019b) Where, when, and how: context-dependent functions of RNA methylation writers, readers, and erasers. Mol Cell 74(4):640–650. https://doi.org/10.1016/j.molcel.2019.04.025

Silonov SA, Mokin YI, Nedelyaev EM, Smirnov EY, Kuznetsova IM, Turoverov KK, Uversky VN, Fonin AV (2023a) On the prevalence and roles of proteins undergoing liquid-liquid phase separation in the biogenesis of PML-bodies. Biomolecules 13(12). https://doi.org/10.3390/biom13121805

Silonov SA, Smirnov EY, Kuznetsova IM, Turoverov KK, Fonin AV (2023b) PML body biogenesis: a delicate balance of interactions. Int J Mol Sci 24(23). https://doi.org/10.3390/ijms242316702

Sirri V, Urcuqui-Inchima S, Roussel P, Hernandez-Verdun D (2008) Nucleolus: the fascinating nuclear body. Histochem Cell Biol 129(1):13–31. https://doi.org/10.1007/s00418-007-0359-6

Sobol M, Krausová A, Yildirim S (2018) Nuclear phosphatidylinositol 4,5-bisphosphate islets contribute to efficient RNA polymerase II-dependent transcription. 131(8). https://doi.org/10.1242/jcs.211094

Spector DL, Fu XD, Maniatis T (1991) Associations between distinct pre-mRNA splicing components and the cell nucleus. EMBO J 10(11):3467–3481. https://doi.org/10.1002/j.1460-2075.1991.tb04911.x

Spector DL, Lamond AI (2011) Nuclear speckles. Cold Spring Harb Perspect Biol 3(2). https://doi.org/10.1101/cshperspect.a000646

Spector DL, O'Keefe RT, Jiménez-García LF (1993) Dynamics of transcription and pre-mRNA splicing within the mammalian cell nucleus. Cold Spring Harb Symp Quant Biol 58:799–805. https://doi.org/10.1101/sqb.1993.058.01.087

Stępiński D (2020) Cajal body dynamics in soybean root meristem cells under chilling stress and recovery. Environ Exp Bot 180:104241. https://doi.org/10.1016/j.envexpbot.2020.104241

Strzelecka M, Trowitzsch S, Weber G, Lührmann R, Oates AC, Neugebauer KM (2010) Coilin-dependent snRNP assembly is essential for zebrafish embryogenesis. Nat Struct Mol Biol 17(4):403–409. https://doi.org/10.1038/nsmb.1783

Stuurman N, Meijne AM, van der Pol AJ, de Jong L, van Driel R, van Renswoude J (1990) The nuclear matrix from cells of different origin. Evidence for a common set of matrix proteins. J Biol Chem 265 (10):5460-5465

Su Y, Maimaitiyiming Y, Wang L, Cheng X, Hsu CH (2021) Modulation of phase separation by RNA: a glimpse on N(6)-Methyladenosine modification. Front Cell Dev Biol 9:786454. https://doi.org/10.3389/fcell.2021.786454

Sun L, Zuo Z, Qiu X, Wang G, Li Q, Qiu J, Peng Q (2023) Recent advances in the interplay between stress granules and m6A RNA modification. Curr Opinion Solid State Mater Sci 27(6):101119. https://doi.org/10.1016/j.cossms.2023.101119

Tapia O, Bengoechea R, Palanca A, Arteaga R, Val-Bernal JF, Tizzano EF, Berciano MT, Lafarga M (2012) Reorganization of Cajal bodies and nucleolar targeting of coilin in motor neurons of type I spinal muscular atrophy. Histochem Cell Biol 137(5):657–667. https://doi.org/10.1007/s00418-012-0921-8

Teixeira D, Parker R (2007) Analysis of P-body assembly in Saccharomyces cerevisiae. Mol Biol Cell 18(6):2274–2287. https://doi.org/10.1091/mbc.e07-03-0199

Turoverov KK, Kuznetsova IM, Fonin AV, Darling AL, Zaslavsky BY, Uversky VN (2019) Stochasticity of biological soft matter: emerging concepts in intrinsically disordered proteins and biological phase separation. Trends Biochem Sci 44(8):716–728. https://doi.org/10.1016/j.tibs.2019.03.005

Urbancokova A, Hornofova T, Novak J, Salajkova SA, Stemberkova Hubackova S, Uvizl A, Buchtova T, Mistrik M, McStay B, Hodny Z, Bartek J, Vasicova P (2024) Topological stress triggers persistent DNA lesions in ribosomal DNA with ensuing formation of PML-nucleolar compartment. elife 12. https://doi.org/10.7554/eLife.91304

Uversky VN (2013) Unusual biophysics of intrinsically disordered proteins. Biochim Biophys Acta 1834(5):932–951. https://doi.org/10.1016/j.bbapap.2012.12.008

Uversky VN (2015) Functional roles of transiently and intrinsically disordered regions within proteins. FEBS J 282(7):1182–1189. https://doi.org/10.1111/febs.13202

Uversky VN (2017a) Intrinsically disordered proteins in overcrowded milieu: membrane-less organelles, phase separation, and intrinsic disorder. Curr Opin Struct Biol 44:18–30. https://doi.org/10.1016/j.sbi.2016.10.015

Uversky VN (2017b) Protein intrinsic disorder-based liquid-liquid phase transitions in biological systems: complex coacervates and membrane-less organelles. Adv Colloid Interf Sci 239:97–114. https://doi.org/10.1016/j.cis.2016.05.012

Uversky VN (2019) Intrinsically disordered proteins and their "mysterious" (meta)physics. Front Phys 7:Article 10 (18 pages). https://doi.org/10.3389/fphy.2019.00010

Uversky VN (2021) Recent Developments in the Field of Intrinsically Disordered Proteins: Intrinsic Disorder–Based Emergence in Cellular Biology in Light of the Physiological and Pathological Liquid–Liquid Phase Transitions. Annu Rev Biophys 50 (1):135–156. doi:https://doi.org/10.1146/annurev-biophys-062920-063704

Uversky VN, Finkelstein AV (2019) Life in phases: intra- and inter- molecular phase transitions in protein solutions. Biomolecules 9(12). https://doi.org/10.3390/biom9120842

Uversky VN, Kuznetsova IM, Turoverov KK, Zaslavsky B (2015) Intrinsically disordered proteins as crucial constituents of cellular aqueous two phase systems and coacervates. FEBS Lett 589(1):15–22. https://doi.org/10.1016/j.febslet.2014.11.028

Valentin GG (1836) Repertorium für Anatomie und Physiologie. Verlag von Veit und comp, Berlin

Verma A, Sumi S, Seervi M (2021) Heat shock proteins-driven stress granule dynamics: yet another avenue for cell survival. Apoptosis 26(7–8):371–384. https://doi.org/10.1007/s10495-021-01678-w

Visa N, Puvion-Dutilleul F, Bachellerie JP, Puvion E (1993) Intranuclear distribution of U1 and U2 snRNAs visualized by high resolution in situ hybridization: revelation of a novel compartment containing U1 but not U2 snRNA in HeLa cells. Eur J Cell Biol 60(2):308–321

Visintin R, Hwang ES, Amon A (1999) Cfi1 prevents premature exit from mitosis by anchoring Cdc14 phosphatase in the nucleolus. Nature 398(6730):818–823. https://doi.org/10.1038/19775

Wagner A, Schosserer M (2022) The epitranscriptome in ageing and stress resistance: a systematic review. Ageing Res Rev 81:101700. https://doi.org/10.1016/j.arr.2022.101700

Wagner R (1835) Einige Bemerkungen und Fragen über das Keimbläschen (vesicular germinativa). Müller's Archiv Anat Physiol Wissenschaft Med:373–377

Wang C, Schmich F, Srivatsa S, Weidner J, Beerenwinkel N, Spang A (2018a) Context-dependent deposition and regulation of mRNAs in P-bodies. elife 7. https://doi.org/10.7554/eLife.29815

Wang K, Liu H, Hu Q, Wang L, Liu J, Zheng Z, Zhang W, Ren J, Zhu F, Liu GH (2022) Epigenetic regulation of aging: implications for interventions of aging and diseases. Signal Transduct Target Ther 7(1):374. https://doi.org/10.1038/s41392-022-01211-8

Wang Y, Hu S-B, Wang M-R, Yao R-W, Wu D, Yang L, Chen L-L (2018b) Genome-wide screening of NEAT1 regulators reveals cross-regulation between paraspeckles and mitochondria. Nat Cell Biol 20(10):1145–1158. https://doi.org/10.1038/s41556-018-0204-2

West JA, Davis CP, Sunwoo H, Simon MD, Sadreyev RI, Wang PI, Tolstorukov MY, Kingston RE (2014) The long noncoding RNAs NEAT1 and MALAT1 bind active chromatin sites. Mol Cell 55(5):791–802. https://doi.org/10.1016/j.molcel.2014.07.012

Wheeler JR, Matheny T, Jain S, Abrisch R, Parker R (2016) Distinct stages in stress granule assembly and disassembly. elife 5. https://doi.org/10.7554/eLife.18413

Wolozin B, Ivanov P (2019) Stress granules and neurodegeneration. Nat Rev Neurosci 20(11):649–666. https://doi.org/10.1038/s41583-019-0222-5

Yang K, Yang J, Yi J (2018) Nucleolar stress: hallmarks, sensing mechanism and diseases. Cell stress 2(6):125–140. https://doi.org/10.15698/cst2018.06.139

Yang P, Mathieu C, Kolaitis RM, Zhang P, Messing J, Yurtsever U, Yang Z, Wu J, Li Y, Pan Q, Yu J, Martin EW, Mittag T, Kim HJ, Taylor JP (2020) G3BP1 is a tunable switch that triggers phase separation to assemble stress granules. Cell 181 (2):325–345 e328. https://doi.org/10.1016/j.cell.2020.03.046

Yao R-W, Xu G, Wang Y, Shan L, Luan P-F, Wang Y, Wu M, Yang L-Z, Xing Y-H, Yang L (2019) Nascent pre-rRNA sorting via phase separation drives the assembly of dense fibrillar components in the human nucleolus. Mol Cell 76 (5):767–783. e711

Youn JY, Dyakov BJA, Zhang J, Knight JDR, Vernon RM, Forman-Kay JD, Gingras AC (2019) Properties of stress granule and P-body proteomes. Mol Cell 76(2):286–294. https://doi.org/10.1016/j.molcel.2019.09.014

Zhang B, Arun G, Mao YS, Lazar Z, Hung G, Bhattacharjee G, Xiao X, Booth CJ, Wu J, Zhang C, Spector DL (2012) The lncRNA Malat1 is dispensable for mouse development but its transcription plays a cis-regulatory role in the adult. Cell Rep 2(1):111–123. https://doi.org/10.1016/j.celrep.2012.06.003

Zhang L, Xia J (2023) N6-Methyladenosine methylation of mRNA in cell senescence. Cell Mol Neurobiol 43(1):27–36. https://doi.org/10.1007/s10571-021-01168-2

Zhang Q, Chen CY, Yedavalli VS, Jeang KT (2013) NEAT1 long noncoding RNA and paraspeckle bodies modulate HIV-1 posttranscriptional expression. mBio 4 (1):e00596–00512. https://doi.org/10.1128/mBio.00596-12

Zhang Y, Lu H (2009) Signaling to p53: ribosomal proteins find their way. Cancer Cell 16(5):369–377. https://doi.org/10.1016/j.ccr.2009.09.024

Zhao Z, Qing Y, Dong L, Han L, Wu D, Li Y, Li W, Xue J, Zhou K, Sun M, Tan B, Chen Z, Shen C, Gao L, Small A, Wang K, Leung K, Zhang Z, Qin X, Deng X, Xia Q, Su R, Chen J (2023) QKI shuttles internal m(7)G-modified transcripts into stress granules and modulates mRNA metabolism. Cell 186 (15):3208–3226 e3227. https://doi.org/10.1016/j.cell.2023.05.047

Chapter 17
Crowding in Anhydrobiosis

Alex Haydon, Charles A. Elder, Rafael S. Demarco, and Michael A. Menze

Abstract Anhydrobiosis, the exceptional adaptation of some organisms to enter a state approaching suspended animation after losing most cellular water, has fascinated scientists and non-scientists since Anthony van Leeuwenhoek described the phenomenon over 300 years ago. Despite progress in elucidating molecular principles enabling 'life at a standstill,' we still lack a comprehensive understanding of the minimum biochemical requirements for desiccation tolerance. Biochemical strategies employed by anhydrobiotic organisms are multifaceted but commonly include the accumulation of intrinsically disordered proteins and compatible osmolytes, investment into antioxidant defense mechanisms, and, more recently recognized, the formation of biomolecular condensates. This chapter explores why water is the solvent capable of sustaining life on Earth and how it promotes biomolecular condensates in cells. We propose four mechanisms by which these biomolecular condensates may facilitate desiccation tolerance: 1.) The selective sequestration hypothesis proposes the formation of 'protective' biomolecular condensates, 2.) The selective target neutralization hypothesis suggests that biomolecular condensates may downregulate metabolic, developmental, and apoptotic pathways during water stress, 3.) The increased viscocapillary effect hypothesis proposes that the preservation of cellular morphology can be maintained during desiccation by a combination of increased viscosity and capillary forces exerted by biomolecular condensates, and 4.) The solvent property hypothesis postulates that water's physicochemical properties inside biomolecular condensates could be modulated to inhibit damage during desiccation. Integrating the known biochemical strategies observed in anhydrobiotes into a comprehensive model of desiccation tolerance may allow us to engineer this trait into water-stress sensitive systems to address societal challenges ranging from crop loss during droughts to stabilization of biomedical relevant cells at room temperature.

Michael A. Menze: Co-corresponding author.

A. Haydon (✉) · C. A. Elder · R. S. Demarco · M. A. Menze (✉)
Department of Biology, University of Louisville, Louisville, KY, USA
e-mail: alex.haydon@louisville.edu; michael.menze@louisville.edu

V. N. Uversky (ed.), *(Macro)Molecular Crowding*, Subcellular Biochemistry 109,
https://doi.org/10.1007/978-3-032-03370-3_17

Keywords Anhydrobiosis · Osmolytes, Biomolecular condensates · Selective sequestration · Selective target neutralization · Viscocapillary effect · Solvent properties

Water

Water is the fundamental solvent of life on Earth, playing a crucial role in maintaining biological systems. One of water's most remarkable properties is its ability to remain liquid across much of Earth's temperature range, providing a suitable medium for biochemical reactions and, depending on the system, it can act as a product, substrate, catalyst, cofactor, intermediate, and chaperone (Brini et al. 2017; Frenkel-Pinter et al. 2021). Unlike most substances, water becomes less dense when it freezes, allowing ice to float and insulate the liquid below, preserving aquatic ecosystems through harsh winters (Clegg and Drost-Hansen 1991; Hampton et al. 2017). Additionally, water's high specific heat capacity allows it to absorb and store large amounts of heat with minimal temperature changes (Brini et al. 2017). This property helps buffer environmental temperature fluctuations, ensuring more stable conditions for life. In aquatic ecosystems, it prevents rapid temperature shifts that could harm marine organisms. At the same time, it dampens climate variations in terrestrial environments by absorbing heat during the day and releasing it at night. At the cellular level, water's resistance to dramatic temperature change is essential for maintaining the precise conditions required for enzymatic activity and metabolic processes, further reinforcing water's role as a critical component of life (Clegg and Drost-Hansen 1991). Water's cohesive and adhesive properties drive capillary action, facilitating plant nutrient transport and animal blood circulation (Steudle 2001). It enables countless biochemical interactions as a universal solvent and reactant, underscoring its indispensability at molecular, cellular, and ecological scales.

Water's robust resistance to heating, melting, and boiling is due to its weak yet immense network of self-interacting hydrogen bonds (Xantheas 2000). When enough thermal energy is removed, the molecular motion of liquid water is reduced such that ice, or rather, a continuous tetrahedral 3-D crystalline lattice, can form. Water near and below 4 °C exists primarily in a tetrahedral ordered state, where the maximum number of four hydrogen bonds occurs in a single water molecule (Clegg and Drost-Hansen 1991). This configuration increases order by decreasing the system's overall motion and free energy. As temperature rises in pure water, hydrogen bonding between water molecules becomes more transient, rapidly breaking and reforming. Instead of forming exclusively tetrahedral networks of water molecules, some amount of liquid water exists in two hydrogen-bonded configurations, in which individual water molecules have one strong donor and one strong acceptor of hydrogen bonds. Therefore, the overall structure of liquid water exists as a blend of tetrahedral networks and ring and chain-like structures, and, on average, each molecule forms just slightly more than three hydrogen bonds (Liu et al. 2017b).

The structure of cellular water is highly dynamic and complex, influenced by the intricate interplay of biomolecules, ions, and hydrophobic interactions. Charged macromolecules, including proteins and nucleic acids, and ions interact with surrounding water molecules through hydrogen bonds. These interactions disrupt the naturally occurring hydrogen-bonding network of bulk water, leading to localized changes in water structure and dynamics (Levy and Onuchic 2004). Conversely, hydrophobic molecules, such as lipids, exhibit the opposite effect by repelling water, forcing surrounding water molecules to adopt a more structured, ordered arrangement. This phenomenon, often referred to as hydrophobic hydration, plays a crucial role in biomolecular self-assembly, including the formation of lipid bilayers and protein folding (Chandler 2005; Disalvo et al. 2008; Liu et al. 2017b). Additionally, intracellular crowding further complicates water organization, as confined spaces and molecular interactions create microenvironments with distinct hydration states. As a result, the structure of cellular water is not uniform but instead exists in a constantly shifting equilibrium, dynamically adapting to regional concentrations of biomolecules and their electrochemical properties (Hansen et al. 2016). This ever-changing nature of cellular water has significant implications for biochemical reactions, molecular diffusion, and overall cellular function. It reinforces the notion that concentrations and conditions are locally defined in the cell.

The compartmentalization of cells and organelles by membranes relies on the polar properties of water. Due to the amphipathic nature of membrane lipids, phospholipids assemble into bilayers in a thermodynamically favorable manner, with the hydrophilic phospholipid heads interacting with surrounding water molecules. Water's high dielectric constant interrupts electrostatic interactions, allowing ions to dissolve in aqueous solutions (Deamer 2017). Additionally, water's ability to form extensive hydrogen-bonded hydration shells around ions further enhances their solubility. Because of these properties, ions remain stabilized in solution and do not associate with lipid membranes' hydrophobic fatty acid tails, making their passage across the membrane energetically unfavorable. As a result, the diffusion of charged molecules across a lipid membrane is up to eight orders of magnitude lower than that of small, uncharged molecules such as oxygen and carbon dioxide. While water is a polar molecule, its small size and ability to transiently form hydrogen bonds allow it to pass through membranes at a moderate rate, often facilitated by aquaporins (Deamer 2017; Vorontsova et al. 2022). The electrochemical properties of water contribute to the stabilization of ion gradients, which are essential for metabolism, cell signaling, and the compartmentalization of biochemical reactions (Chakrabarti and Deamer 1992). Additionally, the extent of hydration surrounding lipid membranes influences membrane fluidity and compactness, affecting the permeability of charged molecules (Eremchev et al. 2023).

Hydration water adjacent to lipid membranes is generally polarized, aligning toward the polar head groups of phospholipids (Disalvo et al. 2008). Depending on the predominant class of phospholipid, such as phosphatidylcholines, glycerophospholipids, or sphingolipids, the incorporation of cholesterol can either disrupt or stabilize the ordered arrangement of water molecules around the head groups, thereby influencing membrane fluidity and permeability. In the case of a sphingolipid-dominant membrane, cholesterol reduces the distance between lipids by facilitating the formation of intermolecular hydrogen bonds between lipids. The tighter packing of lipids further enhances the orientation of water molecules towards the polar head groups, which reinforces the intermolecular interactions between lipids (Orlikowska-Rzeznik et al. 2024). While elevated cholesterol levels support membrane integrity during cellular stress, the cell's immediate response to changing hydration levels can be mediated by detecting membrane architecture changes driven by fluctuations in cell volume and molecular crowding. Watson et al. observed a rapid increase in calcium signaling in mouse chondrocytes immediately following hyperosmotic stress. They speculated that increased osmotic pressure alters membrane tension, activating mechanosensitive calcium channels. This influx of calcium then affects the structure of nearby water molecules, stabilizing the membrane and acting as a secondary signaling mechanism to initiate cellular stress responses (Watson et al. 2023). Beyond its role in stress responses, hydration water influences key membrane-associated processes. The interaction between the polar lipid headgroups and surrounding water molecules affects the translocation of biomolecules, the binding affinity of membrane-associated proteins, and the function of membrane-embedded proteins (Eremchev et al. 2023). These intricate relationships highlight water's indispensable role in maintaining membrane stability, facilitating signaling, and regulating biochemical processes.

Within cells, the structure of water is greatly influenced by the proximity of both hydrophilic and hydrophobic macromolecules (Hansen et al. 2016). *Bulk water* consists of water molecules hydrogen bonded to other water molecules, forming a dynamic and fluid network. Because hydrogen bonds between water molecules are relatively weak, bulk water is highly mobile, with no fixed orientation toward an attractive force. As a result, it does not maintain a persistent dipole moment, as molecular rotations rapidly randomize dipole orientations (Ball 2017; Chen et al. 2008). Bulk water is osmotically responsive and easily exits the cell during hyperosmotic stress. In contrast, *hydration water*, which closely associates with ions, proteins, and other macromolecules, is less affected by osmotic forces due to its retention through hydrogen bonding and other intermolecular interactions (Ball 2017; Chen et al. 2008).

The attractive forces at biomolecular surfaces cause hydration water to be relatively osmotically unresponsive (Cameron et al. 1988). Depending on cell type and stress conditions, the proportion of osmotically unresponsive water can range from 32% to 91% of total cellular water (Fullerton et al. 2006). It should be noted that the importance of osmotically unresponsive water remains debated, as the rapid exchange of individual water molecules between the hydration and bulk phases makes it challenging to quantify experimentally (Halle 2004). Water molecules surrounding a charged surface align their dipole moment according to electrostatic interactions, with hydrogen atoms facing negatively charged interfaces (e.g., carboxyl or phosphate groups) and oxygen atoms facing positively charged regions (e.g., amine groups). Hydration water includes a 'bound' or primary hydration layer, in which water molecules are directly hydrogen-bonded to the macromolecule's surface. Subsequent hydration layers undergoing weaker intermolecular interactions typically extend 2–4 water layers outward, though this can vary depending on the biomolecule and environmental conditions (Chen et al. 2008; Romero-Perez et al. 2023). With increasing distance from a given macromolecule, the dipole alignment of water molecules and their degree of order gradually decreases. As cellular temperature falls (but remains above freezing), water molecules move more slowly, leading to stronger hydrogen bonding around macromolecules and increased macromolecular hydration. This phenomenon results in more hydration water than observed at warmer temperatures (Watson et al. 2023). Hydration water surrounding biomolecules can mediate binding interactions and facilitate electron and proton transfer (Chaplin 2006). Proton hopping, also known as the Grotthuss mechanism, involves proton transfer via hydrogen bond flipping along a continuous network of water molecules (de Grotthuss 1806). These water "wires" are commonly found in structured hydrogen-bonded networks, including hydrophilic protein cavities and bulk water pathways. Furthermore, proton transfer is faster when the water molecules within the wire are more strongly hydrogen-bonded (Chaplin 2006).

The term *vicinal water* is often used to describe water molecules whose structure is altered by proximity to a nearby interface. While this definition excludes the primary hydration layer, it includes subsequent hydration layers and water within the definitive boundary of bulk and hydration water (Drost-Hansen and Singleton 1995). When water solvates ions, structural changes in the surrounding water network typically extend no more than three molecular layers (Drost-Hansen and Singleton 1995). In contrast, at a water-air interface, the hydrophobic nature of air can disrupt the hydrogen-bonding network of water molecules for up to 15 layers (Fig. 17.1) (Clegg and Drost-Hansen 1991). Cellular water with more bent, weak, or unstable hydrogen bonding allows for greater protein mobility. In contrast, more structured water with stronger hydrogen bonding restricts protein movement (Chaplin 2006). It has been shown that vicinal water is induced by all solid surfaces regardless of surface charge, and significant structural changes of vicinal water occur at temperature intervals of about 15 °C from approximately 14–62 °C (Clegg and Drost-Hansen 1991). Lipid membranes, protein complexes, and biomolecular condensates thus likely perturb vicinal water due to their interfacial area and surface chemistry. Whether water molecules are adjacent to a hydrophobic or hydrophilic surface may also significantly affect the structure of water and water's preference for dissolving specific molecules. Water molecules at hydrophobic lipid membranes form a contact angle greater than 65 degrees and thus create a more open and less dense hydrogen-bonded network. In comparison, water molecules near hydrophilic surfaces form a contact angle of less than 30 degrees and condense into a more compact structure. Denser water near hydrophilic surfaces prefers to solubilize divalent ions, while less dense water near hydrophobic surfaces prefers to solubilize monovalent ions (Vogler 1998). The estimated concentration of macromolecules in hydrated cells is 200–350 mg/ml, a volume fraction of about 40% (Munder et al. 2016). A fully hydrated cell, therefore, exists in a gel or liquid crystalline state (Chaplin 2006). Water is the most abundant molecule in cells, making up 60–70% of the cellular mass. Life on Earth and the shape and function of biomolecules have evolved with water as the only solvent. Without water, most proteins would not

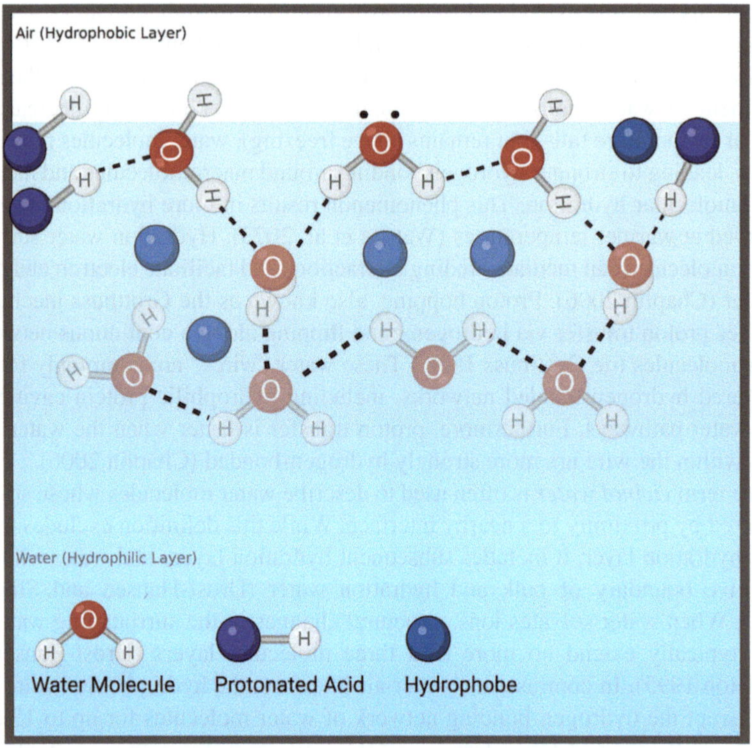

Fig. 17.1 Impact of the air interface on the structure of water. The air-water interface significantly disrupts the hydrogen-bonding network of water due to the hydrophobic nature of air. The interaction between water's hydrogen atoms, lone electron pairs, and the air phase at the interface reduces the potential for hydrogen bond formation among water molecules. As a result, non-polar molecules, known as hydrophobes, tend to accumulate near the air-water interface. Similarly, protonated acids, which are less polar than their deprotonated forms, also concentrate at the interface (Brini et al. 2017). (Created with BioRender)

form their functional secondary and tertiary structures, and it is necessary for cells to sense water fluctuations and to respond to these changes quickly.

Although cells are composed of up to 70% water, they are remarkably crowded environments. Romero-Perez et al. (2023) predict that 10–20 layers of water exist among proteins in the yeast *Saccharomyces cerevisiae* under hydrated conditions. The high concentration of macromolecules, resulting in molecular crowding, profoundly influences the physical properties of proteins, protein complexes, and nucleic acids, causing their behavior in cells to differ significantly from *in vitro* observations. This phenomenon, known as molecular crowding, arises from the dense intracellular environment where macromolecules restrict diffusion, alter reaction kinetics, and stabilize proteins and biomolecular interactions compared to dilute solutions. Molecular crowding increases the excluded volume, the space surrounding a molecule inaccessible to other molecules due to steric hindrance, thereby limiting the volume available for diffusion (Romero-Perez et al. 2023). In addition to excluded volume, preferential exclusion occurs when specific solutes, such as osmolytes or polymers, are excluded from the immediate vicinity of macromolecules due to steric and thermodynamic factors. This phenomenon is associated with enhanced structural stability of macromolecules in crowded environments (Lee and Timasheff 1981).

Crowding can promote the formation of macromolecular complexes through an entropically driven process. While the assembly of monomers into a complex reduces their local configurational entropy, the system's overall entropy can increase. This increase in entropy may occur because the formed complex occupies less excluded volume than individual monomers, effectively "freeing up" space for water, crowding agents, and other macromolecules to access. The resulting increase in solvent and crowding-agent entropy compensates for the loss of monomer entropy, making biomolecular complex assembly thermodynamically favorable (Alfano et al. 2024). In terms of reaction rates in crowded environments, two key hypotheses have been proposed. For reactions with low Michaelis constants (high reaction probabilities), where most encounters between reactants result in a reaction, molecular crowding diminishes reaction rates due to reduced diffusion, which limits how quickly reactants can meet. Conversely, for reactions with high Michaelis constants (low reaction probabilities), where reactions rarely occur on the first encounter, crowding can increase reaction rates by promoting multiple encounters (re-collisions), thereby raising the likelihood of a successful reaction (Kim and Yethiraj 2009).

Anhydrobiosis, the ability of some organisms to survive extreme dehydration by entering a reversible ametabolic state, represents a profound challenge to the water-dependent nature of life (Crowe et al. 1992). In this state, cells must preserve structural integrity, macromolecular organization, and biochemical potential despite a nearly complete water loss. Successfully entering and exiting the anhydrobiotic state is especially complex to conceptualize in the context of macromolecular crowding, where densely packed proteins, nucleic acids, and other solutes rely on hydration to maintain proper spacing, conformational flexibility, and functional interactions. Without water, crowded intracellular environments are at risk of protein denaturation, aggregation, and loss of compartmentalization. Anhydrobiotic organisms counteract these threats through molecular strategies such as accumulating protective sugars like trehalose or sucrose, expressing intrinsically disordered late embryogenesis abundant (LEA) proteins, and forming glassy or highly viscous matrices that stabilize biomolecular structures in the dry state (Boothby and Pielak 2017). These adaptations help preserve the delicate balance of forces observed in the crowded cytoplasm before drying, enabling survival through desiccation and rehydration.

Anhydrobiosis

Organisms are said to be anhydrobiotic when they can survive the near absolute loss of cellular water. Dissecting the word's roots, anhydrobiosis translates to "life without water". The exact cellular water content at which an organism is considered anhydrobiotic is not well defined. However, a common definition is the ability to survive a reduced water content below 0.1 g H_2O per g dry mass. This definition was partly chosen because it is considered the minimum amount of cellular water needed to form a monolayer between macromolecules (Farrant 2008; Romero-Perez et al. 2023). Under anhydrobiotic conditions, metabolic arrest is essential for survival and unavoidable in the absence of bulk water (Olgenblum et al. 2024). In addition to metabolic arrest, most anhydrobiotes require the accumulation of some osmolytes before cellular water loss (Janis 2023). Osmolytes can be amino acids, sugars, polyols, and other charged small molecules that primarily function to 1.) increase intracellular osmolarity during water stress. 2.) protect lipids, proteins, and nucleotides from damage, and 3.) modulate the viscosity and diffusivity of the cellular environment (Romero-Perez et al. 2023). Furthermore, anhydrobiotic organisms often produce specialized proteins known as anhydrobiosis-related intrinsically disordered (ARID) proteins, which mostly lack ordered secondary structure motifs in the hydrated state but gain secondary structure motifs during cellular water loss (Janis 2023). The accumulation of osmolytes and ARID proteins is often developmentally or environmentally regulated, as occurs during diapause or quiescence, respectively, and cells must be able to sense changes in cellular water content at the onset and during anhydrobiosis.

Desiccation tolerance is defined as the ability of an organism to survive significant cellular and tissue water loss and to return to regular activity after rehydration (Jenks and Wood 2008). Many organisms, such as desert cacti and camels, may survive water stress by decreasing water loss during drought and increasing water storage and uptake when available. For example, cacti and other xerophilic plants have fewer and smaller stomata, which reduces evaporative loss. Furthermore, many xerophilic plants undergo crassulacean acid metabolism (CAM), in which stomata are only opened at night when conditions are cooler and more humid (Males and Griffiths 2017). However, during these circumstances, cellular water loss does not occur. These organisms are not desiccation tolerant since they resist dehydration. The distinction between desiccation-tolerant organisms and organisms that can survive water stress without losing cellular water is essential because desiccation-tolerant organisms must have acquired adaptations that protect against increased molecular crowding, protein aggregation, membrane instability, and irreversible cellular damage (Romero-Perez et al. 2023).

Anhydrobiosis in Organisms

Though rare, desiccation-tolerant organisms are diverse and reside in all six kingdoms of life. The ubiquity of anhydrobiosis in nature demonstrates many unique, homologous, and analogous adaptations to desiccation (Alpert 2000). Desiccation-resistant species may rely on adaptations that solely minimize water loss, such as the African killifish embryo, which encases its cells in protective cuticles to reduce evaporative water loss (Hibshman et al. 2020; Polačik and Podrabsky 2015).

Desiccation-tolerant organisms generally have developmental and physiological adaptations to survive intracellular water loss, such as the embryonic diapause cyst of the brine shrimp *Artemia franciscana*. Diapause cysts comprise an outer cuticle constructed primarily of chitin, a middle alveolar layer, and an innermost fibrous layer. This chitin shell may help to precondition the organism to desiccation by slowing the drying rate (Abatzopoulos et al. 2002). Analogously, cyanobacterial species from the orders Nostocales and Stigonematales form spore-like structures called akinetes. Peptidoglycan crosslinking of the cell wall occurs to create a super dense, compact shell. Glycogen and cyanophicin, an amino acid-based biopolymer that stores nitrogen and energetic carbon sources, also accumulate in the dehydrated cell (Garg and Maldener 2021). The free-swimming amoeba of the genus *Acanthamoeba* also forms desiccated cysts composed of a proteinaceous outer wall and a cellulose inner wall. About 70% of *Acanthamoeba* cysts survive desiccation for 2 to 20 years (Sriram et al. 2008). Evolutionary pressure from desiccation has driven a diverse yet often convergent set of adaptations across a wide range of organisms. Ultimately, desiccation-tolerant organisms must contend with a shared set of cellular threats: oxidative stress, membrane destabilization, and the deterioration of vital proteins and nucleotides (Alpert 2000).

The ability of microbes to resist intracellular water loss is noteworthy. Bacteria may form biofilms with algae, fungi, and archaea species in xeric or arid habitats. The main component of biofilms is the extracellular polymeric substance (EPS), which is a sticky polymer complex that either forms a covalently attached capsule to the bacteria's cellular surface or is secreted and generally contributes to the microbial biofilm (Meredith et al. 2018). Due to its hygroscopic and hydrophilic nature, the EPS absorbs moisture from the atmosphere, preventing water loss and triggering rapid water absorption if water becomes available after drying. *Pseudomonas putida* produces the anionic EPS polymer, alginate, which can retain several times its mass in water, and other xeric bacterial species in cold deserts upregulate genes for producing the EPS components rhamnose, xylose, and mannose. *Escherichia coli, Pantoea stewartii,* and *Acinetobacter calcoaceticus* with mutated EPS genes exhibited a sixfold reduction in survival following desiccation stress (Meredith et al. 2018). Fungi are regarded as more desiccation resistant than bacteria in many ways. Many fungal species have hyphae extending into air-filled soil pockets to access nutrients and water. In a study on the microbial composition in dry grasslands, the fungal community was largely unaffected by long dry periods followed by rehydration (Barnard et al. 2013). *S. cerevisiae* may favor beta-oxidation of fatty acids to produce metabolic water during desiccation. Fat oxidation generates 110 g of water per 100 g of fat, a ratio higher than for the oxidation of carbohydrates (~60 g H_2O) or proteins (~40 g H_2O) (Robison et al. 2024).

The hallmark of bacterial anhydrobiosis is the reversible formation of desiccated, dormant endospores, which occur most notably in the genera *Bacillus* and *Clostridium*. Endospores are differentiated cells that exist in a dehydrated, dormant state and are like akinetes but are generally more resistant to additional stresses, such as UV radiation, and extreme temperatures and pressures. Bacterial endospores contain trehalose, glycogen, and triglycerides as carbon stores and potentially as water replacement molecules (see below). Cell membranes also accumulate a higher proportion of negatively charged phospholipids and saturated fatty acids, which help maintain the cell membrane in the liquid-crystalline phase and prevent the formation of non-bilayer structures (Lebre et al. 2017). Additionally, dipicolinic acid and small-acid soluble proteins (SASPs) accumulate in the dehydrated cellular core, shielding DNA from harmful reactive oxygen species (ROS). Some SASPs may also be associated with the cell walls of spores. SASPs are often intrinsically disordered proteins (IDPs) that gain defined secondary structure motifs when water is removed. IDPs are arguably a fundamental aspect of desiccation-tolerance across all of life and exemplify the similarity of strategies to survive extreme drying in simple and more complex organisms (Dahl and Fordice 2011). The survival potential of bacterial endospores is remarkable. This potential is most adeptly demonstrated by the species *Virbacillus arcticus*, which can survive hundreds to thousands of years in permafrost soils (McDonald et al. 2024). *S. cerevisiae* also forms dehydrated spores, which differentiate from dividing yeast cells. In the case of *S. cerevisiae*, desiccation-tolerance depends upon pre-conditioning by starvation. Like many desiccation-tolerant bacteria, plants, and animals, *S. cerevisiae* alters cell membrane components, produces trehalose and IDPs known as hydrophilins, which are predicted to protect intracellular components during water loss (Robison et al. 2024).

Many bacteria have mechanisms to survive osmotic stress. Like fungi and plants, bacteria are better equipped than animal cells to handle osmotic pressure changes because their cell walls can withstand higher turgor pressure (Record et al. 1998a). The pathogenic *Salmonella enterica* and *Cronobacter sakazakii* can survive in preserved foods with high salt concentrations, such as peanut butter and fruit conserves (Lebre et al. 2017). *Halorhabdus utahensis* and *Halorhodospira halophila*, the bacterial species with the lowest known water activity required for cell division, survive hypersaline conditions in the Great Salt Lake (Lebre et al. 2017). In general, bacteria survive hyperosmotic stress by accumulating salts and small molecules to restore osmotic equilibrium, reducing protein aggregation caused by ROS and Maillard reactions, and restructuring the cell membrane to decrease permeability (Grzyb and Sklodowska 2022). Specific solutes upregulated during osmotic stress function exclusively as either osmoprotectants or compatible osmolytes. For example, the divalent cation putrescine was upregulated during osmotic stress in *E. coli*, but did not affect cytoplasmic osmolarity. Putrescine, which binds to nucleic acids, is likely acting as a protectant of nucleic acids. In contrast, a large portion of potassium during hyperosmotic stress is unbound to macromolecules and thus acts primarily as an osmolyte (Record et al. 1998a). *E. coli* have osmotically regulated porin proteins in the cell wall, which allow the diffusion of water and small molecules. Many bacteria and fungi import inorganic ions, like potassium, to match the extracellular osmotic pressure (Lebre et al. 2017). The desiccation-tolerant yeast species *S. cerevisiae* expresses a potassium transporter, Trk2, as well as the glycerol importer Stl1p and exporter Fps1p, which are essential for desiccation-tolerance (Rapoport et al. 2019). *E. coli* and other bacterial species first upregulate potassium uptake and increase trehalose and glutamate biosynthesis to reduce the osmotic pressure imbalance between the cytosol and the extracellular fluid (Record et al. 1998a). However, the secondary response often involves replacing charged ions, like potassium, with compatible osmolytes like proline and glycine betaine, which are less likely to interact deleteriously with biomolecules (Lebre et al. 2017). Glycine betaine accumulates in large concentrations in *E. coli* and acts as an osmoprotectant and osmolyte by preventing the aggregation of biomolecules and reducing the osmotic pressure gradient (Record Jr. et al. 1998b). During hyperosmotic stress in *E. coli*, biomolecular crowding increases markedly as the intracellular volume occupied by biomolecules rises from 30% to 50% in response to an external osmotic pressure increase from 0.1 OsM to 1.0 OsM (Record et al. 1998a). Increased biopolymer proximity under crowded conditions excludes ionic osmolytes, such as potassium, from their surfaces, reducing the likelihood of harmful aggregate formation between osmolytes and biomolecules. However, the increased molecular crowding is accompanied by a dramatic loss in bulk cytoplasmic water, which is necessary for cellular growth (Record et al. 1998a). Record et al. proposed that glycine betaine, due to its bulky nature relative to potassium, is more easily excluded from the hydration layer of biomolecules. In this model, decreased hydration water leads to increased bulk water to prevent the harmful interaction of biomolecules with ionic osmolytes. Therefore, glycine betaine accumulation increases bulk cellular water, which allows *E. coli* to continue cellular growth during hyperosmotic stress (Record et al. 1998b).

Desiccation-tolerance is widespread among plants. Some species of lichens, bryophytes, ferns, and flowering plants can survive desiccation in the adult phase. Generally, it is most common in plants with no vascular tissue (Alpert 2000). Desiccation-tolerance in the adult phase is relatively rare in angiosperms and non-existent in gymnosperms and trees, partly due to the need for water to maintain vascular structure (Ievinsh 2023). Plants face an added threat of oxidative damage due to the presence of photosynthetic pigments and electron transport complexes. As drying ensues, these complexes shut down at rates that are independent of each other, which leads to the generation of free radicals. Desiccation tolerant plants display physical adaptations to avoid excess light absorption, such as rolling and curling leaves. Pigments are also reversibly degraded, and antioxidants are upregulated (Ievinsh 2023). The angiosperm resurrection plant, *Eragrostiella brachyphylla,* upregulates peroxidase, degrades photosynthetic pigments, and increases the production of the osmolytes sucrose and proline, which all aid in preventing oxidative damage and maintaining cellular structure during desiccation (Yathisha et al. 2020). Desiccation-tolerance is present in almost all plants in the embryonic phase, and seeds are considered the canonical example of desiccation-tolerance in plants. Moisture content in seeds decreases as development progresses, and reduced moisture and increased temperature are positively correlated with increased viability and resilience in 90% of plants (Ievinsh 2023). LEA proteins (see below) protect seed cells during molecular crowding conditions and halt development until imbibition. Some LEA proteins may also sense water during rehydration (Field et al. 2023).

Anhydrobiosis is rare in animals. However, desiccation-tolerance is found across diverse taxa, including tardigrades, nematodes, rotifers, larvae of the sleeping chironomid, and brine shrimp embryos (Clegg 2005; Erkut et al. 2013; Guidetti et al.

2011; Lapinski and Tunnacliffe 2003; Shaikhutdinov and Gusev 2022). Many cellular mechanisms for animals to survive desiccation are strikingly similar to those observed in lower organisms. Anhydrobiotic animals enter a state of metabolic arrest, upregulate the import and biosynthesis of osmolytes and non-reducing sugars, and produce specialized proteins that are typically disordered in the hydrated state (Hibshman et al. 2020). Tardigrades and some nematodes are anhydrobiotic in both larvae and adult stages. The nematode *Caenorhabditis elegans* enters a larval diapause state known as dauer in response to desiccation, whereas tardigrades undergo a quiescent tun state that, unlike diapause, is independent of development and can occur at any life stage (Erkut et al. 2013; Guidetti et al. 2011). Desiccation tolerant nematodes and brine shrimp express intrinsically disordered LEA proteins during desiccation (Hand et al. 2011). LEA proteins, which were first discovered in cotton seeds in the 1980s and are ubiquitous in plants, exhibit a multitude of protective mechanisms in crowded cells (Hernandez-Sanchez et al. 2022). Many LEA-like proteins are also found in metazoans that are not strictly anhydrobiotic but may still exhibit a response to dehydration. Tardigrades also express several classes of IDPs which are unrelated to LEA proteins but are functionally similar. Several heat-stable proteins expressed by tardigrades in the tun state, are unique to tardigrades and commonly referred to as tardigrade-specific intrinsically disordered proteins (TDPs) (Hibshman et al. 2020).

The embryos of the brine shrimp, *A. franciscana,* are an excellent model organism for studying desiccation-tolerance in animals due to their small size, relatively fast reproductive cycle, and importance in aquaculture as an industrial fish food source. However, many studies of *A. franciscana* are limited to *in vitro* and *in vivo* studies in other model organisms due to the difficulty of conducting reliable experiments on desiccated cysts (Abatzopoulos et al. 2002). Encysted embryos of *A. franciscana* enter a developmentally regulated diapause when exposed to environmental cues associated with the upcoming winter (Clegg et al. 1999). This state of inactivity is characterized by a near-absolute arrest of metabolic activity within a tightly sealed embryonic cyst. An assortment of ARID proteins and the non-reducing sugar trehalose are abundant in this life-history stage. ARID proteins, which include LEA and several small heat shock proteins (sHsps), contribute to a cellular environment that tolerates stressful events such as desiccation, anoxia, pH shifts, temperature change, and osmotic challenges (Clegg et al. 1999). Prolonged cold and desiccation, cues associated with the upcoming spring weather, trigger the end of diapause. However, if hydrated cysts are exposed to anoxia during early development, they enter a period of environmentally induced metabolic arrest termed quiescence. *A. franciscana* embryos can survive anoxia-induced quiescence for over 6 years in the fully hydrated state (Podrabsky and Hand 2015).

The Role of Non-reducing Sugars in Anhydrobiosis

Maillard reactions, also known as browning reactions, occur when the carbonyl group of a sugar reacts with the primary amine or amino group of a protein. Furthermore, the amines of nucleic acids may be modified. Maillard reactions are commonly associated with cooking certain foods at high temperatures. However, the likelihood of their occurrence is also predicted to increase under anhydrobiotic conditions (Billi and Potts 2002). Just as increased temperature accelerates the kinetics of Maillard reactions, molecular crowding also increases the probability of Maillard reactions occurring due to the diminished spacing between biomolecules, potentially leading to a cascade of irreversible aggregation. Maillard reactions form covalent bonds between sugars, nucleotides, and proteins, leading to extensive and unwanted polymerization (Lebre et al. 2017). Loss of membrane integrity due to desiccation and damaging collisions that trigger Maillard reactions exacerbates respiratory chain complex damage, resulting in elevated ROS levels. Furthermore, transport proteins damaged by the same effects leads to the release of iron and ROS via the Fenton reaction (Lebre et al. 2017). Sugars that participate in Maillard reactions are reducing sugars because of the requirement of a free carbonyl group for the reaction to occur, and non-reducing sugars must be hydrolyzed before they react with proteins. Examples of reducing sugars include glucose, fructose, galactose, ribose, xylose, and mannose (Yahia et al. 2019). Non-reducing sugars, like trehalose, sucrose, and raffinose, are often synthesized in place of reducing sugars during anhydrobiosis to prevent harmful Maillard reactions from occurring. Non-reducing sugars play a critical role in anhydrobiosis by acting as water-replacement molecules and forming a glassy intracellular state that halts molecular movement.

Trehalose, a disaccharide formed by a 1,1-glycosidic linkage of two D-glucose molecules, is one of the most stable sugars in nature and is very resistant to hydrolysis (Richards et al. 2002). Beyond resisting Maillard reactions, trehalose provides at least three additional protective benefits during anhydrobiosis. First, trehalose causes vitrification of the cellular cytoplasm and the formation of a semi-solid matrix known as a glassy state. The glassy state decreases cellular diffusivity and slows potentially unwanted chemical reactions, including those involving free radicals. Second, due to its numerous hydroxyl groups and hydrophilicity, trehalose may mimic water during desiccation. Trehalose's capacity to vitrify the cellular matrix and mimic the hydrophilic nature of water during drying also stabilizes and protects cellular and organelle membranes during stress. Third, when metabolism resumes after rehydration, trehalose is the primary energy source in some species (Crowe et al. 1998; Crowe 2002; Tapia and Koshland 2014).

Water Replacement

In desiccated cysts of the brine shrimp, trehalose comprises up to 15% of the total dry weight, while non-desiccated embryos have virtually none (Clegg 1965). Sucrose, functionally analogous to trehalose and seemingly ubiquitous in desiccation-tolerant plant tissues, can account for as much as 50% of the total dry mass in certain seeds (Crowe et al. 1997). This substantial amount of both trehalose and sucrose during desiccation may be due to these sugar's ability to form stabilizing hydrogen bonds with biomolecules in place of water (Crowe 2002). Trehalose also has an abnormally high flexibility in the glycosidic bond between the two

D-glucose molecules, which may allow it to hydrogen bond with a broad spectrum of polar group configurations (Richards et al. 2002). Proteins, nucleotides, and membranes acquire and retain their native structure due to the folding and shielding of their hydrophobic constituents from water and other polar molecules (Vogler 1998). When water is removed below 0.3 g H_2O/g dry mass, the hydrophobic effects, which maintain cellular structure, drastically decline in desiccation-sensitive organisms (Hoekstra et al. 2001). In desiccation-tolerant organisms, non-reducing disaccharides, as well as some tri and tetra saccharides, may be upregulated in large quantities as a water replacement strategy due to their propensity to form hydrogen bonds with other polar groups and structurally reinforce hydrophobic interactions (Hoekstra et al. 2001). Clegg et al. measured the relaxation times of water molecules in the desiccated cysts of *A. franciscana* using NMR. They observed that at 0.25 g H_2O/g dry mass, water protons' T1 and T2 relaxation times sharply increased (Clegg et al. 1982). If the water replacement hypothesis is disregarded, the proton relaxation time of water would be expected to show a linear decrease as cellular bulk water diminishes and the intracellular concentration of biomolecules, which rely on water interactions to maintain their structure, increases. Clegg et al. attributed the sharp increase in water proton T1 and T2 to water replacement molecules such as trehalose. Trehalose may displace the primary hydration water layer surrounding biomolecules and forms protective hydrogen bonds at low enough water levels. Furthermore, the high concentrations of trehalose and glycerol accumulated before desiccation may help stabilize cyst volume, which would otherwise diminish as water is lost (Clegg et al. 1982).

As mentioned, water molecules maintain the liquid-crystalline state of membranes through interactions with the polar head groups (Lebre et al. 2017). The more fluid and biologically active liquid-crystalline phase may transition into the more rigid gel phase if water molecules are not present to prevent membrane fusion and denser packing of acyl chains (Hoekstra et al. 2001). At the onset of drying, amphiphile molecules may rush to interact with the polar and non-polar components of phospholipids preferentially. While this may help to maintain the fluidity and structure of the cell membranes during desiccation, it is incomplete without the protective properties of water replacement molecules (Hoekstra et al. 2001). Furthermore, amphiphilic molecules may enter the cytoplasm during rehydration at a rate that causes transient leakage and membrane failure during rehydration. Regulation of the rehydration rate by desiccation-tolerant organisms and potentially additional molecular mechanisms are necessary to prevent membrane perturbations caused by amphiphilic interactions during water movement across membranes. Hoekstra et al. propose that some stress-related proteins like sHsps and LEAs, which have amphiphilic properties, may be involved in this process (Hoekstra et al. 2001).

Compatible osmolytes and ARID proteins may exert a protective force on biomolecules during intermediate drying. This mechanism maintains the integrity of the primary hydration layer and minimizes nonspecific interactions by preferentially excluding some solutes from the protein interface, leading to an energetically favorable biomolecular conformation with reduced excluded volume (Lee and

Timasheff 1981). In other words, compatible osmolytes, such as trehalose, are preferentially excluded from interacting with the surfaces of proteins due to their hydrophilicity and preferentially form hydrogen bonds with water molecules instead. At the same time, the chemical potential of water increases near the macromolecules' surface due to this exclusion, which favors decreased excluded volume. Bulkier molecules, such as polyethylene glycols (PEGs), may also be sterically excluded from interacting with proteins due to their large center of mass, contributing to an increase in the chemical potential of water near the surface of macromolecules (Romero-Perez et al. 2023). During desiccation-induced crowding, the preferential exclusion zone around biomolecules expands as available water decreases. With fewer hydration opportunities, proteins energetically favor volume reduction and self-association to minimize unfavorable interactions with excluded solutes (Hoekstra et al. 2001). In addition to the function of preferentially excluding solutes from the surface of proteins, elongated and unfolded IDPs like LEA proteins may form protective "molecular shields" around biomolecules during desiccation, which deter deleterious reactions (Hernandez-Sanchez et al. 2022). However, when desiccation becomes so extreme that even the primary hydration layer is lost, interactions between proteins and solutes become unavoidable. Therefore, the water replacement hypothesis predicts that solutes with chemical properties like water form stabilizing hydrogen bonds in place of the primary hydration layer water molecules during anhydrobiosis, thereby preserving the native structure of proteins and preventing their denaturation. The water replacement hypothesis proposes a direct protective effect of hydrogen bonding to compatible osmolytes during severe desiccation (Hernandez-Sanchez et al. 2022; Olgenblum et al. 2024). When water levels reach the threshold of 0.3 g H_2O/g dry mass, the hydration water begins to be removed, and the present hydrophilic molecules may replace the primary hydration water layer, forming protective hydrogen bonds with proteins and membranes. This process is reversible, and when cells rehydrate, water molecules displace the interacting compounds, reforming the primary hydration layer (Hoekstra et al. 2001).

Cytoplasmic vitrification and water replacement are mechanisms that contribute to desiccation-tolerance and should not be considered mutually exclusive. Without cellular vitrification, harmful biochemical reactions, fusion of membranes, and loss of protein structure could accumulate and inevitably lead to cell death. However, evidence suggests that many other factors, including water replacement, are involved in desiccation-tolerance. For example, some desiccation-sensitive seeds lose viability at water levels well above 0.1 g H_20/g dry mass, but still form glasses when dried below this water content. Hoekstra. et al., suggest that vitrification may be a mechanism for long-term desiccation survival (Hoekstra et al. 2001). A comparison of sorbitol to the polyol 1,2,3-hexanetriol indicated that the protective properties were significantly reduced after removing three hydroxyl groups from sorbitol (Brom et al. 2023). Additional evidence supporting the water replacement hypothesis is that in a glass-forming solution of 100 mg/ml interleukin-1 receptor antagonist, proteins were only protected from damage after lyophilization when there was more than 5% sucrose w/v present, even though all the solutions within a range of 0–10% w/v formed a glassy state (Crowe et al. 1998).

The Glassy State

It is hypothesized that desiccation-tolerant organisms form glasses in the cytoplasm during drying (Buitink et al. 1998; Buitink and Leprince 2004; Crowe et al. 1998; Sun and Leopold 1997). The glassy state is a metastable, non-crystalline phase between the liquid and solid states. It retains the disordered, amorphous molecular arrangement characteristic of liquids, but exhibits mechanical rigidity and solid-like behavior due to significantly reduced molecular mobility and extremely high viscosity. In this state, materials do not undergo long-range molecular diffusion over biologically relevant timescales. Although structurally disordered, glasses respond to mechanical stress like solids and are best characterized by their viscoelastic properties and extremely low flow rates rather than a well-defined melting point (Bhandari and Howes 1999). A liquid in the glass-transition state can have a flow rate of 30 μm per 100 years (Franks et al. 1991). Therefore, desiccated cells that enter the glass state undergo molecular movement at such a slow rate that they are practically motionless (Bhandari and Howes 1999).

The restriction of molecular motion in the glassy state is crucial for preventing harmful biochemical reactions, such as those triggered by Maillard reactions and ROS (Lebre et al. 2017). It is important to note that although the intracellular environment is highly crowded in the fully hydrated state, it still behaves as a cohesive, gel-like liquid (Kwapiszewska et al. 2020; Luby-Phelps 2000). Larger macromolecules may take up to 40% of the cellular volume and thus decrease the overall flow by forming a gel-like matrix (Kwapiszewska et al. 2020). The overall flow of the cytosol can be characterized by its macroviscosity, which is estimated to range from 10,000 to 70,000 centipoise, comparable to that of many gel-like liquids (Kwapiszewska et al. 2020; Luby-Phelps 2000). However, smaller molecules may be less constricted in localized areas of the gel-like matrix and may experience a much higher flow rate. Microviscosity reflects the flow behavior of smaller cytosolic molecules, which is highly heterogeneous and strongly influenced by the local concentration of surrounding biomolecules (Szymanski et al. 2006). Estimates of the microviscosity in hydrated cells are not much greater than in pure water, with estimates ranging from 2–5 centipoise (Dijksterhuis et al. 2007; Kwapiszewska et al. 2020; Szymanski et al. 2006). Therefore, the formation of a glass state during anhydrobiosis deters the rapid diffusion of molecules at nanoscales (Bhandari and Howes 1999). Hydrated fission yeast cells, which are rod-shaped, formed spheres when their cell walls were removed because the cytoplasm remained in the liquid phase. However, energy-depleted and acidified fission yeast cells remained rod-shaped after removing the cell wall, implying that the cytoplasm had transitioned to a state with more solid-like properties (Munder et al. 2016). If a cell remains in the liquid phase after dehydration, the diffusivity rate is high enough to significantly increase the probability of deleterious biochemical reactions (Olgenblum et al. 2024).

Certain sugars and ARID proteins help maintain the cytoplasm in a glassy state during drying. In convective drying, desiccation can lead to cell death if biomolecular diffusivity remains too high due to the persistence of a liquid phase (Hincha et al.

2021). In freeze-drying, cell death may occur from crystallizing damage as the liquid cytoplasm transitions into a crystalline-solid state (Ishiguro and Rubinsky 1994). Vitrifying agents maintain cellular vitrification by affecting the glass transition temperature of the cellular solution. The glass transition temperature is the temperature at which a glass melts to become an amorphous liquid (Bhandari and Howes 1999; Buitink and Leprince 2004; Olgenblum et al. 2024). All amorphous liquids have a glass transition temperature, which varies significantly between pure compounds and mixtures. Mixtures with higher glass transition temperatures maintain a glassy state for a larger range of temperatures (Bhandari and Howes 1999; Buitink and Leprince 2004; Olgenblum et al. 2024). Trehalose enhances desiccation tolerance by increasing the glass transition temperature of the cytoplasm, enabling cells to remain in a motionless, non-reactive, glassy state even at relatively high ambient temperatures during the final drying stage (Richards et al. 2002).

The synergistic or additive effects of ARID proteins and desiccation-tolerance-related sugars, and how they may promote cellular vitrification, is not well understood, and, the types, as well as the concentrations of protective saccharides, vary widely amongst anhydrobiotic organisms (Buitink and Leprince 2004; Janis et al. 2018). While most desiccation-tolerant metazoans rely on trehalose for anhydrobiosis, several desiccation-tolerant tardigrade species, including *Echiniscus granulatus* and *Echiniscus testudo*, produce very low quantities of trehalose and do not upregulate trehalose during drying (Hibshman et al. 2020). This finding implies that other compounds, such as TDPs, may contribute to vitrification. *Af*LEA1 from brine shrimp embryos significantly enhanced the protection of lactate dehydrogenase in *Drosophila melanogaster* Kc167 cell lysates, and ectopic expression increased cytoplasmic viscosity in these cells during water stress (Janis et al. 2022). The protection of lactate dehydrogenase may result from *Af*LEA1 acting as a molecular chaperone, reducing cytoplasmic diffusivity, or both (Janis et al. 2022). Desiccation-tolerant plant seeds upregulate the import and biosynthesis of many different protective sugars like sucrose, trehalose, raffinose, and stachyose, and the ratio of these sugars is vital for the formation of a glass state at a specific temperature (Buitink and Leprince 2004). A mixture of 85% w/w sucrose and 15% w/w raffinose, similar to the sugar ratio found in desiccation-tolerant *Zea mays* (corn) seeds, led to glass formation at temperatures above freezing. In contrast, a mixture of 75% w/w glucose and 25% w/w sucrose, which reflects the typical sugar composition in desiccation-sensitive seeds, only formed a glass at temperatures below 0 °C (Buitink and Leprince 2004). This finding demonstrates both the complexity and specificity of the macromolecular composition needed for forming the glass state during anhydrobiosis in plant seeds. Cellular vitrification can also be induced by a global or localized drop in pH, which presumably affects the solubility of specific biomolecules and interactions with water molecules (Munder et al. 2016). The lowering of cellular pH often results from the metabolic arrest during desiccation-induced dormancy. Upon exposure to anoxia, the brine shrimp encysted embryo, which is still hydrated, releases protons from intracellulat stores, dropping the cytosolic pH from approximately ~7.8 to 6.3 (Hand 1998). These organisms may remain viable in a hydrated and anoxic state for many years (Clegg 2011; Munder et al. 2016).

Macromolecular Adaptations to Anhydrobiosis

A suite of proteins that lack a well-defined secondary structure, known as ARID proteins, contribute to cellular survival during desiccation. These proteins are either constitutively expressed or upregulated at the onset of water loss and include LEA proteins, TDPs, hydrophilins, sHsps, and prion-like proteins. Many ARID proteins are disordered in the hydrated state, but some can undergo a liquid-liquid phase separation (LLPS) once their concentration surpasses a critical threshold. During dehydration, intracellular crowding and solute concentration increase, inevitably driving some ARID proteins to separate into a two- or multi-phase system. These dynamic transitions may be crucial in stabilizing cellular structures, preserving macromolecular function, and organizing biochemical activity during water loss.

Late Embryogenesis Abundant Proteins

LEA proteins are a class of IDPs found in all three domains of life and were first characterized in cotton seeds by Dure et al. in 1981 (Dure and Galau 1981; Hand et al. 2011). LEA proteins are expressed in virtually all plant species and are commonly associated with desiccation-tolerance and the developmental regulation of seeds. However, they are also expressed in some animal species within the phyla Nematoda, Tardigrada, Arthropoda, and Rotifera (Hand et al. 2018). *A. franciscana,* a species of brine shrimp, expresses LEA proteins only during the anhydrobiotic cyst phase of its life cycle in response to photoperiod and temperature cues associated with the autumn season (Hand et al. 2018; Podrabsky and Hand 2015). LEA proteins are thought to aid in survival during desiccation by protecting lipid membrane integrity, sequestering metal ions, enhancing cellular vitrification, and preventing aggregation of proteins through molecular shielding and sequestration of biomolecules in gel-like membraneless organelles (MLO) (Belott et al. 2020; Chakrabortee et al. 2012; French-Pacheco et al. 2018; Goyal et al. 2005; Hincha et al. 2021; Lo Sterzo et al. 2025; Tolleter et al. 2010).

LEA proteins can be categorized into Groups 1–6, seed maturation proteins (SMPs), and dehydrins, but different authors have developed varying grouping schemes. We are referring to the scheme developed by Tunnacliffe and Wise. Groups 1 and 6 are found in *A. franciscana* in addition to Group 3 LEA proteins. All other known LEA proteins in animal species belong to Group 3 (Belott et al. 2020; Hernandez-Sanchez et al. 2022; Tunnacliffe and Wise 2007).

LEA proteins are believed to have evolved in an ancestral species of green algae (Chlorophyta and Streptophyta) and likely existed as a pre-adaptation to land colonization. However, LEA-like proteins and other IDPs are also found in fungi and prokaryotes, especially extremophile prokaryotes.

LEA Proteins—Structure and Function

LEA proteins and IDPs are characterized by a lack of defined secondary structure in the hydrated state and may gain alpha-helical or beta sheets as water is removed. *Af*LEA1 comprises 42% random coils in the hydrated state but gains 85% alpha-helical structures near complete desiccation (Janis et al. 2022). For comparison, *Afr*LEA6 comprises 90% random coils during the hydrated state and adopts mostly alpha-helical structure as dehydration proceeds (LeBlanc et al. 2019). LEA proteins are suspected to protect cellular structure during desiccation in various ways. The alpha-helical sections of some LEA proteins insert themselves directly into lipid membranes via amphipathic interactions and increase the space between lipid heads. This interaction sustains lipid membrane fluidity, stabilizes the membrane curvature, and prevents lipid aggregation. Disordered LEA proteins that attach to lipid membranes and lack alpha-helical structure do not improve lipid membrane stability, suggesting that alpha-helical formation is necessary for the protection of lipid membranes during desiccation (Chen et al. 2020; Kovacs et al. 2008). LEA proteins are generally very hydrophilic and often contain an overrepresentation of amino acids like glycine, alanine, serine, and threonine (Campos et al. 2013). Typical LEA proteins usually contain no cysteine and low amounts of tryptophan. Due to their overall polar composition, they are suspected to mimic the hydrogen bonding of water, which stabilizes the structure of cellular membranes. Besides dehydration, LEA proteins have experimentally been shown to adopt conformational change when exposed to extreme temperatures, increased ion concentration, increased osmolarity, and changes in membrane fluidity. It has been hypothesized that posttranslational modifications and molecular crowding may also trigger a more ordered state. Other conformational changes include decreased hydrodynamic radius and increased self-associative interactions (for review see (Hernandez-Sanchez et al. 2022)).

LEA proteins have been observed to localize to specific subcellular compartments. *Afr*LEA3m is residing in the mitochondrion, presumably the mitochondrial matrix. *Afr*LEA3m was experimentally shown to preserve the respiratory capacity of mitochondria during freezing. The capacity for oxidative phosphorylation was unchanged in mitochondria isolated from *A. franciscana* subject to water stress, whereas the respiratory control ratio of mammalian mitochondria lacking *Afr*LEA3m decreased (Menze et al. 2009). LEA proteins in bdelloid rotifers, *Ar*LEA1A and *Ar*LEA1B, have a hydrophobic leader sequence and colocalize in the endoplasmic reticulum (Tripathi et al. 2012). Other LEA proteins are known to localize to the nucleus, chloroplast, ER, plasma membrane, vacuole, peroxisome, mitochondrion, and chloroplast in plants (Hernandez-Sanchez et al. 2022).

Oligomerization of LEA proteins in plants and animals has been commonly observed under certain stress events. Soybean PM1 and *At*HIRD11 in *Arabidopsis thaliana* were both reported to oligomerize in the presence of copper ions (Hara et al. 2013; Liu et al. 2017a). Even though LEA proteins generally gain order when water is removed, many LEA proteins may be considerably disordered in the desiccated state, allowing LEA protein complexes to interplay between monomeric and oligomeric states. For example, several group 1 and group 6 LEA proteins from *A. thaliana* (LEA15, 16, 17, 20, and 35) only experience a 20–25% increase in helicity upon drying (Knox-Brown et al. 2020). Oligomerization of some LEA proteins may occur in the disordered state, and copper binding of *At*LEA 4–5 in *A. thaliana* was said to happen in 'fuzzy complexes' in the presence of copper (French-Pacheco et al. 2018).

Analysis of LEA genes in the potato plant showed their distribution across most chromosomes, indicating tandem and whole genome duplications of LEA protein encoding DNA over evolutionary time. Plant LEA genes generally have few introns and may have adapted a high rate of duplication events to increase diversity, function, and translation efficiency (Chen et al. 2020). Several highly similar Group 1 LEA proteins in *A. franciscana* are hypothesized to arise from atlernative splicing of exons within the same gene. There are still many questions about the level of diversity versus redundancy of function across LEA proteins and whether transcription levels or concentration of LEA proteins affect their molecular function(s).

*sHsps and Anhydrobiosis—Insights from p26 (***A. franciscana***)*

Small heat shock proteins (sHsps) are members of the protein quality control system and maintain protein homeostasis. sHsps, termed "holdases", are ATP-independent chaperones and act as the first line of defense against stress events by binding misfolded and aggregated proteins to prevent further irreversible misfolding (Reinle et al. 2022). sHsps are usually upregulated during various forms of stress, including anhydrobiosis (Reinle et al. 2022). When metabolism becomes unobservable in the anhydrobiotic state, and intracellular ATP levels have crashed, cells become especially reliant on the ATP-independent protective capabilities of sHsps (King and MacRae 2012). Some sHsps, like p26 in *A. franciscana*, are developmentally regulated and are essential for surviving desiccation during diapause. sHsps prevent the irreversible aggregation of proteins during desiccation, and in several *in vitro* experiments, p26 protected citrate synthase from heat-induced aggregation and insulin from dithiothreitol-induced aggregation significantly more than BSA and IgG (Sun et al. 2006; Sun and Macrae 2005; Sun et al. 2004; Wu and MacRae 2010). The protein p26 from *A. franciscana* is unique compared to other sHsps because of its highly intrinsically disordered N- and C-terminal domains, which increase its overall propensity to undergo LLPS (Brangwynne et al. 2015; Qiu et al. 2006; Sun and Macrae 2005). Several experiments on p26 also investigated the chaperone abilities of truncated mutants of the protein to understand how the N- and C-terminal domains may affect protein function (Sun et al. 2004, 2006; Sun and Macrae 2005; Wu and MacRae 2010). The alpha-crystallin domain (ACD) is highly conserved across all sHsps, and experiments verified that the ACD and the N-terminal domain of p26 are essential for its chaperone activity. However, when removing portions of the highly intrinsically disordered C-terminal domain, the

Table 17.1 Truncation effects on p26 chaperone activity

Publication	Deleted residues	Oligomer mass (kDa)	Chaperone activity
Sun et al. (2004)	1–36	110	Moderate
Sun et al. (2004)	1–60	58	Low
Sun et al. (2004)	153–192	220	Moderate
Sun et al. (2004)	183–192	500	High
Sun et al. (2004)	1–60; 153–192	62	Low
Sun et al. (2005)	8–29	443	Moderate
Sun et al. (2005)	36–45	300	Low
Sun et al. (2005)	169–186	300	Moderate

Deletion experiments targeting various regions of the small heat shock protein p26 were conducted to assess the impact on its ability to prevent the denaturation of stress-sensitive proteins. Removal of residues 183–192 at the C-terminal tail resulted in minimal loss of chaperone activity, suggesting that this region may serve a secondary function. This finding supports the hypothesis that the intrinsically disordered C-terminal domain may contribute to functions beyond protein chaperoning.

protein retained moderate to high chaperoning capabilities (Table 17.1). This finding may indicate a separate function for p26, in addition to molecular chaperoning, that may be more related to the increased solubility of the protein provided by the increased intrinsic disorder in the C-terminal domain. Indeed, it is likely that sHsps, which are categorized as ARID proteins, like p26, possess additional functional roles related to protection from desiccation-induced molecular crowding.

In addition to molecular chaperoning, some sHsps may regulate apoptosis primarily by sequestering or interfering with apoptotic signaling molecules (Takayama et al. 2003). Inhibition of apoptosis is essential for survival during diapause (Wu and MacRae 2010). The disruption of the electron transport chain due to metabolic stress may increase membrane-damaging ROS. Electron transport chain dysfunction, in combination with molecular crowding effects, thus leads to ROS accumulation, which may damage the mitochondrial membrane and cause the release of apoptotic factors like cytochrome *c* (Redza-Dutordoir and Averill-Bates 2016). However, apoptotic cascades and regulation may differ among desiccation-tolerant and sensitive organisms, and cytochrome c failed to increase caspase activity in cell extracts from *A. franciscana* (Menze and Hand 2007). Several studies have shown that mammalian cells transfected with p26 were protected from apoptosis induced by heat shock, drying and rehydration, and the protein kinase C inhibitor, staurosporine (Ma et al. 2005; Villeneuve et al. 2006). Villeneuve et al. also showed that activation of caspase-3 was lower in heat-shocked human embryonic kidney cells (H293) ectopically expressing p26 than in untransfected controls (Villeneuve et al. 2006). Interference with the apoptotic signaling pathways by sHsps has been observed in several other systems, including human Hsp27. Hsp27 inhibited apoptosis in human leukemic cells by interacting with cytochrome *c*, an early apoptotic signal released from the mitochondria, eventually leading to caspase activation and programmed cell death in mammals (Garrido et al. 1999). Hsp27 has also been shown to slow apoptosis downstream in the apoptotic cascade by interacting with procaspase-3 (Pandey et al. 2000). Ma et al. hypothesized that p26 conferred protection to H293 cells during rehydration after convective drying through inhibition of apoptosis. They deduced that neither molecular chaperoning nor membrane protection was likely to convey the observed protection, given the low expression levels in the transfected cells (Ma et al. 2005).

sHsps may also confer protection to cells during molecular crowding by protecting the fluidity and integrity of lipid membranes. Like some LEA proteins, protection may be provided by inserting regions of the sHsps into the membrane through hydrophobic and hydrophilic interactions with the phospholipids (Torok et al. 2001). Hsp17 from the thermophilic cyanobacterium *Thermosynechococcus vulcanus* stabilizes lipid membranes by directly inserting the protein into the membrane, thus allowing the membrane to retain its liquid-crystalline state (Torok et al. 2001). Hsp12 in the corn smut, *Ustilago maydis*, also showed protective properties towards lipid membranes. Furthermore, Hsp12 protects and remodels membranes during anhydrobiotic sporulation and the pathogenetic development stage and appears to remodel larger lipid vesicles into smaller endocytosis vesicles (Mitra et al. 2023).

Hydrophilins

Hydrophilins are a broad class of proteins with a hydrophilicity index of >1.0 and more than 6% glycine residues (Garay-Arroyo et al. 2000). Many LEA proteins are categorized as hydrophilins, except for some atypical group 5 LEA proteins, which are comparatively more hydrophobic (Battaglia et al. 2008). Hydrophilins are strongly associated with water-stress pathways in bacteria, fungi, plants, and some metazoan species (Garay-Arroyo et al. 2000). Cold-regulated proteins (COR proteins) are hydrophilins that are mechanistically similar to LEA proteins and may help to stabilize membranes and proteins during freezing stress (Thomashow 1999). Many hydrophilins are ribosomal proteins and may offer protective mechanisms to ribosomes during desiccation stress. RMF, a hydrophilin found in *E. coli*, is essential for assembling 100S ribosomes, specialized ribosomes generated by bacteria under stress conditions. Without RMF, the 30S and 70S ribosomal subunits cannot adopt the conformation required for dimerization (Yoshida et al. 2021). It should be noted that not all hydrophilins are necessary for desiccation-tolerance. Of 12 characterized hydrophilins in yeast, only five were found to be required for desiccation-tolerance (Lopez-Martinez et al. 2012). One of these proteins, STF2p, is associated with the regulatory peptide inhibitory factor 1 (INF1), which stabilizes the F_1F_0-ATP synthase complex and prevents ATP hydrolysis (Battaglia et al. 2008; Lopez-Martinez et al. 2012).

The high content of structure-breaking glycine residues, combined with the very low hydrophobicity of hydrophilins, enables hypersensitive intrinsically disordered regions, which remain in a constant state of conformational change (Uversky 2013). This structural instability allows the protein to be in tune with shifting cellular energetics primarily conveyed through water dynamics (Uversky 2013). Upon desiccation and macromolecular crowding, the IDP regions of hydrophilins often adopt functional secondary structure motifs or form biomolecular condensates and, thus, may be considered sensors of macromolecular crowding (Meneses-Reyes et al. 2024). While the conformational variability of IDPs and hydrophilins often gives rise to promiscuity in function and binding partners, hydrophilins may also be highly specific in protein interactions, as demonstrated by RMF in *E. coli* and STF2p in *S. cerevisiae* (Battaglia et al. 2008; Uversky 2013). The diversity of hydrophilins demonstrates the plentitude of functions they may possess and the specificity in protein interaction some maintain (Battaglia et al. 2008).

Tardigrade Disordered Proteins

Tardigrades express an array of ARID proteins during the desiccation-tolerant tun state, known as TDPs (Hibshman et al. 2020). CAHS proteins, a type of TDPs, are hypothesized to function similarly to LEA proteins but are evolutionarily unique to tardigrades (Tanaka et al. 2022). Heterotardigrades lack CAHS proteins but possess structurally similar heat-insoluble TDPs which are necessary for desiccation-tolerance (Tanaka et al. 2022). Other TDPs include the mitochondrial-abundant heat-soluble (MAHS) proteins and the secretory-abundant heat-soluble (SAHS) proteins (Lim et al. 2024). MAHS proteins are specific to mitochondrial protection during desiccation and, like CAHS proteins, are highly disordered (Lim et al. 2024). SAHS are secreted extracellularly and are more structured than other TDPs, forming β-barrels with large fatty acid binding cavities. SAHS proteins exhibit protective capabilities to membranes and enzymes during desiccation (Lim et al. 2024). The β-barrel cavity in some SAHS proteins collapses during severe desiccation, forming a dense fibrous network. Lim et al. hypothesized that during desiccation, SAHS proteins might act like a cushioned "packing material", a process mechanistically similar to the molecular shielding hypothesis of LEA proteins, preventing the harmful collisions of desiccation-sensitive biomolecules (Lim et al. 2024).

Some CAHS proteins, like LEA proteins, can adopt alpha-helical structures during desiccation, and many CAHS proteins rely on an alpha-helical domain for oligomerization (Yagi-Utsumi et al. 2021). Tardigrades rely on CAHS proteins to induce cytoplasmic immobilization and safeguard enzymes during desiccation, independently of trehalose (Sanchez-Martinez et al. 2023). Atomic force and transmission electron microscopy provided visual evidence that CAHS proteins oligomerize and form condensates. At higher concentrations, which occur during severe desiccation, CAHS proteins form hydrogels *in vivo* and *in vitro* (Yagi-Utsumi et al. 2021). Hydrogels comprise transiently, non-covalent, cross-linked polymers suspended in an elastic and gel-like solid, often capable of absorbing a significant amount of water (Tanaka et al. 2022). Like the vitrified state, hydrogels are a non-crystalline amorphous solid, but differ from glasses in that they generally retain more water and have increased molecular mobility (Tanaka et al. 2022). Yagi-Utsumi et al. hypothesized that CAHS proteins form fibrous networks within condensates, which functionally protect biomolecules through structural support and retention of water, given the typical properties of hydrogels. Additionally, they proposed that under extreme desiccation, the CAHS protein hydrogel network may expand, potentially offering protective stabilization to larger biomolecular complexes, RNAs, and membranes (Yagi-Utsumi et al. 2021). Yagi-Utsumi et al. claim that the vitrification hypothesis may not apply to CAHS proteins, given these structural observations (Yagi-Utsumi et al. 2021), and Malki et al. observed no evidence of vitrification of CAHS proteins via NMR-spectroscopy (Malki et al. 2022). It is generally understood that a significant benefit of vitrification in other desiccation-tolerant organisms is the immobilization of the cytoplasm, and the question arises as to how tardigrades can survive desiccation in a hydrogel state with larger diffusivity rates. However, Malik et al. observed that despite greater diffusivity, soluble proteins were immobilized in their functional conformations within cavities formed by fibrous CAHS protein networks (Malki et al. 2022). Despite greater hydration levels in hydrogels, Malik et al. state that the water levels within CAHS protein networks are too low to support the native conformation of sequestered proteins and thus hypothesized that the hydrophilic IDP regions of CAHS proteins may aid in solubilizing proteins (Malki et al. 2022). This hypothesis has many commonalities with the water replacement hypothesis proposed for trehalose. The actual mechanism for the gelation of CAHS proteins and their constituents is still debated. Using thermogravimetric analysis, Sanchez-Martinez et al. showed that regardless of gelation, the specific CAHS D class of proteins retained no more water than non-TDPs in control cells (Sanchez-Martinez et al. 2023). However, CAHS D proteins and the gel-forming protein, gelatin, were bound more tightly to water than the non-gelling control protein, lysozyme, which may be a mechanism for stabilizing water-protein interactions. Sanchez-Martinez affirmed that their experiment disproved the hypothesis that gelating CAHS proteins serve as water retention reservoirs and proposed that CAHS protein hydrogels may provide desiccation-tolerance by providing a temporary stress-dependent cytoskeletal matrix (Sanchez-Martinez et al. 2023). However, the authors pointed out that CAHS D proteins did not form condensates during drying in contrast to the behavior of other CAHS protein classes, and it is yet to be determined how biomolecular condensation of CAHS proteins mechanistically affects desiccation-tolerance in tardigrades (Sanchez-Martinez et al. 2023).

Role of Biomolecular Condensates in Anhydrobiosis

The finding that some ARID proteins can form biomolecular condensates led us to develop four hypotheses about their protective mechanism. These can be summarized as (1) selective sequestration, (2) selective target neutralization, (3) increased viscocapillary effect, and (4) modified solvent property hypotheses (Fig. 17.2).

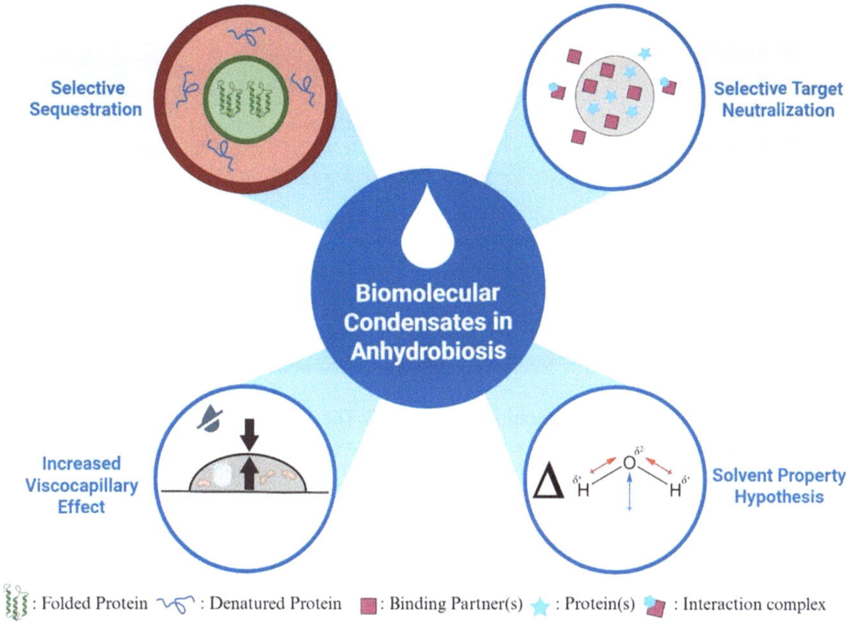

Fig. 17.2 The quadrumvirate of protective hypotheses assigned to biomolecular condensates in anhydrobiosis. Four different hypotheses may offer insights into the protective mechanisms conveyed by biomolecular condensates. Please see the text for additional information. (Created with BioRender)

The Selective Sequestration Hypothesis

Sequestration of biomolecules in biomolecular condensates may occur for several reasons. The selective sequestration hypothesis proposes that a subset of IDPs, specifically ARID proteins, can protect biomacromolecules during desiccation by recruiting and compartmentalizing labile biomolecules into phase-transitioned biomolecular condensates (Janis 2023). Partitioning selected proteins into phase-separated ARID condensates may be based on the physical properties of the partitioned proteins (Fig. 17.3) (Belott et al. 2020; Pak et al. 2016). The group 1 LEA protein for *A. franciscana*, *Af*LEA1, was predicted and then shown to undergo phase separation in the presence of RNA. It could sequester these molecules, potentially modifying their translation kinetics (Janis et al. 2022).

The selective sequestration hypothesis states that partitioning biomolecules into condensates may provide a molecular refuge to protect them from damaging conditions such as protein denaturation, ionic imbalance, or oxidative stress during water loss. Importantly, protection does not require direct binding to the protected biomolecule but may result from passive partitioning into a less hostile environment. Upon rehydration, condensate dissolution leads to the controlled release of critical players, such as RNA, chaperones, ribosomes, or enzymes, enabling the rapid

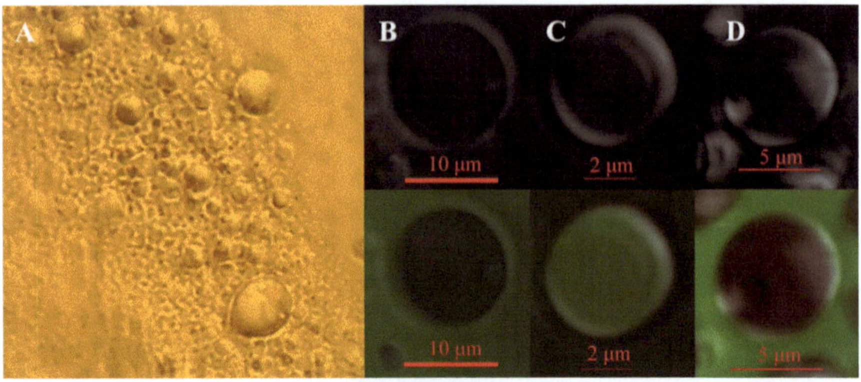

Fig. 17.3 Liquid-liquid phase separation of *Afr*LEA6. The ARID protein *Afr*LEA6 (*A. franciscana)* forms coacervates *in vitro* and selectively sequesters green fluorescent protein (GFP) based on surface charge. (**a**). *Afr*LEA6 (0.17 mg/ml) separates from the solution into a liquid phase in a buffer mimicking the intracellular solute composition of encysted brine shrimp embryos. (**b**) Slightly Negative GFP (stGFP-7) is excluded from the droplet. (**c**) Positive GFP (pGFP+36) is incorporated into the droplet. (D) Highly negative GFP (nGFP-30) is excluded. (Reproduced from Belott et al. 2020)

resumption of metabolism and repair pathways. Thus, ARID proteins may act as stress-dependent molecular scaffolds, protecting the minimal required fraction of essential biomolecules long enough to sustain recovery without necessitating protein degradation or specific binding events, kickstarting cellular repair pathways.

The Selective Target Neutralization Hypothesis

In addition to their proposed role in protecting fragile macromolecules during desiccation, biomolecular condensates formed or promoted by ARID proteins may actively suppress metabolic, developmental, and apoptotic pathways under environmental stress, contributing to the organism's ability to endure prolonged dormancy. This idea is captured in the selective target neutralization hypothesis, which proposes that some phase-separated condensates serve as compartments that temporarily inactivate biomolecules whose unregulated activity would be deleterious under stress or during cellular recovery.

Anhydrobiotic organisms frequently enter states of metabolic depression before the loss of intracellular water. This metabolic depression includes programmed dormancy states like diapause, as seen in *A. franciscana* cysts, and environmentally induced quiescence, which occur in response to unfavorable environmental conditions such as dehydration or anoxia (Hand et al. 2016; MacRae 2016). During quiescence, fully hydrated *A. franciscana* cysts exhibit a pronounced drop in cytosolic pH from ~7.8 to 6.3 (Hand 1998), a shift that may facilitate the nucleation of phase-separated condensates. These condensates are hypothesized to selectively recruit metabolic enzymes, transcription factors, RNAs, or key substrates, temporarily restricting metabolic activity and developmental progression until favorable conditions return. Proteinaceous condensates formed by ARID proteins could selectively recruit or exclude proteins based on their biophysical properties and may also sequester small molecules such as ions. Previous studies have indicated the partitioning of hydroxide ions into phase-separated droplets (Dai et al. 2023). This result is consistent with the idea that negatively charged glycolytic intermediates could be selectively partitioned into ARID protein condensates, downregulating metabolic fluxes. Such a mechanism could explain how early developmental stages in some desiccation-tolerant organisms, like encysted embryos of *Artemia*, orthodox seed plants, and dauer larvae in nematodes, suspend growth in anticipation of environmental change.

The selective neutralization model encompasses developmental inhibition and the suppression of cell death pathways. Emerging evidence suggests that phase-separating proteins can inhibit apoptosis by sequestering pro-death signaling molecules. In one study, the engineered protein DHR81, designed using an ARID protein as a template, formed condensates that recruited the apoptotic effector caspase-7, resulting in significantly reduced apoptosis in camptothecin-treated HT-1080 cells (ATCC, CCL-121) (Veling et al. 2022). By analogy, native ARID protein condensates may sequester programmed cell death-related proteins during desiccation or early rehydration, delaying cell death pathways and allowing sufficient time for molecular repair processes. Inhibiting apoptosis may buy time for ATP-dependent heat shock proteins to refold damaged proteins or for cells to reinstate core functions, thereby increasing post-desiccation survival. This dual functionality, protection, and suppression would reflect a highly adaptive use of phase separation to regulate the proteome, metabolome, and developmental trajectory in response to environmental stress.

The Increased Viscocapillary Effect Hypothesis

The viscocapillary hypothesis proposes that the preservation of cellular morphology during desiccation may be mediated by a combination of elevated cytoplasmic viscosity and capillary forces exerted by phase-transitioned biomolecular condensates that wet and mechanically stabilize intracellular membranes. These synergistic biophysical forces, e.g., macromolecular crowding, polymer entanglement, and condensate-membrane interactions, may provide cohesive structural support that prevents collapse or deformation during extreme water loss. These forces could underlie the enhanced structural preservation observed following desiccation and rehydration in *Drosophila* Kc167 cells ectopically expressing ARID proteins such as *Afr*LEA6 or *Af*LEA1.1 (Belott et al. 2020; Janis et al. 2022). As dehydration proceeds, the progressive rise in molecular crowding promotes phase transitions of macromolecules and amplifies the viscosity of the cytosol. The elevated effects of molecular crowding may foster the formation of multiple distinct condensates across varying hydration states during desiccation (Romero-Perez et al. 2023). Notably, these condensates could nucleate and coalesce preferentially in regions around cellular membranes (Mangiarotti et al. 2023).

Phase-separated compartments can exhibit liquid-like behavior, where the dense phase is markedly more viscous than the surrounding dilute phase. The internal viscosity of proteinaceous condensates has been quantified *in vitro* through particle tracking microrheology. Reported viscosities for such condensates span a wide range, from approximately 100 to over 5,000,000 centipoises, depending on their molecular composition (Fisher and Obermeyer 2024). It is critical to recognize that biomolecular condensates and the cytosol behave as non-Newtonian fluids. Unlike Newtonian fluids, their flow characteristics are not linear with applied stress, leading to substantial differences between macroviscosity and microviscosity (Szymanski et al. 2006). Macroviscosity reflects the bulk resistance of the fluid to flow, whereas microviscosity pertains to the localized frictional environment experienced by diffusing molecules (Shah et al. 2020). Importantly, microviscosity is dependent on molecular size. Smaller molecules exhibit higher mobility than larger ones at a given macroviscosity. In the context of anhydrobiosis, the viscosity of condensates is expected to rise exponentially due to increased macromolecular crowding and entanglement, as predicted by power-law models of concentrated polymer solutions (Uematsu et al. 2003). This dramatic increase in viscosity may be functionally significant. Certain condensates have been shown to interact with and wet membrane surfaces, generating electrocapillary forces capable of deforming membranes spontaneously, manifesting in biological processes such as endocytosis, membrane budding, and tubulation.

The Solvent Property Hypothesis

The solvent property hypothesis proposes that increasing macromolecular crowding and the activity of IDPs and metabolites during desiccation could modulate intracellular water's physicochemical properties. These solvent property changes, including alterations in dipolarity, hydrogen bonding, and dielectric behavior, are hypothesized to potentially contribute to the remarkable capability of anhydrobiotic organisms to prevent excessive cellular damage during the loss of water.

While it is well established that water stress impacts the structure and function of cellular components such as proteins and other biomacromolecules, desiccation presents a double-edged sword. Not only can biomolecules be structurally changed during water loss, but the properties of water itself are altered as the concentrations of cellular constituents increase during drying. Support for this idea comes from studies of polymer-based aqueous two-phase systems, which are often used as simplified analogs of biomolecular condensates (Aumiller Jr. and Keating 2017; Sakuta et al. 2019; Titus et al. 2020). These systems show that solvent properties such as dipolarity/polarizability (π^*), hydrogen bond acceptor basicity (β), and hydrogen bond donor acidity (α) differ significantly between the two coexisting phases

(Madeira et al. 2010, 2021; Titus et al. 2022). These solvent changes could influence solute partitioning and the material properties of each phase.

Similarly, aqueous solutions of some proteins, including human heat shock protein 6B (HSPB6) and several group 2 LEA proteins (e.g., K_n dehydrins), have been shown to alter the solvent characteristics of water (Ferreira et al. 2018a, b). These proteins affect the same key parameters (π^*, β, α) and, notably, induce greater solvent restructuring than many nonionic polymers (Zaslavsky and Uversky 2018). Such protein-driven changes in water behavior are believed to play a functional role in biomolecular phase separation, enabling IDPs to form biomolecular condensates. A comprehensive review concluded that solute-induced changes in water's ability to donate or accept hydrogen bonds and participate in dipole-based interactions may drive LLPS (Zaslavsky and Uversky 2018). These effects may underlie the unique functionality of IDPs in desiccation tolerance.

Moreover, various bulk physicochemical properties of aqueous solutions, such as water activity, osmotic coefficient, relative permittivity, surface tension, relative viscosity, and solubility, can be predicted using a linear combination of just two solvent descriptors: π^* and α (Ferreira et al. 2016). These parameters are sensitive to solute identity and concentration, reinforcing that water becomes structurally and functionally distinct in crowded or protein-rich environments. Compared with ordered protein domains, IDPs and IDRs exhibit an enhanced capacity to interact with water and modulate its properties (Bokor et al. 2005; Tompa et al. 2012, 2016, 2017; Zaslavsky and Uversky 2018). This unique feature may explain why IDPs are frequently found in biomolecular condensates. These water-altering effects may facilitate LLPS and contribute directly to structural stabilization, molecular compartmentalization, and cellular survival during drying. Future research should aim to directly quantify solvent property changes within cells undergoing desiccation and explore how these changes correlate with phase behavior and biological preservation at distinct hydration levels.

Conclusion and Future Direction

Water's indispensable role as the only solvent of life on Earth is underscored by its unique physical properties, including high specific heat capacity, cohesive and adhesive forces, and its ability to form extensive hydrogen-bonded networks. These properties not only support the maintenance of cellular structure and biochemical reactions but also profoundly influence membrane stability, molecular crowding, and metabolic regulation. The complexity of water structure in cells, including bulk water, hydration water, and primary hydration water, underscores the critical importance of localized water dynamics for biomolecular function and stress responses. Within this framework, the survival of anhydrobiotic organisms under extreme desiccation represents a remarkable biological feat, requiring coordinated structural, biochemical, and physiological, adaptations that preserve life in the near absence of water (Fig. 17.4).

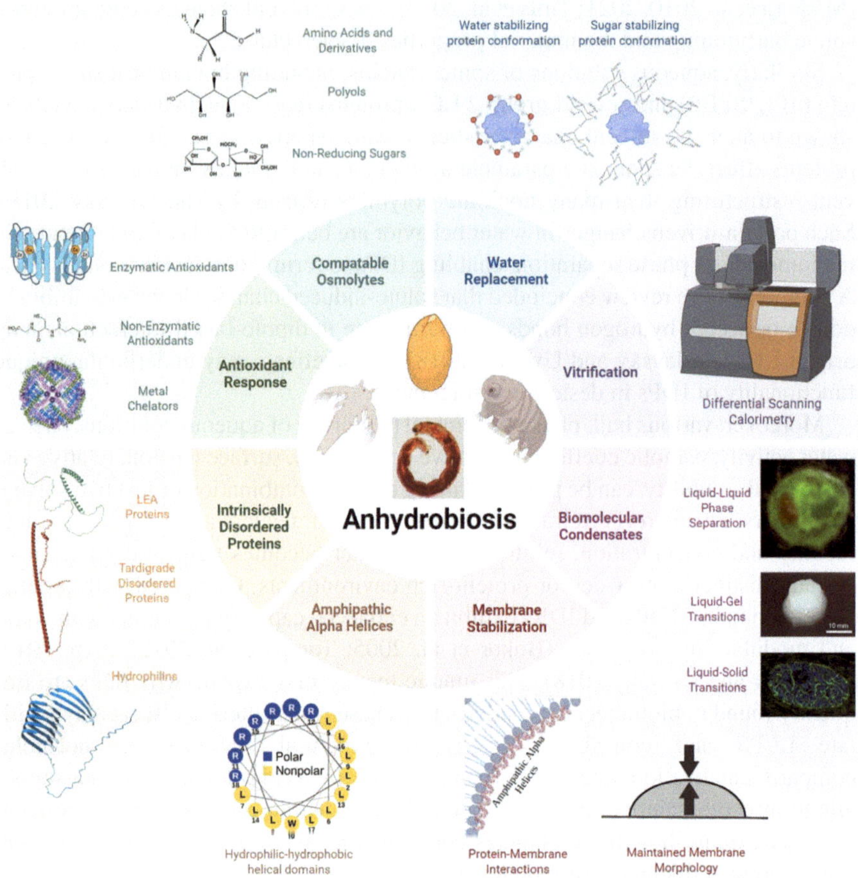

Fig. 17.4 Mechanisms and adaptations operational in anhydrobiotic organisms. Our current understanding of the enigmatic phenomenon of anhydrobiosis includes water replacement by compatible osmolytes, intracellular vitrification, formation of biomolecular condensates, membrane stabilization, the upregulation of antioxidant defense systems, and the accumulation of intrinsically disordered proteins. Anhydrobiotes may rely on these protective mechanisms in varying capacities or may not utilize all the mechanisms shown in the figure. (Created with BioRender)

According to our current knowledge, the extraordinary resilience of desiccation-tolerant organisms results from an intricate interplay between protective sugars, IDPs, and dynamic phase transitions, including cytoplasmic vitrification. We propose several additional mechanisms in which biomacromolecular phase transitions and the resulting biomolecular condensates could help facilitate the survival of desiccation. Taken together, these mechanisms stabilize macromolecular assemblies, mitigate harmful reactions, and maintain essential cellular architecture during extreme dehydration and rehydration. Despite significant advances in our understanding of the molecular underpinnings enabling anhydrobiosis, many mechanistic questions remain unresolved. The interplay between molecular crowding, solvent properties, and biomolecular interactions during drying continues to emerge as a critical determinant of desiccation-tolerance.

Future research elucidating the role of biomacromolecular phase transitions in stress tolerance is warranted. Additionally, synthetic biology approaches may enable the engineering of desiccation-tolerant traits into sensitive species, with profound implications for agriculture, biopreservation, and basic research. A deeper understanding of how biological systems manipulate water dynamics to endure extreme stress may illuminate the fundamental principles of life and expand our capacity to preserve it under otherwise inhospitable conditions. For example, given the well-defined genetic and molecular biology tools available for the precise organ and tissue-manipulation in *D. melanogaster* (Beller and Oliver 2006; Zirin et al. 2024), understanding how and if the expression of specific anhydrobiosis-related genes in this organism (or within a subset of its organs) could increase its tolerance to desiccation would allow for a better understanding of the components necessary for anhydrobiosis to occur in higher organisms. In the past, a few attempts using *Drosophila* cells were performed to understand how the expression of LEA proteins from *A. franciscana* can impact their tolerance to desiccation (Marunde et al. 2013; Rashed et al. 2019). It will be interesting to test whether expressing desiccation-related genes found in *P. vanderplanki*, *A. franciscana*, or other anhydrobiosis-competent invertebrates could increase cold and desiccation resistance in *D. melanogaster*.

References

Abatzopoulos TJ, Beardmore JA, Clegg JS, Sorgeloos P (2002) Artemia basic & applied biology. Kluwer Academic Publishers

Alfano C, Fichou Y, Huber K, Weiss M, Spruijt E, Ebbinghaus S, De Luca G, Morando MA, Vetri V, Temussi PA, Pastore A (2024) Molecular crowding: the history and development of a scientific paradigm. Chem Rev 124(6):3186–3219. https://doi.org/10.1021/acs.chemrev.3c00615

Alpert P (2000) The discovery, scope and puzzle of desiccation tolerance in plants, vol 151

Aumiller WM Jr, Keating CD (2017) Experimental models for dynamic compartmentalization of biomolecules in liquid organelles: reversible formation and partitioning in aqueous biphasic systems. Adv Colloid Interf Sci 239:75–87. https://doi.org/10.1016/j.cis.2016.06.011

Ball P (2017) Water is an active matrix of life for cell and molecular biology. Proc Natl Acad Sci USA 114(51):13327–13335. https://doi.org/10.1073/pnas.1703781114

Barnard RL, Osborne CA, Firestone MK (2013) Responses of soil bacterial and fungal communities to extreme desiccation and rewetting. ISME J 7(11):2229–2241. https://doi.org/10.1038/ismej.2013.104

Battaglia M, Olvera-Carrillo Y, Garciarrubio A, Campos F, Covarrubias AA (2008) The enigmatic LEA proteins and other hydrophilins. Plant Physiol 148(1):6–24. https://doi.org/10.1104/pp.108.120725

Beller M, Oliver B (2006) One hundred years of high-throughput *drosophila* research. Chromosom Res 14(4):349–362. https://doi.org/10.1007/s10577-006-1065-2

Belott C, Janis B, Menze MA (2020) Liquid-liquid phase separation promotes animal desiccation tolerance. Proc Natl Acad Sci USA 117(44):27676–27684. https://doi.org/10.1073/pnas.2014463117

Bhandari BR, Howes T (1999) Implication of glass transition for the drying and stability of dried foods. J Food Eng 40(1–2):71–79

Billi D, Potts M (2002) Life and death of dried prokaryotes. Res Microbiol 153(1):7–12. https://doi.org/10.1016/s0923-2508(01)01279-7

Bokor M, Csizmok V, Kovacs D, Banki P, Friedrich P, Tompa P, Tompa K (2005) NMR relaxation studies on the hydrate layer of intrinsically unstructured proteins. Biophys J 88(3):2030–2037. https://doi.org/10.1529/biophysj.104.051912

Boothby TC, Pielak GJ (2017) Intrinsically disordered proteins and desiccation tolerance: elucidating functional and mechanistic underpinnings of Anhydrobiosis. BioEssays 39(11). https://doi.org/10.1002/bies.201700119

Brangwynne CP, Tompa P, Pappu RV (2015) Polymer physics of intracellular phase transitions. Nat Phys 11:899–904

Brini E, Fennell CJ, Fernandez-Serra M, Hribar-Lee B, Luksic M, Dill KA (2017) How water's properties are encoded in its molecular structure and energies. Chem Rev 117(19):12385–12414. https://doi.org/10.1021/acs.chemrev.7b00259

Brom JA, Petrikis RG, Pielak GJ (2023) How sugars protect dry protein structure. Biochemistry 62(5):1044–1052. https://doi.org/10.1021/acs.biochem.2c00692

Buitink J, Leprince O (2004) Glass formation in plant anhydrobiotes: survival in the dry state. Cryobiology 48(3):215–228. https://doi.org/10.1016/j.cryobiol.2004.02.011

Buitink, J., Claessens, M. M., Hemminga, M. A., & Hoekstra, F. A. (1998). Influence of water content and temperature on molecular mobility and intracellular glasses in seeds and pollen. Plant Physiol, 118(2), 531–541. ://doi.org/https://doi.org/10.1104/pp.118.2.531

Cameron IL, Ord VA, Fullerton GD (1988) Water of hydration in the intra- and extra-cellular environment of human erythrocytes. Biochem Cell Biol 66(11):1186–1199. https://doi.org/10.1139/o88-136

Campos F, Cuevas-Velazquez C, Fares MA, Reyes JL, Covarrubias AA (2013) Group 1 LEA proteins, an ancestral plant protein group, are also present in other eukaryotes, and in the archeae and bacteria domains. Mol Gen Genomics 288(10):503–517. https://doi.org/10.1007/s00438-013-0768-2

Chakrabarti AC, Deamer DW (1992) Permeability of lipid bilayers to amino acids and phosphate. Biochim Biophys Acta 1111(2):171–177. https://doi.org/10.1016/0005-2736(92)90308-9

Chakrabortee S, Tripathi R, Watson M, Schierle GS, Kurniawan DP, Kaminski CF, Wise MJ, Tunnacliffe A (2012) Intrinsically disordered proteins as molecular shields. Mol BioSyst 8(1):210–219. https://doi.org/10.1039/c1mb05263b

Chandler D (2005) Interfaces and the driving force of hydrophobic assembly. Nature 437(7059):640–647. https://doi.org/10.1038/nature04162

Chaplin M (2006) Do we underestimate the importance of water in cell biology? Nat Rev Mol Cell Biol 7(11):861–866. https://doi.org/10.1038/nrm2021

Chen X, Weber I, Harrison RW (2008) Hydration water and bulk water in proteins have distinct properties in radial distributions calculated from 105 atomic resolution crystal structures. J Phys Chem B 112(38):12073–12080. https://doi.org/10.1021/jp802795a

Chen L, Sun Y, Liu Y, Zou Y, Huang J, Zheng Y, Liu G (2020) The N-terminal region of soybean PM1 protein protects liposomes during freeze-thaw. Int J Mol Sci 21(15). https://doi.org/10.3390/ijms21155552

Clegg JS (1965) The origin of Trehalose and its significance during the formation of encysted dormant embryos of *Artemia Salina*. Comp Biochem Physiol 14:135–143. https://doi.org/10.1016/0010-406x(65)90014-9

Clegg JS (2005) Desiccation tolerance in encysted embryos of the animal extremophile, *artemia*. Integr Comp Biol 45(5):715–724. https://doi.org/10.1093/icb/45.5.715

Clegg JS (2011) Stress-related proteins compared in diapause and in activated, anoxic encysted embryos of the animal extremophile, Artemia franciscana. J Insect Physiol 57(5):660–664. https://doi.org/10.1016/j.jinsphys.2010.11.023

Clegg JS, Drost-Hansen W (1991) Phylogenetic and biochemical. Perspectives 1

Clegg JS, Seitz P, Seitz W, Hazlewood CF (1982) Cellular responses to extreme water loss: the water-replacement hypothesis. Cryobiology 19(3):306–316. https://doi.org/10.1016/0011-2240(82)90159-6

Clegg JS, Willsie JK, Jackson S (1999) Adaptive significance of a small heat shock/α-Crystallin protein (p26) in encysted embryos of the brine shrimp. Artemia franciscana American Zoologist 39(6):836–847

Crowe LM (2002) Lessons from nature: the role of sugars in anhydrobiosis. Comp Biochem Physiol A Mol Integr Physiol 131(3):505–513. https://doi.org/10.1016/s1095-6433(01)00503-7

Crowe JH, Hoekstra FA, Crowe LM (1992) Anhydrobiosis. Annu Rev Physiol 54:579–599. https://doi.org/10.1146/annurev.ph.54.030192.003051

Crowe JH, Crowe LM, Carpenter JE (1997) Anhydrobiosis: cellular adaptation to extreme dehydration. Oxford University Press

Crowe JH, Carpenter JF, Crowe LM (1998) The role of vitrification in anhydrobiosis. Annu Rev Physiol 60:73–103. https://doi.org/10.1146/annurev.physiol.60.1.73

Dahl JL, Fordice D (2011) Small acid-soluble proteins with intrinsic disorder are required for UV resistance in *Myxococcus xanthus* spores. J Bacteriol 193(12):3042–3048. https://doi.org/10.1128/JB.00293-11

Dai Y, Chamberlayne CF, Messina MS, Chang CJ, Zare RN, You L, Chilkoti A (2023) Interface of biomolecular condensates modulates redox reactions. Chem 9(6):1594–1609. https://doi.org/10.1016/j.chempr.2023.04.001

de Grotthuss CJT (1806) Sur la décomposition de l'eau et des corps qu'elle tient en dissolution à l'aide de l'électricité galvanique. Ann Chim 58:54–73

Deamer D (2017) The role of lipid membranes in life's origin. Life (Basel) 7(1). https://doi.org/10.3390/life7010005

Dijksterhuis J, Nijsse J, Hoekstra FA, Golovina EA (2007) High viscosity and anisotropy characterize the cytoplasm of fungal dormant stress-resistant spores. Eukaryot Cell 6(2):157–170. https://doi.org/10.1128/EC.00247-06

Disalvo EA, Lairion F, Martini F, Tymczyszyn E, Frias M, Almaleck H, Gordillo GJ (2008) Structural and functional properties of hydration and confined water in membrane interfaces. Biochim Biophys Acta 1778(12):2655–2670. https://doi.org/10.1016/j.bbamem.2008.08.025

Drost-Hansen W, Singleton JL (1995) Our aqueous heritage: evidence for vicinal water in cells, vol 4. JAI Press Inc.

Dure L, Galau GA (1981) Developmental biochemistry of cottonseed embryogenesis and germination: XIII. Regulation of biosynthesis of principal storage proteins. Plant Physiol 68(1):187–194. https://doi.org/10.1104/pp.68.1.187

Eremchev M, Roesel D, Poojari CS, Roux A, Hub JS, Roke S (2023) Passive transport of ca(2+) ions through lipid bilayers imaged by widefield second harmonic microscopy. Biophys J 122(4):624–631. https://doi.org/10.1016/j.bpj.2023.01.018

Erkut C, Vasilj A, Boland S, Habermann B, Shevchenko A, Kurzchalia TV (2013) Molecular strategies of the *Caenorhabditis elegans* dauer larva to survive extreme desiccation. PLoS One 8(12):e82473. https://doi.org/10.1371/journal.pone.0082473

Farrant JM (2008) Mechanisms of desiccation tolerance in angiosperm resurrection plants. Blackwell Ltd

Ferreira LA, Loureiro JA, Gomes J, Uversky VN, Madeira PP, Zaslavsky BY (2016) Why physicochemical properties of aqueous solutions of various compounds are linearly interrelated. J Mol Liq 221:116–123

Ferreira LA, Gusev NB, Uversky VN, Zaslavsky BY (2018a) Effect of human heat shock protein HspB6 on the solvent features of water in aqueous solutions. J Biomol Struct Dyn 36(6):1520–1528. https://doi.org/10.1080/07391102.2017.1328316

Ferreira LA, Walczyk Mooradally A, Zaslavsky B, Uversky VN, Graether SP (2018b) Effect of an intrinsically disordered plant stress protein on the properties of water. Biophys J 115(9):1696–1706. https://doi.org/10.1016/j.bpj.2018.09.014

Field S, Jang GJ, Dean C, Strader LC, Rhee SY (2023) Plants use molecular mechanisms mediated by biomolecular condensates to integrate environmental cues with development. Plant Cell 35(9):3173–3186. https://doi.org/10.1093/plcell/koad062

Fisher RS, Obermeyer AC (2024) Viscoelasticity of globular protein-based biomolecular condensates. Chem Sci 15(47):19795–19804. https://doi.org/10.1039/d4sc03564j

Franks F, Hatley RHM, Mathias S (1991) Materials science and the production of shelf stable biologics. BioPharm 55(4):38–42

French-Pacheco L, Cuevas-Velazquez CL, Rivillas-Acevedo L, Covarrubias AA, Amero C (2018) Metal-binding polymorphism in late embryogenesis abundant protein AtLEA4-5, an intrinsically disordered protein. PeerJ 6:e4930. https://doi.org/10.7717/peerj.4930

Frenkel-Pinter M, Rajaei V, Glass JB, Hud NV, Williams LD (2021) Water and life: the medium is the message. J Mol Evol 89(1–2):2–11. https://doi.org/10.1007/s00239-020-09978-6

Fullerton GD, Kanal KM, Cameron IL (2006) On the osmotically unresponsive water compartment in cells. Cell Biol Int 30(1):74–77. https://doi.org/10.1016/j.cellbi.2005.09.007

Garay-Arroyo A, Colmenero-Flores JM, Garciarrubio A, Covarrubias AA (2000) Highly hydrophilic proteins in prokaryotes and eukaryotes are common during conditions of water deficit. J Biol Chem 275(8):5668–5674. https://doi.org/10.1074/jbc.275.8.5668

Garg R, Maldener I (2021) The formation of spore-like Akinetes: A survival strategy of filamentous cyanobacteria. Microb Physiol 31(3):296–305. https://doi.org/10.1159/000517443

Garrido C, Bruey JM, Fromentin A, Hammann A, Arrigo AP, Solary E (1999) HSP27 inhibits cytochrome c-dependent activation of procaspase-9. FASEB J 13(14):2061–2070. https://doi.org/10.1096/fasebj.13.14.2061

Goyal K, Walton LJ, Tunnacliffe A (2005) LEA proteins prevent protein aggregation due to water stress. Biochem J 388(Pt 1):151–157. https://doi.org/10.1042/BJ20041931

Grzyb T, Sklodowska A (2022) Introduction to bacterial Anhydrobiosis: A general perspective and the mechanisms of desiccation-associated damage. Microorganisms 10(2). https://doi.org/10.3390/microorganisms10020432

Guidetti R, Altiero T, Rebecchi L (2011) On dormancy strategies in tardigrades. J Insect Physiol 57(5):567–576. https://doi.org/10.1016/j.jinsphys.2011.03.003

Halle B (2004) Protein hydration dynamics in solution: a critical survey. Philos Trans R Soc Lond B Biol Sci 359(1448), 1207–1223; discussion 1223–1204, 1323–1208. https://doi.org/10.1098/rstb.2004.1499

Hampton SE, Galloway AW, Powers SM, Ozersky T, Woo KH, Batt RD, Labou SG, O'Reilly CM, Sharma S, Lottig NR, Stanley EH, North RL, Stockwell JD, Adrian R, Weyhenmeyer GA, Arvola L, Baulch HM, Bertani I, Bowman LL, Jr.,…Xenopoulos, M. A. (2017) Ecology under lake ice. Ecol Lett 20(1):98–111. https://doi.org/10.1111/ele.12699

Hand SC (1998) Quiescence in *Artemia franciscana* embryos: reversible arrest of metabolism and gene expression at low oxygen levels. J Exp Biol 201(Pt 8):1233–1242. https://doi.org/10.1242/jeb.201.8.1233

Hand SC, Menze MA, Toner M, Boswell L, Moore D (2011) LEA proteins during water stress: not just for plants anymore. Annu Rev Physiol 73:115–134. https://doi.org/10.1146/annurev-physiol-012110-142203

Hand SC, Denlinger DL, Podrabsky JE, Roy R (2016) Mechanisms of animal diapause: recent developments from nematodes, crustaceans, insects, and fish. Am J Physiol Regul Integr Comp Physiol 310(11):R1193–R1211. https://doi.org/10.1152/ajpregu.00250.2015

Hand SC, Moore DS, Patil Y (2018) Challenges during diapause and anhydrobiosis: mitochondrial bioenergetics and desiccation tolerance. IUBMB Life 70(12):1251–1259. https://doi.org/10.1002/iub.1953

Hansen MM, Meijer LH, Spruijt E, Maas RJ, Rosquelles MV, Groen J, Heus HA, Huck WT (2016) Macromolecular crowding creates heterogeneous environments of gene expression in picolitre droplets. Nat Nanotechnol 11(2):191–197. https://doi.org/10.1038/nnano.2015.243

Hara M, Kondo M, Kato T (2013) A KS-type dehydrin and its related domains reduce cu-promoted radical generation and the histidine residues contribute to the radical-reducing activities. J Exp Bot 64(6):1615–1624. https://doi.org/10.1093/jxb/ert016

Hernandez-Sanchez IE, Maruri-Lopez I, Martinez-Martinez C, Janis B, Jimenez-Bremont JF, Covarrubias AA, Menze MA, Graether SP, Thalhammer A (2022) LEAfing through literature: late embryogenesis abundant proteins coming of age-achievements and perspectives. J Exp Bot 73(19):6525–6546. https://doi.org/10.1093/jxb/erac293

Hibshman JD, Clegg JS, Goldstein B (2020) Mechanisms of desiccation tolerance: themes and variations in brine shrimp, roundworms, and tardigrades. Front Physiol 11:592016. https://doi.org/10.3389/fphys.2020.592016

Hincha DK, Zuther E, Popova AV (2021) Stabilization of dry sucrose glasses by four LEA_4 proteins from *Arabidopsis thaliana*. Biomolecules 11(5). https://doi.org/10.3390/biom11050615

Hoekstra FA, Golovina EA, Buitink J (2001) Mechanisms of plant desiccation tolerance. Trends Plant Sci 6(9):431–438. https://doi.org/10.1016/s1360-1385(01)02052-0

Ievinsh G (2023) Water content of plant tissues: so simple that almost forgotten? Plants (Basel) 12(6). https://doi.org/10.3390/plants12061238

Ishiguro H, Rubinsky B (1994) Mechanical interactions between ice crystals and red blood cells during directional solidification. Cryobiology 31(5):483–500. https://doi.org/10.1006/cryo.1994.1059

Janis BM (2023) Liquid-liquid phase in anhydrobiosis

Janis B, Belott C, Menze MA (2018) Role of intrinsic disorder in animal desiccation tolerance. Proteomics 18(21–22):e1800067. https://doi.org/10.1002/pmic.201800067

Janis B, Belott C, Brockman T, Menze MA (2022) Functional and conformational plasticity of an animal group 1 LEA protein. Biomolecules 12(3). https://doi.org/10.3390/biom12030425

Jenks MA, Wood AJ (2008) Plant desiccation tolerance. Wiley-Blackwell

Kim JS, Yethiraj A (2009) Effect of macromolecular crowding on reaction rates: a computational and theoretical study. Biophys J 96(4):1333–1340. https://doi.org/10.1016/j.bpj.2008.11.030

King AM, MacRae TH (2012) The small heat shock protein p26 aids development of encysting *Artemia* embryos, prevents spontaneous diapause termination and protects against stress. PLoS One 7(8):e43723. https://doi.org/10.1371/journal.pone.0043723

Knox-Brown P, Rindfleisch T, Gunther A, Balow K, Bremer A, Walther D, Miettinen MS, Hincha DK, Thalhammer A (2020) Similar yet different-structural and functional diversity among *Arabidopsis thaliana* LEA_4 proteins. Int J Mol Sci 21(8). https://doi.org/10.3390/ijms21082794

Kovacs D, Kalmar E, Torok Z, Tompa P (2008) Chaperone activity of ERD10 and ERD14, two disordered stress-related plant proteins. Plant Physiol 147(1):381–390. https://doi.org/10.1104/pp.108.118208

Kwapiszewska K, Szczepanski K, Kalwarczyk T, Michalska B, Patalas-Krawczyk P, Szymanski J, Andryszewski T, Iwan M, Duszynski J, Holyst R (2020) Nanoscale viscosity of cytoplasm is conserved in human cell lines. J Phys Chem Lett 11(16):6914–6920. https://doi.org/10.1021/acs.jpclett.0c01748

Lapinski J, Tunnacliffe A (2003) Anhydrobiosis without trehalose in bdelloid rotifers. FEBS Lett 553(3):387–390. https://doi.org/10.1016/s0014-5793(03)01062-7

LeBlanc BM, Le MT, Janis B, Menze MA, Hand SC (2019) Structural properties and cellular expression of AfrLEA6, a group 6 late embryogenesis abundant protein from embryos of *Artemia franciscana*. Cell Stress Chaperones 24(5):979–990. https://doi.org/10.1007/s12192-019-01025-8

Lebre PH, De Maayer P, Cowan DA (2017) Xerotolerant bacteria: surviving through a dry spell. Nat Rev Microbiol 15(5):285–296. https://doi.org/10.1038/nrmicro.2017.16

Lee JC, Timasheff SN (1981) The stabilization of proteins by sucrose. J Biol Chem 256(14):7193–7201. https://www.ncbi.nlm.nih.gov/pubmed/7251592

Levy Y, Onuchic JN (2004) Water and proteins: a love-hate relationship. Proc Natl Acad Sci USA 101(10):3325–3326. https://doi.org/10.1073/pnas.0400157101

Lim S, Reilly CB, Barghouti Z, Marelli B, Way JC, Silver PA (2024) Tardigrade secretory proteins protect biological structures from desiccation. Commun Biol 7(1):633. https://doi.org/10.1038/s42003-024-06336-w

Liu G, Liu K, Gao Y, Zheng Y (2017a) Involvement of C-terminal Histidines in soybean PM1 protein Oligomerization and Cu2+ binding. Plant Cell Physiol 58(6):1018–1029. https://doi.org/10.1093/pcp/pcx046

Liu J, He X, Zhang JZH (2017b) Structure of liquid water - a dynamical mixture of tetrahedral and 'ring-and-chain' like structures. Phys Chem Chem Phys 19(19):11931–11936. https://doi.org/10.1039/c7cp00667e

Lo Sterzo M, Iuso D, Palazzese L, Moncada M, Boffa F, Scudieri A, Gioia L, Czernik M, Loi P (2025) Exogenous LEA proteins expression enhances cold tolerance in mammalian cells by reducing oxidative stress. Sci Rep 15(1):3351. https://doi.org/10.1038/s41598-025-86499-6

Lopez-Martinez G, Rodriguez-Porrata B, Margalef-Catala M, Cordero-Otero R (2012) The STF2p hydrophilin from *Saccharomyces cerevisiae* is required for dehydration stress tolerance. PLoS One 7(3):e33324. https://doi.org/10.1371/journal.pone.0033324

Luby-Phelps, K. (2000). Cytoarchitecture and physical properties of cytoplasm: volume, viscosity, diffusion, intracellular surface area. Int Rev Cytol, 192, 189–221. https://doi.org/10.1016/s0074-7696(08)60527-6, Elsevier

Ma X, Jamil K, Macrae TH, Clegg JS, Russell JM, Villeneuve TS, Euloth M, Sun Y, Crowe JH, Tablin F, Oliver AE (2005) A small stress protein acts synergistically with trehalose to confer desiccation tolerance on mammalian cells. Cryobiology 51(1):15–28. https://doi.org/10.1016/j.cryobiol.2005.04.007

MacRae TH (2016) Stress tolerance during diapause and quiescence of the brine shrimp. Artemia Cell Stress Chaperones 21(1):9–18. https://doi.org/10.1007/s12192-015-0635-7

Madeira PP, Reis CA, Rodrigues AE, Mikheeva LM, Zaslavsky BY (2010) Solvent properties governing solute partitioning in polymer/polymer aqueous two-phase systems: nonionic compounds. J Phys Chem B 114(1):457–462. https://doi.org/10.1021/jp907346s

Madeira PP, Titus AR, Ferreira LA, Belgovskiy AI, Mann EK, Mann JA Jr, Meyer WV, Smart AE, Uversky VN, Zaslavsky BY (2021) Hydrogen bond arrangement is shown to differ in coexisting phases of aqueous two-phase systems. Biomolecules 11(12). https://doi.org/10.3390/biom11121787

Males J, Griffiths H (2017) Stomatal biology of CAM plants. Plant Physiol 174(2):550–560. https://doi.org/10.1104/pp.17.00114

Malki A, Teulon JM, Camacho-Zarco AR, Chen SW, Adamski W, Maurin D, Salvi N, Pellequer JL, Blackledge M (2022) Intrinsically disordered tardigrade proteins self-assemble into fibrous gels in response to environmental stress. Angew Chem Int Ed Engl 61(1):e202109961. https://doi.org/10.1002/anie.202109961

Mangiarotti A, Chen N, Zhao Z, Lipowsky R, Dimova R (2023) Wetting and complex remodeling of membranes by biomolecular condensates. Nat Commun 14(1):2809. https://doi.org/10.1038/s41467-023-37955-2

Marunde MR, Samarajeewa DA, Anderson J, Li S, Hand SC, Menze MA (2013) Improved tolerance to salt and water stress in *Drosophila melanogaster* cells conferred by late embryogenesis abundant protein. J Insect Physiol 59(4):377–386. https://doi.org/10.1016/j.jinsphys.2013.01.004

McDonald MD, Owusu-Ansah C, Ellenbogen JB, Malone ZD, Ricketts MP, Frolking SE, Ernakovich JG, Ibba M, Bagby SC, Weissman JL (2024) What is microbial dormancy? Trends Microbiol 32(2):142–150. https://doi.org/10.1016/j.tim.2023.08.006

Meneses-Reyes GI, Rodriguez-Bustos DL, Cuevas-Velazquez CL (2024) Macromolecular crowding sensing during osmotic stress in plants. Trends Biochem Sci 49(6):480–493. https://doi.org/10.1016/j.tibs.2024.02.002

Menze MA, Hand SC (2007) Caspase activity during cell stasis: avoidance of apoptosis in an invertebrate extremophile, *Artemia franciscana*. Am J Physiol Regul Integr Comp Physiol 292(5):R2039–R2047. https://doi.org/10.1152/ajpregu.00659.2006

Menze MA, Boswell L, Toner M, Hand SC (2009) Occurrence of mitochondria-targeted late embryogenesis abundant (LEA) gene in animals increases organelle resistance to water stress. J Biol Chem 284(16):10714–10719. https://doi.org/10.1074/jbc.C900001200

Meredith HR, Andreani V, Ma HR, Lopatkin AJ, Lee AJ, Anderson DJ, Batt G, You L (2018) Applying ecological resistance and resilience to dissect bacterial antibiotic responses. Sci Adv 4(12):eaau1873. https://doi.org/10.1126/sciadv.aau1873

Mitra A, Bhakta K, Kar A, Roy A, Mohid SA, Ghosh A, Ghosh A (2023) Insight into the biochemical and cell biological function of an intrinsically unstructured heat shock protein, Hsp12 of *Ustilago maydis*. Mol Plant Pathol 24(9):1063–1077. https://doi.org/10.1111/mpp.13350

Munder MC, Midtvedt D, Franzmann T, Nuske E, Otto O, Herbig M, Ulbricht E, Muller P, Taubenberger A, Maharana S, Malinovska L, Richter D, Guck J, Zaburdaev V, Alberti S (2016) A pH-driven transition of the cytoplasm from a fluid- to a solid-like state promotes entry into dormancy. elife 5. https://doi.org/10.7554/eLife.09347

Olgenblum GI, Hutcheson BO, Pielak GJ, Harries D (2024) Protecting proteins from desiccation stress using molecular glasses and gels. Chem Rev 124(9):5668–5694. https://doi.org/10.1021/acs.chemrev.3c00752

Orlikowska-Rzeznik H, Versluis J, Bakker HJ, Piatkowski L (2024) Cholesterol changes interfacial water alignment in model cell membranes. J Am Chem Soc 146(19):13151–13162. https://doi.org/10.1021/jacs.4c00474

Pak CW, Kosno M, Holehouse AS, Padrick SB, Mittal A, Ali R, Yunus AA, Liu DR, Pappu RV, Rosen MK (2016) Sequence determinants of intracellular phase separation by complex Coacervation of a disordered protein. Mol Cell 63(1):72–85. https://doi.org/10.1016/j.molcel.2016.05.042

Pandey P, Farber R, Nakazawa A, Kumar S, Bharti A, Nalin C, Weichselbaum R, Kufe D, Kharbanda S (2000) Hsp27 functions as a negative regulator of cytochrome c-dependent activation of procaspase-3. Oncogene 19(16):1975–1981. https://doi.org/10.1038/sj.onc.1203531

Podrabsky JE, Hand SC (2015) Physiological strategies during animal diapause: lessons from brine shrimp and annual killifish. J Exp Biol 218(Pt 12):1897–1906. https://doi.org/10.1242/jeb.116194

Polačik M, Podrabsky JE (2015) Temporary environments. Springer

Qiu Z, Bossier P, Wang X, Bojikova-Fournier S, MacRae TH (2006) Diversity, structure, and expression of the gene for p26, a small heat shock protein from *Artemia*. Genomics 88(2):230–240. https://doi.org/10.1016/j.ygeno.2006.02.008

Rapoport A, Golovina EA, Gervais P, Dupont S, Beney L (2019) Anhydrobiosis: inside yeast cells. Biotechnol Adv 37(1):51–67. https://doi.org/10.1016/j.biotechadv.2018.11.003

Rashed MZ, Belott CJ, Janis BR, Menze MA, Williams SJ (2019) New insights into anhydrobiosis using cellular dielectrophoresis-based characterization. Biomicrofluidics 13(6):064113. https://doi.org/10.1063/1.5126810

Record MT Jr, Courtenay ES, Cayley DS, Guttman HJ (1998a) Responses of *E. Coli* to osmotic stress: large changes in amounts of cytoplasmic solutes and water. Trends Biochem Sci 23(4):143–148. https://doi.org/10.1016/s0968-0004(98)01196-7

Record MT Jr, Courtenay ES, Cayley S, Guttman HJ (1998b) Biophysical compensation mechanisms buffering *E. Coli* protein-nucleic acid interactions against changing environments. Trends Biochem Sci 23(5):190–194. https://doi.org/10.1016/s0968-0004(98)01207-9

Redza-Dutordoir M, Averill-Bates DA (2016) Activation of apoptosis signalling pathways by reactive oxygen species. Biochim Biophys Acta 1863(12):2977–2992. https://doi.org/10.1016/j.bbamcr.2016.09.012

Reinle K, Mogk A, Bukau B (2022) The diverse functions of small heat shock proteins in the Proteostasis network. J Mol Biol 434(1):167157. https://doi.org/10.1016/j.jmb.2021.167157

Richards AB, Krakowka S, Dexter LB, Schmid H, Wolterbeek AP, Waalkens-Berendsen DH, Shigoyuki A, Kurimoto M (2002) Trehalose: a review of properties, history of use and human tolerance, and results of multiple safety studies. Food Chem Toxicol 40(7):871–898. https://doi.org/10.1016/s0278-6915(02)00011-x

Robison ZL, Ren Q, Zhang Z (2024) How to survive without water: A short lesson on the desiccation tolerance of budding yeast. Int J Mol Sci 25(14). https://doi.org/10.3390/ijms25147514

Romero-Perez PS, Dorone Y, Flores E, Sukenik S, Boeynaems S (2023) When phased without water: biophysics of cellular desiccation, from biomolecules to condensates. Chem Rev 123(14):9010–9035. https://doi.org/10.1021/acs.chemrev.2c00659

Sakuta H, Fujimoto T, Yamana Y, Hoda Y, Tsumoto K, Yoshikawa K (2019) Aqueous/aqueous micro phase separation: construction of an artificial model of cellular assembly. Front Chem 7:44. https://doi.org/10.3389/fchem.2019.00044

Sanchez-Martinez S, Ramirez JF, Meese EK, Childs CA, Boothby TC (2023) The tardigrade protein CAHS D interacts with, but does not retain, water in hydrated and desiccated systems. Sci Rep 13(1):10449. https://doi.org/10.1038/s41598-023-37485-3

Shah M, Corbett D, Lanzaro A, Roche A, Sibanda N, Davis P, Uddin S, van der Walle CF, Curtis R, Pluen A (2020) Micro- and macro-viscosity relations in high concentration antibody solutions. Eur J Pharm Biopharm 153:211–221. https://doi.org/10.1016/j.ejpb.2020.06.007

Shaikhutdinov N, Gusev O (2022) Chironomid midges (*Diptera*) provide insights into genome evolution in extreme environments. Curr Opin Insect Sci 49:101–107. https://doi.org/10.1016/j.cois.2021.12.009

Sriram R, Shoff M, Booton G, Fuerst P, Visvesvara GS (2008) Survival of *Acanthamoeba* cysts after desiccation for more than 20 years. J Clin Microbiol 46(12):4045–4048. https://doi.org/10.1128/JCM.01903-08

Steudle E (2001) The cohesion-tension mechanism and the Acquisition of Water by plant roots. Annu Rev Plant Physiol Plant Mol Biol 52:847–875. https://doi.org/10.1146/annurev.arplant.52.1.847

Sun WQ, Leopold A (1997) Cytoplasmic vitrification and survival of anhydrobiotic organisms. Comp Biochem Physiol A Physiol 117(3):327–333

Sun Y, Macrae TH (2005) Characterization of novel sequence motifs within N- and C-terminal extensions of p26, a small heat shock protein from *Artemia franciscana*. FEBS J 272(20):5230–5243. https://doi.org/10.1111/j.1742-4658.2005.04920.x

Sun Y, Mansour M, Crack JA, Gass GL, Macrae TH (2004) Oligomerization, chaperone activity, and nuclear localization of p26, a small heat shock protein from *Artemia franciscana*. J Biol Chem 279(38):39999–40006. https://doi.org/10.1074/jbc.m406999200

Sun Y, Bojikova-Fournier S, Macrae TH (2006) Structural and functional roles for β-strand 7 in the α-crystallin domain of p26, a polydisperse small heat shock protein from *Artemia franciscana*. FEBS J 273(5):1020–1034. https://doi.org/10.1111/j.1742-4658.2006.05129.x

Szymanski J, Patkowski A, Wilk A, Garstecki P, Holyst R (2006) Diffusion and viscosity in a crowded environment: from nano- to macroscale. J Phys Chem B 110(51):25593–25597. https://doi.org/10.1021/jp0666784

Takayama S, Reed JC, Homma S (2003) Heat-shock proteins as regulators of apoptosis. Oncogene 22(56):9041–9047. https://doi.org/10.1038/sj.onc.1207114

Tanaka A, Nakano T, Watanabe K, Masuda K, Honda G, Kamata S, Yasui R, Kozuka-Hata H, Watanabe C, Chinen T, Kitagawa D, Sawai S, Oyama M, Yanagisawa M, Kunieda T (2022) Stress-dependent cell stiffening by tardigrade tolerance proteins that reversibly form a filamentous network and gel. PLoS Biol 20(9):e3001780. https://doi.org/10.1371/journal.pbio.3001780

Tapia H, Koshland DE (2014) Trehalose is a versatile and long-lived chaperone for desiccation tolerance. Curr Biol 24(23):2758–2766. https://doi.org/10.1016/j.cub.2014.10.005

Thomashow MF (1999) PLANT COLD ACCLIMATION: freezing tolerance genes and regulatory mechanisms. Annu Rev Plant Physiol Plant Mol Biol 50:571–599. https://doi.org/10.1146/annurev.arplant.50.1.571

Titus AR, Ferreira LA, Belgovskiy AI, Kooijman EE, Mann EK, Mann JA Jr, Meyer WV, Smart AE, Uversky VN, Zaslavsky BY (2020) Interfacial tension and mechanism of liquid-liquid phase separation in aqueous media. Phys Chem Chem Phys 22(8):4574–4580. https://doi.org/10.1039/c9cp05810a

Titus AR, Madeira PP, Ferreira LA, Chernyak VY, Uversky VN, Zaslavsky BY (2022) Mechanism of phase separation in aqueous two-phase systems. Int J Mol Sci 23(22). https://doi.org/10.3390/ijms232214366

Tolleter D, Hincha DK, Macherel D (2010) A mitochondrial late embryogenesis abundant protein stabilizes model membranes in the dry state. Biochim Biophys Acta 1798(10):1926–1933. https://doi.org/10.1016/j.bbamem.2010.06.029

Tompa K, Bokor M, Tompa P (2012) Wide-line NMR and protein hydration. Methods Mol Biol 895:167–196. https://doi.org/10.1007/978-1-61779-927-3_13

Tompa P, Han KH, Bokor M, Kamasa P, Tantos A, Fritz B, Kim DH, Lee C, Verebelyi T, Tompa K (2016) Wide-line NMR and DSC studies on intrinsically disordered p53 transactivation domain and its helically pre-structured segment. BMB Rep 49(9):497–501. https://doi.org/10.5483/bmbrep.2016.49.9.037

Tompa K, Bokor M, Agner D, Ivan D, Kovacs D, Verebelyi T, Tompa P (2017) Hydrogen mobility and protein-water interactions in proteins in the solid state. ChemPhysChem 18(6):677–682. https://doi.org/10.1002/cphc.201601136

Torok Z, Goloubinoff P, Horvath I, Tsvetkova NM, Glatz A, Balogh G, Varvasovszki V, Los DA, Vierling E, Crowe JH, Vigh L (2001) *Synechocystis* HSP17 is an amphitropic protein that stabilizes heat-stressed membranes and binds denatured proteins for subsequent chaperone-mediated refolding. Proc Natl Acad Sci USA 98(6):3098–3103. https://doi.org/10.1073/pnas.051619498

Tripathi R, Boschetti C, McGee B, Tunnacliffe A (2012) Trafficking of bdelloid rotifer late embryogenesis abundant proteins. J Exp Biol 215(Pt 16):2786–2794. https://doi.org/10.1242/jeb.071647

Tunnacliffe A, Wise MJ (2007) The continuing conundrum of the LEA proteins. Naturwissenschaften 94(10):791–812. https://doi.org/10.1007/s00114-007-0254-y

Uematsu T, Svanberg C, Nyden M, Jacobsson P (2003) Power laws in polymer solution dynamics. Phys Rev E Stat Nonlinear Soft Matter Phys 68(5 Pt 1):051803. https://doi.org/10.1103/PhysRevE.68.051803

Uversky VN (2013) Unusual biophysics of intrinsically disordered proteins. Biochim Biophys Acta 1834(5):932–951. https://doi.org/10.1016/j.bbapap.2012.12.008

Veling MT, Nguyen DT, Thadani NN, Oster ME, Rollins NJ, Brock KP, Bethel NP, Lim S, Baker D, Way JC, Marks DS, Chang RL, Silver PA (2022) Natural and designed proteins inspired by Extremotolerant organisms can form condensates and attenuate apoptosis in human cells. ACS Synth Biol 11(3):1292–1302. https://doi.org/10.1021/acssynbio.1c00572

Villeneuve TS, Ma X, Sun Y, Oulton MM, Oliver AE, MacRae TH (2006) Inhibition of apoptosis by p26: implications for small heat shock protein function during *Artemia development*. Cell Stress Chaperones 11(1):71–80. https://doi.org/10.1379/csc-154r.1

Vogler EA (1998) Structure and reactivity of water at biomaterial surfaces. Adv Colloid Interf Sci 74:69–117. https://doi.org/10.1016/s0001-8686(97)00040-7

Vorontsova I, Vallmitjana A, Torrado B, Schilling TF, Hall JE, Gratton E, Malacrida L (2022) In vivo macromolecular crowding is differentially modulated by aquaporin 0 in zebrafish lens: insights from a nanoenvironment sensor and spectral imaging. Sci Adv 8(7):eabj4833. https://doi.org/10.1126/sciadv.abj4833

Watson JL, Seinkmane E, Styles CT, Mihut A, Kruger LK, McNally KE, Planelles-Herrero VJ, Dudek M, McCall PM, Barbiero S, Vanden Oever M, Peak-Chew SY, Porebski BT, Zeng A, Rzechorzek NM, Wong DCS, Beale AD, Stangherlin A, Riggi M et al (2023) Macromolecular condensation buffers intracellular water potential. Nature 623(7988):842–852. https://doi.org/10.1038/s41586-023-06626-z

Wu Y, MacRae TH (2010) Truncation attenuates molecular chaperoning and apoptosis inhibition by p26, a small heat shock protein from *Artemia franciscana*. Biochem Cell Biol 88(6):937–946. https://doi.org/10.1139/O10-143

Xantheas SS (2000) Cooperativity and hydrogen bonding network in water clusters. Chem Phys 258(2–3):225–231

Yagi-Utsumi M, Aoki K, Watanabe H, Song C, Nishimura S, Satoh T, Yanaka S, Ganser C, Tanaka S, Schnapka V, Goh EW, Furutani Y, Murata K, Uchihashi T, Arakawa K, Kato K (2021) Desiccation-induced fibrous condensation of CAHS protein from an anhydrobiotic tardigrade. Sci Rep 11(1):21328. https://doi.org/10.1038/s41598-021-00724-6

Yahia EM, Carillo-López A, Bello-Perez LA (2019) Carbohydrates. Elsevier

Yathisha NS, Barbara P, Gugi B, Yogendra K, Jogaiah S, Azeddine D, Sharatchandra RG (2020) Vegetative desiccation tolerance in *Eragrostiella brachyphylla*: biochemical and physiological responses. Heliyon 6(9):e04948. https://doi.org/10.1016/j.heliyon.2020.e04948

Yoshida H, Nakayama H, Maki Y, Ueta M, Wada C, Wada A (2021) Functional sites of ribosome modulation factor (RMF) involved in the formation of 100S ribosome. Front Mol Biosci 8:661691. https://doi.org/10.3389/fmolb.2021.661691

Zaslavsky BY, Uversky VN (2018) In aqua Veritas: the indispensable yet mostly ignored role of water in phase separation and membrane-less organelles. Biochemistry 57(17):2437–2451. https://doi.org/10.1021/acs.biochem.7b01215

Zirin J, Jusiak B, Lopes R, Ewen-Campen B, Bosch JA, Risbeck A, Forman C, Villalta C, Hu Y, Perrimon N (2024) Expanding the *drosophila* toolkit for dual control of gene expression. elife 12. https://doi.org/10.7554/eLife.94073

Chapter 18
Crowding and in-Cell Crystallization

Altijana Hromić-Jahjefendić, Vladimir N. Uversky, and Hana Krnjić

Abstract Macromolecular crowding has significant impact on protein behavior in cellular environments, having influence on their structure, folding, and interactions. This process comes from the high concentration of macromolecules, including proteins, nucleic acids, and carbohydrates, which occupy a substantial fraction of intracellular space. Ficoll 70 and Dextran, synthetic crowding agents, are utilized by researchers to mimic these conditions *in vitro*. Various studies demonstrate that different crowding agents can enhance protein thermal stability and promote structural organization by stabilizing folded states and compacting denatured forms. Additionally, crowding agents like PEG have been found to exert stronger stabilizing effects on proteins compared to others like Dextran. Understanding the effects of crowding agents is important and crucial for elucidating protein dynamics in cellular contexts and has crucial implications for both fundamental biology and therapeutic applications. Continued research into the mechanisms by which these agents influence protein behavior will provide better insights into cellular biochemistry and may lead to novel strategies for drug development.

Keywords Macromolecular crowding · Protein interactions · Crowding agents · Crystallization · Protein stability

A. Hromić-Jahjefendić (✉) · H. Krnjić
Department of Genetics and Bioengineering, Faculty of Engineering and Natural Sciences, International University of Sarajevo, Sarajevo, Bosnia and Herzegovina
e-mail: ahromic@ius.edu.ba

V. N. Uversky
Department of Molecular Medicine and USF Health Byrd Alzheimer's Research Institute, Morsani College of Medicine, University of South Florida, Tampa, FL, USA

V. N. Uversky (ed.), *(Macro)Molecular Crowding*, Subcellular Biochemistry 109,
https://doi.org/10.1007/978-3-032-03370-3_18

Introduction

The term 'macromolecular crowding' was initially utilized to describe "the effect of a high overall weight per volume concentration of macromolecules on the equilibria and kinetics of macromolecular reactions occurring in a highly volume-occupied or 'crowded' solution, regardless of whether or not any single species of reactant or product was concentrated" (Rivas and Minton 2016). The degree of congestion varies according to the type of cell. The concentration of macromolecules in eukaryotic cells is about half that of *E. coli*. The quantities in the mitochondria and nucleus are higher than those seen in bacteria (Pittas et al. 2023).

Crowding effects are defined by the interactions of molecules in solution. Simple crowding theories, which imply that all molecules are hard spheres, indicate that the intracellular environment would stabilize proteins and their complexes via hard-core repulsions, while shape can also play a role (Berg 1990). Understanding macromolecular crowding's impact on proteins is critical for both basic biology and human health. It is critical to understand the features of proteins where they function in cells, especially how crowding impacts liquid-liquid phase separation (Leslie 2021). Recognizing crowding effects on protein complexes is critical because protein networks convert chemical signals into physiological reactions that preserve homeostasis (Cong et al. 2019). Two-thirds of missense mutations have an impact on protein-protein interactions, which are important for human health, yet little is known about protein-protein interactions in cells (Sahni et al. 2015; Stadmiller and Pielak 2021). Additionally, protein aggregation is linked to a specific group of disorders. Crowding by synthetic polymers has a substantial influence on the industrial and medicinal applications mentioned above. In biology, many studies comprehend macromolecular crowding use synthetic polymers, although cells are packed with proteins and nucleic acids. The fact that the composition of the cellular interior depends on many variables complicates efforts to understand protein biophysics in cells (Speer et al. 2022).

Cell crystallization, or protein crystallization in live cells, has been observed unexpectedly often as a natural assembly process in recent decades, and accumulating data suggests that this phenomenon is equally applicable to recombinant proteins (Schönherr et al. 2018). The crystalline state of biological macromolecules is required for X-ray crystallography, a powerful approach for acquiring structural information on proteins at the atomic level. Given that aggregation usually impairs native protein function and, consequently, cell viability, the difficulties in crystal formation may be the result of a negative evolutionary pressure that inhibits spontaneous intracellular protein crystallization (Doye et al. 2004). For this reason, even though crystal formation in living cells has been known for over a century, it was mainly regarded as an unusual occurrence. As seen in several types of human cataracts and anemia, the aberrant crystalline state of proteins that are normally soluble within the cell was recognized as a distinctive pathogenic hallmark (Vekilov et al. 2002). Another example of aberrant protein assembly is the cellular crystallization of recombinant proteins during gene expression. Despite being first reported in

1996 (Fan et al. 1996), two further occurrences were not found for 14 years. It was thought that the living cell's congested environment would inhibit the formation of properly organized crystals. Furthermore, cellular crystals only have modest diffraction capabilities because of their small size, which was thought to be inevitably constrained by the outside dimensions of the cell (Doye and Poon 2006). This is especially true for X-rays generated by the synchrotron sources that were accessible at the time. Therefore, for many years, cellular crystals were not taken into consideration for structural biology, which prevented a methodical examination of this phenomena. Nowadays, a sizable number of protein microcrystals have been found in living cells. The related structural details were clarified multiple times. It is increasingly acknowledged that crystals produced in cells are good structural biology targets.

It is still unclear, though, if the phenomenon of in-cellular crystallization is limited to a small number of proteins that have evolved to give a particular function through native crystal formation, supplemented by a small number of unintentional crystallization events (Schönherr et al. 2018).

Impact of Crowding on Crystallization Kinetics

As was previously indicated, Ficoll 70, Dextran 70, and sucrose are inert, highly soluble, and have an average weight of about 70 kDa, making them suitable for simulating an intracellular congested environment *in vitro* (Zimmerman and Minton 1993). BSA and ovalbumin were similarly included as crowding agents in the experiment to prevent aggregation during lysozyme refolding in the work published by den Berg et al. (van den Berg et al. 1999). Because of the potential interactions these proteins may have with lysozyme, these chemicals were proposed to monitor any effects.

However, the kind of crowders determines the rate of crystallization kinetics. It was demonstrated that the effects of monomeric and polymeric crowders on crystal solubility and catalytic performance varied. Ferreira et al. (2020) investigated how crowding affected the lysozyme's crystal structure. They discovered that the effects of monomeric and polymeric crowders on the solubility of lysozyme crystals are opposite. Whereas sucrose (monomeric) raises solubility and lowers saturation temperature, Ficoll 70 (polymeric) decreases solubility and raises saturation temperature. Both effects show a linear relationship with crowder concentration. Activity coefficients (γ) calculated from solubility data reveal that $\gamma < 1$ for Ficoll 70 suggests formation of lysozyme aggregates, while $\gamma > 1$ for sucrose indicates volume exclusion effects, with $\gamma < 1$ suggesting protein-small molecule interactions (Minton 1983). However, the same group (Ferreira et al. 2020) proposed that the catalytic efficiency of lysozyme is affected differently by monomeric and polymeric crowders. While sucrose raises the Michaelis constant (*Km*), decreasing enzyme efficiency, Ficoll 70 lowers it. Nevertheless, the catalytic turnover (*V*) of the enzyme is unaffected by either crowder. The enzyme-substrate complex assembly processes

($k1$, k—1) are impacted by these crowding effects, while the turnover rate ($k2$) is unaffected. Alterations in thermodynamic activity, which impact $k1$ but not the other rate constants, are the main factor determining the catalytic efficiency, and variations in viscosity between sucrose and Ficoll-70 cannot account for the ***Km*** alterations (Ferreira et al. 2020).

Ferreira et al. confirmed that neither monomeric (sucrose) nor polymeric (Ficoll 70) crowders significantly alter the structural makeup of lysozyme when examining their effects on crystal structure. With RMSD values (0.17 Å for Ficoll 70 and 0.55 Å for sucrose) well below expected variation limits, X-ray crystallographic examination of lysozyme crystals made with and without crowders reveals few structural changes. The conclusion that these crowding agents have little effect on the globular structure of lysozyme is supported by the fact that, even while the space group changes with sucrose, the lysozyme structure essentially stays the same (Ferreira et al. 2020).

Ferreira et al.'s experimental results support the idea that macromolecular steric repulsion plays a part in protein activity, balancing traditional theories with fresh information from folding stability research. The main macromolecular crowding (MC) effect from 300 g/L Ficoll 70 is found to be volume exclusion, which causes a fixed entropy variation (~1 kcal/mol) (Ferreira et al. 2020). Conditions such as pH and ionic strength affect enthalpy changes, which are more noticeable at high ionic strengths. Volume exclusion is validated by control tests using monomeric additives, which demonstrate conflicting effects on crystal solubility. Any crystallizable protein can be quantified using this crystallization-based technique, which forecasts changes in intermolecular collision rates without compromising structural integrity. These results imply that biological events at low protein levels may be catalyzed by steric repulsions in cells (Ferreira et al. 2020).

Amyloid or amyloid-like fibril protein aggregation has been linked to pathological illnesses such as Alzheimer's disease and Type II diabetes, according to another study by White et al. (2010). In a living cell, however, other physicochemical characteristics regulate protein self-assembly, one of the most interesting of which is the impact of macromolecular crowding as the cellular environment comprises numerous volume-excluding macromolecules such as proteins, polysaccharides, and lipids. The dynamics of all biochemical processes that are essential to the functioning of living systems might be disrupted by nonspecific impacts in addition to interactions within this environment (Hall and Dobson 2006). Numerous factors, such as sequence, monomer conformation, nucleation rate, effective protein concentration, intermolecular interactions, diffusive properties, and the activity of species like chaperone molecules, influence the process of protein aggregation (Pedersen and Otzen 2008; Cellmer et al. 2008). According to White et al., the notion that macromolecular crowding is essential to biological events, especially in living systems, is supported by the capacity to greatly affect reaction kinetics (White et al. 2010). This further emphasizes its significance for protein aggregation, which is consistent with previous research on the topic (Ellis and Minton 2006).

Crowding, LLPS and Protein Crystallization

Process where a homogeneous solution separates into two distinct liquid phases (Uversky 2023) is known as liquid-liquid phase separation (LLPS). This mechanism can be observed in the cell between macromolecules which spontaneously form dense liquid droplets, distinguishing them from the more dilute phase (Fahim et al. 2024). Such interactions allow the formation of cellular compartments which are not enclosed by a lipid membrane so called membrane-less organelles (MLOs) and biomolecular condensates (BMCs) such as P-bodies, stress granules, Cajal bodies, nucleoli, etc. (Uversky 2023). MLOs are found both in the nucleus, as well as in the cytoplasm and exist transiently (Mitrea and Kriwacki 2016). They are composed of proteins containing intrinsically disordered regions (IDRs) or modular domains and nucleic acids (mainly RNA). They are also known to exhibit a variety of functions including gene expression regulation, RNA metabolism, ribosome biogenesis, and stress response (Mitrea and Kriwacki 2016; Berchtold et al. 2018). Any disruption in LLPS and MLOs could lead to cellular disfunction, tumors, or neurodegenerative diseases (Gomes and Shorter 2019; Zheng et al. 2023).

Temperature, pH, and ionic strength as environmental conditions have been shown to affect liquid-liquid phase separation (Le Ferrand et al. 2019; Gomes and Shorter 2019; Mudogo et al. 2020; Uversky 2023). The effect of macromolecular crowding, although yet not fully understood, is recognized as another factor influencing LLPS and in turn MLO formation (Sehgal et al. 2020). Studies focused on researching MLOs and LLPS *in vitro* make use of synthetic crowders to simulate in cell environment. They have shown that crowding can induce or even enhance phase separation through various interactions (Kuznetsova et al. 2015; André and Spruijt 2020).

The inner region of membrane-less organelles (MLOs) is more crowded than their exterior. That is due to the phase separation and subsequent increase in macromolecular concentration which leads to inhomogeneous crowding throughout the cell (André and Spruijt 2020). The increase in local concentration of proteins due to preceding LLPS serves to drive biochemical reactions by providing distinct microenvironments where reactions can take place (Mitrea and Kriwacki 2016). Additionally, by concentrating biomolecules beyond their solubility limits, LLPS can facilitate localized supersaturation which could represent an important event in protein crystallization (Mudogo et al. 2020). In order to lower the energy barrier for nucleation, high protein concentration must be obtained. That makes the crystallization process thermodynamically favorable (Mudogo et al. 2020; Liu et al. 2023). LLPS creates supersaturated regions of proteins inside dense liquid droplets also known as liquid dense clusters (LDCs), which are an important pre-step for the nucleation (Zhang et al. 2013). Unlike the classical nucleation mechanism which involves direct nucleation from a supersaturated solution, a two-step nucleation mechanism is proposed for protein crystallization (Zhang et al. 2013). In this model, a metastable intermediate phase exists between the liquid phase and the final crystals, facilitating the necessary supersaturation for nucleation to occur (Fahim et al.

2024). LLPS metastable intermediate region allows the formation of metastable intermediates, promoting nucleation (Zhang et al. 2013). As previously mentioned, LLPS can be induced by crowding agents promoting protein-protein interactions and simulating the macromolecular crowding typically found in living cells, thus enhancing the nucleation process and crystal formation (Mudogo et al. 2020).

Still, metastable intermediate phase has been shown to not be the only prerequisite for crystallization, and more systematic research is needed, as mechanisms underlaying liquid-liquid phase separation, its influence on protein crystal formation, and how crowding affects them, remain not fully understood (Mudogo et al. 2020).

Naturally Occurring Protein Crystals and Their Functions

Protein crystallization, combined with X-ray diffraction, forms the foundation of X-ray crystallography, a powerful technique for determining the three-dimensional structure of proteins (Dale et al. 2003; Ilari and Savino 2008). With the advancement of X-ray crystallography in pursuit of visualizing the protein structure, crystallization of proteins *in vitro* became one of the most important processes deployed in structural biology (Giegé 2013). Despite the efforts to optimize crystal growth it is still an unpredictable trial-and-error process (Timofeev and Samygina 2023). To crystallize proteins *in vitro*, they must be purified carefully, crystallization conditions must be balanced to achieve crystalline structure (instead of amorphous), and environmental conditions such as pH, temperature, and ionic strength should not disrupt the native state of the proteins (McPherson 1990; Durbin and Feher 1996). Given the challenges of protein crystallization *in vitro*, it is not surprising that naturally occurring protein crystals within cells are rare (Schönherr et al. 2015). While efforts have been made to understand and optimize *in vitro* crystallization, *in vivo* crystals were somewhat disregarded and seen as atypical events associated with pathogenic protein aggregation and disease (Doye and Poon 2006). Cataracts, hemoglobin C disease, and the formation of Charcot-Leyden crystals (CLCs) are notable examples of disorders resulting from abnormal *in vivo* protein crystallization (Pande et al. 2001; Doye and Poon 2006; Persson et al. 2019). Normally, the protein concentration in the cell is kept below the solubility threshold, making the nucleation process thermodynamically unfavorable (McPherson 2004). This may be a consequence of "evolutionary negative design" preventing protein crystallization to increase the viability of cells, and a reason why crystallizing proteins *in vitro* remains challenging (Doye et al. 2004; Doye and Poon 2006). However, the increasing amount of recorded native protein crystals reveals that the process is utilized by the cells if it confers an advantageous function (Schönherr et al. 2015). Thanks to the recent technology advancements and utilization of X-ray free electron laser and micro-focused synchrotron radiation which allowed studying of *in vivo* protein.

Jonathan P.K. Doye and Wilson C.K. Poon (Doye and Poon 2006) and more recently Modogo et al. (2020) provided a compendium of the protein *in vivo* crystals observed in different cell systems. Clearly, numerous examples of protein crystallization *in vivo* represent important illustrations of the diversity of biological self-assembly that ranges from the simple cases of oligomer formation to protein fibrillation, assembly of large proteinaceous machines and viral capsids. Abundance and diversity of reported cases of the protein *in vivo* crystals and a wide spectrum of functional implementations of such protein crystals also indicate that *in cellulo* or *in vivo* formation of protein crystals represents a natural phenomenon (Doye and Poon 2006; Koopmann et al. 2012; Duszenko et al. 2015; Schönherr et al. 2018).

Since the first observation of the naturally occurring protein crystals in the nineteenth century (Giegé 2013) until the present day, native protein crystals have been shown to have a variety of functions, such as storage and secretion, protection and stabilization, and solid-state catalysts (Doye and Poon 2006; Schönherr et al. 2018). In fact, packing proteins into dense crystalline structures represents an efficient way to store them (e.g. storage of proteins in seeds) (Doye and Poon 2006; Nasim and Qureshi 2022). In most non-legume seeds, these crystalline structures represent one of the three major parts of the protein storage vesicles, (i) an amorphous matrix, (ii) 'crystalloids' embedded in the matrix containing proteins packed in lattice structure, and (iii) crystals of the salts of phytic or oxalic acids (Gifford et al. 1982; Collada 1993; Müntz 1998; Jiang et al. 2000). Examples of such *in cellulo* crystallizing plant proteins include hemp plant edestin (Hall 1950) and tobacco seed globin (Crowfoot and Fankuchen 1938).

Another important case of natively crystallizing proteins is given by assembly of specific proteins into secretory granules, which represent an essential means for storing a particular protein in concentrated form ready for release from the cell in response to some external signal, with some such proteins destined for secretion being found in the crystalline form (Arvan and Castle 1998). Environmental triggers can then direct a rapid release of these proteins from the crystal. One of the most prominent example of this process is insulin, granules of which are typically 200–300 nm in size, and contain insulin in a crystal form (Dodson and Steiner 1998; Doye and Poon 2006). Another example is given by secretory granules of white blood cells, granulocytes (Doye and Poon 2006), that store some proteins (e.g., the eosinophil major basic protein (EMBP) (Giembycz and Lindsay 1999) and Charcot–Leyden crystal protein (Leonidas et al. 1995), which can be released to combat infection, in crystal form.

Oviparous animals take advantage of this phenomenon as well by storing yolk proteins in crystals in order to deliver the nutrients to oocytes (Papassideri et al. 2007). Some illustrative examples of the well-established natively crystallizing proteins from this category include vitellin yolk protein crystals found in oocytes of bullfrog (Massover 1971), leopard frogs *Rana pipiens* (Ward 1962), and a yellow fever mosquito *A. aegypti* (Snigirevskaya et al. 1997), as well as lipovitellin of bony fish *Polypterus bichir Amia calva*, and *Lepisosteus osseus* (Lange et al. 1982).

There are multiple cases of crystalline secretory granules in protists (Hausmann 1978). For example, in *Paramecium tetraurelia*, secretory granule known as trichocyst represents a structure that is of the order of 3–4 μm in size and contain at least three different families of closely related polypeptides, each localized to different regions of the trichocyst (Vayssié et al. 2000). Trichocysts, which are insoluble needle-shaped objects left behind by *Paramecium* retreating from certain predators by swimming backwards, are secretory vesicles whose contents have crystalline organization (Bannister 1972; Sperling et al. 1987). Another Ciliate, *Tetrahymena thermophila*, contains many hundreds of needle-like, 0.2 × 1 μm secretory granules, which are docked at the plasma membrane. The content of these granules is released synchronously when cells are massively stimulated. This happened because the cargo of granules is spring-loaded, where the polypeptides that comprise the dense core of these granules are organized as a crystal that undergoes a roughly seven-fold expansion following exocytosis (Turkewitz 2004); (Anderer and Hausmann 1977); (Verbsky and Turkewitz 2024).

Naturally in cell crystallized proteins can be found in bacteria as well. The insecticidal Cry3A toxin produced by *Bacillus thuringiensis* are known to be stored in the bacterial cells in crystalline form. These crystal are naturally produced during sporulation as a mechanism to store the toxin in a concentrated, space-efficient manner (Kalman et al. 1995). In 2014, an X-ray crystal structure of the Cry3A toxin naturally crystallized within the *Bacillus thuringiensis* cells was solved at 2.9 Å resolution by Sawayama et al., thereby representing a unique case of a crystalline protein structure as it exists in a living cell (Sawaya 2014).

Besides being used in storage and secretory and secretory granules, *in cellulo* protein crystallization has several other functional implementations. One of them is given by the utilization of the crystal form of the host-synthesized proteins to generate a special form of encapsulation of some insect viruses (Doye and Poon 2006). Here, during the late stages of infection, cytoplasmic and nuclear polyhedrosis viruses, granulosis viruses, and entomopoxviruses force infected cells to express large quantities of polyhedrin, granulin, and spheroidin, respectively. These host proteins then crystallize around the virions to provide a protective environment for the viruses after death of the insect larvae, thereby helping these viruses to remain intact for periods of years in the soil (Vaughn 1978). Although not completely related, the cell envelope of many prokaryotes use crystalline arrays of S-layer proteins for added stability (Sára and Sleytr 2000).

Natural protein crystals can also be used as plugs for sealing sites of cell damage, as in the case of Woronin body (a hexagonal platelet ~5 μm across consisting of a primarily crystalline HEX-1 surrounded by a membrane) that prevents cytoplasmic bleeding in damaged filamentous fungi (Jedd and Chua 2000) and *Neurospora crassa* (Yuan and Jedd 2003). Another example of the crystalline cellular plugs is given by the phloem sieve elements in plants (Van Bel 2003), where the characteristic P-protein crystalloids found in sieve tubes of legumes (Fabaceae) can undergo a reversible transformation to a dispersed state that can plug the sieve tubes in response to the presence of Ca^{2+} (Knoblauch 2001).

There are also reported cases of using some *in cellulo* crystallized proteins in solid state catalysts. This includes peroxisomes, which are the membrane-bound organelles found in eukaryotic cells that are responsible for breakdown of lipids and alcohol catalyzed by enzymes that are typically assembled into regular crystals (Doye and Poon 2006). Among the crystallized enzymes found within the peroxisomes are catalase from the embryonic leaves (cotyledons) of sunflower (*Helianthus annuus*) (Heinze et al. 2000), urate oxidase from the rat liver (Tsukada et al. 1966), and alcohol oxidase from the methylotrophic yeast *Hansenula polymorpha* (Vonck and Van Bruggen 1992).

In the viviparous cockroach *Diploptera punctata* on the other hand, the nutrients in the form of "milk" crystals are stored in the brood sac. This is used to supply embryos until they develop (Banerjee et al. 2016). Proteins that encapsulate viruses are crystallized during unfavorable environmental conditions ensuring protection and stabilization (Chiu et al. 2012).

All those examples represent only a few of the ones documented. Although the unifying mechanisms behind it are yet to be discovered, *in vivo* protein crystallization may be universal to all species. The recent discovery of Charcot-Leyden crystals (CLCs) as suitable pharmacological targets further emphasizes the importance of focusing future research on *in vivo* protein crystallization (Ueki et al. 2018).

Crystallopathies: Diseases Associated with Protein Crystallization *In Vivo*

No good deed goes unpunished. Similar to all other forms of protein self-assembly, which are known to be good, bad, and ugly, some of the naturally occurring protein crystals were shown to be associated with diseases from the category of "crystallopathies" (Mulay and Anders 2016). Although the detailed consideration of this interesting phenomenon is outside the scopes of this chapter, some illustrative examples of such crystallopathies are briefly considered below.

Eosinophilic inflammation is characterized by the presence of the Charcot-Leyden crystals (CLCs), slender bipyramidal hexagonal crystals (~ 100 μm in maximum length), in tissue and body fluids (Ueki et al. 2019). For the first time, CLCs were described in 1853 by a French neurologist Jean-Martin Charcot (1825–1893) who found tiny crystals in the cardiac blood and spleen of a patient who died from leukemia (Charot 1853), whereas in 1872, a German internist Ernst Viktor von Leyden (1832–1910) described colorless crystals in the sputum of asthma patients (Leyden 1872). It was pointed out that presence of CLCs is associated with lytic eosinophils; i.e., eosinophils that have undergone a process called extracellular DNA trap cell death (EETosis) leading to the release of granules and DNA by eosinophils into the extracellular space (Samter 1947; Persson et al. 2019; Ueki et al. 2018, 2019). EETosis is linked to multiple eosinophilic diseases, such as allergic bronchopulmonary aspergillosis (Muniz et al. 2018); (Omokawa et al. 2018), asthma (Mukherjee et al. 2018; Choi et al. 2018), chronic obstructive pulmonary diseases (Echevarría et al. 2017), eosinophilic chronic rhinosinusitis (chronic rhinosinusitis with nasal polyps (CRSwNP)) (Ueki 2016; Ueki et al. 2016; Persson et al. 2019), eosinophilic cellulitis (Wells syndrome) (Wouters et al. 2015), eosinophilic otitis (Ueki et al. 2016; Ohta et al. 2018), and hypereosinophilic syndrome (Ueki et al. 2013, 2018). CLCs are also found in parasitic infections, such as ascariasis (Pereira et al. 1999), infestation with other helminths, including pork tapeworm and filarial worms (Mehta and Furuta 2015), as well as *Toxocara*, *Capilliria*, or *Fasciola* (Bhatia and Sarin 1994; Arora et al. 1997; Hartleb and Januszewski 2001; Kaplan et al. 2001; Leone et al. 2006; Misra et al. 2009), liver inflammatory pseudotumor or parasitic granuloma (Ji et al. 2000; Ryan et al. 2001), fungal-sensitization-associated asthma (Figueiredo and Neves 2018), and parasitic pneumonia (pulmonary paragonimiasis) (Singh et al. 1986; Cheepsattayakorn and Cheepsattayakorn 2014). Furthermore, it was indicated that CLCs can be found in lymphiod organs, e.g., in the bone marrow of patients with acute myeloid leukemia (Vermeersch et al. 2007; Khrizman et al. 2010; Radujkovic et al. 2011; Manny and Ellis 2012), as well as in stroma of solid tumors (Reichman et al. 2016). CLCs, which are human specific lesions as they have not been found in non-human species, contain crystals of the galectin-10, a galectin from the superfamily of S-type lectins (Leonidas et al. 1995; Acharya and Ackerman 2014; Su 2018). During pathological type 2 immunity in mice also triggers the formation of protein crystals, which instead of galectin-10 contain two closely related proteins, Ym1 (Chil3) and Ym2 (Chil4), which are the members of the family of chitinase-like proteins (CLPs) (Guo et al. 2000; Aegerter et al. 2021).

Spontaneous crystallization of mutant γD crystallin was linked to the development of human genetic cataracts (Pande et al. 2001). Although members of all three families of mammalian lens proteins, α-, β-, and γ-crystallins, are essential for maintaining lens transparency, the specific feature of the γ-crystallins that makes them different from the α- and β-crystallins is the presence of attractive interactions between the γ-crystallins, whereas interactions between β-crystallins are repulsive (Tardieu 1992). The presence of such attractive interactions between the γ-crystallins reduces the osmotic pressure in the lens but also increases the risks of the diminishing the homogeneity of the lens leading to cataract due to the increased susceptibility of γ-crystallins to aggregation and phase separation (Pande et al. 1995; Asherie et al. 1998; Albert et al. 2008). Although these attractive interactions do not affect solubility of the wild types of γ-crystallins that preserve solubility for many years at high concentrations and with little turnover (Albert et al. 2008), familiar mutations in these proteins dramatically affect their solubility. In line with these considerations, the Arg-58 to His (R58H) mutant linked to the aculeiform cataract (Héon et al. 1999) and the Arg-36 to Ser (R36S) mutant linked to another form of congenital cataract (Kmoch et al. 2000) were shown to be much less soluble than the wild-type human γD crystallin protein were shown to be more prone to crystallize than the wild-type protein (Pande et al. 2001). Furthermore, crystals of the R36S mutant were extracted from the eye of a young patient (Kmoch et al. 2000).

Finally, mutated human hemoglobin known as hemoglobin C (HbC), where a single mutation at the sixth amino acid position of the β-subunit replaces the negatively charged glutamic acid with a positively charged lysine (β6 Glu → Lys) (Hunt and Ingram 1958), when in the oxygenated (R) liganded state, is known to form crystals inside red blood cells of patients with both homozygous CC (HbC–HbC) and heterozygous CS (HbC–HbS, with HbS being another Hb variant with the β6

Glu → Val substitution) forms of sickle cell anemia (Hirsch et al. 1985; Lawrence et al. 1991). Systematic analysis of the physical mechanisms behind the HbC crystallization revealed that the high crystallization propensity of the R-state HbC can be attributed to the increased hydrophobicity resulting from the conformational changes that accompany the HbC β6 mutation (Vekilov et al. 2002). The stronger net intermolecular attraction between the liganded HbC in comparison with the liganded HbS and HbA was also reflected in lower HbC concentrations requited for liquid-liquid phase separation nd in the ease of crystal formation at the surface of the liquid-liquid interface (Chen et al. 2004). It was also shown that the HbC crystals grow by the attachment of single molecules to the suitable sites on the crystal surface (Feeling-Taylor et al. 2004).

Recent Advances and Challenges

Novel developments in protein crystallization and macromolecular crowding have greatly improved the knowledge of how proteins behave in biological settings. One important recent advancement was made at the Tokyo Institute of Technology. Researchers introduced a novel cell-free protein crystallization (CFPC) technique that enables the quick generation of high-quality protein crystals. This was obtained from unstable proteins that were previously challenging to study using traditional methods. By combining aspects of in-cell protein crystallization (ICPC) with conventional *in vitro* crystallization, this hybrid technique allows scientists to easily crystallize proteins without requiring laborious purification procedures (Abe et al. 2022).

Apart from CFPC, there has been a growing interest in the natural protein crystallization that happens in live cells. This process—often referred to as in-cell crystallization—has been seen more frequently, as researchers investigate its significance for both native and recombinant proteins. Recent developments in synchrotron radiation and X-ray free-electron lasers have allowed researchers to examine micrometer-sized crystals generated in biological surroundings. This analysis has given researchers new information about the structure and function of proteins (Schönherr et al. 2018). Also, these findings imply that living cells can function as efficient protein crystallization chambers, which gives the opportunity to investigate proteins under conditions that closely resemble their native settings.

Understanding the mechanisms underlying intracellular protein crystallization is still difficult. Some proteins may combine unintentionally, making it difficult to identify their biological significance, even though they seem to have evolved to be optimal for crystal formation. The need for additional research into the processes regulating crystal formation within cells is highlighted by the regulatory differences between native and inadvertently crystallizing proteins (Schönherr et al. 2018). Aggregation can jeopardize protein performance and cell viability; the congested nature of biological settings presents intrinsic challenges for attaining ordered crystal formation. Considering these obstacles, enhanced protein crystallization methods have a wide range of possible uses. Improved knowledge of the connection between protein structure and function, for example, can guide the development of medications and treatment plans for illnesses linked to protein aggregation and misfolding. Additionally, by streamlining the production procedures for therapeutic proteins, improvements in crystallization techniques may make it easier to manufacture pharmacological substances (Sun et al. 2022). It is obvious that these discoveries will be essential to the advancement of structural biology.

Conclusions

The importance of crowding agents in influencing protein structure, folding, and interactions in cellular contexts has been highlighted by research on macromolecular crowding and its impact on protein behavior. By changing the conformational states of proteins and improving their functional capacities, crowding agents have a major impact on their dynamics.

Crowding agents change how proteins interact with one another as well as with individual proteins. This is crucial for understanding biological processes. The higher concentrations of macromolecules in crowded environment can raise the effective concentrations of reactants. That on the other hand, can facilitate enzyme reactions and promote biologically relevant structural changes.

Overall, macromolecular crowding profoundly affects the structure and function of proteins, through the action of different crowding agents, greatly altering their behavior. Further research into these topics will improve our knowledge of cellular biochemistry and could result in novel approaches to biotechnological and therapeutic uses. We can gain a greater understanding of the significance of crowding agents in preserving cellular homeostasis and supporting vital biological functions by concentrating on their function.Compliance with Ethical StandardsAuthors declare no conflict of interest.

Funding source: Not applicable.

References

Abe S, Tanaka J, Kojima M (2022) Cell-free protein crystallization for nanocrystal structure determination | Scientific reports. https://www.nature.com/articles/s41598-022-19681-9. Accessed 9 Dec 2024

Acharya KR, Ackerman SJ (2014) Eosinophil granule proteins: form and function. J Biol Chem 289:17406–17415. https://doi.org/10.1074/jbc.R113.546218

Aegerter H, Smole U, Heyndrickx I et al (2021) Charcot-Leyden crystals and other protein crystals driving type 2 immunity and allergy. Curr Opin Immunol 72:72–78. https://doi.org/10.1016/j.coi.2021.03.013

Albert DM, Miller JW, Azar DT (2008) Albert & Jakobiec's principles and practice of ophthalmology, 3rd [rev.] ed. Saunders Elsevier, Philadelphia

Anderer R, Hausmann K (1977) Properties and structure of isolated extrusive organelles. J Ultrastruct Res 60:21–26. https://doi.org/10.1016/S0022-5320(77)80037-3
André AAM, Spruijt E (2020) Liquid-liquid phase separation in crowded environments. Int J Mol Sci 21:5908. https://doi.org/10.3390/ijms21165908
Arora VK, Singh N, Bhatia A (1997) Charcot-Leyden crystals in fine needle aspiration cytology. Acta Cytol 41:409–412. https://doi.org/10.1159/000332532
Arvan P, Castle D (1998) Sorting and storage during secretory granule biogenesis: looking backward and looking forward. Biochem J 332(Pt 3):593–610. https://doi.org/10.1042/bj3320593
Asherie N, Pande J, Lomakin A et al (1998) Oligomerization and phase separation in globular protein solutions. Biophys Chem 75:213–227. https://doi.org/10.1016/S0301-4622(98)00208-7
Banerjee S, Coussens NP, Gallat F-X et al (2016) Structure of a heterogeneous, glycosylated, lipid-bound, in vivo-grown protein crystal at atomic resolution from the viviparous cockroach Diploptera punctata. IUCrJ 3:282–293. https://doi.org/10.1107/S2052252516008903
Bannister LH (1972) The structure of trichocysts in Paramecium caudatum. J Cell Sci 11:899–929. https://doi.org/10.1242/jcs.11.3.899
Berchtold D, Battich N, Pelkmans L (2018) A systems-level study reveals regulators of membrane-less organelles in human cells. Mol Cell 72:1035–1049.e5. https://doi.org/10.1016/j.molcel.2018.10.036
Berg OG (1990) The influence of macromolecular crowding on thermodynamic activity: solubility and dimerization constants for spherical and dumbbell-shaped molecules in a hard-sphere mixture. Biopolymers 30:1027–1037. https://doi.org/10.1002/bip.360301104
Bhatia V, Sarin SK (1994) Hepatic visceral larva migrans: evolution of the lesion, diagnosis, and role of high-dose albendazole therapy. Am J Gastroenterol 89:624–627
Cellmer T, Henry ER, Hofrichter J, Eaton WA (2008) Measuring internal friction of an ultrafast-folding protein. Proc Natl Acad Sci 105:18320–18325. https://doi.org/10.1073/pnas.0806154105
Charot J (1853) Observation de leucocythemie. CR Soc Biol 5:44–50
Cheepsattayakorn A, Cheepsattayakorn R (2014) Parasitic pneumonia and lung involvement. Biomed Res Int 2014:874021. https://doi.org/10.1155/2014/874021
Chen Q, Vekilov PG, Nagel RL, Hirsch RE (2004) Liquid-liquid phase separation in Hemoglobins: distinct aggregation mechanisms of the β6 mutants. Biophys J 86:1702–1712
Chiu E, Coulibaly F, Metcalf P (2012) Insect virus polyhedra, infectious protein crystals that contain virus particles. Curr Opin Struct Biol 22:234–240. https://doi.org/10.1016/j.sbi.2012.02.003
Choi Y, Le Pham D, Lee D-H et al (2018) Biological function of eosinophil extracellular traps in patients with severe eosinophilic asthma. Exp Mol Med 50:1–8. https://doi.org/10.1038/s12276-018-0136-8
Collada C (1993) Development of protein bodies in cotyledons of Fagus sylvatica. Physiol Plant 89:354–359
Cong Q, Anishchenko I, Ovchinnikov S, Baker D (2019) Protein interaction networks revealed by proteome coevolution. Science 365:185–189. https://doi.org/10.1126/science.aaw6718
Crowfoot D, Fankuchen I (1938) Molecular weight of a tobacco seed globulin. Nature 141:522–523. https://doi.org/10.1038/141522a0
Dale GE, Oefner C, D'Arcy A (2003) The protein as a variable in protein crystallization. J Struct Biol 142:88–97. https://doi.org/10.1016/S1047-8477(03)00041-8
Dodson G, Steiner D (1998) The role of assembly in insulin's biosynthesis. Curr Opin Struct Biol 8:189–194. https://doi.org/10.1016/s0959-440x(98)80037-7
Doye JPK, Poon WCK (2006) Protein crystallization in vivo. Curr Opin Colloid Interface Sci 11:40–46. https://doi.org/10.1016/j.cocis.2005.10.002
Doye JPK, Louis AA, Vendruscolo M (2004) Inhibition of protein crystallization by evolutionary negative design. Phys Biol 1:P9–P13. https://doi.org/10.1088/1478-3967/1/1/P02
Durbin SD, Feher G (1996) PROTEIN CRYSTALLIZATION. Annu Rev Phys Chem 47:171–204. https://doi.org/10.1146/annurev.physchem.47.1.171
Duszenko M, Redecke L, Mudogo CN et al (2015) In vivo protein crystallization in combination with highly brilliant radiation sources offers novel opportunities for the structural analysis

of post-translationally modified eukaryotic proteins. Acta Crystallogr F Struct Biol Commun 71:929–937. https://doi.org/10.1107/S2053230X15011450
Echevarría LU, Leimgruber C, González JG et al (2017) Evidence of eosinophil extracellular trap cell death in COPD: does it represent the trigger that switches on the disease? COPD 12:885–896. https://doi.org/10.2147/COPD.S115969
Ellis RJ, Minton AP (2006) Protein aggregation in crowded environments. Biol Chem 387:485–497. https://doi.org/10.1515/BC.2006.064
Fahim A, Pham J, Thomas S, Annunziata O (2024) Boosting protein crystallization from liquid-liquid phase separation by increasing metastability gap. J Mol Liq 398:124164. https://doi.org/10.1016/j.molliq.2024.124164
Fan GY, Maldonado F, Zhang Y et al (1996) In vivo calcineurin crystals formed using the baculovirus expression system. Microsc Res Tech 34:77–86. https://doi.org/10.1002/(SICI)1097-0029(19960501)34:1<77::AID-JEMT11>3.0.CO;2-M
Feeling-Taylor AR, Yau S-T, Petsev DN et al (2004) Crystallization mechanisms of hemoglobin C in the R state. Biophys J 87:2621–2629. https://doi.org/10.1529/biophysj.104.039743
Ferreira C, Pinto MF, Macedo-Ribeiro S et al (2020) Protein crystals as a key for deciphering macromolecular crowding effects on biological reactions. Phys Chem Chem Phys 22:16143–16149. https://doi.org/10.1039/D0CP02469D
Figueiredo RT, Neves JS (2018) Eosinophils in fungal diseases: an overview. J Leukoc Biol 104:49–60. https://doi.org/10.1002/JLB.4MR1117-473R
Giegé R (2013) A historical perspective on protein crystallization from 1840 to the present day. FEBS J 280:6456–6497. https://doi.org/10.1111/febs.12580
Giembycz MA, Lindsay MA (1999) Pharmacology of the eosinophil. Pharmacol Rev 51:213–340
Gifford DJ, Greenwood JS, Bewley JD (1982) Deposition of matrix and crystalloid storage proteins during protein body development in the endosperm of *Ricinus communis* L. cv. Hale Seeds Plant Physiol 69:1471–1478. https://doi.org/10.1104/pp.69.6.1471
Gomes E, Shorter J (2019) The molecular language of membraneless organelles. J Biol Chem 294:7115–7127. https://doi.org/10.1074/jbc.TM118.001192
Guo L, Johnson RS, Schuh JCL (2000) Biochemical characterization of endogenously formed eosinophilic crystals in the lungs of mice *. J Biol Chem 275:8032–8037. https://doi.org/10.1074/jbc.275.11.8032
Hall CE (1950) ELECTRON MICROSCOPY OF CRYSTALLINE EDESTIN. J Biol Chem 185:45–51. https://doi.org/10.1016/S0021-9258(18)56391-2
Hall D, Dobson CM (2006) Expanding to fill the gap: a possible role for inert biopolymers in regulating the extent of the 'macromolecular crowding' effect. FEBS Lett 580:2584–2590. https://doi.org/10.1016/j.febslet.2006.04.005
Hartleb M, Januszewski K (2001) Severe hepatic involvement in visceral larva migrans. Eur J Gastroenterol Hepatol 13:1245–1249. https://doi.org/10.1097/00042737-200110000-00020
Hausmann K (1978) Extrusive organelles in Protists*. In: Bourne GH, Danielli JF, Jeon KW (eds) International review of cytology. Academic, pp 197–276
Heinze M, Reichelt R, Kleff S, Eising R (2000) High resolution scanning electron microscopy of protein inclusions (cores) purified from peroxisomes of sunflower (Helianthus annuus L.) cotyledons. Cryst Res Technol 35:877–886. https://doi.org/10.1002/1521-4079(200007)35:6/7<877::AID-CRAT877>3.0.CO;2-S
Héon E, Priston M, Schorderet DF et al (1999) The γ-Crystallins and human cataracts: a puzzle made clearer. Am J Hum Genet 65:1261–1267
Hirsch RE, Raventos-Suarez C, Olson JA, Nagel RL (1985) Ligand state of Intraerythrocytic circulating HbC crystals in homozygote CC patients. Blood 66:775–777. https://doi.org/10.1182/blood.V66.4.775.775
Hunt JA, Ingram VM (1958) Allelomorphism and the chemical differences of the human Hæmoglobins A, S and C. Nature 181:1062–1063. https://doi.org/10.1038/1811062a0
Ilari A, Savino C (2008) Protein structure determination by X-ray crystallography. In: Keith JM (ed) Bioinformatics: data, sequence analysis and evolution. Humana Press, Totowa, pp 63–87
Jedd G, Chua NH (2000) A new self-assembled peroxisomal vesicle required for efficient resealing of the plasma membrane. Nat Cell Biol 2:226–231. https://doi.org/10.1038/35008652

Ji X-L, Shen M-S, Yin T (2000) Liver inflammatory pseudotumor or parasitic granuloma? World J Gastroenterol 6:458–460. https://doi.org/10.3748/wjg.v6.i3.458

Jiang L, Phillips TE, Rogers SW, Rogers JC (2000) Biogenesis of the protein storage vacuole crystalloid. J Cell Biol 150:755–770. https://doi.org/10.1083/jcb.150.4.755

Kalman S, Kiehne KL, Cooper N et al (1995) Enhanced production of insecticidal proteins in bacillus thuringiensis strains carrying an additional crystal protein gene in their chromosomes. Appl Environ Microbiol 61:3063–3068

Kaplan KJ, Goodman ZD, Ishak KG (2001) Eosinophilic granuloma of the liver: a characteristic lesion with relationship to visceral larva migrans. Am J Surg Pathol 25:1316–1321. https://doi.org/10.1097/00000478-200110000-00014

Khrizman P, Altman JK, Mohtashamian A et al (2010) Charcot–Leyden crystals associated with acute myeloid leukemia: Case report and literature review. Leuk Res 34:e336–e338. https://doi.org/10.1016/j.leukres.2010.08.014

Kmoch S, Brynda J, Asfaw B et al (2000) Link between a novel human γD-crystallin allele and a unique cataract phenotype explained by protein crystallography. Hum Mol Genet 9:1779–1786. https://doi.org/10.1093/hmg/9.12.1779

Knoblauch (2001) Reversible calcium-regulated stopcocks in legume sieve tubes [W] | The Plant Cell | Oxford Academic. https://academic.oup.com/plcell/article-abstract/13/5/1221/6009440?redirectedFrom=fulltext. Accessed 25 Dec 2024

Koopmann R, Cupelli K, Redecke L et al (2012) In vivo protein crystallization opens new routes in structural biology. Nat Methods 9:259–262. https://doi.org/10.1038/nmeth.1859

Kuznetsova IM, Zaslavsky BY, Breydo L et al (2015) Beyond the excluded volume effects: mechanistic complexity of the crowded milieu. Molecules 20:1377–1409. https://doi.org/10.3390/molecules20011377

Lange RH, Grodziński Z, Kilarski W (1982) Yolk-platelet crystals in three ancient bony fishes: Polypterus bichir (polypteri), Amia calva L., and Lepisosteus osseus (L.) (Holostei). Cell Tissue Res 222:159–165. https://doi.org/10.1007/BF00218296

Lawrence C, Fabry ME, Nagel RL (1991) The unique red cell heterogeneity of SC disease: crystal formation, dense reticulocytes, and unusual morphology. Blood 78:2104–2112. https://doi.org/10.1182/blood.V78.8.2104.2104

Le Ferrand H, Duchamp M, Gabryelczyk B et al (2019) Time-resolved observations of liquid–liquid phase separation at the nanoscale using in situ liquid transmission electron microscopy. J Am Chem Soc 141:7202–7210. https://doi.org/10.1021/jacs.9b03083

Leone N, Baronio M, Todros L et al (2006) Hepatic involvement in larva migrans of Toxocara canis: report of a case with pathological and radiological findings. Dig Liver Dis 38:511–514. https://doi.org/10.1016/j.dld.2005.07.007

Leonidas DD, Elbert BL, Zhou Z et al (1995) Crystal structure of human Charcot–Leyden crystal protein, an eosinophil lysophospholipase, identifies it as a new member of the carbohydrate-binding family of galectins. Structure 3:1379–1393. https://doi.org/10.1016/S0969-2126(01)00275-1

Leslie M (2021) Separation anxiety. Science 371:336–338. https://doi.org/10.1126/science.371.6527.336

Leyden E (1872) Zur Kenntniss des Bronchial-Asthma. Archiv f pathol Anat 54:324–344. https://doi.org/10.1007/BF01997025

Liu H, Zhao Y, Sun J (2023) Heterogeneous nucleation in protein crystallization. Biomimetics 8:68. https://doi.org/10.3390/biomimetics8010068

Manny JS, Ellis LR (2012) Acute myeloid leukemia with Charcot-Leyden crystals. Blood 120:503. https://doi.org/10.1182/blood-2011-11-392969

Massover WH (1971) Intramitochondrial yolk-crystals of frog oocytes: I. Formation of yolk-crystal inclusions by mitochondria during bullfrog oogenesis. J Cell Biol 48:266–279. https://doi.org/10.1083/jcb.48.2.266

Mcpherson A (1990) Current approaches to macromolecular crystallization. Eur J Biochem 189:1–23. https://doi.org/10.1111/j.1432-1033.1990.tb15454.x

McPherson A (2004) Introduction to protein crystallization. Methods 34:254–265. https://doi.org/10.1016/j.ymeth.2004.03.019
Mehta P, Furuta GT (2015) Eosinophils in gastrointestinal disorders: eosinophilic gastrointestinal diseases, celiac disease, inflammatory bowel diseases, and parasitic infections. Immunol Allergy Clin N Am 35:413–437. https://doi.org/10.1016/j.iac.2015.04.003
Minton AP (1983) The effect of volume occupancy upon the thermodynamic activity of proteins: some biochemical consequences. Mol Cell Biochem 55:119–140. https://doi.org/10.1007/BF00673707
Misra V, Debnath S, Misra SP, Singh PA (2009) Significance of Charcot Leyden crystals in hepatic aspirates. J Cytol 26:77–79. https://doi.org/10.4103/0970-9371.55227
Mitrea DM, Kriwacki RW (2016) Phase separation in biology; functional organization of a higher order. Cell Commun Signal 14:1. https://doi.org/10.1186/s12964-015-0125-7
Mudogo CN, Falke S, Brognaro H et al (2020) Protein phase separation and determinants of in cell crystallization. Traffic 21:220–230. https://doi.org/10.1111/tra.12711
Mukherjee M, Bulir DC, Radford K et al (2018) Sputum autoantibodies in patients with severe eosinophilic asthma. J Allergy Clin Immunol 141:1269–1279. https://doi.org/10.1016/j.jaci.2017.06.033
Mulay SR, Anders H-J (2016) Crystallopathies. N Engl J Med 374:2465–2476. https://doi.org/10.1056/NEJMra1601611
Muniz VS, Silva JC, Braga YAV et al (2018) Eosinophils release extracellular DNA traps in response to aspergillus fumigatus. J Allergy Clin Immunol 141:571–585.e7. https://doi.org/10.1016/j.jaci.2017.07.048
Müntz K (1998) Deposition of storage proteins. Plant Mol Biol 38:77–99. https://doi.org/10.1023/A:1006020208380
Nasim F, Qureshi IA (2022) Chapter 22 - role of structural biology methods in drug discovery. In: Tripathi T, Dubey VK (eds) Advances in protein molecular and structural biology methods. Academic, pp 357–371
Ohta N, Ueki S, Konno Y et al (2018) ETosis-derived DNA trap production in middle ear effusion is a common feature of eosinophilic otitis media. Allergol Int 67:414–416. https://doi.org/10.1016/j.alit.2017.11.007
Omokawa A, Ueki S, Kikuchi Y et al (2018) Mucus plugging in allergic bronchopulmonary aspergillosis: implication of the eosinophil DNA traps. Allergol Int 67:280–282. https://doi.org/10.1016/j.alit.2017.08.002
Pande J, Lomakin A, Fine B et al (1995) Oxidation of gamma II-crystallin solutions yields dimers with a high phase separation temperature. Proc Natl Acad Sci 92:1067–1071. https://doi.org/10.1073/pnas.92.4.1067
Pande A, Pande J, Asherie N et al (2001) Crystal cataracts: human genetic cataract caused by protein crystallization. Proc Natl Acad Sci 98:6116–6120. https://doi.org/10.1073/pnas.101124798
Papassideri IS, Trougakos IP, Leonard KR, Margaritis LH (2007) Crystalline yolk spheroids in Drosophila melanogaster oocyte: freeze fracture and two-dimensional reconstruction analysis. J Insect Physiol 53:370–376. https://doi.org/10.1016/j.jinsphys.2006.12.011
Pedersen JS, Otzen DE (2008) Amyloid-a state in many guises: survival of the fittest fibril fold. Protein Sci 17:2–10. https://doi.org/10.1110/ps.073127808
Pereira FEL, Musso C, Castelo JS (1999) Pathology of pyogenic liver abscess in children. Pediatr Dev Pathol 2:537–543. https://doi.org/10.1007/s100249900159
Persson EK, Verstraete K, Heyndrickx I et al (2019) Protein crystallization promotes type 2 immunity and is reversible by antibody treatment. Science 364:eaaw4295. https://doi.org/10.1126/science.aaw4295
Pittas T, Zuo W, Boersma AJ (2023) Cell wall damage increases macromolecular crowding effects in the Escherichia coli cytoplasm. iScience 26:106367. https://doi.org/10.1016/j.isci.2023.106367
Radujkovic A, Bellos F, Andrulis M et al (2011) Charcot-Leyden crystals and bone marrow necrosis in acute myeloid leukemia. Eur J Haematol 86:451–452. https://doi.org/10.1111/j.1600-0609.2011.01594.x

Reichman H, Karo-Atar D, Munitz A (2016) Emerging roles for eosinophils in the tumor microenvironment. Trends Cancer 2:664–675. https://doi.org/10.1016/j.trecan.2016.10.002
Rivas G, Minton AP (2016) Macromolecular crowding in vitro, in vivo, and in between. Trends Biochem Sci 41:970–981. https://doi.org/10.1016/j.tibs.2016.08.013
Ryan RS, Al-Hashimi H, Lee MJ (2001) Hepatic abscesses in elderly patients mimicking metastatic disease. Ir J Med Sci 170:251–253. https://doi.org/10.1007/BF03167790
Sahni N, Yi S, Taipale M et al (2015) Widespread macromolecular interaction perturbations in human genetic disorders. Cell 161:647–660. https://doi.org/10.1016/j.cell.2015.04.013
Samter M (1947) Charcot-Leyden crystals: a study of the conditions necessary for their formation. J Allergy 18:221–230. https://doi.org/10.1016/0021-8707(47)90045-2
Sára M, Sleytr UB (2000) S-Layer Proteins. https://journals.asm.org/doi/epub/10.1128/jb.182.4.859-868.2000. Accessed 25 Dec 2024
Sawaya MR (2014) Protein crystal structure obtained at 2.9 Å resolution from injecting bacterial cells into an X-ray free-electron laser beam. https://www.pnas.org/doi/epdf/10.1073/pnas.1413456111. Accessed 25 Dec 2024
Schönherr R, Klinge M, Rudolph JM et al (2015) Real-time investigation of dynamic protein crystallization in living cells. Struct Dyn 2:041712. https://doi.org/10.1063/1.4921591
Schönherr R, Rudolph JM, Redecke L (2018) Protein crystallization in living cells. Biol Chem 399:751–772. https://doi.org/10.1515/hsz-2018-0158
Sehgal PB, Westley J, Lerea KM et al (2020) Biomolecular condensates in cell biology and virology: phase-separated membraneless organelles (MLOs). Anal Biochem 597:113691. https://doi.org/10.1016/j.ab.2020.113691
Singh TS, Mutum SS, Razaque MA (1986) Pulmonary paragonimiasis: clinical features, diagnosis and treatment of 39 cases in Manipur. Trans R Soc Trop Med Hyg 80:967–971. https://doi.org/10.1016/0035-9203(86)90275-0
Snigirevskaya ES, Hays AR, Raikhel AS (1997) Secretory and internalization pathways of mosquito yolk protein precursors. Cell Tissue Res 290:129–142. https://doi.org/10.1007/s004410050915
Speer SL, Stewart CJ, Sapir L et al (2022) Macromolecular crowding is more than hard-core repulsions. Annu Rev Biophys 51:267–300. https://doi.org/10.1146/annurev-biophys-091321-071829
Sperling L, Tardieu A, Gulik-Krzywicki T (1987) The crystal lattice of paramecium trichocysts before and after exocytosis by X-ray diffraction and freeze-fracture electron microscopy. J Cell Biol 105:1649–1662. https://doi.org/10.1083/jcb.105.4.1649
Stadmiller SS, Pielak GJ (2021) Protein-complex stability in cells and in vitro under crowded conditions. Curr Opin Struct Biol 66:183–192. https://doi.org/10.1016/j.sbi.2020.10.024
Su J (2018) A brief history of Charcot-Leyden crystal protein/Galectin-10 research. Molecules 23:2931. https://doi.org/10.3390/molecules23112931
Sun M, Yang H, Miao X et al (2022) Macromolecular protein crystallisation with biotemplate of live cells. Sci Rep 12:3005. https://doi.org/10.1038/s41598-022-06999-7
Tardieu A (1992) Protein interactions in the calf eye lens: interactions between β-crystallins are repulsive whereas in γ-crystallins they are attractive. Eur Biophys J. https://link.springer.com/article/10.1007/BF00195438. Accessed 25 Dec 2024
Timofeev V, Samygina V (2023) Protein crystallography: achievements and challenges. Crystals 13:71. https://doi.org/10.3390/cryst13010071
Tsukada H, Mochizuki Y, Fujiwara S (1966) The nucleoids of rat liver cell microbodies: Fine structure and enzymes. J Cell Biol 28:449–460. https://doi.org/10.1083/jcb.28.3.449
Turkewitz AP (2004) Out with a bang! Tetrahymena as a model system to study secretory granule biogenesis. Traffic 5:63–68. https://doi.org/10.1046/j.1600-0854.2003.00155.x
Ueki S (2016) Eosinophil ETosis and DNA traps: a new look at eosinophilic inflammation | Current allergy and asthma reports. https://link.springer.com/article/10.1007/s11882-016-0634-5. Accessed 25 Dec 2024
Ueki S, Melo RCN, Ghiran I et al (2013) Eosinophil extracellular DNA trap cell death mediates lytic release of free secretion-competent eosinophil granules in humans. Blood 121:2074–2083. https://doi.org/10.1182/blood-2012-05-432088

Ueki S, Konno Y, Takeda M et al (2016) Eosinophil extracellular trap cell death–derived DNA traps: their presence in secretions and functional attributes. J Allergy Clin Immunol 137:258–267. https://doi.org/10.1016/j.jaci.2015.04.041
Ueki S, Tokunaga T, Melo RCN et al (2018) Charcot-Leyden crystal formation is closely associated with eosinophil extracellular trap cell death. Blood 132:2183–2187. https://doi.org/10.1182/blood-2018-04-842260
Ueki S, Miyabe Y, Yamamoto Y et al (2019) Charcot-Leyden crystals in eosinophilic inflammation: active cytolysis leads to crystal formation. Curr Allergy Asthma Rep 19:35. https://doi.org/10.1007/s11882-019-0868-0
Uversky VN (ed) (2023) Biological liquid-liquid phase separation, biomolecular condensates, and Membraneless organelles. MDPI – Multidisciplinary Digital Publishing Institute
Van Bel AJE (2003) The phloem, a miracle of ingenuity. Plant Cell Environ 26:125–149. https://doi.org/10.1046/j.1365-3040.2003.00963.x
van den Berg B, Ellis RJ, Dobson CM (1999) Effects of macromolecular crowding on protein folding and aggregation. EMBO J 18:6927–6933. https://doi.org/10.1093/emboj/18.24.6927
Vaughn JL (1978) Virus-insect relationships. Bull Entomol Soc Am 24:441. https://doi.org/10.1093/besa/24.4.441
Vayssié L, Skouri F, Sperling L, Cohen J (2000) Molecular genetics of regulated secretion in paramecium. Biochimie 82:269–288. https://doi.org/10.1016/S0300-9084(00)00201-7
Vekilov PG, Feeling-Taylor AR, Petsev DN et al (2002) Intermolecular interactions, nucleation, and thermodynamics of crystallization of hemoglobin C. Biophys J 83:1147–1156. https://doi.org/10.1016/S0006-3495(02)75238-7
Verbsky JW, Turkewitz AP (2024) Proteolytic processing and Ca2+-binding activity of dense-core vesicle polypeptides in Tetrahymena. https://www.molbiolcell.org/doi/epdf/10.1091/mbc.9.2.497. Accessed 25 Dec 2024
Vermeersch P, Zachee P, Brusselmans C (2007) Acute myeloid leukemia with bone marrow necrosis and Charcot Leyden crystals. Am J Hematol 82:1029–1029. https://doi.org/10.1002/ajh.20907
Vonck J, van Bruggen EFJ (1992) Architecture of peroxisomal alcohol oxidase crystals from the methylotrophic yeast Hansenula polymorpha as deduced by electron microscopy. https://journals.asm.org/doi/epdf/10.1128/jb.174.16.5391-5399.1992. Accessed 25 Dec 2024
Ward RT (1962) The origin of protein and fatty yolk in Rana Pipiens: II. Electron microscopical and cytochemical observations of young and mature oocytes. J Cell Biol 14:309–341. https://doi.org/10.1083/jcb.14.2.309
White DA, Buell AK, Knowles TPJ et al (2010) Protein aggregation in crowded environments. J Am Chem Soc 132:5170–5175. https://doi.org/10.1021/ja909997e
Wouters J, Waelkens E, Vandoninck S et al (2015) Mass spectrometry of flame figures. Acta Derm Venereol 95:734–735. https://doi.org/10.2340/00015555-2050
Yuan P, Jedd G (2003) A HEX-1 crystal lattice required for Woronin body function in Neurospora crassa. Nat Struct Mol Biol. https://www.nature.com/articles/nsb910. Accessed 25 Dec 2024
Zhang F, Roosen-Runge F, Sauter A et al (2013) The role of cluster formation and metastable liquid—liquid phase separation in protein crystallization. Faraday Discuss 159:313–325. https://doi.org/10.1039/C2FD20021J
Zheng L-W, Liu C-C, Yu K-D (2023) Phase separations in oncogenesis, tumor progressions and metastasis: a glance from hallmarks of cancer. J Hematol Oncol 16:123. https://doi.org/10.1186/s13045-023-01522-5
Zimmerman SB, Minton AP (1993) Macromolecular crowding: biochemical, biophysical, and physiological consequences. Annu Rev Biophys Biomol Struct 22:27–65. https://doi.org/10.1146/annurev.bb.22.060193.000331

Chapter 19
Macromolecular Crowding for Reparative Medicine and Drug Discovery Applications

Giulia Giuffredi, Andrea Rossoni, Laura Trujillo Cubillo, and Dimitrios I. Zeugolis

Abstract Cell-based reparative medicine and drug discovery utilise the innate sophistication of cells to build supramolecular assemblies. Unfortunately, bereft of their optimal tissue context, during *in vitro* culture, this ability is compromised. Herein, we argue that macromolecular crowding is the missing link in the development of tissue facsimiles for reparative medicine and drug discovery purposes.

Keywords Macromolecular crowding · Extracellular matrix · Tissue engineering · Reparative medicine · Drug discovery · *In vitro* disease models

Introduction

Cell-based reparative medicine and drug discovery harness the innate ability of cells to generate tissues (Blache et al. 2020). In reparative medicine, the clinical use of scaffold and scaffold-free cell-based therapies has demonstrated safety, efficacy, and efficiency for a wide spectrum of applications (e.g. reparative medicine of cardiovascular (Miyagawa et al. 2017; Kainuma et al. 2021), orthopaedics (Sato et al. 2019; Hamahashi et al. 2022), and soft (da Cruz et al. 2018; Green et al. 1979) tissues). The secreted extracellular matrix (ECM), acting as a supportive shield for the transplanted cells and a reservoir for trophic / bioactive factors that promote tissue repair and regeneration, is responsible for the efficacy of tissue engineered cell

Authors Giulia Giuffredi, Andrea Rossoni, Laura Trujillo Cubillo have equally contributed to this chapter.

G. Giuffredi · A. Rossoni · L. T. Cubillo · D. I. Zeugolis (✉)
Regenerative, Modular & Developmental Engineering Laboratory (REMODEL), Charles Institute of Dermatology, Conway Institute of Biomolecular & Biomedical Research and School of Mechanical & Materials Engineering, University College Dublin (UCD), Dublin, Ireland
e-mail: dimitrios.zevgolis@ucd.ie

V. N. Uversky (ed.), *(Macro)Molecular Crowding*, Subcellular Biochemistry 109,
https://doi.org/10.1007/978-3-032-03370-3_19

therapies (Daneshmandi et al. 2020; Li et al. 2019; Ferreira et al. 2018). In drug discovery, cell-based disease models have been shown to effectively mimic the degenerative mechanisms of several diseases (e.g. neurodegenerative diseases (Zhao et al. 2020), fibrosis (Xu et al. 2020), and cancer (Ji et al. 2020)), offering not only greater fidelity to human pathophysiology, but also cost advantages over animal models (Lin and Sheltzer 2020; Mak et al. 2014). In this regard, ECM allows for the development of *in vitro* pathophysiology models that more accurately capture disease initiation and progression, as well as proportional to the *in vivo* setting pharmacokinetics, toxicity and effectiveness (Kai et al. 2019; Cox 2021; Sleeboom et al. 2024; Bansaccal et al. 2023). Regrettably, long cultivation times are needed to create ECM-rich constructs *in vitro* (up to 196 days for a tissue engineered blood vessel (L'Heureux et al. 2006) and up to 56 days for the development of a systemic sclerosis fibrosis model (Corriveau et al. 2009)), leading to phenotypic drift, cell senescence, and significant costs tied to production (Darling and Athanasiou 2005; Cigognini et al. 2013; Hunsberger et al. 2018; Wiese et al. 2019).

Macromolecular crowding (MMC) is a biophysical phenomenon that is founded on the excluded volume effect theory, according to which two molecules cannot occupy the same space at the same time (Ellis 2001; Minton 2005). Although several studies have investigated the role of MMC in, for example, RNA formation, dynamics, and functionality (Zacco et al. 2024), DNA compaction and extension (Collette et al. 2023), and protein aggregation (Siddiqui and Naeem 2023), it is still largely under-investigated in eukaryotic cell cultures. Collagen biosynthesis can be used as indicative example (Fig. 19.1). Here, in traditional eukaryotic cell cultures, when the collagen biosynthesis is conducted in the absence of MMC agents, the proteinases and the procollagen are diffused in the large available volume, resulting in the slow procollagen conversion to collagen (Miyahara et al. 1982; Goldberg and Sherr 1973; Limeback and Sodek 1979). The introduction of MMC agents into the

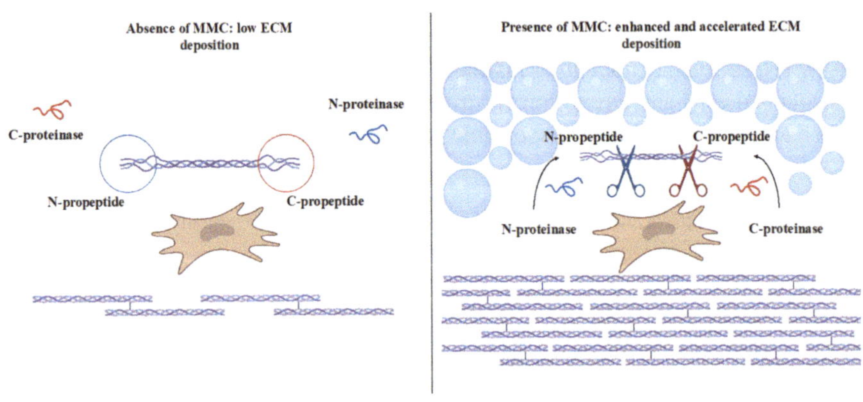

Fig. 19.1 In the absence of MMC (left panel), the procollagen and the proteinases are diffused in the large available volume of the culture medium, leading to limited collagen deposition (left). In the presence of MMC (right panel), the diffusion of procollagen and proteinases is restricted and their interaction is promoted, leading to enhanced and accelerated collagen deposition

culture medium induces volume exclusion, which reduces diffusion and promotes molecular interactions (Kuznetsova et al. 2015), facilitating the cleavage of the non-helical ends of procollagen by the respective proteinases and consequently accelerates collagen deposition (Lareu et al. 2007). Herein, we investigate the use of MMC in eukaryotic cell culture to create ECM-rich tissue surrogates for reparative medicine and drug discovery purposes.

Macromolecular Crowding in Eukaryotic Cell Culture

Physiologically, intracellular and extracellular biological and biochemical reactions occur under highly crowded conditions (Kuznetsova et al. 2014), whilst typical cell culture is deprived of macromolecules (Tsiapalis and Zeugolis 2021). This specific physiological gap MMC aspires to bridge in eukaryotic cell culture through the addition of high molecular weight macromolecules (either through the addition of single (Garnica-Galvez et al. 2021) or mixed (Prewitz et al. 2015) macromolecular species). The effectiveness of the process depends on the physical properties of the MMC agents used (e.g. negative charge and polydispersity, i.e., the heterogeneity of sizes of molecules or particles in a mixture). For example, negatively charged MMC agents have been shown to favour protein assemblies by promoting stronger electrostatic repulsion (Monterroso et al. 2016). Polydisperse MMC agents, due to their size heterogeneity, maximise the excluded volume effect by more effectively occupying available volume (Phillip and Schreiber 2013). In a similar manner, the negatively charged and highly polydispersed carrageenan has been shown to induce the highest ECM deposition in the shortest period of time (Satyam et al. 2014). Subsequent studies demonstrated that artificial polydispersity (i.e. appropriate cocktails of different concentrations and molecular weights dextran sulphate or Ficoll®) can induce similar to carrageenan ECM deposition (Gaspar et al. 2019). It should be noted that although MMC agents are frequently considered as inert macromolecules (Hall and Dobson 2006), in eukaryotic cell culture, the chemistry of the MMC agents has been shown to affect cell phenotype. For example, the sulphated polysaccharide carrageenan has been shown to induce chondrogenesis and osteogenesis in mesenchymal stromal cell cultures (Cigognini et al. 2016; De Pieri et al. 2020; Graceffa and Zeugolis 2019). Interestingly, the sulphated polysaccharide dextran sulphate has been shown to induce osteogenic differentiation of mesenchymal stromal cells (Wan et al. 2022), but to also promote the trans-differentiation of human corneal fibroblasts towards myofibroblast lineage, whilst carrageenan maintained their native phenotype (Kumar et al. 2015). One should note that the non-sulphated polysaccharide hyaluronic acid has also been shown to increase osteogenesis and chondrogenesis (Garnica-Galvez et al. 2023), whilst Ficoll® (a neutral, highly branched synthetic polysaccharide formed by the co-polymerisation of sucrose and epichlorohydrin) has been shown to increase adipogenesis (Ang et al. 2014). For reasons still unclear, MMC agents have been shown to differentially affect cell and orientation of the deposited ECM. Indeed, Ficoll® has been shown to promote actin cytoskeleton and ECM alignment in mesenchymal stromal cell (Zeiger et al. 2012) and corneal fibroblast (Gurdal and Zeugolis 2024) cultures, whilst carrageenan has been shown to induce globular ECM deposition in both permanently differentiated (Ryan et al. 2023a; Ryan et al. 2021) and mesenchymal stromal (Ryan et al. 2023b) cell cultures, even when anisotropic substrates were used. Evidently, MMC in eukaryotic cell culture is a lot more than volume exclusion.

Macromolecular Crowding in Reparative Medicine

The ability of MMC to increase and accelerate ECM deposition has the potential to transform reparative medicine therapies, either in the form of tissue engineered medicines or in the form of decellularised cell derived matrices that can be used for a diverse range of clinical indications (Table 19.1). To put this into perspective, one should consider that MMC has been used effectively (as judged by up to 120-fold increase in the ECM deposition without affecting basic cellular functions) in a wide range of cell types (e.g. dermal fibroblasts (Djalali-Cuevas et al. 2024; Guillaumin et al. 2024), tenocytes (Djalali-Cuevas et al. 2024; Tsiapalis et al. 2021), mesenchymal stromal cells (Cigognini et al. 2016; Garnica-Galvez et al. 2023)). Further, MMC allows for the accelerated development of three-dimensional tissue moduli using only a fraction of cells that traditional approaches use. Indeed, in the absence

Table 19.1 Indicative examples of application of MMC for the development of scaffold-based and scaffold-free tissue surrogates for reparative medicine applications

Clinical indication	Strategy and outcome
Wound healing	Human adipose-derived stromal cells were cultured for 10 days with 75 μg/ml carrageenan. In preclinical setting, the developed tissue engineered medicine resulted in accelerated wound closure, collagen-rich neotissue formation and reduced scar area (De Pieri et al. 2022)
Corneal healing	Human corneal fibroblasts were cultured for 6 days with 100 μg/ml dextran sulphate or 75 μg/ml carrageenan. The use of carrageenan increased tissue-specific ECM deposition and did not influence gene / protein expression i*n vitro,* whilst the addition of dextran sulphate induced myofibroblast differentiation (Kumar et al. 2015)
Bone	Human dermal fibroblasts were cultured for 6 days with fucoidan (1, 10, 50, 100, 250 μg/ml) or dextran sulphate (100 μg/ml) and decellularised. CDMs were seeded with bone marrow mesenchymal stem cells and cultured for 21 days. Fucoidan increased ECM deposition compared to dextran sulphate and control, CDMs obtained with fucoidan promoted bone marrow mesenchymal stem cells osteogenic differentiation (Yoo et al. 2022)
Kidney	Porcine kidney-derived ECM were reconstituted under the presence / absence of Ficoll® 400 kDa (6, 12, 25 mg/ml). ECM reconstitution under Ficoll® increased fibril thickness in a concentration dependent manner; reconstituted scaffolds showed improved *in vitro* morphogenesis of human umbilical vein endothelial cells and murine kidney stem cells (Magno et al. 2017)

of MMC, the production of micro-tissues with 10–15 μm thickness required 28 days and 612,000 cells/cm^2 (Yamamoto et al. 2017; Masumoto et al. 2014; Takagi et al. 2010; Sekine et al. 2011; Okura et al. 2010; Sekine et al. 2013; Ishigami et al. 2018; Sumide et al. 2006), whilst when using carrageenan as MMC agent, 340 μm thick micro-tissues were produced after 10 days in culture using only 50,000 cells/cm^2 (De Pieri et al. 2022). Decellularised cell-derived matrices, due to their multifunctionality, have the potential to transform the cell culture technologies and reparative medicine sectors (Fitzpatrick and McDevitt 2015; Assunção et al. 2020; Chan et al. 2021). Further, in comparison to the tissue engineered medicines, decellularised cell derived matrices, being freeze-dried, have reduced requirements for storage and shipping. In cell culture technologies sector, MMC-produced decellularised cell-derived matrices have been shown to allow for stable cell propagation over lengthy culture periods / cell passages (Satyam et al. 2020; Peng et al. 2012), opening new horizons in *ex vivo* cell culture. In reparative medicine sector, decellularised mesenchymal stromal cell derived matrices that were produced under MMC conditions (carrageenan was used as MMC agent) were shown to have increased retained sulphated glycosaminoglycan content, which in turn improved retention of the endogenous or adsorption of the exogenous bone morphogenic protein 2, leading to increased mineralised tissue formation in an ectopic bone formation model (Fok et al. 2023). Similarly, when Ficoll® 400 kDa was used in the development of decellularised mesenchymal stromal cell derived spheroid matrices, an increased retention of the vascular endothelial growth factor was observed that resulted in an increased intrinsic revascularisation in a subcutaneous nude mouse model (Chiang et al. 2021). Despite these promising preliminary results, more detailed and clinical indication-specific studies are required to evaluate the potential of MMC utilization in the reparative medicine.

Macromolecular Crowding in Drug Discovery

Drug discovery is a long and costly (10–15 years and US$ 1–2 billion, respectively, per molecule) process that is more often associated with failure than success, considering that 90% of molecules that would enter into the clinical trial phase will fail during phase I, II, or III (Sun et al. 2022; Dowden and Munro 2019). This extremely high number of ineffective drugs being tested in clinical trials highlights a critical issue in the drug development pipeline, namely inadequate *in vitro*, *ex vivo*, and *in vivo* models that fail to imitate the targeted diseased tissue composition and accurately predict a molecule's performance with respect to efficacy, efficiency, specificity, and/or localised or systemic toxicity (Cruz-Acuna et al. 2021; Khalil et al. 2020; Yadav et al. 2021) Although the role of ECM in disease manifestation, progression and establishment has been well-documented for a range of pathophysiologies (Theocharis et al. 2019; Chakravarti et al. 2022), traditional two-dimensional and three-dimensional *in vitro* models are ECM-poor. Again, this technological gap MMC aspires to bridge by creating *in vitro* disease models that will not only imitate disease progression, but also accurately predict in *in vitro* setting the *in vivo* performance of the molecular therapies under investigation. Considering that the primary outcome of incorporating MMC into cell culture is the increased collagen deposition, it is understandable that MMC has been used extensively to develop organotypic models for fibrosis and cancer (Table 19.2). In the context of fibrosis, the addition of 500 kDa dextran sulphate or a cocktail containing 70 kDa Ficoll® and 400 kDa Ficoll® in lung fibroblasts with TGFβ1 generated, almost fifteen years ago, the 'Scar-in-a-jar' fibrosis model (Chen et al. 2009), which not only imitated the high ECM content that is encountered in fibrotic conditions, but also outperformed in sensitivity the state-of-the-art fibroplasia models, using less cells (50,000 cells per well, as opposed to 100,000 cells per well) and shorter period of time (2–6 days, as opposed to 8–14 days). Since then, several modifications of the model have been proposed, and numerous molecules have been assessed (Coentro et al. 2021; Puerta Cavanzo et al. 2021; Alghamdi et al. 2020; Rønnow et al. 2020; Fan et al. 2019; Coentro et al. 2022; Good et al. 2019). It is worth noting that the model has been even used to elucidate the antifibrotic mechanism of molecules in clinical trials (Decato et al. 2022). In oncological context, decellularised cell derived matrices produced under MMC conditions not only closely imitated the composition and structure of cancerous tissues, but also when they were recellularised with cancer cell lines, a significantly increased doxorubicin resistance was observed, as compared to the two-dimensional non-MMC and three-dimensional scaffold-based control groups (Rubi-Sans et al. 2021; Shologu et al. 2022). Although the *in vivo* predictability of MMC-based *in vitro* models has yet to be assessed, these preliminary data are very positive indeed, and we anticipate this technology to be soon expanded into other pathophysiologies.

Conclusions

In eukaryotic cell culture, MMC dramatically increases and accelerates ECM deposition, in a MMC agent / cocktail physicochemical properties (e.g. negative charge, polydispersity index) dependent manner. This increased and enhanced ECM deposition has the potential to revolutionise reparative medicine and drug discovery, as multifunctional healthy and diseased, respectively, tissue facsimiles can now be realised using only a fraction of the cells and time that traditional approaches require. Considering the very short period of time required for the development of such tissue surrogates, we anticipate that MMC will also be used in personalised medicine applications. Although we are at early stages of understanding and implementing MMC in eukaryotic cell culture, the preliminary data are positive. We anticipate, as our knowledge expands, MMC will become the new *status quo* in the field.

Acknowledgements This work has received funding from the European Research Council (ERC) under the European Union's Horizon 2020 research and innovation programme, grant agreement

Table 19.2 Indicative examples of MMC application in the development of *in vitro* pathophysiology models for drug discovery purposes

Pathophysiology	Strategy and outcome
Breast cancer	Human adult mammary fibroblasts, human adult dermal fibroblasts and human adult lung fibroblasts were cultured for 10 days using carrageenan (100 μg/ml) to obtain cell-derived matrices. Following decellularisation with different techniques (NP40, NH_4OH, DOC and SDS-EDTA) and lyophilisation, the decellularised matrices were recellularised with a triple-negative breast cancer cell line (MDA-MB-231) and treated with doxorubicin (100 nm). The MDA-MB-231 seeded on human adult mammary fibroblasts matrices decellularised using NP40 showed a significant increase in chemoresistance against doxorubicin compared to the monolayer culture and non-crowded control (Shologu et al. 2022)
Colorectal cancer	Human adipose derived stem cells were cultured for 10 days on poly-L-lactic acid microcarriers and a cocktail of Ficoll® 70 (37.5 mg/ml) and Ficoll® 400 (25 mg/ml). The cell-derived matrix was decellularised and the microcarriers were removed using 0.5% Triton 100X in 25 mM NH_4OH, 30 μg/ml DNAse I and 0.5 M Na_3PO_4. Two colorectal cancer cell lines (HT29 and SW480) were seeded on top of the decellularised matrices and cultured for 7 days before treatment with doxorubicin (0, 0.5, 1, 5 μM). The cancer cells exhibited a significant increase in chemoresistance against doxorubicin compared to the monolayer culture and culture in collagen gels (Rubi-Sans et al. 2021)
Clubfoot fibrosis	Following MMC optimisation using different cocktails of Ficoll® 70 kDa and 400 kDa or polyvinylpyrrolidone 40 kDa, primary human fibroblasts from the medial and lateral side of clubfoot were cultured for 7 days with a cocktail of 37.50 mg/ml Ficoll® 70 kDa and 25 mg/ml Ficoll® 400 kDa and with Minoxidil, a lysis hydroxylase inhibitor, at concentrations of 0.25, 0.5, 0.75 mM. Minoxidil treatment at 0.25, 0.5, 0.75 mM inhibited ECM deposition and downregulated lysyl hydroxylase encoding genes PLOD1, PLOD2 and PLOD3 (Doubková et al. 2024)
Skin fibrosis	Primary human adult fibroblasts were cultured for 4 days using a cocktail of Ficoll® 70 (18.75 mg/ml) and Ficoll® 400 (12.5 mg/ml), polyvinylpyrrolidone 40 kDa (21.5 mg/ml), polyvinylpyrrolidone 360 kDa (11.34 mg/ml), neutral dextran 670 (100 μg/ml), dextran sulphate 500 (100 μg/ml) and carrageenan (100 μg/ml) in normal (10% FBS) and starved (0.5% FBS) conditions to screen for the optimal cell culture condition that induced the maximum ECM deposition without affecting the normal fibrotic response. Subsequently, the primary human fibroblasts were cultured for 4 days in starved with the optimal crowder, polyvinylpyrrolidone 40 kDa, and TGFβ1 to induce fibrosis, plus 4 different antifibrotic drugs: Galunisertib (0.5, 1, 2.5 μM), Omipalisib (0.01, 0.1, 1 μM), Imatinib (0.1, 0.5, 1 μM) and Nintenadib (0.1, 0.5, 1 μM). The results obtained are in line with published data obtained without MMC (Puerta Cavanzo et al. 2021)

No. 866126 and Science Foundation Ireland, grant agreement No. 19/FFP/6982. The authors would like to acknowledge the significant contribution of Dr. Oonagh Dwane (University of Galway, Ireland) in the writing of all grants.

Author Contribution Giulia Giuffredi, Andrea Rossoni and Laura Trujillo Cubillo wrote the first draft. Dimitrios I. Zeugolis edited the manuscript. All authors have approved the final version of the manuscript.

Conflicts of Interest The authors declare no conflicts of interest.

Artificial Intelligence Statement Artificial intelligence was not used.

References

Alghamdi M, Al-Eitan L, Stevenson A, Chaudhari N, Hortin N, Wallace H, Danielsen P, Manzur M, Wood F, Fear M (2020) Secreted factors from keloid keratinocytes modulate collagen deposition by fibroblasts from normal and fibrotic tissue: A pilot study. Biomedicine 8(7):200

Ang X, Lee M, Blocki A, Chen C, Ong L, Asada H, Sheppard A, Raghunath M (2014) Macromolecular crowding amplifies adipogenesis of human bone marrow-derived mesenchymal stem cells by enhancing the pro-adipogenic microenvironment. Tissue Eng Part A 20(5–6):966–981

Assunção M, Dehghan-Baniani D, Yiu C, Später T, Beyer S, Blocki A (2020) Cell-derived extracellular matrix for tissue engineering and regenerative medicine. Front Bioeng Biotechnol 8:602009

Bansaccal N, Vieugue P, Sarate R, Song Y, Minguijon E, Miroshnikova YA, Zeuschner D, Collin A, Allard J, Engelman D (2023) The extracellular matrix dictates regional competence for tumour initiation. Nature 623(7988):828–835

Blache U, Stevens MM, Gentleman E (2020) Harnessing the secreted extracellular matrix to engineer tissues. Nat Biomed Eng 4(4):357–363. https://doi.org/10.1038/s41551-019-0500-6

Chakravarti S, Enzo E, Monteiro R, de Barros M, Maffezzoni MBR, Pellegrini G (2022) Genetic disorders of the extracellular matrix: From cell and gene therapy to future applications in regenerative medicine. Annu Rev Genomics Hum Genet 23:193–222. https://doi.org/10.1146/annurev-genom-083117-021702

Chan W, Yu F, Le Q, Chen S, Yee M, Choudhury D (2021) Towards biomanufacturing of cell-derived matrices. Int J Mol Sci 22(21):11929

Chen CZ, Peng YX, Wang ZB, Fish PV, Kaar JL, Koepsel RR, Russell AJ, Lareu RR, Raghunath M (2009) The Scar-in-a-Jar: Studying potential antifibrotic compounds from the epigenetic to extracellular level in a single well. Br J Pharmacol 158(5):1196–1209. https://doi.org/10.1111/j.1476-5381.2009.00387.x

Chiang CE, Fang YQ, Ho CT, Assunção M, Lin SJ, Wang YC, Blocki A, Huang CC (2021) Bioactive decellularized extracellular matrix derived from 3D stem cell spheroids under macromolecular crowding serves as a scaffold for tissue engineering. Adv Healthc Mater 10(11):2100024

Cigognini D, Gaspar D, Kumar P, Satyam A, Alagesan S, Sanz-Nogues C, Griffin M, O'Brien T, Pandit A, Zeugolis DI (2016) Macromolecular crowding meets oxygen tension in human mesenchymal stem cell culture – a step closer to physiologically relevant in vitro organogenesis. Sci Rep 6:30746. https://doi.org/10.1038/srep30746

Cigognini D, Lomas A, Kumar P, Satyam A, English A, Azeem A, Pandit A, Zeugolis D (2013) Engineering in vitro microenvironments for cell based therapies and drug discovery. Drug Discov Today 18(21–22):1099–1108. https://doi.org/10.1016/j.drudis.2013.06.007

Coentro J, di Nubila A, May U, Prince S, Zwaagstra J, Järvinen T, Zeugolis D (2022) Dual drug delivery collagen vehicles for modulation of skin fibrosisin vitro. Biomed Mater 17(2):025017

Coentro JQ, May U, Prince S, Zwaagstra J, Ritvos O, Järvinen TA, Zeugolis DI (2021) Adapting the Scar-in-a-Jar to skin fibrosis and screening traditional and contemporary anti-fibrotic therapies. Front Bioeng Biotechnol 9:756399

Collette D, Dunlap D, Finzi L (2023) Macromolecular crowding and DNA: Bridging the gap between in vitro and in vivo. Int J Mol Sci 24(24):17502. https://doi.org/10.3390/ijms242417502

Corriveau MP, Boufaied I, Lessard J, Chabaud S, Senecal JL, Grodzicky T, Chartier S, Raymond Y, Moulin VJ (2009) The fibrotic phenotype of systemic sclerosis fibroblasts varies with disease duration and severity of skin involvement: Reconstitution of skin fibrosis development using a tissue engineering approach. J Pathol 217(4):534–542. https://doi.org/10.1002/path.2482

Cox TR (2021) The matrix in cancer. Nat Rev Cancer 21(4):217–238

Cruz-Acuna R, Vunjak-Novakovic G, Burdick JA, Rustgi AK (2021) Emerging technologies provide insights on cancer extracellular matrix biology and therapeutics. iScience 24(5):102475. https://doi.org/10.1016/j.isci.2021.102475

da Cruz L, Fynes K, Georgiadis O, Kerby J, Luo YH, Ahmado A, Vernon A, Daniels JT, Nommiste B, Hasan SM, Gooljar SB, Carr AF, Vugler A, Ramsden CM, Bictash M, Fenster M, Steer J, Harbinson T, Wilbrey A, Tufail A, Feng G, Whitlock M, Robson AG, Holder GE, Sagoo MS, Loudon PT, Whiting P, Coffey PJ (2018) Phase 1 clinical study of an embryonic stem cell-derived retinal pigment epithelium patch in age-related macular degeneration. Nat Biotechnol 36(4):328–337. https://doi.org/10.1038/nbt.4114

Daneshmandi L, Shah S, Jafari T, Bhattacharjee M, Momah D, Saveh-Shemshaki N, Lo KW, Laurencin CT (2020) Emergence of the stem cell secretome in regenerative engineering. Trends Biotechnol 38(12):1373–1384. https://doi.org/10.1016/j.tibtech.2020.04.013

Darling EM, Athanasiou KA (2005) Rapid phenotypic changes in passaged articular chondrocyte subpopulations. J Orthop Res 23(2):425–432. https://doi.org/10.1016/j.orthres.2004.08.008

De Pieri A, Korntner SH, Capella-Monsonis H, Tsiapalis D, Kostjuk SV, Churbanov S, Timashev P, Gorelov A, Rochev Y, Zeugolis DI (2022) Macromolecular crowding transforms regenerative medicine by enabling the accelerated development of functional and truly three-dimensional cell assembled micro tissues. Biomaterials 287:121674

De Pieri A, Rana S, Korntner S, Zeugolis DI (2020) Seaweed polysaccharides as macromolecular crowding agents. Int J Biol Macromol 164:434–446. https://doi.org/10.1016/j.ijbiomac.2020.07.087

Decato B, Leeming D, Sand J, Fischer A, Du S, Palmer S, Karsdal M, Luo Y, Minnich A (2022) LPA1 antagonist BMS-986020 changes collagen dynamics and exerts antifibrotic effects in vitro and in patients with idiopathic pulmonary fibrosis. Respir Res 23(1):61

Djalali-Cuevas A, Rettel M, Stein F, Savitski M, Kearns S, Kelly J, Biggs M, Skoufos I, Tzora A, Prassinos N (2024) Macromolecular crowding in human tenocyte and skin fibroblast cultures: A comparative analysis. Materials today Bio 25:100977

Doubková M, Knitlová J, Vondrášek D, Eckhardt A, Novotný Ts, Ošt'ádal M, Filová E, Bačáková L (2024) Harnessing the biomimetic effect of macromolecular crowding in the cell-derived model of clubfoot fibrosis. Biomacromolecules 25 (10):6485–6502

Dowden H, Munro J (2019) Trends in clinical success rates and therapeutic focus. Nat Rev Drug Discov 18(7):495–496. https://doi.org/10.1038/d41573-019-00074-z

Ellis RJ (2001) Macromolecular crowding: Obvious but underappreciated. Trends Biochem Sci 26(10):597–604. https://doi.org/10.1016/s0968-0004(01)01938-7

Fan C, Lim L, Loh S, Ying Lim K, Upton Z, Leavesley D (2019) Application of "macromolecular crowding" in vitro to investigate the naphthoquinones shikonin, naphthazarin and related analogues for the treatment of dermal scars. Chem Biol Interact 310:108747

Ferreira JR, Teixeira GQ, Santos SG, Barbosa MA, Almeida-Porada G, Goncalves RM (2018) Mesenchymal stromal cell secretome: Influencing therapeutic potential by cellular preconditioning. Front Immunol 9:2837. https://doi.org/10.3389/fimmu.2018.02837

Fitzpatrick L, McDevitt T (2015) Cell-derived matrices for tissue engineering and regenerative medicine applications. Biomater Sci 3(1):12–24

Fok SW, Gresham RC, Ryan W, Osipov B, Bahney C, Leach JK (2023) Macromolecular crowding and decellularization method increase the growth factor binding potential of cell-secreted extracellular matrices. Front Bioeng Biotechnol 11:1091157

Garnica-Galvez S, Korntner S, Skoufos I, Tzora A, Diakakis N, Prassinos N, Zeugolis D (2021) Hyaluronic acid as macromolecular crowder in equine adipose-derived stem cell cultures. Cells 10(4):859

Garnica-Galvez S, Skoufos I, Tzora A, Diakakis N, Prassinos N, Zeugolis DI (2023) Macromolecular crowding in equine bone marrow mesenchymal stromal cell cultures using single and double hyaluronic acid macromolecules. Acta Biomater 170:111–123

Gaspar D, Fuller KP, Zeugolis DI (2019) Polydispersity and negative charge are key modulators of extracellular matrix deposition under macromolecular crowding conditions. Acta Biomater 88:197–210. https://doi.org/10.1016/j.actbio.2019.02.050

Goldberg B, Sherr CJ (1973) Secretion and extracellular processing of procollagen by cultured human fibroblasts. Proc Natl Acad Sci USA 70(2):361–365. https://doi.org/10.1073/pnas.70.2.361

Good R, Eley J, Gower E, Butt G, Blanchard A, Fisher A, Nanthakumar C (2019) A high content, phenotypic 'scar-in-a-jar' assay for rapid quantification of collagen fibrillogenesis using disease-derived pulmonary fibroblasts. BMC Biomed Eng 1:14

Graceffa V, Zeugolis D (2019) Carrageenan enhances chondrogenesis and osteogenesis in human bone marrow stem cell culture. Eur Cell Mater 37:310–332

Green H, Kehinde O, Thomas J (1979) Growth of cultured human epidermal cells into multiple epithelia suitable for grafting. Proc Natl Acad Sci USA 76(11):5665–5668. https://doi.org/10.1073/pnas.76.11.5665

Guillaumin S, Gurdal M, Zeugolis DI (2024) Gums as macromolecular crowding agents in human skin fibroblast cultures. Life 14(4):435

Gurdal M, Zeugolis D (2024) Macromolecular crowding agent dependent extracellular matrix deposition and growth factor retention in human corneal fibroblast cultures. Exp Eye Res 250:110162

Hall D, Dobson CM (2006) Expanding to fill the gap: A possible role for inert biopolymers in regulating the extent of the 'macromolecular crowding'effect. FEBS Lett 580(11):2584–2590

Hamahashi K, Toyoda E, Ishihara M, Mitani G, Takagaki T, Kaneshiro N, Maehara M, Takahashi T, Okada E, Watanabe A, Nakamura Y, Kato R, Matoba R, Takagi T, Akutsu H, Umezawa A, Kobayashi H, Akamatsu T, Yamato M, Okano T, Watanabe M, Sato M (2022) Polydactyly-derived allogeneic chondrocyte cell-sheet transplantation with high tibial osteotomy as regenerative therapy for knee osteoarthritis. NPJ Regen Med 7(1):71. https://doi.org/10.1038/s41536-022-00272-1

Hunsberger JG, Shupe T, Atala A (2018) An industry-driven roadmap for manufacturing in regenerative medicine. Stem Cells Transl Med 7(8):564–568. https://doi.org/10.1002/sctm.18-0060

Ishigami M, Masumoto H, Ikuno T, Aoki T, Kawatou M, Minakata K, Ikeda T, Sakata R, Yamashita JK, Minatoya K (2018) Human iPS cell-derived cardiac tissue sheets for functional restoration of infarcted porcine hearts. PLoS One 13(8):e0201650

Ji W, Zhang W, Wang X, Shi Y, Yang F, Xie H, Zhou W, Wang S, Guan X (2020) c-myc regulates the sensitivity of breast cancer cells to palbociclib via c-myc/miR-29b-3p/CDK6 axis. Cell Death Dis 11(9):760. https://doi.org/10.1038/s41419-020-02980-2

Kai F, Drain AP, Weaver VM (2019) The extracellular matrix modulates the metastatic journey. Dev Cell 49(3):332–346. https://doi.org/10.1016/j.devcel.2019.03.026

Kainuma S, Miyagawa S, Toda K, Yoshikawa Y, Hata H, Yoshioka D, Kawamura T, Kawamura A, Kashiyama N, Ito Y, Iseoka H, Ueno T, Kuratani T, Nakamoto K, Sera F, Ohtani T, Yamada T, Sakata Y, Sawa Y (2021) Long-term outcomes of autologous skeletal myoblast cell-sheet transplantation for end-stage ischemic cardiomyopathy. Mol Ther 29(4):1425–1438. https://doi.org/10.1016/j.ymthe.2021.01.004

Khalil AS, Jaenisch R, Mooney DJ (2020) Engineered tissues and strategies to overcome challenges in drug development. Adv Drug Deliv Rev 158:116–139. https://doi.org/10.1016/j.addr.2020.09.012

Kumar P, Satyam A, Fan X, Rochev Y, Rodriguez BJ, Gorelov A, Joshi L, Raghunath M, Pandit A, Zeugolis DI (2015) Accelerated development of supramolecular corneal stromal-like assemblies from corneal fibroblasts in the presence of macromolecular crowders. Tissue Eng Part C Methods 21(7):660–670

Kuznetsova IM, Turoverov KK, Uversky VN (2014) What macromolecular crowding can do to a protein. Int J Mol Sci 15(12):23090–23140. https://doi.org/10.3390/ijms151223090
Kuznetsova IM, Zaslavsky BY, Breydo L, Turoverov KK, Uversky VN (2015) Beyond the excluded volume effects: Mechanistic complexity of the crowded milieu. Molecules 20(1):1377–1409. https://doi.org/10.3390/molecules20011377
L'Heureux N, Dusserre N, Konig G, Victor B, Keire P, Wight TN, Chronos NA, Kyles AE, Gregory CR, Hoyt G, Robbins RC, McAllister TN (2006) Human tissue-engineered blood vessels for adult arterial revascularization. Nat Med 12(3):361–365. https://doi.org/10.1038/nm1364
Lareu R, Subramhanya K, Peng Y, Benny P, Chen C, Wang Z, Rajagopalan R, Raghunath M (2007) Collagen matrix deposition is dramatically enhanced in vitro when crowded with charged macromolecules: The biological relevance of the excluded volume effect. FEBS Lett 581(14):2709–2714
Li M, Ma J, Gao Y, Yang L (2019) Cell sheet technology: A promising strategy in regenerative medicine. Cytotherapy 21(1):3–16. https://doi.org/10.1016/j.jcyt.2018.10.013
Limeback HF, Sodek J (1979) Procollagen synthesis and processing in periodontal ligament in vivo and in vitro. A comparative study using slab-gel fluorography. Eur J Biochem 100 (2):541–550. https://doi.org/10.1111/j.1432-1033.1979.tb04200.x
Lin A, Sheltzer JM (2020) Discovering and validating cancer genetic dependencies: Approaches and pitfalls. Nat Rev Genet 21(11):671–682. https://doi.org/10.1038/s41576-020-0247-7
Magno V, Friedrichs J, Weber HM, Prewitz MC, Tsurkan MV, Werner C (2017) Macromolecular crowding for tailoring tissue-derived fibrillated matrices. Acta Biomater 55:109–119
Mak IW, Evaniew N, Ghert M (2014) Lost in translation: Animal models and clinical trials in cancer treatment. Am J Transl Res 6(2):114–118
Masumoto H, Ikuno T, Takeda M, Fukushima H, Marui A, Katayama S, Shimizu T, Ikeda T, Okano T, Sakata R (2014) Human iPS cell-engineered cardiac tissue sheets with cardiomyocytes and vascular cells for cardiac regeneration. Sci Rep 4(1):6716
Minton AP (2005) Influence of macromolecular crowding upon the stability and state of association of proteins: Predictions and observations. J Pharm Sci 94(8):1668–1675. https://doi.org/10.1002/jps.20417
Miyagawa S, Domae K, Yoshikawa Y, Fukushima S, Nakamura T, Saito A, Sakata Y, Hamada S, Toda K, Pak K, Takeuchi M, Sawa Y (2017) Phase I clinical trial of autologous stem cell-sheet transplantation therapy for treating cardiomyopathy. J Am Heart Assoc 6(4):e003918. https://doi.org/10.1161/JAHA.116.003918
Miyahara M, Njieha FK, Prockop DJ (1982) Formation of collagen fibrils in vitro by cleavage of procollagen with procollagen proteinases. J Biol Chem 257(14).8442–8448
Monterroso B, Reija B, Jiménez M, Zorrilla S, Rivas G (2016) Charged molecules modulate the volume exclusion effects exerted by crowders on FtsZ polymerization. PLoS One 11(2):e0149060
Okura H, Matsuyama A, Lee C-M, Saga A, Kakuta-Yamamoto A, Nagao A, Sougawa N, Sekiya N, Takekita K, Shudo Y (2010) Cardiomyoblast-like cells differentiated from human adipose tissue-derived mesenchymal stem cells improve left ventricular dysfunction and survival in a rat myocardial infarction model. Tissue Eng Part C Methods 16(3):417–425
Peng Y, Bocker M, Holm J, Toh W, Hughes C, Kidwai F, Lajoie G, Cao T, Lyko F, Raghunath M (2012) Human fibroblast matrices bio-assembled under macromolecular crowding support stable propagation of human embryonic stem cells. J Tissue Eng Regen Med 6(10):e74–e86
Phillip Y, Schreiber G (2013) Formation of protein complexes in crowded environments - From in vitro to in vivo. FEBS Lett 587(8):1046–1052
Prewitz MC, Stissel A, Friedrichs J, Traber N, Vogler S, Bornhauser M, Werner C (2015) Extracellular matrix deposition of bone marrow stroma enhanced by macromolecular crowding. Biomaterials 73:60–69. https://doi.org/10.1016/j.biomaterials.2015.09.014
Puerta Cavanzo N, Bigaeva E, Boersema M, Olinga P, Bank RA (2021) Macromolecular crowding as a tool to screen anti-fibrotic drugs: The Scar-in-a-Jar system revisited. Front Med 7:615774
Rønnow S, Dabbagh R, Genovese F, Nanthakumar C, Barrett V, Good R, Brockbank S, Cruwys S, Jessen H, Sorensen G, Karsdal M, Leeming D, Sand J (2020) Prolonged Scar-in-a-Jar: An

in vitro screening tool for anti-fibrotic therapies using biomarkers of extracellular matrix synthesis. Respir Res 21(1):108

Rubi-Sans G, Nyga A, Rebollo E, Perez-Amodio S, Otero J, Navajas D, Mateos-Timoneda MA, Engel E (2021) Development of cell-derived matrices for three-dimensional in vitro cancer cell models. ACS Appl Mater Interfaces 13(37):44108–44123. https://doi.org/10.1021/acsami.1c13630

Ryan C, Pugliese E, Shologu N, Gaspar D, Rooney P, Islam M, O'Riordan A, Biggs M, Griffin M, Zeugolis D (2021) A combined physicochemical approach towards human tenocyte phenotype maintenance. Mater Today Bio 12:100130

Ryan CNM, Pugliese E, Shologu N, Gaspar D, Rooney P, Islam MN, O'Riordan A, Biggs MJ, Griffin MD, Zeugolis DI (2023a) Physicochemical cues are not potent regulators of human dermal fibroblast trans-differentiation. Biomater Biosyst 11:100079. https://doi.org/10.1016/j.bbiosy.2023.100079

Ryan CNM, Pugliese E, Shologu N, Gaspar D, Rooney P, Islam MN, O'Riordan A, Biggs MJ, Griffin MD, Zeugolis DI (2023b) The synergistic effect of physicochemical in vitro microenvironment modulators in human bone marrow stem cell cultures. Biomater Adv 144:213196. https://doi.org/10.1016/j.bioadv.2022.213196

Sato M, Yamato M, Mitani G, Takagaki T, Hamahashi K, Nakamura Y, Ishihara M, Matoba R, Kobayashi H, Okano T, Mochida J, Watanabe M (2019) Combined surgery and chondrocyte cell-sheet transplantation improves clinical and structural outcomes in knee osteoarthritis. NPJ Regen Med 4:4. https://doi.org/10.1038/s41536-019-0069-4

Satyam A, Kumar P, Fan X, Gorelov A, Rochev Y, Joshi L, Peinado H, Lyden D, Thomas B, Rodriguez B, Raghunath M, Pandit A, Zeugolis D (2014) Macromolecular crowding meets tissue engineering by self-assembly: A paradigm shift in regenerative medicine. Adv Mater 26(19):3024–3034. https://doi.org/10.1002/adma.201304428

Satyam A, Tsokos MG, Tresback JS, Zeugolis DI, Tsokos GC (2020) Cell-derived extracellular matrix-rich biomimetic substrate supports podocyte proliferation, differentiation, and maintenance of native phenotype. Adv Funct Mater 30(44):1908752

Sekine W, Haraguchi Y, Shimizu T, Umezawa A, Okano T (2011) Thickness limitation and cell viability of multi-layered cell sheets and overcoming the diffusion limit by a porous-membrane culture insert. J Biochip Tissue Chip 1:007

Sekine W, Haraguchi Y, Shimizu T, Yamato M, Umezawa A (2013) Okano T (2013) Chondrocyte differentiation of human endometrial gland-derived MSCs in layered cell sheets. ScientificWorldJournal 1:359109

Shologu N, Gurdal M, Szegezdi E, FitzGerald U, Zeugolis DI (2022) Macromolecular crowding in the development of a three-dimensional organotypic human breast cancer model. Biomaterials 287:121642. https://doi.org/10.1016/j.biomaterials.2022.121642

Siddiqui GA, Naeem A (2023) Connecting the dots: Macromolecular crowding and protein aggregation. J Fluoresc 33(1):1–11. https://doi.org/10.1007/s10895-022-03082-2

Sleeboom JJ, van Tienderen GS, Schenke-Layland K, van der Laan LJ, Khalil AA, Verstegen MM (2024) The extracellular matrix as hallmark of cancer and metastasis: From biomechanics to therapeutic targets. Sci Transl Med 16 (728):eadg3840

Sumide T, Nishida K, Yamato M, Ide T, Hayashida Y, Watanabe K, Yang J, Kohno C, Kikuchi A, Maeda N (2006) Functional human corneal endothelial cell sheets harvested from temperature-responsive culture surfaces. FASEB J 20(2):392–394

Sun D, Gao W, Hu H, Zhou S (2022) Why 90% of clinical drug development fails and how to improve it? Acta Pharm Sin B 12(7):3049–3062. https://doi.org/10.1016/j.apsb.2022.02.002

Takagi R, Murakami D, Kondo M, Ohki T, Sasaki R, Mizutani M, Yamato M, Nishida K, Namiki H, Yamamoto M (2010) Fabrication of human oral mucosal epithelial cell sheets for treatment of esophageal ulceration by endoscopic submucosal dissection. Gastrointest Endosc 72(6):1253–1259

Theocharis AD, Manou D, Karamanos NK (2019) The extracellular matrix as a multitasking player in disease. FEBS J 286(15):2830–2869. https://doi.org/10.1111/febs.14818

Tsiapalis D, Kearns S, Kelly JL, Zeugolis DI (2021) Growth factor and macromolecular crowding supplementation in human tenocyte culture. Biomater Biosyst 1:100009

Tsiapalis D, Zeugolis DI (2021) It is time to crowd your cell culture media - Physicochemical considerations with biological consequences. Biomaterials 275:120943. https://doi.org/10.1016/j.biomaterials.2021.120943

Wan HY, Shin RLY, Chen JCH, Assuncao M, Wang D, Nilsson SK, Tuan RS, Blocki A (2022) Dextran sulfate-amplified extracellular matrix deposition promotes osteogenic differentiation of mesenchymal stem cells. Acta Biomater 140:163–177. https://doi.org/10.1016/j.actbio.2021.11.049

Wiese DM, Ruttan CC, Wood CA, Ford BN, Braid LR (2019) Accumulating transcriptome drift precedes cell aging in human umbilical cord-derived mesenchymal stromal cells serially cultured to replicative senescence. Stem Cells Transl Med 8(9):945–958. https://doi.org/10.1002/sctm.18-0246

Xu H, Zhao Q, Song N, Yan Z, Lin R, Wu S, Jiang L, Hong S, Xie J, Zhou H, Wang R, Jiang X (2020) AdipoR1/AdipoR2 dual agonist recovers nonalcoholic steatohepatitis and related fibrosis via endoplasmic reticulum-mitochondria axis. Nat Commun 11(1):5807. https://doi.org/10.1038/s41467-020-19668-y

Yadav J, El Hassani M, Sodhi J, Lauschke VM, Hartman JH, Russell LE (2021) Recent developments in in vitro and in vivo models for improved translation of preclinical pharmacokinetics and pharmacodynamics data. Drug Metab Rev 53(2):207–233. https://doi.org/10.1080/03602532.2021.1922435

Yamamoto K, Yamato M, Morino T, Sugiyama H, Takagi R, Yaguchi Y, Okano T, Kojima H (2017) Middle ear mucosal regeneration by tissue-engineered cell sheet transplantation. NPJ Regen Med 2(1):6

Yoo Y-I, Ko K-W, Cha S-G, Park S-Y, Woo J, Han DK (2022) Highly effective induction of cell-derived extracellular matrix by macromolecular crowding for osteogenic differentiation of mesenchymal stem cells. J Ind Eng Chem 107:391–400

Zacco E, Broglia L, Kurihara M, Monti M, Gustincich S, Pastore A, Plath K, Nagakawa S, Cerase A, Sanchez de Groot N, Tartaglia GG (2024) RNA: The unsuspected conductor in the orchestra of macromolecular crowding. Chem Rev 124(8):4734–4777. https://doi.org/10.1021/acs.chemrev.3c00575

Zeiger AS, Loe FC, Li R, Raghunath M, Van Vliet KJ (2012) Macromolecular crowding directs extracellular matrix organization and mesenchymal stem cell behavior. PLoS One 7(5):e37904

Zhao J, Fu Y, Yamazaki Y, Ren Y, Davis MD, Liu CC, Lu W, Wang X, Chen K, Cherukuri Y, Jia L, Martens YA, Job L, Shue F, Nguyen TT, Younkin SG, Graff-Radford NR, Wszolek ZK, Brafman DA, Asmann YW, Ertekin-Taner N, Kanekiyo T, Bu G (2020) APOE4 exacerbates synapse loss and neurodegeneration in Alzheimer's disease patient iPSC-derived cerebral organoids. Nat Commun 11(1):5540. https://doi.org/10.1038/s41467-020-19264-0

Chapter 20
Simulating Crowding *In Vitro*: Not an Elusive Goal Any Longer

Piero Andrea Temussi and Matteo Levantino

Abstract A few decades ago, it became clear that most biophysical studies were performed on proteins under conditions not corresponding to those found in vivo, thus limiting the generality of at least some of the conclusions. The main difference was identified in the absence, in the dilute solutions typically used, of large amounts of other macromolecules. Solution conditions similar to those found in vivo were dubbed as crowded. The main consequences of crowding were identified as the influence of crowding on protein stability and on alterations of reaction conditions due to changes in protein activity. Initial experimental work to assess the effect of crowding on protein stability was dominated by theoretical estimations that suggested an increase of the unfolding temperatures between 5 and 20 °C. The present chapter examines some of the techniques typically used in these studies in a critical way, exploring the most common methodologies employed to study proteins under crowded conditions. One of the critical problems is the choice of the crowders. There are two categories of commonly used macromolecular crowders, synthetic polymers and common proteins. Synthetic crowders often do not interfere with optical measurements of proteins. When proteins are used as crowder it is necessary to label the protein under study. We demonstrate that studies of protein stability in crowded solutions provide more valuable information when observing the behaviour of marginally stable proteins.

Keywords Confinement · Methodologies · Molecular crowding · Synchrotron

P. A. Temussi (✉)
Dipartimento di Scienze Chimiche, University "Federico II" of Naples, Naples, Italy
e-mail: temussi@unina.it

M. Levantino
European Synchrotron Radiation Facility, Grenoble, France

V. N. Uversky (ed.), *(Macro)Molecular Crowding*, Subcellular Biochemistry 109,
https://doi.org/10.1007/978-3-032-03370-3_20

Introduction

Crowding, the theme of this book, has been introduced in previous chapters. Nonetheless, it is appropriate to spend very few words to summarize the origins of the term.

While experimental studies on proteins are usually performed in dilute solutions, the liquid inside cells contains many other macromolecules and accordingly it is now called a "crowded" environment. The concept of crowding was introduced by A. G. Ogston (Laurent and Ogston 1963; Ogston 1970) but was widely accepted only after Allen Minton rationalized it clearly (Minton 1980, 1983, 2000). To quantify the extent of crowding in cells it has been generally assumed that all proteins inside the cell contribute to it. Following measurements by Zimmerman and Trach (1991) the concentration of macromolecules in *E. coli* is often assumed to range from 0.20 g/mL to 0.40 g/mL, that corresponds roughly to a range of occupancy of 20% to 40% of the cellular volume. As a consequence, the volume available to a given protein is excluded, *i.e.* it becomes inaccessible. This observation implies that protein conformations of larger volume would be disfavoured with respect to conformations of smaller volume. The belief, at the basis of the concept of crowding, that all macromolecules of the cell are free in solution, amounts to regarding any (prokaryotic) cell as a bag full of proteins. Such a view is probably too simplistic, as pointed out by James Clegg (1984) and Paul Srere (2000), mainly because at least some proteins are part of immobile complexes or confined in smaller compartments or embedded into the membrane.

Describing methods to study crowding is not strictly equivalent to describing technical approaches. During the first years after the concept of crowding was explicitly formulated by Allen Minton, experimental studies of crowding *in vitro* were characterized by a few topics rather than by specific techniques. The two main aspects, thought to be affected by the presence of crowders were protein stability and protein activity. In this chapter, we will review some of the most crucial effects and discuss the techniques able to provide useful information.

Techniques

Even if the emphasis is on themes, it is appropriate to summarize the main techniques used in *in vitro* studies on crowding.

The technical methods used in these studies are essentially the same previously used in all studies on proteins performed in dilute solutions, only with special care in choosing the crowders and the proteins under study. Structural features of proteins or their interaction with other molecules are investigated with several biophysical methods, mainly various types of spectroscopies. The techniques used for these studies are various spectroscopies such as circular dichroism (CD) (Greenfield 2006a, b), other optical spectroscopies (Serrano et al. 2012; Keiderling and Xu

2002; Royer 1995) and nuclear magnetic resonance (NMR) (Dyson and Wright 2004).

Among these techniques, the spectroscopy most used to check protein stability is probably CD spectroscopy, particularly when protein stability is checked by measuring protein unfolding.

Circular dichroism (CD) refers to the difference in the absorption between left- and right-handed polarized light, where left-hand circular (LHC) and right-hand circular (RHC) polarized light are the two possible states of the spin angular momentum of a photon. CD spectroscopy has many applications but probably the most important one is the detection of the secondary structure of proteins. Usually, commercial instruments for CD spectroscopy detect absorption bands comprised between 1000 to 190 nm. Nucleic acids and proteins have absorption bands in this range, but most sugars are transparent in this region. Along with traditional CD, Synchrotron radiation circular dichroism (SRCD) spectroscopy is CD carried out at a synchrotron source. This possibility extends the frontiers of applications of conventional CD spectroscopy by exploiting the high flux of a synchrotron source that enables collection of data at lower wavelengths and detection of spectra with higher signal-to-noise levels. The technique has also the advantage to permit measurements in the presence of other absorbing components (buffers, salts, lipids and detergents) thus reducing the problem of the background. By this global technique it has been possible to gain important static and dynamic structural information on proteins in solution.

Techniques less used in general to study unfolding are NMR spectroscopy and fluorescence. However, NMR is unique in giving residue specific information and fluorescence, particularly single-molecule fluorescence detection in combination with Förster resonance energy transfer (FRET), is very good to study labeled proteins even in very crowded solutions.

Protein Stability

In *in vitro* studies, a common way to measure protein stability is to monitor protein unfolding. Techniques use different reporters of unfolding, depending on whether they measure global properties of the whole protein or they monitor properties of single residues. In the case of CD spectroscopy, a typical global technique, the contributing signals come essentially from the hydrophobic core, *i.e.* the part of the protein totally involved in unfolding. Global techniques provide little information on unfolding from regions only partially structured, often far away from the hydrophobic core. Techniques that yield information on site-specific unfolding, mainly NMR and, to a lesser extent, fluorescence may be a rich source of information. However, this information, when taken as representative of the stability of the whole protein, may be misleading. We will examine the influence of crowders on protein stability in the light of these different points of view.

When we study a protein in a crowded solution, signals from crowders may interfere with those coming from the protein under study. This may be particularly dramatic and prevent the observation if we use another protein as a crowder. If we want to avoid the interference of the crowder's signals there are two possible strategies: 1) use crowders that give no signal for a particular spectroscopy; or 2) label the protein to make it the only visible molecule. Luckily, many synthetic polymers commonly used as crowders are invisible to some spectroscopies. This is the case, for instance, of PEG, Ficoll and some other saccharides when studied by CD spectroscopy.

A word of caveat: most of the early experimental studies on crowding were directed at measuring the influence of crowding on protein stability. On the basis of the model proposed by Minton (1980), in which unfolded forms were depicted as very large spheres, large increments of protein stability in crowded solutions were expected. On the contrary, studies of thermal stability showed in general only a modest increase in unfolding temperature upon addition of high amounts of crowders (Theillet et al. 2014). We will examine all possible reasons for this discrepancy.

The Proper Choice of Crowders In Vitro

Which molecules have been used to simulate crowding in living cells? By and large, the most commonly used crowders have been different types of macromolecules, although there are several studies in which the molecules used as crowders were small molecular weight molecules. In some cases, the small crowders were used as a simple comparison to polymers of which they were considered the corresponding monomers. All instances of small molecules appear as a contradiction of the original proposal by Allen Minton (1980). The extreme case is the proposal, put forward by a theoretician (Sharp 2015), that water is "the" ideal crowder. This proposal is a bit opinable, as it is certainly far away from the initial proposals by Ogston (1970) and by Minton (2000).

There are two categories of macromolecular crowders, synthetic polymers of various nature and common proteins. The synthetic polymers most used as crowders are all very soluble in water and, generally, do not interact strongly with any protein. The most common crowders are Ficoll, dextran and polyethylene glycol (PEG). Ficoll, a trademark owned by General Electric companies, is a neutral, highly branched, hydrophilic polysaccharide. Ficoll radii range from 2 to 7 nm. It is commonly used to adjust the density and viscosity of solutions containing biomolecules. It is formed by the co-polymerisation of sucrose and epichlorohydrin. Due to its many hydroxyl groups, Ficoll is highly soluble in water. Dextran is a glucan whose chains have lengths generally comprised between 3 and 2000 kDa. It is a complex branched polysaccharide derived from the condensation of glucose, originally discovered by Louis Pasteur as a microbial product in wine. Dextran chains are of varying lengths (from 3 to 2000 kilodaltons). Commercial dextran is now produced from sucrose using lactic acid bacteria of the family lactobacillus. In most studies of the effect of crowding on protein stability it was assumed that saccharides like Ficoll and dextran do not interact significantly with the protein under study, but at least in one study the presence of even weak interactions with the protein (Yfh1) was excluded by a careful NMR study (Alfano et al. 2017).

Polyethylene glycol (PEG) is a synthetic polyether compound corresponding to the monomer ethylene glycol, used in many applications, including the preparation of protein crystals for x-ray crystallography. In the field of crowding, it was originally considered as completely neutral with respect to proteins, but recently there have been several cases in which it was suspected of weak interactions with the protein under study. In the quoted paper by Alfano et al. (2017) it was shown by NMR that it did not interact with Yfh1 (the protein under study) but it was suspected to alter significantly the activity of water.

All three quoted synthetic crowders can be considered "invisible" for most experimental crowding studies because they have no absorption in the range of frequencies normally explored in CD spectroscopy and other optical spectroscopies commonly used in experimental studies of proteins.

Optical Spectroscopies and "Invisible" Crowders

One of the first studies on the influence of crowding on protein stability was performed by Allen Minton himself and collaborators (Sasahara et al., 2003), using dextran as a crowder. The one used by Sasahara et al. (2003) was a commercial sample of average molecular weight 35 kDa purchased from Sigma Chemical Co. (St. Louis, MO) and used without further purification.

Sasahara et al. (2003) studied the influence of dextran on the stability of two proteins, lysozyme and cytochrome C. Thermal unfolding of hen egg-white lysozyme was measured in solutions at different protein concentrations, containing several concentrations of dextran, from 100 gl^{-1} to 300 gl^{-1} at pH 2.0 using near-UV and far-UV CD spectroscopy. The curves of folded protein population *vs* temperature, fitted to a two-state model by a non-linear, least-squares method yielded the transition temperature (T_m), enthalpy change ($\Delta H_u(T_m)$), and free energy change ($\Delta G_u(T)$) of the unfolding transition. The increases in $T_{m,}$ at almost constant $\Delta H_u(T_m)$ values, in the presence of added dextran, varied from 0.3 K at 100 gl^{-1} of dextran to ca. 2.5 K at 300 gl^{-1} of dextran. In the case of cytochrome c, the possible influence of dextran on the stability was studied simply starting from the unfolded state induced by acid at pH 2.0. On addition of dextran, the CD spectrum shifted from that characteristic of the unfolded state to that characteristic of the salt-induced molten globule state that is more compact, a result suggesting that the molten globule state is stabilized in crowded environments.

Dextran has not been used much as a crowding agent but in an interesting article of 2019 it was used together with polyethylene glycol (PEG) to have a comparison between crowding agents with different polarity (Köhn and Kovermann 2019).

These authors made a detailed analysis of the influence of the mentioned macromolecular crowders on the stability of *Bacillus subtilis* Cold shock protein B (BsCspB) during both thermal and chemical denaturation. The dextran used was Dex20, technical grade (Pharmacosmos). They noted that, contrary to what observed in many classical studies of chemical unfolding, the description of thermal unfolding by means of a single parameter was not possible. According to Köhn and Kovermann (2019) the change in thermodynamic stability could not be judged solely by the change in temperature midpoint (T_m), nor by the change in enthalpy (ΔH), or by the difference in the heat capacity (ΔCp), but the effect could be not linear and thus the consequence of changes in any one, two, or all of these three parameters combined together (i.e. giving a cooperative effect). Thermal unfolding of BsCspB, monitored by CD spectroscopy in the presence of 10, 20, and 30% (w/v) PEG8 or Dex20, showed a pronounced increase in T_m in the case of Dex20 (2.8, 5.5 and 8.4 K respectively) but only marginal ones for PEG8. At the end of their comprehensive investigation of the chemically and thermally induced unfolding of BsCspB, Köhn and Kovermann (2019) concluded that they found that unfolding in crowded solutions depends only on the weight per volume fraction of the crowder, and not on the type of macromolecule used nor on its size.

Among polysaccharides, the crowder most largely employed is probably Ficoll. One of the first studies using Ficoll as a crowder (Stagg et al. 2007) apparently confirmed the prediction of Minton (1980) on the entity of stabilization, as measured by the increase of ΔT_m. Stagg et al. (2007) chose to study the effect of Ficoll 70 on the stability of Desulfovibrio desulfuricans apoflavodoxin, a small single domain protein of modest stability, using far-UV CD spectroscopy to assess stability. The authors found that, contrary to previous studies on the effect of crowding on protein stability, their study was in full agreement with the predictions of Minton (1980): addition of Ficoll 70 produces an increase in the stability of the protein that is proportional to the concentration of the crowder. The increment of ΔT_m observed in the range of Ficoll concentration from 100 mg/ml to 400 mg/ml varied from 4 K to 20 K, exactly as predicted by Minton (1980). This was one of the very few cases in which the gain in stability was so conspicuous. We believe that the main reason for it is probably that the authors used a protein marginally stable: the influence of crowding is clearer and more appreciable on a protein of modest stability, whereas it is difficult to detect any influence on stable proteins, as was the case with lysozyme and cytochrome c. (Sasahara et al., 2003).

A similar study, using CD spectroscopy, was performed by Wang et al. (2010a, b) on a larger protein, the dimeric creatine kinase (CK), relatively more stable than the apoflavodoxin selected by Stagg et al. (2007). Wang et al. (2010a, b) found a trend qualitatively similar but with lower increments of the ΔT_m. In the range of crowder concentration from 30 mg/ml to 260 mg/ml the ΔT_m. increased from 2 °C to 8.5 °C. The authors analysed changes of secondary and tertiary structure of CK from 25 °C to 80 °C in the presence of Ficoll 70 in a range of concentrations and found that the unfolding of CK in the presence of the crowder is highly cooperative, as indicated by the quality of the fitting based on a two-state model and the consistency of the transition curves measured by CD spectra and those measured by fluorescence spectra. The authors claimed that their results provided strong evidence that macromolecular crowding enhances protein stability. In their experiments, they observed that the crowder induces a decreased volume of both the folded and the thermally unfolded CK. The conclusion of the authors is that enhancement of thermal stability of CK in the presence of Ficoll 70 proves the theory that macromolecular crowding is caused by an excluded volume effect.

Ficoll was also chosen as a crowder in the paper by Sanfelice et al. (2013) aimed at comparing effects of crowding and confinement using the marginally stable protein Yfh1. The measurement of stability of Yfh1 in solutions crowded by Ficoll, by CD spectroscopy resulted remarkable for its accuracy because it was measured on the basis of the whole stability curve (Becktel and Schellman 1987). The term protein stability may in fact refer to several different thermodynamic parameters (T_m, ΔH, etc.) that can be proportional to stability. The temperature dependence of a spectroscopic parameter during temperature unfolding is typically used to derive by least-squares fitting two parameters of the process, *i.e.* T_m and ΔH_m, which is instead insensitive to the magnitude of the difference in heat capacity (ΔC_p). It would instead be useful to be able to measure the whole stability curve of a protein to judge the effect of crowding on its stability. When both cold and heat unfolding can be observed, it is possible to use the temperature dependence of the spectroscopic parameters directly to derive the thermodynamic parameters T_m, ΔH_m and ΔC_p by fitting the data using the modified Gibbs-Helmholtz equation (Szyperski et al. 2006). The corresponding curve is known as the stability curve of the protein (Becktel and Schellman 1987). In the case of Yfh1, the protein studied by Sanfelice et al. (2013), it was indeed possible to observe cold denaturation above the water freezing temperature (Pastore et al. 2007) and thus the measurement of the influence of crowding is particularly accurate. Sanfelice et al. (2013) measured full stability curves of Yfh1 in the range of Ficoll concentrations from 3.75% w/w to 15% w/w and found that the presence of Ficoll did lead to an increased stabilisation of Yfh1 (corresponding to a ΔT_m of 5 C at 15%). However, the effects corresponding to Ficoll values higher than 15% were not linear and were excluded from the full stability curve measurement. It is worth noticing that Yfh1 is a marginally stable pro tein, a circumstance in part related to the possibility of observing cold denaturation above water freezing. (Pastore et al. 2019).

At approximately the same time, it was published a paper using Ficoll to study its influence on the stability of three well known, fairly stable proteins, *i.e.* on RNase-A, lysozyme and calcium-loaded (holo) α-Lactalbumin (Mittal and Singh, 2013). The authors measured unfolding of each protein at four different pH values (7, 6, 5,4) in a range of Ficoll concentration from 100 g/l to 400 g/l. Temperature unfolding of the three proteins was monitored by following the changes in absorbance at frequencies at which Ficoll does not absorb: 287 nm for RNase-A, 300 nm for lysozyme and 295 nm for α-LA as a function of temperature. At pH 7 all three proteins have a rather high maximum stability and the increase observed with addition of Ficoll is negligible, ranging from 0.1 °C to 0.3 °C. The only sizeable effects of the crowder were observed for RNase-A at pH 4, when the T_m in absence of crowder drops from 63.7 °C to 56.8 °C. Addition to the solution of RNase-A at pH 4 of 400 g/l of Ficoll causes an increase of T_m of 4 °C. The authors concluded that the structural and thermodynamic influence of macromolecular crowding are strictly dependent on the nature of the protein-crowder couple and that the effect on protein stability depends mainly on the structural features of the denatured state (Mittal and Singh, 2013). Therefore, it is difficult to generalize the effect of macromolecular crowding on the basis of studies on specific proteins. Surprisingly, these authors did not perceive the importance of the intrinsic stability of the proteins, in the absence of crowders.

Ficoll has also been used as a control in vitro for a study of stability in cell. The protein used in this *in cell* study is a variant of the superoxydismutase 1 (SOD1), designed to observe cold denaturation above water freezing. This protein variant (named $SOD1^{barrel}$) was designed by truncating a metal-binding loop and an electrostatic loop of the wild protein, changes that hampered dimerization and suppressed catalytical activity, leaving almost unchanged the shape of the protein. $SOD1^{barrel}$ I35A showed a fully resolved NMR spectrum in mammalian cells. The unfolding temperatures of the variant in PBS buffer were 0.64 °C for cold denaturation and 35.6 °C for heat denaturation. Addition of Ficoll (at 100 mg/ml) led these temperatures to −0.62 °C and to 38.5 °C respectively, *i.e* to an appreciable increase in stability. The technique used for the measurement of unfolding was ^{15}N NMR. For quantification of the populations of folded and unfolded forms the authors used the volumes of the C-terminal Q153 cross peak, which are well separated and insensitive to relaxation effects induced by temperature or viscosity. However, as we have recently shown (Puglisi et al. 2021), the choice of single residues to measure unfolding can be risky, particularly when the residue is far from the hydrophobic core and/or affected by a different dynamic behaviour. The importance of the choice of the protein in studying the influence of crowding on protein stability does not exhaust the causes of the discrepancy between the expected effect of molecular crowding and experimental findings. One of the causes most strongly put forward for many years was also the relevance of enthalpic effects with respect to entropic effects, as seen in the original formulation by Allen Minton (1980). In some cases, the antithesis between entropic origins and enthalpic causes was put to extremes, almost disallowing the actual relevance of entropic causes. It would probably be wiser to try to assess the relevance of entropic and enthalpic causes individually to be able to dissect the specific contributions.

A study in this direction is that described by Alfano et al. (2017), in which the mentioned protein Yfh1 was studied in solutions crowded by Ficoll or by PEG. Although it is almost impossible to find crowders with no enthalpic contribution, it is possible to check carefully the extent of interactions by examining them experimentally. Alfano et al. (2017) before measuring by CD spectroscopy the stability curve of Yfh1 in the presence of crowders, examined solutions containing Yfh1 and the crowder by NMR spectroscopy, whose parameters are very sensitive to even weak interactions and found no influence of those concentrated crowders on the Yfh1 structure. In the absence of significant enthalpic contributions, the authors showed that protein stability is indeed affected by volume exclusion. In addition, they found that the effect on the protein is more pronounced when the crowder's size is closer to that of the protein.

Proteins Made Visible in Complex Solutions: Labelled Proteins

An alternative to the use of "invisible" crowders is to make the studied protein individually visible by means of specific labelling of one or more residues. The spectroscopies mainly used to look at labelled proteins are fluorescence spectroscopy and NMR spectroscopy.

In fluorescence, the method best suited for these studies is single-molecule fluorescence detection in combination with Förster resonance energy transfer (FRET). It allows to probe suitably labeled proteins even in the presence of very large concentrations of unlabeled solutes, *i.e.* the situation typical of crowding studies.

One paradigmatic application of this approach regards Intrinsically Disordered Proteins (IDP). We have seen that the influence of crowding on protein stability is best seen when the protein under study is marginally stable. The case of IDPs can be seen as the extreme situation of marginally stable proteins. Soranno et al. (2014) studied several IDPs representative of observed typical sequences in combination with a range of sizes of crowding agents, represented mainly by different molecular weights of polyethylene glycol (PEG). In particular, Soranno et al. (2014) investigated the effect of crowding on four different IDPs that span the N- and C-terminal segments of two proteins: human prothymosin-α (ProTα-N and -C), the activator for thyroid hormones and retinoid receptors (ACTR), and the N-terminal domain of HIV-1 integrase (IN). To probe the dimensions of the IDPs, *i.e.* the intramolecular distance distributions of the IDPs, the authors attached to cysteine residues at suitable positions the donor and acceptor fluorophores Alexa Fluor 488 and Alexa Fluor 594, with excitation at 488 and 594 respectively. They observed a compaction of IDPs not only when the concentration of the crowder was increased, but also when the size of the crowding agents was increased. The authors interpreted these results by suggesting that the excluded volume between overlapping biopolymers can be an essential contribution to the physics governing the crowded cellular medium.

Wang et al. (2018) used a labelled kinase to go one step further in examining the possible consequences of crowding in vivo. They reasoned that crowding can be dynamic: as cell volume changes, for instance in the course of the cell cycle, water moves in or out through membrane channels, and crowding changes accordingly. Protein stability would be altered as a result of these changes, but the extent of changes is not known. To test the extent of change Wang et al. (2018) looked at the response of a FRET-labelled kinase when volume changes in live cells were induced by osmotic changes. They examined both the folded and unfolded states of the kinase by changing the temperature of the media surrounding the cell. Their data showed that crowding compacts the structures of its unfolded ensemble and stabilizes the already folded protein.

One very interesting FRET study, performed by Davis et al. (2020) employed AcGFP1, a green fluorescent protein as the donor (D) fused to the N-terminus of two proteins and mCherry as the acceptor (A) cloned to the C-terminus of the proteins. Using the donor fluorescence-acceptor fluorescence ratio (*D*/*A*) to report on the end-to-end distance of the FRET construct it was possible to monitor thermal denaturation.

Davis et al. (2020) wanted to explain the contradictory behaviour of two proteins when studied *in vivo* with respect to their behaviour *in vitro*. Yeast PGK and *Borrelia burgdorferi* VlsE have opposite trends in stability when moved from *in vitro* into cells. In sodium phosphate buffer, PGK and VlsE have similar melting temperatures, $T_m = 39 \pm 1$ °C and $T_m = 40 \pm 1$ °C, respectively. However, inside U-2 OS cells PGK is stabilized, $T_m = 44 \pm 2$ °C, (Ebbinghaus et al. 2010; Dhar et al. 2011) and VlsE is destabilized, $T_m = 35 \pm 2$ °C. To mimic *in vivo* conditions, Davis et al. (2020) tried several combinations of Ficoll (as an entropic crowder) and Pierce IP lysis buffer, used in lysis of whole mammalian cells, optimized to minimize disruption of protein complexes, a sticking agent, because these mixtures can generate a stabilization landscape. In the case of the two proteins of their study, VlsE is known to be more sensitive to sticking interactions. Thus, the component of lysis buffer may be more effective in denaturing VlsE and screening it from the stabilizing influence of Ficoll. For PGK and VlsE, the mixture that reproduced the in-cell stability of both proteins simultaneously was 150 mg/ml Ficoll and 60% lysis buffer. The authors proposed this combination as an *in vitro* mimic of crowding and nonspecific interactions in cells.

One of the main consequences of labelling the protein under study is that it opens the possibility of using proteins as crowders. In turn, this opening allowed a better understanding of the relative importance of enthalpy with respect to the original entropic forces controlling crowding, essentially because any protein cannot have negligible interactions with another protein.

Enthalpy Versus Entropy

In 2011, the Pielak group explicitly described crowding effects in terms of an antinomy between entropic and enthalpic effects. Wang et al. (2011) said that while experiments using synthetic polymers as crowders show stabilizing effects on protein stability, they found that protein crowders can be mildly destabilizing. This competition yields a tunable stability, a situation that is crucial in the understanding of the role of proteins in living systems.

Among the techniques that make use of labelled proteins for studies in crowded environments, the one mostly used is certainly NMR spectroscopy and a large number of these studies were performed in the Pielak laboratory. Pielak et al. (2009) described in a review the possible use of NMR in several biological environments characterized by crowding. In their words: "Most proteins operate inside cells where the concentration of macromolecules can be >300 mg/mL." They mentioned fluorine resonance and ^{15}N filtered NMR spectra as two of the most relevant methods. In subsequent years the Pielak laboratory published the great majority of NMR studies on proteins in crowded solutions.

One of the first aspects investigated in the Pielak laboratory about reproducing crowding conditions *in vitro* was the influence of typical crowders on protein diffusion. Wang and Pielak (2010) found that the influence of synthetic crowders was essentially different from that of proteins used as crowders. Synthetic macromolecules induced negative deviation from the Stokes laws whereas protein crowders cause positive deviation. In addition, synthetic polymers affect translation more than rotation, but proteins reduce rotational diffusion more than translational diffusion. Therefore, Wang and Pielak (2010) suggested that synthetic polymers may not be suitable mimics of the intracellular environment. According to them, weak, nonspecific, noncovalent chemical interactions between proteins might play the fundamental impact in protein diffusion in cells.

Wang et al. (2010a, b) studied a ^{15}N labelled chymotrypsin inhibitor 2 (CI2) in solutions crowded by poly(vinylpyrrolidone)s (PVPs) of varying molecular weights. To quantify the stability of this small protein the authors used NMR-detected amide ^{1}H exchange. At a PVP concentration of 100 g/L, Wang et al. (2020) found that the data are consistent with the effect of hard particle exclusion. However, the data showed also that the crowder stabilizes the CI2 protein by weakly binding to its folded state. They concluded that the role of native-state binding and other soft interactions should be seriously considered in studies of macromolecular crowding.

Miklos et al. (2010) used a very large synthetic polymer as an upper limit to study macromolecular crowding. This polymer microgel composed of poly(N-isopropylacrylamide-co-acrylic acid) [p-NIPAm-co-AAc], was used to study the stability of a globular test protein, chymotrypsin inhibitor 2 (CI2). By absorbing a large amount of water, the polymer yields spherical particles of 300 nm radii that exclude large amounts of solution volume. They chose this very large crowding agent because it is pharmaceutically relevant, and because its ability to take up water provides a model for volume exclusion by a molecule much larger than the test protein. As a result, the authors found no change in protein backbone dynamics, indicating the absence of strong crowder-protein interactions and only mild stabilization of CI2 at a volume occupancy of 70%, which approaches the occupancy of closely packing spheres. They concluded that large microgels that interact weakly with proteins do not strongly influence protein stability because they constitute an upper size limit on crowding effects.

As a crucial test on the relevance of crowding with respect to protein stability, Schlesinger et al. (2011) investigated the ability of macromolecular crowding in refolding an unfolded protein. They chose as a test protein the immunoglobulin G binding domain of protein L (ProtL) from the mesophile Streptococcus magnus. Substituting seven of its lysines with glutamic acids lowered the stability, causing the majority of the protein molecules to be in the denatured state in dilute buffer. The authors demonstrated that 84% of the molecules of the variant protein was in the denatured state in dilute buffer at room temperature, compared with 0.1% for the wild-type protein. They used in-cell NMR spectroscopy, particularly ^{19}F NMR resonance, to show that the cytoplasm of *Escherichia coli* does not overcome even this modest (~1 kcal/mol) free-energy deficit. Altogether, the data were consistent with the idea that nonspecific interactions between cytoplasmic components can overcome the excluded-volume effect.

Like other groups, also the Pielak group investigated intrinsically disordered proteins (IDPs). Wang et al. (2012a, b) tried to understand whether crowding influenced also diffusion. They compared the behaviour of globular protein chymotrypsin inhibitor 2 (CI2, 7.4 kDa) with that of the disordered protein α-synuclein (14 kDa). They found that, in dilute solution, the diffusion of CI2 is faster than that of α-synuclein, but in crowded solutions; α-synuclein diffuses faster than CI2, in spite of the fact that it is larger. Wang et al. (2012a, b) concluded that shape is a key parameter determining protein diffusion under crowded conditions, a property that had not been considered explicitly among the parameters affected by crowding.

The Pielak group elaborated again on the relative importance, in crowding, of entropic and enthalpic forces. Wang et al. (2012a, b) reasoned that efforts to understand crowding effects on protein stability had focused on hard-core repulsions, regarded as inherently entropic and stabilizing. They stated instead that, having assessed these phenomena by measuring the temperature dependence of amide proton exchange by NMR, it was possible to compare the entropic and enthalpic contributions to the stability of ubiquitin. According to their data, the contribution of soft chemical interactions is large and, in many cases, dominates the contribution from hardcore repulsions. Therefore, they concluded that both chemical interactions and hard-core repulsions should be considered when assessing the effects of crowding.

Similar conclusions were also reached in a comparison of the influence on stability of one chymotrypsin inhibitor 2 (CI2) of Ficoll and its monomer, sucrose (Benton et al. (2012). In spite of the enormous difference in molecular weight, Ficoll and sucrose have a similar stabilizing effect on CI2. Furthermore, the stabilization is driven by enthalpy, not entropy.

Finally, the Pielak's group extended their considerations from *in vitro* crowding studies to a more realistic approach using an actual cytoplasm (Sarkar et al. (2013). They used a reductionist approach: *Escherichia coli* cytoplasm was dialyzed, lyophilized, and resuspended at two concentrations to study by NMR the stability of the globular protein chymotrypsin inhibitor 2 (CI2) on the basis of amide proton exchange. The cytosol destabilized CI2, and the destabilization increased with increasing cytosol concentration, thus showing that the cytoplasm interacts favourably with CI2, and these interactions overcome the destabilizing hard-core repulsions.

Recently, Stringer et al. (2023) investigated a protein with an intrinsically disordered region adjacent to a folded domain, the RNA-binding domain of the SARS-CoV-2 nucleocapsid protein, using PEG as crowder and single-molecule Förster resonance energy transfer as the main technique. The authors found a different behaviour for PEG samples of different molecular weight. High-molecular-weight

PEG promoted a refolding of the disordered tail, but low-molecular-weight PEG induced chain expansion. Their data can be explained by accounting for two opposing contributions: favourable weak interactions between the protein and crowder and entropic excluded volume interactions.

Experiments Versus Predictions

Summarizing what mentioned above, experimental studies on the influence of crowding on protein stability expected to find, for unfolding midpoints, increments between 5 and 20 °C. However, the values of ΔT_m found experimentally were, in general, much lower.

The reasons for this discrepancy are several but can be roughly summarized in three categories: 1) non-appropriate choice of the protein chosen to study the effect of crowding, 2) the choice of the volume and shape of conformations of the unfolded ensemble, 3) the contribution of forces different from the simple entropic forces of volume exclusion. Let's examine them in more detail.

(1) The proper choice of the protein is essential to detect visible effects. A stable protein may be a poor choice, because the change in stability due to external causes corresponds to a minor variation of a large value, *i.e.* it will be a small percentual value. Some take a typical prototype protein like RNase T1, a protein composed of 104 amino acids, with a $\Delta G_{20} = 9$ kcal/mol, as a marginally stable protein because the 9 kcal is a relatively small value with respect to stabilizing and destabilizing forces of the order of 300 kcal/mol (Pace et al. 1996). However, even if we consider a protein with a $\Delta G_{20} = 5$ kcal/mol, we find that the population of unfolded conformations in equilibrium with the folded form at room temperature is ca. 0.02%.

    As calculated by Pastore et al. (2019), it is clear that, when the difference in free energy between folded and unfolded species is higher than 2 kcal/mole, only the folded species is practically detectable in solution, whereas the folded and unfolded species coexist between 2 and 0 kcal/mole. Therefore, one cannot expect to see great influence of crowding when studying stable proteins. (Pastore et al. 2019). It is clear that when we study the stability of a protein that has a folded population of 99.99% it is difficult to assess an increase towards 100% population.

    In addition, the use of an increase of ΔT_m as a measure of stability is problematic when used on a stable protein. The value of T_m is not necessarily proportional to changes in free energy. In some cases, to an increase in thermodynamic stability may correspond a very little increase in T_m or even a decrease. To overcome problems of this kind, Alfano et al. (2017) proposed a new parameter to measure protein stability, *i.e.* they proposed the use of the area of the stability curve comprised between the cold unfolding and the heat unfolding temperatures. This parameter is very sensitive to even small changes in

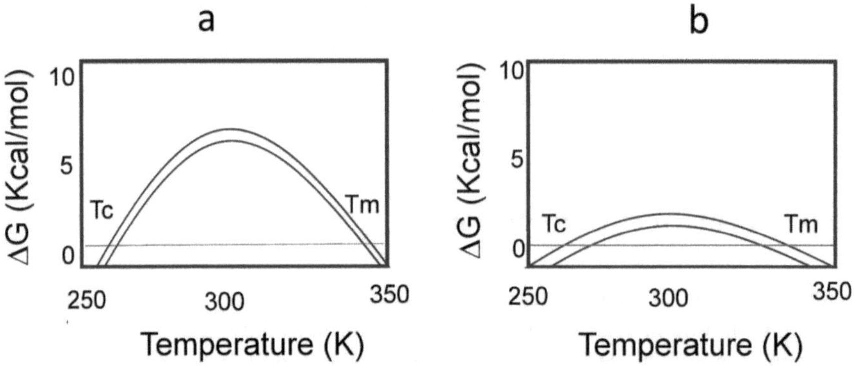

Fig. 20.1 Cartoons showing the influence of stability change on the size of ΔT_c and ΔT_m. (**a**) External change of stability for a protein with intrinsic stability between 5 and 10 kcal/mol. The values of ΔT_c and ΔT_m are small. (**b**) The same external change of stability for a protein with intrinsic stability between 2 and 0 kcal/mol. The values of ΔT_c and ΔT_m are large

protein stability (Alfano et al. 2017) and can be used to illustrate the difficulty of using the ΔT_m criterion to judge the crowding influence on the stability of proteins. As an example, we can compare the changes of ΔT_m caused by the same increment of the area under the stability curve on two proteins of different free energy difference between folded and unfolded states, *i.e.* 10 kcal/mole and 1 kcal/mole respectively. The consequences are very different. For the protein with a difference in free energy of 10 kcal/mole (Fig. 20.1a) the relative increment of area is accompanied by very small changes in both cold and heat denaturation temperatures. On the other hand, for the protein with a difference in free energy of 1 kcal/mole (Fig. 20.1b) we observe large changes in ΔT_ms.

(2) Another important aspect to interpret the disagreement between predictions and experiments on the influence of crowding on protein stability is the estimate of protein volumes. In the original view of Minton (2000), the volume of proteins becomes one order of magnitude larger with respect to that of the folded one upon denaturation. This view originated from the estimated value of small angle scattering (SAXS) data and from a simplistic estimate of the shape of the unfolded form as a regular sphere. It is true that many of the folded species have a shape resembling an irregular sphere or an ellipsoid, as measured by x-ray crystallography. However, there is no correlation with the shape of unfolded species. The shape of unfolded species is essentially unknown, also because the unfolded state is not representable as a single species but as an ensemble of species. Politou and Temussi (2015) have shown that, by assuming that unfolded shapes are irregular, it is possible to explain small experimental changes in melting temperatures.

(3) Last but not least, the cause invoked by the largest number of studies is probably the contribution of enthalpic forces alongside the original entropic forces. The laboratory of Pielak has shown in a series of studies that even weak enthalpic forces between crowders and the protein under study could decrease the

stabilization induced by entropic forces and counterbalance them. In the view of the Pielak laboratory, enthalpic forces are the main forces playing a role in crowding. This view was very important in the search for a mimic of *in vivo* environments.

The studies of the Pielak lab have shed light on the relevance of soft interactions *in vivo*. However, it is probably wrong to describe those interactions as an antithesis of entropic interactions. Both contributions are important *in vivo* and it would be essential to understand them independently. Although it is difficult to find crowders that have only entropic or only enthalpic interactions, there are actual situations close to these limiting conditions. In some cases, synthetic crowders have indeed negligible enthalpic interactions with some proteins (Alfano et al. 2017). This is so because many of the synthetic crowders commonly used are neutral and with surface not similar to those of proteins. For instance, Alfano et al. (2017) found, by careful NMR analysis, that Yfh1, the protein they studied in crowded solutions, did not interact in a measurable way with Ficoll or PEG. Therefore, the effects of crowders found on the Yfh1 stability could be safely attributed to entropic forces.

On the other hand, studies in which the crowders are proteins use generally a single protein as crowder and thus enhance the effect of enthalpic forces, inevitably present between two proteins, *i.e.* between the crowder and the protein. For instance, the molar ratio between the concentrations of the two proteins can be as high as 20. In a typical experiment, Wang et al. (2010a, b) used a 20 mM lysozyme crowder to study a 1 mM CI2 protein. It is clear that a concentration 20 times larger can enhance the role of even very small forces. In cell, a similar global concentration is achieved with the contribution of many proteins. It is conceivable that the forces exerted by each protein will be in part offset by those exerted by other proteins.

Thermodynamic Activity

The two main fields affected by crowding can be roughly quoted as "protein stability" and "thermodynamic activity", but they are intrinsically different since protein stability is a circumscribed and well defined topic whereas activity is much more diverse and cannot be easily restricted because the main influence of crowding is inevitably linked mainly to the reactions involving biological macromolecules and it is more difficult to treat reactions in which one of the reagents is a small molecular weight molecule.

As far as methods of experimental works are concerned, there is little difference with respect to studies on protein stability in crowded solutions. Most experimental studies imply the use of "invisible" crowders or deal with labelled proteins.

As recalled above, Minton predicted that the presence of crowders could increase the melting temperature of a protein (ΔT_m) by values up to 20 K (Minton 1980). Soon after, the same author used similar statistical considerations to predict the consequences of crowding on the thermodynamic activity of reacting proteins

(Minton,1983). It is not surprising that the majority of examples quoted by Minton in this introductory article regard biochemical reactions among macromolecules.

According to Minton (1983), the knowledge of crowding effects was rudimentary at the time. Nevertheless, simple hard-particle models were able to yield even quantitative information about protein activity in crowded solutions. The most successful application of the quoted models was the thermodynamic analysis of gelation of sickle hemoglobin (Minton 1977; Ferrone et al. 1980), a case in which the non-ideal behavior was attributed almost entirely to volume exclusion.

In a review of a few years later, Zimmerman and Minton (1993) summarized the early applications of crowding to biochemical reactions. Admittedly, they concentrated on "[..] the steric-repulsive interactions deriving from the fundamentally impenetrable nature of molecules, as these interactions are always present in addition to any other interactions that may or may not be present." These authors focused attention on thermodynamic interpretations of experimental data. In particular, they focused on two approaches to calculate thermodynamic activity in the presence of crowders. The method introduced by Ogson to calculate the probability that a randomly selected point in the solution of composition {c} can accommodate a probe molecule of species *i*, was referred by Zimmerman and Minton as "available volume theory", or AVT. They called the second approximate method "scaled particle theory" or SPT. According to the SPT the logarithm of the activity coefficient of species *i* may be expressed as a finite series in powers of R_i, the radius of the probe particle.

Before analyzing specific examples of experimental studies, Zimmerman and Minton made the following predictions of equilibrium excluded-volume analysis, where ϕ denotes the fraction of the total volume occupied by macromolecules:

(1) For self- and hetero-association reactions, the nonideal correction factor Γ increases monotonically with increasing ϕ, and the value of $d\Gamma/d\phi$ increases monotonically with increasing ϕ.
(2) For fixed ϕ, the value of Γ increases with the degree of association.
(3) Increasing volume occupancy tends to favor compact or globular conformations (or assemblies) relative to highly anisometric conformations (or assemblies).
(4) For fixed ϕ, the value of Γ for association reactions increases with decreasing molecular weight of background species.
(5) Volume occupancy by macromolecules has a greater influence on association equilibria in three dimensions than the same degree of volume (area) occupancy has upon the corresponding reaction equilibria in two dimensions.
(6) For fixed ϕ, the effect of crowding upon associations between macromolecules is expected to be much greater than upon associations between a macromolecule and a small molecule.
(7) Finally, and most importantly, under conditions of macromolecular volume occupancy comparable to those found in vivo ($0.1 < \phi < 0.5$), macromolecular association constants are predicted to exceed those in dilute solution by as much as several orders of magnitude.

All studies reported by Zimmerman and Minton show an increase of association tendency in the presence of crowders, although the effects described are often qualitative in nature.

Association Among Macromolecules

A large proportion of studies of reactions in crowded solutions concerns association among macromolecules, partly because they are at the core of many diseases and partly because macromolecules are naturally more influenced than small molecules by crowding conditions.

Here are two significant examples, one qualitative and another quantitative.

Munishkina et al. (2004) used thioflavin T (ThT) fluorescence and CD spectroscopy to observe the influence of several crowders (dextran, Ficoll and PEG) on the fibrillation of alpha-synuclein. They found an acceleration of fibrillation in all cases and estimated an increase of approximately ten-fold with PEG, ca. three-fold with Ficoll and ca. five-fold with proteins. The same group found that crowding (plus Zn ions) was able to force nonamyloidogenic beta-synuclein to fibrillate (Yamin et al. 2005). Similarly, del Alamo et al. (2005) found that the addition of Ficoll could accelerate HIV capsid assembly ca. ten-fold.

In the past, the only really quantitative study on the influence of PEG on association reactions was performed by Kozer et al. (2007) when dealing with the peculiarities of PEG as a crowder.

To illustrate the potential complexity of the interaction between PEG and proteins Kozer et al. (2007) studied the interaction in a range of concentrations, from dilute to semidilute to concentrated solutions. They monitored the interaction of two proteins, TEM1-b-lactamase (TEM) and the b-lactamase inhibitor protein (BLIP), in solutions containing crowding agents of different molecular weight, from monomers (ethylene glycol, glycerol, or sucrose) to polymers like polyethylene glycols (PEGs) of different molecular weights (0.2 to 8 kDa) and Ficoll. They found, in all solutions, an inverse linear relation between viscosity and the translational diffusion of the proteins, according to the well-known relation of Stokes-Einstein. They found that deviations from this relation of association rates were related to the three distinct regimes of polymer solutions: dilute, semidilute, and concentrated. In the dilute regime PEGs interfere with protein association by introducing a repulsive force originating from preferential hydration. When one goes to the semidilute regime, it is possible to observe faster association rates due to the depletion interaction, which causes an attraction between the two proteins. In the concentrated regime, PEGs again slow down the association between the two proteins, an effect that depends on their mass concentration.

Mixed Cases

An important contribution to studies of biochemical reactions in crowded environments was published by the Pielak laboratory (Stadmiller et al. 2020).

They used the NMR lineshape analysis of protein SH3 labelled with ^{19}F to quantify the equilibrium thermodynamics and kinetics of binding with two proline-rich peptides in solutions crowded with other proteins. Their results highlight the importance of using proteins to mimic crowded environments. They demonstrated that the results obtained with protein cosolutes do not resemble those obtained with the small-molecule controls, sucrose, glucose, or urea. However, the protein−peptide complex destabilization in protein cosolutes differs from studies in which synthetic polymers were used to study crowding (Kozer et al. 2007). This aspect may reflect the abnormal enhancement of enthalpic interactions when using very high concentrations of a single protein as crowder.

An illustration of the wide spectrum and complexity of the influence of crowding on reactions in biological systems is furnished by a paper by Nussinov et al. (2021). These authors examined how cells receive, communicate, and respond to signals. They described how efficient regulation of assemblies, in a crowded environment, can transmit effectively at micrometer scales, and what is the role of the cell organization. According to them, cells can navigate long-range diffusion-controlled signaling via association/dissociation of spatially proximal entities.

As mentioned above, when quoting crowding effects on conformation, a recently published paper (Stringer et al. 2023) investigated a protein with an intrinsically disordered region adjacent to a folded domain, the RNA-binding domain of the SARS-CoV-2 nucleocapsid protein, using PEG as crowder and single-molecule Förster resonance energy transfer as the main technique. They characterized the interaction between this protein and RNA in the presence of PEG. For all PEG molecules tested, they observed an increase in the binding affinity, but with a trend that is not monotonic as a function of the PEG molecular weight. This points to the role of nonspecific protein–PEG interactions on binding in addition to the entropic effects due to crowding.

The diversity of the fields in which activity is influenced by crowding is well summarized by a recent review (Perillo et al. 2023). These authors claim that in several situations, such as in cells, food and other biotechnological environments, owing to the presence of macromolecular crowding, the thermodynamic activity of macromolecules can be several orders of magnitude higher than in dilute solutions. In particular, the diffusion rates are directly affected by the volume exclusion induced by the crowders. Immiscible liquid phases, which can arise by liquid-liquid phase separation induced by macromolecular crowding, may lead to a dynamic confinement of reactants. In addition, owing to crowding conditions, membranes and macromolecules provide large surfaces that can perturb the solvent, causing its immobilisation by adsorption and also affecting the solvent viscosity.

Changes in the water structure can also alter the interaction between enzyme and substrate, and, in the case of hydrolytic enzymes, where water is one of the substrates, it also affects the reaction mechanism.

Emerging Methods

Traditional methods for studying molecular crowding often face limitations in accurately capturing the complexity of cellular environments. Advanced X-ray based techniques are emerging as powerful tools for investigating molecular crowding in life sciences, offering high resolution and non-destructive analysis capabilities. Advancements in microfocussing and coherence of X-ray beams coupled with improved data collection methods allow the study of larger, more complex biological assemblies in crowded environments with unprecedented precision.

Among the available methods, X-ray solution scattering (both static and time-resolved) and X-ray photon correlation spectroscopy (XPCS) are particularly useful and have the highest potential to yield relevant new insights in molecular crowding studies. X-ray solution scattering (XSS) both in the so-called small-angle X-ray scattering (SAXS) and wide-angle X-ray scattering (WAXS) regions is a versatile technique for characterising the size, shape, and interactions of macromolecules in solution. SAXS is particularly well-suited for studying molecular crowding due to its ability to probe the overall structural properties of biomolecules within complex cellular environments. It can also be used to derive valuable information about the spatial organisation, packing density and the interaction potential between macromolecules in extreme conditions (Peters et al. 2023). Through SAXS based studies it has been for example shown that molecular crowding can stabilise RNA folding (Kilburn et al. 2010, 2013) or enhance enzyme activity by making their structure more compact (Akabayov et al. 2013). Extending scattering data collection in the WAXS region, more detailed structural information can be obtained. It is for example possible to track how the relative position of different protein subunits or domains is altered in crowding conditions. WAXS data suggest that when samples at protein concentrations similar to the ones observed in vivo are investigated, intramolecular equilibrium breathing motions may be significantly reduced in amplitude or even inhibited (Makowski et al. 2008). A further particularly appealing possibility offered by XSS methods is that of performing time-resolved investigation with high temporal resolution in crowding conditions or even on samples containing entire cells. Indeed, if a macromolecule conformational change can be triggered by a laser pulse, it is possible to probe the induced structural changes as a function of time by monitoring the sample with short X-ray pulses (Pounot et al. 2023). Time-resolved X-ray scattering (TR-XSS) has been successfully used in a first proof-of-principle experiment to track the R-T quaternary structure of human haemoglobin within intact red blood cells (Spilotros et al. 2012). Functionally relevant conformational change of photosensors and other light sensitive proteins in the cellular environment could potentially be investigated with TR-XSS with nanosecond (or better) time-resolution, thus allowing to observe how crowding affects the dynamics of protein function with a directly structural sensitive technique.

Another promising advanced X-ray technique relevant for crowding studies is XPCS. By analysing the fluctuations in X-ray scattering patterns caused by the motion of macromolecules within cellular environments, XPCS provides insights into the dynamics of molecular crowding phenomena such as macromolecular diffusion, transport, and phase transitions. While molecular diffusion can be studied with well-known techniques including single-molecule fluorescence and high-speed single particle imaging, these techniques have limited spatial resolution (hundreds of nanometres at best). XPCS is thus more suited to study molecular diffusion as it is based on the use of the scattering of X-ray at sub-nanometer wavelengths. Moreover, XPCS can be used to obtain insights into the viscoelastic properties of crowded biological systems. XPCS has been for example used to show that the viscosity of highly concentrated BSA solutions is significantly lower than what is expected based on established models (Ross and Minton 1977) and rheology measurements (Otto et al. 2024). XPCS can also be used to investigate the kinetics and thermodynamics of phase transitions, including liquid-liquid phase separation, gelation, and protein aggregation processes (Girelli et al. 2021; Moron et al. 2022; Reiser et al. 2022). These studies have provided first insights into the role of molecular crowding in driving phase transitions and self-assembly phenomena, which are implicated in cellular organisation, biomaterial formation, and disease pathology. With the advent of nearly diffraction-limited storage ring synchrotrons and X-ray free-electron laser, XPCS now benefits from higher beam coherence and time-resolution (Möller et al. 2019; Chushkin et al. 2022) and could potentially become an extremely powerful tool for studying how molecular crowding influences a wide range of biological processes, from macromolecular diffusion and phase behaviour to protein dynamics and function (Perakis and Gutt 2020).

Concluding Remarks

This chapter was conceived originally as a simple account of methods currently in use in studies involving the concept of crowding. The history of methods used to study the relevance of crowding is entangled with the view of actual conditions in *in-vivo* studies. At the time when nearly all biochemical and biophysical studies were performed in solutions where in addition to a dilute macromolecule there were only a dilute buffer and small amounts of salt, the introduction of the concept of crowding by Allen Minton was a turning point to advance biophysical studies closer to conditions in vivo.

Although in the beginning the possible applications of the concept were limited to protein stability and activity, in the last few years the field has expanded to a much wider horizon. At the same time, the methodologies used to study the effects of crowding have now been enriched by the new possibilities offered by new advanced techniques, including methods linked to the use of synchrotron. Together, these advancements offer new perspectives to approach a theme of increasing interest in Biology.

References

Akabayov SR, Akabayov B, Richardson CC, Wagner G (2013) Molecular crowding enhanced ATPase activity of the RNA helicase eIF4A correlates with compaction of its quaternary structure and association with eIF4G. J Am Chem Soc 135(27):10040–10047

Alfano C, Sanfelice D, Martin SR, Pastore A, Temussi PA (2017) An optimized strategy to measure protein stability highlights differences between cold and hot unfolded states. Nat Commun 8:15428

Becktel WJ, Schellman JA (1987) Protein stability curves. Biopolymers 26:1859–1877

Benton LA, Smith AE, Young GB, Pielak GJ (2012) Unexpected effects of macromolecular crowding on protein stability. Biochemistry 51(49):9773–9775

Chushkin Y, Gulotta A, Roosen-Runge F, Pal A, Stradner A, Schurtenberger P (2022) Probing cage relaxation in concentrated protein solutions by X-ray photon correlation spectroscopy. Phys Rev Lett 129(23):238001

Clegg JS (1984) Properties and metabolism of the aqueous cytoplasm and its boundaries. Am J Phys 246:R133–R151

Davis CM, Deutsch J, Gruebele M (2020) An in vitro mimic of in-cell solvation for protein folding studies. Protein Sci 29(4):1060–1068

del Alamo M, Rivas G, Mateu MG (2005 Nov) Effect of macromolecular crowding agents on human immunodeficiency virus type 1 capsid protein assembly in vitro. J Virol 79(22):14271–14281

Dhar A, Girdhar K, Singh D, Gelman H, Ebbinghaus S, Gruebele M (2011) Protein stability and folding kinetics in the nucleus and endoplasmic reticulum of eucaryotic cells. Biophys J 101:421–430

Dyson HJ, Wright PE (2004) Unfolded proteins and protein folding studied by NMR. Chem Rev 104(8):3607–3622

Ebbinghaus S, Dhar A, McDonald JD, Gruebele M (2010) Protein folding stability and dynamics imaged in a living cell. Nat Methods 7:319–323

Ferrone FA, Hofrichter J, Sunshine HR, Eaton WA (1980) Kinetic studies on photolysis-induced gelation of sickle cell hemoglobin suggest a new mechanism. Biophys J 32(1):361–380

Girelli A, Rahmann H, Begam N, Ragulskaya A, Reiser M, Chandran S, Westermeier F, Sprung M, Zhang F, Gutt C, Schreiber F (2021) Microscopic dynamics of liquid-liquid phase separation and domain coarsening in a protein solution revealed by X-ray photon correlation spectroscopy. Phys Rev Lett 126(13):138004

Greenfield NJ (2006a) Analysis of the kinetics of folding of proteins and peptides using circular dichroism. Nat Protoc 1(6):2891–2899

Greenfield NJ (2006b) Analysis of the kinetics of folding of proteins and peptides using circular dichroism. Nat Protoc 1:2876–2890

Keiderling TA, Xu Q (2002) Unfolded peptides and proteins studied with infrared absorption and vibrational circular dichroism spectra. Adv Protein Chem 62:111–161

Kilburn D, Roh JH, Guo L, Briber RM, Woodson SA (2010) Molecular crowding stabilizes folded RNA structure by the excluded volume effect. J Am Chem Soc 132(25):8690–8696

Kilburn D, Roh JH, Behrouzi R, Briber RM, Woodson SA (2013) Crowders perturb the entropy of RNA energy landscapes to favor folding. J Am Chem Soc 135(27):10055–10063

Köhn B, Kovermann M (2019) Macromolecular crowding tunes protein stability by manipulating solvent accessibility. Chembiochem 20(6):759–763

Kozer N, Kuttner YY, Haran G, Schreiber G (2007) Protein-protein association in polymer solutions: from dilute to semidilute to concentrated. Biophys J 92(6):2139–2149

Laurent TC, Ogston AG (1963) The interaction between polysaccharides and other macromolecules. 4. The osmotic pressure of mixtures of serum albumin and hyaluronic acid. Biochem J 89:249–253

Makowski L, Rodi DJ, Mandava S, Minh DD, Gore DB, Fischetti RF (2008) Molecular crowding inhibits intramolecular breathing motions in proteins. J Mol Biol 375(2):529–546

Miklos AC, Li C, Sharaf NG, Pielak GJ (2010) Volume exclusion and soft interaction effects on protein stability under crowded conditions. Biochemistry 49(33):6984–6991

Minton AP (1977) Non-ideality and the thermodynamics of sickle-cell hemoglobin gelation. J Mol Biol 110(1):89–103

Minton AP (1980) Excluded volume as a determinant of protein structure and stability. Biophys J 32:77–79

Minton AP (1983) The effect of volume occupancy upon the thermodynamic activity of proteins: some biochemical consequences. Mol Cell Biochem 55:119–140

Minton AP (2000) Effect of a concentrated "inert" macromolecular cosolute on the stability of a globular protein with respect to denaturation by heat and by chaotropes: a statistical-thermodynamic model. Biophys J 78:101–109

Möller J, Sprung M, Madsen A, Gutt C (2019) X-ray photon correlation spectroscopy of protein dynamics at nearly diffraction-limited storage rings. IUCrJ 6(Pt 5):794–803

Moron M, Al-Masoodi A, Lovato C, Reiser M, Randolph L, Surmeier G, Bolle J, Westermeier F, Sprung M, Winter R, Paulus M, Gutt C (2022) Gelation dynamics upon pressure-induced liquid-liquid phase separation in a water-lysozyme solution. J Phys Chem B 126(22):4160–4167

Munishkina LA, Cooper EM, Uversky VN, Fink AL (2004) The effect of macromolecular crowding on protein aggregation and amyloid fibril formation. J Mol Recognit 17(5):456–464

Ogston AG (1970) On the interaction of solute molecules with porous networks. J Phys Chem 74:668–669

Otto F, Dallari F, Westermeier F, Wieland D, Parak W, Lehmkühler F, Schulz F (2024) The dynamics of PEG-coated nanoparticles in concentrated protein solutions up to the molecular crowding range, vol 5. Aggregate

Pace CN, Shirley BA, McNutt M, Gajiwala K (1996 Jan) Forces contributing to the conformational stability of proteins. FASEB J 10(1):75–83

Pastore A, Martin SR, Politou A, Kondapalli KC, Stemmler T, Temussi PA (2007) Unbiased cold denaturation: low- and high-temperature unfolding of yeast frataxin under physiological conditions. J Am Chem Soc 129(17):5374–5375

Pastore A, Martin SR, Temussi PA (2019) Generalized view of protein folding: in Medio stat Virtus. J Am Chem Soc 141(6):2194–2200

Perakis F, Gutt C (2020) Towards molecular movies with X-ray photon correlation spectroscopy. Phys Chem Chem Phys 22(35):19443–19453

Perillo MA, Burgos I, Clop EM, Sanchez JM, Nolan V (2023) The role of water in reactions catalysed by hydrolases under conditions of molecular crowding. Biophys Rev 15(4):639–660

Peters J, Oliva R, Caliò A, Oger P, Winter R (2023) Effects of crowding and Cosolutes on biomolecular function at extreme environmental conditions. Chem Rev 123(23):13441–13488

Pielak GJ, Li C, Miklos AC, Schlesinger AP, Slade KM, Wang GF, Zigoneanu IG (2009) Protein nuclear magnetic resonance under physiological conditions. Biochemistry 48(2):226–234

Politou A, Temussi PA (2015) Revisiting a dogma: the effect of volume exclusion in molecular crowding. Curr Opin Struct Biol 30:1–6

Pounot K, Schirò G, Levantino M (2023) Tracking the structural dynamics of proteins with time-resolved X-ray solution scattering. Curr Opin Struct Biol 82:102661

Puglisi R, Karunanithy G, Hansen DF, Pastore A, Temussi PA (2021) The anatomy of unfolding of Yfh1 is revealed by site-specific fold stability analysis measured by 2D NMR spectroscopy. Commun Chem 4(1):127

Reiser M, Girelli A, Ragulskaya A, Das S, Berkowicz S, Bin M, Ladd-Parada M, Filianina M, Poggemann HF, Begam N, Akhundzadeh MS, Timmermann S, Randolph L, Chushkin Y, Seydel T, Boesenberg U, Hallmann J, Möller J, Rodriguez-Fernandez A, Rosca R, Schaffer R, Scholz M, Shayduk R, Zozulya A, Madsen A, Schreiber F, Zhang F, Perakis F, Gutt C (2022) Resolving molecular diffusion and aggregation of antibody proteins with megahertz X-ray free-electron laser pulses. Nat Commun 13(1):5528

Ross PD, Minton AP (1977) Hard quasispherical model for the viscosity of hemoglobin solutions. Biochem Biophys Res Commun 76(4):971–976

Royer CA (1995) Fluorescence Spectroscopy. In: Shirley BA (ed) Protein stability and folding. Methods in molecular biology™, vol 40. Humana Press

Sanfelice D, Politou A, Martin SR, De Los RP, Temussi P, Pastore A (2013) The effect of crowding and confinement: a comparison of Yfh1 stability in different environments. Phys Biol 10(4):045002

Sarkar M, Smith AE, Pielak GJ (2013) Impact of reconstituted cytosol on protein stability. Proc Natl Acad Sci USA 110(48):19342–19347

Schlesinger AP, Wang Y, Tadeo X, Millet O, Pielak GJ (2011) Macromolecular crowding fails to fold a globular protein in cells. J Am Chem Soc 133(21):8082–8085

Serrano AL, Waegele MM, Gai F (2012 Feb) Spectroscopic studies of protein folding: linear and nonlinear methods. Protein Sci 21(2):157–170

Sharp KA (2015) Analysis of the size dependence of macromolecular crowding shows that smaller is better. Proc Natl Acad Sci USA 112:7990–7995

Soranno A, Koenig I, Borgia MB, Hofmann H, Zosel F, Nettels D, Schuler B (2014) Single-molecule spectroscopy reveals polymer effects of disordered proteins in crowded environments. Proc Natl Acad Sci USA 111(13):4874–4879

Spilotros A, Levantino M, Schirò G, Cammarata M, Wulff M, Cupane A (2012) Probing in cell protein structural changes with time-resolved X-ray scattering. Soft Matter 8. https://doi.org/10.1039/C2SM25676B

Srere PA (2000) Macromolecular interactions: tracing the roots. Trends Biochem Sci 25:150–153

Stagg L, Zhang SQ, Cheung MS, Wittung-Stafshede P (2007) Molecular crowding enhances native structure and stability of alpha/beta protein flavodoxin. Proc Natl Acad Sci USA 104(48):18976–18981

Stringer MA, Cubuk J, Incicco JJ, Roy D, Hall KB, Stuchell-Brereton MD, Soranno A (2023) Excluded volume and weak interactions in crowded solutions modulate conformations and RNA binding of an intrinsically disordered tail. J Phys Chem B 127(26):5837–5849

Szyperski T et al (2006) Combined NMR-observation of cold denaturation in supercooled water and heat denaturation enables accurate measurement of deltaC(p) of protein unfolding. Eur Biophys J 35:363–366

Theillet FX, Binolfi A, Frembgen-Kesner T, Hingorani K, Sarkar M, Kyne C, Li C, Crowley PB, Gierasch L, Pielak GJ, Elcock AH, Gershenson A, Selenko P (2014) Physicochemical properties of cells and their effects on intrinsically disordered proteins (IDPs). Chem Rev 114(13):6661–6714

Wang Y, He H, Li S (2010a May) Effect of Ficoll 70 on thermal stability and structure of creatine kinase. Biochemistry (Mosc) 75(5):648–654

Wang Y, Li C, Pielak GJ (2010b) Effects of proteins on protein diffusion. J Am Chem Soc 132(27):9392–9397

Wang Y, Benton LA, Singh V, Pielak GJ (2012a) Disordered protein diffusion under crowded conditions. J Phys Chem Lett 3(18):2703–2706

Wang Y, Sarkar M, Smith AE, Krois AS, Pielak GJ (2012b) Macromolecular crowding and protein stability. J Am Chem Soc 134(40):16614–16618

Wang Y, Sukenik S, Davis CM, Gruebele M (2018) Cell volume controls protein stability and compactness of the unfolded state. J Phys Chem B 122(49):11762–11770

Yamin G, Munishkina LA, Karymov MA, Lyubchenko YL, Uversky VN, Fink AL (2005) Forcing nonamyloidogenic beta-synuclein to fibrillate. Biochemistry 44(25):9096–9107

Zimmerman SB, Trach SO (1991) Estimation of macromolecule concentrations and excluded volume effects for the cytoplasm of Escherichia coli. J Mol Biol 222:599–620

Chapter 21
Molecular Crowding by Computational Approaches

Orkid Coskuner-Weber, Mert Koca, and Vladimir N. Uversky

Abstract Molecular crowding plays a crucial role in biological and medicinal systems, impacting the structure, behavior, and function of biomolecules within the densely packed environments of cells. This chapter provides an overview of the implications of molecular crowding, exploring how the high concentration of macromolecules such as proteins, nucleic acids, and other biological entities impacts biochemical reactions and cellular processes. The discussion highlights the challenges associated with experimental studies of molecular crowding, including challenges in creating accurate in vitro models, controlling concentrations, and isolating crowding effects from other interactions. To address these challenges, the chapter emphasizes the importance of computational techniques. Various computational approaches, including molecular dynamics simulations, Monte Carlo simulations, Brownian dynamics, lattice-models, finite element analysis, coarse-grained modeling, quantum mechanics/molecular mechanics simulations, and multi-scale modeling, are discussed in detail. Each of these techniques contributes unique insights into the molecular-level impacts of crowding, enhancing our understanding of biophysical processes critical for therapeutic development and biological function. The chapter discusses also quantum computing, machine learning and classical simulations hybrid approaches for future directions around molecular crowding studies.

Keywords Molecular dynamics (MD) simulations · Quantum computing · Machine learning · Simulations hybrid approaches · Monte Carlo simulations · Brownian dynamics simulations · Lattice-based models · Coarse-grained models · Finite element analysis · Multi-scale modeling

---

O. Coskuner-Weber (✉) · M. Koca
Molecular Biotechnology, Turkish-German University, Istanbul, Turkey
e-mail: weber@tau.edu.tr

V. N. Uversky (✉)
Department of Molecular Medicine and USF Health Byrd Alzheimer's Research Institute, Morsani College of Medicine, University of South Florida, Tampa, FL, USA
e-mail: vuversky@usf.edu

V. N. Uversky (ed.), *(Macro)Molecular Crowding*, Subcellular Biochemistry 109,
https://doi.org/10.1007/978-3-032-03370-3_21

Introduction

Molecular crowding studies in biology and medicine focus on understanding how the high concentration of macromolecules within cellular environments impacts the behavior, structure, and function of biomolecules (Zhou et al. 2008). In a cellular environment, there is a dense population of nucleic acids, proteins, and other biological macromolecules, leading to what is defined as a "crowded" environment (Kuznetsova et al. 2014). Crowding can have various aspects in biochemistry and biophysics (Burg 2000). For instance, it can stabilize specific protein conformations and affect folding pathways, and even impact protein stability due to the exclusion volume effect, which limits the space available for molecules to unfold (Christiansen et al. 2013). Furthermore, the dynamics of biochemical reactions can change significantly under crowded conditions, often accelerating reactions or altering kinetic parameters because molecules are in closer proximity (Weiss 2014). Additionally, crowded environments can enhance or inhibit interactions between macromolecules, affecting processes such as protein complex formation and enzyme activity (Mittal et al. 2015). Understanding molecular crowding is crucial in the development of therapeutics as the effective concentration and behavior of drugs can differ in crowded physiological environments compared to dilute laboratory conditions (Alfano et al. 2024).

Even though experiments are very useful in the studies of molecular crowding, these studies face several limitations due to the complexity of cellular conditions and accurately measuring the effects of crowding (Nakano et al. 2014). Creating accurate in vitro models that mimic the highly crowded and heterogeneous environment of a cell is challenging (Rodenhizer et al. 2018). The variability and complexity of cellular environments are difficult to replicate in laboratory settings (Nicholson et al. 2004). It is also very difficult to control and measure the concentrations of multiple macromolecules in experimental setups (Minton 2001). Moreover, many molecular phenomena affected by crowding, such as diffusion processes, occur on short time scales that can be challenging to capture with conventional experimental tools (Hall and Hoshino 2010). In addition, the presence of a multitude of interacting species makes it difficult to isolate the effects of crowding from other interactions (Rivas and Minton 2016). Unfortunately, NMR and fluorescence spectroscopy may not always be sensitive or specific enough to detect changes in molecular behavior due to crowding (Alfano et al. 2024).

Computational studies can complement experiments. In fact, computational approaches play a crucial role in molecular crowding studies by providing insights that are difficult to obtain through experimental studies alone (Musiani and Giorgetti 2017). Specifically, computational models can simulate crowded cellular environments, helping to predict how macromolecules behave under varying conditions (Feig et al. 2017). Moreover, algorithms and simulations can assess how molecular crowding effects biochemical reactions, including changes in equilibria and reaction rates, which may then be validated by experiments (Schnell and Turner 2004). Simulations provide insights at an atomic level into how crowding influences the structure, folding and interactions of proteins and nucleic acids and we can even define transient states and intermediate conformations with such simulations (Grassmann et al. 2024). Furthermore, computational tools can predict how crowding impacts molecular recognition and binding affinities, providing insights into protein and protein interactions or protein-DNA interactions (Rakers et al. 2015; Peters et al. 2023).

Different computational approaches and techniques are used to investigate molecular crowding and each technique/approach offers unique insights into how crowded environments impact biological systems. In this chapter, we describe molecular dynamics simulations, Monte Carlo simulations, Brownian dynamics simulations, lattice-based models, coarse-grained models, finite element analysis, and multi-scale modeling in the studies of molecular crowding in biology and medicine.

Molecular Dynamics (MD) Simulations

MD simulations are a computational technique used to simulate the physical movements of atoms and molecules over time (Badar et al. 2022). These simulations are based on classical mechanics, where Newton's laws of motion are utilized to calculate the trajectory of atoms (Allison et al. 2011). The positions, velocities, and accelerations of particles are calculated iteratively over small time steps (Adcock and McCammon 2006). A set of parameters known as a force field is used to estimate the forces acting between particles (Wise and Coskuner 2014). These force fields include bonded and non-bonded interactions (Allison et al. 2011). Bonds, angles, and dihedrals define the bonded terms while van der Waals forces and electrostatics define non-bonded interactions. The potential energy of the system is calculated using a force field, and the gradients of this potential energy determine the forces acting on each particle (Ponder and Case 2003). The simulation starts with an initial configuration that is usually based on X-ray or NMR spectroscopy measurements (Robustelli et al. 2010). The simulation progresses in discrete time steps, usually on the order of femtoseconds, enabling the observation of molecular behavior over nanoseconds to microseconds (Schlick et al. 1997). Periodic boundary conditions are usually applied for simulating a bulk environment, where particles leaving one side of the simulation box re-enter on the opposite side, which reduces edge effects (Allison et al. 2011). Thermostats and barostats are used for controlling the temperature and pressure of the system, which enables the simulation of real-world conditions (Allison et al. 2011).

MD simulations are useful in molecular crowding studies as these simulations offer detailed insights into how the dense cellular environment impact molecular behavior at an atomic level (Feig et al. 2017). Using this technique, we can simulate systems containing a high density of macromolecules, providing a good model of the intracellular environment (Feig et al. 2017). In crowded conditions, this technique helps to determine how proteins fold and unfold, exploring the stability of different conformational states due to excluded volume effects (Samiotakis et al. 2009). MD simulations also help in mapping folding pathways that might differ from those in dilute conditions (Cheung and Thirumalai 2007). Using such simulations, we can study how crowding affects the kinetics and thermodynamics of reactions (Zhou and Qin 2013). Such investigations help to understand changes in reaction rates and pathways in a cellular context. More importantly, MD simulations can provide insights into how molecular crowding modulates interactions between proteins, nucleic acids, and other macromolecules (Fig. 21.1) (Pastore et al. 2024). Such studies can be crucial for investigating processes like signaling or metabolic pathways. In fact, MD simulations have been utilized to demonstrate how molecular crowding stabilizes protein conformations that might otherwise be unstable in dilute solution (Mittal et al. 2015). For instance, the compactness of proteins and their diffusion coefficients can be altered in crowded environments (Harada et al. 2012). Moreover, MD simulations are used to simulate the cytoplasm, which is a highly crowded environment, for gaining insights into enzyme behavior that can differ from in vitro conditions (Huggins et al. 2019). Such knowledge provides better predictions of enzyme dynamics under physiological conditions. Furthermore, MD

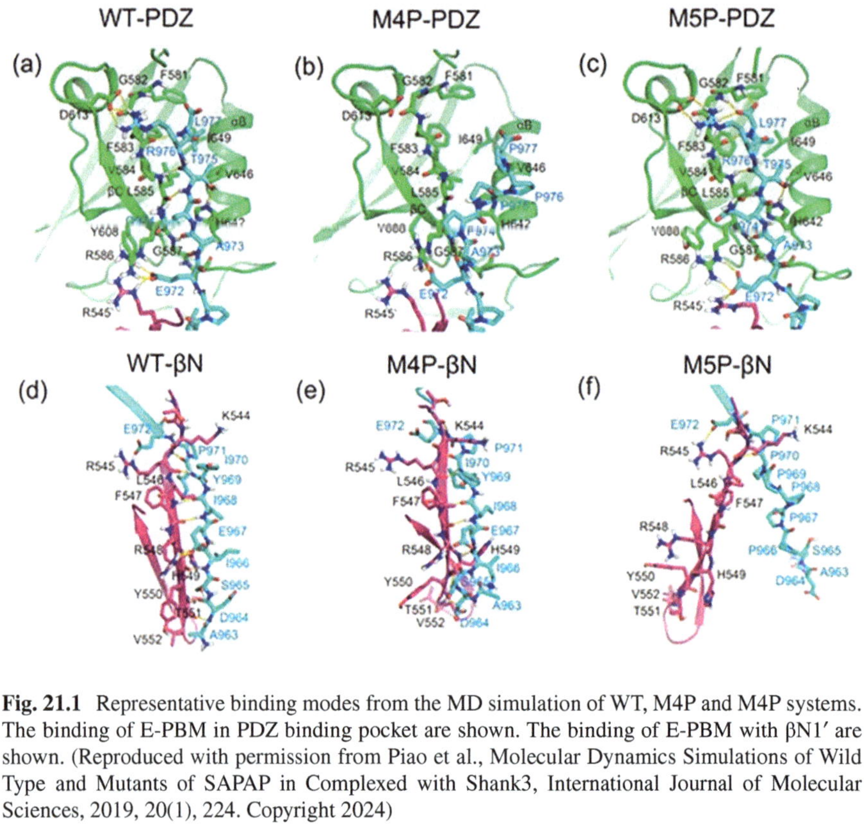

Fig. 21.1 Representative binding modes from the MD simulation of WT, M4P and M4P systems. The binding of E-PBM in PDZ binding pocket are shown. The binding of E-PBM with βN1′ are shown. (Reproduced with permission from Piao et al., Molecular Dynamics Simulations of Wild Type and Mutants of SAPAP in Complexed with Shank3, International Journal of Molecular Sciences, 2019, 20(1), 224. Copyright 2024)

simulations have been used to study how crowding impacts the structure and dynamics of DNA and RNA (Verdian Doghaei et al. 2015). Specifically, how crowding influences the binding and unbinding of nucleic acids with proteins or small molecules is an active area of research. Moreover, these simulations are used to study how drugs interact with their targets in crowded cellular environments, leading to the redesign of drugs for better efficiency in vivo (Muratov et al. 2021). All in all, MD simulations provide atomistic detail and dynamic insights that are difficult to capture through experiments alone (Allison et al. 2011). Therefore, this technique has an impact on our understanding of molecular crowding effects in biology and medicine.

On the other hand, we have to mention that MD simulations can be computationally expensive, especially for large size systems and long simulation times (Hollingsworth and Dror 2018). Furthermore, as we have mentioned in earlier studies, the reliability of results depends heavily on the accuracy of the chosen force field (Mandaci et al. 2020; Caliskan et al. 2021). In addition, MD simulations capture fast molecular motions, however, longer time-scale processes such as large conformational changes or protein folding often remain challenging (Lane et al. 2013).

Monte Carlo (MC) Simulations

MC simulations are a statistical method used to simulate and analyze complex systems and processes that involve significant uncertainty and variability (Coskuner and Deiters 2007). Named after the famous Monte Carlo casino due to the element of chance involved, this technique uses randomness to solve problems that might be deterministic in their nature (Gobet 2016). At the core of MC simulations is the concept of random sampling (Gobet 2016). By running numerous simulations, we compute statistical averages of the outcomes, which provide estimates of expected values, variances, and probability distributions (Bonate 2001). MC methods are employed for simulating molecular systems, such as predicting thermodynamic properties or studying molecular interactions in crowded environments (Vitalis and Pappu 2009). By employing randomness and statistical analysis, we gain insights into complex problems.

As mentioned above, MC simulations are useful in studying molecular crowding and they provide an effective way to simulate complex systems with many interacting components. This is very useful especially when conventional analytical approaches become challenging. MC simulations, for example, help to explore high-dimensional conformational spaces of biomolecules, which is important when studying how crowding impacts stability and interaction (Mohanty 2020). By relying on random sampling, these simulations can mimic the realities of molecular interactions in crowded environments. Such studies account for randomness and fluctuations that occur in biological systems. MC is usually computationally more efficient than MD simulations for certain types of analyses, especially those involving equilibria and sampling trajectories over different energy landscape (Paquet and Viktor 2015). In fact, MC simulations have been used to study protein stability in crowded environments. For instance, the folding and binding of proteins differ in the presence of crowding agents by MC simulations (Prytkova et al. 2016). By sampling different spatial configurations, we can study how crowding impacts the likelihood of aggregation or misfolding (Morriss-Andrews and Shea 2015). Furthermore, using MC simulations, one can model enzyme kinetics under crowded conditions (Anand et al. 2019). Such simulations help to understand the impacts of molecular crowding on reaction rates. For instance, such simulations have illustrated how crowding alters the reactivity and diffusion of substrates, leading to enhanced or inhibited enzymatic reactivity (Balcells et al. 2014). In the studies of nucleic acids, these simulations are useful to investigate how the binding of proteins to DNA is affected by the presence of other macromolecules (Bombelli et al. 2008). By simulating crowded environments, we can analyze changes in binding affinities of nucleoprotein complexes. Furthermore, MC simulations have been applied to model the release profiles of drugs within crowded biological environments (Bairagee et al. 2024). Overall, the ability to simulate complex systems with a high degree of variability makes MC simulations extremely useful in gaining insights into molecular crowding phenomena and their implications in biology and medicine.

We have to mention here that depending on the complexity of the model and the number of simulations required, MC simulations can be computationally expensive and time-consuming. The accuracy of the results relies heavily on the quality of input probability distributions used to present uncertainties (Janssen 2013). While accuracy improves with more iterations, gathering sufficient samples for convergence can be challenging, particularly in high-dimensional spaces (Janssen 2013).

Brownian Dynamics (BD) Simulations

BD simulations are a tool used to study the motion of particles suspended in a fluid, particularly in biological systems (Huber and McCammon 2019). These simulations build on the principles of Brownian motion, which describes the random movement of particles in a fluid due to collisions with the molecules of the surrounding medium (Chen and Kim 2004). Brownian motion is the random movement of particles resulting from the thermal agitation of molecules in a fluid (Philipse 2018). This motion is characterized by a series of stochastic movements and is impacted by temperature and viscosity (Philipse 2018). The dynamics of a particle undergoing Brownian motion can be described by the Langevin equation, which incorporates both deterministic forces such as potential energy gradients and stochastic forces which account for thermal fluctuations (Sung 2018). The random motion can be further characterized by the diffusion constant, which describes how quickly a particle diffuses through a medium (Dickinson 1985). The Langevin equation is discretized to evolve the position and velocity of the particles over time (Schuss 2013). Usually, this involves converting continuous time dynamics into discrete time steps (Schuss 2013). These simulations proceed in small time increments, where the new position of particles is calculated based on their previous position, deterministic forces, and a random force sampled from a Gaussian distribution (Schuss 2013). These simulations can incorporate different types of particle interactions, such as hard-sphere collisions, steric repulsions, or these can be dictated by potentials (Karimnejad et al. 2022). BD simulations help to study the crowding of macromolecules related motion and interactions, mimicking the crowded environments found in living cells (Długosz and Trylska 2011). Through their ability to capture both deterministic and stochastic processes, BD simulations enhance our understanding of various biological phenomena. BD simulations are useful in the studies of molecular crowding due to their ability to simulate the random motion of particles subjected to thermal fluctuations in crowded environments (Ando et al. 2016). In fact, BD simulations simulate how particles behave in environments with high concentrations of macromolecules and capture thereby the effects of steric hindrance and excluded volume that characterize crowded cellular conditions (Fig. 21.2) (Muñiz-Chicharro et al. 2023). Moreover, BD simulations are effective in evaluating how molecular crowding affects diffusion rates, which is critical for understanding processes such as enzyme kinetics, molecular interactions, and transport mechanisms within cells (Weiss 2014). These simulations can often be computationally more efficient than detailed MD simulations, especially when we study systems where long-range interactions are important but precise atomic-details are less critical (Ozboyaci et al. 2016). Specifically, BD simulations have been conducted to study how molecular crowding affects the kinetics of enzymatic reactions (Echeverria and Kapral 2015). In the studies of protein dynamics, BD simulations can help to understand how folding pathways and stability are affected when proteins are in crowded cellular environments (Basak et al. 2019). It was shown that crowding can stabilize certain intermediate states of protein folding, affecting the overall folding landscape (Rojnuckarin et al. 1998). Moreover, these simulations are useful for studying how the presence of crowding agents affects the

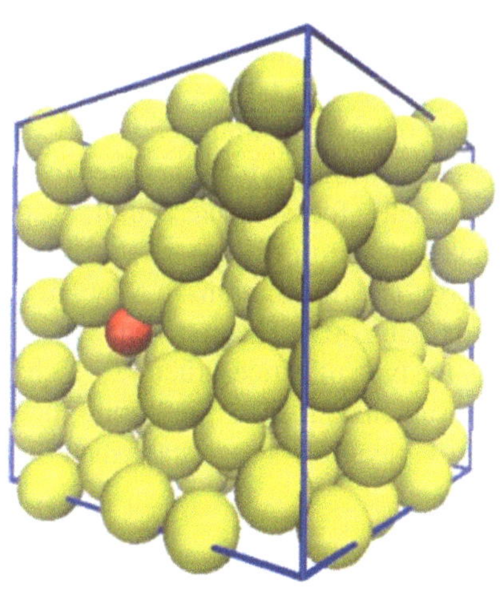

Fig. 21.2 Snapshot of one of the performed dynamics. A protein (in red) diffuses among dextran molecules (in yellow) which act as crowding agents. (Reproduced with permission from Blanco et al., Brownian Dynamics Computational Model of Protein Diffusion in Crowded Media with Dextran Macromolecules as Obstacles, Entropy, 2017, 19(3), 105. Copyright 2024)

binding dynamics of proteins to DNA or RNA (Tabaka et al. 2014). For instance, the diffusion and binding processes of transcription factors in a crowded chromatin environment was simulated by BD simulations (Shim et al. 2020). This provided insights into gene regulation mechanisms. In addition, these simulations can model the self-assembly of lipid membranes or protein complexes in crowded environments, providing insights into how crowding affects the efficiency and kinetics of assembly processes (Mathur et al. 2024). BD simulations were used to study how the presence of crowding agents can facilitate or hinder the formation of lipid bilayers (Alfano et al. 2024). Finally, BD simulations have been conducted to study how nanoparticles diffusing through a crowded environments interact other biomolecules (Al-Obaidi and Florence 2015). Such studies help to understand the effectiveness of drug delivery systems in vivo.

However, while BD simulations can incorporate various interactions as described above, these simulations often simplify complex interactions. This may result in not capturing all dynamics accurately. Furthermore, these simulations often focus on relatively short time scales due to the stochastic nature of the dynamics (Hänggi and Marchesoni 2005). Some processes that occur over longer time scales may require enhanced sampling techniques. For a system with a large number of particles or complex interactions, these simulations can become computationally expensive (Hänggi and Marchesoni 2005).

Lattice-Based Models

Lattice-based models are computational frameworks commonly used in biological modeling to study systems whose components can be represented on a discrete grid or lattice (Bittig and Uhrmacher 2010). These models are significant for their capability to simplify complex interactions within a system while still providing insights into various physical, chemical, and biological processes (Van Liedekerke et al. 2015). At the core of lattice-based models is the grid—a lattice—where each point or site of the lattice represents a possible configuration or state of the simulated system (Dallon 2007). Each point is connected to its neighboring points, enabling interactions to be simulated locally (Nikolić et al. 2018). Unlike continuous models that deal with varying quantities over a continuum, lattice models discretize space and usually focus on interactions at specific points. Usually lattice-based models are rooted in the principles of statistical mechanics and use partition functions, free energy, and distributions to calculate the thermodynamic properties of systems (Cao and Chen 2005). The first step in these simulations is to define the lattice, i.e., selecting the type of lattice structure which can be square, triangular or cubic for example, and dimensions for representing the system. Next, interaction rules between adjacent lattice points are established, which can range from simple nearest-neighbor interactions to more complex, long-range interactions. Lattice models can be explored using techniques like Monte Carlo simulations for sampling configurations and estimating thermodynamic properties, or lattice dynamics can be conducted for examining time evolution and state changes (Skolnick and Kolinski 1991). In MC simulations, for example, an updating algorithm such as Metropolis algorithm or cluster update method could be used to sample states by probabilistically accepting or rejecting change based on energy consideration. Lattice models are applied to study processes such as protein folding, molecular crowding, and the dynamics of cellular processes. These models greatly simplify complex systems by reducing the degrees of freedom and focusing on local interactions. They are particularly useful as these models help to study large populations of interacting particles.

These models are useful in studying molecular crowding due to their ability to simplify complex interactions and facilitate the exploration of how crowding affects molecular behavior in confined environments. Specifically, lattice models provide a discrete representation of space, enabling the simulations of crowded environments where macromolecules interact. By focusing on nearest-neighbor interactions, lattice models can effectively capture the essential effects of steric hindrance and excluded volume that arise in crowded environments (Li et al. 2021). These approaches employ statistical mechanics principles to derive thermodynamic properties and phase behavior. This helps to predict how crowding affects reaction kinetics and molecular stability. These models can be more computationally efficient than MD simulations or BD simulations especially when scaling up to study larger systems or longer timescales. Lattice models of proteins, such as the HP (hydrophobic-polar) model have been used to study how proteins fold in crowded environments (Takahashi et al. 2005). These models can reveal how different folding pathways are impacted by the presence of other macromolecules, highlighting the stability and aggregation tendencies under crowding (Visser et al. 2024). These models were employed to simulate enzymatic reactions in crowded media. For instance, lattice model studies have shown that crowding alters the diffusion of substrates toward enzymes (Weilandt and Hatzimanikatis 2019). Lattice random walk models can describe the diffusion of particles in crowded systems (Abbaspour and Klumpp 2021). By simulating how molecules diffuse through a lattice filled with other molecules, we can understand how crowding affects diffusion coefficients and spatial distribution.

However, we should mention here that the discrete nature of lattice models can introduce artifacts that may not accurately reflect continuous systems (Park and Levitt 1995). Furthermore, by focusing on local interactions, these models may overlook long-range interactions or spatial heterogeneity present in real systems (Choi et al. 2019). Depending on the system and nature of interactions, lattice models may become complex and require high computational resources as the size of the system increases.

Coarse-Grained (CG) Models

CG models are a simplification technique used in the studies of complex molecular systems by reducing the level of detail (Noid 2013). Rather than considering every atom in a molecule, CG models aggregate groups of atoms into larger units or "beads" (Ingólfsson et al. 2014). This approach allows us to simulate larger systems or longer time scales that would be computationally expensive or impossible to achieve with all-atom models (Ingólfsson et al. 2014). Specifically, CG involves simplifying a molecular system by grouping atoms into larger structural units (Takada et al. 2015). Each bead in the model represents a cluster of atoms or a functional group (Takada et al. 2015). This reduces the overall number of particles without significant loss of important physical properties (Takada et al. 2015). The interactions between CG beads are typically represented by effective potentials that account for the contributions of the multiple atoms in each bead (Marrink et al. 2023). This approach helps to preserve the system's thermodynamics and kinetics while simplifying the representation (Marrink et al. 2023). CG models can capture collective phenomena and dynamics that occur over large scales and longer time frames (Pak and Voth 2018). This makes them suitable for studying processes that are challenging to simulate using detailed atomistic models. The first step involves deciding how to group atoms into beads based on their interactions and structural coherence. This may be determined by functional groups, secondary structure elements, or geometric considerations (Wassenaar et al. 2014). Once beads are defined, the next step involves defining the potentials that govern interactions between beads (Zhou et al. 2007). This can be achieved through various methods, including fitting to experimental data or higher-resolution simulations. CG models can be used with various simulation techniques such as MD or MC (Izvekov and Voth 2005). CG MD simulations can efficiently model the system dynamics over longer time scales. CG MC simulations can be employed to explore conformational space and to calculate thermodynamic properties (Gniewek and Kolinski 2010). CG models are widely used in the studies of proteins, nucleic acids, and lipid membranes. For instance, these models help simulate protein folding and dynamics over long time scales (Derreumaux and Mousseau 2007). CG is instrumental in simulating lipid bilayers and membrane proteins, enabling to investigate the effects of membrane composition on protein dynamics and ligand binding (Cascella and Vanni 2015). By reducing the number of beads to be simulated, CG models can simulate larger systems and longer simulation time scales. These models can help to focus on essential interactions and/or collective behaviors.

CG models have been useful in the studies of molecular crowding due to their ability to simplify complex systems while capturing essential physical behaviors and interactions (Grassmann et al. 2024). As mentioned above, these models reduce the number of degrees of freedom by grouping atoms into larger beads or particles. This simplification enables us to simulate larger systems and explore how molecular crowding affects molecular interactions without the computational burden of an all-atom approach. Furthermore, CG models facilitate the studies of processes that occur over larger timescales, which are critical in molecular crowding where diffusion, binding, and conformational changes can take significant time to unfold (Bhattacherjee et al. 2016). These models are adept at capturing collective behaviors in crowded environments, such as diffusion coefficients, reaction kinetics, and assembly processes, which may be obscured in more detailed models (Bhattacherjee et al. 2016).

CG models have been widely used to study the folding of proteins in crowded environments, such as the effects of macromolecular crowding on the folding kinetics of proteins such as chaperonins (Hyeon and Thirumalai 2011). These simulations have shown that crowding can either stabilize or destabilize intermediates during folding pathways (Zhang et al. 2009). CG models have been also conducted in the studies of enzyme-substrate interactions and how crowding agents affect such interactions (Echeverria and Kapral 2015). In studies of DNA and RNA, CG models are used to study how crowding impacts the binding dynamics of proteins and nucleic acids (Hong et al. 2020). For instance, simulations of transcription factor binding to DNA showed that crowding can impact binding affinities by altering the available conformations of both the DNA and proteins (Guardiani et al. 2014). Moreover, CG models of lipid membranes can help study how crowding impacts the dynamics of membrane proteins and their interactions within a crowded membrane environment (Heo et al. 2022). For instance, simulations explored how crowded lipid systems alter the lateral diffusion of proteins and lipid rafts' formation (Guigas and Weiss 2016). An interesting area of application includes CG modeling in self-assembly processes of structures such as vesicles or micelles and how these processes can be impacted by crowding conditions (Musiani and Giorgetti 2017). An example is provided by the simulations of molecular crowding modulations of the kinetics of self-assembly that leads to different morphologies (Grime et al. 2016). Another interesting area of research involves the application of CG models in the studies of cellular processes, such as the diffusion of signaling molecules and how crowding impacts these movements (Xie et al. 2014). For instance, CG models were applied to study the distribution and transport of signaling molecules in a crowded cytoplasm for understanding intracellular signaling pathways (Ridgway et al. 2008).

Despite, we have to mention here, the primary drawback is that CG inevitably loses some atomistic details, which in turn may be critical for studying certain biological phenomena, such as specific bond formations (Cascella and Vanni 2015).

Finite Element Analysis (FEA)

FEA is a numerical method used for solving complex problems (Srirekha and Bashetty 2010). FEA helps in understanding how structures behave under different physical conditions, such as heat, forces, and fluid flow (Okereke and Keates 2018). The main idea of FEA is to break down a complex system which we call the "domain" into smaller, simpler parts called finite elements (Givoli and Keller 1989). This process is known as discretization (Givoli and Keller 1989). These elements can be in the form of 1D line elements, 2D surface elements (triangles or quadrilaterals), or 3D volume elements (tetrahedra or hexahedra) (Cohen and Pernet 2017). These elements are connected at points called nodes. The physical behavior of the entire domain can be expressed as a sum of the behaviors of these elements (Argyris et al. 1979). For each finite element, equations based on physical laws such as equilibrium equations or heat conduction equations are formulated (George et al. 1997). These equations relate the properties of elements and the applied loads to the resulting displacements or temperatures. The stiffness matrix or mass matrix is defined for each element based on the equations and the system's properties (Friswell and Mottershead 1995). The global stiffness matrix of the entire system is assembled from the individual element stiffness matrices while applying the transformations for each element's position (Friswell and Mottershead 1995). Conditions such as fixed supports or external loads are applied to the global system (Friswell and Mottershead 1995). The system of equations derived from the assembled stiffness matrix is solved using numerical methods, such as Gaussian elimination or iterative solvers. This process yields the displacements or temperatures at each node. Coupled with computational fluid dynamics (CFD), FEA can help to study how fluids interact with structures (Sun et al. 2010). In fact, FEA can be applied to a wide range of biological problems incorporating various complex geometries and boundary conditions (Dumont et al. 2009). By refining the mesh, which is the number of elements and their distribution, FEA may yield highly accurate solutions.

FEA Is a useful computational approach in studying molecular crowding, particularly in applications where the mechanical properties and structural dynamics of biological molecules are of interest (Kolston 2000). However, we should note that FEA is more associated with continuum mechanics and structural analysis, while molecular crowding studies often use MC or MD to capture molecular interactions. However, FEA can analyze the mechanical properties of biomolecules, such as proteins and their aggregates in crowded environments, helping us to understand how crowding impacts stability and deformation (Solernou et al. 2018). Interestingly, FEA can model how cellular structures such as membranes or the cytoskeleton respond to external forces, providing insights into the mechanical resilience of cells under crowded conditions (Iakovliev et al. 2018). Moreover, FEA can be employed for biomechanical analyses and can help to understand the impact of molecular crowding on cellular processes, such as the movement of cytoskeletal elements and their interactions in crowded cellular environments (Johnson et al. 2021). FEA has been successfully used to study the mechanical stability of proteins and how their structures respond to the steric hindrance provided by other macromolecules in crowded cellular conditions (Sedeh et al. 2018). For instance, modeling of force distribution within a protein structure highlighted regions at risk of unfolding or denaturation due to crowding (Solernou et al. 2018). Furthermore, FEA can model how the mechanical properties of a cell change with crowded intracellular environments (Kroupa et al. 2023). FEA simulations were conducted for understanding the deformation of red blood cells in crowded capillary network to assess how crowding impact blood flow and cellular interactions (Cupelli et al. 2013). Moreover, FEA is used to analyze the mechanical properties of lipid membranes in crowded environments (Kwak et al. 2010). FEA simulations on the effect of crowding on membrane fluctuation and stability revealed insights into how protein-lipid interactions are altered (Feng and Klug 2006). Finally, FEA was applied to understand the mechanics of vesicle formation and fusion processes in crowded cellular environments (Haspinger et al. 2021). While FEA is usually associated with mechanical systems, its application in studying molecular crowding is emerging as a useful computational approach; specifically, in understanding how structural and mechanical characteristics of biomolecules and cellular systems change under crowded conditions. The combination of FEA with MD can lead to a more comprehensive understanding of molecular crowding phenomena in biology and medicine.

FEA faces certain limitations, and these are related to the analysis which can require significant computational resources, particularly for large and complex models with millions of elements. Furthermore, one needs to be careful with the mesh since the accuracy of the results heavily depends on the quality of the mesh (Burkhart et al. 2013).

Quantum Mechanics/Molecular Mechanics (QM/MM) Studies

QM/MM is a computational technique that combines the principles of quantum mechanics (QM) and molecular mechanics (MM) to study complex systems especially in biochemistry (Allison et al. 2011). This hybrid approach is useful for studying molecular systems where both quantum effects and classical interactions play significant roles, such as in enzymatic reactions and biomolecular interactions (Monard and Merz 1999). QM provides a detailed description of the behavior of electrons and the atomic nuclei in a molecule (Allison et al. 2011). It is important for accurately capturing electronic structure, bond formation, and spectral properties. However, QM calculations can be computationally intensive for large systems. MM is a less computationally demanding technique that treats atoms as classical particles, using force fields to approximate interactions as described above (Allison et al. 2011). This allows for the simulation of larger systems over longer timescales. QM/MM combines both methods by treating a small, chemically interesting region of the system (the QM region) using quantum mechanics, while the rest of the system (the MM region) is treated with classical mechanics (Senn and Thiel 2009). This allows the exploration of complex systems where electronic effects are important while still benefitting from the efficiency of classical mechanics. The QM region contains the atoms involved in a chemical reaction or electronic transition, where quantum effects are significant, such as the reactive site of an enzyme. The MM region consists of the rest of the molecular system that interacts with the QM region but can be adequately treated classically. The definition of the boundary between QM and MM regions is critical (Shurki and Warshel 2003). This often involves identifying which atoms or groups of atoms are directly involved in the electronic interactions and should be included in the QM region (Shaik et al. 2010). It has to be carefully ensured that the delineation minimizes boundary effects while capturing important interactions between the regions (Haji Dehabadi et al. 2024). The QM region is usually treated using methods like Density Functional Theory (DFT) or Hartree-Fock (HF) (Allison et al. 2011). The MM region employs a classical force field and to connect the two regions, the QM and MM regions are coupled through a set of interaction terms (Allison et al. 2011). These interaction terms can include electrostatic interactions whereby the electrostatic potential of the MM atoms acting on the QM region is considered or by van der Waals interactions between QM and MM regions that are typically calculated to account for steric effects (Tzeliou et al. 2022). On the other hand, the effect of the QM region on the MM atoms is also considered in some studies, allowing for polarization of MM atoms by the QM electron density (Ryde 2016). Special techniques are used to address potential artifacts that arise at the QM/MM interface (Csizi and Reiher 2023). In some approaches, additional "link" atoms and these are often hydrogen atoms are used to saturate the bonds at the boundary to prevent unphysical interactions (Morzan et al. 2018). Furthermore, geometric constraints can be applied to maintain the connectivity of the MM and QM regions or solvent effects can be integrated using continuum solvation models for the MM region (Gkionis et al. 2016). QM/MM is a useful approach for studying enzymatic mechanisms because it allows us to model the reactions with quantum accuracy while also considering the larger protein environment (Van Der Kamp and Mulholland 2013). Furthermore, QM/MM is helpful in understanding how drug candidates interact with specific targets at the atomic level, considering electronic effects important in binding (Omer et al. 2015).

QM/MM is also useful in studying molecular crowding, although direct applications in this field are somewhat limited compared to other areas like enzymatic reactions (Sinnecker and Neese 2006). As mentioned above, QM applies quantum mechanical precision to specific areas where chemical reactions occur, such as enzyme active sites, while maintaining a simplified classical model for the surrounding crowded environment. This allows studies on how crowding impacts reaction mechanisms at these localized sites (Zhang et al. 2020). This method can be employed to study how the electronic properties and reactivity of a molecule change due to the presence of crowding agents, using the MM model to simulate the crowding environment and its physical constraints. Specifically, QM/MM helps to elucidate changes in interaction strengths and binding affinities due to crowding, especially for reactions that depend on specific electronic interactions at the site of interest (Gherman et al. 2004). QM/MM can be used to understand how the crowded cellular environment impacts enzyme reaction pathways and kinetics (Velázquez-Libera et al. 2020). By modeling the enzyme's active site with QM and the surrounding crowding macromolecules with MM, insights into catalytic activity can be obtained. Furthermore, QM/MM assists in understanding how molecular crowding affects the folding energy landscape of small peptides or protein domains (Bock et al. 2023). This also involves studying how macromolecular interactions stabilize or destabilize certain conformations. In metabolic pathways, where substrates are channeled between enzyme complexes, QM/MM can be useful in studying how crowding impacts the efficiency of such processes by focusing on critical bond-making breaking stages within active sites, influenced by the surrounding crowded environment (Knop et al. 2023). Moreover, studies investigating the effects of crowding on DNA or RNA stabilization, folding, and interactions may deploy QM/MM to analyze base-pair formation where electronic detail is important (Šponer et al. 2018). Finally, in pharmaceutical research, understanding how drugs interact in crowded environments can inform more realistic in vivo predictions (Huggins et al. 2019). QM/MM is used to simulate how crowding impacts drug binding to protein targets at the quantum level (Haldar et al. 2018).

While the QM/MM approach isa powerful computational tool for studying complex systems, it comes with certain limitations. The interface between QM and MM regions can introduce artifacts if not treated properly (Watanabe and Cui 2019). The use of link atoms or other boundary methods can sometimes lead to spurious effects (König et al. 2005). The accuracy of the MM region heavily depends on the chosen force field's quality (Akbayrak et al. 2020). Many QM/MM implementations are static, meaning they focus on a singe point or limited number of molecular geometries. Simulating time-dependent processes accurately requires advanced techniques like ab initio MD simulations, which are computationally intensive.

Multi-Scale Modeling

Multi-scale modeling is a computational approach used to study complex systems by integrating information and models across multiple spatial and/or temporal scales (Wang et al. 2024). This method is useful in biology and medicine where phenomena at different scales can interact to impact system behaviour (Sloot and Hoekstra 2010). Spatial scales can range from atomic (Angstroms) to macroscopic (meters). Multi-scale models bridge these gaps by combining methods like quantum mechanics for atomic level precision with continuum mechanics for large-scale behaviors (Horstemeyer 2009). Temporal scales can range from femtoseconds to seconds or longer. Different computational techniques are used to address processes occurring over these varying timescales. Multi-scale modeling often involves coupling different levels of theory, using techniques like coarse-graining, where fine-scale details are averaged out to allow for efficient simulation of large systems (Dada and Mendes 2011). In hierarchical modeling, different levels of description are connected hierarchically (Schaffer and Ideker 2021). For instance, results from atomistic simulations might inform parameters in a mesoscale model, which in turn informs a macroscale model (Rhodes et al. 2013). On the other hand, in concurrent coupling, different scales are simulated simultaneously, with data exchange occurring in real time (Qu et al. 2011). This approach is more complex but enables dynamic interaction between scales. However, in coarse-grained modeling, we focus on simplifications of a detailed model by averaging or grouping variables that focus on relevant large-scale behavior (Fig. 21.3) (Jebahi et al. 2016).

Multi-scale modeling is extremely useful in studies of molecular crowding because it can address the complexities arising from the dense and diverse cellular

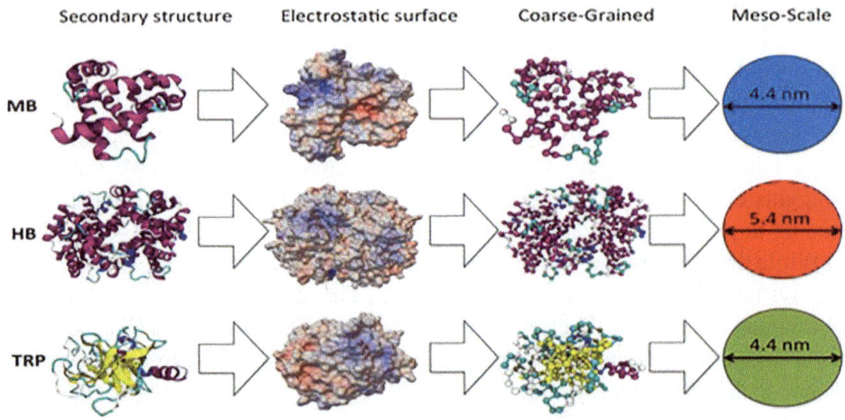

Fig. 21.3 Models of the myoglobin (MB), hemoglobin (HB), and trypsin (TRP) proteins at different scale-levels. The Coarse-Grained model, and the Meso-Scale representation are depicted. (Reproduced with permission from Tavanti et al., Multiscale Molecular Dynamics Simulation of Multiple Protein Adsorption on Gold Nanoparticles, International Journal of Molecular Science, 2019, 20(14), 3539. Copyright 2024)

environment at different levels of detail (Tozzini 2010). Molecular crowding has significant effects on biochemical reactions, molecular interactions, and cellular functions, and these effects can be best understood by integrating phenomena across various scales. At the molecular or atomistic level, multi-scale models help understand how crowding impacts molecular conformations and interactions, such as protein folding and ligand binding (Heo et al. 2022). At the mesoscopic level, these models can describe the diffusion and transport properties impacted by crowding, which are crucial for understanding cellular processes including signaling and metabolism (Długosz and Trylska 2011). On the other hand, on a macroscopic scale, multi-scale modeling can incorporate the cumulative effects of crowding on cell behavior and function, such as changes in metabolism and gene expression (Zangooei et al. 2021). Multi-scale models can help to study how crowding agents impact protein folding pathways (Ostrowska et al. 2019). Atomistic simulations provide insights onto secondary structure stability, while mesoscopic models simulate how these proteins fold or misfold in crowded environments (Ghosh et al. 2024). These models can be used to study how molecular crowding affects enzyme reaction rates. Detailed quantum mechanical models can be used for active sites, while larger-scale simulations assess how crowding alters substrate diffusion and enzyme orientation. Furthermore, crowding effects on RNA folding can be studied by combining detailed atomistic simulations with coarser, larger-scale models that describe the impact of other biomolecules and macromolecular crowding agents in the cellular environment (Feig and Sugita 2019). Moreover, in signal transduction pathways, multi-scale models can help map how molecular crowding affects protein-protein interactions and protein localization, by collecting data from detailed molecular interactions and broader-scale cellular models (Cheng et al. 2012). In addition, multi-scale modeling helps in understanding how drugs diffuse through and act within crowded cellular environments, optimizing drug design and predicting efficacy by considering crowding-induced changes at atomic, molecular, and cellular levels (Amaro and Mulholland 2018). Finally, multi-scale modeling helps in evaluating the effects of crowding on metabolic pathways (Widmer and Stelling 2018). Atomistic models can be used to study the effects on enzyme activity, which are then integrated into larger-scale models to predict overall metabolix flux under crowded conditions (Coveney and Fowler 2005). In short, this modeling approach can be adapted to focus on different aspects of crowding by adjusting the levels of detail at which each component is described.

On the other hand, developing multi-scale models can be extremely complex, requiring significant expertise in different areas. Though generally more efficient than single-scale models, multi-scale models can still be computationally intensive, especially when scales are coupled concurrently (Southern et al. 2008).

Future Directions

A hybrid approach that combines quantum computing, machine learning, and classical simulations can significantly enhance studies of molecular crowding. Such an integration leverages the strengths of each method to address the complex interactions and phenomena present in crowded environments. Quantum computers can solve certain problems faster than classical computers, especially those involving combinatorial optimization and complex quantum systems (Montiel Ross 2020). Using quantum algorithms such as the Quantum Approximate Optimization Algorithm (QAOA) could help to identify optimal configurations of crowded molecular systems more efficiently. Classical MD or MC simulations can be used to model the bulk behavior of molecules in crowded environments. By employing quantum computing in conjunction with these approaches, we could more rapidly study relevant configurations and dynamic properties, which also improves the efficiency of simulations. Quantum computing has the potential to perform electronic structure calculations more accurately than classical methods through quantum algorithms that simulate quantum systems (Motta and Rice 2022). This can provide detailed information about the interactions between biomolecules in a crowded environment. Machine learning models can be trained on the results of quantum calculations to predict molecular properties and interactions in complex systems, which in turn would help to turn computational results into practical models for predicting behavior in crowded conditions. In crowded environments, the energy landscape can become highly complex, making it challenging to sample configurational space adequately. Quantum MC methods may enable better sampling of such landscapes. Machine learning techniques, such as reinforcement learning or generative models, can assist in sampling strategies by predicting likely configurations and reaction pathways. As new data about molecular crowding effects are generated via quantum simulations or experimental data, machine learning algorithms can adaptively refine their predictions, leading to better-informed models over time. We can develop new multi-scale modeling techniques by coupling quantum computing, classical simulations, and machine learning. For this purpose, we could use quantum mechanics in specific regions where electronic behavior is critical, such as the active site of an enzyme. We could employ classical simulations for the bulk environment and implement machine learning to connect these two levels, which could ensure smooth transitions between the two.

Conclusions

In conclusion, understanding molecular crowding is essential for elucidating the complex behavior of biomolecules in physiological environments. This chapter demonstrates that while experimental approaches provide valuable insights, they are often limited by the inricate nature of cellular conditions. Computational approaches serve as a powerful complement to experiments, enabling us to model crowded environments and predict the resulting behavior of of macromolecules at an atomic level. Techniques such as molecular dynamics and Monte Carlo simulations offer crucial insights into protein folding, interactions, and reaction kinetics under crowded conditions, while approached like Brownian dynamics further enhance our understanding of particle behavior within biological fluids. Additionally, finite element analysis enables the study of mechanical properties and stress distributions in crowded environments, enabling studies of how structural deformations can impact molecular interactions. The incorporation of quantum mechanics/molecular mechanics methods provides a hybrid strategy for addressing specific electronic interactions within macromolecules while treating the bulk components classically. This enables more accurate simulations of reactions and interactions under crowded conditions. Multi-scale modeling techniques enable a comprehensive approach, integrating information across various spatial and temporal dimensions. Therefore, these capture the complex dynamics of molecular crowding effectively. Coarse-grained simulations simplify molecular representations to focus on larger-scale behaviors while maintaining essential interactions.

Looking ahead, the integration of quantum computing into molecular crowding studies hold significant promise. Quantum computers have the potential to revolutionize the way we simulate and understand complex molecular systems. This approach offers enhanced computational power for exploring vast conformational spaces and intricate electronic properties. As these technologies advance, the synergy between classical simulations, quantum methods, and machine learning will likely lead to breakthroughs in our understanding of molecular crowding. This will ultimately facilitate the development of targeted therapeutics or foster deeper insights into fundamental biological processes, marking a new era in the study of molecular interactions in crowded environments.

References

Abbaspour L, Klumpp S (2021) Enhanced diffusion of a tracer particle in a lattice model of a crowded active system. Phys Rev E 103(5):052601. https://doi.org/10.1103/PhysRevE.103.052601

Adcock SA, McCammon JA (2006) Molecular dynamics: survey of methods for simulating the activity of proteins. Chem Rev 106(5):1589–1615. https://doi.org/10.1021/cr040426m

Akbayrak IY, Caglayan SI, Ozcan Z, Uversky VN, Coskuner-Weber O (2020) Current challenges and limitations in the studies of intrinsically disordered proteins in neurodegenerative diseases by computer simulations. Curr Alzheimer Res 17(9):805–818. https://doi.org/10.2174/1567205017666201109094908

Alfano C, Fichou Y, Huber K, Weiss M, Spruijt E, Ebbinghaus S, De Luca G, Morando MA, Vetri V, Temussi PA, Pastore A (2024) Molecular crowding: the history and development of a scientific paradigm. Chem Rev 124(6):3186–3219. https://doi.org/10.1021/acs.chemrev.3c00615

Allison TC, Coskuner O, Gonzalez CA (eds) (2011) Metallic systems: a quantum chemist's perspective, 0th edn. CRC Press

Al-Obaidi H, Florence AT (2015) Nanoparticle delivery and particle diffusion in confined and complex environments. J Drug Deliv Sci Technol 30:266–277. https://doi.org/10.1016/j.jddst.2015.06.017

Amaro RE, Mulholland AJ (2018) Multiscale methods in drug design bridge chemical and biological complexity in the search for cures. Nat Rev Chem 2(4):0148. https://doi.org/10.1038/s41570-018-0148

Anand R, Agrawal M, Mattaparthi VSK, Swaminathan R, Santra SB (2019) Consequences of heterogeneous crowding on an enzymatic reaction: a residence time Monte Carlo approach. ACS Omega 4(1):727–736. https://doi.org/10.1021/acsomega.8b02863

Ando T, Yu I, Feig M, Sugita Y (2016) Thermodynamics of macromolecular association in heterogeneous crowding environments: theoretical and simulation studies with a simplified model. J Phys Chem B 120(46):11856–11865. https://doi.org/10.1021/acs.jpcb.6b06243

Argyris JH, Balmer H, Doltsinis JST, Dunne PC, Haase M, Kleiber M, Malejannakis GA, Mlejnek H-P, Müller M, Scharpf DW (1979) Finite element method — the natural approach. Comput Methods Appl Mech Eng 17–18:1–106. https://doi.org/10.1016/0045-7825(79)90083-5

Badar MS, Shamsi S, Ahmed J, Alam MA (2022) Molecular dynamics simulations: concept, methods, and applications. In: Rezaei N (ed) Transdisciplinarity. Springer, Cham, pp 131–151

Bairagee D, Panchawat S, Jain N, Pingali S (2024) Controlling the drug release rate and targeted drug delivery to the desired site by molecular simulation. In: Malviya R, Sundram S, Meenakshi DU (eds) Drug delivery systems using quantum computing, 1st edn. Wiley, pp 353–387

Balcells C, Pastor I, Vilaseca E, Madurga S, Cascante M, Mas F (2014) Macromolecular crowding effect upon *in vitro* enzyme kinetics: mixed activation–diffusion control of the oxidation of NADH by pyruvate catalyzed by lactate dehydrogenase. J Phys Chem B 118(15):4062–4068. https://doi.org/10.1021/jp4118858

Basak S, Sengupta S, Chattopadhyay K (2019) Understanding biochemical processes in the presence of sub-diffusive behavior of biomolecules in solution and living cells. Biophys Rev 11(6):851–872. https://doi.org/10.1007/s12551-019-00580-9

Bhattacherjee A, Krepel D, Levy Y (2016) Coarse-grained models for studying protein diffusion along DNA. WIREs Comput Mol Sci 6(5):515–531. https://doi.org/10.1002/wcms.1262

Bittig AT, Uhrmacher AM (2010) Spatial modeling in cell biology at multiple levels. In: Proceedings of the 2010 Winter simulation conference. IEEE, Baltimore, MD, USA, pp 608–619

Bock LV, Gabrielli S, Kolář MH, Grubmüller H (2023) Simulation of complex biomolecular systems: the ribosome challenge. Annu Rev Biophys 52(1):361–390. https://doi.org/10.1146/annurev-biophys-111622-091147

Bombelli FB, Gambinossi F, Lagi M, Berti D, Caminati G, Brown T, Sciortino F, Nordén B, Baglioni P (2008) DNA closed nanostructures: a structural and Monte Carlo simulation study. J Phys Chem B 112(48):15283–15294. https://doi.org/10.1021/jp804544u

Bonate PL (2001) A brief introduction to Monte Carlo simulation. Clin Pharmacokinet 40(1):15–22. https://doi.org/10.2165/00003088-200140010-00002

Burg MB (2000) Macromolecular crowding as a cell VolumeSensor. Cell Physiol Biochem 10(5–6):251–256. https://doi.org/10.1159/000016371

Burkhart TA, Andrews DM, Dunning CE (2013) Finite element modeling mesh quality, energy balance and validation methods: a review with recommendations associated with the modeling of bone tissue. J Biomech 46(9):1477–1488. https://doi.org/10.1016/j.jbiomech.2013.03.022

Caliskan M, Mandaci SY, Uversky VN, Coskuner-Weber O (2021) Secondary structure dependence of amyloid-β(1-40) on simulation techniques and force field parameters. Chem Biol Drug Des 97(5):1100–1108. https://doi.org/10.1111/cbdd.13830

Cao S, Chen S-J (2005) Predicting RNA folding thermodynamics with a reduced chain representation model. RNA 11(12):1884–1897. https://doi.org/10.1261/rna.2109105

Cascella M, Vanni S (2015) Toward accurate coarse-graining approaches for protein and membrane simulations. In: Springborg M, Joswig J-O (eds) Chemical modelling. Royal Society of Chemistry, Cambridge, pp 1–52

Chen JC, Kim AS (2004) Brownian dynamics, molecular dynamics, and Monte Carlo modeling of colloidal systems. Adv Colloid Interf Sci 112(1–3):159–173. https://doi.org/10.1016/j.cis.2004.10.001

Cheng TMK, Gulati S, Agius R, Bates PA (2012) Understanding cancer mechanisms through network dynamics. Brief Funct Genomics 11(6):543–560. https://doi.org/10.1093/bfgp/els025

Cheung MS, Thirumalai D (2007) Effects of crowding and confinement on the structures of the transition state Ensemble in Proteins. J Phys Chem B 111(28):8250–8257. https://doi.org/10.1021/jp068201y

Choi J-M, Dar F, Pappu RV (2019) LASSI: a lattice model for simulating phase transitions of multivalent proteins. PLoS Comput Biol 15(10):e1007028. https://doi.org/10.1371/journal.pcbi.1007028

Christiansen A, Wang Q, Cheung MS, Wittung-Stafshede P (2013) Effects of macromolecular crowding agents on protein folding in vitro and in silico. Biophys Rev 5(2):137–145. https://doi.org/10.1007/s12551-013-0108-0

Cohen G, Pernet S (2017) Definition of different types of finite elements. In: Finite element and discontinuous Galerkin methods for transient wave equations. Springer, Dordrecht, pp 39–93

Coskuner O, Deiters UK (2007) Hydrophobic interactions of xenon by Monte Carlo simulations. Z Phys Chem 221(6):785–799. https://doi.org/10.1524/zpch.2007.221.6.785

Coveney PV, Fowler PW (2005) Modelling biological complexity: a physical scientist's perspective. J R Soc Interface 2(4):267–280. https://doi.org/10.1098/rsif.2005.0045

Csizi K, Reiher M (2023) Universal QM/MM approaches for general nanoscale applications. WIREs Comput Mol Sci 13(4):e1656. https://doi.org/10.1002/wcms.1656

Cupelli C, Borchardt T, Steiner T, Paust N, Zengerle R, Santer M (2013) Leukocyte enrichment based on a modified pinched flow fractionation approach. Microfluid Nanofluid 14(3–4):551–563. https://doi.org/10.1007/s10404-012-1073-9

Dada JO, Mendes P (2011) Multi-scale modelling and simulation in systems biology. Integr Biol 3(2):86. https://doi.org/10.1039/c0ib00075b

Dallon JC (2007) Models with lattice-free center-based cells interacting with continuum environment variables. In: Anderson ARA, Chaplain MAJ, Rejniak KA (eds) Single-cell-based models in biology and medicine. Birkhäuser Basel, Basel, pp 197–219

Derreumaux P, Mousseau N (2007) Coarse-grained protein molecular dynamics simulations. J Chem Phys 126(2):025101. https://doi.org/10.1063/1.2408414

Dickinson E (1985) Brownian dynamic with hydrodynamic interactions: the application to protein diffusional problems. Chem Soc Rev 14(4):421. https://doi.org/10.1039/cs9851400421

Długosz M, Trylska J (2011) Diffusion in crowded biological environments: applications of Brownian dynamics. BMC Biophys 4(1):3. https://doi.org/10.1186/2046-1682-4-3

Dumont ER, Grosse IR, Slater GJ (2009) Requirements for comparing the performance of finite element models of biological structures. J Theor Biol 256(1):96–103. https://doi.org/10.1016/j.jtbi.2008.08.017
Echeverria C, Kapral R (2015) Enzyme kinetics and transport in a system crowded by mobile macromolecules. Phys Chem Chem Phys 17(43):29243–29250. https://doi.org/10.1039/C5CP05056A
Feig M, Sugita Y (2019) Whole-cell models and simulations in molecular detail. Annu Rev Cell Dev Biol 35(1):191–211. https://doi.org/10.1146/annurev-cellbio-100617-062542
Feig M, Yu I, Wang P, Nawrocki G, Sugita Y (2017) Crowding in cellular environments at an atomistic level from computer simulations. J Phys Chem B 121(34):8009–8025. https://doi.org/10.1021/acs.jpcb.7b03570
Feng F, Klug WS (2006) Finite element modeling of lipid bilayer membranes. J Comput Phys 220(1):394–408. https://doi.org/10.1016/j.jcp.2006.05.023
Friswell MI, Mottershead JE (1995) Finite element modelling. In: Finite element model updating in structural dynamics. Springer, Dordrecht, pp 7–35
George AZ, Robinson BA, Dash ZV, Trease V (1997) Summary of the models and methods for the FEHM application-a finite-element heat- and mass-transfer code
Gherman BF, Goldberg SD, Cornish VW, Friesner RA (2004) Mixed quantum mechanical/molecular mechanical (QM/MM) study of the Deacylation reaction in a penicillin binding protein (PBP) versus in a class C β-lactamase. J Am Chem Soc 126(24):7652–7664. https://doi.org/10.1021/ja036879a
Ghosh D, Biswas A, Radhakrishna M (2024) Advanced computational approaches to understand protein aggregation. Biophys Rev 5(2):021302. https://doi.org/10.1063/5.0180691
Givoli D, Keller JB (1989) A finite element method for large domains. Comput Methods Appl Mech Eng 76(1):41–66. https://doi.org/10.1016/0045-7825(89)90140-0
Gkionis K, Kruse H, Šponer J (2016) Derivation of reliable geometries in QM calculations of DNA structures: explicit solvent QM/MM and restrained implicit solvent QM optimizations of G-Quadruplexes. J Chem Theory Comput 12(4):2000–2016. https://doi.org/10.1021/acs.jctc.5b01025
Gniewek P, Kolinski A (2010) Coarse-grained Monte Carlo simulations of mucus: structure, dynamics, and thermodynamics. Biophys J 99(11):3507–3516. https://doi.org/10.1016/j.bpj.2010.09.047
Gobet E (2016) Monte-Carlo methods and stochastic processes: from linear to non-linear, 1st edn. Chapman and Hall/CRC
Grassmann G, Miotto M, Desantis F, Di Rienzo L, Tartaglia GG, Pastore A, Ruocco G, Monti M, Milanetti E (2024) Computational approaches to predict protein-protein interactions in crowded cellular environments. Chem Rev 124(7):3932–3977. https://doi.org/10.1021/acs.chemrev.3c00550
Grime JMA, Dama JF, Ganser-Pornillos BK, Woodward CL, Jensen GJ, Yeager M, Voth GA (2016) Coarse-grained simulation reveals key features of HIV-1 capsid self-assembly. Nat Commun 7(1):11568. https://doi.org/10.1038/ncomms11568
Guardiani C, Cencini M, Cecconi F (2014) Coarse-grained modeling of protein unspecifically bound to DNA. Phys Biol 11(2):026003. https://doi.org/10.1088/1478-3975/11/2/026003
Guigas G, Weiss M (2016) Effects of protein crowding on membrane systems. Biochimica et Biophysica Acta (BBA) Biomembr 1858(10):2441–2450. https://doi.org/10.1016/j.bbamem.2015.12.021
Haji Dehabadi M, Saidi H, Zafari F, Irani M (2024) Assessing the accuracy and efficacy of multiscale computational methods in predicting reaction mechanisms and kinetics of SN2 reactions and Claisen rearrangement. Sci Rep 14(1):16791. https://doi.org/10.1038/s41598-024-67468-x
Haldar S, Comitani F, Saladino G, Woods C, Van Der Kamp MW, Mulholland AJ, Gervasio FL (2018) A Multiscale Simulation Approach to Modeling Drug–Protein Binding Kinetics. J Chem Theory Comput 14(11):6093–6101. https://doi.org/10.1021/acs.jctc.8b00687

Hall D, Hoshino M (2010) Effects of macromolecular crowding on intracellular diffusion from a single particle perspective. Biophys Rev 2(1):39–53. https://doi.org/10.1007/s12551-010-0029-0
Hänggi P, Marchesoni F (2005) Introduction: 100years of Brownian motion. Chaos: an interdisciplinary. J Nonlinear Sci 15(2):026101. https://doi.org/10.1063/1.1895505
Harada R, Sugita Y, Feig M (2012) Protein crowding affects hydration structure and dynamics. J Am Chem Soc 134(10):4842–4849. https://doi.org/10.1021/ja211115q
Haspinger DC, Klinge S, Holzapfel GA (2021) Numerical analysis of the impact of cytoskeletal actin filament density alterations onto the diffusive vesicle-mediated cell transport. PLoS Comput Biol 17(5):e1008784. https://doi.org/10.1371/journal.pcbi.1008784
Heo L, Sugita Y, Feig M (2022) Protein assembly and crowding simulations. Curr Opin Struct Biol 73:102340. https://doi.org/10.1016/j.sbi.2022.102340
Hollingsworth SA, Dror RO (2018) Molecular dynamics simulation for all. Neuron 99(6):1129–1143. https://doi.org/10.1016/j.neuron.2018.08.011
Hong F, Schreck JS, Šulc P (2020) Understanding DNA interactions in crowded environments with a coarse-grained model. Nucleic Acids Res 48(19):10726–10738. https://doi.org/10.1093/nar/gkaa854
Horstemeyer MF (2009) Multiscale modeling: a review. In: Leszczynski J, Shukla MK (eds) Practical aspects of computational chemistry. Springer, Dordrecht, pp 87–135
Huber GA, McCammon JA (2019) Brownian dynamics simulations of biological molecules. Trends Chem 1(8):727–738. https://doi.org/10.1016/j.trechm.2019.07.008
Huggins DJ, Biggin PC, Dämgen MA, Essex JW, Harris SA, Henchman RH, Khalid S, Kuzmanic A, Laughton CA, Michel J, Mulholland AJ, Rosta E, Sansom MSP, Van Der Kamp MW (2019) Biomolecular simulations: from dynamics and mechanisms to computational assays of biological activity. WIREs Comput Mol Sci 9(3):e1393. https://doi.org/10.1002/wcms.1393
Hyeon C, Thirumalai D (2011) Capturing the essence of folding and functions of biomolecules using coarse-grained models. Nat Commun 2(1):487. https://doi.org/10.1038/ncomms1481
Iakovliev A, Dasmahapatra S, Bhaskar A (2018) Structural stability of a Mitotic Spindle: parametric Finite element approach. https://doi.org/10.13140/RG.2.2.16352.12808
Ingólfsson HI, Lopez CA, Uusitalo JJ, De Jong DH, Gopal SM, Periole X, Marrink SJ (2014) The power of coarse graining in biomolecular simulations. WIREs Comput Mol Sci 4(3):225–248. https://doi.org/10.1002/wcms.1169
Izvekov S, Voth GA (2005) A multiscale coarse-graining method for biomolecular systems. J Phys Chem B 109(7):2469–2473. https://doi.org/10.1021/jp044629q
Janssen H (2013) Monte-Carlo based uncertainty analysis: sampling efficiency and sampling convergence. Reliab Eng Syst Saf 109:123–132. https://doi.org/10.1016/j.ress.2012.08.003
Jebahi M, Dau F, Charles J-L, Iordanoff I (2016) Multiscale modeling of complex dynamic problems: an overview and recent developments. Arch Computat Methods Eng 23(1):101–138. https://doi.org/10.1007/s11831-014-9136-6
Johnson ME, Chen A, Faeder JR, Henning P, Moraru II, Meier-Schellersheim M, Murphy RF, Prüstel T, Theriot JA, Uhrmacher AM (2021) Quantifying the roles of space and stochasticity in computer simulations for cell biology and cellular biochemistry. MBoC 32(2):186–210. https://doi.org/10.1091/mbc.E20-08-0530
Karimnejad S, Delouei AA, Basagaoglu H, Nazari M, Shahmardan MM, Falcucci G, Lauricella M, Succi S (2022) A review on contact and collision methods for multi-body hydrodynamic problems in complex flows. https://doi.org/10.48550/ARXIV.2211.11728
Knop J-M, Mukherjee S, Jaworek MW, Kriegler S, Manisegaran M, Fetahaj Z, Ostermeier L, Oliva R, Gault S, Cockell CS, Winter R (2023) Life in multi-extreme environments: brines, osmotic and hydrostatic pressure—a physicochemical view. Chem Rev 123(1):73–104. https://doi.org/10.1021/acs.chemrev.2c00491
Kolston PJ (2000) Finite-element modelling: a new tool for the biologist. Philos Trans R Soc London, Ser A 358(1766):611–631. https://doi.org/10.1098/rsta.2000.0548

König PH, Hoffmann M, Frauenheim T, Cui Q (2005) A critical evaluation of different QM/MM frontier treatments with SCC-DFTB as the QM method. J Phys Chem B 109(18):9082–9095. https://doi.org/10.1021/jp0442347

Kroupa KR, Gangi LR, Zimmerman BK, Hung CT, Ateshian GA (2023) Superficial zone chondrocytes can get compacted under physiological loading: a multiscale finite element analysis. Acta Biomater 163:248–258. https://doi.org/10.1016/j.actbio.2022.10.013

Kuznetsova IM, Turoverov KK, Uversky VN (2014) What macromolecular crowding can do to a protein. Int J Mol Sci 15(12):23090–23140. https://doi.org/10.3390/ijms151223090

Kwak KJ, Valincius G, Liao W-C, Hu X, Wen X, Lee A, Yu B, Vanderah DJ, Lu W, Lee LJ (2010) Formation and finite element analysis of tethered bilayer lipid structures. Langmuir 26(23):18199–18208. https://doi.org/10.1021/la1021802

Lane TJ, Shukla D, Beauchamp KA, Pande VS (2013) To milliseconds and beyond: challenges in the simulation of protein folding. Curr Opin Struct Biol 23(1):58–65. https://doi.org/10.1016/j.sbi.2012.11.002

Li S, Dutta B, Cannon S, Daymude JJ, Avinery R, Aydin E, Richa AW, Goldman DI, Randall D (2021) Programming active cohesive granular matter with mechanically induced phase changes. Sci Adv 7(17):eabe8494. https://doi.org/10.1126/sciadv.abe8494

Mandaci SY, Caliskan M, Sariaslan MF, Uversky VN, Coskuner-Weber O (2020) Epitope region identification challenges of intrinsically disordered proteins in neurodegenerative diseases: secondary structure dependence of α-synuclein on simulation techniques and force field parameters. Chem Biol Drug Des cbdd.13662. https://doi.org/10.1111/cbdd.13662

Marrink SJ, Monticelli L, Melo MN, Alessandri R, Tieleman DP, Souza PCT (2023) Two decades of martini: better beads, broader scope. WIREs Comput Mol Sci 13(1):e1620. https://doi.org/10.1002/wcms.1620

Mathur A, Ghosh R, Nunes-Alves A (2024) Recent progress in modeling and simulation of biomolecular crowding and condensation inside cells

Minton AP (2001) The influence of macromolecular crowding and macromolecular confinement on biochemical reactions in physiological media. J Biol Chem 276(14):10577–10580. https://doi.org/10.1074/jbc.R100005200

Mittal S, Chowhan RK, Singh LR (2015) Macromolecular crowding: macromolecules friend or foe. Biochim Biophys Acta Gen Subj 1850(9):1822–1831. https://doi.org/10.1016/j.bbagen.2015.05.002

Mohanty S (2020) Aggregation and coacervation with Monte Carlo simulations. In: Progress in molecular biology and translational science. Elsevier, pp 505–520

Monard G, Merz KM (1999) Combined quantum mechanical/molecular mechanical methodologies applied to biomolecular systems. Acc Chem Res 32(10):904–911. https://doi.org/10.1021/ar970218z

Montiel Ross OH (2020) A review of quantum-inspired metaheuristics: going from classical computers to real quantum computers. IEEE Access 8:814–838. https://doi.org/10.1109/ACCESS.2019.2962155

Morriss-Andrews A, Shea J-E (2015) Computational studies of protein aggregation: methods and applications. Annu Rev Phys Chem 66(1):643–666. https://doi.org/10.1146/annurev-physchem-040513-103738

Morzan UN, Alonso De Armiño DJ, Foglia NO, Ramírez F, González Lebrero MC, Scherlis DA, Estrin DA (2018) Spectroscopy in complex environments from QM–MM simulations. Chem Rev 118(7):4071–4113. https://doi.org/10.1021/acs.chemrev.8b00026

Motta M, Rice JE (2022) Emerging quantum computing algorithms for quantum chemistry. WIREs Comput Mol Sci 12(3):e1580. https://doi.org/10.1002/wcms.1580

Muñiz-Chicharro A, Votapka LW, Amaro RE, Wade RC (2023) Brownian dynamics simulations of biomolecular diffusional association processes. WIREs Comput Mol Sci 13(3):e1649. https://doi.org/10.1002/wcms.1649

Muratov EN, Amaro R, Andrade CH, Brown N, Ekins S, Fourches D, Isayev O, Kozakov D, Medina-Franco JL, Merz KM, Oprea TI, Poroikov V, Schneider G, Todd MH, Varnek A,

Winkler DA, Zakharov AV, Cherkasov A, Tropsha A (2021) A critical overview of computational approaches employed for COVID-19 drug discovery. Chem Soc Rev 50(16):9121–9151. https://doi.org/10.1039/D0CS01065K
Musiani F, Giorgetti A (2017) Protein aggregation and molecular crowding. In: International review of cell and molecular biology. Elsevier, pp 49–77
Nakano S, Miyoshi D, Sugimoto N (2014) Effects of molecular crowding on the structures, interactions, and functions of nucleic acids. Chem Rev 114(5):2733–2758. https://doi.org/10.1021/cr400113m
Nicholson JK, Holmes E, Lindon JC, Wilson ID (2004) The challenges of modeling mammalian biocomplexity. Nat Biotechnol 22(10):1268–1274. https://doi.org/10.1038/nbt1015
Nikolić M, Karavelić E, Ibrahimbegovic A, Miščević P (2018) Lattice element models and their peculiarities. Arch Computat Methods Eng 25(3):753–784. https://doi.org/10.1007/s11831-017-9210-y
Noid WG (2013) Perspective: coarse-grained models for biomolecular systems. J Chem Phys 139(9):090901. https://doi.org/10.1063/1.4818908
Okereke M, Keates S (2018) Finite Element Applications. Springer International Publishing, Cham
Omer A, Suryanarayanan V, Selvaraj C, Singh SK, Singh P (2015) Explicit drug re-positioning. In: Advances in protein chemistry and structural biology. Elsevier, pp 89–112
Ostrowska N, Feig M, Trylska J (2019) Modeling crowded environment in molecular simulations. Front Mol Biosci 6:86. https://doi.org/10.3389/fmolb.2019.00086
Ozboyaci M, Kokh DB, Corni S, Wade RC (2016) Modeling and simulation of protein–surface interactions: achievements and challenges. Q Rev Biophys 49:e4. https://doi.org/10.1017/S0033583515000256
Pak AJ, Voth GA (2018) Advances in coarse-grained modeling of macromolecular complexes. Curr Opin Struct Biol 52:119–126. https://doi.org/10.1016/j.sbi.2018.11.005
Paquet E, Viktor HL (2015) Molecular dynamics, Monte Carlo simulations, and Langevin dynamics: a computational review. Biomed Res Int 2015:1–18. https://doi.org/10.1155/2015/183918
Park BH, Levitt M (1995) The complexity and accuracy of discrete state models of protein structure. J Mol Biol 249(2):493–507. https://doi.org/10.1006/jmbi.1995.0311
Pastore A, Rivas Caballero G, Temussi PA (2024) Introduction: molecular crowding. Chem Rev 124(11):6697–6699. https://doi.org/10.1021/acs.chemrev.4c00287
Peters J, Oliva R, Caliò A, Oger P, Winter R (2023) Effects of crowding and cosolutes on biomolecular function at extreme environmental conditions. Chem Rev 123(23):13441–13488. https://doi.org/10.1021/acs.chemrev.3c00432
Philipse AP (2018) Brownian motion: elements of colloid dynamics. Springer, Cham
Ponder JW, Case DA (2003) Force fields for protein simulations. In: Advances in protein chemistry. Elsevier, pp 27–85
Prytkova V, Heyden M, Khago D, Freites JA, Butts CT, Martin RW, Tobias DJ (2016) Multi-conformation Monte Carlo: a method for introducing flexibility in efficient simulations of many-protein systems. J Phys Chem B 120(33):8115–8126. https://doi.org/10.1021/acs.jpcb.6b00827
Qu Z, Garfinkel A, Weiss JN, Nivala M (2011) Multi-scale modeling in biology: how to bridge the gaps between scales? Prog Biophys Mol Biol 107(1):21–31. https://doi.org/10.1016/j.pbiomolbio.2011.06.004
Rakers C, Bermudez M, Keller BG, Mortier J, Wolber G (2015) Computational close up on protein–protein interactions: how to unravel the invisible using molecular dynamics simulations? WIREs Comput Mol Sci 5(5):345–359. https://doi.org/10.1002/wcms.1222
Rhodes NR, Tschopp MA, Solanki KN (2013) Quantifying the energetics and length scales of carbon segregation to α-Fe symmetric tilt grain boundaries using atomistic simulations. Modelling Simul Mater Sci Eng 21(3):035009. https://doi.org/10.1088/0965-0393/21/3/035009
Ridgway D, Broderick G, Lopez-Campistrous A, Ruaini M, Winter P, Hamilton M, Boulanger P, Kovalenko A, Ellison MJ (2008) Coarse-grained molecular simulation of diffusion and

reaction kinetics in a crowded virtual cytoplasm. Biophys J 94(10):3748–3759. https://doi.org/10.1529/biophysj.107.116053
Rivas G, Minton AP (2016) Macromolecular crowding in vitro, in vivo, and in between. Trends Biochem Sci 41(11):970–981. https://doi.org/10.1016/j.tibs.2016.08.013
Robustelli P, Kohlhoff K, Cavalli A, Vendruscolo M (2010) Using NMR chemical shifts as structural restraints in molecular dynamics simulations of proteins. Structure 18(8):923–933. https://doi.org/10.1016/j.str.2010.04.016
Rodenhizer D, Dean T, D'Arcangelo E, McGuigan AP (2018) The current landscape of 3D in vitro tumor models: what cancer hallmarks are accessible for drug discovery? Adv Healthc Mater 7(8):1701174. https://doi.org/10.1002/adhm.201701174
Rojnuckarin A, Kim S, Subramaniam S (1998) Brownian dynamics simulations of protein folding: access to milliseconds time scale and beyond. Proc Natl Acad Sci USA 95(8):4288–4292. https://doi.org/10.1073/pnas.95.8.4288
Ryde U (2016) QM/MM calculations on proteins. In: Methods in enzymology. Elsevier, pp 119–158
Samiotakis A, Wittung-Stafshede P, Cheung MS (2009) Folding, stability and shape of proteins in crowded environments: experimental and computational approaches. IJMS 10(2):572–588. https://doi.org/10.3390/ijms10020572
Schaffer LV, Ideker T (2021) Mapping the multiscale structure of biological systems. Cell Syst 12(6):622–635. https://doi.org/10.1016/j.cels.2021.05.012
Schlick T, Barth E, Mandziuk M (1997) Biomolecular dynamics at long timesteps: bridging the timescale gap between simulation and experimentation. Annu Rev Biophys Biomol Struct 26(1):181–222. https://doi.org/10.1146/annurev.biophys.26.1.181
Schnell S, Turner TE (2004) Reaction kinetics in intracellular environments with macromolecular crowding: simulations and rate laws. Prog Biophys Mol Biol 85(2–3):235–260. https://doi.org/10.1016/j.pbiomolbio.2004.01.012
Schuss Z (2013) Brownian dynamics at boundaries and interfaces. In: Physics, chemistry, and biology. Springer, New York
Sedeh RS, Yun G, Lee JY, Bathe K-J, Kim D-N (2018) A framework of finite element procedures for the analysis of proteins. Comput Struct 196:24–35. https://doi.org/10.1016/j.compstruc.2017.10.015
Senn HM, Thiel W (2009) QM/MM methods for biomolecular systems. Angew Chem Int Ed 48(7):1198–1229. https://doi.org/10.1002/anie.200802019
Shaik S, Cohen S, Wang Y, Chen H, Kumar D, Thiel W (2010) P450 enzymes: their structure, reactivity, and selectivity—modeled by QM/MM calculations. Chem Rev 110(2):949–1017. https://doi.org/10.1021/cr900121s
Shim AR, Nap RJ, Huang K, Almassalha LM, Matusda H, Backman V, Szleifer I (2020) Dynamic crowding regulates transcription. Biophys J 118(9):2117–2129. https://doi.org/10.1016/j.bpj.2019.11.007
Shurki A, Warshel A (2003) Structure/function correlations of proteins using MM, QM/MM, and related approaches: methods, concepts, pitfalls, and current progress. In: Advances in protein chemistry. Elsevier, pp 249–313
Sinnecker S, Neese F (2006) QM/MM calculations with DFT for taking into account protein effects on the EPR and optical spectra of metalloproteins. Plastocyanin as a case study. J Comput Chem 27(12):1463–1475. https://doi.org/10.1002/jcc.20426
Skolnick J, Kolinski A (1991) Dynamic Monte Carlo simulations of a new lattice model of globular protein folding, structure and dynamics. J Mol Biol 221(2):499–531. https://doi.org/10.1016/0022-2836(91)80070-B
Sloot PMA, Hoekstra AG (2010) Multi-scale modelling in computational biomedicine. Brief Bioinform 11(1):142–152. https://doi.org/10.1093/bib/bbp038
Solernou A, Hanson BS, Richardson RA, Welch R, Read DJ, Harlen OG, Harris SA (2018) Fluctuating finite element analysis (FFEA): a continuum mechanics software tool for mesoscale simulation of biomolecules. PLoS Comput Biol 14(3):e1005897. https://doi.org/10.1371/journal.pcbi.1005897

Southern J, Pitt-Francis J, Whiteley J, Stokeley D, Kobashi H, Nobes R, Kadooka Y, Gavaghan D (2008) Multi-scale computational modelling in biology and physiology. Prog Biophys Mol Biol 96(1–3):60–89. https://doi.org/10.1016/j.pbiomolbio.2007.07.019

Šponer J, Bussi G, Krepl M, Banáš P, Bottaro S, Cunha RA, Gil-Ley A, Pinamonti G, Poblete S, Jurečka P, Walter NG, Otyepka M (2018) RNA structural dynamics as captured by molecular simulations: a comprehensive overview. Chem Rev 118(8):4177–4338. https://doi.org/10.1021/acs.chemrev.7b00427

Srirekha A, Bashetty K (2010) Infinite to finite: an overview of finite element analysis. Indian J Dent Res 21(3):425. https://doi.org/10.4103/0970-9290.70813

Sun Z, Chew JW, Hills NJ, Volkov KN, Barnes CJ (2010) Efficient finite element analysis/computational fluid dynamics thermal coupling for engineering applications. J Turbomach 132(3):031016. https://doi.org/10.1115/1.3147105

Sung W (2018) Brownian motions. In: Statistical physics for biological matter. Springer, Dordrecht, pp 241–268

Tabaka M, Kalwarczyk T, Szymanski J, Hou S, Holyst R (2014) The effect of macromolecular crowding on mobility of biomolecules, association kinetics, and gene expression in living cells. Front Phys 2. https://doi.org/10.3389/fphy.2014.00054

Takada S, Kanada R, Tan C, Terakawa T, Li W, Kenzaki H (2015) Modeling structural dynamics of biomolecular complexes by coarse-grained molecular simulations. Acc Chem Res 48(12):3026–3035. https://doi.org/10.1021/acs.accounts.5b00338

Takahashi K, Arjunan SNV, Tomita M (2005) Space in systems biology of signaling pathways—towards intracellular molecular crowding in silico. FEBS Lett 579(8):1783–1788. https://doi.org/10.1016/j.febslet.2005.01.072

Tozzini V (2010) Multiscale modeling of proteins. Acc Chem Res 43(2):220–230. https://doi.org/10.1021/ar9001476

Tzeliou CE, Mermigki MA, Tzeli D (2022) Review on the QM/MM methodologies and their application to Metalloproteins. Molecules 27(9):2660. https://doi.org/10.3390/molecules27092660

Van Der Kamp MW, Mulholland AJ (2013) Combined quantum mechanics/molecular mechanics (QM/MM) methods in computational enzymology. Biochemistry 52(16):2708–2728. https://doi.org/10.1021/bi400215w

Van Liedekerke P, Palm MM, Jagiella N, Drasdo D (2015) Simulating tissue mechanics with agent-based models: concepts, perspectives and some novel results. Comp Part Mech 2(4):401–444 https://doi.org/10.1007/s40571-015-0082-3

Velázquez-Libera JL, Caballero J, Tuñón I, Hernández-Rodríguez EW, Ruiz-Pernía JJ (2020) On the nature of the enzyme–substrate complex and the reaction mechanism in human arginase I. A combined molecular dynamics and QM/MM study. ACS Catal 10(15):8321–8333. https://doi.org/10.1021/acscatal.0c00981

Verdian Doghaei A, Housaindokht MR, Bozorgmehr MR (2015) Molecular crowding effects on conformation and stability of G-quadruplex DNA structure: insights from molecular dynamics simulation. J Theor Biol 364:103–112. https://doi.org/10.1016/j.jtbi.2014.09.015

Visser BS, Lipiński WP, Spruijt E (2024) The role of biomolecular condensates in protein aggregation. Nat Rev Chem 8(9):686–700. https://doi.org/10.1038/s41570-024-00635-w

Vitalis A, Pappu RV (2009) Chapter 3: methods for Monte Carlo simulations of biomacromolecules. In: Annual reports in computational chemistry. Elsevier, pp 49–76

Wang H, Yan H, Rong C, Yuan Y, Jiang F, Han Z, Sui H, Jin D, Li Y (2024) Multi-scale simulation of complex systems: a perspective of integrating knowledge and data. ACM Comput Surv 56(12):1–38. https://doi.org/10.1145/3654662

Wassenaar TA, Pluhackova K, Böckmann RA, Marrink SJ, Tieleman DP (2014) Going backward: a flexible geometric approach to reverse transformation from coarse grained to atomistic models. J Chem Theory Comput 10(2):676–690. https://doi.org/10.1021/ct400617g

Watanabe HC, Cui Q (2019) Quantitative analysis of QM/MM boundary artifacts and correction in adaptive QM/MM simulations. J Chem Theory Comput 15(7):3917–3928. https://doi.org/10.1021/acs.jctc.9b00180

Weilandt DR, Hatzimanikatis V (2019) Particle-based simulation reveals macromolecular crowding effects on the Michaelis-Menten mechanism. Biophys J 117(2):355–368. https://doi.org/10.1016/j.bpj.2019.06.017

Weiss M (2014) Crowding, diffusion, and biochemical reactions. In: International review of cell and molecular biology. Elsevier, pp 383–417

Widmer LA, Stelling J (2018) Bridging intracellular scales by mechanistic computational models. Curr Opin Biotechnol 52:17–24. https://doi.org/10.1016/j.copbio.2018.02.005

Wise O, Coskuner O (2014) New force field parameters for metalloproteins I: divalent copper ion centers including three histidine residues and an oxygen-ligated amino acid residue. J Comput Chem 35(17):1278–1289. https://doi.org/10.1002/jcc.23622

Xie Z-R, Chen J, Wu Y (2014) A coarse-grained model for the simulations of biomolecular interactions in cellular environments. J Chem Phys 140(5):054112. https://doi.org/10.1063/1.4863992

Zangooei MH, Margolis R, Hoyt K (2021) Multiscale computational modeling of cancer growth using features derived from microCT images. Sci Rep 11(1):18524. https://doi.org/10.1038/s41598-021-97966-1

Zhang J, Li W, Wang J, Qin M, Wu L, Yan Z, Xu W, Zuo G, Wang W (2009) Protein folding simulations: from coarse-grained model to all-atom model. IUBMB Life 61(6):627–643. https://doi.org/10.1002/iub.223

Zhang Y, Ma Y, Zhang K, Wang C-K, Lin L, Fan J (2020) Insights on aggregation induced room temperature phosphorescence properties: a QM/MM study. J Lumin 221:117046. https://doi.org/10.1016/j.jlumin.2020.117046

Zhou H-X, Qin S (2013) Simulation and modeling of crowding effects on the thermodynamic and kinetic properties of proteins with atomic details. Biophys Rev 5(2):207–215. https://doi.org/10.1007/s12551-013-0101-7

Zhou J, Thorpe IF, Izvekov S, Voth GA (2007) Coarse-grained peptide modeling using a systematic multiscale approach. Biophys J 92(12):4289–4303. https://doi.org/10.1529/biophysj.106.094425

Zhou H-X, Rivas G, Minton AP (2008) Macromolecular crowding and confinement: biochemical, biophysical, and potential physiological consequences. Annu Rev Biophys 37(1):375–397. https://doi.org/10.1146/annurev.biophys.37.032807.125817

Chapter 22
Modeling Facilitated Diffusion of Proteins in Crowded Environment

Shrawan Kumar Choudhary, Sangeeta, Pinki Dey, and Arnab Bhattacherjee

Abstract The interaction between proteins and DNA is fundamental to numerous enzymatic processes, relying on the precise recognition of specific DNA bases by DNA-binding proteins (DBPs). This interaction occurs within the densely packed environment of the cell nucleus, which contains high concentrations of biomolecules. Consequently, the limited intracellular space can impact the diffusion and association of large macromolecules. This scenario prompts the question of how the complex milieu of the cell nucleus influences the ability of DBPs to locate and bind to their target DNA sites within a large polymeric DNA substrate. In this work, we review the *in silico* approaches used to investigate the facilitated diffusion of DBPs along DNA and the influence of crowded environments on this process. We illustrate that the effects of crowding differ significantly from those observed in protein-protein associations and cannot be solely attributed to volume exclusion principles. Additionally, we demonstrate how the physicochemical characteristics of crowding agents impact the facilitated diffusion of DBPs, crucial for understanding the protein's search for its DNA binding motif within the densely packed nuclear environment.

Keywords Macromolecular crowding · Protein-DNA interactions · Facilitated diffusion · Depletion layer · Coarse-grained simulation

---

Authors Shrawan Kumar Choudhary, Sangeeta have equally contributed to this chapter.

S. K. Choudhary · Sangeeta · A. Bhattacherjee (✉)
School of Computational and Integrative Sciences, Jawaharlal Nehru University, New Delhi, India
e-mail: arnab@jnu.ac.in

P. Dey
Centre for Healthy Futures (CHEF), Torrens University, Surry Hill Campus, Sydney, Australia

V. N. Uversky (ed.), *(Macro)Molecular Crowding*, Subcellular Biochemistry 109,
https://doi.org/10.1007/978-3-032-03370-3_22

Introduction

DNA-binding proteins (DBPs) swiftly seek out and bind to their respective DNA sites to facilitate crucial biological processes such as DNA repair, transcription, and DNA replication (Watson and Crick 1953). However, the challenge arises when proteins need to search for specific DNA-binding sites, given the significant difference in size between the binding site (typically 12 base pairs) (Janin et al. 2007) and the entire genome (comprising billions of base pairs). Considering the substantial discrepancy in sizes, the likelihood of DBPs reaching their target site via three-dimensional collisions is exceedingly low. Consequently, most interactions occur with non-specific DNA, compelling the protein to undergo multiple rounds of dissociation and re-association through 3D diffusion. Presumably, this process takes a considerable amount of time to reach the target DNA site, and the associated kinetics are expected to precede with a rate constant, $k_a = 4\pi DR$, where D represents the diffusion coefficient, and R is the sum of the radii of the interacting partners (Von Hippel and Berg 1989) assuming it a 3D diffusion-controlled reaction.

Interestingly, in 1960, Riggs et al. discovered in an *in vitro* experiment that the association rate of LacI to the lac operator site can exceed that of 3D diffusion-controlled reactions by two orders of magnitude (Riggs et al. 1970). Subsequently, the idea of rapid kinetics in target DNA site recognition was recognized as a biological imperative (Richter and Eigen 1974; Berg and Blomberg 1976; Berg et al. 1981). This spurred a flurry of experimental and theoretical studies seeking an explanation for why and how nature surpasses the diffusion limit in the target search kinetics of DBPs.

After almost three decades, in 1989, Berg and Von Hippel proposed a model suggesting that the specific binding of a protein to its target DNA site involves two stages. Firstly, there is a non-specific association between the protein and DNA, aiding in sequestering the protein to the region of space where the target is likely to be found, leading to a one-dimensional facilitated diffusion (FD) process (Von Hippel and Berg 1989). Secondly, formation of the specific protein-DNA complex upon arrival of the protein at its cognate site. The complex formation is triggered by short-ranged yet specific interactions between the amino acids side chains of protein and nucleo-bases of the DNA molecule. In contrast, the protein engages with the DNA through nonspecific interactions (primarily electrostatic interactions) during its one-dimensional facilitated diffusion characterized by a random walk along the DNA chain. Despite its apparent simplicity, characterizing the underlying molecular mechanisms of facilitated DNA search by DBPs is a complex challenge due to the transient and elusive nature of non-specific protein-DNA interactions. Several experimental and computational approaches have been developed to provide insights into these processes. Among *in silico* techniques, coarse-grained (CG) molecular simulations have proven instrumental in revealing critical insights. For instance, they capture the helical motion of DBPs as they navigate the DNA major grooves during linear diffusion (sliding), as well as various other search modes (hopping, intersegmental transfer, and 3D diffusion), highlighting their dependence on salt concentration (Givaty and Levy 2009). Moreover, the role of conformational heterogeneity of the searching protein has been thoroughly investigated (Bhattacherjee and Levy 2014a, b; Krepel et al. 2016; Krepel and Levy 2016, 2017; Bhattacherjee et al. 2016). These studies suggest that the transition from non-specific to specific protein-DNA interactions often involves conformational changes and water molecule expulsion, factors that can significantly impact target search kinetics (Mirny et al. 2009; Marcovitz and Levy 2011; Cuculis et al. 2016). Recently, a similar coarse-grained approach has been extensively employed to probe the role of DNA, including groove geometry, helical twist, curvature, and DNA topology, in the DNA target search problem by DBPs (Bhattacherjee and Levy 2014a, b). Thus, while understanding of the molecular interactions that underpin the DNA search by proteins has improved greatly in recent years, a number of fundamental issues remain poorly understood. Particularly majority of the observations stem from experiments conducted either at the single-molecule level or in diluted buffers. Consequently, these studies overlook the influence of the crowded nuclear environment on facilitated diffusion of the proteins on DNA.

The cellular environment constitutes a dynamic and intricate amalgamation of various biomolecules, encompassing proteins, nucleic acids, and cytoskeletal elements, among others. In the context of a typical *E. coli* cell, these macromolecules occupy a substantial portion of the total volume, ranging from 10% to 40% (Zimmerman and Trach 1991; Ellis 2001). This elevated concentration creates a "crowded" milieu within the cell, where diverse macromolecules coexist, delineating boundaries and influencing the behavior of other soluble biomolecules. Although the impacts of a crowded cellular environment have been extensively explored in the realms of protein-protein association and protein folding kinetics and thermodynamics (Hall and Minton 2003; Rivas et al. 2004; Zhou 2004a, b; Minton 2005, 2006 Ellis and Minton 2006; Zhou et al. 2008), its influence on the target search dynamics of DBPs has only recently garnered attention (Liu and Luo 2014; Liu and Luo 2015). For instance, an experimental study by Akabayov et al. reported the formation of a more compact DNA polymerase due to the presence of macromolecular crowders, showcasing the induction of conformational changes in the protein-DNA complex formation. Other studies investigating the DNA translocation efficiency of Human Uracil DNA Glycosylase (hUNG) (Rodriguez et al. 2017) and human 8–oxoguanine DNA glycosylase (hOGG1) (Cravens et al. 2015) state that crowders enhance the translocation efficiency of an enzyme by preventing its escape into the bulk solution. The diffusion of the hUNG protein further increases in the presence of its N-terminal disordered tail (Rodriguez et al. 2017; Dey and Bhattacherjee 2019a). In addition to the crowders in solution, approximately 50% of the DNA surface is consistently occupied by various proteins, known as 'genomic crowders'. An *in vivo* study on Lac repressor confirms that the protein spends 90% of the time non-specifically diffusing along the DNA, thus occupying multiple DNA base pairs. These DNA-bound proteins can also act as potential crowding agents, hindering the sliding track of other proteins searching for their regulatory sites on the DNA surface. A study by Shvets et al. indeed reported that the presence of such 'genomic crowders' on DNA significantly influences the target search dynamics of

DBPs (Flyvbjerg et al. 2006). A set of *in silico* studies, parallel to these experimental investigations, were conducted to unravel the underlying molecular details.

In this chapter, we review the computational modeling approaches that investigate the facilitated diffusion of DBPs in a crowded cellular milieu, as previously presented by Dey et al. (Dey and Bhattacherjee 2018, 2019b, 2020) and other researchers (Krepel et al. 2016; Krepel and Levy 2016, 2017). We delve into the intricacies of the crowded cellular environment, particularly focusing on the impact of 'inert' crowding agents on the target search process of DBPs. Furthermore, we explore the effects of physical properties of macromolecular crowders, including size, mobility, and shape, on the diffusion exponent of searching proteins. Additionally, we discuss the mechanisms and the consequences of the presence of both 'interacting' bulk and 'genomic crowders' on facilitated diffusion and the target search efficiency of DBPs.

Model Description

To explore the relatively long timescales of protein movement on DNA by molecular dynamics simulations, we used a reduced description of protein and DNA model. The coarse-grained (CG) approach revolutionizes the simulation of large biomeolecular processes, offering time scales previously unattainable with all-atom models. Similarly, a CG approach was adopted for the crowder molecules. These models are then subjected to molecular dynamics simulations using our in-house computer code to investigate how crowding influences the way DBPs search for their targets.

Protein Model

The protein structure is modelled via a CG representation, where a single amino acid residue is described by a single bead positioned precisely at its respective C_α coordinates. This C_α model effectively delineates the protein backbone, capturing essential secondary structural elements like α-helices and β-sheets. The protein's energetic modeling utilizes a potential energy function (Eq. 22.1), which accounts for the energy contributions stemming from fluctuations in bonds, angles, and torsional angles around their respective mean values (bonded interactions). A native topology-based model was used to represent native contact interactions (Clementi et al. 2000). This model excludes non-native interactions and uses the Lennard-Jones potential. Sap-1 protein was used as a case of study (PDB ID: 1 BC8).

$$E_{pot}^{pro} = E_{bond}^{pro} + E_{bend}^{pro} + E_{torsion}^{pro} + E_{native}^{pro} + E_{non-native}^{pro} \tag{22.1}$$

DNA Model

We generated the detailed atomic structure of the linear B-DNA with the desired sequence using the w3DNA web server (Zheng et al. 2009). To represent DNA, 3SPN.2 (Three Site Per Nucleotide) CG model was utilized, developed by Pablo et al. (Hinckley et al. 2013). In this model, each nucleotide is coarse-grained into 3 beads positioned at the centre of mass of the sugar, bases and phosphate. The energetics of DNA are modeled using Pablo's 3SPN.2, which accurately describes structural properties such as base rise, duplex width, and major and minor groove widths. This model not only captures the correct persistence length for both dsDNA and ssDNA but also predicts DNA melting temperatures for various sequences and compositions consistent with experimental data. It considers the impacts of sequence, temperature, and ionic strength, successfully capturing the correct hybridization rate constant. These attributes make the 3SPN.2 model an ideal choice for exploring the molecular basis of protein-DNA interactions. The energy function employed to model DNA is expressed as:

$$E_{pot}^{DNA} = E_{bond}^{DNA} + E_{bend}^{DNA} + E_{torsion}^{DNA} + E_{ev}^{DNA} + E_{bstk}^{DNA} + E_{bp}^{DNA} + E_{cstk}^{DNA} + E_{elec}^{DNA} \quad (22.2)$$

We employed the Debye-Hückel potential to describe the electrostatic interactions between the interacting biomolecules. The amino acids with positive and negative charges (Arg, Lys, Asp, and Glu) and the DNA's phosphate groups were charged in the model.

$$E_{elec} = \sum_{n_{elec}}^{i<j} \frac{q_i q_j e^{-\frac{r_{ij}}{\lambda_D}}}{4\pi\epsilon_0\epsilon(T,C) r_{ij}} \quad (22.3)$$

q_i and q_j denote charges on i and j bead and r_{ij} is the separation between them. n_{elec} represents the count of electrostatic interactions. ϵ_0 and ϵ are the dielectric permittivity of the vacuum and solution respectively. Debye screening length, λ_D is defined as

$$\lambda_D = \sqrt{\frac{\epsilon_0\epsilon(T,C)}{2\beta N_A e_c^2 I}} \quad (22.4)$$

I and e_c denotes the ionic strength of the solution and the elementary charge. N_A is the Avogadro's number and $\beta = \frac{1}{k_B T}$ denotes inverse thermal energy. The dielectric permittivity of the solution is denoted by $\epsilon(T, C)$, which is a function of molarity of NaCl and temperature T. This can be approximated as

$$\varepsilon(T,C)=\varepsilon(T)a(C) \tag{22.5}$$

where

$$\varepsilon(T)=249.4-0.788T/K+7.20\times10^{-4}(T/K)^2 \tag{22.6}$$

$$a(C)=1.00-0.2551C/M+5.151\times10^{-2}(C/M)^2-6.889\times10^{-3}(C/M)^3 \tag{22.7}$$

Protein-DNA Interaction

The searching protein and the DNA interact non-specifically using two ways. First, the charged residues of the protein electrostatically interact with the phosphate beads of DNA. Second, excluded volume interactions between the two biomolecules influence their dynamics. The Debye-Hückel potential Eq. 22.3 models electrostatic interactions, and excluded volume interaction is represented by Eq. 22.8. The parameters for protein-DNA excluded volume interactions are determined using combining rules, with a geometric mean for energy parameters (Berthelot rule) and an arithmetic mean for size parameters (Lorentz rule).

$$\varepsilon_{ij}=\sqrt{\varepsilon_{ii}\varepsilon_{jj}}\,(\text{Bertherlot rule})$$

$$\sigma_{ij}=\frac{\sigma_{ii}+\sigma_{jj}}{2}(\text{Lorentz rule})$$

The associated potential energy function is given by

$$E_{ij}=\sum_{ij}\begin{cases}\varepsilon_{ij}\left[5\left(\frac{\sigma_{ij}}{r_{ij}}\right)^{12}-6\left(\frac{\sigma_{ij}}{r_{ij}}\right)^{10}\right] & r<r_c\\ 0 & r\geq r_c\end{cases} \tag{22.8}$$

σ_{ij} and r_{ij} are the average site diameter and inter-site separation, respectively. The cutoff distance, r_c, is set at 15 Å, allowing protein-DNA interaction only during sliding motion.

Crowder Model

Crowders in the simulations are represented as randomly distributed uncharged spheres, and their volume fraction (Φ), is determined based on the radius of each crowder (R_{cr}), the total number of crowders (N_c), and the simulation box dimensions (L_x, L_y, L_z). In our investigation, we employ two categories of crowders: (1) Inert crowders, exerting solely excluded volume effects, and (2) Crowders capable of enthalpic interactions with the protein, mimicking other proteins within the cellular environment. Inert crowders interact with other biomolecules using a Lennard-Jones potential which can be represented by:

$$E_{ij} = \sum_{i<j} \begin{cases} k_{ev}\left[\left(\frac{\sigma_{ij}}{r_{ij}}\right)^{12} - 2\left(\frac{\sigma_{ij}}{r_{ij}}\right)^{6} + 1\right] & r < r_c \\ 0 & r \geq r_c \end{cases} \tag{22.9}$$

where k_{ev} = 1.0 kJ/mol, which is the energy parameter, σ_{ij} and r_{ij} is the mean site diameter and separation between the sites *i* and *j* respectively. r_c denotes the cut off distance and is set at $2^{\frac{1}{6}}\sigma_{ij}$ Å.

To model crowders capable of engaging in both repulsive and short-ranged attractive interactions, we employed a pairwise potential represented by the following form (Putzel et al. 2014):

$$U_r = \begin{cases} U_{large} - \left(\left(U_{large} + \varepsilon\right)/\delta\right)\left(r - r_t + \delta\right) & r < r_t \\ -\varepsilon \exp\left[-\left(r - r_t\right)/\lambda\right] & r \geq r_t \end{cases} \tag{22.10}$$

r_t represents summation of the hardcore radius of interacting beads and crowders. The first part of the equation accounts for the repulsive interaction, while the second part contributes to an attractive interaction. The hardcore interaction approximation is upheld by a large finite force denoted by $\frac{U_{large}}{\delta}$, where ϵ quantifies the strength of the attractive interaction (measured in units of k_BT) acting within the characteristic range λ of the crowders interacting with other molecules. In this context, U_{large} is set to 167.36 kJ/mol, δ is 1 Å, and the characteristic range λ is 5 Å. Any attractive interaction beyond ($r - r_t > 6\lambda$) is disregarded.

Protein, DNA and Crowders were placed inside a box of dimension 150 Å × 150 Å × 400 Å. A 100 bp DNA molecule was aligned with the Z-axis, placed at the center of the box throughout the simulation. Protein, DNA and crowders were allowed to diffuse freely. The dynamics of the protein–DNA system is simulated with the Langevin equation:

$$m_i \dot{v}_i = F_i - \gamma m_i v_i + R_i(t) \tag{22.11}$$

where m_i, v_i are the mass and velocity of the particle i and γ is the friction coefficient and set to 0.01 in this study. F_i represents the deterministic force applied to the particle i. R_i denotes a stochastic variable drawn from a Gaussian distribution with zero mean and variance:

$$< R_i(t) R_i(t') \geq 2\Gamma\delta(t-t') \tag{22.12}$$

where Γ is the strength of the random force to be determined and the δ function is the Dirac delta function defined by $\delta(t-t') = \{\begin{matrix} +\infty, & t = t' \\ 0, & t \neq t' \end{matrix}$.

Following parameters are used to characterize the mechanistic intricacies of facilitated diffusion.

Identification of Various Search Modes

In order to differentiate the search modes of the protein, i.e., sliding, hopping and 3D diffusion, three different parameters were used. These are: (1) orientation of the protein's recognition region towards DNA (θ) (2) percentage of recognition region of protein in contact with the DNA major groove and (3) distance between the centre of mass of protein and the nearest DNA base pair center (R). The red colored patch in Fig. 22.1a denotes the recognition region of the protein. In our study, we assumed that the protein undergoes '3D diffusion' when the center of mass of its recognition region is more than 25 Å away from the nearest base pair center. A snapshot is categorized as 'sliding' if at least 70% of the recognition region remained in contact with the DNA major groove, if the recognition region's center of mass is within 25 Å from the nearest DNA base pair and orientation angle (θ) is < 25°. Furthermore, a snapshot can be defined as 'hopping' if the recognition region is less than 25 Å from the nearest DNA base pair but did not meet any of the criteria for sliding dynamics.

Calculation of Diffusion Coefficient

Diffusion coefficient for both 1D (sliding and hopping) and 3D search regimes of the protein can be estimated by measuring the linear behaviour of the mean square displacement (MSD) of the protein along the Z-axis of the DNA over time (Fig. 22.1b) which can be calculated by following equation:

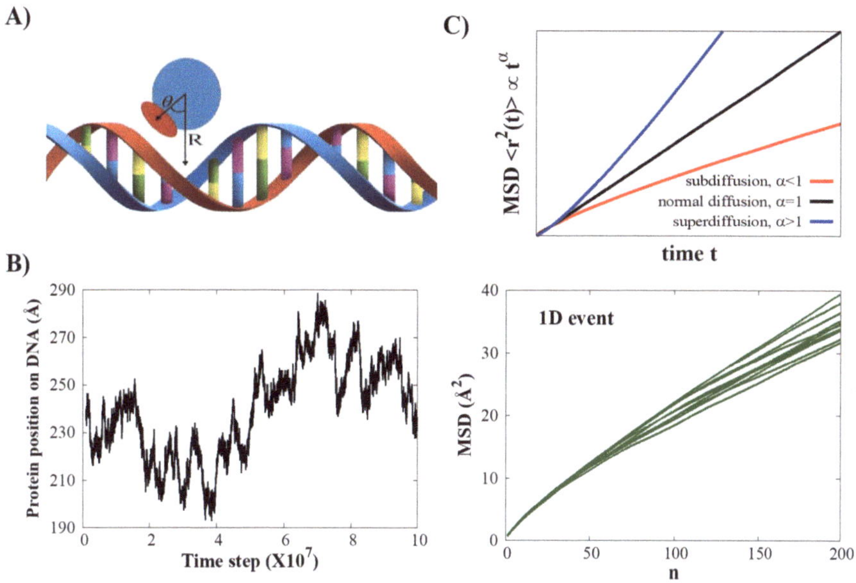

Fig. 22.1 (**a**) A schematic diagram illustrates the parameters utilized for characterizing different searching modes undertaken by a protein, including sliding, hopping, and 3D diffusion. In this representation, R signifies the distance between the center of the recognition region (depicted by a red ellipse) and the nearest DNA base pair's center. Additionally, θ represents the orientation angle θ. (**b**) Calculation of the one-dimensional diffusion coefficient from Mean Squared Displacement (MSD) profiles for protein diffusion along DNA. (**c**) Mean square displacement in normal and anomalous diffusion

$$\mathrm{MSD}(n,N) = \sum_{N-n}^{i=1} \frac{\left(Z_{i+n} - Z_i\right)^2}{N-n} = 2D_1 n\Delta t \tag{22.13}$$

In the given eq. N denotes the number of measured time steps in the measurement window given by n that ranges from 1 to N. D_1 is the 1D diffusion coefficient and Δt gives the time interval between two successive steps. Hence, the linear slope of MSD(n,N) at n will be $2D_1n\Delta$t. The MSD(n,N) at n below a cutoff n_c will be linear with a slope of $2D_1\Delta$t. Similarly, we estimate the 3D diffusion coefficient when the protein performs 3D diffusion in the bulk medium.

Alternate Approach to Calculate Diffusion Coefficient

The equation for the one-dimensional mean square displacement (MSD) of a diffusing Brownian particle is expressed as follows:

$$< r^2(t) \geq 2D_1 t \tag{22.14}$$

Here, D_1 denotes the one-dimensional diffusion coefficient, which is interconnected with the solvent molecules' viscosity and the hydrodynamic radius of the Brownian particle. In a complex medium, the MSD of a diffusing particle adheres to a power law (Fig. 22.1c).

$$< r^2(t) \geq 2\Gamma t^{\alpha} \tag{22.15}$$

In this context, the constant Γ remains unchanged over time. In the case of normal diffusion, the exponent α takes a value of 1. Anomalous diffusion occurs when α is not equal to 1. A value of α less than 1 denotes anomalous sub-diffusion, while α greater than 1 indicates anomalous super-diffusion shown in given figure. So the effective diffusion coefficient can be calculated from this given equation:

$$D_1^{eff}(t) = \Gamma t^{\alpha-1} \tag{22.16}$$

Results and Discussion

Characterization of Depletion Layer in the Presence of Crowders

In vitro investigations conducted in crowded environments have revealed significant alterations in the structure, dynamics, and intermolecular relationships of biomolecules compared to dilute conditions, commonly attributed to excluded volume effects caused by crowding. Despite the expectation that high crowder concentrations would hinder the movement of diffusing species, recent experimental findings have shown unexpected facilitation of DBPs binding to their target DNA sites and enhancement of their enzymatic function (Cravens et al. 2015; Krepel and Levy 2016). This observation poses challenges for an explanation solely based on excluded volume interactions of the crowders. The apparent inconsistency strongly suggests the involvement of several other factors that manifest only in the presence of high crowder concentrations (Kuznetsova et al. 2015).

Crowding involving large molecules can restrict some of the three-dimensional volume accessible to smaller searching proteins, a phenomenon known as the 'exclusion effect'. Computational investigations, from our group, into how crowding concentration influences protein-DNA interactions using molecular dynamics simulations have revealed, a clear separation between crowders and DNA molecules, forming a depletion layer between them. To quantify this separation, we calculated the average distance between each DNA base pair and the nearest crowder molecule, denoted by l_d (see Fig. 22.2a). l_d serves as an approximation of the width

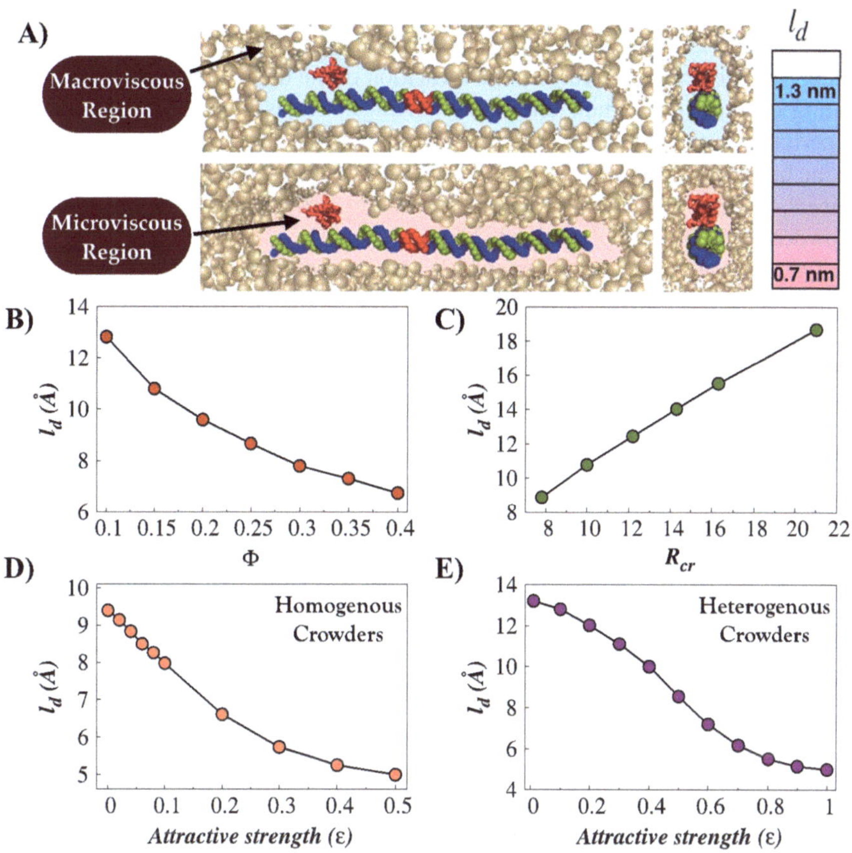

Fig. 22.2 (**a**) A schematic depiction is presented illustrating the formation of the depletion layer surrounding the DNA molecule. The formation of the depletion region arises from the dynamic interplay between DNA and crowder molecules, without any imposed artificial conditions. As the volume fraction of crowders (Φ) increases, the depletion width diminishes (blue colored region to pink colored region). Depletion layer width, l_d (Å) around the DNA molecule as a function of (B) Φ, (**c**) Crowder size (R_{cr}), (**d**) protein-crowder interaction strength, ϵ (in the presence of uniformly sized crowders), € protein-crowder interaction strength, ϵ (in the presence of heterogeneous crowders)

of the depletion region separating the crowder cloud from the DNA surface. The dynamics of the crowder and DNA molecules restrict their proximity to a limited extent, resulting in the formation of a depletion volume in between. This depletion layer arises from entropic factors, with no artificial constraints imposed. Fig. 22.2b illustrates the width of the depletion region as a function of changes in the crowder concentration, represented by variations in crowder fraction (Φ), while keeping the size of the crowder constant. Φ ranged from 0.1 to 0.4 (equivalent to 0.003–0.15% w/v). As Φ increases, l_d decreases from approximately 1.28 nm to 0.68 nm. Certain experimental studies (Kozer et al. 2007; Cravens et al. 2015) have reported l_d to be around 0.9 nm for 5% w/v of PEG600 crowders. A recent Fluorescence correlation spectroscopy (FCS) experiment on ligand binding affinity to G-quadruplex DNA by Clovis and Sen has also reported the formation of the depletion layer, thereby providing experimental validation for our findings (Clovis et al. 2023). An another experimental study by Zosel et al. has presented the decrease in the depletion width in the presence of varying crowder concentration, where they have investigated the impact of depletion interactions on the binding between disordered proteins in crowded environments (Zosel et al. 2020).

The relationship between depletion Width and crowder concentration becomes notably intricate in semi-dilute and concentrated solutions, leading to psi condensation (Cravens et al. 2015). Consequently, higher concentrations of crowder molecules are often disregarded when assessing their impacts on the width of the depletion region. The depletion region observed in our investigations aligns with the dimensions of the third (approximately 1 nm) and second (approximately 0.7 nm) hydration shells reported for B-DNA molecules (Chuprina et al. 1991; Egli et al. 1998). This analogy implies the formation of a depletion volume under *in vivo* conditions due to 'hydration interaction', wherein water molecules arrange to create a 'spine of hydration' around the DNA (Drew et al. 1981). The presence of such hydration shells around B-DNA molecules is crucial in regulating processes involving interactions with transcription factors and ligands (Schwabe 1997; Jayaram and Jain 2004; Privalov et al. 2007). Upon increasing the crowder fraction (Φ), beyond a threshold limit, the width of the depletion layer diminishes. Recent experimental findings (Cravens et al. 2015) also support that beyond a certain threshold concentration, crowder molecules such as PEG polymers tend to overlap, forming a non-covalent polymeric network with an average mesh size. Both the protein and DNA become embedded within this mesh. The size of this mesh varies with the Radius of Gyration (R_g) of the polymeric crowders. Additionally, R_g decreases with increasing crowder concentration, leading to a reduction in the width of the depletion region surrounding the DNA molecule. Furthermore, as the crowder molecules are excluded from the depleted volume, this results in the presence of two distinct viscosity regimes: microscopic and macroscopic viscosity as a result of the depletion layer formation (Fig. 22.2a). The microviscosity within the depletion zone experiences minimal change with variations in bulk crowder concentration. Conversely, the presence of crowders increases the macroscopic viscosity in the bulk solution. Consequently, when exploring its DNA cognate site, the searching protein encounters markedly different viscous environments: one while diffusing three-dimensionally in the bulk and another when migrating one-dimensionally along the DNA contour.

Impact of Crowder Properties on Depletion Layer

Since the investigation of crowder molecules primarily revolves around their volume occupancy or excluded volume component, extensive studies have been conducted to explore the effects of crowder molecule size on various reaction equilibria (Sharp 2015; Shin et al. 2015; Acosta et al. 2017). These studies encompass investigations into the impact of crowder size across a spectrum of processes ranging from RNA folding to protein fibrillation (Munishkina et al. 2004; Truong Co and Li 2013). The cellular milieu constitutes a heterogeneous mixture comprising proteins, cytoskeletal elements, nucleic acids, and cellular organelles, all serving as macromolecular crowders. Generally, crowding agents reduce protein's diffusivity by elevating the bulk viscosity. This phenomenon is particularly pronounced with smaller crowders, as reported by Sharp and Batra (Sarkar et al. 2014; Acosta et al. 2017). However, in certain instances, larger crowders have been noted to enhance biochemical processes, as reported in some studies (Kolesov et al. 2007; Shin et al. 2015), while others have found no significant size dependency (Wang et al. 2012; Truong Co and Li 2013). One aspect of interest is the formation of depletion layer around the DNA molecule in the presence of different sizes of the crowders. We investigated the influence of crowder size, ranging from 7.8 Å to 21 Å, on depletion layer formation while maintaining constant crowder concentration (Fig. 22.2c). Larger crowders induce the formation of broader depletion layers reducing the available space for crowder molecules to diffuse.

The theoretical framework of crowder actions, as per the Asakura-Oosawa model (AOM) (Asakura and Oosawa 1954), has shown significant discrepancies when applied to recent experiments involving protein-based crowding agents. Experiments demonstrating enhanced facilitated diffusion of the protein in the presence of synthetic crowding agents such as PEG have erroneously attributed the observed effects solely to 'volume exclusion' by the crowding agents, overlooking the potentially active role of these agents. Recent calorimetric experiments indeed suggest that commonly used synthetic agents like glucose, Dextran, and PEG can substantially contribute to both entropic destabilization and enthalpic stabilization of the proteins (Senske et al. 2014) . Given these insights and biomolecules' presence in *in vivo* conditions that can engage in non-specific attractive interactions with DBPs, there is a pressing need to investigate the role of an interacting crowded environment in facilitating DBP's diffusion. We examined the fundamental distinctions between hard spheres, entropic crowders as described in the AOM, and enthalpic protein-based crowding agents, investigating how the enthalpic contribution of crowders influences the facilitated diffusion of proteins. The width of the depletion layer was estimated for the protein-DNA system in the presence of interacting crowding agents exhibiting the varying affinity towards other biomolecules. To comprehend the influence of nonspecific crowder interactions on the microenvironment affecting the target search dynamics of DBPs, the strength of nonspecific interactions ($\frac{\varepsilon}{k_B T}$)

of the crowders was varied gradually from 0 to 0.5. Fig. 22.2d illustrates its direct impact on the depletion region's width (l_d), which reduces from approximately 9.5 to 5.0 Å with a rise in interaction strength, ϵ. This range of depletion width, l_d, aligns well with the experimental observations addressing the depletion widths of PEG600 (Bouvier et al. 2011). The intracellular environment is highly heterogeneous consisting the macromolecules of different sizes. The heterogeneous crowded environment is modeled by incorporating crowder molecules with different radii ranging from 7.8 Å to 21 Å. These crowder sizes correspond to the dimensions of PEG crowders used in experimental studies (Sabirov et al. 1993). The variation in the depletion region's width (l_d) around the DNA surface in the presence of heterogeneous crowder sizes, relative to ϵ, follows a similar trend (Fig. 22.2e) independent of crowder physiology.

Facilitated Diffusion: Protein Search in the Microviscous and Macroviscous Region

As previously mentioned and supported experimentally, DBPs nonspecifically scan DNA at a faster pace by employing a facilitated diffusion mechanism, which involves a weighted combination of one-dimensional (1D) and three-dimensional (3D) diffusion of the DBPs (Von Hippel and Berg 1989; Mirny et al. 2009). During 1D diffusion, protein slides along the DNA's helical pitch ('sliding'). Alternatively, it can engage in a series of association, dissociation and re-association events of short life spans along the DNA contour ('hopping'). Moreover, the search process can be facilitated through intersegmental 'jumps/transfer' by proteins between nearby DNA segments, allowing them to skip the scanning of several DNA bases (Sheinman and Kafri 2009; Esadze and Iwahara 2014). In three-dimensional diffusion (3D), the protein diffuses randomly in the bulk. The advantage of combining 1D and 3D diffusion lies in the rapid diffusion of the protein in 3D, enabling it to swiftly reach various DNA segments. Meanwhile, in the one-dimensional mode, the searching protein can closely inspect the nucleobases of DNA to precisely identify specific DNA sequences. However, relying solely on either 1D or 3D diffusion entails a trade-off between search speed and sequence precision, resulting in slower recognition of target DNA sites. Experimental evidence of facilitated diffusion in crowded environments extends beyond the realm of DBPs. It includes observations such as the human DNA glycosylases' translocation inside the highly dense PEG solutions (Bouvier et al. 2011) and lac repressor's optimized regulation kinetics in *E. coli* (Bailly et al. 1989).

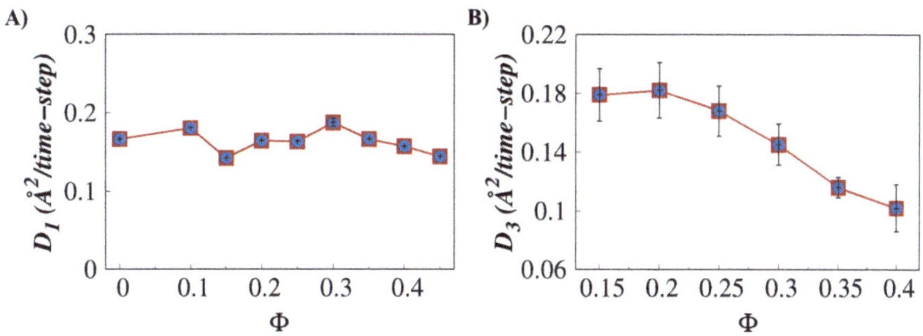

Fig. 22.3 Variation of (**a**) 1D diffusion coefficient (D_1) and (**b**) 3D diffusion coefficient (D_3) of the searching protein relative to crowder concentration (Φ)

Impact of Crowder Concentration

As explained above, the presence of crowders leads to the formation of a depletion layer, creating two distinct viscous environments for the searching protein. Dey and Bhattacherjee investigated the dynamics of the searching protein in these two distinct viscous media using an *in silico* approach (as described in the methods section). The crowders considered in this study are inert. Crowder concentration was varied, and Fig. 22.3a illustrates the consistent behavior of the robust one-dimensional diffusion constant, D_1, regardless of the crowder concentration (Φ). This indicates that the influence of bulk crowders on the 1D target search process is minimal. However, the protein can utilize a combination of one-dimensional (1D) and three-dimensional (3D) modes to reach the target site, consistent with recent site transfer experiments suggesting that the nonspecific diffusion of protein on the DNA surface is independent of bulk crowder concentration. Our study elucidates the mechanistic details behind this observation. For nonspecific scanning of the DNA surface, protein closely scans the DNA bases utilizing the 1D search mechanism. Thus, the protein typically resides within the depletion region given that its size aligns with the depletion width, which typically ranges from 7 to 12 Å. While diffusing away from the DNA surface, the protein experiences minimal electrostatic attraction from the DNA surface. However, upon entering the depletion region, its dynamics are governed by two opposing effects. As depicted in Fig. 22.2b, the depletion region narrows with increasing Φ, compelling the searching protein to remain close to the DNA surface. This process enhances the electrostatic attraction between the searching protein and DNA molecule, promoting sliding dynamics along the DNA major grooves. Secondly, increasing Φ leads to a rise in excluded volume energies between the protein and crowders due to the crowders' opposing roles. This results in frequent collisions that impede the protein's smooth motion along the DNA surface, necessitating more hopping dynamics. However, the attempts by crowder molecules to promote both sliding and hopping dynamics by reducing the depletion layer width and enhancing protein-crowder crosstalk with increasing Φ counterbalance each other, eventually showing no effects on the 1D search mechanism of DBPs. On the other hand, the effect of crowder molecules in the bulk solution is distinct. In the bulk solution, the protein resides far from the DNA surface and search the conformation space utilizing 3D diffusion mechanism. Figure 22.3b shows that with an increase in Φ, the 3D diffusion coefficient decreases, indicating reduced protein dynamics in the bulk solution with increasing Φ. This also highlights the presence of two different viscosities: one in the bulk (macroscopic viscosity), which is sensitive to changes in Φ, and the other being a constant microscopic viscosity inside the depletion region, independent of Φ. Despite the narrowing of the depletion layer at high crowder concentrations, interactions between DBPs and crowders balance out the effects, allowing DBPs to navigate the DNA surface effectively. DBP can readily diffuse on the DNA surface through a constant microscopic viscosity unaffected by bulk crowders.

Impact of Crowder Properties

Dey and Bhattacherjee conducted a study to evaluate the diffusive behavior of DBPs in response to various attributes of crowding agents, including size (R_{cr}), mass (M_{cr}), and shapes. To investigate the search protein's diffusivity, we calculated the anomalous diffusion exponents corresponding to both one-dimensional (1D) and three-dimensional (3D) diffusion, denoted as α_1 and α_3, respectively (detail in methods section). The observations reported in Fig. 22.4a, c, e reveal that the 1D diffusion of proteins (α_1) along the DNA contour remains unaffected by crowder size, mobility, and shape. However, the 3D diffusion exponent, α_3, of proteins is notably influenced by crowder properties. Further explanation in this regard is provided in the subsequent discussion.

Impact of Crowder Size

The α_3 exhibits non-monotonic behavior with respect to crowder size (R_{cr}), featuring a minimum at $R_{Cr} \approx 12$ Å (Fig. 22.4a). The variation of α_3 spans approximately 0.88 to 0.94 (from pronounced to minimal sub-diffusive regime), consistent with the protein diffusion coefficient in the cytoplasm of *E. coli* (Banks and Fradin 2005). Interestingly, the minimum observed for α_3 aligns with a crowder size ($R_{cr} \approx 12$ Å) matching the radius of gyration of the searching protein ($R_g \approx 12$ Å for Sap-1 used in our study). To gain insight into this, we examined the dynamics of the protein within 3D diffusion space, particularly focusing on its interactions with crowder molecules while diffusing in the bulk. We measured the protein-crowder crosstalk (E_{ev}^{3D}) arising from collisions between crowder molecules and the protein in the 3D search regime varying with the crowder size (Fig. 22.4b). E_{ev}^{3D} exhibits non-monotonic behavior with increasing crowder size, reaching a minimum value at $R_{cr} \approx 12$ Å. This can be explained by two opposing effects: one effect arises from the reduction in the crowder population due to an increase in crowder size, which is necessary to uphold the crowder volume fraction. This reduction decreases the

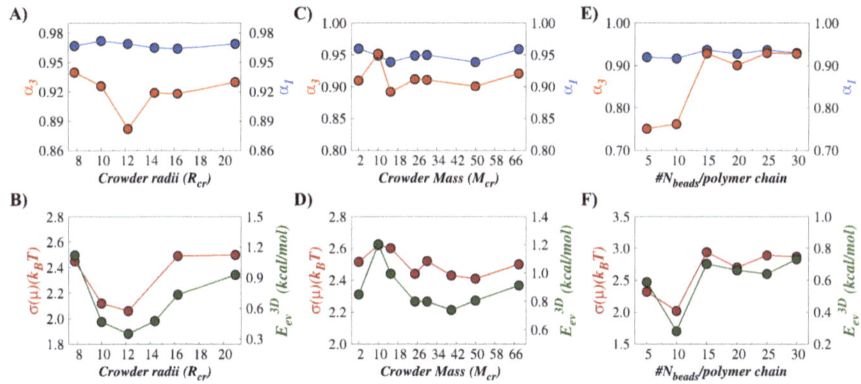

Fig. 22.4 The influence of crowder properties on the diffusion of the searching protein. (**a**) The changes in 3D diffusion exponent (α_3) (orange color) and 1D diffusion exponent (α_1) (blue color) relative to crowder radii (R_{cr}). (**b**) Estimation of the roughness of the potential energy landscape, $\sigma(\mu)$ (red color) and the protein-crowder excluded volume interactions, E_{ev}^{3D} (green color) relative to R_{cr}. (**c**) Variation in α_3 and α_1 & (**d**) $\sigma(\mu)$ and E_{ev}^{3D} relative to crowder mass (M_{cr}). € Variation in α_3 and α_1 & (F) $\sigma(\mu)$ and E_{ev}^{3D} relative to polymeric crowder length

frequency of interactions between crowders and the protein molecule, leading to a decrease in E_{ev}^{3D} until $R_{cr} \approx 12$ Å. However, the increase in depletion layer width with increasing crowder size (Fig. 22.2c) reduces the available volume for the crowders resulting in enhanced effective crowder density. Consequently, during three-dimensional diffusion, the protein molecule encounters crowders more frequently, leading to a higher E_{ev}^{3D}. Thus, the observed variation in α_3 suggests the fastest diffusivity of the searching protein at $R_{cr} \approx 12$ Å (Dey and Bhattacherjee 2019b). Furthermore, the ruggedness of the potential energy landscape (σ) varies with crowder size (Fig. 22.4b), indicating the direct impact of protein-crowder crosstalk in regulating the ruggedness of the associated potential energy landscape ($\sigma(\mu)$). Higher the ruggedness of the energy landscape, the protein diffusion will be more hindered. These results clearly suggest that encounters between the crowders and protein in the 3D search regime regulate protein diffusivity. However, the search mechanism in the 1D regime is notably different. The searching protein employs 1D dynamics (sliding and hopping) to scan the DNA surface within the depletion region. With increasing width of the depletion layer (l_d), the excluded volume interaction between the protein and crowder in 1D dynamics, E_{ev}^{1D}, decreases with increasing crowder size (Dey and Bhattacherjee 2019b). Despite changes in l_d and E_{ev}^{1D}, α_1 remains independent of crowder size. For smaller crowders, the frequency of encounters between crowder and protein is higher (resulting in high E_{ev}^{1D}), promoting hopping, while a narrow l_d forces the protein to slide. Conversely, with larger crowder sizes, the frequency of encounters between crowder and protein decreases, favoring protein sliding motion, but the physiological salt concentration promotes hopping. The opposing actions of these two factors cancel each other out, resulting in a constant 1D diffusion, α_1, at various R_{cr} (Fig. 22.4a). Moreover, the motion of the searching protein within the depletion layer encounters constant microscopic viscosity, which significantly differs from the bulk media.

Impact of Crowder Mass

In the heterogeneous cellular environment, macromolecules coexist with varying masses and mobilities. It was observed that higher crowder mass correlates with lower crowder mobility and vice versa. Next, we investigated the impact of crowder mass on both the one-dimensional (1D) and three-dimensional (3D) search regimes, focusing on crowder mobility at $R_{Cr} = 21$ Å. Figure 22.4c depicts the observed values of the 1D and 3D exponents (α_1 and α_3 respectively) as a function of increasing crowder mass (decreasing crowder mobility). The influence of crowder mass on α_1 is minimal, mirroring the trend observed for crowder size. In contrast, a non-monotonic trend is observed for 3D diffusion behavior (α_3), with a maximum observed at $M_{Cr} \approx 10$ kDa. Notably, this crowder mass aligns with the mass of the searching protein utilized in the study by Dey et al. (≈10.23 kDa) (Dey and Bhattacherjee 2019b). This phenomenon arises due to protein-crowder interactions during the 3D diffusion of the protein, reflected by the maximum protein-crowder crosstalks (E_{ev}^{3D}) at $M_{Cr} \approx 10$ kDa (Fig. 22.4d). When the mass of the crowder differs from that of the protein molecule, there is a notable difference in the diffusivity between the protein and the crowding agents. Crowders with high mobility diffuse faster compared to proteins; thus, within a certain time range, the relatively slow-moving protein encounters only a few crowder molecules. Conversely, heavier crowders move so slowly compared to the searching protein's mobility that the protein experiences an environment of static crowders. The diffusing protein encounters only those crowder molecules that happen to be along its trajectory. The roughness of the associated potential energy landscape ($\sigma(\mu)$) impedes the free motion of the diffusing protein (Fig. 22.4d).

Impact of Crowder Shape

As mentioned, the intracellular environment is highly heterogeneous, comprising macromolecules of diverse sizes and shapes; for instance, cytoplasmic proteins often exhibit non-globular shapes. Many eukaryotic proteins contain intrinsically disordered regions spanning over 30 amino acids, which tend to fold synergistically upon interacting with their partners to adopt specific secondary structures. Dey and Bhattacherjee utilized random coil polymers of varying lengths to investigate the role of proteins with intrinsically disordered regions as crowding agents (Dey and Bhattacherjee 2019b). The impact of these crowders on the diffusion of the searching protein in both one-dimensional (1D) and three-dimensional (3D) regimes is illustrated in Fig. 22.4e. Consistent with prior findings, contrasting diffusivity characteristics of the searching protein in 1D and 3D regimes are observed. A marginal alteration is noted in the 1D diffusion exponent (α_1) of the searching protein with varying polymer lengths, as the depletion region lacks crowders during 1D search. However, the 3D diffusion exponent (α_3) of the protein varies from pronounced sub-diffusion (~0.75) to minimal sub-diffusion (~0.92) as a function of polymer lengths. With an increase in polymer length (15 beads per polymer), the associated α_3 value rises by approximately 18% compared to a polymer length of 5 beads, remaining constant upon further lengthening. The excluded volume interaction energy, represented by crowder-protein crosstalks (E_{ev}^{3D}), reaches a minimum at a polymer length of 10, subsequently increasing with longer crowder polymers (Fig. 22.4f). A similar pattern is observed for the associated roughness of the potential energy landscape (Fig. 22.4f). In the presence of large polymeric crowders, the searching protein may become ensnared while diffusing in the bulk media within the mesh-like network formed by the lengthy crowders. Previous studies by Kozer et al. have suggested that the mesh created by polymeric crowders is a primary contributor to the anomalous diffusion observed in tracer proteins (Kozer et al. 2007).

Interacting Crowders

Early investigations proposed that in the presence of crowding agents, the repeated collisions between protein and DNA serve as the molecular foundation for the tighter association of protein-DNA complexes (Golding and Cox 2006; Guigas and Weiss 2008). Further investigations on diffusion in crowded media supported this notion, suggesting that bulk crowders prevent the searchinh protein's dissociation from the DNA surface, promoting one-dimensional diffusion of the protein along the DNA contour over three-dimensional diffusion. This observation aligns with the Asakura-Oosawa model of crowder action, where the excluded volume interactions with the DNA molecule lead to a preferentially depleted volume around the DNA surface, facilitating independent translational and rotational diffusion of the protein due to the presence of a constant microviscosity. This microviscosity is much lower than that of the bulk region (macroviscous region). However, Asakura-Oosawa model is inconsistent with observations involving protein-based cytoplasmic crowders. These crowders exert nonspecific attractive forces on the other biomolecules while also excluding the available volume for them. Notably, experiments reporting enhanced facilitated diffusion of the searching protein in the presence of synthetic crowding agents like PEG, attributed solely to 'volume exclusion' effect by these agents, have misinterpreted the inert nature of the crowding agents. Recent calorimetric experiments have indicated that commonly used synthetic agents, e.g., PEG, glucose, and Dextran can significantly contribute to the entropic destabilization and enthalpic stabilization of proteins. Given the findings and the occurrence of biomolecules in *in vivo* conditions that can mediate attractive non-specific interactions with DBPs highlights the need to investigate the role of interacting crowded environments in the facilitated diffusion of DBPs. Dey and Bhattacherjee addressed the biophysics of the enthalpic contribution of crowders influencing the facilitated diffusion (Dey and Bhattacherjee 2020).

In the context of attractive interactions occurring between crowders and proteins, Fig. 22.5a illustrates the 3D and 1D diffusion coefficients (D_3 & D_1 respectively) of the protein relative to diffusivity measurements conducted without crowders (D_0), varying with nonspecific crowder interactions strength, ϵ. The increasing adhesive

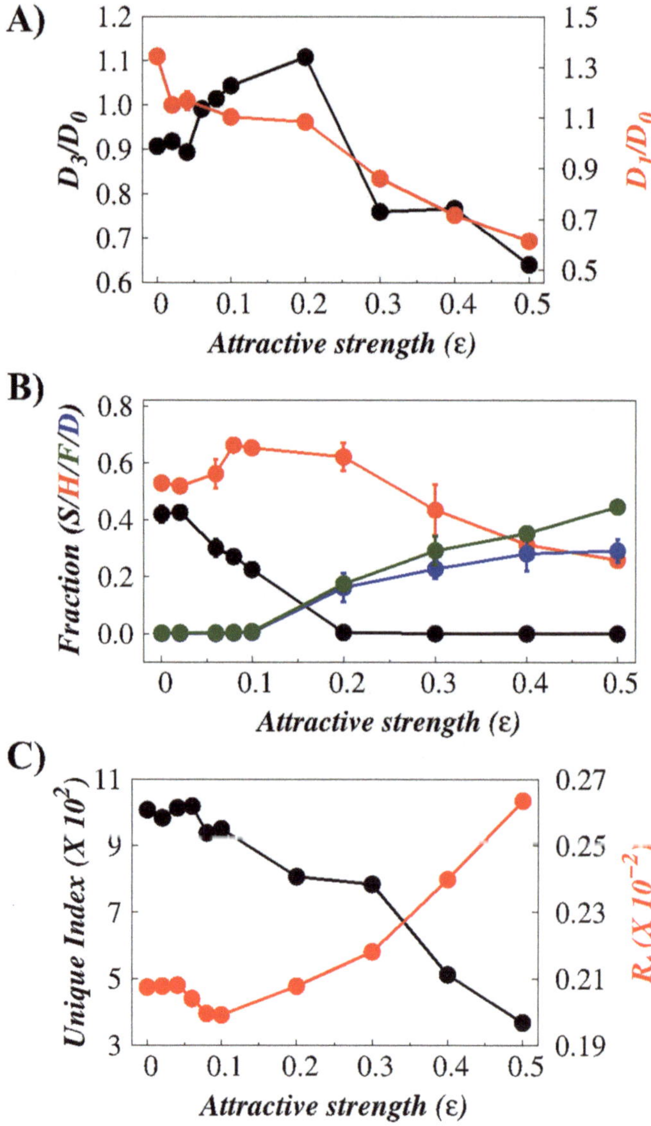

Fig. 22.5 (**a**) The variation of D_1 and D_3 with respect to diffusion coefficient when crowder molecules are absent (D_0) in relation to non-specific protein-crowder interactions (ϵ). The error values are narrower than the size of individual data points. (**b**) The variation in the various search modes propensities of the protein, Sliding (S), Hopping (H), Floating (F), and three-dimensional diffusion (D). (**c**) The overall changes in the count of distinct interacting crowders (unique index) and the transfer rate (R_t) of proteins (illustrated by the red line) between individual crowders relative to ϵ

nature of crowder molecules with increasing ϵ restrains the searching protein via strong nonspecific attractive forces, impeding its free 3D diffusion (black color, Fig. 22.5a). Conversely, the 1D diffusion coefficient of the searching protein displays a robust non-monotonic dependence on interaction strength, ϵ, reaching a peak at an intermediate value of ϵ, $\epsilon = 0.2$, as shown in Fig. 22.5a (red color). Enhanced attractive nonspecific interactions between crowder and protein compel the crowder molecule to take out the protein from the depletion layer into the macroviscous region, where protein movement is restricted to 3D diffusion. Conversely, a small ϵ minimizes interference from crowding agents during protein translocation through the depletion layer under the influence of strong electrostatics. Strong electrostatics keep the protein in close proximity to the DNA, where it diffuses slowly along the DNA helical pitch to read the DNA bases closely. The high propensity for sliding at a small ϵ value, as depicted in Fig. 22.5b, supports this argument.

To investigate the acceleration of the protein search process during 1D dynamics and the involvement of crowder molecules, Dey and Bhattacherjee monitored the motion of the crowders interacting with the diffusing protein across various nonspecific attractive protein-crowder interaction strengths, ϵ (Fig. 22.5c). The unique indices of such crowders decrease with rising protein-crowder interaction strength, ϵ (Fig. 22.5c). Simultaneously, the transfer rate (R_t) of the protein molecule, influenced by the average time spent in the close proximity to the nearest crowder molecule, escalates. At low interaction strength, ϵ, the protein interacts nonspecifically and weakly with the crowding agents having affinity towards the protein, allowing interaction with numerous crowders while diffusing autonomously. Conversely, at high ϵ, robust protein-crowder interactions restrict interactions to neighboring crowder molecules (as indicated by the decline in unique crowder indices in Fig. 22.5c), albeit for brief periods. The protein molecule experiences a strong attraction from the surrounding crowders, compelling it to shuttle among them, resulting in a high transfer rate (R_t). At intermediate ϵ, the protein employs a blend of independent and crowder-mediated search dynamics, facilitating longer hopping events and consequently faster protein diffusion.

The distance between centre of mass of the recognition region of the searching protein and the nearest DNA base pairs ($R_{protein-DNA}$), under the influence of crowding agents with varying affinities (as illustrated in Fig. 22.6), indicates distinct modes of protein-DNA interaction. In the presence of solely repulsive crowding agents ($\epsilon = 0$), the protein predominantly engages in 1D diffusion along the DNA. Below a threshold of $R_{protein-DNA} < 8$ Å, strong electrostatic forces orient the protein, facilitating thorough interaction with the DNA in a sliding mode (Fig. 22.6a, b). The snapshot presented in Fig. 22.6c highlights the position of the protein on DNA surface. Conversely, within the range of 8–15 Å, i.e., 8 Å $< R_{protein-DNA} < 15$ Å, the protein executes small, iterative jumps along the DNA contour, influenced by moderate DNA electrostatic interactions (−8 to −3 kcal/mol), enabling temporary dissociation from the DNA surface before reengaging (Fig. 22.6d, e). With increasing non-specific crowder interactions, the propensity for sliding diminishes, favoring hopping and 3D diffusion, due to the stronger non-specific attraction between the protein and crowders, pulling the protein away from the DNA surface (Fig. 22.5c). Moreover, the protein assumes an another search mode approximately 15–25 Å away from the DNA bases, where the $E^{el}_{protein-DNA}$ ranges between −3 to

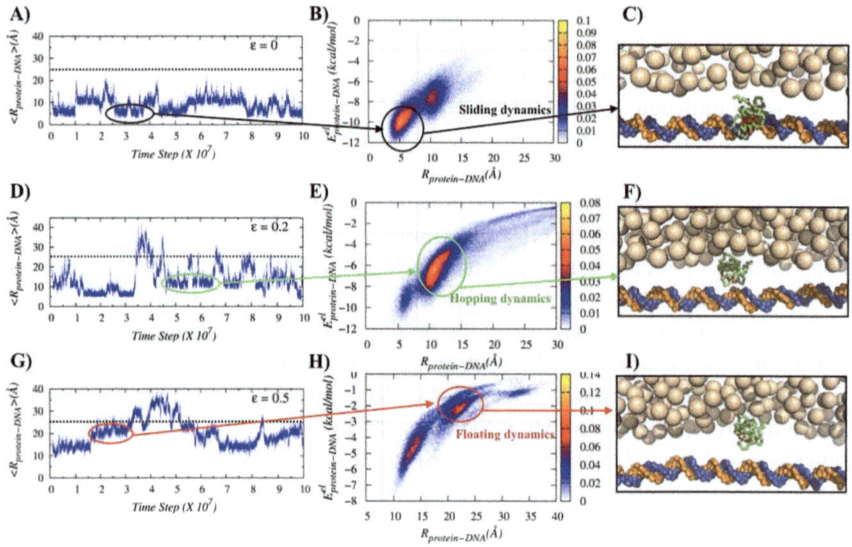

Fig. 22.6 The structural foundation underlying the enhanced facilitated diffusion of DBPs in the presence of crowders. Horizontally, the figures present (1) the variation of the distances between the center of recognition region of the searching protein and the center of the closest DNA base pair (bp) with simulation time, (2) $E^{el}_{protein-DNA} - R_{protein-DNA}$ contour plots presenting the differences in various search modes, and (3) snapshots showing the protein position during these search modes for different interaction strength, ϵ (**a–c**) 0.0, (**d–f**) 0.2, and (**g–i**) 0.5. For clear representation, the crowder molecules are not shown in the back and front of the DNA. (Adapted from Biophysical Journal, 2020, 118: 505–517)

−1 kcal/mol (Fig. 22.6h). This level of electrostatic energy is inadequate for protein hopping (i.e., −8 to −3 kcal/mol) (Fig. 22.6e, h), indicating a floating dynamics mediated by crowder interaction. Our observations suggest that protein-crowder interactions act as a driving force, facilitating protein transport along the DNA through a crowder-free depletion zone. The proportion of various search modes relative to the strength of non-specific crowder interactions (ϵ) is depicted in Fig. 22.5b. As non-specific crowder interaction strength increases, the propensity for sliding decreases, favoring hopping, floating, and three-dimensional diffusion. This phenomenon may offer a strategy to expedite the search process by bypassing pseudo-cognate sites, albeit with the caveat that optimal kinetics require a delicate balance between hopping and floating dynamics.

Genomic Crowders

In addition to the crowders in solution, ∼ 50% of the DNA surface is consistently occupied by diverse array of proteins, referred to as "genomic crowders." These proteins are distributed throughout the genome, covering approximately 20–50% of it (Hedglin and O'Brien 2008). An *in vivo* study performed on Lac repressor confirms that the protein spend a significant portion, around 90%, of their time engaged in non-specific diffusion along the DNA molecule, thereby occupying multiple DNA base pairs. LacI copies are bound to the DNA surface nonspecifically at various sites that are not the targets sites but shares a varying degree of sequence similarities with the target DNA site. The "genomic crowders" can range from slow-moving nucleosome-bound histone pairs to highly dynamic transcription factors and repair enzymes, all of which play pivotal roles in essential cellular processes such as gene transcription, DNA repair, and artificial genome editing. Moreover, these DNA-bound proteins can also serve as potential obstacles, hindering the sliding motion of other proteins searching for their respective regulatory sites on the DNA surface (Marcovitz and Levy 2013; Krepel and Levy 2016). Studies performed by Shvets et al. have highlighted the significant influence of "genomic crowders" on the target search dynamics of DBPs (Brackley et al. 2013; Marcovitz and Levy 2013; Shvets et al. 2016; Shvets and Kolomeisky 2016). The effectiveness of the search process is dependent upon the size and spatial distribution of these crowders or obstacles along the DNA track. Few *in silico* and analytical model based studies suggest that the presence of the roadblocks prompts the searching protein to adopt a hopping mechanism as an alternative strategy to bypass these obstacles, thereby facilitating faster diffusion along the DNA (Hedglin and O'Brien 2011; Krepel and Levy 2016; Shvets and Kolomeisky 2016; Koslover et al. 2017). Interestingly, the presence of static "genomic crowders" are found to impede the one-dimensional diffusion of searching proteins, in contrast to mobile crowders on DNA (Shvets and Kolomeisky 2016). However, studies examining the effects of roadblocks on facilitated diffusion of DBPs have reported minimal differences or none at all in the translocation of the searching protein to reach their target DNA sites (Brackley et al. 2013; Krepel and Levy 2016). The existing ambiguity necessitates a deeper understanding, particularly regarding the potential nonspecific interactions between crowders and diffuser proteins. Does such interaction hinder the diffusion of nearby proteins, analogous to the halting effect observed with pseudo-cognate sites? This question is addressed through an investigation of facilitated diffusion in the presence of interacting "genomic crowders."

To access target search dynamics of DBPs, Dey and Bhattacherjee estimated the 1D diffusion coefficient, D_1, of the protein with respect to the diffusivity measured in the absence of crowders as a function of the varying nonspecific crowder interaction strengths (ϵ). An illustration of the "genome crowders" alongside the target site is provided in Fig. 22.7a. Figure 22.7b presents a non-monotonic relationship between D_1 and ϵ. The maximum was observed at interacting strength of value 0.06 ($\epsilon = 0.06$). The observed D_1 was found to be approximately a 12% higher compared to the scenarios where no genomic crowders were present. Examination of the ruggedness of the potential energy landscape (σ_{μ}^{eff}) revealed minimal ruggedness at ϵ = 0.06, facilitating rapid protein diffusion along the DNA. This observation is further supported by contour plots, as depicted in Fig. 22.7c–e, where the spots denote the protein's positions during simulations. Interestingly, these positions of the proteins in all these plots do not precisely align with genomic crowder locations,

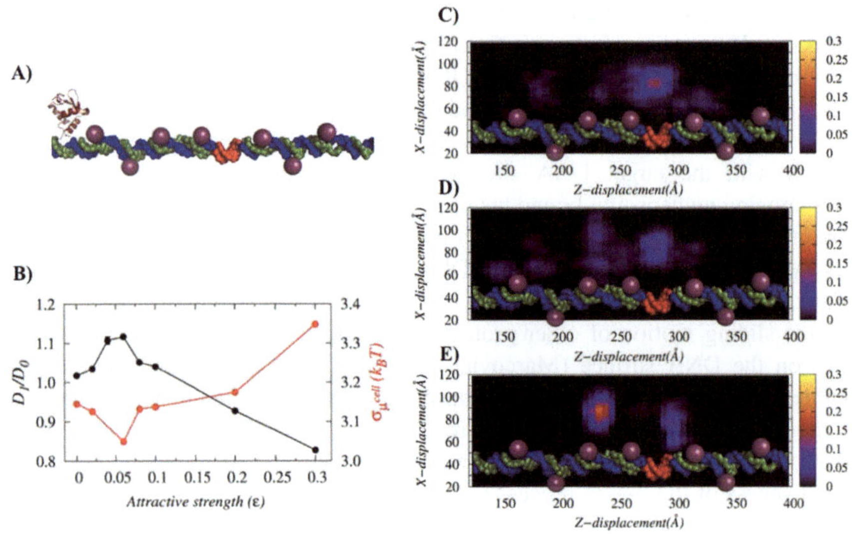

Fig. 22.7 Impact of the genomic crowders on target search dynamics of DBPs. (**a**) Illustrated depiction of protein's (in orange) target search process in the presence of DNA bound crowders (represented as colored spheres). (**b**) Graphical representation of the change in D_1 (black lines), with respect to the diffusion coefficient in the absence of crowder molecules (D_0), and the change in the complexity of the potential energy landscape (σ_μ^{eff} , red lines) based on ϵ values. (**c–e**) The contour plots illustrating the probability distribution of the searching protein positions on the X-Z plane for different crowder interaction strength $\epsilon = 0$, 0.06, and 0.5, respectively. The intensity of the spot correlates with the likelihood of finding the protein at that specific position. (Adapted from Biophysical Journal, 2020, 118: 505–517)

instead they are more commonly found between two consecutive genomic crowders. The connection between protein diffusion and this spatial arrangement warrants further investigation.

A comprehensive structural analysis indicates that nonspecific interactions with crowders enable adjacent or closely spaced genomic crowders to concurrently interact with the nearby searching proteins, forming transient bridged complexes (refer to Fig. 22.8b). Such interactions on linear DNA require deformation in the DNA conformation to bring the interacting crowders into spatial proximity. Dey and Bhattacherjee defined an angle θ formed between two adjacent genomic crowders to quantify this DNA deformation (Dey and Bhattacherjee 2020). Figure 22.8a shows the DNA deformation (θ) relative to the nonspecific protein crowder interaction strength, ϵ. The findings suggest that moderate ϵ values minimize the kink angle θ, facilitating protein jumping between crowders via local DNA conformational deformation and subsequent bridged complex formation. Conversely, for crowders that are purely repulsive, the formation of such bridged complexes is not allowed, while strongly attractive genomic crowders (high ϵ) tightly bind the searching protein, limiting its interaction with other crowders for bridged complex formation. This is evident from the reduction in DNA deformation (θ). It is noteworthy that cytoplasmic crowder interaction strengths can vary widely, potentially

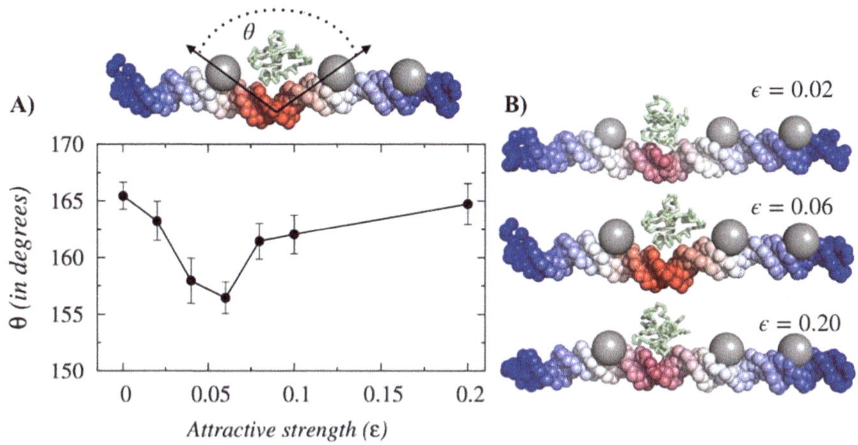

Fig. 22.8 Crowder-induced bends in the DNA structure lead to a transient bridged complex formation. (**a**) The degree of DNA deformation, defined by θ varies with the crowder affinity ϵ. This deformation occurs when searching protein interacts attractively with nearby genomic crowders. These crowders tend to come close to the interacting protein, leading to localized bends in the DNA conformation. (**b–d**) Illustrative representation of the DNA bend formed at three different crowder attraction strength ($\epsilon = 0.02$, 0.06, and 0.2). A color shift from blue to red denotes the increasing degree of bending. (Adapted from Biophysical Journal, 2020, 118: 505–517)

offsetting facilitated diffusion via transient bridged complexes observed with moderately interacting crowders, thereby impeding protein target search dynamics. Additionally, factors such as crowder positions, mobility, and size critically influence the efficiency of DBP target search (Krepel and Levy 2016; Shvets et al. 2016; Shvets and Kolomeisky 2016). Nonetheless, specific DBPs possess alternative mechanisms to navigate around genomic crowders (Bauer et al. 2015) by promoting the formation of DNA loops and coils, facilitating protein target search processes (van den Broek et al. 2008; Lomholt et al. 2009).

Conclusion

The interaction between proteins and DNA plays a pivotal role in nearly all enzymatic processes, hinging on the precise recognition of DBPs. This necessitates the rapid and efficient association of DBPs to their target sites, where fast kinetics is achieved through a weighted proportion of one and three-dimensional diffusion of the DBPs, termed as facilitated diffusion. This protein-DNA interaction unfolds within the cell nucleus, characterized by high concentrations of biomolecules. Consequently, the intracellular space is limited, impacting the diffusion and association of large macromolecules. This raises a fundamental question: How does the crowded environment of the cellular milieu influence the facilitated diffusion of DBPs on a large polymeric DNA substrate?

In this chapter, we present a suite of computational models to elucidate the mechanistic intricacies of protein-DNA interaction under crowded conditions using molecular dynamics simulations. These aspects include: the role of a crowded cellular environment with 'inert' crowding agents on the target search dynamics of the DBPs, the impacts of physiological properties of macromolecular crowders such as size, mobility, and shape on the diffusion exponent of the searching proteins, the consequences of the presence of 'interacting' bulk crowders and 'genomic crowders' on the facilitated diffusion and the target search efficiency of DBPs. Our findings reveal that the presence of cellular crowders leads to a distinct spatial separation between the crowder molecules and the DNA surface, resulting in the formation of a depletion layer. Notably, we quantitatively characterized this separation for the first time by computing the average distance between each crowder molecule and its nearest DNA index. The dimensions of the depletion region observed in our simulations correspond to those of the second (approximately 0.7 nm) and third (approximately 1 nm) hydration shells reported for a B-DNA molecule (Drew et al. 1981; Chuprina et al. 1991; Egli et al. 1998; Jayaram and Jain 2004; Laage et al. 2017). Experimental evidence has corroborated this observation and indicated that the dynamics of the protein remain unaltered when it searches within the depleted region (Clovis et al. 2023). Our findings regarding the impact of crowder concentration and their physiological characteristics indicate that the width of the depletion layer varies with the concentration of crowders and their properties. Specifically, a higher concentration of crowders correlates with a narrower depletion layer when the crowders are of the same size. Conversely, larger crowders lead to a wider depletion region around the DNA at constant volume fraction of the crowders in the system. Additionally, when interacting crowders exhibit a high affinity for the searching protein, the depletion region is reduced at a constant volume fraction, regardless of whether the crowders are homogeneous or heterogeneous. The formation of the depletion layer gives rise to two different viscous regimes: microviscous and macroviscous. Within the microviscous regime, crowder molecules are excluded, enabling the protein to closely and nonspecifically scan the DNA surface without encountering hindrance from the crowders. Conversely, the presence of bulk crowders elevates the macroviscosity, potentially leading to interactions between the searching protein and these crowders along its trajectory. The findings indicate that the concentration of inert crowders does not affect the nonspecific target search process of proteins. However, the presence of interacting crowders, which mimic the numerous copies of other DBPs within a cell, markedly influences protein dynamics. It is noteworthy that while the nonspecific interactions between the crowder molecules and the searching protein are individually weak, collectively they exert a significant effect on protein-DNA recognition. Our observations present a disparity between the physical laws governing the diffusive search process of the searching proteins in 1D and 3D regimes with respect to the physical properties of crowders. Our another finding highlights that the 3D diffusivity of the protein is maximally affected when the physical properties of the crowder molecules closely resemble those of the searching protein. Through precise selection of crowder molecules with desired size, mobility, and shape, one can manipulate the roughness of the associated potential energy landscape to achieve an optimal balance in the speed-stability paradox.

Another crucial aspect of the crowded cellular milieu is its heterogeneity. The majority of crowding agents are proteins, which interact among themselves, leading to intricate crosstalk. This crosstalk is particularly significant when investigating the facilitated diffusion ofDBPs through crowded media. Our findings highlight that the presence of interacting crowders, mimicking the numerous copies of other DBPs within a cell, profoundly influences protein dynamics. Typically, DBPs function akin to miniature molecular motors, utilizing DNA as a pathway for movement. Nevertheless, when bulk crowders are present, the protein switches to a floating mode, moving along the DNA contour in one dimension. This contrasts with the typical 1D hopping and sliding dynamics, where the protein diffusion is solely governed by protein-DNA electrostatics. In floating dynamics, the protein still performs 1D movement on DNA but diffuses slightly off from the DNA surface where the ruggedness of the binding energy landscape is minimal. The interplay between weak DNA electrostatics and moderate crowder interaction results in optimal target search kinetics. In addition to bulk crowders, genomic DNA can also be crowded with already bound DBPs. Analogous to congested roads, the high density of DBPs nonspecifically bound to DNA acts as obstacles to the diffusing protein along the DNA track. Interestingly, our results suggest that, contrary to traffic congestion on roads, molecular crowding on genomic DNA can enhance the migration of proteins diffusing along the DNA track in search of their binding sites. This phenomenon is further validated by experimental studies. These findings corroborate earlier investigations conducted by Krepel et al. (Krepel et al. 2016; Krepel and Levy 2017), who reported that crowding expedites search kinetics by constraining protein mobility along DNA regions between crowding particles, particularly evident at moderate salt concentrations. However, at high salt concentrations, crowding may impede 3D diffusion and promote hopping events, thereby impacting the overall search process. Moreover, their research revealed that crowding, particularly with larger particles, could enhance the efficiency of intersegmental transfer, a pivotal aspect of the monkey-bar mechanism. Crowding exerts influence on the dynamics of DBPs by modulating their various search modes, such as sliding, hopping, and 3D diffusion. This study underscores that crowding may not invariably decelerate search kinetics but could instead expedite specific search modes contingent upon salt concentrations and crowding particle size.

While the preceding discussion offers a comprehensive overview of crowder action on the facilitated diffusion of DBPs), several intriguing questions persist and demand further investigation. For example, histone proteins play a pivotal role in the compaction of chromatin fiber by forming the fundamental structural unit, known as the nucleosome. In doing so, they obstruct DNA regulatory sites for other proteins. How do DBPs navigate such obstacles on genomic DNA to access their specific sites? Although our recent study has elucidated the underlying mechanism of nucleosome invasion by DBPs (Mondal et al. 2021, 2022; Kumar Mishra and Bhattacherjee 2023; Mishra and Bhattacherjee 2023), facilitated diffusion on multi-nucleosome systems, especially in the presence of nucleosome remodeling complexes, remains unexplored. Additionally, comprehending the self-crowding of the DNA molecule poses an interesting challenge. Chromatin fiber organizes in a hierarchical manner, suggesting compartmentalization with varying degrees of compaction and DNA density. Understanding facilitated diffusion in such an organizational context could potentially unravel the gene expression pattern in euchromatin and heterochromatin.

Acknowledgement We express our gratitude for the financial assistance provided by DBT India (BT/PR46247/BID/7/1015/2023), DST India (CRG/2023/000636), and the DBT CoE research grant. A.B. gratefully acknowledges support from the Alexandar von Humboldt Foundation, Germany. S.K.C. thanks the University Grant Commission for the financial support. Sangeeta acknowledges the financial support from the Ministry of Human Resource Development (MHRD), Govt. of India for Prime Minister's Research.

References

Acosta LC, Perez Goncalves GM, Pielak GJ, Gorensek-Benitez AH (2017) Large cosolutes, small cosolutes, and dihydrofolate reductase activity. Protein Sci 26:2417–2425

Asakura S, Oosawa F (1954) On interaction between two bodies immersed in a solution of macromolecules. J Chem Phys 22:1255–1256

Bailly V, Verly WG, O'Connor T, Laval J (1989) Mechanism of DNA strand nicking at apurinic/apyrimidinic sites by Escherichia coli [formamidopyrimidine] DNA glycosylase. Biochem J J 262:581–589

Banks DS, Fradin C (2005) Anomalous diffusion of proteins due to molecular crowding. Biophys J 89:2960–2971

Bauer M, Rasmussen ES, Lomholt MA, Metzler R (2015) Real sequence effects on the search dynamics of transcription factors on DNA. Sci Rep 5:10072

Berg OG, Blomberg C (1976) Association kinetics with coupled diffusional flows. Special application to the lac repressor–operator system. Biophys Chem 4:367–381

Berg OG, Winter RB, Von Hippel PH (1981) Diffusion-driven mechanisms of protein translocation on nucleic acids. 1. Models and theory. Biochemistry 20:6929–6948

Bhattacherjee A, Levy Y (2014a) Search by proteins for their DNA target site: 1. The effect of DNA conformation on protein sliding. Nucleic Acids Res 42:12404–12414

Bhattacherjee A, Levy Y (2014b) Search by proteins for their DNA target site: 2. The effect of DNA conformation on the dynamics of multidomain proteins. Nucleic Acids Res 42:12415–12424

Bhattacherjee A, Krepel D, Levy Y (2016) Coarse-grained models for studying protein diffusion along DNA. WIREs Comput Mol Sci 6:515–531

Bouvier B, Zakrzewska K, Lavery R (2011) Protein-DNA recognition triggered by a DNA conformational switch. Angew Chemie Int Ed 50:6516–6518

Brackley CA, Cates ME, Marenduzzo D (2013) Intracellular facilitated diffusion: searchers, crowders, and blockers. Phys Rev Lett 111:108101

Chuprina VP, Heinemann U, Nurislamov AA, Zielenkiewicz P, Dickerson RE, Saenger W (1991) Molecular dynamics simulation of the hydration Shell of a B-DNA Decamer reveals two Main types of minor-groove hydration depending on groove width. Proc Natl Acad Sci 88:593–597

Clementi C, Nymeyer H, Onuchic JN (2000) Topological and energetic factors: what determines the structural details of the transition state ensemble and "en-route" intermediates for protein folding? An investigation for small globular proteins. J Mol Biol 298:937–953

Clovis NS, Alam P, Chand AK, Sardana D, Khan MF, Sen S (2023) Molecular crowders modulate ligand binding affinity to G-quadruplex DNA by decelerating ligand association. J Photochem Photobiol A Chem 437:114432

Cravens SL, Schonhoft JD, Rowland MM, Rodriguez AA, Anderson BG, Stivers JT (2015) Molecular crowding enhances facilitated diffusion of two human DNA glycosylases. Nucleic Acids Res 43:4087–4097

Cuculis L, Abil Z, Zhao H, Schroeder CM (2016) TALE proteins search DNA using a rotationally decoupled mechanism. Nat Chem Biol 12:831–837

Dey P, Bhattacherjee A (2018) Role of macromolecular crowding on the intracellular diffusion of DNA binding proteins. Sci Rep 8:844

Dey P, Bhattacherjee A (2019a) Mechanism of facilitated diffusion of DNA repair proteins in crowded environment: case study with human uracil DNA glycosylase. J Phys Chem B 123:10354–10364

Dey P, Bhattacherjee A (2019b) Disparity in anomalous diffusion of proteins searching for their target DNA sites in a crowded medium is controlled by the size, shape and mobility of macromolecular crowders. Soft Matter 15:1960–1969

Dey P, Bhattacherjee A (2020) Structural basis of enhanced facilitated diffusion of DNA-binding protein in crowded cellular milieu. Biophys J 118:505–517

Drew HR, Wing RM, Takano T, Broka C, Tanaka S, Itakura K, Dickerson RE (1981) Structure of a B-DNA dodecamer: conformation and dynamics. Proc Natl Acad Sci 78:2179–2183

Egli M, Tereshko V, Teplova M, Minasov G, Joachimiak A, Sanishvili R, Weeks CM, Miller R, Maier MA, An HY, Cook PD, Manoharan M (1998) X-ray crystallographic analysis of the hydration of A-form and B-form DNA at atomic-resolution. Biopolymers 48:234–252

Ellis RJ (2001) Macromolecular crowding : obvious but underappreciated. Trends Biochem Sci 26:597–604

Ellis RJ, Minton AP (2006) Protein aggregation in crowded environments. Biol Chem 387:485–497

Esadze A, Iwahara J (2014) Stopped-flow fluorescence kinetic study of protein sliding and intersegment transfer in the target DNA search process. J Mol Biol 426:230–244

Flyvbjerg H, Keatch SA, Dryden DTF (2006) Strong physical constraints on sequence-specific target location by proteins on DNA molecules. Nucleic Acids Res 34:2550–2557

Givaty O, Levy Y (2009) Protein sliding along DNA : dynamics and structural characterization. J Mol Biol 385:1087–1097

Golding I, Cox EC (2006) Physical nature of bacterial cytoplasm. Phys Rev Lett 96:98102

Guigas G, Weiss M (2008) Sampling the cell with anomalous diffusion – the discovery of slowness. Biophys J 94:90–94

Hall D, Minton AP (2003) Macromolecular crowding: qualitative and semiquantitative successes, quantitative challenges. Biochim Biophys Acta 1649:127–139

Hedglin M, O'Brien PJ (2008) Human alkyladenine DNA glycosylase employs a processive search for DNA damage. Biochemistry 47:11434–11445

Hedglin M, O'Brien P (2011) Hopping enables a DNA repair glycosylase to search both strands and bypass a bound protein. ACS Chem Biol 5:427–436

Hinckley DM, Freeman GS, Whitmer JK, de Pablo JJ (2013) An experimentally-informed coarse-grained 3-site-per-nucleotide model of DNA: structure, thermodynamics, and dynamics of hybridization. J Chem Phys 139:144903

Janin J, Rodier F, Chakrabarti P, Bahadurb RP (2007) Macromolecular recognition in the protein data Bank. Acta Crystallogr D Biol Crystallogr 63:1–8

Jayaram B, Jain T (2004) The role of water in protein-DNA recognition. Annu Rev Biophys Biomol Struct 33:343–361

Kolesov G, Wunderlich Z, Laikova ON, Gelfand MS, Mirny LA (2007) How gene order is influenced by the biophysics of transcription regulation. Proc Natl Acad Sci 104:13948–13953

Koslover EF, De La Rosa MD, Spakowitz AJ (2017) Crowding and hopping in a protein's diffusive transport on DNA. J Phys A Math Theor 50:74005

Kozer N, Kuttner YY, Haran G, Schreiber G (2007) Protein-protein association in polymer solutions: from dilute to semidilute to concentrated. Biophys J 92:2139–2149. https://doi.org/10.1529/biophysj.106.097717

Krepel D, Levy Y (2016) Protein diffusion along DNA : on the effect of roadblocks and crowders. J Phys A Math Theor 49:494003

Krepel D, Levy Y (2017) Intersegmental transfer of proteins between DNA regions in the presence of crowding. Phys Chem Chem Phys 19:30562–30569

Krepel D, Gomez D, Klumpp S, Levy Y (2016) Mechanism of facilitated diffusion during a DNA search in crowded environments. J Phys Chem B 120:11113–11122

Kumar Mishra S, Bhattacherjee A (2023) Understanding the target search by multiple transcription factors on Nucleosomal DNA. ChemPhysChem 24:e202200644

Kuznetsova IM, Zaslavsky BY, Breydo L, Turoverov KK, Uversky VN (2015) Beyond the excluded volume effects: mechanistic complexity of the crowded milieu. Molecules 20:1377–1409

Laage D, Elsaesser T, Hynes JT (2017) Water dynamics in the hydration shells of biomolecules. Chem Rev 117:10694–10725

Liu L, Luo K (2014) Molecular crowding effect on dynamics of DNA-binding proteins search for their targets molecular crowding effect on dynamics of DNA-binding proteins search. J Chem Phys 141:225102

Liu L, Luo K (2015) DNA-binding protein searches for its target : non-monotonic dependence of the search time on the density of roadblocks bound on the DNA chain. J Chem Phys 142:125101

Lomholt MA, Van Den BB, Kalisch S-MJ, Wuite GJL, Metzler R (2009) Facilitated diffusion with DNA coiling. Proc Natl Acad Sci 106:8204–8208

Marcovitz A, Levy Y (2011) Frustration in protein–DNA binding influences conformational switching and target search kinetics. Proc Natl Acad Sci 108:17957 LP – 17962

Marcovitz A, Levy Y (2013) Obstacles may facilitate and direct DNA search by proteins. Biophys J 104:2042–2050

Minton AP (2005) Models for excluded volume interaction between an unfolded protein and rigid macromolecular Cosolutes : macromolecular crowding and protein stability revisited. Biophys J 88:971–985

Minton AP (2006) How can biochemical reactions within cells differ from those in test tubes? J Cell Sci 119:2863–2869

Mirny L, Slutsky M, Wunderlich Z, Tafvizi A, Leith J, Kosmrlj A (2009) How a protein searches for its site on DNA: the mechanism of facilitated diffusion. J Phys A Math Theor 42. https://doi.org/10.1088/1751-8113/42/43/434013

Mishra SK, Bhattacherjee A (2023) How do nucleosome dynamics regulate protein search on DNA? J Phys Chem B 127:5702–5717

Mondal A, Mishra SK, Bhattacherjee A (2021) Kinetic origin of nucleosome invasion by pioneer transcription factors. Biophys J 120:5219–5230

Mondal A, Mishra SK, Bhattacherjee A (2022) Nucleosome breathing facilitates cooperative binding of pluripotency factors Sox2 and Oct4 to DNA. Biophys J 121:4526–4542

Munishkina L, Cooper EM, Uversky VN, Fink AL (2004) The effect of macromolecular crowding on protein aggregation and amyloid fibril formation. J Mol Recognit 17:456–464

Privalov PL, Dragan AI, Crane-robinson C, Breslauer KJ, Remeta DP, Minetti CASA (2007) What drives proteins into the major or minor grooves of DNA? J Mol Biol 365:1–9

Putzel GG, Tagliazucchi M, Szleifer I (2014) Nonmonotonic diffusion of particles among larger attractive crowding spheres. Phys Rev Lett 113:138302

Richter PH, Eigen M (1974) Diffusion controlled reaction rates in spheroidal geometry. Application to repressor–operator association and membrane bound enzymes. Biophys Chem 2:255–263

Riggs AD, Bourgeois S, Cohn M (1970) The lac repressor-operator interaction. 3. Kinetic studies J Mol Biol 53:401–417

Rivas G, Ferrone F, Herzfeld J (2004) Life in a crowded world. EMBO Rep 5:23–27. https://doi.org/10.1038/sj.embor.7400056

Rodriguez G, Esadze A, Weiser BP, Schonhoft JD, Cole PA, Stivers JT (2017) Disordered N-terminal domain of human uracil DNA glycosylase (hUNG2) enhances DNA translocation. ACS Chem Bol 12:2260–2263
Sabirov RZ, Krasilnikov OV, Ternovsky VI, Merzliak PG (1993) Relation between ionic channel conductance and conductivity of media containing different nonelectrolytes. A novel method of pore size determination. Gen Physiol Biophys 12:95–111
Sarkar M, Lu J, Pielak GJ (2014) Protein Crowder charge and protein stability. Biochemistry 53:1601–1606
Schwabe JWR (1997) The role of water in protein—DNA interactions. Curr Opin Struct Biol 7:126–134
Senske M, To L, Born B, Havenith M, Herrmann C, Ebbinghaus S (2014) Protein stabilization by macromolecular crowding through enthalpy rather than entropy. J Am Chem Soc 136:9036–9041
Sharp KA (2015) Analysis of the size dependence of macromolecular crowding shows that smaller is better. Proc Natl Acad Sci 112:7990–7995
Sheinman M, Kafri Y (2009) The effects of intersegmental transfers on target location by proteins. Phys Biol 6:16003
Shin J, Cherstvy AG, Metzler R (2015) Self-subdiffusion in solutions of star-shaped crowders: non-monotonic effects of inter-particle interactions. New J Phys 17
Shvets A, Kolomeisky AB (2016) Crowding on DNA in protein search for targets. J Phys Chem Lett 7:2502–2506
Shvets A, Kochugaeva M, Kolomeisky AB (2016) Role of static and dynamic obstacles in the protein search for targets on DNA. J Phys Chem B 120:5802–5809
Truong Co N, Hu C-K, Li MS (2013) Dual effect of crowders on fibrillation kinetics of polypeptide chains revealed by lattice models. J Chem Phys 138:185101
van den Broek B, Lomholt MA, Kalisch S-MJ, Metzler R, Wuite GJL (2008) How DNA coiling enhances target localization by proteins. Proc Natl Acad Sci 105:15738–15742
Von Hippel PH, Berg OG (1989) Facilitated target location in biological systems*. J Biol Chem 264:675–678
Wang Y, Sarkar M, Smith AE, Krois AS, Pielak GJ (2012) Macromolecular crowding and protein stability. J Am Chem Soc 134:16614–16618
Watson JD, Crick F (1953) Molecular structure of nucleic acids. Nature 171:737–738
Zheng G, Jun LX, Olson WK (2009) Web 3DNA – a web server for the analysis, reconstruction, and visualization of three-dimensional nucleic-acid structures. Nucleic Acids Res 37:240–246
Zhou H-X (2004a) Loops, linkages, rings, catenanes, cages, and crowders: entropy-based strategies for stabilizing proteins. Acc Chem Res 37:123–130
Zhou H-X (2004b) Polymer models of protein stability, folding, and interactions. Biochemistry 43:2141–2154
Zhou H-X, Rivas G, Minton AP (2008) Macromolecular crowding and confinement: biochemical, biophysical, and potential physiological consequences. Annu Rev Biophys 37:375–397
Zimmerman SB, Trach SO (1991) Estimation of macromolecule concentrations and excluded volume effects for the cytoplasm of Escherichia coli. J Mol Biol 222:599–620
Zosel F, Soranno A, Buholzer KJ, Nettels D, Schuler B (2020) {D}epletion interactions modulate the binding between disordered proteins in crowded environments. Proc Natl Acad Sci USA 117:13480–13489

Chapter 23
The Hidden Influences of Macromolecular Crowding: A New Frontier in Cellular Biology and Medicine

Isabella V. Gimón, Conner Sandefur, and Santiago Schnell

Abstract Macromolecular crowding, the dense packing of macromolecules within cells, significantly influences enzyme kinetics, cell signaling, liquid-liquid phase separation, and disease progression. This review explores the multifaceted effects of crowding on osmotic balance, intracellular compartmentalization, and metabolic efficiency. We highlight the exclusion volume effect's role in protein folding, stability, and osmotic homeostasis. Crowding also modulates enzyme activity and kinetics, contributing to metabolic regulation, and plays a crucial role in forming membraneless organelles. Dysregulation of macromolecular crowding is implicated in neurodegenerative diseases, cancer, and type 2 diabetes, highlighting its importance for therapeutic development. Future research should focus on elucidating crowding mechanisms, developing in vivo models, and exploring therapeutic interventions targeting macromolecular crowding.

Keywords Macromolecular crowding · Cell biology · Cell signaling · Liquid-liquid phase separation · Enzyme kinetics · Protein aggregation · Disease · Therapeutics

I. V. Gimón
Integrated Biomedical Sciences Program, University of Notre Dame, Indiana, USA

C. Sandefur
Sandefur Modeling, Pittsboro, NC, USA

S. Schnell (✉)
Department of Biological Sciences, and Department of Applied & Computational Mathematics & Statistics, University of Notre Dame, Notre Dame, IN, USA
e-mail: santiago.schnell@nd.edu

V. N. Uversky (ed.), *(Macro)Molecular Crowding*, Subcellular Biochemistry 109,
https://doi.org/10.1007/978-3-032-03370-3_23

Introduction

For most of us, being in a crowded environment is an unpleasant experience. Imagine navigating a packed house party, searching for a friend across the room. The crowd influences your trajectory, and even the chance to encounter you have might along the way.

Similarly, the crowded and heterogenous intracellular environment profoundly impacts the behavior of individual biomolecules. Linus Pauling's research on protein structure began to acknowledge the purpose behind the intracellular environment (Nakada 2009). This paved the way for Britton Chance and colleagues to analyze crowding effects on mitochondrial oxidative phosphorylation in the 1970s. However, the concept of macromolecular crowding emerged in the 1980s, it was not until the 1990s, with a clearer understanding of macromolecular thermodynamics (Minton 1983), that the field truly gained traction. As research in crowding progressed into the 90 s, scientists relied on experimental techniques, such as NMR to study macromolecular crowding *in vitro* (Ellis 2001). Since then, researchers have relied on in vitro and in silico models to analyze the different effects brought by macromolecular crowding.

Two main methodologies are used to study crowding effects on biomolecules:

1. **Inert crowders:** These substances, such as polyethylene glycol or Ficoll, occupy space without directly interacting with the target molecule. This approach isolates the effects of volume exclusion, a fundamental aspect of crowding.
2. **Complex crowding agents:** Substances like bovine serum albumin or lysozyme can mimic the diverse chemical environment inside cells. These agents not only occupy space but also engage in weak interactions with the target molecule, providing a more physiologically relevant context.

Both methodologies have revealed that to truly understand cellular physiology, it is essential to move beyond the study of isolated molecules or simple networks and consider the complex, crowded environment in which they function. This crowded environment, filled with proteins, nucleic acids, carbohydrates, lipids and other macromolecules, profoundly influences the chemistry, physics, and ultimately, the function of the cell.

The following sections delve into the integral role macromolecular crowding plays in many fundamental cellular processes.

Macromolecular Crowding and Exclusion Volume Effect

Macromolecular crowding plays a role in maintaining osmotic balance, a critical factor for cell survival and function. The exclusion volume effect, a fundamental consequence of crowding, regulates the availability of water molecules within the

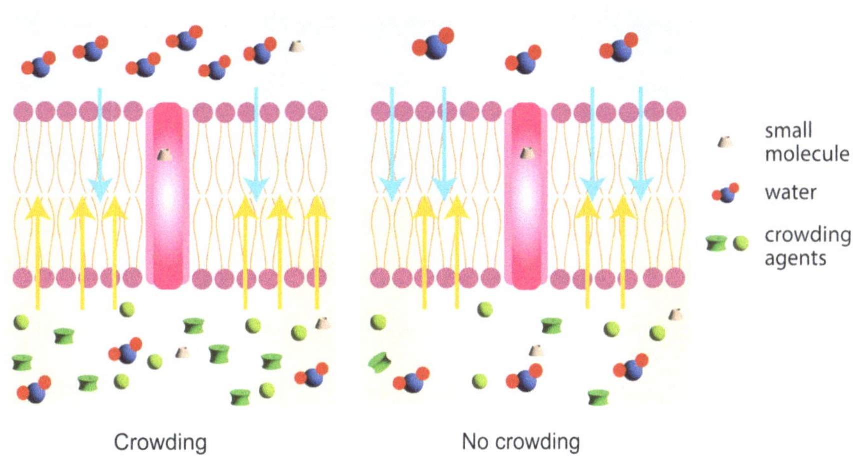

Fig. 23.1 Schematic representation of the impact of molecular crowding on osmotic pressure and water transport across the cell membrane. A crowded intracellular environment (left) increases osmotic pressure, driving water influx, while a less crowded environment (right) results in lower osmotic pressure

intracellular environment. This, in turn, influences water thermodynamics, driving processes like protein folding and stability.

For example, during the folding of ribonuclease A an enzyme that cleaves RNA, the exclusion volume effect reduces the number of water molecules interacting with ribonuclease A's interior as it assumes its native structure (Neira and Rico 1997). This decrease in water entropy around ribonuclease A increases osmotic pressure, counteracting the pressure exerted by metabolites, ions, and other solutes, ultimately achieving osmotic homeostasis within the cell.

Moreover, the excluded volume effect ensures proper distribution of water, supporting optimal membrane and organelle function (Fig. 23.1). In mitochondria, with their double-membrane structure housing essential molecules for the citric acid cycle and electron.

transport chain, the excluded volume effect maintains hydration of the inner membrane.

This is crucial for optimizing the efficiency of metabolic reactions, including ATP production, within the mitochondrial matrix (Ikizawa et al. 2023).

Metabolic Activity and Macromolecular Crowding

The crowded intracellular environment significantly influences enzyme interactions, activity, and kinetics. Volume exclusion, a consequence of crowding, increases effective enzyme concentrations, promoting the formation of enzyme-substrate complexes (Fig. 23.2). An early study on glyceraldehyde-3-phosphate

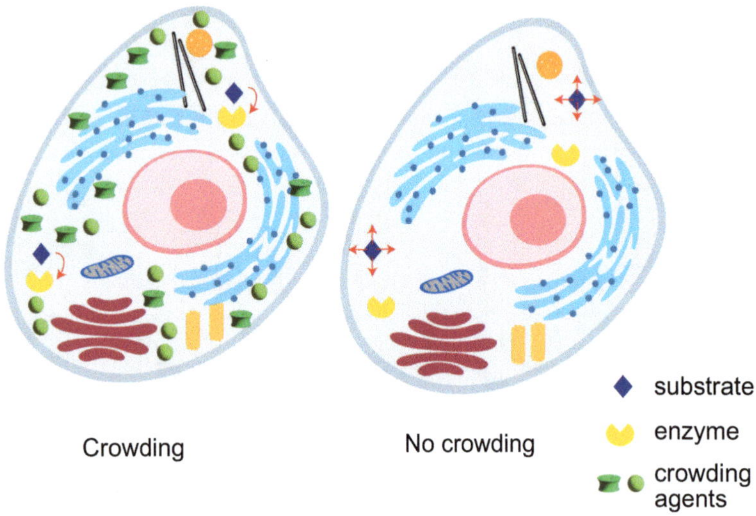

Fig. 23.2 Schematic comparison of enzyme-substrate complex formation in crowded (left) versus non-crowded (right) cellular environments. Increased proximity in crowded conditions enhances the likelihood of complex formation

dehydrogenase in high crowding conditions (300 mg/mL) demonstrated a shift towards tetramer formation over the monomeric (Minton and Wilf 1981). Similarly, Rohwer and colleagues found that the phosphoenolpyruvate:carbohydrate phosphotransferase system, responsible for glucose uptake, exhibited enhanced stability and catalytic activity in crowded environments with high PEG concentrations (Rohwer et al. 1998).

Macromolecular crowding can also influence enzyme conformation and structural dynamics, with effects depending on the specific crowding agent used. For instance, studies with alpha-chymotrypsin have shown that PEG increased the enzyme's catalytic rate constant, k_{cat}, while dextran decreases the limiting rate V and increases the Michaelis constant K_M (Pastor et al. 2011; Verma et al. 2011). Understanding the reasons behind these shifts in kinetic parameter is crucial for elucidating enzyme behavior under physiological conditions. To accurately quantify these effects within crowded environments, researchers have proposed modified approaches to account for reactant interactions with macromolecules (Kuthan, 2001; Minton & Wilf, 1981; Grima and Schnell, 2006). Calculating the equilibrium constant, for example, requires the consideration of a non-ideal correction factor that reflects the influence of crowding agents (Giddings 1970; Minton and Wilf 1981).

Most cellular processes rely on the diffusion, movement and interaction of molecules, leading to reactions. In the past two decades, scientists have demonstrated a crossover between different diffusion types in environments with high macromolecular content (Dix and Verkman 2008). The confined cellular environment leads to enhanced collision rates, reduced diffusion, and higher molecular concentrations.

One consequence of limited diffusion is the emergence of fractal-like kinetics in chemical reactions (Schnell and Turner 2004). This phenomenological kinetics describe a time-dependent rate under limited diffusion conditions. Compared to a Michaelis-Menten reaction, a reaction exhibiting fractal-like kinetics would initially have a higher rate but then slow down over time, eventually becoming slower than the Michaelis-Menten rate (Vasilescu et al. 2013; Mourão et al. 2014).

Fractal-like kinetics describe situations where reaction rates vary over time, influenced by factors like enzyme concentration and reactant availability. This phenomenon can enhance energy efficiency, as seen in glycolytic reactions. For example, the glycolytic enzymes hexokinase and phosphofructokinase, which phosphorylate glucose and fructose-6-phosphate respectively, exhibit fractal-like kinetics (Chandel et al. 2021). Macromolecular crowding prevents the free diffusion of glucose and fructose-6-phosphate in the cytoplasm, enabling efficient rerouting and directing glucose towards glycolysis.

Macromolecular Crowding and Intracellular Compartmentalization

Macromolecular crowding, through its influence on fractal-like kinetics and volume elusion effect, contributes to efficient cellular compartmentalization (Mourão et al. 2014; Stroberg and Schnell 2017). This compartmentalization minimizes waste and ensures the proper localization of molecules. A prominent example is chromatin, the DNA-protein-RNA complex. Studies have shown that chromatin exhibits fractal-like organization, partly due to the crowded environment (Bancaud et al. 2009). This architecture is hypothesized to optimize chromosome packing while maintaining accessibility for essential genomic processes (Bancaud et al. 2012).

Within cells, most biochemical reactions, regardless of their kinetics, are organized into membraneless compartments. Many of these compartments are liquid-like structures that form through liquid-liquid phase separation (Shin & Brangwynne, 2017). Examples of these membraneless structures includes stress granules, which help regulate mRNA translation during cellular stress (Protter and Parker 2016), and Cajal bodies, nuclear regions involved in modifying and processing small nuclear RNAs (Gall 2000).

Macromolecular crowding significantly influences the molecular interactions involved that drive liquid-liquid phase separation. The increased local concentration of crowding agents can promote the formation of condensed phases (Fig. 23.3). In intrinsically disordered proteins and intrinsically disordered regions, crowding can induce the formation of foldons, protein segments that fold independently of the overall protein structure (André et al. 2023) . For example, the C-terminal domain of histone H1 adopts a secondary structure in the presence of PEG and Ficoll-70 (Roque et al. 2007) and PEG also induces liquid-like condensate formation in nucleophosmin (André et al. 2023).

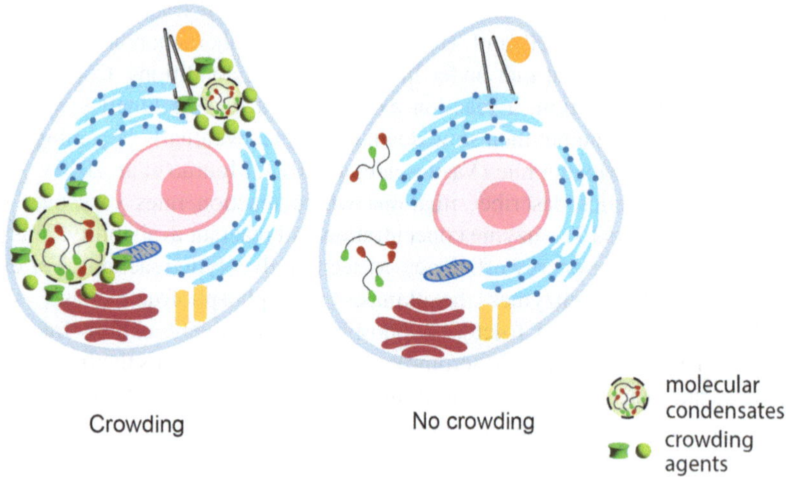

Fig. 23.3 Schematic comparison of liquid-liquid phase separation in crowded (left) and non-crowded (right) environments. Crowding increases the local concentration of proteins, promoting the formation of robust protein-rich droplets (molecular condensates). In contrast, less crowded conditions result in fewer and smaller molecular condensates

The dynamic nature of liquid-liquid phase separation means that different crowding agents can have distinct effects, an important consideration when mimicking the intracellular environment in vitro. Polymers like PEG and Ficoll may induce reversible effects or condensate disassembly, while smaller molecules like trehalose can exert osmotic effects that alter the size and hydration of the liquid phase. For instance, the intrinsically disordered protein Fused in Sarcoma, involved in membraneless organelle formation forms smaller condensates with PEG and larger ones with bovine serum albumin (Nguemaha and Zhou 2018). This difference likely stems from PEG's primarily induction of excluded volume effects, while bovine serum albumin can facilitate specific Fused in Sarcoma interactions.

Environmental factors within the cell, such as pH, can also significantly alter the dynamics of crowding agents and liquid-liquid phase separation. Changes in intracellular pH can induce conformational changes in macromolecules, promoting or inhibiting liquid-liquid phase separation. This is achieved by altering the ionization state of amino acids, modifying protein structure, and affecting hydrogen bonding within and between crowding agents (Xu et al. 2023).

Macromolecular Crowding and Cell Signaling

Crowding significantly influences the efficiency of signaling pathways. To transmit messages, signaling molecules must physically interact with cellular components. Crowding can enhance signal transduction by spatially organizing signaling molecules into transient cluster, both within the cell membrane and through the cytoplasm (Fig. 23.4) (Nussinov et al. 2021). This increases the likelihood of productive interactions.

The cell membrane, a highly crowded space (Goñi 2014), presents unique challenges in understanding the consequences of crowding due to its non-uniform and asymmetry of the lipid composition (Lorent et al. 2020). Nevertheless, we are gaining a better insight into how membrane crowding impacts cell physiology. For example, ion channels cluster into dynamic compartments within the membrane, with the sizes and shape of the clusters influenced by the density of nearby proteins and the complexity of the lipid mixture (Duncan et al. 2017). Neurotransmitter receptors, such as AMPA receptors essential for excitatory synaptic transmission, reside in the postsynaptic density. Within this membrane structure, a dynamic density of crowders and clustering of scaffold proteins keep these signal transducing AMPA receptors within the synapse (Li et al. 2016). Recent advancements in crowding sensors offer promising tools for further quantifying the impact of membrane crowding on cell signaling (Löwe et al. 2023).

Cytoplasmic diffusion of molecules is 5–20 times slower than in saline solutions (Takahashi et al. 2005) with larger particles experiencing greater diffusion restriction due to volume exclusion. This reduced diffusion can paradoxically enhance signaling efficiency through molecular "channeling", where intermediates or products from one reaction are directly transferred to the next enzyme in the pathway (Miles et al. 1999). This channeling allows for localized organization of reactions, as exemplified by purine biosynthesis, where the physical associate of purine

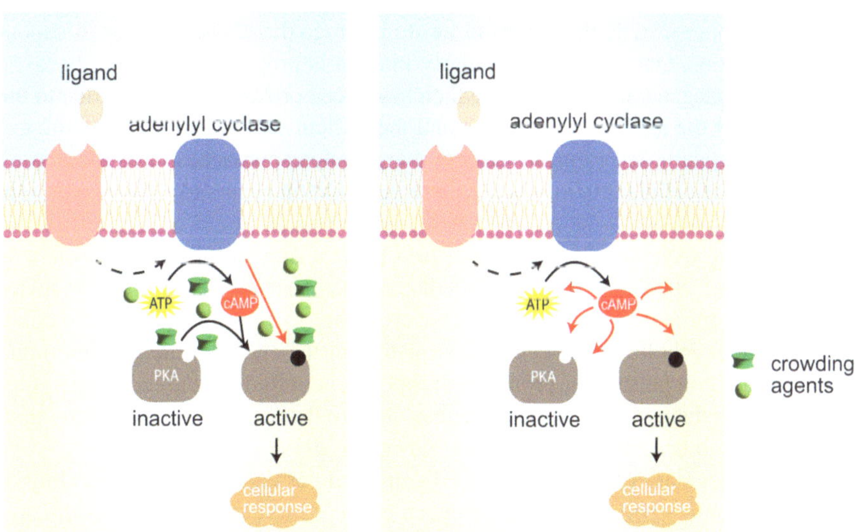

Fig. 23.4 Schematic illustration of signal transduction in crowded (left) and non-crowded (right) intracellular environments. Crowding facilitates continuous signaling due to frequent molecular interactions, while less crowded conditions result in dispersed and fragmented signaling

nucleotide phosphorylase and adenylosuccinate synthase reduces intermediates in competing pathways and increases overall efficiency (Wojtyś et al. 2021).

Cellular responses to osmotic stress also highlight the role of crowding in cell signaling. When faced with high extracellular solute concentrations, cells can shrink due to water expulsion, leading to reduced cell signaling and vesicle-mediated protein transport. Miermont et al. (2013) observed a gradual decrease in the high-osmolarity glycerol pathway with cell volume reduction, leading to a loss of homeostatic control. This effect is reversible through a regulatory volume increase, which dilutes crowding agents, enhancing cell responsiveness and nutrient uptake (Alhuthali et al. 2021). Conversely, regulatory volume decrease activates signaling pathways that promote greater protein-protein association, such as the mitogen-activated protein kinase pathway, leading to spatially organized phosphorylation events (Zhang et al. 2002).

Beyond individual cells, crowding impacts intercellular communication. Tunneling nanotubes, membrane connections between cells, facilitate cell-to-cell communication (Loewenstein and Kanno 1966). A recent theoretical model suggests that diverse membrane shapes, like those required for tunneling nanotubes, can be generated by adjusting the size of the space, and the size and density of proteins within it (Liese and Carlson 2021).

Macromolecular Crowding and Disease

As our understanding of the relationship of macromolecular crowding with cellular processes deepens, so do the opportunities to leverage that knowledge for therapeutic interventions. One area of particularly interest is protein aggregation diseases, such as neurodegenerative diseases, which have been proven challenging due to the complexity of the aggregation process and the difficulty of identifying specific targets for therapeutic intervention. Recent therapeutics approaches have focused on modulating the cellular environment where aggregation occurs to influence protein behavior.

Tauroursodeoxycholic Acid, a drug used to treat amyotrophic lateral sclerosis (Albanese et al. 2022) and Huntington's disease (Keene et al. 2002), exerts its therapeutic effect by altering macromolecular crowding to improve cellular function. Tauroursodeoxycholic Acid decreases crowding in the endoplasmic reticulum and other cellular compartments, promoting proper protein folding and reducing aggregation. This reduction in protein aggregates lead to decreased cellular toxicity and, in turn, reduced neuron loss in patients (Kusaczuk 2019).

Macromolecular crowding can also disrupt cellular communication, leading to impaired signaling pathways. This has been observed in the context of insulin signaling in type 2 diabetes (Zamora and Villena 2019). Increased intracellular crowding, often associated with elevated lipid levels in individuals with type 2 diabetes (Qi and Wang 2023), disrupts the spatial organization and diffusion of proteins involved in insulin signaling (Mika and Poolman 2011). For example, this disruption can lead to the phosphorylation of serine residues on the insulin receptor substrate by protein kinase C, negatively regulating insulin receptor substrate proteins and inhibiting the translocation of glucose transporter 4 to the cell membrane. This results in the hallmark of type 2 diabetes: increased blood glucose levels due to decreased glucose uptake (Lee et al. 2008).

Studies have also revealed that changes in macromolecular crowding within the tumor microenvironment impact cellular metabolism. Changes exhibited in the intracellular environment of cancer cells have been shown to restrict proper molecular diffusion and extracellular matrix remodeling. These deviations are reflected in the tumor microenvironment through anomalies in the cell metabolism (Barthes et al. 2014). In this context, cancer cells can undergo a metabolic switch, converting glucose to lactate even in the presence of oxygen. This switch, known as the Warburg effect, promotes tumor progression by directly influencing glycolytic flux and lactate production (Liberti and Locasale 2016). Additionally, the relocation of glucose transporter 1, a glucose transporter, leads to increased glucose uptake in cancer cells (Shin & Koo, 2021). Crowding-induced impairment of oxidative phosphorylation has also been observed, further altering mitochondrial metabolism in cancer cells (Debska-Vielhaber et al. 2021).

Conclusion

Macromolecular crowding is a fundamental aspect of cellular biology, influencing a wide range of processes from enzyme kinetics to cell signaling. The excluded volume effect, a key consequence of crowding, drives processes like protein folding and stability, and contributes to osmotic homeostasis. Crowding also modulates enzyme activity and kinetics, impacting metabolic efficiency. Furthermore, it plays a crucial role in the formation of membraneless organelles through liquid liquid phase separation.

Crowding influences cell signaling pathways, both within and between cells. Dysregulation of crowding has been linked to various diseases, including neurodegenerative disorders, type 2 diabetes, and cancer. Understanding the complex interplay between macromolecular crowding and cellular processes is essential for developing novel therapeutic strategies for these diseases.

We have come a long way in our understanding of macromolecular crowding. Our advancements in super-resolution microscopy, fluorescence resonance energy transfer, and single-molecules techniques have provided greater insight towards crowding behavior and function. Moving forward, we are making greater efforts in perfecting the current models used to study crowding. Using more *in vivo* models to analyze crowding will help answer important unanswered questions. First, how does a biological system influence crowding at the cellular level? Is there an observable difference in the type of crowding we see between healthy versus unhealthy individuals? If so, can changes in crowding be used as a hallmark for diagnosis?

Moreover, *in vivo* models could aid researchers in providing drugs whose mechanism of action is to maintain proper crowding organization.

Future research in this field should focus on further elucidating the mechanisms by which crowding influences cellular processes, developing more sophisticated in vivo and in vitro models, and exploring the potential of targeting macromolecular crowding for therapeutic interventions. As our knowledge of this fascinating field continues to grow, we can anticipate new and exciting discoveries that will shed light on the intricate workings of the cell.

Acknowledgements IBL thanks support from the National Science Foundation Graduate Research Fellowship Program (Fellow No. 2023359072).

References

Albanese A, Ludolph AC, McDermott CJ et al (2022) Tauroursodeoxycholic acid in patients with amyotrophic lateral sclerosis: the TUDCA-ALS trial protocol. Front Neurol 13:1009113. https://doi.org/10.3389/fneur.2022.1009113

Alhuthali S, Kotidis P, Kontoravdi C (2021) Osmolality effects on CHO cell growth, cell volume, antibody productivity and glycosylation. Int J Mol Sci 22:3290. https://doi.org/10.3390/ijms22073290

André AAM, Yewdall NA, Spruijt E (2023) Crowding-induced phase separation and gelling by co-condensation of PEG in NPM1-rRNA condensates. Biophys J 122:397–407. https://doi.org/10.1016/j.bpj.2022.12.001

Bancaud A, Huet S, Daigle N et al (2009) Molecular crowding affects diffusion and binding of nuclear proteins in heterochromatin and reveals the fractal organization of chromatin. EMBO J 28:3785–3798. https://doi.org/10.1038/emboj.2009.340

Bancaud A, Lavelle C, Huet S, Ellenberg J (2012) A fractal model for nuclear organization: current evidence and biological implications. Nucleic Acids Res 40:8783–8792. https://doi.org/10.1093/nar/gks586

Barthes J, Özçelik H, Hindié M et al (2014) Cell microenvironment engineering and monitoring for tissue engineering and regenerative medicine: the recent advances. Biomed Res Int 2014:921905. https://doi.org/10.1155/2014/921905

Chandel V, Sharma PP, Nayar SA et al (2021) In silico identification of potential inhibitor for TP53-induced glycolysis and apoptosis regulator in head and neck squamous cell carcinoma 3 Biotech 11:117. https://doi.org/10.1007/s13205-021-02665-3

Debska-Vielhaber G, Miller I, Peeva V et al (2021) Impairment of mitochondrial oxidative phosphorylation in skin fibroblasts of SALS and FALS patients is rescued by in vitro treatment with ROS scavengers. Exp Neurol 339:113620. https://doi.org/10.1016/j.expneurol.2021.113620

Dix JA, Verkman AS (2008) Crowding effects on diffusion in solutions and cells. Annu Rev Biophys 37:247–263. https://doi.org/10.1146/annurev.biophys.37.032807.125824

Duncan AL, Reddy T, Koldsø H et al (2017) Protein crowding and lipid complexity influence the nanoscale dynamic organization of ion channels in cell membranes. Sci Rep 7:16647. https://doi.org/10.1038/s41598-017-16865-6

Ellis RJ (2001) Macromolecular crowding: obvious but underappreciated. Trends Biochem Sci 26:597–604. https://doi.org/10.1016/S0968-0004(01)01938-7

Gall JG (2000) Cajal bodies: the first 100 years. Annu Rev Cell Dev Biol 16:273–300. https://doi.org/10.1146/annurev.cellbio.16.1.273

Giddings JC (1970) Effect of membranes and other porous networks on the equilibrium and rate constants of macromolecular reactions. J Phys Chem 74:1368–1370. https://doi.org/10.1021/j100701a037

Goñi FM (2014) The basic structure and dynamics of cell membranes: an update of the Singer–Nicolson model. Biochim Biophys Acta 1838:1467–1476. https://doi.org/10.1016/j.bbamem.2014.01.006

Grima R, Schnell S (2006) How reaction kinetics with time-dependent rate coefficients differ from generalized mass action. ChemPhysChem 7:1422–1424. https://doi.org/10.1002/cphc.200600081

Ikizawa T, Ikeda K, Arita M et al (2023) Mitochondria directly sense osmotic stress to trigger rapid metabolic remodeling via regulation of pyruvate dehydrogenase phosphorylation. J Biol Chem 299:102837. https://doi.org/10.1016/j.jbc.2022.102837

Keene CD, Rodrigues CMP, Eich T et al (2002) Tauroursodeoxycholic acid, a bile acid, is neuroprotective in a transgenic animal model of Huntington's disease. Proc Natl Acad Sci USA 99:10671–10676. https://doi.org/10.1073/pnas.162362299

Kusaczuk M (2019) Tauroursodeoxycholate-bile acid with chaperoning activity: molecular and cellular effects and therapeutic perspectives. Cells 8:1471. https://doi.org/10.3390/cells8121471

Kuthan H (2001) Self-organisation and orderly processes by individual protein complexes in the bacterial cell. Prog Biophys Mol Biol 75:1–17. https://doi.org/10.1016/s0079-6107(00)00023-7

Lee S, Lynn EG, Kim JA, Quon MJ (2008) Protein kinase C-ζ phosphorylates insulin receptor substrate-1, −3, and −4 but not −2: isoform specific determinants of specificity in insulin signaling. Endocrinology 149:2451–2458. https://doi.org/10.1210/en.2007-1595

Li TP, Song Y, MacGillavry HD et al (2016) Protein crowding within the postsynaptic density can impede the escape of membrane proteins. J Neurosci 36:4276. https://doi.org/10.1523/JNEUROSCI.3154-15.2016

Liberti MV, Locasale JW (2016) The Warburg effect: how does it benefit cancer cells? Trends Biochem Sci 41:211–218. https://doi.org/10.1016/j.tibs.2015.12.001

Liese S, Carlson A (2021) Membrane shape remodeling by protein crowding. Biophys J 120:2482–2489. https://doi.org/10.1016/j.bpj.2021.04.029

Loewenstein WR, Kanno (1966) Intercellular communication and the control of tissue growth: lack of communication between cancer cells. Nature 209:1248–1249. https://doi.org/10.1038/2091248a0

Lorent JH, Levental KR, Ganesan L et al (2020) Plasma membranes are asymmetric in lipid unsaturation, packing and protein shape. Nat Chem Biol 16:644–652. https://doi.org/10.1038/s41589-020-0529-6

Löwe M, Hänsch S, Hachani E et al (2023) Probing macromolecular crowding at the lipid membrane interface with genetically-encoded sensors. Protein Sci 32:e4797. https://doi.org/10.1002/pro.4797

Miermont A, Waharte F, Hu S et al (2013) Severe osmotic compression triggers a slowdown of intracellular signaling, which can be explained by molecular crowding. Proc Natl Acad Sci USA 110:5725–5730. https://doi.org/10.1073/pnas.1215367110

Mika JT, Poolman B (2011) Macromolecule diffusion and confinement in prokaryotic cells. Curr Opin Biotechnol 22:117–126. https://doi.org/10.1016/j.copbio.2010.09.009

Miles EW, Rhee S, Davies DR (1999) The molecular basis of substrate channeling. J Biol Chem 274:12193–12196. https://doi.org/10.1074/jbc.274.18.12193

Minton AP (1983) The effect of volume occupancy upon the thermodynamic activity of proteins: some biochemical consequences. Mol Cell Biochem 55:119–140. https://doi.org/10.1007/BF00673707

Minton AP, Wilf J (1981) Effect of macromolecular crowding upon the structure and function of an enzyme: glyceraldehyde-3-phosphate dehydrogenase. Biochemistry 20:4821–4826. https://doi.org/10.1021/bi00520a003

Mourão M, Kreitman D, Schnell S (2014) Unravelling the impact of obstacles in diffusion and kinetics of an enzyme catalysed reaction. Phys Chem Chem Phys 16:4492–4503. https://doi.org/10.1039/c3cp52417e

Nakada T (2009) Neuroscience of water molecules: a salute to professor Linus Carl Pauling. Cytotechnology 59:145–152. https://doi.org/10.1007/s10616-009-9216-x

Neira JL, Rico M (1997) Folding studies on ribonuclease A, a model protein. Fold Des 2(1):R1–R11. https://doi.org/10.1016/S1359-0278(97)00001-1

Nguemaha V, Zhou HX (2018) Liquid-liquid phase separation of patchy particles illuminates diverse effects of regulatory components on protein droplet formation. Sci Rep 8:6728. https://doi.org/10.1038/s41598-018-25132-1

Nussinov R, Tsai CJ, Jang H (2021) Signaling in the crowded cell. Curr Opin Struct Biol 71:43–50. https://doi.org/10.1016/j.sbi.2021.05.009

Pastor I, Vilaseca E, Madurga S et al (2011) Effect of crowding by dextrans on the hydrolysis of N-succinyl-l-phenyl- ala-p-nitroanilide catalyzed by α-chymotrypsin. J Phys Chem B 115:1115–1121. https://doi.org/10.1021/jp105296c

Protter DSW, Parker R (2016) Principles and properties of stress granules. Trends Cell Biol 26:668–679. https://doi.org/10.1016/j.tcb.2016.05.004

Qi Y, Wang X (2023) The role of gut microbiota in high-fat-diet-induced diabetes: lessons from animal models and humans. Nutrients 15:922. https://doi.org/10.3390/nu15040922

Rohwer JM, Bader R, Westerhoff HV, Postma PW (1998) Limits to inducer exclusion: inhibition of the bacterial phosphotransferase system by glycerol kinase. Mol Microbiol 29:641–652. https://doi.org/10.1046/j.1365-2958.1998.00963.x

Roque A, Ponte I, Suau P (2007) Macromolecular crowding induces a molten globule state in the C-terminal domain of histone H1. Biophys J 93:2170–2177. https://doi.org/10.1529/biophysj.107.104513

Schnell S, Turner TE (2004) Reaction kinetics in intracellular environments with macromolecular crowding: simulations and rate laws. Prog Biophys Mol Biol 85:235–260. https://doi.org/10.1016/j.pbiomolbio.2004.01.012

Shin E, Koo JS (2021) Glucose metabolism and glucose transporters in breast cancer. Front Cell Dev Biol 9:728759. https://doi.org/10.3389/fcell.2021.728759

Shin Y, Brangwynne CP (2017) Liquid phase condensation in cell physiology and disease. Science 357:eaaf4382. https://doi.org/10.1126/science.aaf4382

Stroberg W, Schnell S (2017) On the origin of non-membrane-bound organelles, and their physiological function. J Theor Biol 434:42–49. https://doi.org/10.1016/j.jtbi.2017.04.006

Takahashi K, Vel Arjunan SN, Tomita M (2005) Space in systems biology of signaling pathways – towards intracellular molecular crowding in silico. FEBS Lett 579:1783–1788. https://doi.org/10.1016/j.febslet.2005.01.072

Vasilescu C, Olteanu M, Flondor P, Calin GA (2013) Fractal-like kinetics of intracellular enzymatic reactions: a chemical framework of endotoxin tolerance and a possible non-specific contribution of macromolecular crowding to cross-tolerance. Theor Biol Med Model 10:55. https://doi.org/10.1186/1742-4682-10-55

Verma S, Jain SL, Sain B (2011) PEG-embedded KBr(3): a recyclable catalyst for multicomponent coupling reaction for the efficient synthesis of functionalized piperidines. Beilstein J Org Chem 7:1334–1341. https://doi.org/10.3762/bjoc.7.157

Wojtyś MI, Jaźwiec R, Kazazić S et al (2021) A comprehensive method for determining cellular uptake of purine nucleoside phosphorylase and adenylosuccinate synthetase inhibitors by H. Pylori. Appl Microbiol Biotechnol 105:7949–7967. https://doi.org/10.1007/s00253-021-11510-9

Xu Z, Wang W, Cao Y, Xue B (2023) Liquid-liquid phase separation: fundamental physical principles, biological implications, and applications in supramolecular materials engineering. Supramol Mater 2:100049. https://doi.org/10.1016/j.supmat.2023.100049

Zamora M, Villena JA (2019) Contribution of impaired insulin signaling to the pathogenesis of diabetic cardiomyopathy. Int J Mol Sci 20:2833. https://doi.org/10.3390/ijms20112833

Zhang W, Liu HT, Tu H (2002) MAPK signal pathways in the regulation of cell proliferation in mammalian cells. Cell Res 12:9–18. https://doi.org/10.1038/sj.cr.7290105

MIX
Papier aus verantwortungsvollen Quellen
Paper from responsible sources
FSC® C105338

If you have any concerns about our products,
you can contact us on
ProductSafety@springernature.com

In case Publisher is established outside the EU,
the EU authorized representative is:
Springer Nature Customer Service Center GmbH
Europaplatz 3, 69115 Heidelberg, Germany

Printed by Libri Plureos GmbH
in Hamburg, Germany